Structural Monitoring with Fiber Optic Technology

Structural Monitoring with Fiber Optic Technology

Raymond M. Measures
Institute for Aerospace Studies
University of Toronto
Downsview, Ontario
Canada

ACADEMIC PRESS
A Harcourt Science and Technology Company
San Diego San Francisco New York Boston
London Sydney Tokyo

This book is printed on acid-free paper. ♾

Academic Press
A Harcourt Science and Technology Company
525 B Street, Suite 1900, San Diego, California 92101-4495, USA
http://www.academicpress.com

Academic Press
Harcourt Place, 32 Jamestown Road, London NW1 7BY, UK

Library of Congress Catalog Card Number: 00-107677

International Standard Book Number: 0-12-487430-4

PRINTED IN THE UNITED STATES OF AMERICA
01 02 03 04 05 QW 9 8 7 6 5 4 3 2 1

Dedication

To my sisters, Sadie and Pearl, who passed away during the preparation of this book and are dearly missed.

Contents

Preface

Structures today are primarily designed for a single purpose, such as supporting a load. Nature, by contrast, almost always creates multi-functional structures. Look at your hand! Your hand can certainly lift weights, but in addition you can assess the weight and the speed with which you are lifting it. You are also aware of any force or heat directed at your hand and can even locate the site of this stress. Although it will be some time before we can duplicate nature, the first step to following nature's paradigm is to build structures with the ability to monitor their status and health.

Fiber optic technology is destined to become one of the core technologies of the twenty-first century. The purpose of this book is to show that this exciting new technology could also have an important role to play in structural monitoring. We explain the merits of this technology and show that it is unrivaled in terms of measurement versatility and convenience when it comes to integrating sensor arrays into practical structures as diverse as bridges, ships, and aircraft. One of its most significant advantages over conventional technology is the ability to use a single strand of optical fiber to replace the hundreds of wires required for measuring a given strain field using strain gauges. The fact that such a highly instrumented optical fiber can readily be embedded within FRP composite material structures, even if pultruded, contributes to the excitement in regard to its potential applications. When it is also appreciated that fiber optic sensors are immune to electrical interference and corrosion and do not create electrical pathways, it can be understood why there is so much interest in this revolutionary technology.

In the case of the construction industry, the condition of large numbers of major structures {bridges, dams, tunnels...} could be monitored by this fiber optic sensing technology from central monitoring stations through telephone lines, cellular phones, or satellite links. It may be possible in the twenty-first century for an engineer, or student, anywhere on the planet to use the Internet to log-on to one of many major structures and see in real-time the structure's response to its environment and loading and then to compare this response to that predicted by a computer model of the structure.

This book is designed to introduce fiber optic structural sensing to those interested in learning what is possible with this new monitoring technology. The book is intended to serve as a primer for understanding the basics of the technology and, most important, to provide an insight into its extraordinary measurement capability.

Chapter 1 provides some background and places fiber optic structural monitoring into the broader context of the emerging field of "smart structures." Chapter 2 establishes the need for structural monitoring with fiber optic technology in many diverse disciplines ranging from civil to aerospace engineering. Chapter 3 introduces the most important concepts and principles of lightwave theory and lays the groundwork for fiber optic technology. Chapter 4 reviews in some detail light sources and detectors as these topics constitute the heart of the optoelectronic systems designed to read fiber optic sensors. Chapter 5 describes the propagation of light in optical fibers and describes the primary fiber optic components referred to in subsequent chapters.

Chapter 6 indicates the merits of fiber optic sensors and provides an overview of the different types of fiber optic sensors. Chapter 7 lays the foundation for understanding the strain and temperature sensitivity of fiber optic sensors and discusses thermal apparent strain and methods of compensating its influence. Chapter 8 is concerned with the installation of fiber optic sensors on, and within, structures and addresses a number of important material integration issues, including the influence of embedded optical fiber sensors on the properties of the host structure.

Chapter 9 focuses on the short gauge-length fiber optic sensor and reviews methods of demodulating (interrogating) fiber Bragg grating and Fabry–Perot interferometric sensors. A rich collection of applications for such sensors is described, including a number of key field site applications. Chapter 10 concentrates on long gauge-length fiber optic sensors, describing methods of demodulation and a broad range of applications. Chapter 11 presents a detailed description of techniques used to serial multiplex and demodulate strings of fiber Bragg grating sensors and illustrates this topic with a number of applications. Chapter 12 is concerned with truly distributed strain and temperature sensing, with an emphasis on techniques involving fiber Bragg gratings. Chapter 13 presents a brief discussion on future prospects and describes a universal fiber Bragg grating sensor demodulation system based on the use of an advanced rapidly scanned, tunable DFB laser. It also provides a brief summary of many of the key points made in the sensing part of the book, that is to say, Chapters 6 through 12.

In a field that is advancing as rapidly as fiber optic structural monitoring, the best we can hope to achieve is provide: (i) a strong foundation that will serve the reader for some years, (ii) a snapshot of the status of the technology today, and (iii) insight into the myriad of ways in which this technology can be used for structural monitoring. I have tried to also steer the reader in the direction for future advances and demonstrate that the technology has an established record of success in many applications, including a number of major field-site applications.

The reader is invited to visit the web site *http:members.home.net/smoft* where they can obtain a brief overview of the book, including color versions of some of the illustrations in the book. It is also planned to provide an update on research and applications of this technology through this web site. A more in-depth overview of the subject material of the book is available directly from the author in the form of an interactive CD-ROM. This presentation is highly animated and visually exciting and intended for people who wish to quickly grasp the most important aspects of this subject.

Raymond M. Measures
Thornhill, Ontario
rm.measures@home.com

Acknowledgments

I express my gratitude to all of my colleagues who directly contributed papers and pictures. Since I have tried to illustrate the subject material of the book as much as possible, the colored photographs were especially appreciated, even though I have not been able to use them in their color format.

I thank all of my students and professionals, past and present, who helped build the Fiber Optic Smart Structures Laboratory at the University of Toronto Institute for Aerospace Studies into one of the pioneering centers for this subject in the world. In particular, I acknowledge those who were awarded the FOSSL Achievement Award: Tom Valis, Beck Mason, Kexing Liu, Serge Melle, Michel LeBlanc, Tino Alavie, Myo Ohn, and Trent Coroy. I appreciate Sharon Huang's outstanding work and help with the students of the FOSSL. I also acknowledge the extraordinary talents of Tino Alavie, who with the solid backing of Rob Maaskant, were able to run fast enough to catch a bridge and subsequently create a very successful engineering company: *ElectroPhotonics Corporation*. Lastly, I thank Sami Rizkalla, the president of ISIS—Canada for our fruitful time together creating a world class center for the activity that is at the heart of this book.

I would like to thank my family for their help and sacrifice in the preparation of this book. In particular, my daughter Karen for her beautiful art work, my younger son Warren for his hours of typing references and filing material, and my older son Jeffrey for some gems of advice. Last, and by no means least, I thank my wife for her help in very many ways, from typing blocks of the manuscript to carefully spell checking the entire document, and for just being there when I needed her.

CHAPTER 1

Introduction

I believe we are at the beginning of a new age of engineering — one that will lead to radical changes in the way we design, build, and operate structures. To date, structures have been built primarily for a single purpose — for example, supporting a load. Nature is a "master builder" and almost always creates multifunctional structures. Look at your hand! It can certainly lift weights, but in addition it permits you to assess: the weight you are lifting, the speed with which you are lifting it, and any acceleration experienced. Your hand can also sense any wind pressure or change of temperature to which it is subjected. If your hand sustains an injury, you certainly know it, and you have a good idea of the location and to some extent the severity of the injury without seeing your hand. Furthermore, your hand can repair itself, unless the injury is very severe. It is also interesting to reflect on the versatility, dexterity, and range of motion possible with the human hand. There are men who can use their hands to punch a 6-inch nail through a beam of timber, while the hands of a neuro-surgeon can undertake the most delicate and precise operations, such as stitching a nerve together. The hand is clearly a remarkable structure and it will probably be some time before we can duplicate what nature has achieved. Fortunately, for most applications it is clearly not necessary. Be this as it may, the first step to following nature's paradigm is to build structures with the ability to monitor their status and health.

Fiber optic technology is destined to become one of the core technologies of the twenty-first century. Its role in communications is already assured, as only its enormous information-carrying capacity will permit the *information currency* of the next millennium to readily flow around the world on the Internet. The purpose of this book is to show that certain aspects of this exciting new technology could have an important role to play in those industrial sectors concerned with structural monitoring. In the case of the construction industry, the condition of large numbers of bridges, dams, tunnels, etc., could be monitored using fiber optic sensing technology and this information transmitted through telephone lines, cellular or digital phones, or satellite links, to central monitoring stations (Fig. 1.1). It may be possible in the twenty-first century for an engineer or a student anywhere on the planet to use the Internet to log on to one of many major structures and see in real time the structure's response to its environment and loading and compare this response to that predicted by a computer model of the structure.

This book is intended to introduce fiber optic structural sensing to those interested in assessing what is possible with this new monitoring

FIGURE 1.1. Future central monitoring station for a number of major bridges.

technology. This is accomplished by an extensive review of the fundamentals of the subject combined with an insight into the extraordinary measurement capability afforded by fiber optic structural sensing. A large number of in-depth examples of how and where this technology has been applied at site locations is provided to complement the discussions of the technology. It is hoped that this will enable the reader to better appreciate the application potential for this technology. An emphasis on applications for the construction industry will be evident, but this reflects on the reality that the civil engineering community have started to seriously embrace this technology as they see its value to the future.

1.1 SMART STRUCTURES

The end of the twentieth century was witness to the merger of several technological disciplines that could eventually revolutionize engineering design philosophy and lead to the creation of intelligence within otherwise inanimate structures. This confluence of material and structural engineering, sensing systems, actuation and control systems, and neural networks sets the stage for a broad class of structures (Measures, 1990), the most important of which are identified in the Venn diagram presented as Fig. 1.2. This new technology will endow a structure with senses and the ability to react to its environment and change its state. The consequences of this paradigm shift for engineered objects is likely to be profound, and we have as much chance of perceiving the future impact of this technology as Bell had when he was working on the telephone, Shockley the transisitor, or Maiman the laser. Gerard Bricogne is credited with the statement, "Mankind is a catalyzing enzyme for the transition from carbon-based to silicon-based intelligence" (Rogers, 1993). A similar concept is expressed by Steven Levy in his book entitled *Artificial Life—the Quest for a New Creation*. Of course it is easy for those with a skeptical mind to look at the embryonic state of the smart structures field today and take serious issue with those who would claim major changes stemming from this technology. However, these same people would probably never have believed that early aircraft that were made of string, wood and paper could lead to the Concorde; or that the vacuum tube would lead to large-scale integrated microelectronics and the marvel of modern television; or that one day a person would be able to carry in a briefcase a computer that is vastly superior to the room-size Univac computer.

The term "smart" has already crept into our everyday language (Coghlan, 1992) and is used to designate:

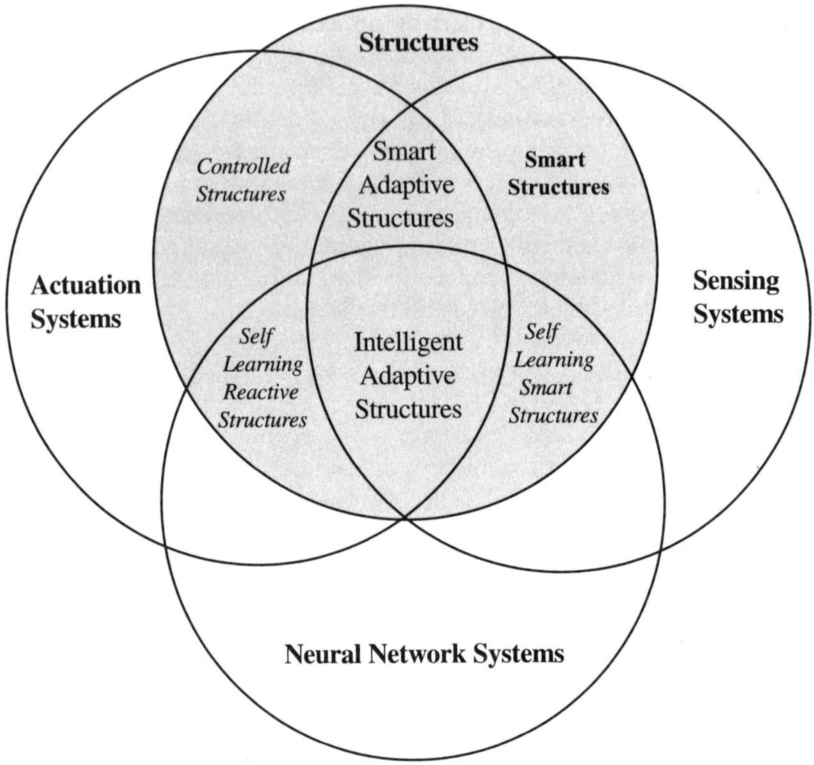

FIGURE 1.2. Venn diagram indicating some of the types of future structures possible by the confluence of the fields of structures, sensing systems, actuation and control systems, and neural networks. From Measures, R.M., (1990), Invited paper "Fiber Optics in Composite Materials – Materials with Nerves of Glass" Int. Congress of Optical Sciences and Engineering, The Hague, *SPIE* **1267**, 241–256.

- A food label that changes color if the product is stored at too high a temperature
- A pair of eyeglasses that serve as sunglasses outdoors but revert to being clear indoors
- A climber's rope that changes color when weakened
- A paint that indicates the aerodynamic pressure distribution over the wing of an airplane
- A highway that automatically monitors and weighs vehicles on the move
- A house with sensors and computers that control temperature, lighting, and security

Clearly, some form of self-diagnosis is a necessary first step in the evolution of smart structures. A structure can hardly be termed "smart" if it is incapable of sensing when it is in trouble and has suffered significant damage. As we shall see later, *structurally integrated fiber optic sensors* (SIFORS) represents one of the most promising means of achieving this self-diagnostic capability. The term structurally integrated will be taken, in this book, to mean that the sensors are either affixed to the surface or embedded within a structure in such a manner that they are treated as part of the structure. A resident sensing system of this form could continuously monitor the state and response of the host structure to imposed loads (thermal and mechanical) and thereby lead to improvements in safety and economics for many industrial sectors. A number of interesting articles have appeared on this subject in the popular press (Troy, 1997; Hamilton, 1996; Rausch, 1995; Williams, 1993; Dotto, 1992; and Shelly, 1989).

An early set of definitions (Measures, 1989) states that a *passive smart structure* possesses a structurally integrated optical microsensor system for determining the state of the structure, while a *reactive smart structure* possesses both a structurally integrated optical microsensor system and an actuator system. Athough a more appropriate name for a structure that only senses its state might be a "sensory structure," as suggested by Wada *et al.* (1990), the term "smart structure" has become fairly well established for this kind of structure. However, for the remainder of this book, we shall use the following definitions:

A *smart structure* possesses a structurally integrated sensing system.

A *smart adaptive structure* possesses a structurally integrated sensing and actuation system. The sensing information is used by the actuation control system to change the structure's state.

An *intelligent structure* possesses a structurally integrated sensing and actuation system and learns from experience.

In general, we will use the terms in the more restrictive sense that the sensing system will be based on fiber optic technology. In the case of "birth to retirement" smart structures, the resident sensing system would monitor the structure through its fabrication phase and possibly improve quality control. The resident sensing system would then keep track of the structure during installation to warn of mishandling, then provide in-service monitoring of its state of health through to the end of its useful life. This built-in monitoring system could track load history, wear and fatigue damage, and excess loadings, and it could schedule detailed inspections, repair, or replacements until the structure is ready to be retired.

The next level of sophistication for smart structures, according to the biological paradigm, involves the engineering equivalent of a combination of nerves and muscle. In a smart adaptive structure, the built-in sensing system would work in concert with its actuation system to modify the structure, or its response to loads, so as to perform certain functions or minimize structural damage. This reaction could involve moving part of the structure or changing its stiffness over a specific region in order to achieve optimum performance.

Shape control has tremendously exciting prospects in terms of adaptive aircraft wings, where ailerons, flaps, articulated parts, and hydraulic systems would all be replaced by an airfoil that modifies its shape to meet various flight conditions, by means of a structurally integrated and distributed sensing and actuation system (Crawley, 1992). The resident sensing system within such a "adaptive smart cambered wing" would provide the necessary information on its shape and aerodynamic load distribution. Shape control of propellers, helicopter rotor blades, aircraft wings, and even wind power generators offers the potential for increased efficiency, improved maneuverability, reduced vibration, and less noise.

A longer life can be expected from reduced vibration, as this also lessens structural fatigue. Similar prospects arise with marine vehicles, ground transportation systems, etc. Preliminary research in this direction is already underway with fins of naval vehicles (Beauchamp *et al.*, 1992) and a new type of helicopter rotor blade that uses adaptive materials in the form of piezoelectric sheets to twist sections of the blade root (Barrett and Stutts, 1997). This permits the pitch of the rotor blade to change as it rotates and represents a significant advance in helicopter design (Fig. 1.3). The solid-state adaptive rotor arrangement used in this model helicopter is indicated in Fig. 1.4.

There are a number of actuation materials, the three most prominent being shape memory alloys, piezoelectric materials, and magnetostrictive elements (Cross, 1992; Barrett and Gross, 1996). A comparison of these is presented in Fig. 1.5 and clearly shows that piezoelectric ceramics have the highest energy density. These materials also possess excellent dynamic characteristics and are relatively low-cost, making them the best suited for flight control applications (Barrett and Stutts, 1997). Active shape control will also be required for any large space structure, as launch weight restrictions make it impractical to build sufficient rigidity into such structures. Flexible robotic manipulators could also benefit from this technology. In earthquake-susceptible zones there is an ongoing need to improve the design of structures to resist the destructive seismic forces and develop technology that could lessen the effects of these events

FIGURE 1.3. Small remotely piloted helicopter fitted with piezoelectric adaptive blades that can be made to change the pitch angle of the blade by application of the appropriate voltage. From Barrett, R., Schliesman, M., and Frye, P., "Design, Development and Testing of a Mini Solid State Adaptive Rotorcraft," Proceedings of the 4th Annual Society of Photo-Optical Instrumentation Engineers, *SPIE*, Annual Symposium on Smart Structures and Materials, San Diego, CA, 3–6 March 1997.

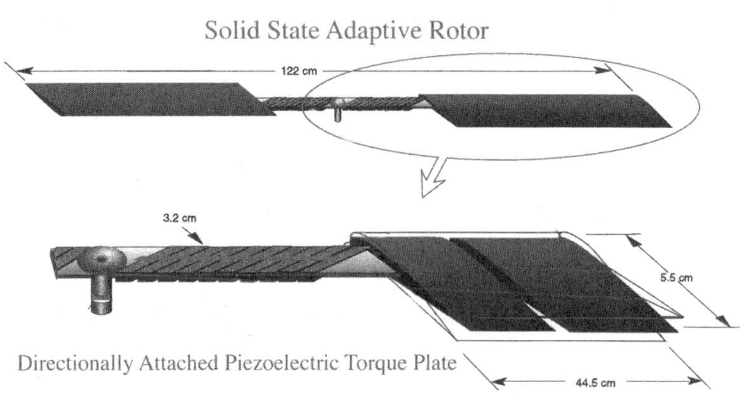

FIGURE 1.4. More detailed view of the solid state adaptive rotor blade used in the helicopter shown in Fig. 1.3. From Barrett, R. and Stutts, J., (1997), "Design and Testing of a 1/12th scale Solid State Adaptive Rotor," *Smart Mater. Structures*, **6**, 491–497.

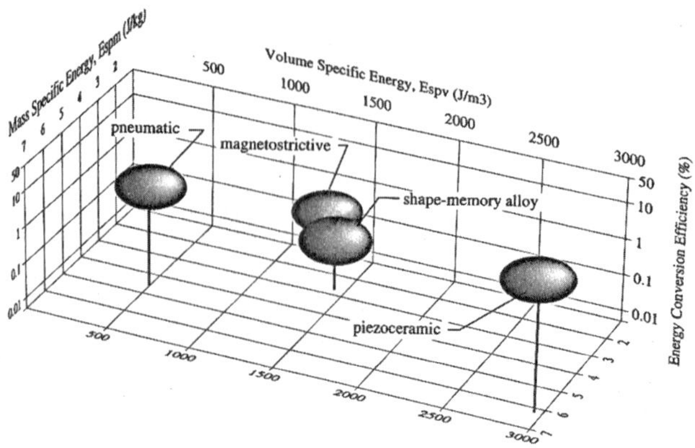

FIGURE 1.5. Comparison of the properties of a number of actuation materials. From Barrett, R. and Stutts J., (1997) "Design and Testing of a 1/12th-Scale Solid State Adaptive Rotor" *Smart Mater. Structures J*, **6**, 491–497.

(Housner *et al.*, 1992). As the technology moves ever closer to the biological paradigm, we may even see structures that can initiate some kind of self-repair (Dry and McMillan, 1996).

More broadly based philosophical discussions on the concept of intelligent materials have been presented by Takagi (1992). According to Takagi, an intelligent material is one that responds to environmental change in the optimum manner and manifests its function according to this change. An example of an intelligent material according to this kind of functional definition would be one that, when exposed to a high temperature or voltage, would modify its molecular composition and state to resist damage from this hostile environment. Wada (1990), on the other hand, introduced a framework that leads to a more practical set of definitions that are more in keeping with those used in this book. Spillman (1992), in his discussion on the evolution of smart structures, refers to an organizational hierarchy that has a parallel to biological structures. He suggests that an optimal smart structure must have at each level of complexity a more limited support structure that is locally self-acting, yet supports the overall purpose of the entity. For example, in the case of an autonomous robot that is able to move about on a set of legs, each leg would have its own smart structure that would sense the position and state of the leg and move it in accordance with the need of moving the robot, taking account of objects detected by a vision system.

It may be that the ultimate goal for human kind is to develop entities that are better able to fulfill the "mission of intelligent life." Although we may never know this goal, our major purpose may be to create, intelligent adaptive structures that are better equiped to fulfill this mission than humankind and so constitute the next stage in the evolutionary process. A truly intelligent entity may not necessarily perform the function for which it was initially primed, but may learn from experience the real purpose of its being. A child can be taught a given way of life and behavior patterns—but the adult may go in a different direction if this fits better with life's experience and a better understanding of the world. Indeed, history is rich in human beings who have transcended their culture and current knowledge base and elevated the human state.

1.2 BRIEF HISTORICAL OVERVIEW OF SMART STRUCTURES

The emergence of any new field is most difficult to map and is inevitably tempered by the knowledge and bias of the person attempting the task. However, there are usually a few widely accepted milestones that can be used to chart the course of events. In the fiber optic structural sensing field, the paper of Butter and Hocker in 1978, which demonstrated the ability of an optical fiber to measure interferometrically the strain experienced by a mechanically loaded structure, could be classified as one of the earliest milestones. The work of Varnham *et al.* (1983) to develop a polarimetric-based fiber optic strain gauge was also important in that it showed how to isolate the sensing region and form a localized sensor with an insensitive lead-in optical fiber. Corke *et al.* (1984) used the same principle to make a localized fiber optic temperature sensor. Valis *et al.* (1989a) pointed out the importance of measuring the "strain tensor" in order to uniquely specify the strain state of a structure, and developed the first practical fiber optic strain rosette (Valis *et al.*, 1992); see Fig. 1.6. This illustration also clearly demonstrates why fiber optic structural sensors are so well suited for integration into materials compared to conventional strain gauge technology. This is even more so when we recognize that in this photograph the fiber optic strain rosette is overexposed and appears much thicker than it is in reality.

Lee and Taylor (1988) demonstrated the feasibility of producing a mirror within an optical fiber by sputtering TiO_2 on the end of an optical fiber and then fusion splicing it to another length of optical fiber. They then repeated the process to form two internal mirrors and produce the first intrinsic fiber-optic Fabry–Perot interferometric sensor. This device was shown to make an optical strain gauge having small dimensions, high sensitivity, and lead-in fiber insensitivity. An extrinsic version of the

FIGURE 1.6. First fiber optic strain-rosette based on three intrinsic Fabry–Perot interferometric sensors (right side) shown against a conventional resistive foil strain rosette (left side). Both are bonded to a small aluminum beam and placed beside a pen to demonstrate the scale.

fiber optic Fabry–Perot interferometric sensor was also developed at about the same time (Murphy *et al.*, 1991c). These sensors have become commercially available from Fiber and Sensor Technologies, Virginia.

A truly major advance in the field was made unwittingly by Hill *et al.* (1978). They discovered how to form Bragg gratings within the core of optical fibers by exposure to light. This invention was turned into a practical device by Meltz *et al.* (1989), who demonstrated that it was possible to make an *intracore Bragg grating* by transverse holographic illumination of a section of GeO_2-doped optical fiber with ultraviolet radiation at about 245 nm. Morey *et al.* (1989) showed that such fiber optic Bragg gratings make excellent optical strain gauges with high sensitivity and lead-in fiber insensitivity. This type of sensor possessed a number of advantages over the Fabry–Perot sensor, including a much greater versatility in gauge length as well as automated method of production during the manufacture of the optical fibers by exposure to an intense flash of UV laser radiation (Dong *et al.*, 1993a,b; Askins, 1994; Friebele *et al.*, 1995). The transverse holographic method of writing fiber Bragg gratings has now largely been superseded by the much less expensive *phase mask technique* (Hill *et al.*, 1993).

A simple passive method of reading fiber Bragg gratings was developed in the author's laboratory (Measures *et al.*, 1992b) and led to the first commercial sensing systems based on fiber Bragg gratings by ElectroPhotonics Corporation, Ontario. These have been installed on a number of major innovative new bridges, including the 13-km-long ocean-spanning Confederation Bridge in Canada. A low-coherence long gauge fiber optic structural sensor developed by Inaudi *et al.* (1994) has been used extensively by the Construction Industry in Switzerland and is now available from Smartech, Switzerland. Sirkis and Haslach (1990) were the first to recognize some of the major issues that need to be addressed before an optical signal, whether it be the change of phase from a Fabry–Perot interferometer or the wavelength of a Bragg grating sensor, can be interpreted into a meaningful strain in the host material. They also formulated a more general phase-strain model than that used by Butter and Hocker (1978).

The development of a simple built-in fiber optic damage assessment system could be of considerable benefit to any user of composite materials where structural integrity is of prime importance. The fracture of embedded optical fibers represents the simplest technique for damage assessment. Early attempts (Hale *et al.*, 1980; Crane and Gagorik, 1984; Hofer, 1987) at implementing this concept led to systems that could only detect severe damage. Measures *et al.* (1989) devised a special etching treatment that enabled the damage sensitivity of optical fibers to be tailored to the point that barely visible impact damage (BVID) could be detected. The ability of such optical fibers to detect impact induced zones of delamination was successfully demonstrated with a grid that was embedded within the aramid fiber/epoxy leading edge of an aircraft (Glossop *et al.*, 1990; LeBlanc and Measures, 1992); see Fig. 1.7.

Multiplexed serial and parallel arrays of fiber optic structural sensors and distributed strain measurements represent the current important area of focus for fiber optic structural sensing development and application. Groups from several Laboratories including the U.S. Naval Research Laboratory (Kersey, 1997), University of Kent, (Jackson *et al.*, 1993), University of Southampton (Volanthen *et al.*, 1996a,b,c), and the author's laboratory (Measures *et al.*, 1998; Duck, 1999) have made significant progress in this area.

One of the most obvious indicators of the crystallization of an identifiable new research field is the appearance of a conference on the topic. The *Fiber Optic Smart Structures and Skins* Conference, organized under the auspices of SPIE and first held in 1988 in Boston (Udd, 1988), could be so classified for fiber optic structural sensing. The Symposium on Active Materials and Adaptive Structures organized by the American

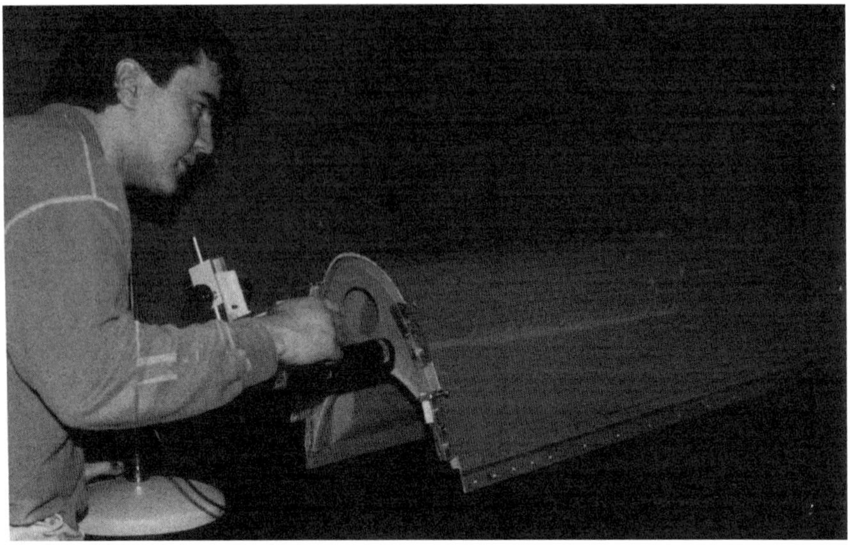

FIGURE 1.7. A section of the FRP composite material leading edge of a deHavilland DASH-8 aircraft wing, instrumented with a damage assessment system involving an embedded two-dimensional grid of 250 specially sensitized optical fibers.

Defense Preparedness Association and held in Alexandria in November 1991 was the first International conference that encompassed all aspects of the Smart Adaptive Structures field. However, a US–Japan Workshop on Smart/Intelligent Materials and Systems was held the previous year in Honolulu (Ahmad *et al.*, 1990).

An even clearer indicator of the recognition of an identifiable new research field is the publication of specialized journals. In 1990, Technomic Publishing, Inc., commenced the *Journal of Intelligent Material Systems and Structures* with Craig Rogers as the Editor-in-Chief; the Institute of Physics commenced the publication of the *Smart Materials and Structures Journal*, with Richard O. Claus, Gareth J. Knowles, and Vijay K. Varadan as Editors-in-Chief, in 1992.

The first major European program in this field was titled Optical Sensing Technologies for Intelligent Composites (OSTIC) and was initiated in 1988 as a BRITE-EURAM project (Michie *et al.*, 1992). The aims of this program were to demonstrate quasi-distributed dynamic strain measurements within a composite material and use of a single sensing optical fiber to simultaneously evaluate both strain and temperature. The former was undertaken by Bertin et Cie, who developed a coherence multiplexed polarimetric sensor (Sansonetti *et al.*, 1992). The latter investigation was primarily undertaken at the University of Strathclyde (Michie *et al.*, 1992).

The BRITE-EURAM II Optical Fiber Sensing Systems for Monitoring of Structures (OSMOS) program involved eight institutions and five industrial partners in five countries and was initiated in December 1992 (Lecot *et al.*, 1993). The objectives of this 3-year program were to demonstrate strain and temperature sensing in realistic structures in both civil engineering and aeronautics. Some emphasis was placed on structural usage and health monitoring for aircraft, while in the case of Civil Engineering highly stressed structural members such as long stay cables, suspension cables, ground anchors, and prestressing tendons were considered.

A major milestone in the development fiber optic structural monitoring was achieved in 1993, with the first integration of a fiber optic structural sensing array into a number of the precast concrete girders of a new, innovative highway bridge (Measures *et al.*, 1995). The significance of this advance was further enhanced by the historic first use of carbon fiber composite material prestressing tendons to replace steel in some of the girders of this bridge, the Beddington Trail Bridge in Calgary (Maaskant *et al.*, 1994). The sensors were intermittently monitored over a 20-month period to track the strain relief experienced by both the carbon fiber and steel tendons. More details including results can be found in Chapter 9, Section 9.3.1.

This first so-called "smart bridge" served as a precursor for the formation of the *Intelligent Sensing for Innovative Structures* (ISIS) Network of Centres of Excellence. This major Canadian initiative was established to develop and apply both advanced composite materials and fiber optic structural sensing to the construction industry. Its work was initially conducted within the five themes indicated in Fig. 1.8. An international workshop to formulate a 5-year sequential research program for implementation of fiber optic sensor technology in construction materials and bridges was held May 1998 under the auspices of the U.S. National Science Foundation, the Federal Highway Administration, and the New Jersey Department of Transportation (Ansari, 1998).

In earthquake-susceptible zones there is an ongoing need to improve the design of structures that can resist destructive seismic forces and develop technology that can lessen the effects of seismic events. A panel on Structural Control Research was established by the U.S. National Science Foundation to facilitate the transmission and dissemination of information on state-of-the-art developments, and plan and prioritize needed research and development, including full-scale testing and demonstrations of new technology (Housner *et al.*, 1992).

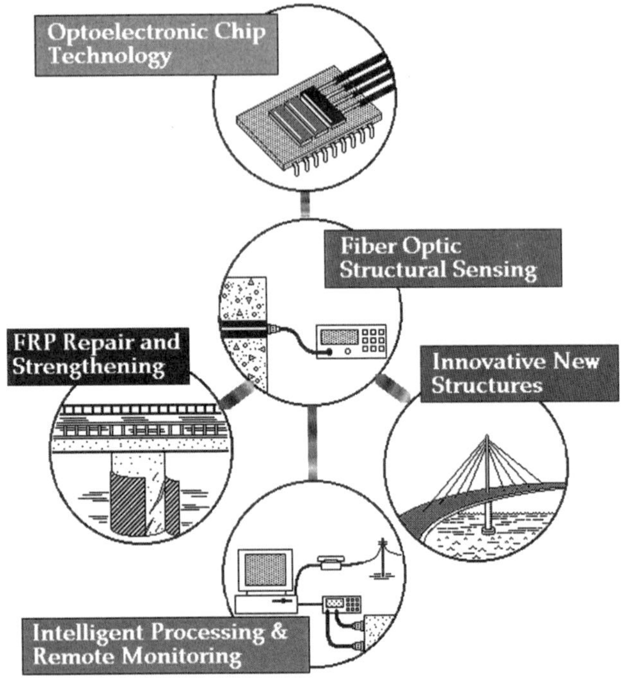

FIGURE 1.8. Five themes that originally constituted the Networks of Centres of Excellence for Intelligent Sensing for Innovative Structures. Used with permission, ISIS Canada.

The fiber optic structural sensing field can be seen to have undergone major changes in the past decade. It has evolved from rather crude laboratory experiments to commercial monitoring systems that rival conventional technology. What makes this technology especially exciting, however, is the new types of measurement not practical by conventional structural monitoring systems. Examples are long-gauge, serial multiplexed, and truly distributed measurement systems. Commensurate with the development of this new monitoring technology has been the emergence of advance composite materials as a challenge to the use of steel in the construction industry and aluminum in the aerospace industry. The marriage of these very diverse technologies can be seen as the leading edge of a new age of multifuctional, smart materials.

CHAPTER 2

Need for Integrated Structural Monitoring

2.1 INTRODUCTION

Over the past decade most industrial sectors have been forced to reduce cost and become more competitive. This strategy is driven by growing competition from all parts of the world as more countries enter the global economy as both producers and users of goods and services. World markets are expanding, and this represents both an exciting challenge and a potential threat to many industrial sectors. Governments are also exposed to this changing world economy and are having to become more fiscally responsible, which means they are constantly looking for ways to cut their costs. Many programs, such as transportation, are consequently under significant pressure to reduce costs and improve their effectiveness. This climate is conductive to the introduction of new technology if it can prove to lower costs and enhance efficiency.

We shall see in this chapter that a number of diverse areas can benefit from integrated structural monitoring in three ways: improved performance; reduced costs; and improved safety. At this point it is probably helpful to define the term *structurally integrated fiber optic remote sensors* (SIFORS). In this book we shall take SIFORS to mean the measurement of "strain," and where necessary "temperature," by a fiber optic sensing system that is built into the structure to be monitored. This generally covers sensors that adhere to the surface of, or are embedded within, the structure. When a structural element of length L is subject to a tensile (compressive) force, it will deform by expanding (contracting) by an amount $\pm\Delta L$ in the loaded direction (Fig. 2.1), where (+) is for expansion and (−) is for contraction. We define the *strain* in the element:

$$\varepsilon \equiv \pm\Delta L/L. \tag{2.1}$$

Thus, the *strain* in an element refers to the internal tensile, or compressive, state of material resulting from local loading. Almost all so-called strain sensors (except for photoelasticity) are actually deformation sensors with a short *gauge length* (or measurement base). For inhomogeneous materials (concrete, fiber reinforced polymers, timber), the microscopic strain field will vary significantly if measured on the scale of the inhomogeneity. For such materials the sensor gauge length should be at least 10 times the granularity of the material. In concrete the gauge length should probably be at least 100 mm, if macroscopic information is required. Strain sensors are best suited to monitor the local condition of the material and should be placed at the critical points of a structure where high strains are expected. For large structures, such as suspension bridges, that require dimensional stability, deformation measurements are important and gauge lengths for the sensors can be in the meters. Here the load state is usually much less than the material failure limit, so local strain measurements are of little interest. An interesting example would be a bridge that sinks at midspan because of ground movement and reduces driver and passenger comfort

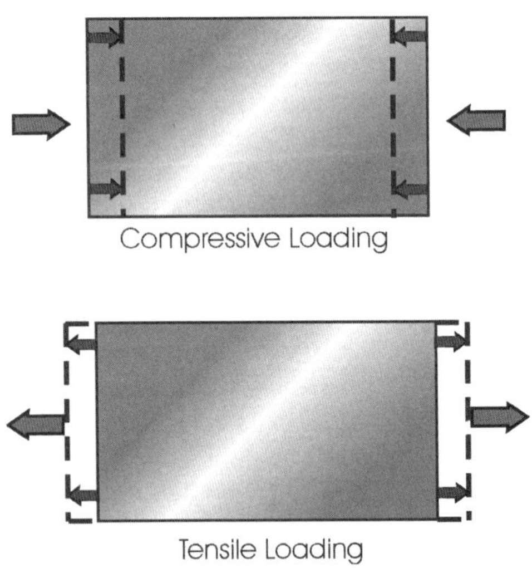

FIGURE 2.1. Schematic illustration of strain.

for vehicles crossing the bridge. This deformation could, however, improve the structural safety by reducing the strain in the bridge's girders (Inaudi *et al.*, 1996c).

We shall see that the desire to introduce new materials, particularly *fiber-reinforced polymer* (FRP), acts as a strong driver for the adoption of SIFORS for many new types of structures, as well as those undergoing repair; this is true across many disciplines, ranging from civil to aerospace engineering. A number of specific areas of potential application for fiber optic structural monitoring are identified and the potential value to be derived from the use this technology is discussed.

2.2 CIVIL ENGINEERING PROBLEMS

The civil engineering infrastructure, including bridges, tunnels, highways, railways, dams, seaports, and airports, represents an enormous financial investment — an investment that has recently run into many problems and is about to undergo a number of radical changes. The major building boom that took place in North America in the 1950s and 1960s has left in its wake a vast number of bridges, highways, and other concrete structures that are now in urgent need of rehabilitation or replacement. Indeed, in the early 1990s, Dunker and Rabbat (1993) indicated that of the nearly half-million highway bridges in the United States, more than 200,000 were deficient. Some indication of the types of processes contributing to the deterioration of bridges is presented as Fig. 2.2. The inset provides two examples of concrete that has been degraded by corrosion of its steel reinforcements. More recently, the Federal Highway Administration has classified 42% of the United States' 578,000 bridges as structurally or functionally deficient — if not obsolete (Page, 1996). The cost for correcting all of the deficient bridges exceeds $90 billion. Similar problems are to be found in other developed countries.

Ever since the Silver Bridge across the Ohio River collapsed in 1967, killing 46 people, extensive records of bridge safety and adequacy have been maintained and today most bridges are inspected every two years. Although a number of people have been killed and injured by bridge collapses each year since 1967 — for example, the Mianus River Bridge collapse killed three people — it is the disruption of traffic caused by bridge closures for substantial repair that is a major concern. The closure of the Williamsburg Bridge between Manhattan and Brooklyn, from April to August 1988, caused 100,000 drivers a day to find alternative routes. Sometimes accidents arise that, in themselves, do not lead to death or

IMPORT

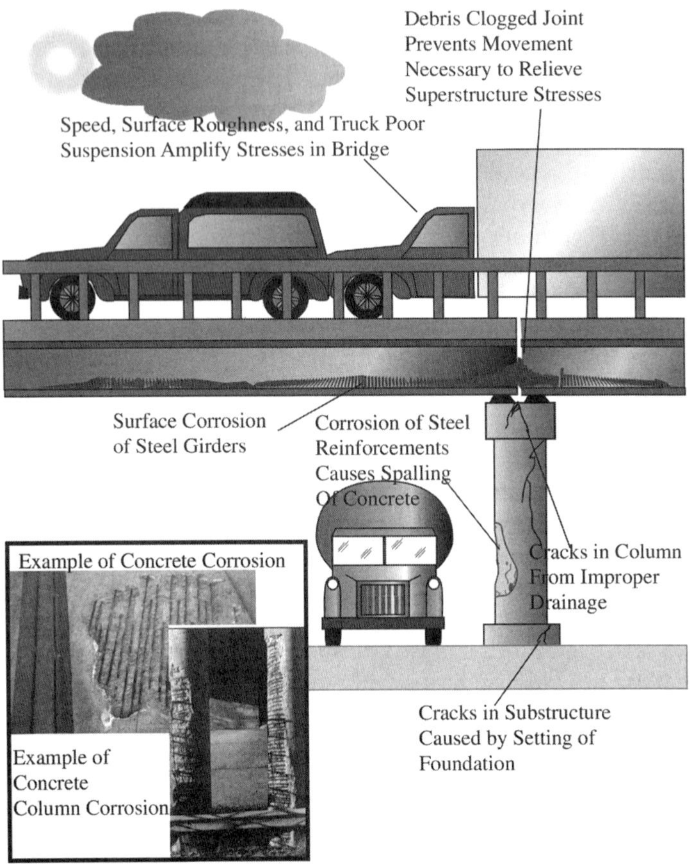

FIGURE 2.2. Deterioration processes for bridges.

injury, but weaken a structure such that a more serious accident subsequently occurs. Such was the case in September 1993, when a train careered into the water off a bridge in Mobile, Alabama, killing 40 people. Minutes earlier the bridge had been struck by an errant barge. If the bridge had been tied into some form of structural monitoring network, the second, more serious incident might have been averted.

Other than earthquakes, the most severe damage to large concrete structures arises from corrosion of the embedded steel reinforcements. The extreme cold climate experienced by concrete structures in northern latitudes accelerates deterioration from corrosion due to the use of deicing salts. Added to this is the considerable increase in traffic volume and the size of truck loads over the past couple of decades. Extensive studies reveal that although excess loading by large trucks contributes to the poor state of bridges, even more important is a lack of regular maintenance. It would seem that when a bridge has a few weak spots and these are pounded by heavy trucks, the deterioration spreads and is accelerated. In an attempt to make bridges safer and more economic, the U.S. government is funding research into use of new materials, such as FRP, that can lower costs of construction and repair and reduce the need for maintenance.

2.3 NEW MATERIALS FOR THE CONSTRUCTION INDUSTRY

The current economic climate has also led to a transition from "construction costs" to "lifetime costs" as considerations in the design of new major constructions. The 13-km-long Confederation Bridge (originally known as the PEI-Fixed Link) and the 407 Electronic Toll Highway in Toronto serve as Canadian examples. This paradigm shift is leading to innovative designs and the use of unconventional materials.

The use of FRP materials is at the forefront of the move toward more imaginative structures that will achieve superior performance, require less maintenance, and lead to lower life cycle costs. The specific advantages of fiber-reinforced polymers over steel are as follows:

- High strength to weight
- High stiffness to weight
- Excellent corrosion resistance, even to salt water
- Tailored to complex shapes
- Carbon and aramid FRP have excellent fatigue characteristics
- Immunity to electromagnetic waves
- Low axial coefficient of thermal expansion (especially CFRP)

Fiber reinforced polymers do, however, also have some disadvantages:

- High cost
- Low modulus of elasticity
- Low failure strain and many failure modes
- High ratio of axial to lateral strength (anchorages require special attention)
- Long-term strength can be less than short-term strength
- Susceptible to UV damage

There are three primary types of fiber used in FRP: carbon, aramid, and glass. The supporting matrix for the fiber is often some form of thermosetting epoxy resin. Each material has its own merits and demerits. For use in concrete structures, *carbon fiber reinforced polymer* (CFRP) is noncorrosive and has the highest tensile modulus of elasticity (typically 65% that of steel). The maximum strain at failure lies between 1.2 and 2%, which requires particular attention when CFRP is used in reinforced concrete (Fig. 2.3).

Two forms of CFRP have been used to date in a number of concrete bridges in Canada (Erki and Rizkalla, 1993a; Rizkalla *et al.*, 1997). Unidirectional carbon fiber composite cable (CFCC) made by Tokyo Rope Manufacturing Co. Ltd., comprises carbon fibers of the polyacrylonitrile type. These are made from roving prepreg, consisting of 12,000 filaments impregnated with resin. The prepreg is twisted to make a fiber core with diameter 3 to 40 mm. The shear strength and heat resistance are low due to the thermosetting matrix (Fig. 2.4). Mitsubishi Kasei, Japan, makes round Leadline rods of CFRP using Dialead coal tar pitch based continuous carbon fiber and an epoxy resin. Rods with diameters that range from 1 to 17 mm are available. The noncorrosive nature and good fatigue properties of CFRP will ensure that concrete structures in which reinforcing steel has been replaced with this material will have increased

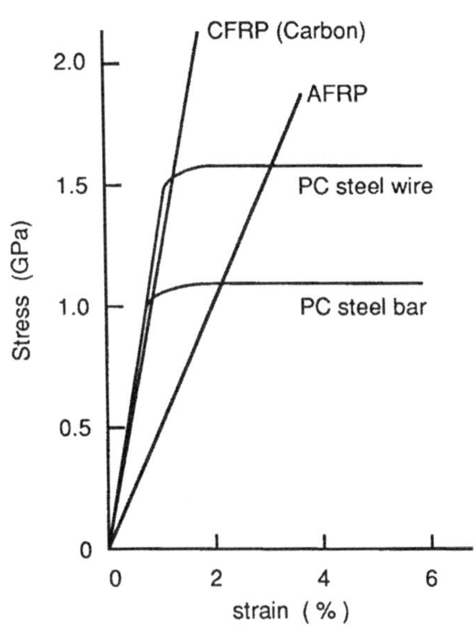

FIGURE 2.3. Stress against strain for steel and FRP. From Mufti, A.A., Erki, M.A., Jaeger, L.G., (Eds), "*Advanced Composite Materials in Bridges and Structures in Japan*" The Canadian Society of Civil Engineers, p. 104.

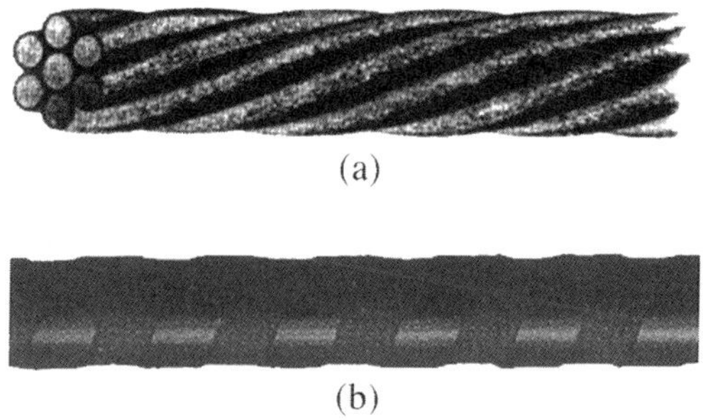

FIGURE 2.4. Representative samples of (a) Tokyo Rope's CFCC, and (b) Mitsubishi Kasei's Leadline.

service life and reduced long-term maintenance cost. The high strength-to-weight ratio of such CFRP may also lead to additional cost savings in terms of lower transportation and site erection costs. As we shall see several new bridges have led the way in utilizing these new materials.

2.4 BRIDGES OF ADVANCED DESIGN

The Beddington Trail Bridge built in 1993 in Calgary was the first bridge to use CFRP prestressing tendons in place of steel in a number of its precast concrete girders (Rizkalla and Tadros, 1994). The Taylor Bridge built in 1997 in Winnipeg across the Assiniboine River is the first bridge in the world to completely replace all of the steel in a number of its precast concrete girders with CFRP prestressing tendons and shear stirrups (Fig. 2.5) (Rizkalla *et al.*, 1997). The first FRP highway bridge in the United States was designed at Lawrence Technological Institute and was built in Southfield, Michigan (Page, 1996). The first two projects will be described in detail in Chapter 9, because of their innovative integration of fiber optic structural sensing systems.

The characteristics of CFRP also make them well suited for long stay cables, suspension bridge cables, masonry structures and ground anchorages for concrete dams, retaining walls, and bridge abutments. The tensile and bond strength of FRP is generally equal or superior to that of steel reinforcements (Erki and Rizkalla, 1993b). However, when FRP is used in prestressed concrete, the main challenge is finding an anchorage system that will permit full development of the FRP strength. FRP reinforcement's high ratio of axial tensile strength to lateral compressive

FIGURE 2.5. Taylor Bridge in Winnipeg. Used with permission, ISIS Canada.

strength (~20 : 1) introduces serious problems with regard to anchorage. A comparison of the strain distribution within various anchorage designs might help lead to an optimized system. Such an undertaking will be shown later to be plausible with a certain kind of fiber optic strain measurement system; see Chapter 12.

The span length of a bridge is a key parameter in terms of aesthetics, technical efficiency, and cost. The longer the span length, the more challenging the engineering and the higher the cost. In general, the longest span lengths are found in suspension bridges. The first was built in 1826 across the Menai Straits and had a span of 177 m; by the end of the twentieth century, the 2000 m span limit had been approached by the Akashi Kaikyo Suspension Bridge (span 1990 m) that was recently built between Honshu and Shiku in Japan. This is one of the world's longest bridges, but was built using conventional engineering and materials and to some extent represents a cumbersome example of traditional civil engineering (Menn and Billington, 1995; Normile and Vizard, 1998). This can be understood, for in the past the price has sometimes been high for introducing innovative design into bridge building. Almost every engineering student is shown the infamous film clip of the wind-driven destruction of the Tacoma Narrows Bridge in Washington State just 4 months after it opened in 1940. This catastrophic bridge failure curtailed new designs in suspension bridges for more than two decades. When the Severn Bridge between England and Wales opened in 1960, its lightweight deck was heralded as a landmark in aerodynamic design, but the bridge began falling apart under the traffic load in 15 years. In 1987 the Alex Fraser Bridge in Vancouver, at the time the longest cable-stayed bridge, developed cracks only 1 year after it opened (O'Neill, 1993b).

Depending upon the ratio of bridge width to span length, suspension bridges are sensitive to vibrations, especially under wind loading. For extremely long spans ($\sim$3000 m), this problem will dominate. Although steel has, to date, been the primary material for suspension bridges, the prospects for new materials such as CFRP look very promising because of their much lighter weight with comparable stiffness. Indeed, a number of bridges that are currently planned will test the use of FRP material for both the cables and the girders (Menn and Billington, 1995). One of the best-known is the so called "Glass Bridge" that is planned for the San Diego area by Frieder Seible, Director of the Charles Lee Powell Structural Research Laboratory (Wilson, 1997). The first stay-cable bridge in the world to replace steel with CFRP (in two of the cables) is Storck's Bridge in Winterthur, Switzerland (Fig. 2.6) (Nellen *et al.*, 1997). Both of the 35-m CFRP cables were instrumented with a combination of fiber optic strain sensors and resistive foil strain gauges. This monitoring system has been in operation for several years and is providing a useful comparison between the steel and CFRP cables.

Clearly, these new materials offer the prospect of designing structures to meet a broader range of performance objectives than is possible with conventional materials. There is, however, a downside to such tailored materials: the difficulty of including them in design codes that provide litigation protection in the event of some failure. With increasing experience it is quite likely that the design codes will soon be modified to reflect the use of these materials. Instrumenting such innovative structures with built-in fiber optic structural sensors, which could ascertain how well the new materials are performing, could go a long way toward alleviating many concerns and could hasten the extensive use of FRP by the construction industry.

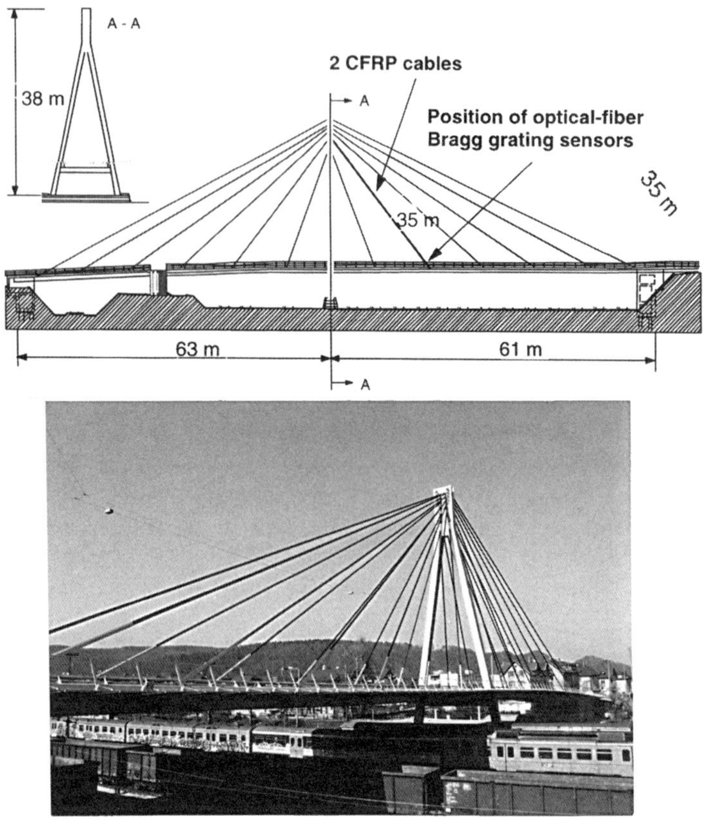

FIGURE 2.6. Storck's Bridge in Winterthur, Switzerland. From Nellen, P.M., Anderegg, P., Brönnimann, R., and Sennshauser, U., "Application of Fiber Optical and Resistance Strain Gauges for Long-Term Surveillance of Civil Engineering Structures," *SPIE*, **3043**, San Diego, CA, March 5–6, 1997, 77-86. In: Smart Structures and Materials 1997, Conference on Smart Systems for Bridges, Structures and Highways. *SPIE*.

2.5 DETECTION OF STRUCTURAL WEAKNESS

Monitoring the static and dynamic response of structures can, in general, provide insight into their structural health by identifying abnormal vibrational patterns and frequencies, or anomalous strain distributions. Previous studies of structural elements have revealed that changes in eigenfrequencies can be used to detect damage such as cracks (Gudmunson, 1982; Cawley and Adams, 1979; Biswas *et al.*, 1990; Natke and Yao, 1990; Salane and Baldwin, 1990). More recently, Kim (1995) affixed 26 accelerometers at various sites on two girders of a plate-girder bridge that carries the I-40 highway over the Rio Grande river. A 2200 lb hydraulic shaker was used to excite the bridge into a single flexural vibration mode while four levels of damage were inflicted on one of the steel girders. Predamage and postdamage accelerometer readings were taken before and after each damaging episode. Kim (1995) developed a robust algorithm for detecting and localizing damage in bridges from changes in mode shapes. The next step would be to develop a method of assessing the severity of the damage as well as its location!

Although fiber optic accelerometers have been developed that could be integrated into concrete structures, it is also possible to measure directly the vibrational modes of such structures with a series of strain sensors strung out along a single optical fiber. Indeed, researchers from Los Alamos Laboratory and Texas A&M University have shown that damage, such as a substantial crack in a steel bridge girder, can be detected by subtle changes in the structure's mode shapes under excitation (Beard, 1995).

Even with the conservatism of civil engineering there have been a number of disastrous incidents, such as the Ronan Point Apartment Block collapse, UK or the Kansas City Hyatt walkway collapse (Wyatt, 1994). These can be directly blamed on poor engineering, although sometimes the introduction of new materials or procedures may be responsible. If these structures had in place some form of health monitoring system, it is possible that these terrible accidents could have been avoided.

In the case of the terrorist bombing of the World Trade Center (Fig. 2.7) in February 1993 (Shuldiner and Retseck, 1993), the initial evaluation of the extent of the damage resulting from the detonation of a 1500 lb explosion in the subterranean parking garage was very uncertain. More than 50,000 employees and visitors fled the 110-story towers and more than 1000 people were injured. Part of the huge cost of this attack (~ $0.5B) was attributed to insurance claims associated with the closure of the complex for an extended period due to the uncertainties of the extent of the structural damage. If some temporary means of quickly checking the structural integrity of the complex had been available, millions of dollars might have been saved.

A useful structural monitoring system should be able to discern critical events, change the scan frequency and the dynamic range, and perform a preliminary evaluation of which data should be stored and which can be discarded. Information that is deemed relevant, such as response intensities and deformation modes should be converted into visual graphics that can quickly be assessed by engineers. Continuous health monitoring of structures reinforced with FRP can increase the

FIGURE 2.7. Simulated illustration of World Trade Center bombing, Februrary 1993.

security factor and avoid the structural overdesign that often abrogates the benefits that should be derived from the use of these new materials.

A favorable outcome for FRPs has, ironically, emerged from the preliminary study on the collapse of highway bridges in the January 1994 Northridge Earthquake. This investigation revealed that the massive inertia of heavy reinforced concrete used in these structures was a major contributing factor to their failure (Chong *et al.*, 1994). Future bridges and other structures built with FRP could be 3 to 5 times lighter, and this will have to be given serious consideration in earthquake-prone zones.

2.6 MEASUREMENT PROSPECTS FOR FIBER OPTIC TECHNOLOGY

In spite of the attractive features of these new materials, there is substantial resistance to switching to them materials without a long trial period.

Indeed, before FRPs are widely used for the civil infrastructure, a number of barriers will have to be scaled. The first hurdle is the high initial cost of these materials. This may not be too serious if the lifetime costs can be shown to be competitive, or superior, to those when using conventional materials. The second hurdle is the very large number of failure and degradation mechanisms associated with FRP composite materials (see Fig. 2.8). This is particularly of concern with regard to the issue of long-term safety. Integrated structural monitoring could overcome these latter concerns.

It would appear that a number of important factors are conspiring to make fiber optic structural sensing a vital new technology for the construction industry in the next decade. There is the move, discussed earlier, to replace steel embedded within concrete structures by FRP materials (Mufti *et al.*, 1991, 1992) and the rapidly increasing use of this material in the rehabilitation and strengthening of existing concrete structures (Meier *et al.*, 1992; Bonacci, 1996; Demers *et al.*, 1996; Pantazopoulou *et al.*, 1996; Alexander and Cheng, 1996) discussed later.

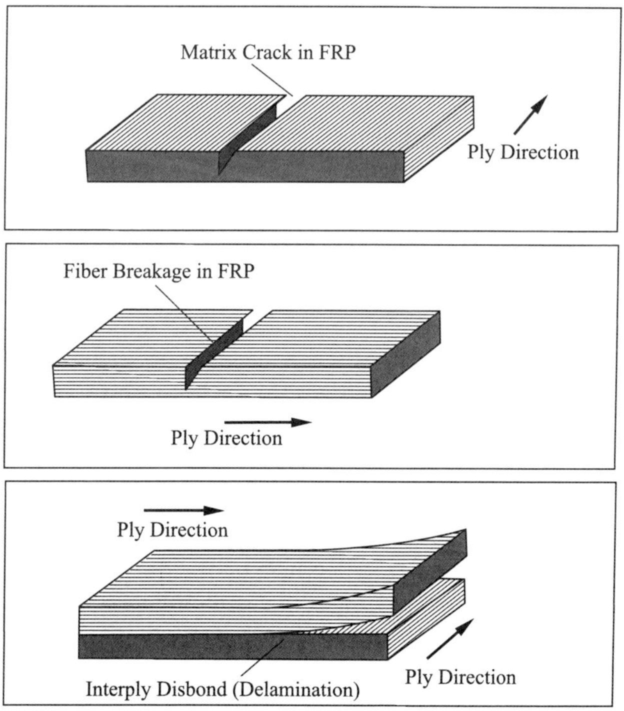

FIGURE 2.8. Schematic of different FRP failure modes.

Aktan *et al.* (1995) provide two good reasons for instrumenting bridges with a structural monitoring system. First, we need to improve our understanding of the actual loading environment and the corresponding bridge responses. Currently, there are significant gaps in our understanding of the actual service loading, especially damage limit loading, experienced by bridges. This becomes even more important as a bridge ages and deteriorates. The second reason for a structural monitoring system would be to determine if such automatically gathered information can reduce the need for visual inspection, provide an objective indication on the state of health of a bridge, and most important, warn of any impending failure. The timely emergence of practical fiber optic structural sensing technology with its concomitant merits makes it a serious contender for this task.

Three kinds of measurement are envisioned for this new kind of monitoring technology:

1. Structural monitoring and damage evaluation.
2. Experimental stress analysis.
3. Management and control of systems and service installations.

The first kind of application involves the measurement of deflection and bending for various kinds of concrete structures, such as beams, columns, arches, and slabs. In the field of experimental stress analysis it would be possible to measure the strain field for complex structures that would be difficult to model. By comparing actual strain fields and deflections with those predicted by computational models, more accurate design factors can be determined, and this could lead to structures that are safer and more economic. The third set of applications could include, for example, the measurement of traffic flow over a bridge, or access to the runways of an airport. This monitoring system could determine the frequency and extent of excess loading episodes due to large or high-speed trucks on a bridge, or hard landings at an airport. This information would help evaluate the impact of such events on the needs for repair and maintenance of such structures.

The benefits of fiber optic structural monitoring are as follows:

1. Assess load history.
2. Evaluate effect of damage on capacity and performance.
3. Assess effectiveness of repairs and maintenance.
4. Check performance against design predictions.
5. Warn of abnormal conditions or behavior.

Conventional strain gauges are used primarily for *point* strain measurements, although a number of them can be used to provide a form of *quasi-distributed* measurement. Fiber optic sensors are extremely well suited (notably fiber Bragg gratings; see Chapters 9, 11, and 12) to undertake, *point*, *quasi-distributed*, and truly *distributed* measurements of strain (Fig. 2.9). For example, "lap joints" have been used to provide continuity of tension and compression in many reinforced concrete structures. Transfer of force between a pair of overlapping rods relies on the bond at the steel/concrete interface. In tension lap joints, cracks can extend rapidly along the joint, sometimes leading to explosive failures. Consequently, research into lap joints has been concerned with identifying factors that influence the magnitude and resistance to such splitting forces.

The strength of such lap joints is often dependent on the strain gradient in the reinforcing steel. Judge *et al.* (1990) investigated this failure process using a large number of resistive foil strain gauges to determine the distribution of bond stresses along tension lap joints (Fig. 2.10) in order to ascertain the relative role played by peak and mean stresses.

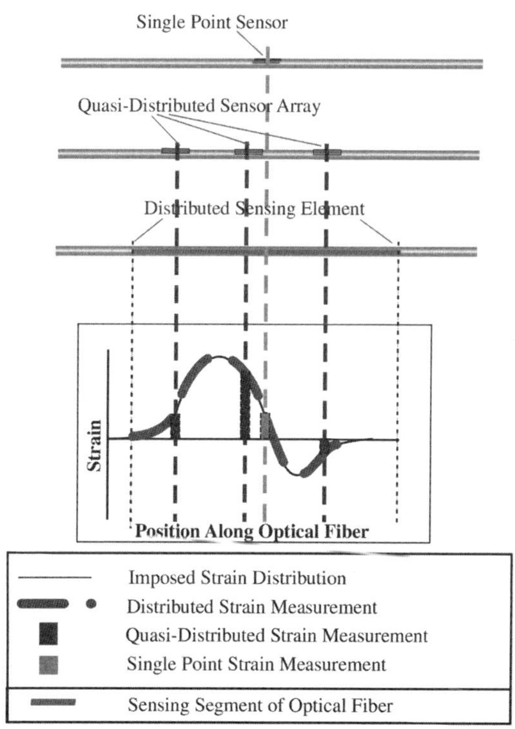

FIGURE 2.9. Comparison of point, quasi-distributed, and distributed strain sensing.

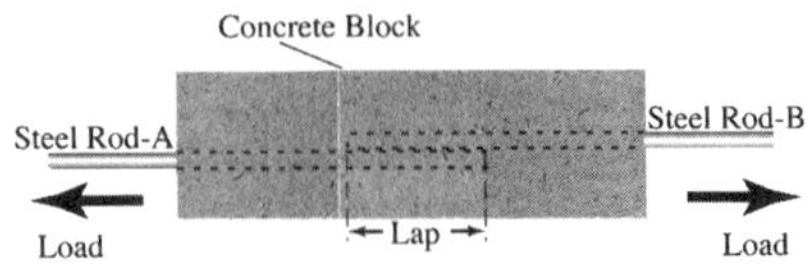

FIGURE 2.10. Schematic of steel reinforcement lap point within concrete specimen. From Judge, R.C.B., Scott, R.H. and Gill, P.A.T., (1990), "Strain and Bond Stress Distributions in Tension Lap Joints in Reinforced Concrete" *Magazine of Concrete Research*, **42**, 5–14, with permission from Thomas Telford Publishing Ltd.

Strain distribution measurements along steel reinforcements have also been undertaken with a large number of conventional resistive foil strain gauges bonded into grooves machined into two half bars (Fig. 2.11). The two halves were then glued together after installation of the strain gauges. They observed, prior to crack formation, steep gradients at the free ends of the rod and approximately, 60% of average strain was developed within one rod diameter of the lap end (Fig. 2.12).

Similar *strain distribution measurements* were undertaken by Scott and Gill (1987). Here the instrumented steel rods were 2.6 m long, with diameters of either 12 or 20 mm, and embedded in prisms of concrete. Each rod contained 80 strain gauges, with a 12.5 mm separation, in the strain gauged concrete zone. In addition, two further gauges were installed at each end of the rod, outside the zone of concrete, making 84 strain gauges in all (inset to Fig. 2.13). When the specimens were loaded hydraulically, the change in the strain distribution associated with the formation of cracks in the concrete can be seen by reference to Fig. 2.13b. Debonding between the reinforcement and the surrounding concrete occurred at the ends of the specimen and sometimes would extend into the strain-gauged region.

The measurement technique of Scott and Gill (1987) was later applied to concrete structures where an investigation of strains in reinforced concrete beam–column connections subject to simulated seismic loading

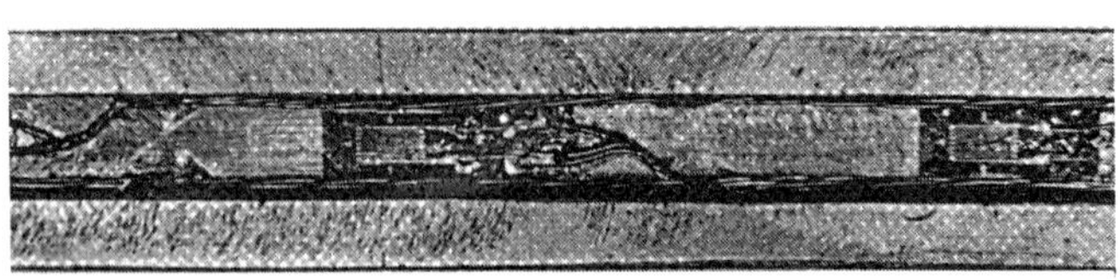

FIGURE 2.11. Set of resistive foil strain gauges incorporated into slot cut in steel rod. From Scott, R.H. and Gill, P.A.T., (1987), "Short-Term Distributions of Strain and Bond Stress along Tension Reinforcement" *The Structural Engineer*, **65B**, 39–48.

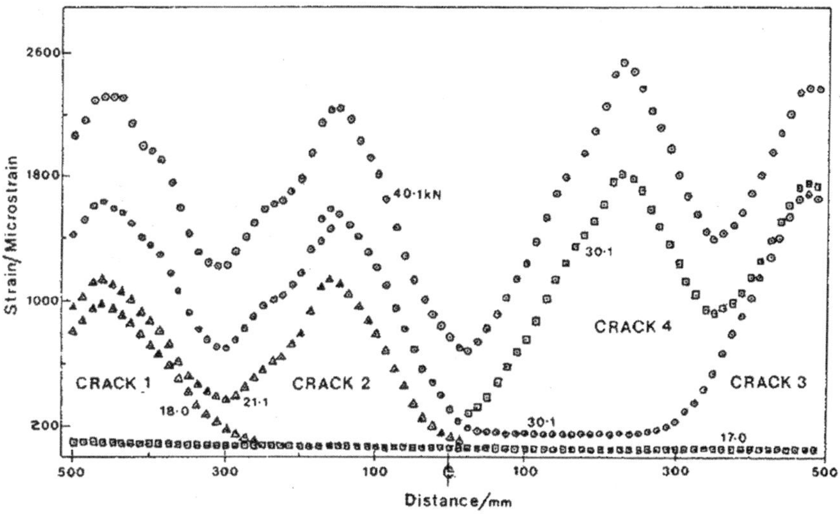

FIGURE 2.12. Strain profiles recorded by strain gauges distributed along a steel reinforcement. From Judge, R.C.B., Scott, R.H. and Gill, P.A.T., 1990, "Strain and Bond Stress Distributions in Tension Lap Joints in Reinforced Concrete" *Magazine of Concrete Research*, **42**, 5–14, with permission from Thomas Telford Publishing Ltd.

was undertaken (Scott, 1991). Figure 2.14 shows a picture of the experimental setup, and Fig. 2.15 displays a representative strain distribution under upward and downward loads. The strain-gauged reinforcement showed clearly the strain distributions giving rise to hingelike failures. There is often a desire to extensively monitor and test in the laboratory new concrete structures. Unfortunately, the complexity of wiring harnesses for an array of strain and temperature sensors acts as a considerable deterrent to such monitoring in the field.

A considerable simplification in the measurement of strain profiles arises when using a fiber optic monitoring system. This is primarily because the optical fiber serves as both the sensor and the conduit for the optical sensing signals. Indeed, a single optical fiber may contain a string of structural sensors that can undertake quasi-distributed strain measurements and in essence replace the equivalent number of foil strain gauges, each with several soldered electrical leads (Fig. 2.16). Furthermore, where FRP reinforcements, or tendons, are being used instead of their steel counterparts, such a string of optical fiber sensors can be integrated with the FRP structural components during pultrusion (Section 8.7; Kalamkarov *et al.*, 1997). This advance both serves to protect the sensors and makes the sensing technology almost transparent to the site

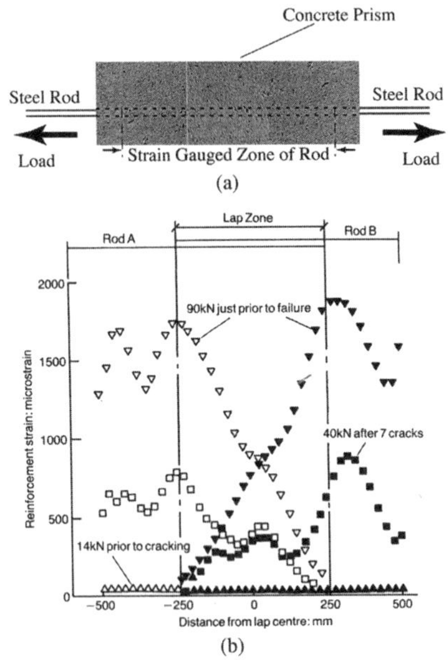

FIGURE 2.13. Strain profiles recorded by strain gauge instrumented steel reinforcement. Part (a) from Scott, R.H. and Gill P.A.T., 1987, "Short-Term Distributions of Strain and Bond Stress along Tension Reinforcement" *The Structural Engineer*, **65B**, 39–48.

construction workers. Optical connectivity to such resident fiber optic sensors, however, poses a challenging problem.

The small diameter of optical fibers (generally less than 250 μm) permits bolts and other components to be instrumented in a manner that does not compromise the integrity or strength of the component. This means that the instrumented component can be a permanent part of a structure to be tested, which is not the case for conventional strain gauge instrumentation. For a resistive foil strain gauge, a section of the bolt is milled away to make a flat footprint for the gauge (Fig. 2.17). By comparison, the optical fiber requires only a fine hole drilled in the center of the bolt. An example of a fiber optic strain sensor instrumented bolt, used to test the emergency sealing of a flange in a high-pressure pipe of the nuclear industry, is presented in Fig. 2.18, courtesy of ElectroPhotonics Corporation. This figure shows a fiber optic sensor instrumented bolt in the top right corner inset, and reveals the hostile installation site in which these bolts functioned perfectly. As indicated previously, the bolts instrumented with the fiber optic sensors were affected so little by this procedure that they were left in place after the pressure test.

FIGURE 2.14. Experimental arrangement for investigating the strain in a concrete beam–column connection. From R. H. Scott, (1991), "A Preliminary Investigation of Strains in Reinforced Concrete Beam Column Connections due to Seismic Loading," *Magazine of Concrete Research*, **43**, 59–64, with permission from Thomas Telford Publishing Ltd.

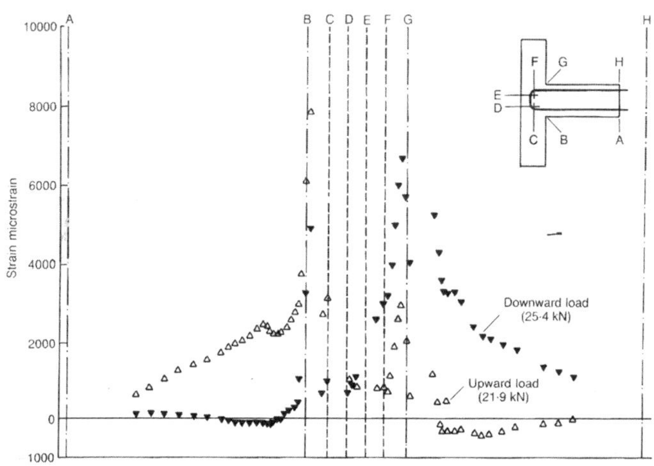

FIGURE 2.15. Representative strain profiles with both up and down loading of a beam–column connection. From Scott, R.H., (1991), "A Preliminary Investigation of Strains in Reinforced Concrete Beam Column Connections due to Seismic Loading," *Magazine of Concrete Research*, **43**, 59–64, with permission from Thomas Telford Publishing Ltd.

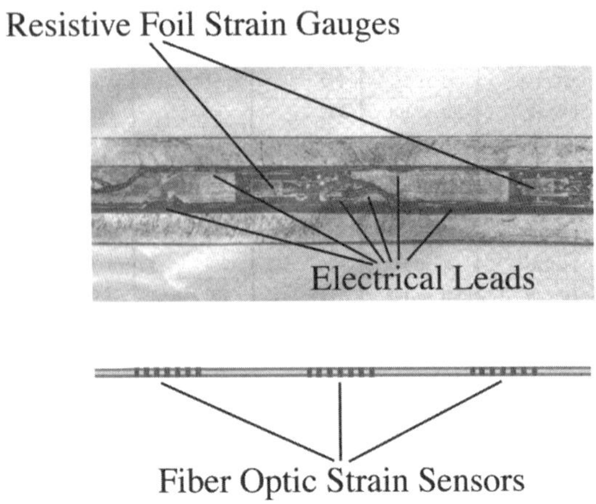

FIGURE 2.16. A comparison of a string of fiber optic strain sensors and an equivalent set of resistive foil gauges.

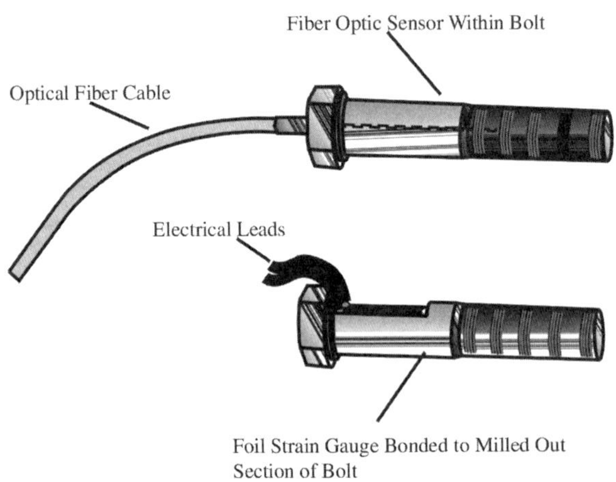

FIGURE 2.17. A fiber optic strain sensor instrumented bolt and one instrumented with a resistive foil strain gauge.

2.7 EARTHQUAKES AND NEW MATERIALS FOR REPAIR

The catastrophic earthquake that devastated Kobe in 1995 was the strongest ever recorded in Japan, say seismologists who measured the intensity of shaking during the quake. Horizontal accelerations in excess of $8\,m\,s^{-2}$ were recorded. This corresponds to about 80% of the earth's

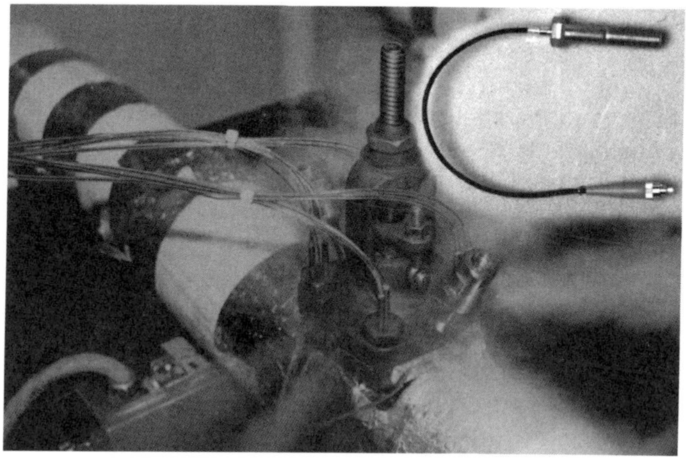

FIGURE 2.18. Photograph of a fiber optic strain sensor instrumented bolt. Used with permission, ElectroPhotonics Solutions Corporation.

gravity and is thought to arise from amplification from seismic waves bouncing off firmer rocks (Hadfield, 1995). In 1994 the city of Northridge near Los Angeles was struck by a 6.7 (on the Richter scale) magnitude earthquake. Although this earthquake did a considerable amount of damage and killed more than 40 people, its impact was relatively small compared to the devastation unleashed by the 6.9 (on the Richter scale) Kobe earthquake, which killed 5470 people (Valery, 1995). Unfortunately, the Northridge earthquake did nothing to relieve the stress building in the fault line under Los Angeles. The release of this energy could deliver an earthquake with 15 times the intensity of the quake that shook Northridge (Mestel, 1995). A substantial buildup of strain on a section of the San Andreas Fault south of San Francisco also continues and is even greater than observed before the Loma Prieta earthquake of October 1989. This earthquake, which was centered about 100 km from the heart of San Francisco, caused extensive fires in the marina district of the city and caused sections of the Bay Bridge and the Nimitz Freeway in Oakland to collapse (Anderson, 1992).

New estimates of seismic risk for people living on the Eastern Seaboard presented at the American Association for the Advancement of Science in March 1995 indicates the likelihood of a damaging earthquake within the next few decades. A disturbing aspect of this news is that in California, the earth's shattered crust quickly absorbs the energy of an earthquake, so its range is less than that of a comparable quake on the eastern side of North America. Consequently, eastern earthquakes tend to be more damaging than western quakes of the same magnitude. Well-designed skyscrapers such as the World Trade Center, built on solid rock with a natural resonance frequency far lower than easily excited by an earthquake, would probably emerge unscathed. However, much of downtown Manhattan is built on silt sediments, which can liquefy in a severe earthquake (Holmes, 1995). Also, few of the bridges and buildings in cities such as Montreal, Ottawa, New York, Boston, and Philadelphia were designed to withstand a strong earthquake.

The earthquakes in Mexico City in 1985, San Francisco in 1989, Los Angeles in 1994, and Kobe in Japan in 1995 taught us how important it is to quickly assess the extent of damage suffered by various structures. This knowledge permits rescue crews to be directed to evacuate first those buildings most likely to collapse with subsequent aftershocks. Another lesson stemming from the Los Angeles earthquake of 1994 was that rupture of underground gas pipelines can cause major damage and loss of life. Fiber optic sensors could provide a faster means of detecting such ruptures and activating fast gas shutoff valves.

A similar problem arose after the terrorist explosion under the World Trade Center in 1992 (Fig. 2.7). It was necessary to know if the tens of thousands of people in the building should be evacuated or not, and to evaluate the extent of the structural damage in order to decide when it was safe for them to return to work. If the building had been equipped with a fiber optic damage assessment system, or if some form of the emergency, or temporary, fiber optic monitoring patch technology discussed later were available, it might have been possible to quickly determine the extent to which the building's structural integrity had been compromised. As it was, the financial loss was greatly increased by the extended period of closure of this huge complex due to uncertainty. With the move to install fiber optic information networks in future buildings, it may be possible to design a dual use for such optical networks so that structural integrity can be addressed, if it should ever be necessary. This may be particularly relevant for the ultratall buildings ($\sim$1 km high) planned for the next century (O'Neill, 1993b).

FRP patching and wrapping represents a promising alternative to retrofitting with steel and concrete jackets for the purpose of rehabilitation and strengthening of existing concrete structures. The new FRP materials are much easier and quicker to install and interrupt traffic less than the conventional technology (Fig. 2.19). Where concrete column supported structures have been damaged by earthquakes, repair by concrete or steel jackets could make the rest of the structure more vulnerable to damage by aftershocks because of the resulting increase in stiffness of the repaired structure. In contrast, FRP rehabilitation and strengthening does not suffer

FIGURE 2.19. Repair and strengthening of a corroded concrete column by FRP material wrap. Used with permission, ISIS Canada.

this problem and can equally be used to repair corrosion damaged concrete structures. The FRP repairs are also immune to corrosion and are more aesthetic.

Meier *et al.* (1993) provided the following reasons for choosing to use FRP rather than bonded steel plates to strengthen a defective concrete structure:

1. Weathering and deicing salts lead to corrosion, which compromises the bond between the steel plates and the concrete structure.
2. FRP is very much lighter than the equivalent steel plate and is much easier to handle at the field site.
3. Because of the weight of the steel plates, their length is generally restricted to about 6 m. Greater lengths require joints, but if these joints are welded, this destroys the adhesive bond.
4. If bonded steel plates are compression loaded, they may fall off in a buckling mode.

They also indicate that FRP sheets can replace steel plates with the overall cost savings stemming from the simplicity of the strengthening method because:

1. They do not corrode.
2. They are easier to handle and can be bonded to the structure by someone in a truck lift, avoiding expensive scaffolding.
3. They are available in endless lengths on bobbins, avoiding joints.
4. CFRP does not fall off when subject to compressive stresses. It also displays outstanding fatigue behavior.

Figure 2.20 provides two cartoon illustrations of some of these advantages. Despite this, reservation exist regarding the performance and durability of FRP repairs. The instrumentation of these FRP rehabilitation and strengthening elements and wraps with integrated fiber optic structural sensing systems could provide a means of continuously monitoring their performance, either intermittently or on-line. The development of

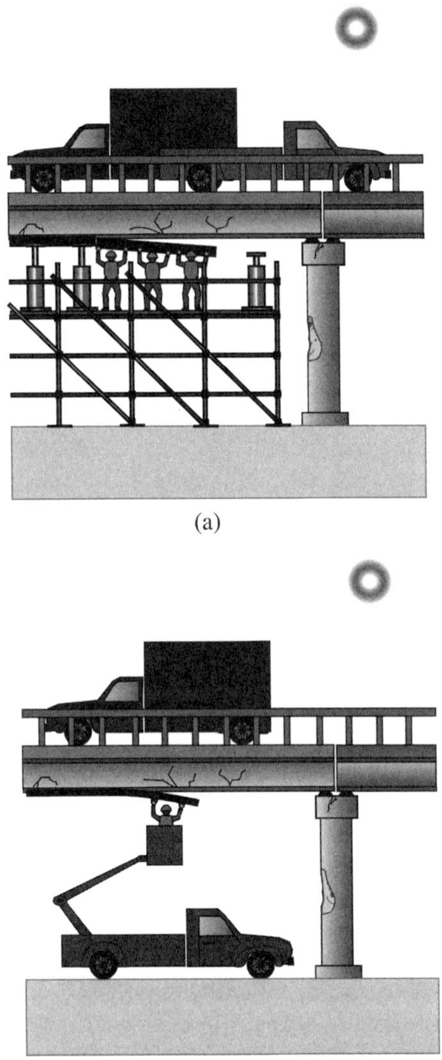

(a)

(b)

FIGURE 2.20. Comparison of bridge repair with steel plates and FRP material.

"smart" FRP rehabilitation and strengthening elements and wraps could also lead to a useful by-product: "temporary fiber optic monitoring patches." These could be used with existing structures after earthquakes, explosions, or impacts, to assess the integrity of the structure and determine the extent of any damage sustained (Fig. 2.21). This technology could also find application when it is necessary to check the response of a bridge to a large truck with a load that is greater than the maximum normally permitted across the bridge. Such incidents often arise when there are detours due to accidents or bridge and road repairs. A schematic summary of the range of measurements possible with a fiber optic structural monitoring system integrated into a bridge is presented as Fig. 2.22.

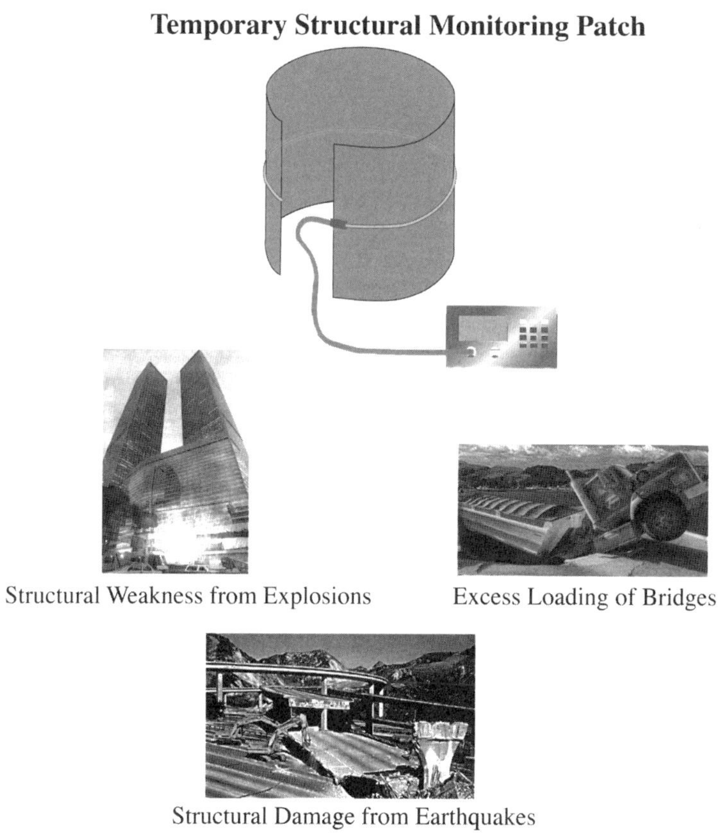

FIGURE 2.21. Application potential for a temporary fiber optic monitoring FRP wrap or patch.

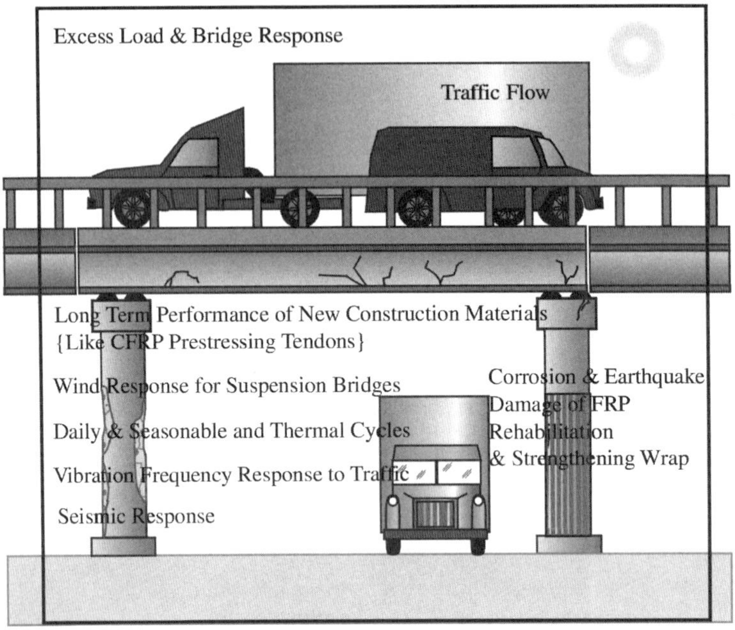

FIGURE 2.22. Overview of application potential for fiber optic integrated structural monitoring in the case of bridges.

2.8 OTHER STRUCTURAL MONITORING APPLICATIONS

Fiber optic sensing systems can be designed to provide thermal as well as mechanical information. Furthermore, the unique ability of an optical fiber to serve as both the sensor and the sensing data channel permits strain and temperature information to be obtained, in principle, from any point along its length. This attribute would allow an optical fiber to provide an early warning of the buildup of a dangerous level of ice on a major electrical power line (Ogawa *et al.*, 1997) during a fierce winter storm (Fig. 2.23) or detect the excess stress at some point along a pipeline arising from an unacceptable ground movement at that location. In both cases a crew could be sent to the correct location to rectify the situation early enough to avoid a disruption of power or the rupture of the pipeline. It is also possible for a single strand of optical fiber woven through one structure, or a group of structures, to warn of the development of a hot spot and provide an indication of its location. This technology could catch fires at a sufficiently early stage that damage and personnel jeopardy are minimized.

FIGURE 2.23. Fiber optic structural monitoring prospects for electrical power lines and oil and gas pipelines.

Monitoring the cure of concrete structures, especially grouted features, is often important in civil engineering, as premature transportation of large concrete girders can cause internal weakness to develop, while extending the period of inactive waiting for the cure to complete is costly in terms of tying up resources such as people and equipment. A disrupted grout created through premature movement or post tensioning can lead to void formation, which can permit moisture, if not water per se, to enter the concrete structure through these narrow channels (Culshaw, 1993). Fiber optic sensors are capable of monitoring both the temperature and the degree of cure for concrete and could therefore indicate when it is appropriate for movement or tensioning of the concrete structure. Concrete curing is always accompanied by the heat of hydration, a process that leads to the hardening of the portland cement. As cooling begins, the outer layers rapidly gain strength and stiffness. For large concrete structures, tensile stresses can build because restraints to the subsequent contraction. Embedded fiber optic sensors could monitor the distribution of both stress and temperature, and this information could be used to control the rate of cooling in order to avoid the build up of excess residual stress within the concrete structure (Méndez *et al.*, 1989).

The rate of destruction of highways and bridges depends more on traffic than on weather, and one large truck can do as much damage as 100,000 automobiles (Newland, 1990). Checking traffic flow and determining truck weight without interruption of the truck's journey is another application for fiber optic sensing technology. Traditionally, trucks are stopped at off-highway operating weight stations distributed along the highways. This is a time-consuming process for the trucker, and if the truck is known to be overweight, the trucker can avoid the sections of the highway that contain these sites. We see that there would be a bonus in instrumenting bridges with integrated fiber optic structural monitoring systems if they could monitor truck weight and distribution of load, and thereby record valuable statistical information on truck traffic and their loads. This would permit the actual load history for an individual bridge to be recorded. With additional equipment, such as video cameras, it would also be possible to identify trucks that use bridges even though they exceed the posted load limit for the bridge.

There is also interest in the prospect of developing a fiber optic sensor that could directly monitor corrosion. Bennett and McLaughlin (1995) developed what amounts to a fiber optic "corrosion fuse." In essence, a length of an optical fiber is held in a tight bend by a wire that is exposed to the same environmental condition as the structure being monitored. At a certain point corrosion causes the wire to break, releasing the optical fiber so it is no longer held in the tight bend (Fig. 2.24). This has the effect of allowing the optical fiber to transmit much more light. The sudden increase in the photodetector's output indicates that some precalibrated degree of corrosion has occurred at that location. It is also possible to use a number of such sensors along a given optical fiber and use *optical time domain reflectometry* (OTDR), or time of flight, to identify which corrosion fuse has blown from the disappearance of its reflected signal and loss of light (Fig. 2.25).

Fuhr and Huston (1998) have been developing a fiber optic sensor that can detect the chemical conditions that promote corrosion of steel

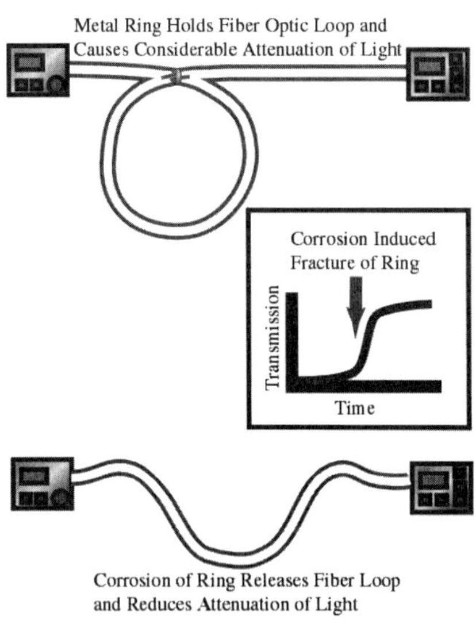

FIGURE 2.24. Principle behind the fiber optic corrosion fuse.

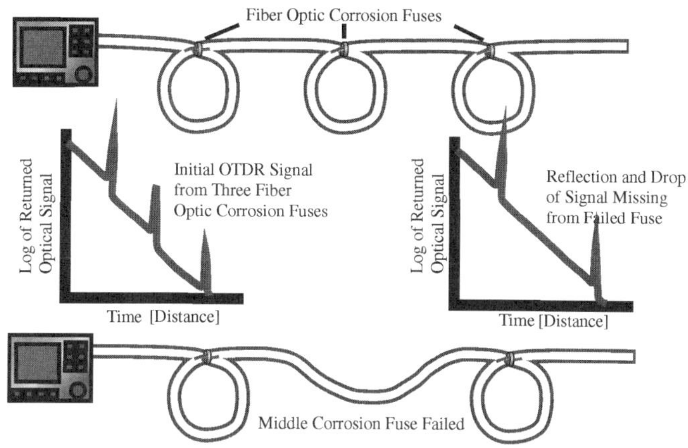

FIGURE 2.25. Optical time domain reflectometry and its application to the fiber optic corrosion fuse.

bars in reinforced concrete. Such sensors could prove especially useful for minimizing costly and dangerous road damage by determining maintenance schedules. They have borrowed a technique from Brian MacCraith of Dublin City University, which involves coating an optical fiber with a ceramic layer that changes color under exposure to alkaline conditions. These color changes can be detected by monitoring the light traveling through the optical fibers, and this in turn permits the alkalinity of the environment to be assessed. Fuhr and Huston (1998) found that they could detect quite early stages of corrosion using this technique.

2.9 WIND POWER AND STRUCTURAL MONITORING

With the growing threat of global warming, the demise of the nuclear power promise, and the eventual depletion of fossil fuels, alternative sources of power generation are becoming a high priority. Wind power is one of the most benign renewable sources of power, and although it is unlikely to satisfy all our needs, it is growing in importance. California's wind farms had reached a capacity in excess of 1500 MW, which represented about 1.5% of the state's electricity, by 1993 (McGowan, 1993). In October 1997, Denmark launched the first phase of its plan to use wind power to generate about half its national consumption by the year 2030. It is planned to operate 500 large turbines generating 750 MW in its shallow coastal waters by the year 2008 (*New Scientist*, 1997). An important step to encouraging wind-power generation in Canada was taken in Alberta in 1988 when the "Small Power Research and Development Act" was passed. This acknowledged a formal role for windpower by giving renewable energy producers the right to generate nearly 18% of the Alberta Interconnected System (Tulley, 1994).

The UKs Central Electricity Generating Board estimates that around 1% of Britain's electricity requirements could be satisfied by this means by 2005, and several windpower farms are under construction or planned. Since windpower increases as the cube of the wind speed, careful siting is very important, and studies have indicated that the minimum wind speed for economic extraction of electrical power is $6.5\,\mathrm{m\,s^{-1}}$. Energy planners have suggested that an area of about $6000\,\mathrm{km^2}$ might be needed to produce 20% of Britain's electricity by wind (Clarke, 1989). The Office of Energy Policy in South Australia has estimated that an original scheme based on a diffuser-augmented wind turbine tested in the 1970s by Northrop Grumman could produce 5.5 times the power of a conventional turbine of the same cross-section (Anderson 1997). The idea was originally abandoned because no material was available at the time that was light and strong enough to withstand the strain generated by the higher wind speed. Structural monitoring might permit such machines to work at the highest efficiency.

Standardization in manufacture and siting have made wind-power farms competitive with conventional power generation today. The introduction of advanced composite material, especially CFRP, makes it practical to build taller, larger-diameter wind-power generators than possible with current technology. Such designs could become even more efficient and reduce the mean cost of power generation by permitting the turbines to make electricity when the wind speed is low or variable.

To minimize cost and maximize efficiency, the blades of wind-power generators need to operate under high loads at the upper part of their wind-speed operating range. Concerns exist regarding the possibility of delamination of CFRP turbine blades, causing catastrophic failure that might damage not only the local generator, but adjacent machines. Since it is possible to monitor the entire strain profile along a blade with a single optical fiber containing a string of strain sensors, the prospect of low-cost monitoring to give due warning of the development of any structural weakness in the blades is quite realistic (Fig. 2.26). Thus, wind-power generators could in the future represent a fertile ground for use of fiber optic structural sensing.

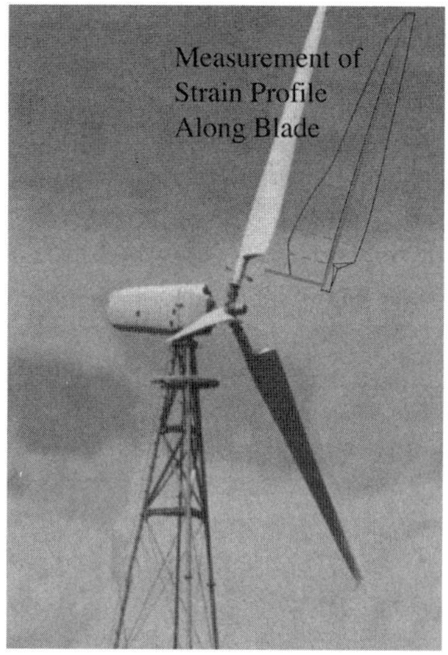

FIGURE 2.26. Structural monitoring for wind power electrical generator.

2.10 MAGNETIC LEVITATION TRAIN MONITORING

High-speed ground transportation systems are highly desirable as they are environmentally more benign than airplanes and also represent a more energy-efficient mode of transportation. Magnetic levitation (maglev) is a revolutionary form of high-speed ground transportation that uses magnetic forces for noncontact suspension, guidance, and propulsion of vehicles at speeds $\sim$500 km h^{-1}. The Linear Express, which is being developed and tested at full scale by the Japanese railway, is magnetically levitated by *electrodynamic suspension* (EDS) by means of the repulsive force generated between superconductive magnets (carried by the vehicle) and currents induced in short-circuited aluminium coils mounted along the guideway (Eastham, 1995). Since the magnitude of the induced current depends on the speed with which the vehicle moves over the guideway coils, the levitation force is speed dependent and the vehicle requires wheels at low speeds (50 to 100 km h^{-1}). The EDS system is characterized by guideway clearances of $\sim$100 mm at high speeds and tends to be dynamically stable.

The German Transrapid 07 involves a partnership between the German government, the German National Railways, Lufthansa Airline, and a number of leading industrial companies such as Messerschmitt

Boelkow Blohm (Najafi and Nassar, 1996). These maglev trains are designed for 400 to 500 km h^{-1}, are capable of climbing 10% grades, and are magnetically levitated by means of an attractive force between the vehicle-borne iron-cored electromagnets and ferromagnetic elements in the guideway. This *electromagnetic suspension* (EMS) system is inherently unstable and must be dynamically stabilized by active feedback control of the magnetic excitation in response to changes in the gap. This gap is 10 to 15 mm, or an order of magnitude smaller than that of the rival EDS, but is nearly independent of speed so no wheels are required at low speed (Eastham, 1995). Both EDS and EMS vehicles have magnetic guidance whereby a lateral displacement generates a strong restoring force toward the center of the guideway. A synchronous long-stator linear motor is used to propel and brake the train. Derailment is not possible because the suspension system wraps around the guideway.

Although the expectations expressed in the 1980s did not materialize because of global recession and the high cost of introducing this new mode of transportation, the Swiss plan to link some of their major cities with a 400 km h^{-1} magnetically levitated train that would run through partially evacuated tunnels because of the rapid increase in energy requirements with speed (*Popular Science*, 1997). Structural stability of the maglev guideways is a key consideration for these trains because of the small clearances for this form of suspension. The presence of large electromagnetic fields make fiber optic sensors a natural choice for monitoring the structural stability of these maglev guideway structures because these sensors are immune to electromagnetic interference.

2.11 AEROSPACE ENGINEERING PROBLEMS

As the passengers of the Aloha Airlines Boeing 737 at Hilo Airport in Hawaii found their seats, none of them could imagine the terrifying ordeal awaiting them. Minutes after takeoff, when the plane was several thousand meters above the ground, a *6-meter* section of the fuselage roof ripped off, sucking out one of the flight attendants. In addition to that fatality, 61 of the 95 passengers were injured. This tragic accident in 1988 riveted world attention on metal fatigue and the "aging aircraft" syndrome. The 737 had been built in 1969 and although it had been regularly inspected, small fatigue cracks had accumulated and been undetected (Friend, 1996). A second, even more deadly incident occurred in February of 1989, when nine people were sucked out through a gaping hole that ripped open the right side of a United Airlines Boeing 747 some six miles above the Pacific Ocean.

These incidents have led to more stringent inspection procedures, which are undertaken at fixed intervals and represent the major part of the aircraft health monitoring program. Checking for cracks is a painstaking task that entails stripping the plane of its seats and internal fittings to reveal the basic structure, and even though visual inspection has undergone some radical advances, the information that can be gathered by this means is still very limited. Ultrasound sensors are often used to detect cracks, but this can take days, or even weeks for a large plane. Within the past few years the concept of *health and usage monitoring* (HUM) has become widely accepted. For this purpose an aircraft can be divided into four systems:

- Aircraft structure
- Hydraulic systems
- Electronics/avionics
- Propulsion systems

HUM for the propulsion system is highly advanced, providing in-flight data on revolutions per minute, vibrations, temperatures, pressure, and rate of fuel consumption. Immediate data analysis of the engine condition provides warnings of any potential malfunction before it can become serious. Initial work on in-flight structural monitoring has been performed with strain gauges bonded to well selected, highly loaded sites on aircraft and helicopters (Boller and Dilger, 1992).

The U.S. Air Force currently conducts an *Aircraft Structural Integrity Program* (ASIP) that provides design and operational inspection guidelines. These guidelines rely on the predictable accumulation of structural damage in response to measurable flight environments, to detect and repair flaws before they become catastrophic. Real-time structural monitoring would permit more effective maintenance as technicians would know from the monitoring system those areas that have either sustained appreciable damage or been excessively loaded. Some current systems use counting accelerometers to track load exceedances for later structural fatigue analysis. The data gathered is suitable for metal structures that suffer fatigue damage from numerous low-stress cycles and can be used to predict the remaining life of the structural component.

Weight is a key factor in the design of most aircraft, especially high-performance aircraft. Weight savings can be directly translated into improved performance, or longer range. Unfortunately, weight reduction normally runs counter to safety assurance, as it lowers the margin of structural safety (Agnes and Silva, 1992). The incorporation of sensing

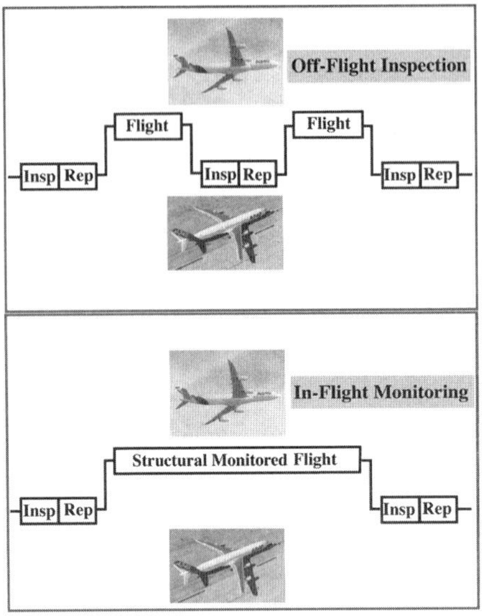

FIGURE 2.27. Schematic illustration of the reduction in airplane downtime resulting from in flight structural monitoring.

systems for real-time health monitoring will allow higher performance and would prevent major failures due to unanticipated rates of damage accumulation between inspection and maintenance checks (Agnes and Silva, 1992). A fast turnaround during maintenance and repair is important for all aircraft to keep operating costs down. This necessitates a fast and thorough inspection, which would be greatly improved in an aircraft with a built-in self-diagnostic capability (Fig. 2.27; Schmidt and Boller, 1992). For military aircraft fast turnaround is even more important to ensure a high level of fleet operation.

2.12 NEW MATERIALS FOR THE AEROSPACE INDUSTRY

Many of the merits spelled out earlier for FRP materials make them valuable for the aerospace industry. Their use in space, in aircraft, and especially in helicopters is increasing each year. Repair patches for aircraft based on boron fibers in a resin matrix are under serious consideration. However, elaborate techniques such as thermography, are needed to reveal regions of debond that could compromise the patch's function. It is possible that a resident fiber optic sensing technique could be developed that would provide such information without taking the aircraft out of service (Section 12.10.2). In addition to their obvious high strength and stiffness-to-weight, FRP materials are less sensitive to low stress cycles, but instead are more vulnerable to the occasional high strain event. This and the fact that FRP structures are designed to strain criteria makes direct strain measurement highly desirable. The inhomogeneity of FRPs ensures that cracks do not propagate as cleanly as they do in homogeneous materials such as metals. FRP materials also display a wider range of electrical characteristics than metals, complicating electrical-based sensors. Their directional structural characteristics lead to nonuniform energy propagation at various levels through the thickness of a structure. Current measurement systems based on acoustic wave interrogation have limited diagnostic capabilities in FRP materials, as the waves propagate in a complex pattern due to the material's inhomogeneity and directional properties. This situation is made even worse when the structure includes holes or steplike features.

Often FRP structures are fabricated in layers (layup or filament-wound), making it relatively easy to embed optical fiber sensors during manufacture. The fact that optical fibers are typically the same size as the thickness of one FRP ply (Fig. 2.28) makes embedment a viable option, providing protection to the sensor and excellent siting for the appropriate measurements. Unfortunately, embedment can lead to a reduction in structural strength without due diligence (Sirkis and Dasgupta, 1995; Jensen and Sirkis, 1995) and accurate strain sensing requires precise knowledge on the depth of the optical fibers relative to the neutral axis of any given structure. Ingress/egress and optical connectivity represent additional issues to be addressed later.

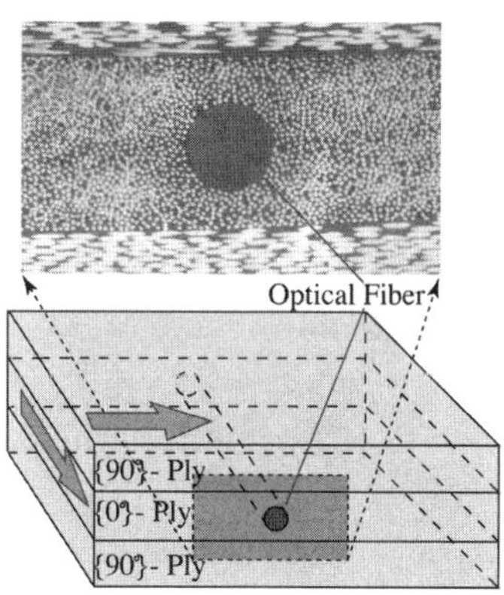

FIGURE 2.28. Embedment of fiber optic structural sensors within multi- ply FRP structure.

2.13 FIBER OPTIC MONITORING OF AIRCRAFT

The concept of self-monitoring structures is clearly very appealing from both an economic and a safety point of view. However, unlike a bridge, the airframe of a commercial or military airplane is extremely complex and can involve hundreds to thousands of parts. Clearly, there is no way they all could, or need, be monitored. There are, nevertheless, a sizable number of highly loaded critical parts that could benefit from SIFORS. This is especially the case for helicopters or propeller-driven aircraft, which are subject to considerable vibrational loading. In-service fiber optic structural monitoring could lead to improved aircraft performance, reduced fuel consumption, and minimal downtime and inspection costs. Furthermore, continuous monitoring of structural loads, especially in high-stress regions such as wing roots of aircraft, could lead to ongoing improvements in design that would avoid the creation of excess strains. It is clear that SIFORS could substantially reduce life-cycle cost as it would provide the means of checking structures quite literally from birth to retirement.

An overview of some of the possible sites for structural monitoring is schematically presented as Fig. 2.29. Translating the concept into everyday practice is, however, a formidable challenge, especially with the requirement that it be economic. Before this technology can become a commercial reality, it has to be proven that the sensing systems will be reliable over the life of the structure, and when damaged they can be repaired, or replaced. It is also crucial that the presence of the sensing system in no

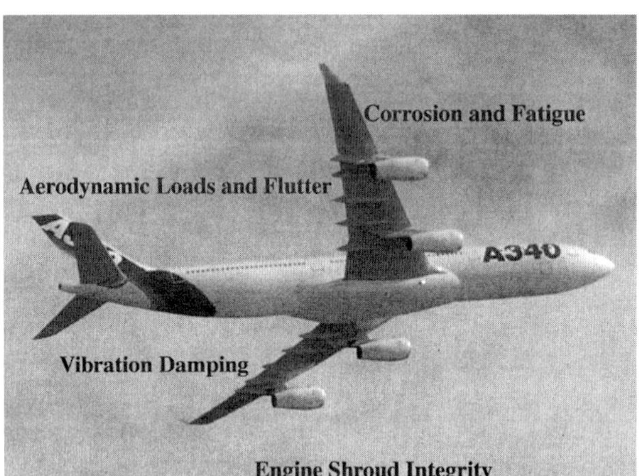

FIGURE 2.29. Schematic illustration of the possible aircraft applications for a fiber optic structural monitoring system.

way compromise the material's structural property or require more expensive procedures for construction of the airplane.

Managing the enormous volume of data generated by a comprehensive sensing system aboard an airplane represents an extraordinary challenge in its own right. This sensing data will have to be processed and analyzed by a computer, classified as to its importance and priority, stored, and, when necessary, brought to the attention of maintenance and repair crews, or in the event of a real-time crisis, presented to the pilot in a dramatic manner. Clearly, false alarms would be extremely unpopular as well as uneconomic, so reliability would have to be guaranteed.

It is possible that expert systems based on neural networks could be used to provide damage assessment and structural health status. SIFORS could provide a variety of responses that include strain maps, vibration modes, and temperature profiles, which could be compared with healthy signatures for specified loading events. The system would learn to detect anomalous responses and their interpretation in terms of structural weakening. This information could be passed to the pilot or stored and passed to maintenance crews after the flight, depending on the perceived danger. The ability of the neural networks to infer possible damage in response to new inputs that the system has not yet been trained to recognize is of crucial importance and crosses the boundary to intelligent systems. Depending on the extensiveness of this structurally integrated sensing system, all critically important individual components could be instrumented and monitored. The prospects of having on call the actual load history of each critical component in an aircraft is certainly daunting, but is not beyond the realm of possibility.

Structurally integrated sensors and actuators form key elements of any smart adaptive structure, and weight considerations have indicated the potential superiority of distributed sensing and control over high authority point actuation (Warkentin *et al.*, 1992). The U.S. Air Force became interested in the prospect of embedding radar antenna directly into the structure of its military aircraft in the late 1970s and coined the phrase "smart skins" to designate this concept in their Project II Forecast. Integrating sensing systems into the airframe of an aircraft is the logical first phase in the evolution of multifunctional materials (Mazur *et al.*, 1988; Measures, 1989; Wada, 1990). In 1995 a team led by Ron Barrett of the University of Alabama built a small revolutionary model aircraft called "Mothra." Instead of conventional aerodynamic control surfaces such as ailerons and flaps, this plane changed the shape of its tail plane and tail wings by means of a distributed piezoelectric actuators (Fig. 1.3). This method of aerodynamic flight control holds the promise of great weight savings and much simpler mechanical structures, eliminating the massive network of hydraulic systems and heavy, powerful motors.

CHAPTER 3

Introduction to Lightwaves

3.1 BACKGROUND AND OVERVIEW

Few of us can remember what life was like without television, telephones, and computers. Electronics pervades our lives in many ways—some obvious, others not so obvious—from the cars we drive to the supermarket we visit. So what is electronics? To appreciate the answer we first need to recognize that all matter can be broken down into atoms, which themselves can be further divided. At the heart of each atom is an extremely small ($\sim 10^{-15}$ m) positively charged nucleus. The nucleus comprises over 99% of the mass of the atom. Swirling around the nucleus, at a relatively large distance (10^{-10} m), are just the right number of negatively charged electrons to make the atom appear electrically neutral (Fig. 3.1). *Electronics* refers to the use and manipulation of free electrons, i.e., electrons that have been sufficiently energized to rip free of their atom and move in the space between atoms under the influence of any applied electric field.

In a similar vein we use the term *photonics* to refer to the use and manipulation of photons, the smallest packets of light. There is an important difference between electrons and photons that we must recognize. Electrons exist independent of their motion and can be found in various states of motion. Photons can exist in only one state: traveling at the speed of light! The term *light* is usually restricted to mean electromagnetic energy with a wavelength (color) that is in the range detected by the human eye. The complete electromagnetic spectrum runs from gamma rays at very short wavelengths to the far infrared at long wavelengths. In this book "light" is often used in the broader sense to mean electromagnetic radiation. It is noteworthy that humans used light long before the existence of the electron was appreciated. For example, the ancient Romans sometimes communicated over large distances with shields polished to serve as mirrors (Fig. 3.2). It is interesting to note that as we

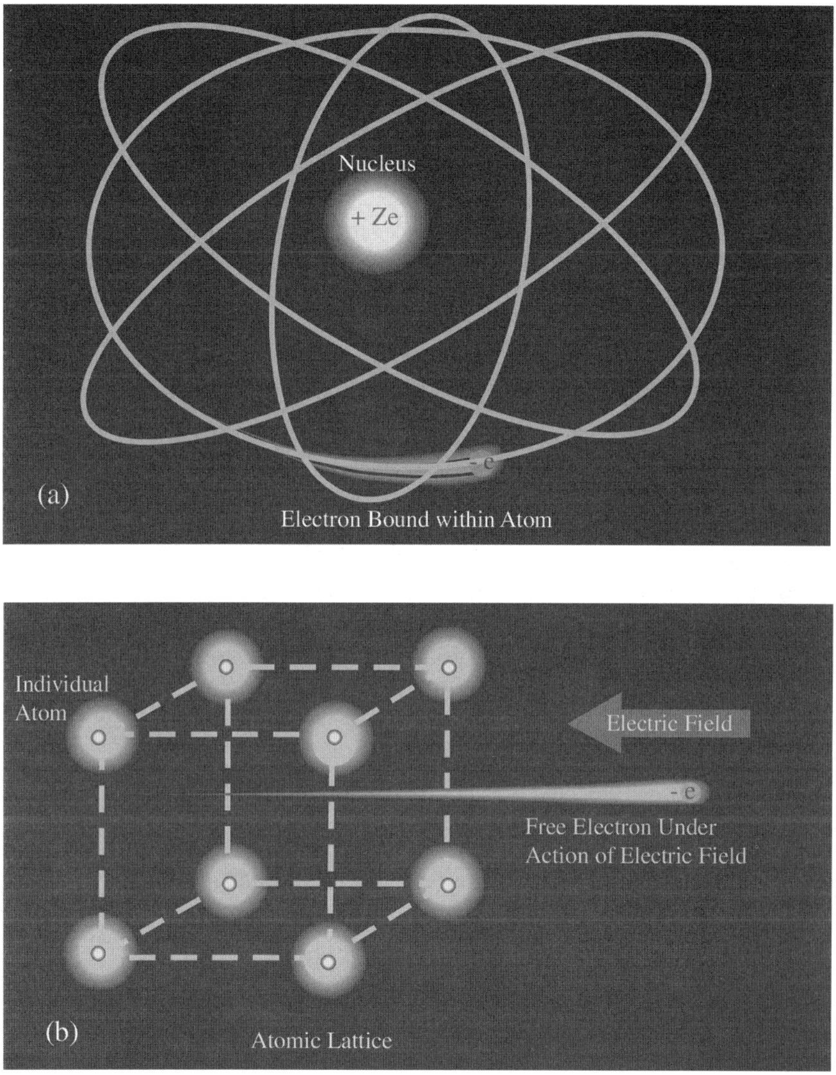

FIGURE 3.1. (a) Bound electron, (b) free electron within atomic lattice moving under the action of an applied electric field.

commence the twenty first century we are witnessing a spectacular resurgence in the use of light, especially for communications.

Two major technological developments have made optical (lightwave) communications so attractive today that it challenges the dominance of wires and electronics. First came the invention of the laser, an extraordinarily powerful source of light, then the invention of the optical fiber, a form of glass so transparent that you could see through a window of this material that is many kilometers thick. The first *laser* was produced in 1960 by Theodore Maiman and used a small crystal of ruby to convert the undisciplined burst of light from a helical flash lamp into a beam of light that can cut through matter (Fig. 3.3). Although the principle of guiding light along an optical fiber was recognized in the early 1960s (Snitzer, 1961), it was the discovery of low-loss silica-glass fiber (Kao and Hockman, 1966; Fig. 3.4) that eventually led to the technology of optical fiber communications. Many subsequent advances have made fiber optic technology so important today that we are witnessing, essentially, the rewiring of the industrialized world.

The original impetus for fiber optics was the insatiable need for telephone voice communications. Fiber optics initially offered longer intervals of transmission across the oceans with lower-weight cables. The higher bandwidth prospect was also extremely important, as one optical fiber could carry the same number of voice channels as hundreds of copper wires. Today, however, it is the need for hundreds of television channels, the explosive rise in the transmission of computer-based information, especially images, and the ever-increasing use of the Internet that is driving the demand for ever higher bandwidth in optical fibers.

Information can be transmitted along an optical fiber in a number of different ways (Fig. 3.5). The simplest involves changing the magnitude of light with time. We call this *analog modulation*. In *digital modulation* a continuous stream of light pulses carry information by the presence, or absence, of pulses in the sequence. This is the optical analogue of digital electronic communication. *Spectral modulation* refers to encoding the information in variations of the wavelength of the light. This usually involves shifting the wavelength of a very narrow spectral width signal in sympathy with the information. If the light signal is modulated at a given frequency (usually in the range of kilohertz to gigahertz), then information can be encoded on this carrier frequency through changes of its frequency. This is termed *frequency modulation*. Finally, the angular orientation (about the axis of the optical fiber) of the plane of the electric field that constitutes the lightwave can be varied to carry information, and this is termed *polarization modulation*.

Although optical fibers represent a near-perfect communication medium, they can be made sensitive to their environment. This allows them to serve as both a sensing element and the transmission conduit for optical sensing signals. In this book we shall restrict our focus to the use of optical fibers for "structural monitoring," or the measurement of strain and temperature. It should be noted in passing, however, that fiber optic

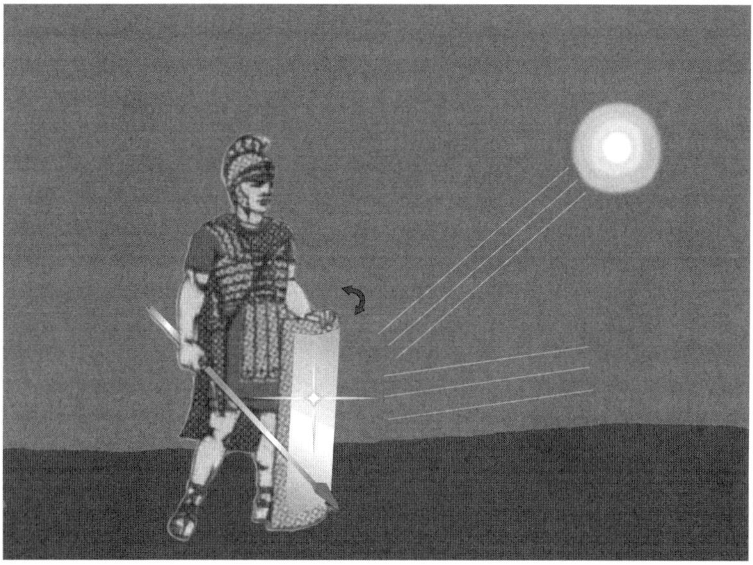

FIGURE 3.2. Roman soldiers communicated using their polished shields to modulate reflected sunlight.

sensors have been made to respond to very many parameters, including strain, temperature, pressure, moisture, hydrogen gas, deformation, acceleration, acoustic waves, electric and magnetic fields, and electric currents (Dakin and Culshaw, 1988, 1997; Udd, 1991).

In this chapter we introduce the basic concepts of lightwave theory. This will prepare us for the next chapter on light sources and detectors

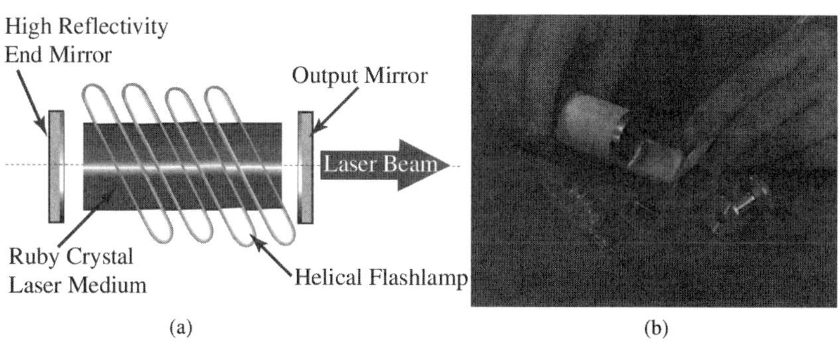

FIGURE 3.3. (a) Schematic of a ruby laser. (b) Original Ruby laser of T. Maiman. From Nehrich Jnr., R.B., Voran, G.I., Dessel, N.F., *"Atomic Light Lasers,"* Sterling Publishing, 1967. Now out of print.

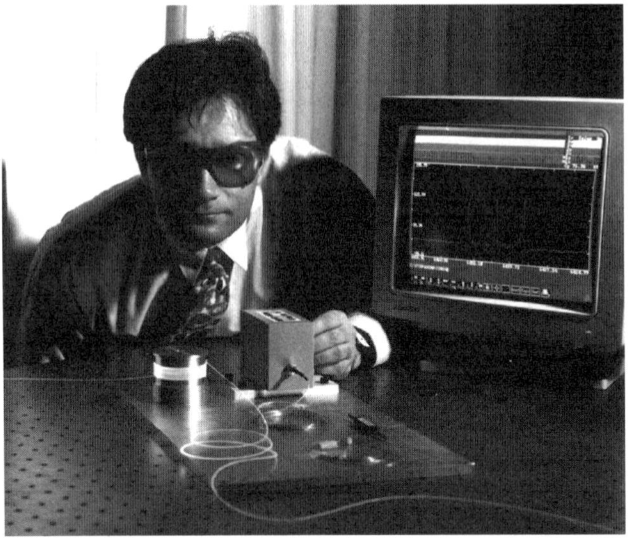

FIGURE 3.4. An illuminated low-loss optical fiber used today in telecommunications.

and the subsequent chapter on optical fibers. First we introduce the electromagnetic wave equation, that describes the propagation of lightwaves. We then discuss the polarization of light and explore interference and interferometers. Finally, we look into the coupling of light between optical fibers and end with a discussion on the new and important topic of Bragg grating reflection.

3.2 ELECTROMAGNETIC RADIATION

Electromagnetic radiation is a form of energy that, once created, propagates through space or matter until it interacts and is converted into some other form of energy. We have all experienced the pleasant feeling of being warmed by the sun. In this instance our skin is absorbing the sun's electromagnetic radiation, here truly in the form of light, and transforming it into an elevation of our skin temperature. The great scientist James Clerk Maxwell, in the mid-1880s, described electromagnetic radiation by a wave equation that was derived from the fundamental set of four equations that bear his name (Saleh and Teich, 1991). These equations describe all electromagnetic phenomena in terms of an electric field (**E**) and a magnetic induction field (**B**). Both are vector quantities possessing

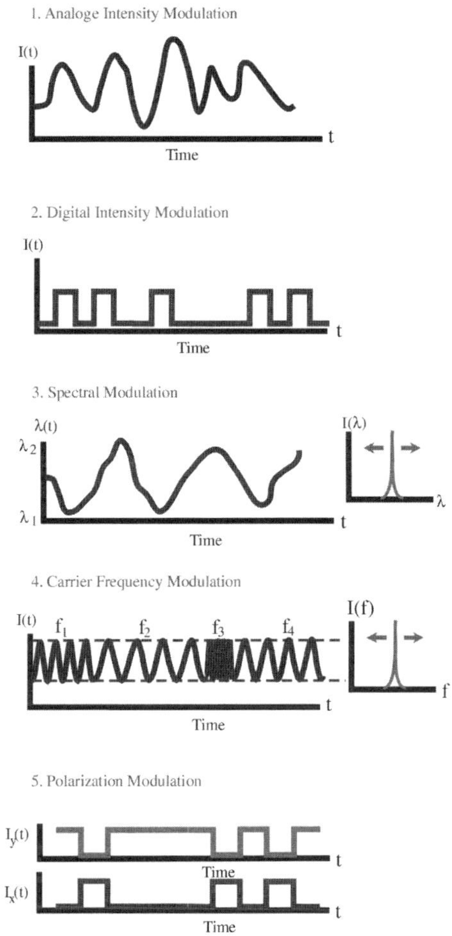

FIGURE 3.5. Different methods of encoding information transmitted by light along an optical fiber.

magnitude and direction. When a free electron, that is to say one that is not bound to an atom, is subject to an electric field **E**, the electron will experience a force in the direction opposite (because the electron is negatively charged) to the field. The energy expended on the free electron, W, by the electric field in moving the electron from position s_1 to s_2 (Fig. 3.6) can be written

$$W = \int_{s_1}^{s_2} (-e)E ds. \tag{3.1}$$

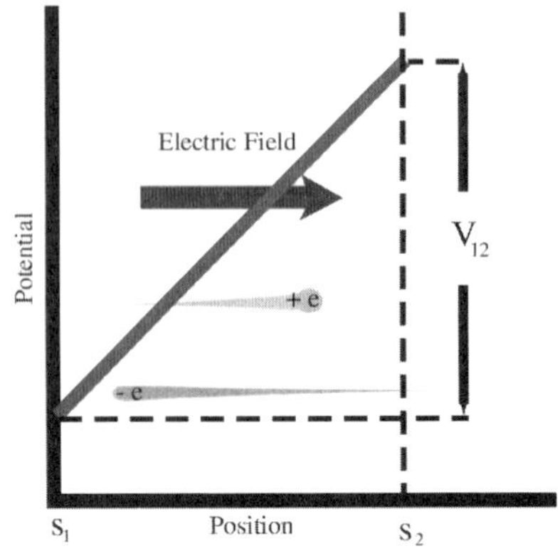

FIGURE 3.6. Electric field created by a difference in potential between two locations and the resulting force acting on a free electrons and a positive ion.

However, the change in voltage, V_{12}, between positions s_1 and s_2 is given by the integral of the field over the distance, i.e.,

$$V_{12} = \int_{s_1}^{s_2} E ds. \tag{3.2}$$

Thus, we can write

$$W = (-e)V_{12}. \tag{3.3}$$

The change in the energy of an electron can thus be expressed in terms of *electron volts* (or *eV*), where 1 eV is the energy acquired by an electron in falling through a voltage difference, or potential, of −1 volt.

An electron that is part of an atom is constrained by the laws of quantum mechanics such that it cannot absorb an arbitrary amount of energy. This can be understood in terms of the quantized model of the atom and the quantization of electromagnetic radiation (Fig. 3.7). According to quantum mechanics, the smallest package of electromagnetic energy at a *frequency* ν is $h\nu$, where h is *Planck's constant* and has the value 6.626×10^{-34} J s. Furthermore, the electron can only exist in certain allowed energy states of the atom ($U_1, U_2, U_3, \ldots$), see Fig. 3.7. For an electron to absorb energy from passing electromagnetic radiation, the radiation must contain a frequency component ν that satisfies the relation

$$\nu = \Delta U/h, \tag{3.4}$$

where ΔU corresponds to the energy difference between two permitted energy states of the atom.

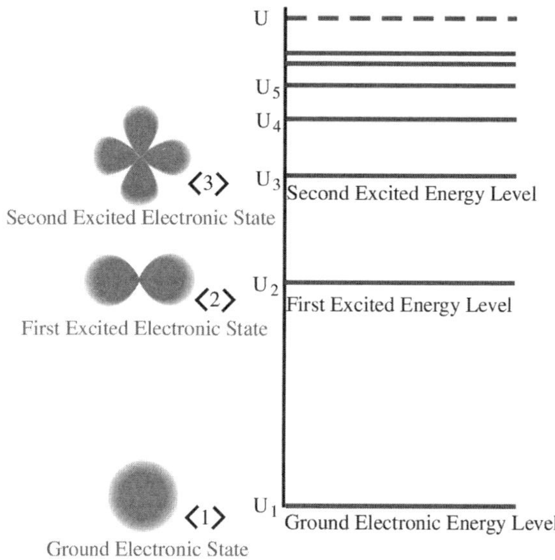

FIGURE 3.7. Energy levels within an atom and some examples of the corresponding electronic states.

The electron tries to always reside in the lowest state but can be excited to one of the higher states by a number of processes, including the absorption of the appropriate amount of electromagnetic radiation (Fig. 3.8). Such an excited electron will then spontaneously return to the *ground* (lowest) state, U_1, by either the spontaneous or stimulated emission of a photon of electromagnetic energy of frequency, v (Fig. 3.9) or by some other relaxation process, such as lattice vibrations if the atom is part of a solid. Stimulated emission will be discussed in Section 4.2.

One of the most common sources of light are incandescent, or hot, bodies, such as the filament within a light bulb or our sun. In these sources electrons are excited into higher states by collisions of atoms possessing large amounts of random kinetic energy due to their elevated temperature. The propagation of this light is governed by the *electromagnetic wave equation* (Saleh and Teich, 1991), which in free space (i.e., a vacuum) takes the form

$$\nabla^2 F = \frac{1}{c^2} \frac{\partial^2 F}{\partial t^2}, \tag{3.5}$$

where c is the velocity of light in free space and $F = (\mathbf{E} \text{ or } \mathbf{B})$ and represents either the electric or the magnetic field, as both satisfy this

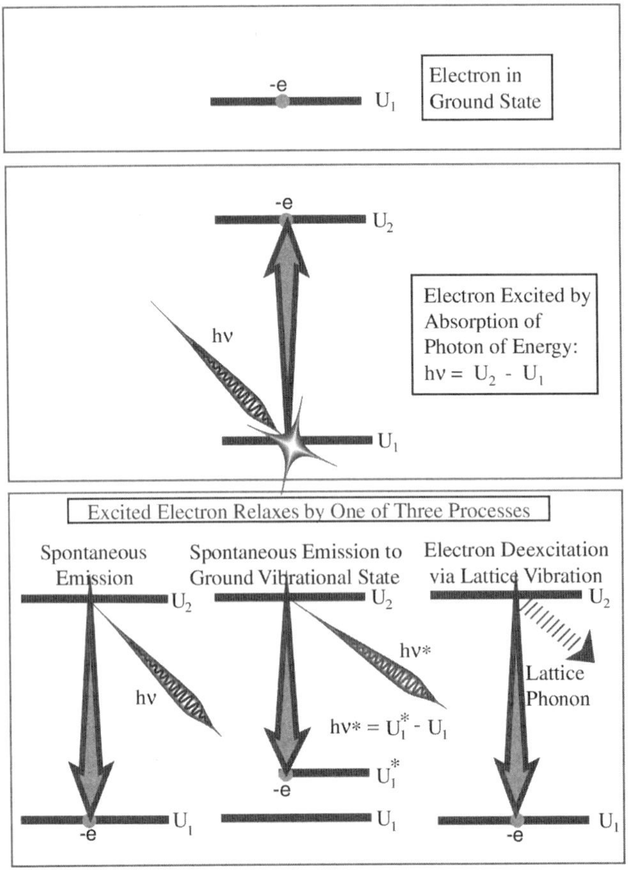

FIGURE 3.8. Excitation of bound electron by the absorption of light of the appropriate wavelength and the subsequent three modes of relaxation.

wave equation. Although **E** and **B** are vectors of the form: (E_x, E_y, E_z) and (B_x, B_y, B_z), respectively, in Cartesian coordinates, each of their components also satisfies (3.5), so we shall use **E** to generally represent either of these components. For a monochromatic (i.e., single frequency) plane wave traveling primarily in the z-direction, the **E**-*field* solution of Eq. (3.5) takes the form

$$(E, x, y, z, t) = E(x, y)e^{i(kz-\omega t)} + E(x, y)e^{-i(kz-\omega t)}, \tag{3.6}$$

where $E(x, y)$, is the *amplitude* of the wave and represents the spatial part of the **E**-field that depends upon its location in the x–y plane, $k = 2\pi/\lambda$ is the is the *free-space propagation constant*, $\omega = 2\pi\nu$, is the *angular frequency*, and λ is the *free- space wavelength* of the wave. The second term on the right side of Eq. (3.6) is the *complex conjugate* (*c.c*) of the first term. The plane wave nature of this wave is evident from the independence of $E(x, y)$ on z; see Fig. 3.9. The magnetic field, **B**, has the same form of solution, except that Maxwell's equations require its direction to be perpendicular to that of the **E**-*field* in an isotropic medium (Saleh and Teich, 1991). If the orientation of the **E**-*field* remains fixed as the wave propagates, then we refer to this as a *linearly polarized wave* (Fig. 3.10). We shall see later that light propagating along an optical fiber can exist in a more complex state of polarization.

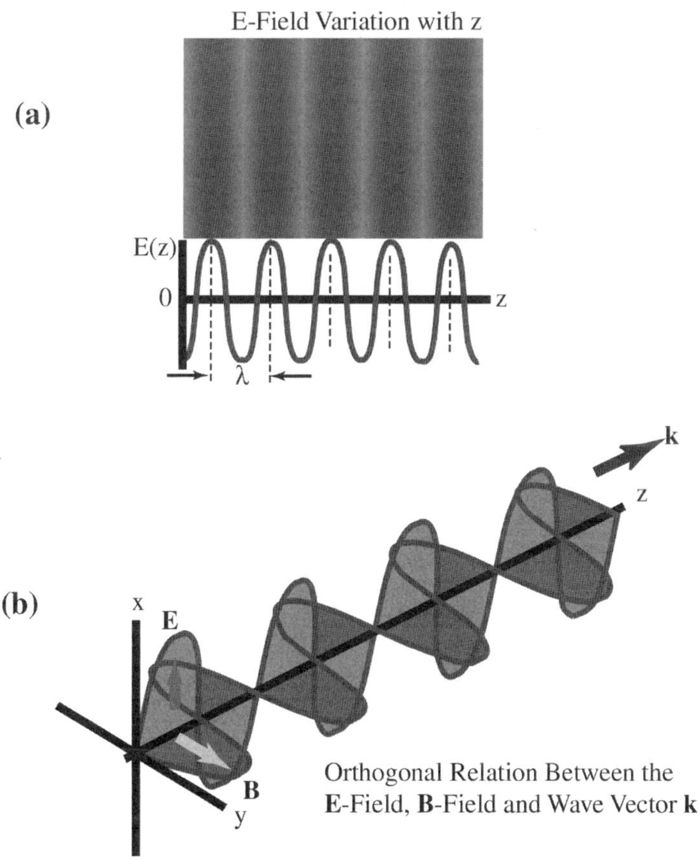

FIGURE 3.9. For a plane, monochromatic electromagnetic wave traveling in the z-direction: (a) presents the **E**-field variation along the direction of propagation, and (b) the orthogonal **E**- and **B**-fields of this plane wave.

FIGURE 3.10. Fixed E-field orientation seen for linearly polarized lightwave.

The relationship between the wavelength and frequency of an electromagnetic wave can be obtained by substitution of the plane wave solution, Eq. (3.6), into the wave equation. This yields

$$k^2 = \omega^2/c^2. \tag{3.7}$$

In the case of a homogeneous isotropic dielectric material, such as glass, we introduce the *general propagation constant*, β, which depends upon the material properties. Note that the real part of the right side of Eq. (3.6) describes the physical value of the **E**-*field*. Under these conditions the plane wave solution takes the form

$$E(x, y, z, t) = R_e : E(x, y)e^{i(\beta z - \omega t)} = E(x, y)\cos\{\beta z - \omega t\}, \tag{3.8}$$

and the *electromagnetic wave equation* for a material in which the *velocity of light* is υ takes the form

$$\nabla^2 F = \frac{1}{\upsilon^2}\frac{\partial^2 F}{\partial t^2}. \tag{3.9}$$

Substitution of (3.8) into (3.9) now yields

$$\beta^2 = \omega^2/\upsilon^2, \tag{3.10}$$

in which case, if we introduce the *index of refraction* of the material,

$$n \equiv c/\upsilon, \tag{3.11}$$

and combining (3.7) with (3.10), we see that

$$\beta = nk. \tag{3.12}$$

As we shall see later, β is an important quantity when considering the propagation of light along an optical fiber. In essence, β describes the change in phase of the wave as it propagates unit distance. Thus, at a distance z, the *phase* of the wave is different from its value at $z = 0$ by $\phi = \beta z$. A change of phase of 2π, or a multiple thereof, returns the **E**-*field* to its value at $z = 0$ (Fig. 3.11), i.e., $E(\phi + 2m\pi) = E(\phi)$ where $m = 0, 1, 2, \ldots$.

As indicated earlier, the electromagnetic radiation spectrum extends from the shortest wavelengths (less than 0.1 nm), referred to as *gamma rays*, through to the longest wavelengths (~100 μm) in the far infrared. Since we are primarily concerned with light in optical fibers, we shall be interested in the near infrared—specifically, 1.3 and 1.5 μm, the wavelengths for which optical fibers have the best transmission characteristics (see Chapter 5 and specifically Fig. 5.15).

The very high frequencies of lightwaves, especially in the visible and near infrared, precludes photodetectors from responding directly to the **E**-field. Instead, photodetectors respond to the time average of the lightwave energy falling upon their detection surface. Since lightwave energy is proportional to the square of its **E**-field (Born and Wolf, 1975), we introduce the term *intensity*, I (or irradiance), defined as the light energy flowing per second across unit area perpendicular to the propagation direction of the wave. This intensity can be shown to be given by the relation

$$I \equiv \varepsilon_0 c \langle \mathbf{E} \cdot \mathbf{E} \rangle \qquad (\mathrm{Wm}^{-2}). \tag{3.13}$$

Note that $\langle \mathbf{E} \cdot \mathbf{E} \rangle$ represents a time average of the product of the **E**-field vector with its complex conjugate over a time interval much longer than the period of the wave, but usually much shorter than times of interest in any experiment. For a harmonic-plane electromagnetic wave of the sort indicated in Eq. (3.6), $\langle \cos^2(\omega t) \rangle = \langle \sin^2(\omega t) \rangle = \frac{1}{2}$, while $\langle \cos(\omega t) \sin(\omega t) \rangle = 0$. Using this in the square of (3.8) leads us to write (3.13) in the form

$$I = \tfrac{1}{2} \varepsilon_0 c E_0^2 \qquad (\mathrm{Wm}^{-2}), \tag{3.14}$$

where we have introduced E_0 to represent the time-independent amplitude of the wave.

A more fundamental quantity in regard to a light source is the *radiance*, $J(\chi, \varphi)$ (or brightness), which is defined as the flow of electromagnetic energy per second per unit solid angle per unit projected area, i.e.

$$J(\chi, \varphi) = \frac{1}{\cos \chi} \cdot \frac{d^2 \mathrm{P}}{dS d\Omega} \qquad (\mathrm{Wm}^{-2}\ \mathrm{str}^{-1}), \tag{3.15}$$

where P is the power (W) radiated by the source, χ is the angle from the normal to the source surface, φ is the angle around that normal, and d is the element of solid angle into which the optical power is radiating, see

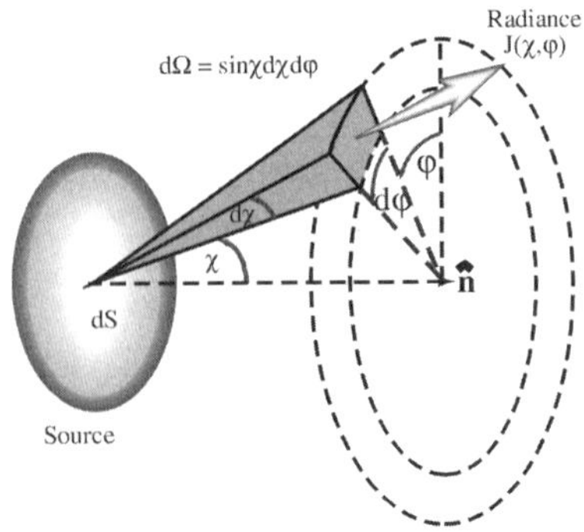

FIGURE 3.11. Radiance from a light source.

Fig. 3.11. In this case the relationship between the radiance of the source and the intensity in the radiated lightwave takes the form

$$I = \int_{\chi=0}^{\pi/2} \int_{\varphi=0}^{2\pi} J(\chi, \varphi) \cos \chi d\Omega \qquad (\mathrm{Wm^{-2}}). \tag{3.16}$$

For isotropic radiation, $J(\chi, \varphi) = J$, and $I = \pi J$. We also need to introduce the *spectral radiance*,

$$J(v) = \frac{dJ}{dv} \qquad (\mathrm{Wm^{-2}\ str^{-1}\ Hz^{-1}}), \tag{3.17a}$$

and the *spectral irradiance*,

$$I(v) = \frac{dI}{dv} \qquad (\mathrm{Wm^{-2}\ Hz^{-1}}). \tag{3.17b}$$

These terms are used to describe the spectral distribution of the lightwave in terms of unit frequency bandwidth (*Hz*). Note that Eq. (3.7) relates the free-space wavelength, λ, and the frequency, v, of light through the equation

$$v\lambda = c \qquad (\mathrm{ms^{-1}}). \tag{3.18}$$

3.3 BIREFRINGENCE AND POLARIZATION

When a plane, monochromatic, harmonic electromagnetic wave propagates in the *z*-direction through an isotropic medium (properties are independent of the direction of propagation), it can be described by Eq. (3.6). By contrast, if the lightwave propagates in the *z*-direction within a *birefringent medium,* then the index of refraction experienced by the wave, and subsequently the propagation velocity, depends on the orientation of the **E**-field. For the situation illustrated in Fig. 3.12, where $n_x \neq n_y$, the velocity of the E_x and E_y components of the lightwave are different. Under these conditions the lightwave is described by a pair of equations that describe what happens to each of the **E**-field components. This can be expressed in the form of a *matrix equation,*

$$\tilde{E}(z,t) = R_e\begin{bmatrix} E_x(z,t) \\ E_y(z,t) \end{bmatrix} = R_e\begin{bmatrix} E_{xo}e^{i(\beta_x z - \omega t + \delta_x)} \\ E_{yo}e^{i(\beta_y z - t + \delta_y)} \end{bmatrix}, \tag{3.19}$$

where the amplitude of the light wave is

$$E_o = (E_{xo,}e^{i\delta_x}, E_{yo,}e^{i\delta_y}, o), \tag{3.20}$$

and

δ_x, δ_y are the respective *phases* of the **E**-*field* at $z = 0$ and $t = 0$,

β_x, β_y are the respective *propagation constants* i.e., $\beta_j = n_j k$ with $(j = x, y)$, and

E_{xo}, E_{yo} are the respective *positive field amplitudes.*

We can simplify (3.19) by introducing the *total phase retardation* ("phase" for short,

$$\phi(z,t) \equiv \beta_x z - \omega t + \delta_x, \tag{3.21}$$

which allows us to write

$$\begin{bmatrix} E_x(z,t) \\ E_y(z,t) \end{bmatrix} = \begin{bmatrix} E_{xo}\cos\phi(z,t) \\ E_{yo}\cos\phi(z,t) + \delta(z) \end{bmatrix}, \tag{3.22}$$

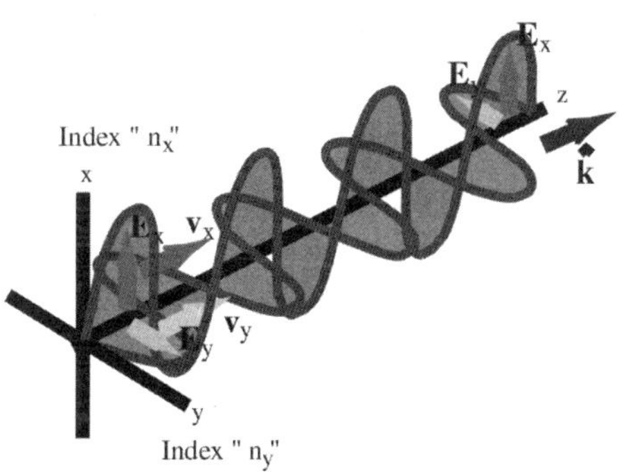

FIGURE 3.12. Birefringence stemming from different indices of refraction for the x- and y-components of the E-field.

where

$$\delta(z) = (\beta_y - \beta_x)z + \delta_y - \delta_x \tag{3.23}$$

is the termed the *differential phase.*

In the case of an "isotropic" medium,

$$\beta_y = \beta_x = \beta \tag{3.24}$$

and we can write

$$\begin{bmatrix} E_x(z,t) \\ E_y(z,t) \end{bmatrix} = \begin{bmatrix} E_{xo}\cos\{\phi(z,t)\} \\ E_{yo}\cos\{\phi(z,t)+\delta\} \end{bmatrix}, \tag{3.25}$$

where the *differential phase* now takes on the value

$$\delta \equiv \delta_y - \delta_x \tag{3.26}$$

and the *phase*

$$\phi(z,t) \equiv \beta z - \omega t + \delta_x. \tag{3.27}$$

The *polarization angle* θ is defined as the ratio of the y- and x-components of the **E**-field (Fig. 3.13),

$$\theta(z,t) \equiv \tan^{-1}\left[R_e\left\{\frac{E_y(z,t)}{E_x(z,t)}\right\}\right], \tag{3.28}$$

which can be written using (3.22) in the form

$$\theta(z,t) = \tan^{-1}\left[\frac{E_{yo}\cos\{\phi(z,t)+\delta\}}{E_{xo}\cos\{\phi(z,t)\}}\right]. \tag{3.29}$$

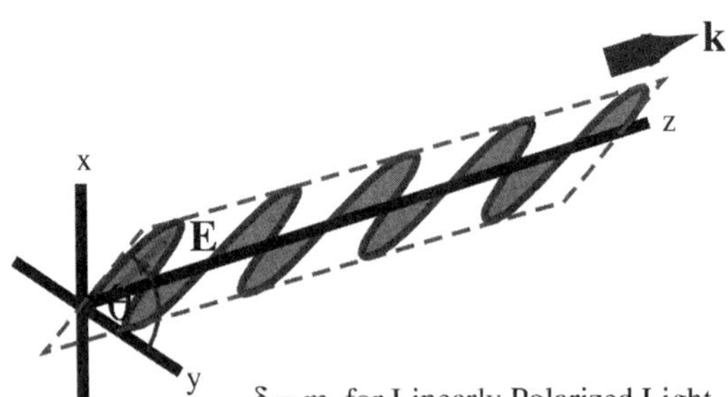

FIGURE 3.13. Linearly polarized light described by the polarization angle θ.

Equation (3.29) is the key to determining the *state of polarization* (SOP), i.e., the relation between the polarization angle θ and the phase ϕ. Indeed, we can write

$$\tan\{\theta(z,t)\} = \gamma[\cos\delta - \sin\delta\tan\{\phi(z,t)\}], \tag{3.30}$$

where we have introduced

$$\gamma \equiv E_{yo}/E_{xo}. \tag{3.31}$$

The most common form of polarized light is termed *linearly polarized light* (LPL). This arises when

$$\delta = m\pi \qquad (m = 0, 1, 2, \ldots). \tag{3.32}$$

Under these circumstances,

$$\tan\theta = \gamma\cos(m\pi) = (-1)^m\gamma, \tag{3.33}$$

and the polarization angle, θ, is seen to be independent of z or t so the plane of the **E**-field is fixed. That is to say, the plane of polarization does not change with time or space (Fig. 3.9).

A *linear polarizer* is a form of optical filter that permits only the component of light with its **E**-field parallel to the transmission axis of the device to be efficiently transmitted; see inset to Fig. 3.14. By contrast, if the plane of polarization of linearly polarized light is orthogonal to the transmission axis of the polarizer, then no light is transmitted. Thus, in Fig. 3.15 we see that the vertical polarizer only transmits the E_x component of the light wave, while the horizontal polarizer blocks the transmission of this component. There are a number of different types of *polarizer*, ranging from a simple *dichroic sheet* of stretched polyvinyl that absorbs the component of **E**-field perpendicular to its transmission axis (Saleh and Teich, 1991), to a length of special polarizing optical fiber that strongly attenuates one **E**-field component relative to its orthogonal component.

Under special circumstances the magnitude of the **E**-field vector does not change, but the plane of the **E**-field is observed to rotate about the direction of propagation. This state of polarization is termed *circularly polarized light* (CPL) and arises when

$$\delta = \pm\left(\frac{2m+1}{2}\right)\pi \qquad (m = 0, 1, 2, \ldots) \tag{3.34}$$

and

$$\gamma = 1, \qquad \text{i.e.,} \qquad E_{yo} = E_{xo}.$$

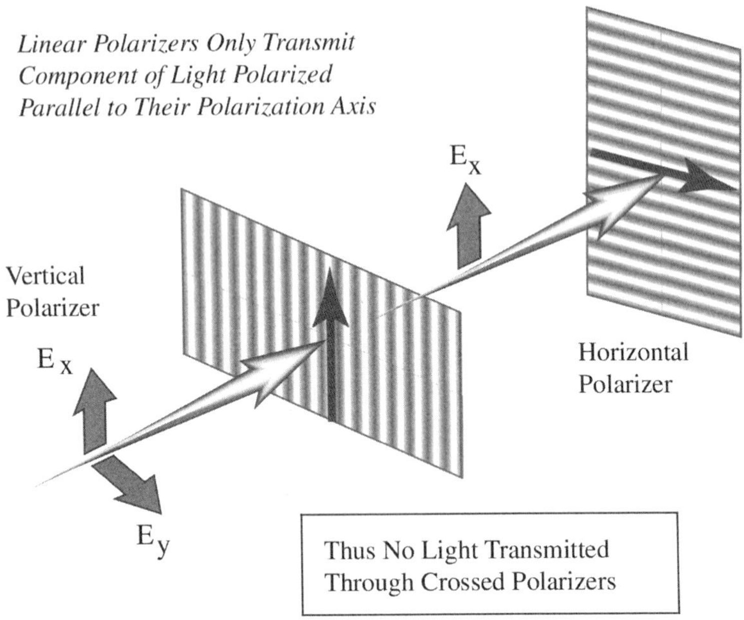

FIGURE 3.14. Light-blocking action of crossed polarizers.

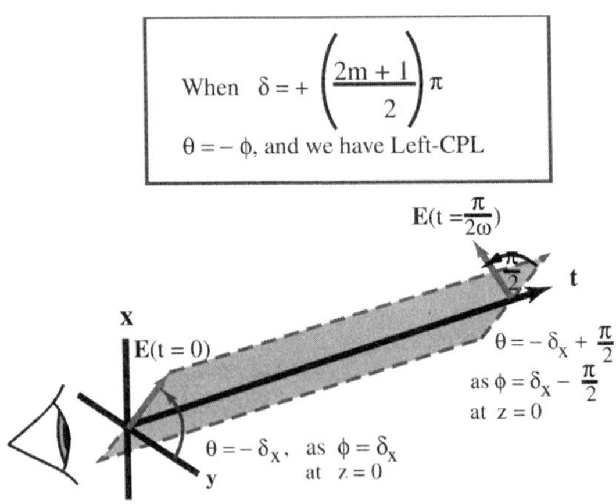

FIGURE 3.15. Rotation of the *E*-field with time for left circularly polarized light.

Under these conditions (3.30) becomes

$$\tan\{\theta(z, t)\} = 1.0[0 - (\pm)\tan\phi(z, t)],$$

which is the same as writing

$$\theta(z, t) = -(\pm)\phi(z, t) \tag{3.35}$$

Thus, for CPL the **E**-field vector direction (specified by θ) rotates with phase angle $\phi(z, t)$. In the case of $\delta = +(2m + 1)\pi/2$, then $\theta(z, t) = -\phi(z, t)$ and we have *left*-CPL, which corresponds to an anticlockwise rotation of **E** as seen by an observer peering into the lightwave (Fig. 3.15). Alternatively, if $\delta = -(2m + 1)\pi/2$, then $\theta(z, t) = \phi(z, t)$, and we have *right*-CPL, which corresponds to a clockwise rotation of **E**, as seen by an observer.

3.4 SUPERPOSITION, COHERENCE, AND INTERFERENCE

When two monochromatic lightwaves overlap in space, they are said to superpose. The *principle of superposition* follows from the linearity of the electromagnetic wave equation (3.5) and states that the **E**-fields (or **B**-fields) of two waves can be summed to constitute a third (resultant) wave. Since the intensity of a lightwave is proportional to the square of the **E**-field, Eq. (3.13), it is clear that the intensity of the two waves may not be the same as the sum of the intensities of the individual waves. Indeed, the harmonic nature of the **E**-field suggests that the intensity of the resultant wave will have strong variations in space and time. This variation in intensity is termed *interference*.

Consider two linearly polarized light waves with the same polarization angle, θ, emanating from the same source and overlapping at a photodetector after each wave has traveled a different path, A and B in Fig. 3.16. The **E**-field components of the two waves at the spacetime coordinate (z, t) can be expressed in the form

$$E_1(z, t) = E_{10}\cos\left[\int_{\mathrm{SAD}} \beta_1 ds - \omega t + \delta_1\right] \tag{3.36}$$

$$E_2(z, t) = E_{20}\cos\left[\int_{\mathrm{SBD}} \beta_2 ds - \omega t + \delta_2\right], \tag{3.37}$$

where we assume a common *angular frequency*, ω, and that

E_{10}, E_{20} are the respective **E**-*field amplitudes*,
β_1, β_2 are the respective *propagation constants* (sometimes termed wave vectors).

We have introduced the *spatial phase retardation*,

$$\phi_\alpha = \int_{\text{path}} \beta_\alpha ds + \delta_\alpha, \tag{3.38}$$

experienced by the waves as they travel over their respective paths (SAD or SBD; Fig. 3.16. In Eq. (3.38) we set $\alpha = 1$ or 2. We also introduce the *temporal phase retardation* $-\omega t$, where the time t is understood to mean the time taken by the lightwave to travel between the source point, S, and the photodetector, D. We also need to introduce δ_1 and δ_2 as the respective *initial phases* for the two waves, since either could be at any point in its cycle at $z = 0$ and $t = 0$.

The intensity at the point D can be determined from (3.13), i.e.,

$$I = \varepsilon_0 c \langle |E_1 + E_2|^2 \rangle. \tag{3.39}$$

This can be expressed in the form

$$I = \varepsilon_0 c \langle \{E_{10}\cos(\phi_1 - \omega t) + E_{20}\cos(\phi_2 - \omega t)\}^2 \rangle, \tag{3.40}$$

which on expanding becomes

$$\begin{aligned} I = \varepsilon_0 c \langle & E_{10}^2 \cos^2(\phi_1 - \omega t) + E_{20}^2 \cos^2(\phi_2 - \omega t) \\ & + 2E_{10}E_{20}\cos(\phi_1 - \omega t)\cos(\phi_2 - \omega t) \rangle. \end{aligned}$$

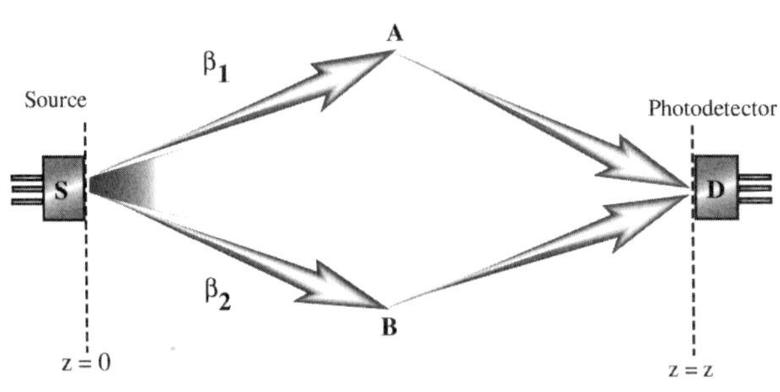

FIGURE 3.16. Interferometric arrangement whereby light travels two paths from source to photodetector.

We can separate the spatial and temporal components of the phase retardation into different factors using simple trigonometric relations

$$I = \varepsilon_0 c\langle E_{10}^2\{\cos\phi_1 \cos\omega t + \sin\phi_1 \sin\omega t\}^2 + \cdots\rangle \tag{3.41}$$

Just as there is no "absolute time," there is also no "absolute phase." Only the difference in phase between the waves has meaning and can be observed by its effect on the photodetection system. We therefore redefine the spatial phase retardations for the two waves and write

$$\phi_1 = \int_{\text{SAD}} \beta_1 ds \tag{3.42}$$

and

$$\phi_2 = \int_{\text{SBD}} \beta_2 ds + \delta, \tag{3.43}$$

where

$$\delta \equiv \delta_1 - \delta_2, \tag{3.44}$$

is the difference between the initial phases for the two lightwaves. If the light source is *coherent*, there is no temporal variation in the initial phase difference between the two waves. That is to say, δ is a constant with no time dependence, in which case, ϕ_1 and ϕ_2 are time independent and can be removed from the time average in Eq. (3.41), i.e.,

$$I = \varepsilon_0 c[E_{10}^2 \cos^2\phi_1\langle\cos^2\omega t\rangle + E_{10}^2 \sin^2\phi_1\langle\sin^2\omega t\rangle + \cdots]. \tag{3.45}$$

As indicated earlier for a harmonic function,

$$\begin{aligned} \langle\cos^2\omega t\rangle &= \langle\sin^2\omega t\rangle = \tfrac{1}{2} \\ \langle\cos\omega t \sin\omega t\rangle &= 0. \end{aligned} \tag{3.46}$$

in which case many of the terms in Eq. (3.45) go to zero when we time average and we are left with the basic *interference equation*

$$I = I_1 + I_2 + 2\sqrt{I_1 I_2}\cos\phi, \tag{3.47}$$

where I_1 and I_2 are the intensities that would be observed at D (Fig. 3.17) if each of the respective lightwaves alone were to be incident upon a photodetector at D, i.e.,

$$I_\alpha \equiv \tfrac{1}{2}\varepsilon_0 E_{\alpha o}^2 \qquad (\alpha = 1, 2). \tag{3.48}$$

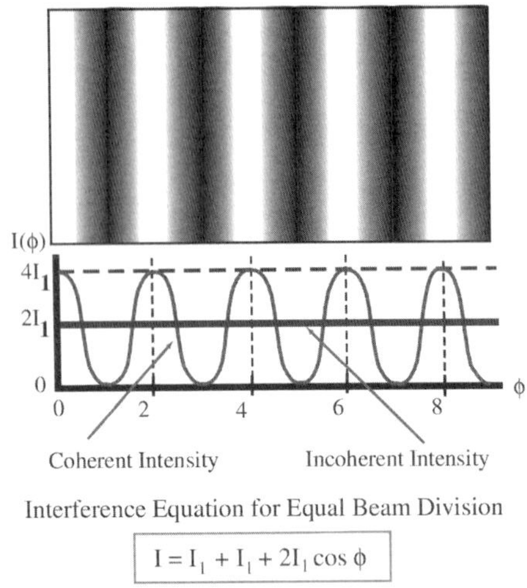

FIGURE 3.17. Interference of two waves of equal intensity.

We see that in the presence of both lightwaves the intensity is not just the sum of the individual intensities. There exists an additional *interference term* that is proportional to $\cos\phi$, where

$$\phi \equiv \int_{\mathrm{SAD}} \beta_1 ds - \int_{\mathrm{SBD}} \beta_2 ds + \delta \tag{3.49}$$

is introduced as the *differential phase* between the two lightwaves. It is clear that the intensity arising from the two waves is strongly modulated.

If the two waves have equal amplitudes (i.e., $E_{10} = E_{20}$) and are *coherent*, then the total intensity can go to zero, when $\phi = \pi$, see Fig. 3.17. Under these circumstances the two waves undergo "complete" *destructive interference*. By contrast the total intensity can be double the sum of the intensities for the separate waves when $\phi = 0$. Here, complete *constructive interference* arises. The term "coherent" is introduced to describe the condition that δ is independent of time. That is to say there is no rapid (in terms of the order of the period of the wave) change in the difference of the initial phases of the waves.

3.5 PARTIAL COHERENCE AND COHERENCE LENGTH

If the two waves are "incoherent," δ changes rapidly over the interval of the time average and we know from experience that there is no interference, i.e.,

$$I = I_1 + I_2.$$

Between the extremes of *incoherence* and *coherence* lies the middle ground of *partial coherence*. To better understand this topic, consider the truly single-frequency cosine lightwave described by

$$E = E_0 \cos\{kz\}$$

and depicted in Fig. 3.18a, where $k = 2\pi/\lambda$, and λ is the free-space wavelength. The spectral content of this wave is a single frequency component, ν, shown in Fig. 3.18b, where $\nu\lambda = c$, with c being the free-space velocity of the wave. The phase of this wave is a continuous function of z, and this would certainly represent light from a coherent source. Since light from an incandescent source has its origin in the bursts of spontaneous emission that follow the thermal excitation of electrons within atoms, it follows that the emitted lightwave will, in reality, suffer a number of sudden phase and amplitude disruptions. This is schematically illustrated in Fig. 3.19. Fourier analysis makes it clear that the more complex the temporal behavior of a wave, the broader its spectrum (Saleh and Teich, 1991; Fig. 3.20).

To establish the relationship between the bandwidth of a light source and its temporal coherence we use *Fourier transform theory*. If a lightwave

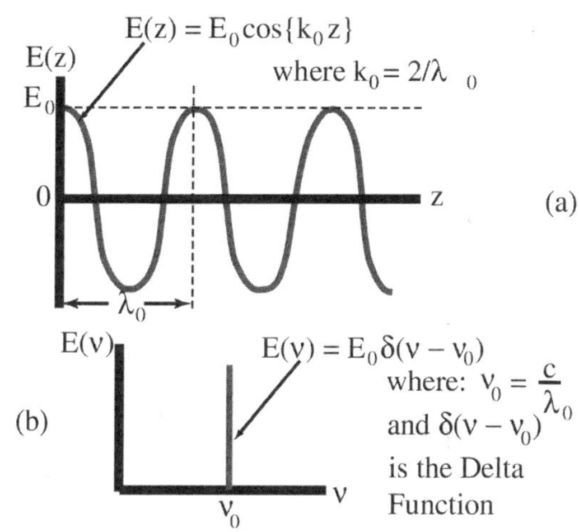

FIGURE 3.18. (a) Monochromatic lightwave and (b) its corresponding delta function spectrum.

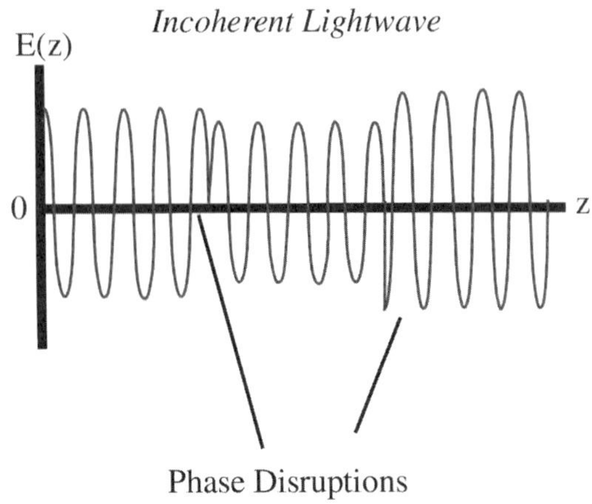

FIGURE 3.19. Illustration of phase interruptions in the E-field of a lightwave generated by spontaneous emission.

in the temporal domain is described by $f(t)$ and its description in the frequency domain

$$s(\omega) = \frac{1}{\sqrt{2\pi}} \int_{-\infty}^{\infty} f(t) e^{i\omega t}\, dt, \tag{3.50}$$

is given by the *Fourier transform* of $f(t)$, then the *inverse Fourier transform* of $s(\omega)$ gives $f(t)$ (Fowles, 1975):

$$f(t) = \frac{1}{\sqrt{2\pi}} \int_{-\infty}^{\infty} s(\omega) e^{-i\omega t}\, d\omega. \tag{3.51}$$

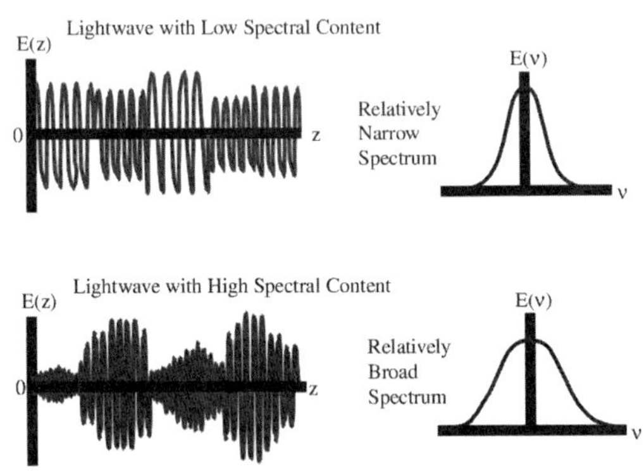

FIGURE 3.20. Schematic illustrations of the **E**-field spectral dependence on the complexity of the temporal behavior of the lightwave. The more complex the temporal variations, the wider the spectrum.

Suppose that $f(t)$ describes a lightwave comprising primarily a single frequency, ω_c, for a finite duration τ_c (Fig. 3.21a). That is to say,

$$\begin{aligned} f(t) &= e^{-i\omega_c t} \qquad \text{for } -\tau/2 < t < \tau/2, \\ f(t) &= 0 \qquad \text{for all other times.} \end{aligned} \tag{3.52}$$

Taking the Fourier transform, of $f(t)$ yields the corresponding *frequency distribution*:

$$s(\omega) = \frac{1}{\sqrt{2\pi}} \int_{\tau_c/2}^{\tau_c/2} e^{i(\omega-\omega_c)t}\, dt = \sqrt{\frac{2}{\pi}}\, \frac{\sin[(\omega - \omega_c)\tau_c/2]}{(\omega - \omega_c)}. \tag{3.53}$$

The form of this frequency distribution for the truncated wave is shown in Fig. 3.21b.

The *power spectrum*, $S(\omega)$, for this finite duration wave is obtained directly from the square of the Fourier transform (Fowles, 1975), i.e.,

$$S(\omega) = |s(\omega)|^2 = \frac{2}{\pi}\, \frac{\sin^2[(\omega - \omega_c)\tau_c/2]}{(\omega - \omega_c)^2}, \tag{3.54}$$

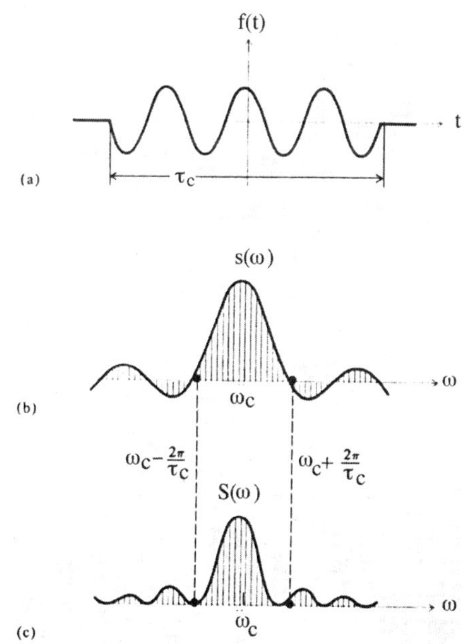

FIGURE 3.21. (a) A lightwave of finite temporal duration, (b) its Fourier transform, and (c) its corresponding power spectrum. From Fowles G.R., (1975), *"Introduction to Modern Optics"*, 2nd edition, p. 72.

and is shown in Fig. 3.21c. We see that the power spectrum has a maximum value at $\omega = \omega_c$ and drops to zero for $\omega = \omega_c \pm 2\pi/\tau_c$. Most of the energy of the wave is contained between these first two minima, so we can claim that the *effective width* of the power spectrum is

$$\Delta\omega = 2\pi/\tau_c. \tag{3.55}$$

Although we have derived this relationship for a specific example, it does allow us to appreciate the more general statement that the *coherence time* of a light source is equal to the inverse of its bandwidth, i.e.,

$$\tau_c = 1/\Delta\nu. \tag{3.56}$$

We can restate this in terms of the linewidth, $\Delta\lambda$, of the source, using (3.18):

$$\tau_c = \frac{\lambda^2}{c\Delta\lambda}. \tag{3.57}$$

The *coherence length* for this source is given by $c\tau_c$, and this takes the form

$$\ell_c = \frac{\lambda^2}{\Delta\lambda}, \tag{3.58}$$

We see that in general the broader the spectrum of a source the shorter will be its coherence length. For a light-emitting diode operating at a wavelength of 1.55 μm with a linewidth of 50 nm, $\ell_c = 48\,\mu\text{m}$. By contrast, for a single-mode laser diode operating at 1.55 μm with a bandwidth of 10 MHz, the coherence length can be substantial; in fact, $\ell_c = 30\,\text{m}$.

3.6 HIGH-COHERENCE INTERFEROMETERS

Most interferometers split the amplitude of a lightwave into two components, sending each component along a different path to a detector. This amplitude division can be accomplished, for example, with a beam splitter that also serves to recombine the two waves after each has experienced a different phase retardation. This is illustrated in the case of a bulk optic *Michelson interferometer* in Fig. 3.22. A collimated beam of light, formed from a point source by a suitable lens, is intercepted by the thin beam splitter (BS) and divided into two beams as shown in Fig. 3.23. One beam is sent to a plane mirror, M_1, while the second beam is sent to another plane mirror, M_2. The back-reflected lighwaves are then recombined at the beam splitter and each is once again divided, part being transmitted and part reflected at 90°.

In the case of a high-coherence source, we can assume a plane, monochromatic lightwave of angular frequency ω. The **E**-field of the

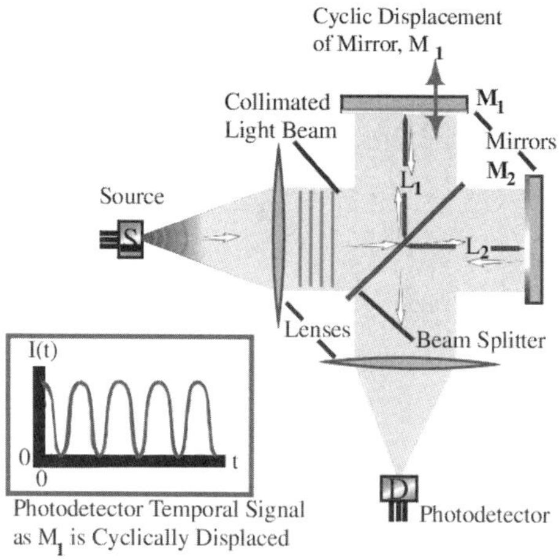

FIGURE 3.22. Bulk optic Michelson interferometer. Inset reveals the interference pattern of light intensity recorded at the photodetector when one mirror is cyclically displaced.

wave reflected from M_1 and arriving back at the beam splitter can be written

$$\mathbf{E}_1 = E_{10} \cos\{\beta L_1 - \omega t + \delta_1\}, \tag{3.59}$$

and the **E**-field for the wave reflected from M_2 and arriving back at the beam splitter can be written

$$\mathbf{E}_2 = E_{20} \cos\{\beta L_2 - \omega t + \delta_2\}. \tag{3.60}$$

L_1 and L_2 represent the total distance covered by the first and second beams in going from the beam splitter to their mirrors and back, respectively. Note that since the distance from the BS to the detector, D, is common to both beams (for a thin enough beam splitter), we do not have to take it into account.

If the two paths are identical, then $L_1 = L_2$, and $\delta_2 = \delta_1$ as the waves recombining at the detector share a common time origin and total constructive interference takes place. This can be seen if, for example, one of the mirrors was to be displaced back and forth a very small amount (on the order of the wavelength of the light), see the inset to Fig. 3.22.

If $L_2 \neq L_1$ then $\delta_2 \neq \delta_1$, as waves arriving at the same instant at D will have originated at different times at BS. In some sense the lightwave can be thought to interference with its past! In which case, if we write $L_2 = L_1 + \Delta L$, then in free space,

$$\delta_2 = \delta_1\left(t = -\frac{\Delta L}{c}\right).$$

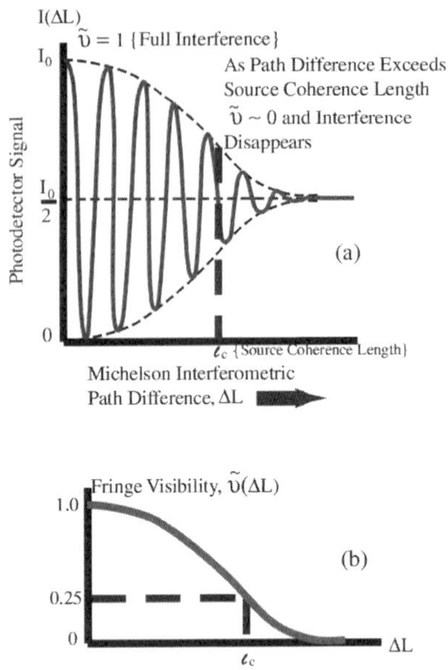

FIGURE 3.23. (a) Variation in the interference depth of modulation with increasing path difference *L*. (b) Interference envelope, or fringe visibility, variation with *L*.

That is to say, the initial (or reference) phase of the wave reflected from M_2 is the same as the initial phase for the wave reflected from M_1 for the *retarded time* $\Delta L/c$. In this case interference will arise only if $\Delta L/c$ is within the *coherence time*, τ_c, of the source, for then $\delta(\equiv \delta_2 - \delta_1)$ is time independent. An illustration of the variation in the interference pattern with mirror path difference ΔL produced by a Michelson interferometer with a 50/50 beam splitter is presented as Fig. 3.23a. The detected intensity is described by the interference Eq (3.47) modified to account for a source that is not perfectly coherent, i.e.,

$$I = \frac{I_0}{2}[1 + \tilde{\upsilon} \cos \phi], \tag{3.61}$$

where I_0 is the *intensity* of the beam before its **E**-field was divided in half by the beam splitter, and ϕ is the *phase difference* given by (3.49). The "visibility" of the fringe pattern under these circumstances is defined by the relation

$$\tilde{\upsilon} = \frac{I_{\text{max}} - I_{\text{min}}}{I_{\text{max}} + I_{\text{min}}}, \tag{3.62}$$

where I_{max} and I_{min} represent the maximum and minimum values of the intensity.

It should be noted that the form of (3.61) is based on the 50/50 division of the **E**-field, not the division of the intensity, by the beam splitter. Also, in the case of a common, uniform medium, $\beta_1 = \beta_2 = \beta$, so $\phi = \beta \Delta L$. A Michelson interferometer with a 50/50 beam splitter can be used to determine the coherence length, $c\tau_c$, of a light source by measuring the *visibility* of the fringes as a function of ΔL (Fig. 3.23b). An often-used definition of *coherence length* is "that value of the mirror path difference for which the fringe visibility drops to 0.25."

The inquisitive reader may ask, "What happens to the light when $\phi = \pi$, and we have a high-coherence source, so $\tilde{v} = 1$, for under these circumstances there is complete destructive interference, and $I = 0$ according to (3.62)?" The simple answer is that we have only been considering what happens to the light that is directed to the photodetector after reflection at the two mirrors. A little thought reveals that light is as likely to be directed back to the source via the beam splitter as it is to be directed toward the photodetector.

Thus when the phase relation is such that *destructive interference* occurs for light traveling toward the photodetector (so the photodetector receives no light), then *constructive interference* arises for the light traveling back towards the source (and all the light goes back to the source). This satisfies the basic law of physics stating that light energy cannot be destroyed unless an *absorbing medium* is included in its path. Although we have not taken it into account, there is an additional contribution to the phase change when light is reflected. It is this phase change that determines which of the two routes experiences constructive interference.

3.7 MULTIPASS FABRY–PEROT INTERFEROMETER

A different type of interferometer is formed when two plane, parallel mirrors are separated by some distance L and aligned with their normals in the direction of propagation of a plane lightwave (see Fig. 3.24). Clearly if the mirror reflection coefficients are not 100%, then some light will be transmitted through this *Fabry–Perot interferometer* (FPI). Furthermore, multireflections are possible, as some of the light is repeatedly sent back and forth through the cavity formed by the mirrors. This type of interferometer lies at the heart of most lasers and its fiber-optic analogue can be used for sensing.

We assume that a plane, monochromatic lightwave of angular frequency ω and amplitude E_0 propagates from the surface of the first

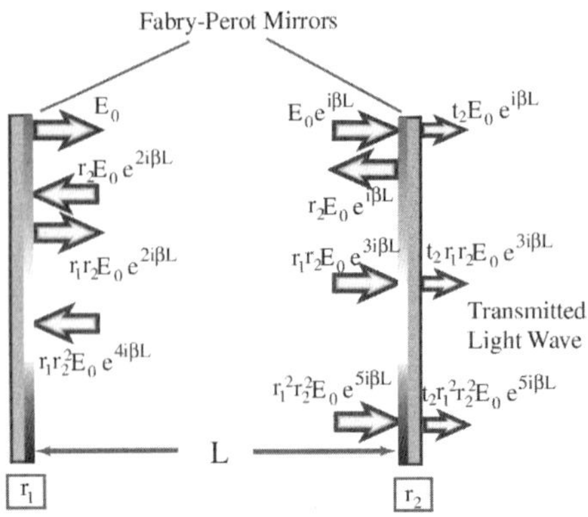

FIGURE 3.24. Illustration of the multiple reflections within a plane mirror Fabry–Perot interferometer, or resonator.

mirror toward the second mirror. The *amplitude reflection coefficients* of the mirrors are taken to be r_1 and r_2, respectively. The **E**-*field* at axial position, z, and time, t, can be expressed in the form

$$E(z, t) = E_0 e^{i(\beta z - \omega t)}, \tag{3.63}$$

where the *propagation constant*

$$\beta = nk = \frac{2\pi n}{\lambda}. \tag{3.64}$$

The z-axis is chosen to coincide with the normal to the center of the first mirror. For simplicity we consider the steady state, so (3.63) is simplified by removing the common temporal factor so that the **E**-*field* reaching the second mirror, at $z = L$, is given by the expression

$$E(L) = E_0 e^{i\beta L}. \tag{3.65}$$

If t_2 is the *transmission coefficient* for the second mirror, then the fractions of this **E**-field transmitted and reflected are $t_2 E_0 e^{i\beta L}$ and $r_2 E_0 e^{i\beta L}$, respectively. The total lightwave **E**-field transmitted through the second mirror is the sum of the **E**-field components transmitted after one, three, five, ... passes through the cavity, i.e.,

$$E_T = E_0 t_2 e^{i\beta L}[1 + r_1 r_2 e^{2i\beta L} + r_1^2 r_2^2 e^{4i\beta L} + \cdots]. \tag{3.66}$$

This is an infinite series that can be written in the closed form

$$E_T = \frac{E_0 t_2 e^{i\beta L}}{1 - r_1 r_2 e^{2i\beta L}}. \tag{3.67}$$

The transmitted intensity, I_T, can be derived from Eq. (3.67) using (3.13) after a little algebra:

$$I_T = \tfrac{1}{2}\varepsilon_0 c|E_T|^2 = I_0 \frac{T_2}{(1-R)^2}\mathscr{A}(R, \beta L). \qquad (3.68)$$

Here we introduce the *Airy function*,

$$\mathscr{A}(R, \beta L) \equiv \frac{1}{1 + \mathscr{F}\ \sin^2(\beta L)}, \qquad (3.69)$$

and the *coefficient of finesse*,

$$\mathscr{F} \equiv \frac{4R}{(1-R)^2}. \qquad (3.70)$$

In expression (3.68), I_0 is the *total intensity* in the FP cavity; $T_2(\equiv |t_2|^2)$ and $R(\equiv r_1 r_2)$ are the intensity *transmission* and *reflection coefficients*, respectively.

Since the Airy function gives rise to the comblike transmitted spectrum schematically illustrated in Fig. 3.25, the FPI is seen to act as a multipeaked filter for light originating within the cavity. This behavior of a Fabry–Perot cavity will be important later in Chapter 5, when we discuss lasers.

The peaks correspond to the condition $\mathscr{A}(R, \beta L) = 1.0$ and arise when $\beta L = m\pi$. This latter condition essentially defines the set of *FPI resonant frequencies*,

$$\nu_m = \frac{mc}{2nL}, \qquad (3.71)$$

where m is an integer.

Light transmitted through the FPI from the left side to the right side experiences a reflection first before it passes through the cavity. This leads

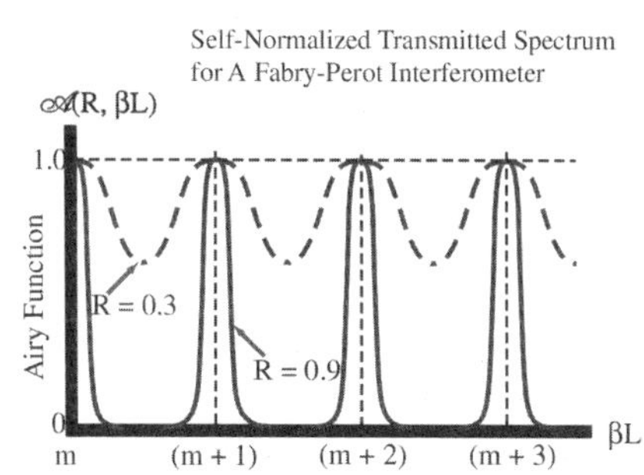

FIGURE 3.25. Airy function corresponds to the self-normalized transmission spectrum of a Fabry–Perot interferometer, since β is inversely proportional to λ.

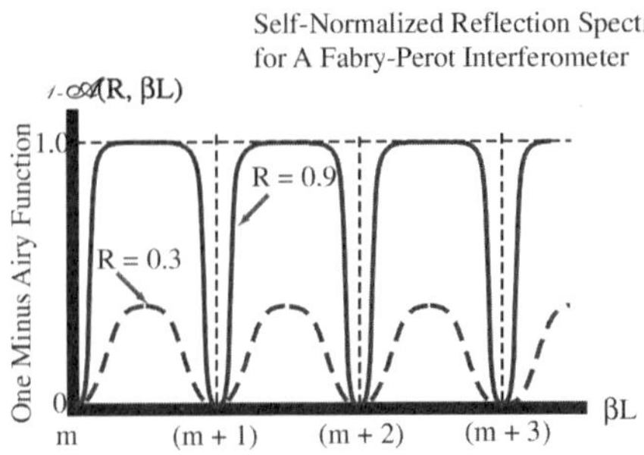

FIGURE 3.26. One minus the Airy function corresponds to the self- normalized reflection spectrum for a Fabry–Perot interferometer, or resonator.

to a comblike transmission spectrum (Hecht and Zajac, 1974) of the form indicated in Fig. 3.26. Power continuity requires that the normalized reflection spectrum be given by one minus the normalized reflection spectrum, as we have not allowed for any loss mechanism. One minus the Airy function is shown in Fig. 3.26.

3.8 LOW-COHERENCE INTERFEROMETRY

Although most interferometers operate with high coherence sources, such as lasers, there are some that work with low coherence sources of light, such as a *light-emitting diode* (LED). Sometimes, cost is the rationale, as LEDs are less expensive than laser diodes. Other times, the scale of the measurement might be so large that it is impractical to use a laser with such long coherence length, that is to say, such a narrow linewidth; see Section 3.5. The basic premise for high coherence interferometers is that the *coherence length* of the light is greater that the size of the interferometer. Thus, in the Michelson interferometer discussed in Section 3.6, we assumed that the two lightwaves reflected from M_1 and M_2, respectively, were still coherent when they recombined. In fact, the actual requirement for interference with good visibility fringes is that

$$\Delta L < \ell_c, \tag{3.72}$$

where L_1 and L_2 are the respective optical paths involving reflections at M_1 and M_2, respectively; $\Delta L \equiv L_1 - L_2$; and ℓ_c is the *source coherence length*. In situations where this condition is difficult, or costly, to satisfy, we can "turn the problem into the solution" and use low-coherence interferometry.

In order to understand this idea let us consider what happens when we illuminate a bulk optic *Michelson interferometer* with a broadband (low coherence) source (Fig. 3.27). If we start with $L_1 \gg L_2 + \ell_c$, then we observe no interference fringes at the photodetector as we scan the mirror M_2 a small distance (a few wavelengths of the light). This is in contrast with what was described in Section 3.6, where the source was assumed to be narrowband and coherent. However, if we continue to translate M_2 to the point where $L_1 = L_2$, then we will observe a set of fringes, as the two paths lie within the coherence length of the source (see inset to Fig. 3.28). If the source coherence length is around 50 μm, then we can probably determine that the two paths match to within 5 μm, or better, by looking for the position of M_2 that gives the peak fringe. If L_1 is a distance we wish to measure, we can do so (to within 5 μm) using this approach, provided L_1 is accurately determined. For a distance L_1 of, say, 1 m, this is equivalent to a *strain resolution* of 5 με (microstrain).

We shall see later that *low-coherence interferometry* can also be used in more complex configurations, such as combinations of fiber optic Fabry–Perot and Michelson interferometers. We now introduce an aid to under-

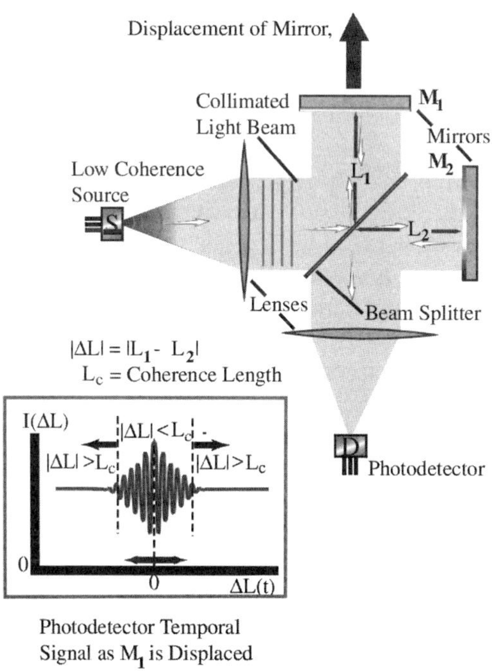

FIGURE 3.27. Schematic illustration of the limited variation in path difference leading to interference in a Michelson interferometer excited by a low-coherence light source.

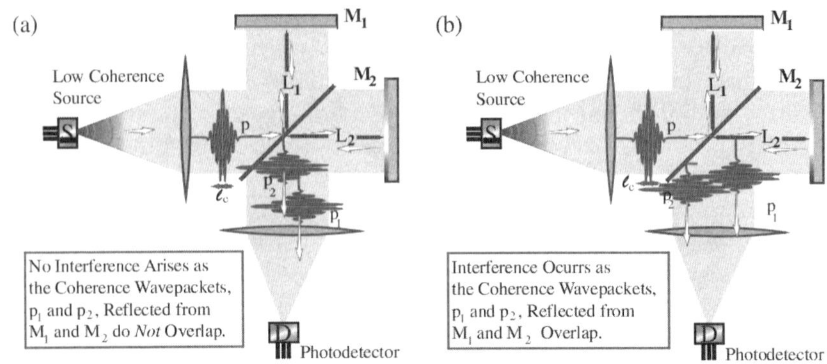

FIGURE 3.28. Explanation of low-coherence interferometry in terms of overlapping coherence wavepackets.

standing low-coherence interferometry. At any point along a lightwave we can visualize a *coherence wavepacket* of length ℓ_c that represents the local region of coherence. In the case of the bulk optic Michelson interferometer, let us identify this coherence wavepacket, p, just prior to the light being divided by the beam splitter. After the two beams have been reflected from their respective mirrors and recombined by the beam splitter, we have two coherence wavepackets, p_1 and p_2, enroute to the photodetector (see Fig. 3.28). If these coherence wavepackets do not overlap (to within ℓ_c), then no interference will be observed at the photodetector. Clearly, they will only overlap if they have traveled the same distance, i.e., $L_1 = L_2$.

One way of tracking changes in the distance of mirror M_1 (i.e., L_1) with time is to repetitively scan M_2 and check for the value of L_2 at which maximum interference is observed. A fast measurement of L_2 is of course required if good frequency response is desired. It is possible to undertake this low-coherence interferometric tracking task with no moving mirrors. This can be accomplished by interposing a *Fizeau interferometer* and a linear photodetector array, a *charge coupled detector* (CCD), in the path of the recombined beam.

We can see how this system works by reference to the coherence wavepackets in Fig. 3.29a. Here we show the paths and coherence wavepackets, p_a and p_b, for two rays that travel through the Michelson interferometer. We have separated the paths in an unrealistic manner in order to more easily follow them. It is apparent that we have set the difference in the mirror separations, ΔL, to exceed the coherence length of the light source, ℓ_c. Thus, the separation between p_{a1} and p_{a2}, and between

p_{b1} and p_{b2}, which corresponds to ΔL, is much greater than the width of the coherence wavepackets, ℓ_c, in which case no interference would be observed in the absence of the Fizeau interferometer.

In Fig. 3.29b, we show the continuation of these two rays as they pass through the Fizeau interferometer and strike the CCD array. A Fizeau interferometer is similar to a Fabry–Perot interferometer, except that there is a linear variation in the separation of the two mirror surfaces along the device. In Fig. 3.29b, the separation of the Fizeau mirror surfaces at location B is just right to compensate for the path difference, ΔL, between M_2 and M_1 of the Michelson interferometer. Consequently, interference is observed by the CCD in the vicinity of B, as the fraction of p_{b2} that is

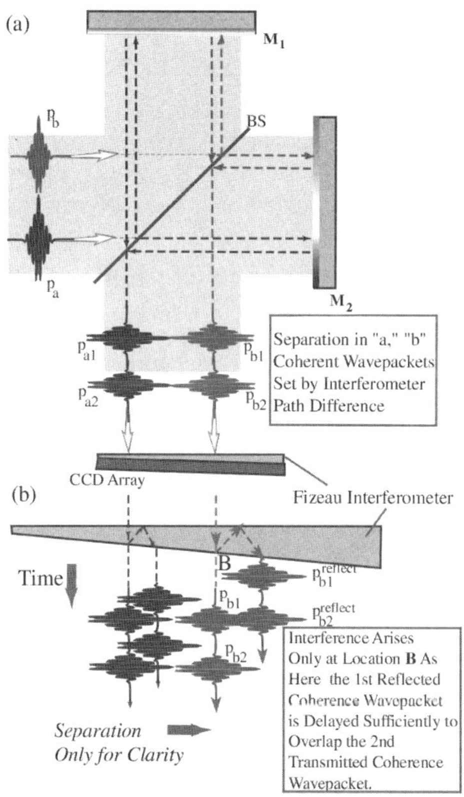

FIGURE 3.29. (a) Low coherence interferometer using a Fizeau etalon and a CCD array for real-time measurements. (b) Illustration of Fizeau interferometer using coherence wavepackets.

reflected from the far side of the Fizeau interferometer, and subsequently makes a double pass through the device, just overlaps the fraction of p_{b1} that is directly transmitted. Since the separation of the mirrors at each position along the Fizeau interferometer is known, we can relate the location of this interference to the separation of the Michelson mirrors, and this can be done extremely rapidly, as the only scanning involved is that of the voltage waveform required to read the photodiode array.

3.9 RADIATION COUPLING BETWEEN OPTICAL FIBERS

As we shall see in Chapter 5, light travels along an isolated optical fiber with little attenuation because its energy is constrained from spreading in the lateral directions beyond the core of the optical fiber. This core region has a higher index of refraction than the surrounding cladding material, and the light rays can be thought of as being confined by continuous internal reflection at the boundary between the high- and low-index regions (Fig. 3.30). Although the energy of the lightwave is confined within the core of the optical fiber, a more rigorous mathematical description requires the existence of a so-called *evanescent* **E**-*field* to extend beyond the core into the cladding region. As long as the optical fiber is isolated, this evanescent **E**-field does not lead to any loss of light energy from the core.

The operative word is "isolated," for when two such optical fibers are brought close together, optical power can be exchanged between their cores, providing the cladding regions are not too large. This will be seen to be important in Chapter 5, as this exchange of light between two optical fibers lies at the heart of a number of devices, including *direction couplers*. What makes this interaction possible is the penetration of each core by the evanescent **E**-field of the light being guided within the other optical fiber

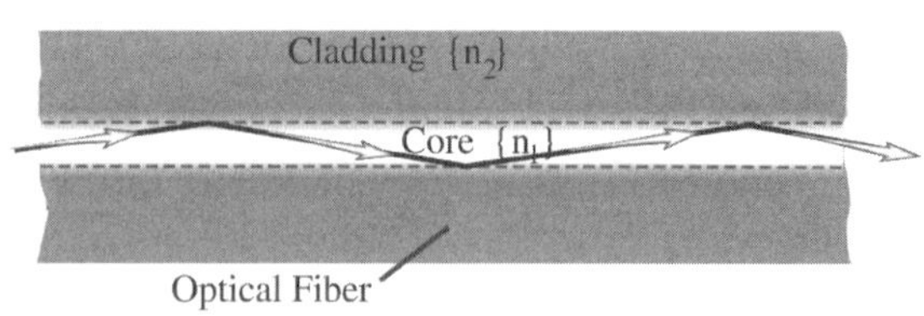

FIGURE 3.30. Core confinement of light by internal reflection at the core–cladding interface.

(Fig. 3.31). We can describe the coupling of light between two optical fibers using a pair of coupled equations.

We consider the propagation of a plane, monochromatic light wave in the z-direction and confined within the core of an optical fiber. We have seen in Eq. (3.6) that the **E**-field for such a lightwave can be expressed in the form

$$E(z, t) = E(0, 0)e^{i(\beta z - \omega t)} + \text{c.c.} \tag{3.73}$$

This can be written in the form

$$E = ae^{i\beta z}, \tag{3.74}$$

if we ignore time variations. We also recognize that this equation can be thought of as the solution of the simple differential equation that describes the axial variation of the **E**-field along an isolated optical fiber, i.e.,

$$\frac{dE}{dz} = i\beta E. \tag{3.75}$$

This equation states that for an isolated optical fiber the only change experienced by the **E**-field as a function of z is a phase change. Now,

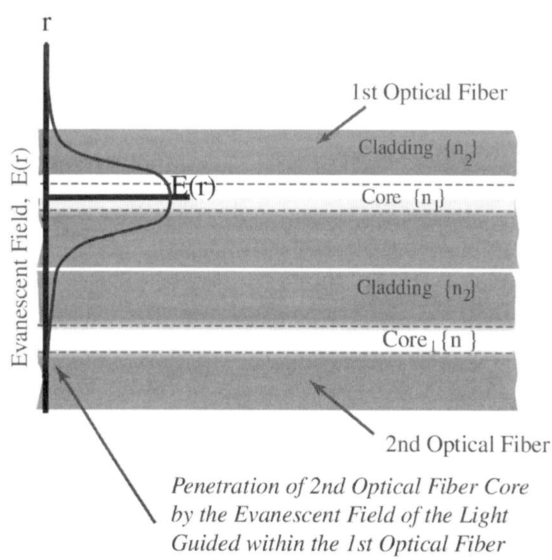

FIGURE 3.31. E-field distribution within one of a pair of closely spaced optical fibers, showing penetration of the second fiber core by the evanescent field of the light being guided within the first optical fiber core.

observation reveals that when the cores of two optical fibers are brought sufficiently close together, there is an exchange of optical power. We can account for this interaction phenomenologically by introducing a *coupling term* in Eq. (3.75) for each optical fiber that is proportional to the **E**-field in the neighboring optical fiber, i.e.,

$$\frac{dE_1}{dz} = i\beta_1 E_1 + iC_{12}E_2 \tag{3.76}$$

$$\frac{dE_2}{dz} = iC_{21}E_1 + i\beta_2 E_2, \tag{3.77}$$

where C_{12} and C_{21} represent the respective *coupling coefficients*. These coefficients depend on the difference in the indices of refraction for the two cores and the degree of penetration of the evanescent **E**-field from one optical fiber into the core of the other optical fiber (Lee, 1986). We assume that we can express the **E**-fields in the first and second optical fibers by the expressions

$$E_1 = a_1 e^{i\beta_1 z} \tag{3.78}$$

and

$$E_2 = a_2 e^{i\beta_2 z}, \tag{3.79}$$

where we have introduced the *amplitude coefficients* a_1 and a_2 as functions of z. Substitution of Eq. (3.78) and (3.79) into the pair of Eqs. (3.76) and (3.77) leads to a pair of coupled first-order differential equations for these *amplitude coefficients* in the form

$$\frac{da_1(z)}{dz} = iC_{12}a_2(z)e^{i\Delta\beta z} \tag{3.80}$$

$$\frac{da_2(z)}{dz} = iC_{21}a_1(z)e^{-i\Delta\beta z}, \tag{3.81}$$

where we have introduced the *phase mismatch*

$$\Delta\beta \equiv \beta_2 - \beta_1, \tag{3.82}$$

and we notice that the rate of change of a_1 is proportional to a_2 and vice versa. If we assume that the amplitude of the lightwave entering the first optical fiber is $a_1(0)$ and that no light enters the second optical fiber at that location, so $a_2(0) = 0$, then the pair of Eqs. (3.80) and (3.81) can be solved by assuming solutions of the form

$$a_1(z) = p_1 e^{i\gamma_1 z} \tag{3.83}$$

and

$$a_2(z) = p_2 e^{i\gamma_2 z}. \tag{3.84}$$

Using these solutions in Eqs. (3.80) and (3.81) leads to the relations

$$i\gamma_1 p_1 e^{i\gamma_1 z} = iC_{12} p_2 e^{i(\gamma_2 + \Delta\beta)z} \tag{3.85}$$

and

$$i\gamma_2 p_2 e^{i\gamma_2 z} = iC_{21} p_1 e^{i(\gamma_1 - \Delta\beta)z}. \tag{3.86}$$

In order for p_1 and p_2 to be independent of z, it is necessary that

$$\gamma_2 = \gamma_1 - \Delta\beta. \tag{3.87}$$

Furthermore, if we divide Eq. (3.85) by (3.86), we obtain the relation

$$\gamma_2 \gamma_1 = C_{12} C_{21}. \tag{3.88}$$

Combining Eq. (3.87) with (3.88) yields a quadratic equation for γ_1, with solutions

$$\gamma_1 = \frac{\Delta\beta}{2} \pm \sqrt{\left(\frac{\Delta\beta}{2}\right)^2 + C^2}, \tag{3.89}$$

where we introduced

$$C^2 \equiv C_{12} C_{21}. \tag{3.90}$$

If we now define

$$\gamma = \sqrt{\left(\frac{\Delta\beta}{2}\right)^2 + C^2}, \tag{3.91}$$

then the *amplitude coefficients* for the lightwave in the second optical fiber take the form

$$a_2(z) = e^{-i\Delta\beta z/2}\{p_2 e^{i\gamma z} + q_2 e^{-i\gamma z}\}. \tag{3.92}$$

Consideration of the boundary condition on the second optical fiber, i.e., $a_2(0) = 0$, requires that $q_2 = -p_2$, so we can write

$$a_2(z) = 2ip_2 e^{-i\Delta\beta z/2} \sin^2 \gamma z. \tag{3.93}$$

Using this expression in (3.81) yields the amplitude coefficient for the lightwave in the first optical fiber:

$$a_1(z) = \frac{2p_2}{C_{21}} e^{i\Delta\beta z/2}\left[\gamma \cos \gamma z - \frac{i\Delta\beta}{2} \sin \gamma z\right]. \tag{3.94}$$

If we now draw on the boundary condition for this optical fiber, i.e., $a_1(0) = a_0$, then (3.94) yields the relation:

$$a_0 = \frac{2p_2\gamma}{C_{21}}, \tag{3.95}$$

which yields the final expressions for the amplitude coefficients for the lightwaves in the first and second optical fibers:

$$a_1(z) = a_0 e^{i\Delta\beta z/2}\left[\cos\gamma z - \frac{i\Delta\beta}{2}\sin\gamma z\right] \tag{3.96}$$

$$a_2(z) = \frac{iC_{21}a_0}{\gamma} e^{-i\Delta\beta z/2}\sin^2\gamma z. \tag{3.97}$$

Since the optical power is proportional to the square of the field, we can write

$$P_1(z) = P_1(0)\left[\cos^2\gamma z + \left\{\frac{\Delta\beta}{2\gamma}\right\}^2 \sin^2\gamma z\right] \tag{3.98}$$

and

$$P_2(z) = P_1(0)\,\frac{|C_{12}|^2}{\gamma^2}\sin^2\gamma z. \tag{3.99}$$

Under *phase-matched* conditions,

$$\beta_1 = \beta_2 = \beta, \tag{3.100}$$

and $\Delta\beta = 0$.

Equations (3.98) and (3.99) subsequently simplify to the form

$$P_1(z) = P_1(0)\cos^2 Cz \tag{3.101}$$

and

$$P_2(z) = P_1(0)\sin^2 Cz. \tag{3.102}$$

leading to the sinusoidal exchange of the total power between the two optical fibers, as indicated in Fig. 3.32, with an *exchange length*

$$z_{\text{exc}} = \pi/2C. \tag{3.103}$$

It is reasonable to ask the question, "why does all the optical power exchange between the fibers? Why does the process not stop when the light power in each optical fiber is the same?" The answer lies in the 90° phase difference between the *driving field* in the first optical fiber and the *driven field* in the second optical fiber. This is expressed by the i in Eq. (3.97) (as $i = e^{i\pi/2}$) for the amplitude coefficient of the **E**-field in the second optical fiber. This 90° phase difference means that the field induced in the

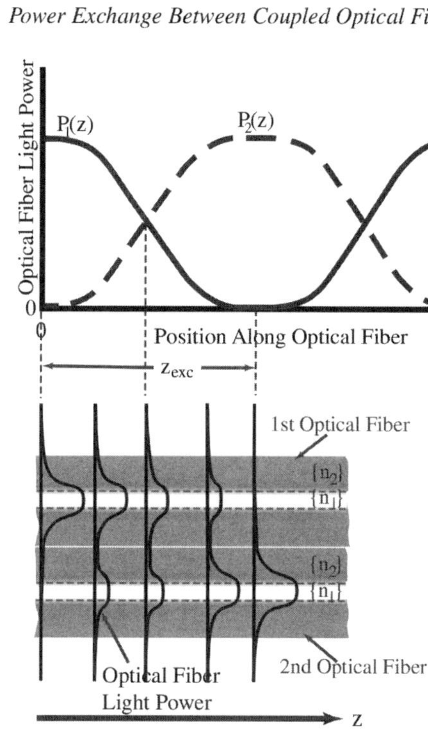

FIGURE 3.32. Light power exchange between coupled optical fibers. Lower sketch indicates changing distribution of the **E**-field across the two optical fibers as the power is exchanged.

first optical fiber by the field building in the second optical fiber will be 180° out of phase with the original field (90° plus 90°) and therefore will undergo destructive interference. Thus, for the phase-matched condition, no field can start to be regenerated in the first optical fiber until all of the field is transferred to the second optical fiber.

3.10 BRAGG GRATING REFLECTION

Another example of radiation mode coupling arises when a lightwave impinges upon a region of periodic variation in the index of refraction. Such regions are often referred to as *Bragg gratings* because the interference that takes place between the incident lightwave and the waves scattered from these periodic variations, or gratings, is similar to the interaction observed by the famous scientist Lawrence Bragg in his work on X-ray scattering from lattice planes in crystals. We shall see in Chapter 4 that Bragg gratings play a role in making narrowband lasers, and from Chapters 6 on that Bragg gratings formed in optical fibers constitute very important sensors for structural monitoring. In each of these instances a strong narrowband back reflection of light is generated when the resonance (or phase match) condition is satisfied. This reflection results from constructive interference between the forward wave and the contrapropagating lightwave that is coupled from the forward lightwave by the grating.

With this in mind we consider a plane monochromatic lightwave propagating in the z-direction through a region in which the index of refraction has a sinusoidal variation, Fig. 3.33, described by the relation

$$n(z) = n_0 + \Delta n \cos(2pz), \tag{3.104}$$

where n_0 is the mean value of the core index of refraction, Δn is the amplitude of the refractive index modulation, and

$$p \equiv \pi/\Lambda, \tag{3.105}$$

where Λ is the period of the refractive index modulation. If we express the lightwave in the form of a harmonic disturbance,

$$E(z, t) = E(z)e^{i\omega t}, \tag{3.106}$$

where ω is the angular frequency, and use (3.106) in the electromagnetic wave equation for the medium (3.9), we arrive at the one-dimensional form of the *Helmholtz wave equation*:

$$\frac{d^2E(z)}{dz^2} = -\frac{\omega^2}{v^2}E(z). \tag{3.107}$$

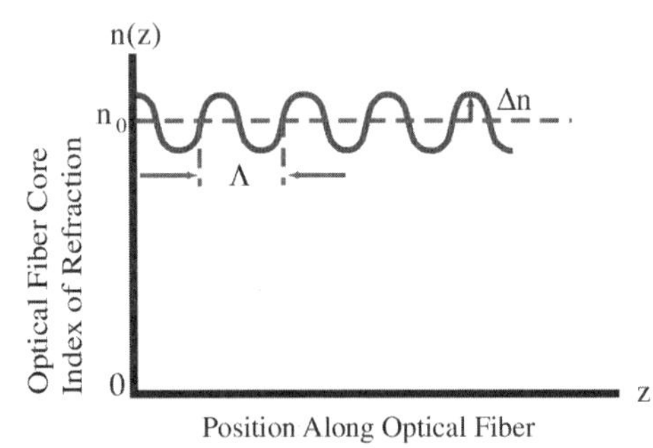

FIGURE 3.33. Axial variation of the core index of refraction that constitutes a fiber Bragg grating.

If we further draw on (3.10) and (3.12) and substitute (3.104) for n, we arrive at the equation

$$\frac{d^2E(z)}{dz^2} = -\{n_0 + \Delta n \cos(2pz)\}^2 k^2 E(z). \tag{3.108}$$

If the perturbation in the index of refraction is small (justified as $\Delta n/n_0 \ll 10^{-2}$), we can expand the square within the curly bracket in (3.108), drop the term proportional to Δn^2, and use

$$2\cos(2pz) = e^{2ipz} + e^{-2ipz} \tag{3.109}$$

to write

$$\frac{d^2E(z)}{dz^2} = -[\beta^2 + \xi\{e^{2ipz} + e^{-2ipz}\}]E(z), \tag{3.110}$$

where we have introduced the *propagation constant*

$$\beta = n_0 k \tag{3.111}$$

and the *coupling factor*

$$\xi = k^2 n_0 \Delta n. \tag{3.112}$$

We now assume plane wave solutions for (3.110) of the form

$$E(z) = a_f(z)e^{-ipz} + a_b(z)e^{ipz}, \tag{3.113}$$

where $a_f(z)$ is the "forward" wave **E**-field amplitude and $a_b(z)$ is the "backward" wave **E**-field amplitude. This solution is substituted into (3.110); terms with the factor e^{-ipz} are collected and yield a second-order differential equation of the form

$$\frac{d^2a_f}{dz^2} - 2ip\,\frac{da_f}{dz} + (\beta^2 - p^2)a_f - \xi a_b = 0. \tag{3.114}$$

Collecting terms with the factor e^{ipz} yields

$$\frac{d^2a_b}{dz^2} + 2ip\,\frac{da_b}{dz} + (\beta^2 - p^2)a_b - \xi a_f = 0. \tag{3.115}$$

Making a further assumption that second derivative terms are negligible (Erdogan, 1997) leads to a pair of first-order coupled differential equations:

$$2ip\,\frac{da_b}{dz} + (\beta^2 - p^2)a_b - \xi a_f = 0 \tag{3.116}$$

and

$$2ip\,\frac{da_{\mathrm{f}}}{dz} - (\beta^2 - p^2)a_{\mathrm{f}} + \xi a_{\mathrm{b}} = 0 \tag{3.117}$$

If we use (3.116) to express $a_{\mathrm{f}}(z)$ in terms of $a_{\mathrm{b}}(z)$ and its z-derivative, we obtain the second-order differential equation

$$\frac{d^2 a_{\mathrm{b}}}{dz^2} - \kappa^2 a_{\mathrm{b}} = 0, \tag{3.118}$$

which is exclusively in terms of $a_{\mathrm{b}}(z)$. This simplification is possible by introducing the *grating coupling coefficient*:

$$\kappa^2 = \frac{\xi^2}{4p^2} - \frac{(\beta^2 - p^2)^2}{4p^2}. \tag{3.119}$$

Note that a second-order differential equation identical in form to (3.118) can be derived for $a_{\mathrm{f}}(z)$. In both cases κ is *real* and we can introduce solutions of the form

$$a_{\mathrm{b}}(z) = b_1 e^{\kappa z} + b_2 e^{-\kappa z}. \tag{3.120}$$

In most situations of interest the incident, forward, wave will impinge upon the Bragg grating at $z = 0$, and as it propagates through the grating a fraction of its power in a narrow band is coupled into the backward wave (Fig. 3.34). The relevant boundary conditions usually take the form $a_{\mathrm{b}}(L_{\mathrm{G}}) = 0$ and $P_{\mathrm{f}}(0) = P_0$, where L_{G} is the length of the Bragg grating, and P_0 is the value of the optical power in the forward lightwave $P_{\mathrm{f}}(z)$ at the start of the grating. Using the first boundary conditions with (3.120) yields

$$b_2 = -b_1 e^{2\kappa L_{\mathrm{G}}}. \tag{3.121}$$

Substitution into (3.120) yields the expression for the backward lightwave amplitude,

$$a_{\mathrm{b}}(z) = 2b_1 e^{\kappa L_{\mathrm{G}}} \sinh\{\kappa(z - L_{\mathrm{G}})\}, \tag{3.122}$$

and substitution of this equation into (3.116) provides an expression for the forward lightwave amplitude:

$$a_{\mathrm{f}}(z) = \frac{2b_1 e^{\kappa L_{\mathrm{G}}}}{\xi}[(\beta^2 - p^2)\sinh\{\kappa(z - L_{\mathrm{G}})\} + 2ip\kappa\cosh\{\kappa(z - L_{\mathrm{G}})\}]. \tag{3.123}$$

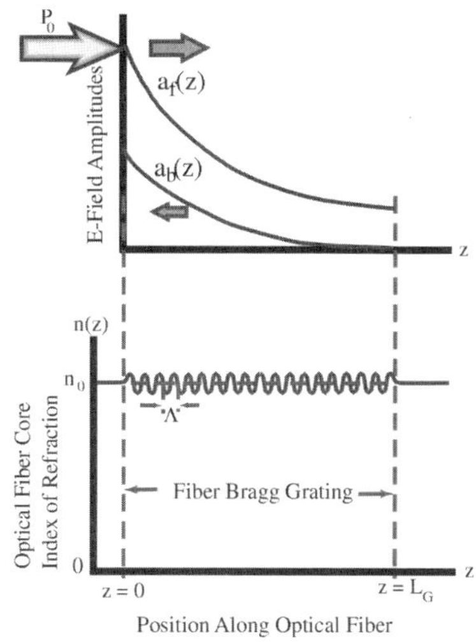

FIGURE 3.34. Forward and backward lightwave *E*-field amplitudes along a fiber Bragg grating.

Then, using (3.14) and the second boundary condition to eliminate b_1 in terms of the incident optical power P_0, we arrive at the optical power in the backward wave at the start of the Bragg grating:

$$P_b(0) = \frac{P_0 \xi^2 \sinh^2\{\kappa L_G\}}{[(\beta^2 - p^2)^2 \sinh^2\{\kappa L_G\} + 4p^2\kappa^2 \cosh^2\{\kappa L_G\}]}. \tag{3.124}$$

The *reflectivity* of the Bragg grating,

$$R(\kappa L_G) \equiv P_b(0)/P_0, \tag{3.125}$$

can be seen to take the form

$$R(\kappa L_G) = \frac{\sinh^2\{\kappa L_G\}}{\left[\frac{(\beta^2 - p^2)^2}{\xi^2} \sinh^2\{\kappa L_G\} + \frac{4p^2\kappa^2}{\xi^2} \cosh^2\{\kappa L_G\}\right]}. \tag{3.126}$$

From (3.119) we see that κ is real, provided

$$\xi^2 \geq (\beta^2 - p^2)^2. \tag{3.127}$$

This is valid for small wavelength displacements from the *Bragg wavelength* λ_B, the wavelength corresponding to the *phase-matched condition*, defined as

$$\beta = p, \qquad \text{or} \qquad 2\pi n_0/\lambda_B = \pi/\Lambda. \tag{3.128}$$

We see that this *phase-matched condition* leads to the important expression for the *Bragg wavelength*, i.e.,

$$\lambda_B = 2n_0\Lambda. \tag{3.129}$$

If the phase-matched condition applies, Eq. (3.119) takes the form

$$\kappa^2 = \frac{\xi^2}{4p^2}, \tag{3.130}$$

and the *grating coupling coefficient* takes on the simple form

$$\kappa = \frac{\pi \Delta n}{\lambda_B}. \tag{3.131}$$

Under these circumstances, we see from (3.126) the peak reflectivity of the Bragg grating,

$$R_{max}(\kappa L_G) = \tanh^2\{\kappa L_G\}, \tag{3.132}$$

where κL_G is termed the *grating strength*. The dependence of the peak reflectivity of the Bragg grating on this parameter is presented in Fig. 3.35, and an indication of its effect on the reflection spectrum is provided in Fig. 3.36.

The reflection spectrum for a Bragg grating can be determined from (3.126). As indicated previously, the peak in the reflection spectrum is given by (3.132) and arises at the Bragg wavelength. For small displace-

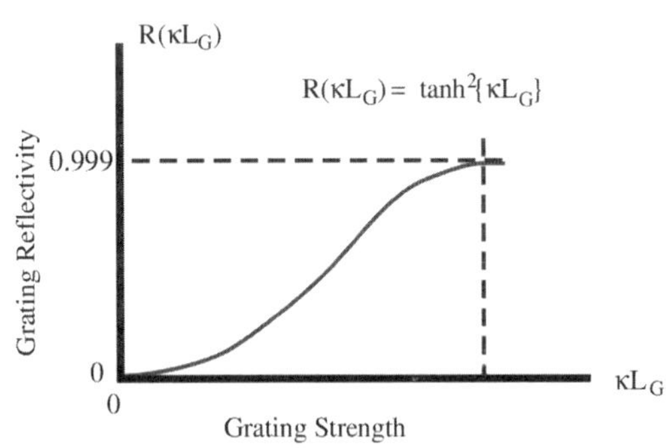

FIGURE 3.35. Variation in the peak reflectivity of a fiber Bragg grating with the strength of the grating.

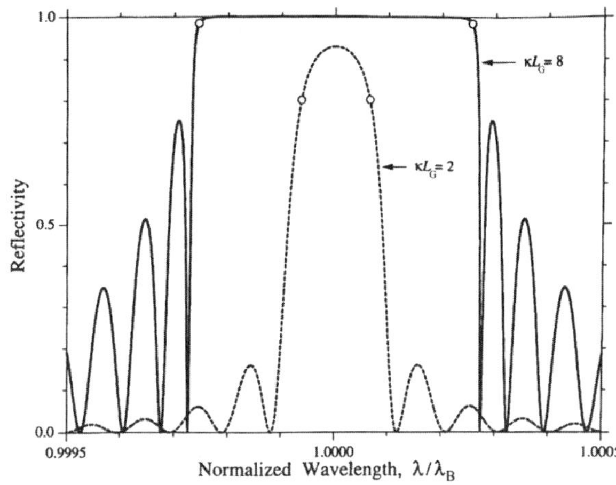

FIGURE 3.36. Fiber Bragg grating reflectivity spectra for two values of the grating strength. From Erdogan, T., "Fiber Grating Spectra" *J. Lightwave Technology*, **15**, 1277–1287 © IEEE.

ments from the Bragg wavelength, expression (3.126) with κ real describes the spectrum. At larger displacements from the Bragg wavelength, $\xi^2 \leq (\beta^2 - p^2)^2$, and according to (3.119), κ becomes imaginary. We therefore introduce

$$\kappa^* \equiv i\kappa = \sqrt{\frac{(\beta^2 - p^2)^2}{4p^2} - \frac{\xi^2}{4p^2}}, \tag{3.133}$$

and the grating reflection spectrum in this part of wavelength space is described by the expression

$$R(\kappa^* L_G) = \frac{\sin^2\{\kappa^* L_G\}}{\left[\frac{(\beta^2 - p^2)^2}{\xi^2}\sin^2\{\kappa^* L_G\} + \frac{4p^2\kappa^{*2}}{\xi^2}\cos^2\{\kappa^* L_G\}\right]}. \tag{3.134}$$

A representative Bragg grating reflection spectrum (measured and calculated) for a 1-mm-long uniform grating with a strength $\kappa L_G = 1.64$ (Erdogan, 1997) is presented in Fig. 3.37. To acquire an indication of the spectral width of Bragg grating reflection spectra, we evaluate the wavelength separation of the two null points on either side of the peak reflection. The first null in $R(\kappa^* L_G)$ arises when

$$\kappa L_G = \pi. \tag{3.135}$$

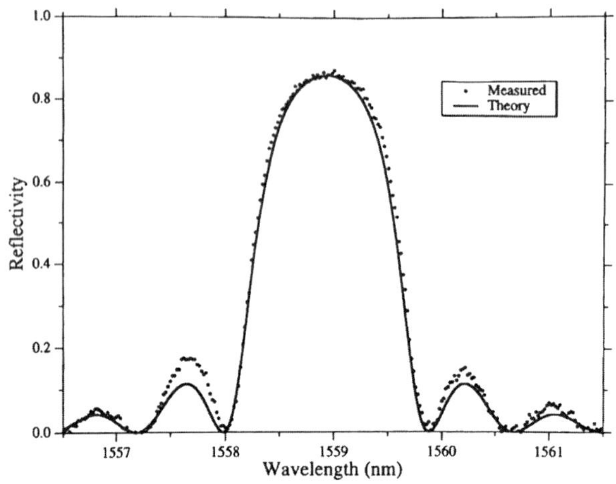

FIGURE 3.37. Comparison of theoretical and experimental grating reflectivity spectra. From Erdogan, T., "Fiber Grating Spectra" *J. Lightwave Technology*, **15**, 1277–1287 © IEEE.

Using (3.105) and (3.112) with (3.133) and recognizing that at the wavelength corresponding to this first null,

$$\beta = \frac{2\pi n_0}{\lambda_B + \Delta\lambda_w}, \tag{3.136}$$

where $\Delta\lambda_w$ is the wavelength displacement from the Bragg wavelength for this null (Fig. 3.38), Eq. (3.135) is approximated by the expression

$$\frac{\pi L_G}{\Lambda}\sqrt{\left\{\frac{\Delta\lambda_w}{\lambda_B}\right\}^2 - \left\{\frac{\Delta n}{2n_0}\right\}^2} = \pi, \tag{3.137}$$

assuming $\Delta\lambda_w \ll \lambda_B$, with some reasonable justification. This relation can be used to ascertain the *half-spectral width* of the Bragg grating reflection spectrum

$$\Delta\lambda_w = \lambda_B\sqrt{\left\{\frac{\Lambda}{L_G}\right\}^2 + \left\{\frac{\Delta n}{2n_0}\right\}^2}. \tag{3.138}$$

We see that this width decreases with longer gratings for a given period and with smaller modulation in the index of refraction for a given mean index. Using (3.131) in (3.132) leads to the expression

$$R_{max}(\kappa L_G) = \tanh^2\left\{\pi\left(\frac{\Delta n}{2n_0}\right)\left(\frac{L_G}{\Lambda}\right)\right\}. \tag{3.139}$$

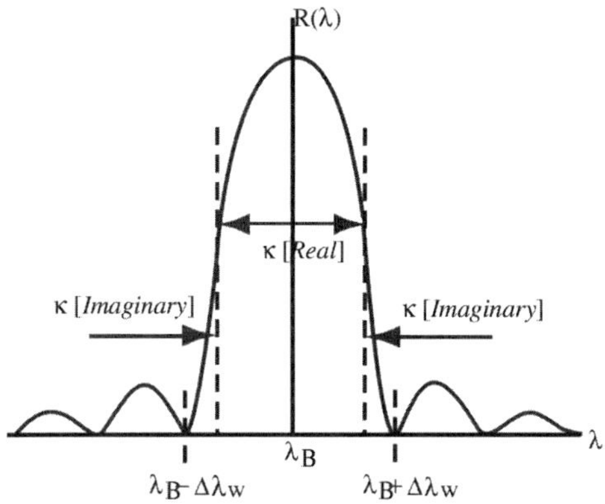

FIGURE 3.38. Fiber Bragg grating reflectivity spectrum indicating the real and imaginary domains of the strength parameter.

We observe that a large grating peak reflectivity requires the grating to be long compared to the period, but the index modulation should be large compared to the mean index. We see that a trade-off may be necessary if it is desirable to have both a spectrally narrow and a high-reflectivity grating. A more accurate form of the relation for the peak reflectivity (Friebele *et al.*, 1995) is

$$R_{max}(\lambda_B, L_G) = \tanh^2\left\{0.8\pi\Delta n\,\frac{L_G}{\lambda_B}\right\}. \tag{3.140}$$

By way of example we shall consider a Bragg grating formed within an optical fiber, in which case the following parameters are reasonable: $n_0 = 1.5$, $\Delta n = 10^{-4}$, $\Lambda = 0.517\,\mu m$, $L_G = 5\,mm$. Then $\lambda_B = 2n_0\Lambda = 1.55\,\mu m$,

$$R_{max} = \tanh^2\left\{\pi\left(\frac{\Delta n}{2n_0}\right)\left(\frac{L_G}{\Lambda}\right)\right\} = 58.8\%$$

and

$$\Delta\lambda_w = \lambda_B\sqrt{\left\{\frac{\Lambda}{L_G}\right\}^2 + \left\{\frac{\Delta n}{2n_0}\right\}^2} = 0.168\ \text{nm}.$$

CHAPTER 4

Light Sources and Detectors

4.1 INTRODUCTION

In this chapter we describe how light is generated and detected and identify the types of light source and photodetector most suitable for use in fiber optic structural monitoring. In general light sources can be divided into those that are *broadband and incoherent*, and those that are *narrowband and coherent*. Incandescent light bulbs fall into the first category, while many lasers are described by the second category. We shall focus on semiconductor sources, as these: are very small, require little power, are rugged, and can be produced with consistent properties at low cost in large numbers. They can be used with semiconductor photodetectors which also have desirable attributes, notably small size, high sensitivity, ruggedness and low cost. These features are germane to making inexpensive, portable sensor demodulation (interrogation) systems.

4.2 LIGHT GENERATION AND GAIN MEDIA

Electromagnetic radiation, or light as we have defined it, is most often generated by the spontaneous transition of an electron from an excited state to some lower energy state within an atom. If the electron was excited by a thermal collisional process, then we have an incandescent source. In such a source the collisions result from the elevated temperature of the atoms. The sun is of course the most powerful (local) incandescent source. In the familiar neon tube, or a helium–neon laser, a few free electrons are accelerated by a voltage established between electrodes in the gas. With sufficient voltage, a number of the electrons gain enough energy to ionize additional atoms and release more free electrons. Most of the electrons suffer collisions before they acquired sufficient energy to ionize atoms and instead excite the bound electrons of the atoms in their path to energetic states from which they decay with the emission of radiation (Fig. 4.1). In this way a fraction of the electrical energy dissipated in the gas is converted to light via the process of "spontaneous emission."

We saw in Chapter 3 that light of the right wavelength can be absorbed, or spontaneously emitted, by the atoms depending on the state of the atom (Fig. 3.8). One of the most important contributions made by Albert Einstein to the quantum theory of matter was the recognition that equilibrium considerations require the existence of a third interaction between light and atoms. This process he termed "stimulated emission," and it is this process that makes possible the laser. Indeed, the name *laser* is this acronym: "light amplification by stimulated emission of radiation." *Stimulated emission* arises when excited atoms are immersed in a radiation field of the appropriate frequency for the transition. An incident photon forces, or *induces*, the atom in the excited state to fall back to the lower state with the emission of a photon that is identical to the stimulating photon (Fig. 4.2). It is important to note that identical means in this instance that the stimulated photon has the same *frequency, direction*, and *phase* as the incident photon.

What makes stimulated emission so important is that it leads to amplification of the light wave if more atoms exist in the higher energy state than the lower energy state of the transition (Fig. 4.3). If we designate the population density of the upper and lower states by N_p and N_q, respectively, then the condition for the existence of such a *population inversion* is

$$\mathcal{N} \equiv N_p - N_q > 0, \tag{4.1}$$

where we have introduced $\mathcal{N}$ to represent the *inverted population density* of the two states. For simplicity we assume no degeneracy of states (i.e., we assume only one state corresponds to a given energy level). Clearly, a positive inverted population density is required for the amplification, or gain, in the radiation field. A medium in which light grows rather than suffer attenuation is termed an "active" or "gain" medium and represents a highly non-equilibrium state that is very rare to find in nature. Thermal equilibrium is the normal state for materials when the temperature is not changing rapidly, (see Fig. 4.3). In thermal equilibrium the ratio of the population densities in the higher and lower states of the transition of interest is described by the *Boltzmann distribution*

$$N_p = N_q e^{-\Delta U/\kappa T}, \tag{4.2}$$

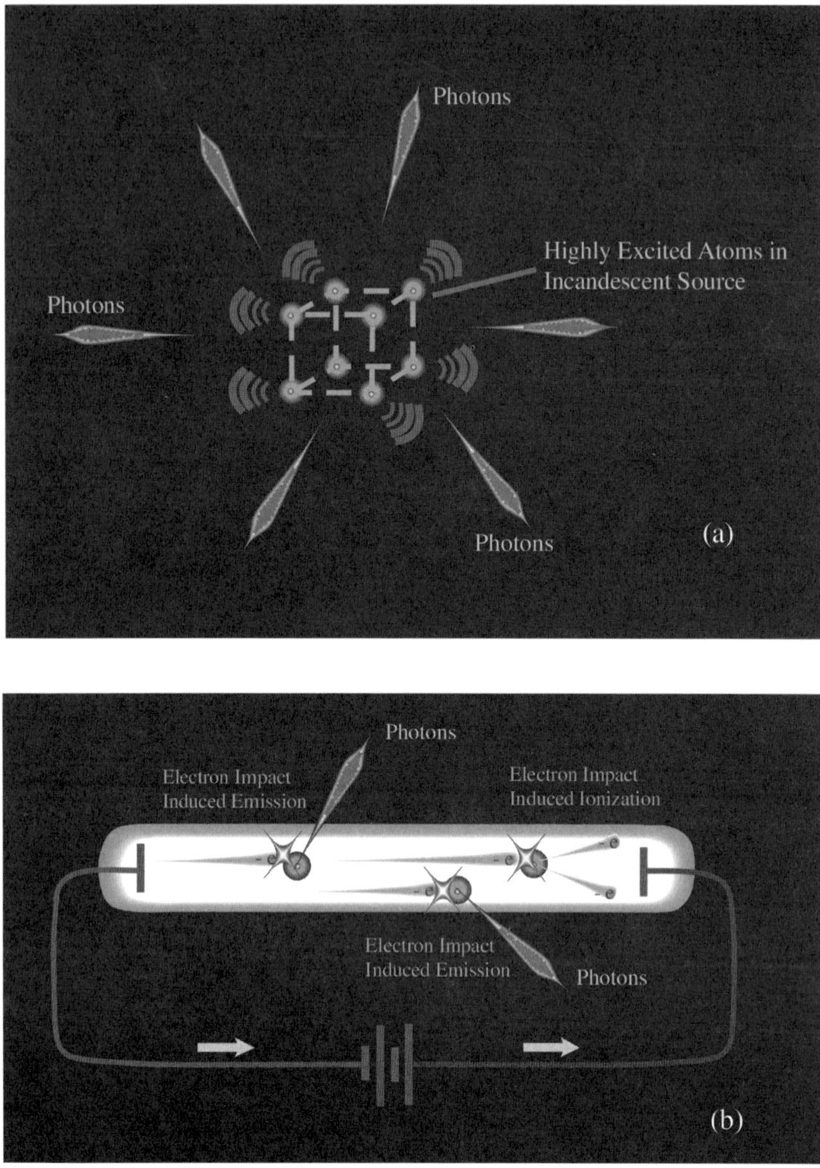

FIGURE 4.1. (a) Spontaneous emission of light from the atoms of a solid heated to incandescence. (b) Electric discharge in low-pressure gas gives rise to spontaneous emission as electrons, accelerated in the applied electric field, collide with atoms and either excite them, or if the electrons acquire sufficient energy, ionize them.

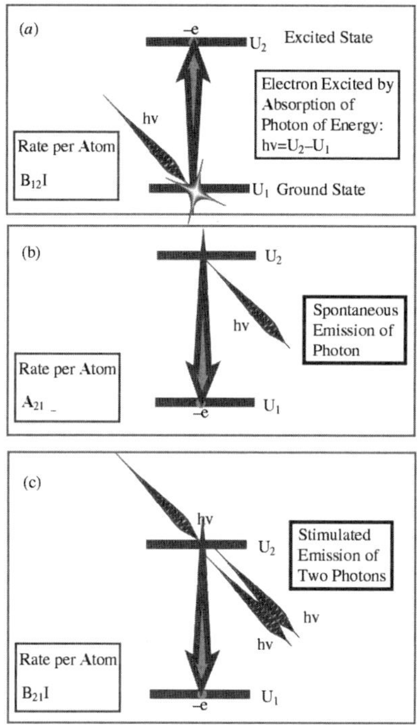

FIGURE 4.2. Three primary radiative process: (a) absorption, (b) spontaneous emission, (c) stimulated emission.

where, $\Delta U \equiv U_p - U_q$, is the energy difference between the higher and lower states, $\kappa = 1.3807 \times 10^{-23}\ \text{JK}^{-1}$, is the *Boltzmann constant*, and T is the temperature in Kelvin (K).

The rate of *absorption* per atom in the lower state by radiation at frequency, ν, is $B_{qp}I(\nu)$, while the rate of *stimulated emission* per atom in the higher state by radiation at frequency, ν, is $B_{pq}I(\nu)$, where $I(\nu)$ is the *spectral intensity* (spectral irradiance in $\text{Wm}^{-2}\ \text{Hz}^{-1}$) of the radiation field, and B_{pq} and B_{qp} are the respective: *stimulated emission* and *absorption* Einstein coefficients for the transition of interest. The *volume rate of absorption* for atoms in the lower state is $N_q B_{qp} I(\nu)\zeta(\nu)d\nu$, while the *volume rate of stimulated emission* from atoms in the higher state is $N_p B_{pq} I(\nu)\zeta(\nu)d\nu$. In these expressions, $N_q\zeta(\nu)d\nu$ and $N_p\zeta(\nu)d\nu$ represent the *number of atoms per unit volume* in the q and p states, respectively, that can interact with radiation at frequency, ν.

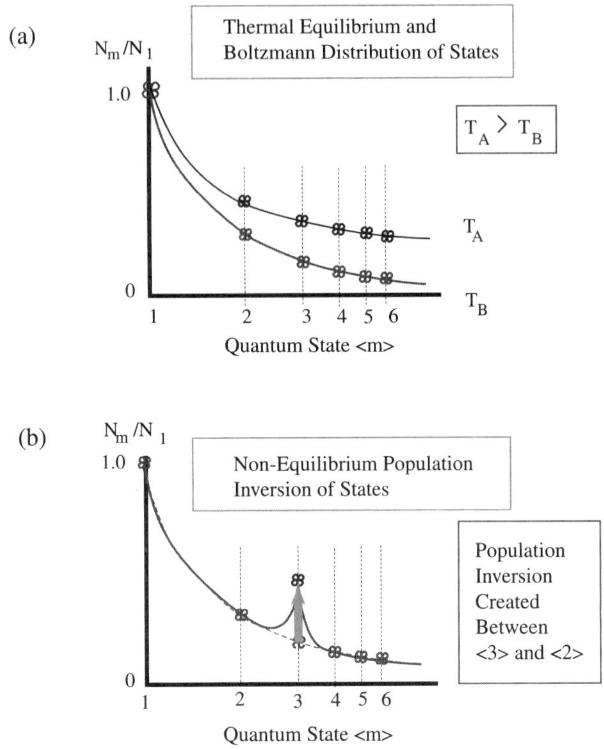

FIGURE 4.3. (a) Thermal equilibrium and the Boltzmann distribution of states. (b) Non-equilibrium of states associated with a population inversion.

Since the photon density is increased by one for each stimulated emission event and decreased by one for each absorption, the net increase in the *spectral photon density*, $\Phi(v)$, of the radiation field is

$$\Delta\Phi(v)dv = N_p B_{pq} I(v)\zeta(v)dv - N_q B_{qp} I(v)\zeta(v)dv \tag{4.3}$$

Since we are neglecting the degeneracy of states

$$B_{pq} = B_{qp}, \tag{4.4}$$

and we can write $B_{pq} = B_{qp} = \mathscr{B}$. It follows that (4.3) can be rewritten

$$\Delta\Phi(v)dv = \mathscr{N}\mathscr{B}I(v)\zeta(v)dv, \tag{4.5}$$

where the *spectral photon density* can in general be related to the *spectral intensity* of the beam by the relation (Saleh and Teich, 1991):

$$I(v) = \Phi(v)hvc. \tag{4.6}$$

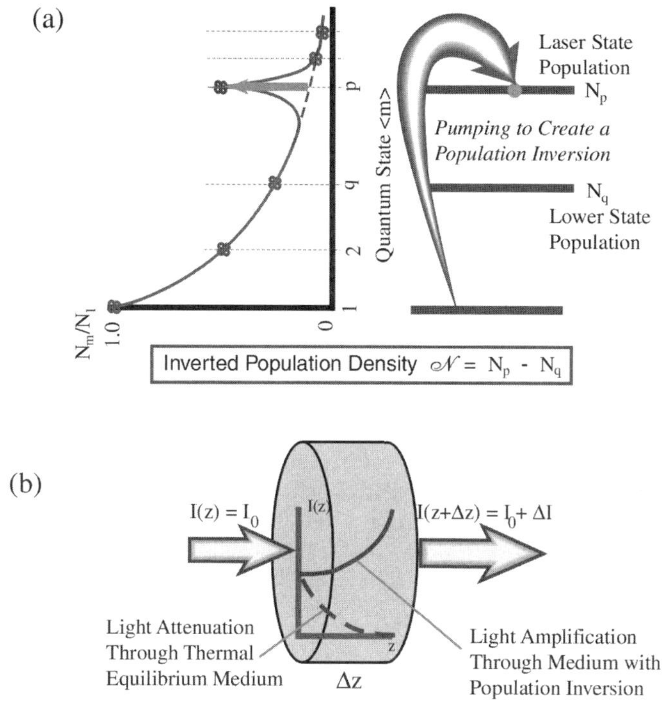

FIGURE 4.4. (a) Creation of an inverted population density between the two states that constitute the laser transition. (b) Exponential attenuation of light in passing through a medium in thermal equilibrium and amplification (exponential growth) of light in the presence of a population inversion.

As indicated in Chapter 3, $h\nu$ corresponds to the *photon energy*. From Fig. 4.4 it follows that the incremental increase in the intensity of the beam of radiation at frequency ν, passing through a thickness Δz of the gain medium

$$\Delta I(\nu) = \Delta\Phi(\nu)h\nu c\Delta z, \tag{4.7}$$

in which case, using (4.7), we can write

$$\Delta I(\nu) = \mathcal{N}\mathcal{B}I(\nu)h\nu\zeta(\nu)c\Delta z. \tag{4.8}$$

If we introduce the *spectral gain coefficient* (gain per unit length) for the active medium

$$g(\nu) = \mathcal{N}\mathcal{B}h\nu\zeta(\nu)c, \tag{4.9}$$

then we can express (4.9) in terms of a radiative transfer equation of the form

$$dI(\nu)/dz = g(\nu)I(\nu). \tag{4.10}$$

The solution of this equation indicates an exponential growth of the radiation field.

$$I(\nu, z) = I(\nu, 0)\exp[g(\nu)z], \tag{4.11}$$

and this is depicted in Fig. 4.4.

A *gain medium* is created by pumping a large amount of energy selectively into the higher state of the transition on which optical gain is to be created. This produces the desired population inversion (Fig. 4.4). In the case of the ruby laser of Maiman, an intense burst of light from a flash lamp was used to create a population inversion in the ruby crystal, Fig. 3.3. This actually arises in two stages. First, a spectrally broadband pulse of the flash lamp radiation is absorbed by a large fraction of the ground-state Cr^{3+} ions in the ruby crystal, which pumps the electrons of these ions into two broad energy bands (Fig. 4.5). Instead of decaying back to the

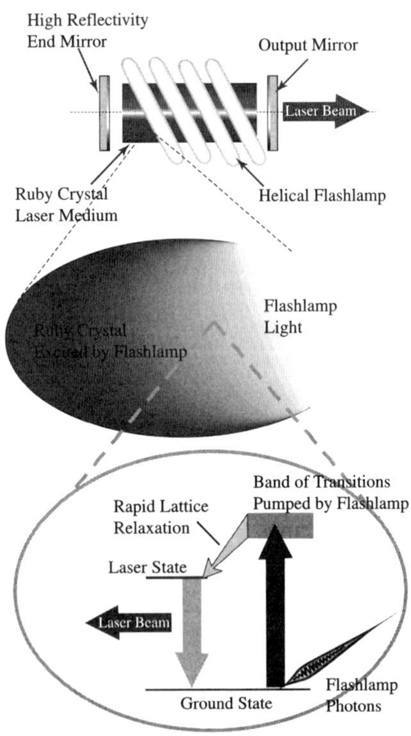

FIGURE 4.5. Schematic illustration of ruby laser and its pumping configuration.

ground state, however, the second stage involves the very rapid transfer, by lattice vibrations, of most of these excited ions to the narrow energy laser state creating a population inversion between this state and the depleted ground state. Ironically, although this was the pumping scheme used for creating the first laser, it is the most difficult because a sizable fraction of the ground-state population has to be excited to create the population inversion. Many lasers today achieve laser action with much less energy as they do not use the heavily populated ground state as the lower state of the laser transition (Siegman, 1986).

The existence of a *population inversion*, i.e., ($\mathcal{N} > 0$) and the concomitant gain in the radiation field is a necessary but not sufficient condition to produce a laser. This is because all light of the appropriate frequency, no matter in what direction it is traveling, will be amplified and this rapidly exhausts the amplification process before substantial growth can be achieved. The crucial step in turning a *gain medium* into a laser is the introduction of a pair of mirrors that are parallel to each other and perpendicular to the desired direction of propagation for the resulting light beam (Fig. 4.6). Usually these mirrors are plane and placed at either end of the gain medium, constituting a *Fabry–Perot cavity*, or interferometer, discussed in Chapter 3.

The length of the gain medium in the desired direction of propagation is made large relative to its length in the transverse direction. Thus, light propagating through the gain medium in the desired direction will be amplified, then reflected back through the gain medium and amplified again. As this process is repeated many times, the lightwave grows substantially and given sufficient gain, we have a laser. We shall study this further in the next section. It should be noted, however, that there are more complex forms of semiconductor lasers that do not comply with this *Fabry–Perot cavity* (FPC) configuration. These have special properties and will be discussed later in Section 4.7.

4.3 FABRY–PEROT CAVITY LASERS

It is fairly obvious from the outset that the creation of a laser really hinges on the relative magnitude of the gain experienced by the light wave in traveling through the gain medium as compared to the loss it suffers at each reflection, assuming no other losses. In order to understand the full influence of the FPC arrangement on the laser's *threshold condition* and the spectral properties of the output beam we consider the simple model of a homogeneous gain medium within an FPC, (Fig. 4.6). A highly reflecting plane mirror with an amplitude reflection coefficient, r_1 is positioned at one end of the gain medium, where $z = 0$, and a partial reflecting plane

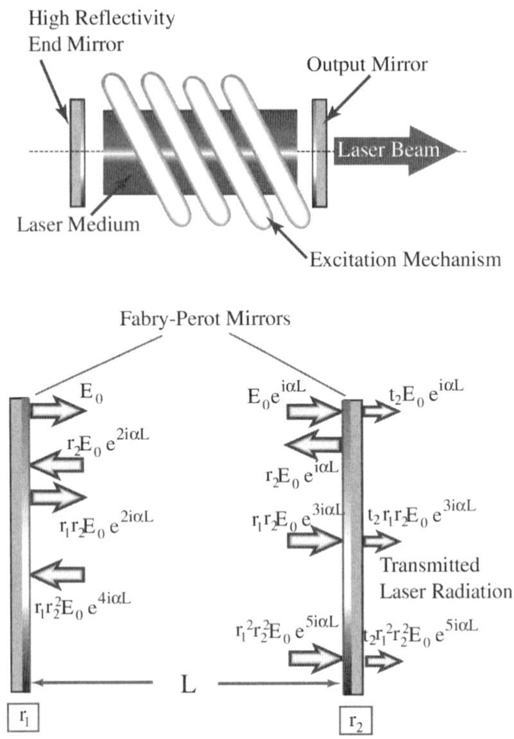

FIGURE 4.6. Fabry–Perot laser and schematic illustration of multiple reflections and gain.

mirror with an amplitude reflection coefficient r_2 is positioned at the other end of the gain medium, at $z = L$. We assume a *monochromatic, plane wave* with no variation in the x and y directions, propagating in the z-direction. Before we proceed, however, we need to know how the intensity of the light wave is related to its **E**-field. For a monochromatic, plane wave, we have seen in Section 3.2 that

$$I = \varepsilon_0 c E^2, \tag{4.12}$$

which means that the gain in the **E**-field is expressed in terms of a new coefficient, ψ, termed the **E**-*field gain coefficient*, where a comparison of Eqs. (4.11) and (4.12) reveal that

$$\psi(\nu) = g(\nu)/2. \tag{4.13}$$

We now introduce the *complex propagation constant*,α, to account for both the phase change and optical gain experienced by the **E**-field at frequency, ν as it propagates through the medium

$$\alpha(\nu) \equiv \beta(\nu) + i\psi(\nu). \tag{4.14}$$

If we assume that the **E**-field amplitude of this plane wave at $z = 0$ is $E_0 e^{-i\omega t}$, Then at $z = L$,

$$E(\nu, L) = E_0 e^{i(\alpha L - \omega t)}. \tag{4.15}$$

Now since the temporal factor, $e^{-i\omega t}$, is common, we can dispense with it, as it plays no significant role in our discussion. The lightwave on reaching the end of the gain medium is partially reflected and partially transmitted. The fraction of the lightwave **E**-field at frequency (ν) that is transmitted through the second output mirror is given by

$$E(\nu, L)_{t1} = (1 - r_2) E_0 e^{i\alpha L}, \tag{4.16}$$

assuming no losses, such as absorption, at the mirror Fig. 4.6. The fraction of the lightwave **E**-field at frequency, (ν) that is reflected

$$E(\nu, L)_{r1} = r_2 E_0 e^{i\alpha L}. \tag{4.17}$$

This part of the original wave makes another pass through the gain medium, suffers a reflection at the other mirror, and after a further pass through the gain medium is incident on the output mirror once again. In this case, the fraction of this wave that is transmitted is given by

$$E(\nu, L)_{t2} = (1 - r_2) r_1 r_2 E_0 e^{3i\alpha L}. \tag{4.18}$$

This process will repeat itself many times, and if we sum all of the transmitted **E**-field components, the net transmitted **E**-field at frequency ν takes the form

$$E(\nu, L) = (1 - r_2)\left\{E_0 e^{i\alpha L} + r_1 r_2 E_0 e^{3i\alpha L} + [r_1 r_2]^2 E_0 e^{5i\alpha L} \ldots\right\}. \tag{4.19}$$

This represents an infinite series that can be written in the simple form

$$E(\nu, L) = (1 - r_2) \frac{E_0 e^{i\alpha(\nu)L}}{[1 - r_1 r_2 E_0 e^{2i\alpha(\nu)L}]}, \tag{4.20}$$

provided, $r_1 r_2 E_0 e^{2i\alpha L} < 1$.

The output of the laser is essentially the intensity transmitted through this partially reflecting *output mirror*. Its value is derived from Eq. (4.20) using, Eq. (4.12), and with some algebra takes the form

$$I(\nu) = I_0 T \frac{\zeta(\nu)}{\{1 - G(\nu)\}^2} \frac{1}{\left[1 + \dfrac{4G(\nu)\sin^2\{\beta(\nu)L\}}{\{1 - G(\nu)\}^2}\right]}, \tag{4.21}$$

where $I(\nu)$ is the *spectral intensity* (spectral irradiance) in $\mathrm{Wm^{-2}\,Hz^{-1}}$, I_0 is the total spectrally integrated intensity, T is the intensity transmission coefficient for the partially reflecting output mirror, $\zeta(\nu)$ is the *line profile function* for the laser transition, $G(\nu)$ is the *double pass gain*, given by

$$G(\nu) \equiv r_1 r_2 \exp[2g(\nu)L], \tag{4.22}$$

and $\beta(\nu)$ is the propagation constant defined by Eq. (3.10), i.e.

$$\beta(\nu) = 2\pi n\nu/c. \tag{4.23}$$

As indicated earlier, the *line profile function*, $\zeta(\nu)$, expresses the probability of spontaneous, or stimulated, emission at a frequency ν relative to that at the line center frequency ν_0, as indicated in Fig. 4.7, and $\zeta(\nu)d\nu$, represents the probability of emission, or absorption, in the frequency interval $(\nu, d\nu)$.

The presence of the gain medium within the Fabry–Perot cavity leads to the observed multi-peaked spectrum shown in Fig. 4.8. The line profile function serves as the envelope for the comb like spectrum generated by the Fabry–Perot cavity. The narrowing of this spectral envelope with

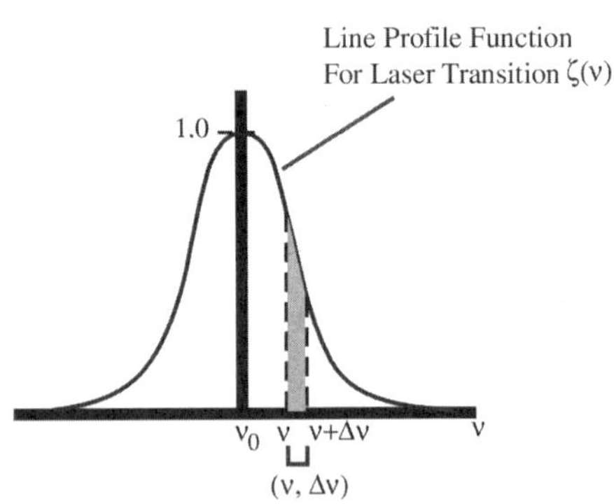

FIGURE 4.7. Line profile function for the laser transition.

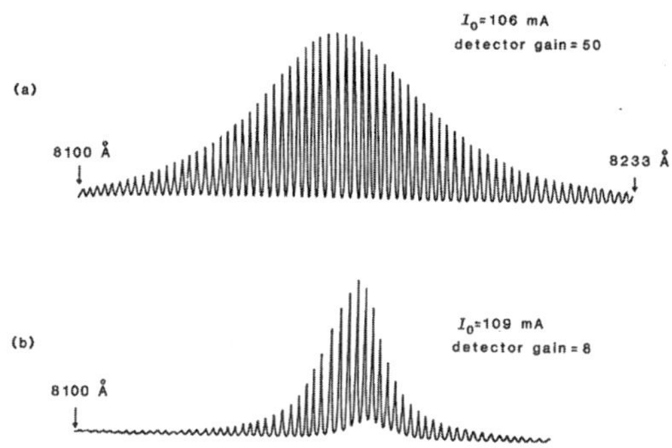

FIGURE 4.8. Amplified spontaneous emission spectra from a GaAs semiconductor laser diode for two values of gain just below the laser threshold. The central modes are seen to become more dominant and the envelope narrows as the threshold for laser oscillation is approached. Note that the "detector" gain is less for (b) and (a), so that the actual light signal for (b) is much larger than for (a). From *"Lasers"* A. E. Siegman, University Science Books, 1986, p. 446.

increasing gain is expressed by the *gain-modified line profile* function, the third factor in the output intensity expression (4.21). That is to say

$$\zeta^*\{\nu, G(\nu)\} = \frac{\zeta(\nu)}{\{1 - G(\nu)\}^2}, \tag{4.24}$$

constitutes the *gain curve* for the active medium, and we see that the more closely ν approaches the *line center frequency* ν_0, the closer $G(\nu)$ is to unity and the larger the value of $\zeta^*\{\nu, G(\nu)\}$. In a gas laser, such as a helium–neon laser the gain curve can be quite narrow, much less than 0.1 nm. By comparison the gain curve in a semiconductor laser can be in excess of 50 nm.

The fourth factor in (4.21)

$$\Omega(\nu, G) = \frac{1}{\left[1 + \dfrac{4G(\nu)\sin^2\{\beta(\nu)L\}}{\{1 - G(\nu)\}^2}\right]}, \tag{4.25}$$

is termed the *gain-modified Airy function* and accounts for gain narrowing of the Fabry–Perot comb like spectral structure evident in Fig. 4.8. We saw in Chapter 3, Fig. 3.25, the *normalized Airy function* reflection spectrum of a Fabry Perot cavity in the "absence" of a gain medium.

In the absence of a gain medium the *frequency separation* of the Fabry–Perot modes shown in Fig. 3.25, is given by the relation

$$\Delta\nu = \frac{c}{2nL}. \tag{4.26}$$

This is changed slightly when a gain medium is present, but to first order it gives a good idea of the spectral separation in the low-loss reflection notches arising from the cavity. Laser action is most likely to arise at these low-loss frequencies, explaining the complex spectral form of the laser output.

As the gain medium is pumped more intensely, the energy grows in the form of the population inversion, and the *double pass gain* approaches unity. It is quite apparent from (4.21) that the *output intensity* of the laser becomes singular when two conditions are simultaneously satisfied

$$\text{(i)}\quad 1 - G(\nu) = 0, \tag{4.27}$$

$$\text{(ii)}\quad \sin^2 \beta(\nu)L = 0. \tag{4.28}$$

The pair of conditions (4.27) and (4.28) can be restated in the form

$$\text{(i)}\quad G \equiv r_1 r_2 e^{2gL} = 1, \tag{4.29}$$

and

$$\text{(ii)}\quad \beta(\nu_m)L = m\pi, \tag{4.30}$$

where m is an integer that defines the *Fabry–Perot cavity resonance mode.*

If we introduce the *cavity loss per pass*

$$\gamma = \frac{1}{2}\log\left\{\frac{1}{r_1 r_2}\right\}, \tag{4.31}$$

we can express the first condition in terms of a critical, or threshold, *population inversion density*, $\mathcal{N}_c$, by requiring that the *gain per pass* at the line centre frequency ν_0 in Eq. (4.9), must just equal the *loss per pass*, i.e.

$$\text{(i)}\quad g_c L = \mathcal{N}_c \mathcal{B} h\nu c L \zeta(\nu_0) = \gamma, \tag{4.32}$$

which can also be expressed in the form

$$\mathcal{N}_c = \frac{\gamma}{\sigma(\nu_0)L} \tag{4.33}$$

where we introduce the *stimulated emission cross-section* at the line center frequency

$$\sigma(\nu_0) \equiv \mathcal{B} h\nu \zeta(\nu_0) c. \tag{4.34}$$

Equation (4.32) essentially states that for laser action to be possible, the single-pass gain at threshold, $g_c L$, must be at least equal the effective single-pass loss through the mirrors, γ. Equation (4.33) restates this requirement in terms of the *critical value* for the *inverted population density*, $\mathcal{N}_c$. The second condition (4.30) gives us the *Fabry–Perot mode frequencies*

$$\nu_m = \frac{mc}{2nL}, \tag{4.35}$$

mentioned in Chapter 3, Eq. (3.71).

In summary, laser action requires that the gain medium is pumped hard enough to at least attain the threshold value for the inverted population density. When this is achieved, laser action is restricted to those frequencies that coincide with the low-loss Fabry–Perot cavity modes. This is schematically illustrated in Fig. 4.9, where it is seen that laser action arises only for those frequencies where the loss gain curve is greater, or matches, the low loss modes (spikes) of the FP cavity.

The repeated reflections of the lightwave within the laser forces it to propagate through an effective gain medium of length much greater than L, the physical length of the gain medium. Because the rate of interaction is proportional to the stimulating intensity those frequencies where the field is high acquire energy more rapidly than those frequencies where the field is weak. Competition for the limited energy in the form of the

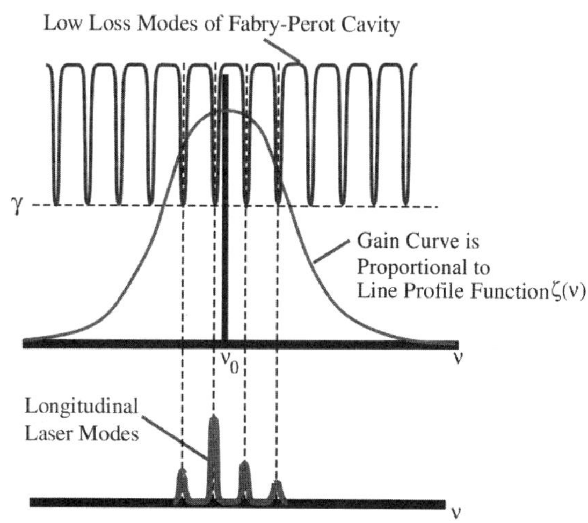

FIGURE 4.9. The laser transition line profile and the Fabry–Perot cavity loss curves, revealing that laser action can only occur on those cavity modes for which the gain exceeds the loss.

inverted population inversion thus leads to the observed narrowing of the spectral features.

Furthermore, as the gain is increased by stronger pumping, the spectrum narrows so that laser action eventually occurs on only one Fabry–Perot mode, usually the mode closest to the peak of the gain curve. This narrowing process can be seen, in Fig. 4.10, for a *semiconductor laser* as its *injection current* is increased from 155 to 176 mA. The injection current is the method of pumping energy into the gain region in this type of laser, and it should be noted that the intensity scale factor is changed from 1 to 100 between these four outputs, so the peak intensity at 176 mA is over 100 times that at 155 mA. The resulting laser radiation is seen to be very

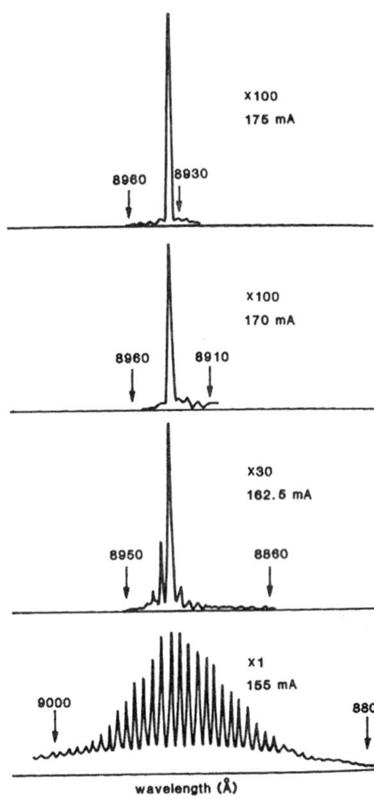

FIGURE 4.10. Single-mode laser action rises up from the multimode amplified spontaneous emission of a GaAs laser diode subject to increasing injection current. Note the changes in the vertical scale in the succession of curves. From Siegman, A.E., "*Lasers*," (1986). University Science Books, p. 464.

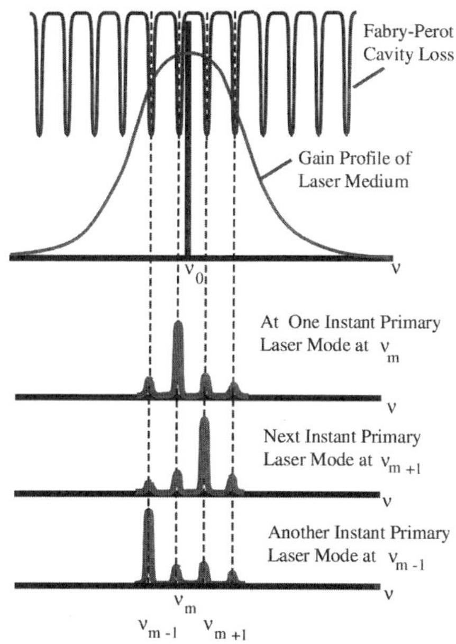

FIGURE 4.11. Mode hopping of longitudinal laser modes arises with spatial hole burning and small fluctuations in the laser conditions.

narrowband. In lasers with a broad and fairly flat gain curve, which particular mode lases can change instant by instant due to small fluctuations in variables like temperature or injection current. The effect of this "mode hopping," as it is called, makes the time average laser spectrum somewhat broader than a single mode, Fig. 4.11. This will become an important issue when we consider tuning a laser for demodulation of certain fiber optic sensors.

In reality, *gain saturation* is the limiting process that prevents the laser intensity from achieving the singular infinite value predicted by Eq. (4.21). Nevertheless, the pair of conditions (4.29) and (4.30) serve as the *necessary* and *sufficient* condition for laser action. Gain saturation is easily understood in terms of the limited amount of energy available for the laser radiation in the form of the population inversion. Each atom that has been pumped into the higher state of the laser transition is available to donate its energy to the laser beam through the stimulated emission process. Once this has occurred, the atom has to be reexcited, and that takes time, so the ultimate laser power that can be attained is limited by the rate of pumping.

4.4 SEMICONDUCTOR RADIATION SOURCES

A number of very different types of radiation source are used with fiber optic sensors. Sources can be classified in a number of ways, so the requirements of the sensor have to be known before the most suitable source can be identified. The spectral requirements of the sensor are often the most important in choosing a source. Light sources are either *broadband* or *narrowband*, and operate in a *pulsed*, or *continuous* manner. As we shall see shortly, the properties of the optical fiber and its sensing element dictate the wavelength of the source, while the kind of sensor and sometimes the method of demodulation (or, interrogation) of the sensor determines if the source should be broadband or narrowband, pulsed or continuous. The power of the source, or to be more exact, the fraction thereof that finds its way into the optical fiber, need only be sufficient to ensure an adequate *signal-to-noise ratio* at the photodetector. Lastly, the source can be specified as being *coherent* or *incoherent*. We shall see later that in many cases the choice is not so stark and often the specific property of the source has to be spelled out carefully.

By and large, *semiconductor* light sources are chosen for fiber optic structural sensors, because they are very small and have low power requirements, which permits compact and highly portable sensor demodulation systems to be produced. Semiconductor sources can be classified as follows:

1. Light emitting diodes (LED)
2. Superluminescent diodes (SLD)
3. Laser diodes (LD)

Before we describe these in detail it is necessary to introduce some of the basic physics of semiconductor materials.

A *semiconductor* is a crystalline or amorphous solid with an *electrical conductivity* that lies between the values of *metals* and *insulators*. Furthermore, its electrical conductivity can be significantly changed by altering its temperature or doping it with special impurities. Electronic devices, such as computer memory chips, are made primarily with silicon (Si), while optoelectronic devices, such as laser diodes, are made of binary, ternary, or quaternary compounds of (III–V) elements of the periodic table. Optoelectronic devices include light sources and detectors and are typically multilayer pin-head size devices that are fabricated by vapour deposition in expensive vacuum machines in extremely clean room facilities, that can cost close to a one billion dollars. First, extremely pure substrates are made from binary compounds like *gallium arsenide* (GaAs) or *indium phosphide* (InP), in the form of very high quality thin discs. Features such as active, or gain, regions (in the case of a laser) are formed from extremely thin layers of materials like GaInAsP that are fabricated on top of these substrates, (Fig. 4.12).

In order to attain a basic understanding of optoelectronic devices we need to introduce the concepts of *energy bandgaps*, *p–n junctions* and *electron–hole recombination*. We saw earlier that the electrons within isolated atoms are constrained to exist in quantized states that are defined in very narrow energy ranges referred to as energy levels, see Fig. 3.7. When a large number of these atoms exists as part of a solid these energy levels are smeared out into energy bands, (Fig. 4.13). The most weakly bound electrons are confined to one of two energy bands. These are termed the *valence* and *conduction* bands. In an isolated atom, bound electrons can only exist in the allowed energy states and are confined to the immediate vicinity of the atom. That is to say the quantum state of a bound electron is highly localized in space, see Fig. 3.7. This is not the case in the *conduction* (C) band of a solid where the electrons are free to move anywhere within the solid under the action of an applied electric field. Electrons within a solid cannot occupy the energy interval between the C- and V-bands. This *forbidden energy zone* is important and termed the *bandgap energy* (E_g). It is important because it represents the smallest *quanta of radiation* ($h\nu$) necessary to excite an electron from the V-band to the C-band. It also plays a role in determining the material properties.

Insulator materials, like glass, have large bandgap energies (>3 eV) so virtually no electrons are likely to be thermally excited from the V-band into the C-band, and these materials exhibit extremely low electrical

FIGURE 4.12. Optoelectronic wafer used in the manufacture of lasers and photodetectors.

conductivity, (Fig. 4.14). We might ask why the electrons in the V-band do not move under the influence of an applied electric field and contribute to the electrical conductivity of the material. The answer to this question comes from the quantized nature of material. In particular, the *Pauli exclusion principle* states that "no two electrons can occupy the same quantum space." In an isolated atom, that is understood to mean the same quantum state. In the case of the solid, all of the constituent atom's ground states merge to fill the physical space of the solid and form the V-band. If all of the states in the V-band are occupied (i.e., filled with electrons) then none of its electrons can move.

When a few of the V-band electrons are excited thermally into the C-band they become free to move and, furthermore, the "holes" left behind in the V-band are also free to move. This can be understood in terms of an analogy. If you fill a box with identical marbles, then none of the marbles are free to move, even though all of the marbles are subject to the force of gravity. If, however, you were to magically make one of the marbles vanish, leaving a hole, then the marble above it would fall into the space created, and this process would be repeated until the hole reached the top of the marbles (we neglect friction).

An observer might say that the hole was like a marble with negative mass moving against gravity. In a similar manner the vacancies (holes) in the V-band act as if they are positively charged and move, under an applied electric field, in the direction opposite to that of the elevated

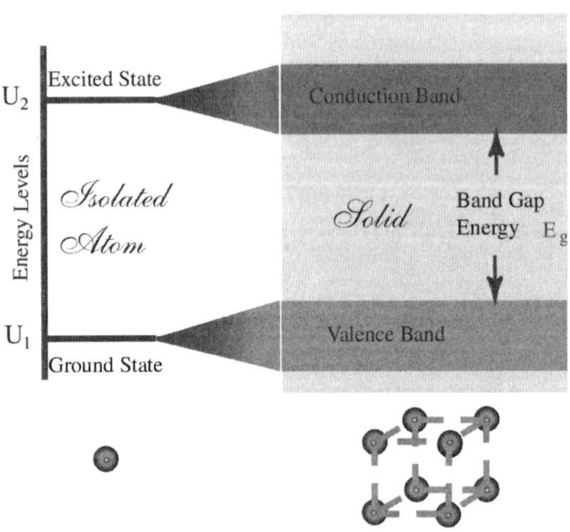

FIGURE 4.13. Difference between the primary energy levels of an isolated atom and the "conduction" and "valence" bands of a solid made of many such atoms.

electron. Indeed, it is the quick separation in space between the excited electron and its hole that prevents this electron from falling back into its hole. When some other excited electron blunders into the hole and drops back into the V-band, *electron–hole recombination*, the excess energy is often released in the form of radiation, (Fig. 4.15). The frequency of the resulting radiation, v, is primarily determined by the *bandgap energy* (E_g), i.e.

$$hv = E_g, \tag{4.36}$$

from which the wavelength, λ, of the electron-hole recombination radiation is given by

$$\lambda = hc/E_g. \tag{4.37}$$

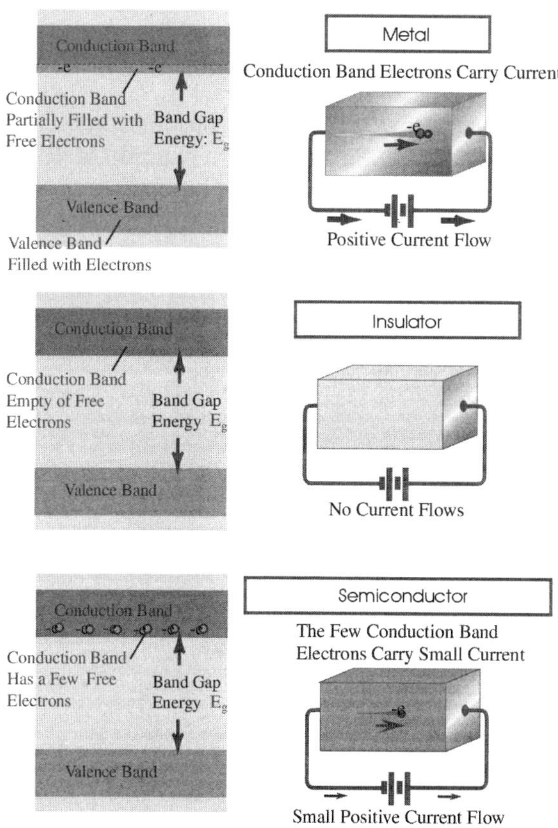

FIGURE 4.14. Some of the most significant differences in terms of electrical properties for metals, insulators, and semiconductor materials.

In a *metal*, the bandgap energy is small so many electrons can thermally be excited to lie within the C-band, and their ability to move under the influence of an applied electric field is what gives rise to the high electrical conductivity that we associate with metals, see Fig. 4.14.

Semiconductor materials (like germanium) fall between the extremes of metals and insulators as a few electrons exist in the C-band and give rise to a low level of electrical conductivity. The inherent conductivity can be greatly increased by suitable doping of a semiconductor material. Doping with a *donor* element contributes electrons to the C-band, making an *n-type* material with a majority of negative carriers. This is because the donor atoms have a lower *electron affinity* than the host atom and are able to donate some of their electrons to the host C-band. In contrast, doping with an *acceptor* element (atoms with a higher *electron affinity* than the host atoms) removes electrons from the V-band, creating holes, (Fig. 4.16), and forming a *p*-type semiconductor with a majority of positive carriers. The few mobile electrons in the *p*-type material and the holes in the *n*-type material are referred to as *minority carriers.*

To create optoelectronic components various combinations of *p*- and *n*-type materials are used. Since the wavelength of the electron–hole recombination radiation is inversely proportional to the *bandgap energy,* Eq. (4.37), specific wavelength requirements necessitate producing semiconductor materials with the appropriate characteristics. In the case of light sources it is very important that the electron–hole recombination proceed via the emission of radiation, as opposed to the released energy

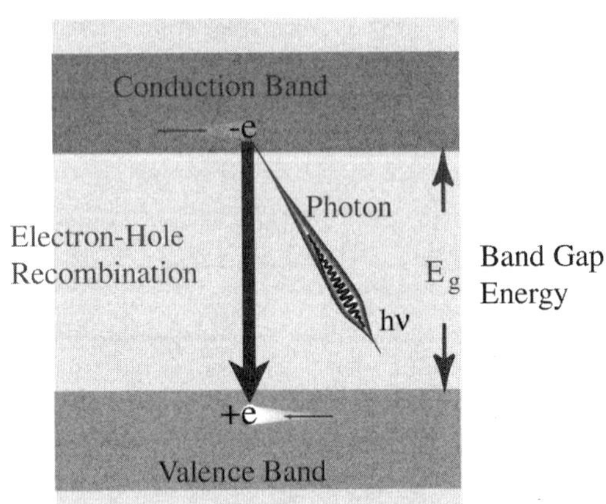

FIGURE 4.15. Electron hole recomdinbation in a solid resulting in the spontaneous emission of a photon of energy $h\nu$ equal to the bandgap energy E_g.

TABLE 4.1 Popular LED and Laser Diode Materials

Material	Bandgap (eV)	Wavelength (μm)	
Binary			
GaP	2.24	0.55	Gallium–Phosphide
AlAs	2.09	0.59	Aluminium–Arsenide
GaAs	1.42	0.87	Gallium–Arsenide
InP	1.33	0.93	Indium–Phosphide
InAs	0.34	3.60	Indium–Arsenide
Ternary			
AlGaAs	1.4–1.6	0.77–0.87	Aluminium–Gallium–Arsenide
Quaternary			
InGaAsP	0.7–1.1	1.10–1.67	Indium–Gallium–Arsenide-Phosphide

going into lattice vibration. Certain compound semiconductor materials made from III and V elements of the periodic table satisfy both the wavelength and light generation efficiency requirements. The most popular semiconductor materials used for light sources and some of their characteristics are provided in Table 4.1.

4.5 LIGHT-EMITTING DIODES

Conversion of an electric current into light within a semiconductor is best accomplished by a *p–n junction*, created when *p*-type and *n*-type materials are produced adjacent to each other. If the *p–n* junction is formed by different doping of the same substrate, it is referred to as a *homojunction*. In contrast, a *p–n* junction formed with different semiconductor materials is termed a *heterojunction*. Although the latter is more costly to produce, its light generation efficiency is far superior to the former, and as a result most semiconductor light sources are based on some form of heterojunction. As soon as a *p–n* junction is formed some of the electrons from the *n*-type material diffuse from the high electron concentration region into *p*-type material while some holes from the *p*-type material diffuse into the *n*-type material. This diffusion of charge carriers leaves unneutralized (negative) acceptor ions in the *p*-type material and unneutralized (positive) donor ions in the *n*-type material. This creates a *depletion layer* with a potential difference (barrier) between the *p*- and *n*-type materials that is just sufficient to terminate the charge flow of the majority carriers across the p–n junction, (Fig. 4.17). As a result, a narrow region on both sides of the junction, called the "depletion layer," becomes almost totally depleted of mobile majority carriers and is therefore very resistive in comparison to the bulk material properties.

If a *forward bias voltage* (positive voltage applied to the p-region of the junction) of the appropriate amount is applied across the junction the barrier potential is reduced to zero and the electrons and holes, from their respective n- and p-type materials, are free to move across the junction and recombine producing light emission (Fig. 4.18). This is the basis of the *light-emitting diode* (LED) where copious amount of light via spontaneous emission is created at high values of injection current. In essence, a high injection current forces large numbers of electrons and holes to the same region of space (the p–n junction) where they can recombine with the emission of radiation, (Fig. 4.18). The LED optical power output, P_0, can be expressed in the form (Saleh and Teich, 1991)

$$P_0 = \eta h\nu J/e, \tag{4.38}$$

where J is the LED injection current, ν is the emitted frequency of light, and e is the electron charge. η represents the product of several efficiencies, including the efficiency of light generation and the efficiency with which the light is transmitted through the device output window. Output powers for an LED are typically a few milliwatts for an injection current in excess of 100 mA. However, as we shall see later, only a fraction of this output power is coupled into the core of a single-mode optical fiber.

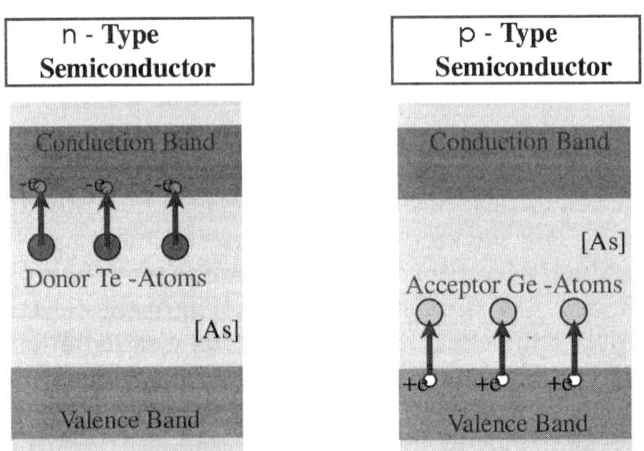

FIGURE 4.16. Schematic illustration of n- and p-type semiconductor material. Donor atoms in n-type material populate the conduction band with negative electron carriers, whereas acceptor atoms create a positive hole carriers in p-type material.

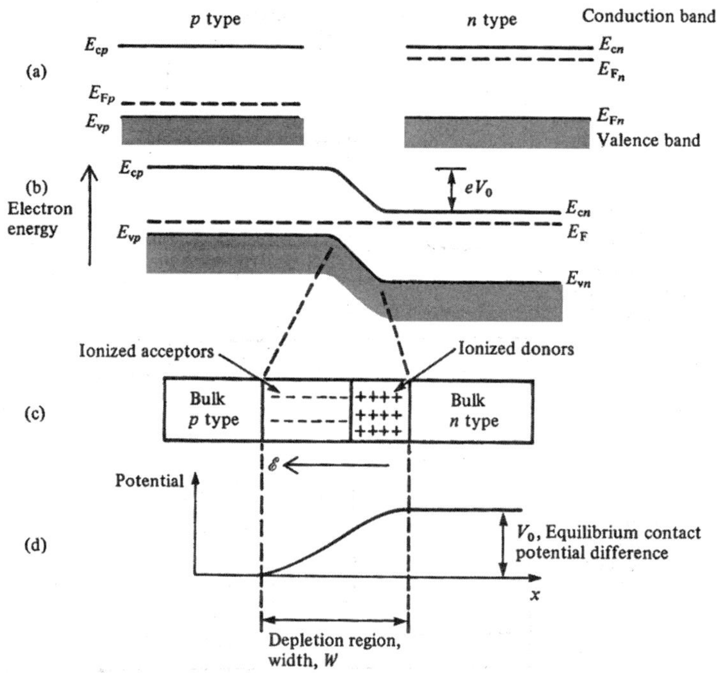

FIGURE 4.17. Schematic representation of the formation of a *p n* junction: (a) separated *p*- and *n*-type materials; (b) energy band distribution after the junction is formed; (c) the space charge layers of ionized impurity atoms within the depletion region W; and (d) the potential distribution at the junction. From *"Optoelectronics, An Introduction,"* 2nd Edn, J. Wilson & J.F.B. Hawkes, Prentice Hall International, 1989, p. 62.

The spectral distribution, or spectrum, emitted by an LED varies with the composition of the device. In general, the *spectral width*, $\Delta\lambda_{FW}$ (full width at half maximum height), of an LED increases as the bandgap energy decreases, (Saleh and Teich, 1991). This means, according to Eq. (4.37), that the longer the wavelength of the emission from an LED, the wider its spectrum. This is illustrated for three InGaAsP LEDs of different composition in Fig. 4.19, (Gowar, 1984). A spectral width of about 100 nm is reasonable for an LED operating at about 1.5 μm. The wide spectral width of LEDs ensures their low coherence and low sensitivity to back reflections. *Low coherence* will be shown to allow a certain type of spectral interferometry that is important for structural monitoring (see Chapter 10) while invulnerability to back reflections avoids the need for expensive optical isolators (see Chapter 5). As we saw earlier, Eq. (4.37), the center

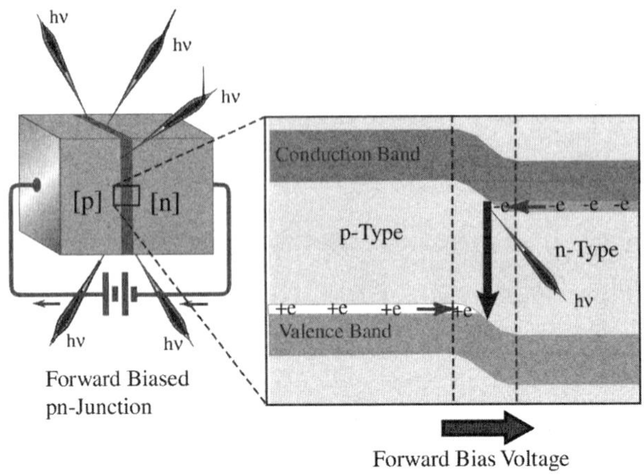

FIGURE 4.18. The application of a forward bias voltage across a *p* n junction drives the electrons (in the conduction bands) and the holes (in the valence band) toward each other at the *p n* junction. This is the principle of the light-emitting diode.

emission wavelength for an LED depends on the bandgap energy of the semiconductor material forming the *p–n* junction; i.e.,

$$\lambda[\mu\text{m}] = \frac{1.24}{E_g[\text{eV}]}. \tag{4.39}$$

LEDs are constructed as *surface-emitting* or *edge-emitting* devices, (Fig. 4.20). Examples of different colored visible LEDs shown in the inset to Fig. 4.19. The *surface-emitting* LED, or SLED, emits light in an isotropic manner from a face of the device that is parallel to the plane of the *p–n* junction. One of the most popular surface-emitting designs is the *Burrus diode* (Dakin and Culshaw, 1988–97), schematically illustrated in Fig. 4.21. This source is easy to use with large core, multimode, optical fiber as it can conveniently be *butt coupled* to the end of the fiber and held in place with epoxy as indicated in Fig. 4.21. The etched well in this device permits light to be collected directly from the junction layer, but even so only a small fraction of the light produced (1 or 2%) is coupled into the optical fiber.

The far-field radiation pattern from a surface-emitting LED is similar to that of a *Lambertian source* where the intensity of light varies as $\cos\theta$, where θ is the angle from the emission plane normal. *Microlenses* are often used to reduce this angular spread and improve the efficiency of light coupled into optical fibers (Dakin and Culshaw, 1988–97; Gowar, 1984). Even with such microlenses, it is difficult to couple much light from a SLED into the very small core of single-mode optical fibers (Senior, 1985).

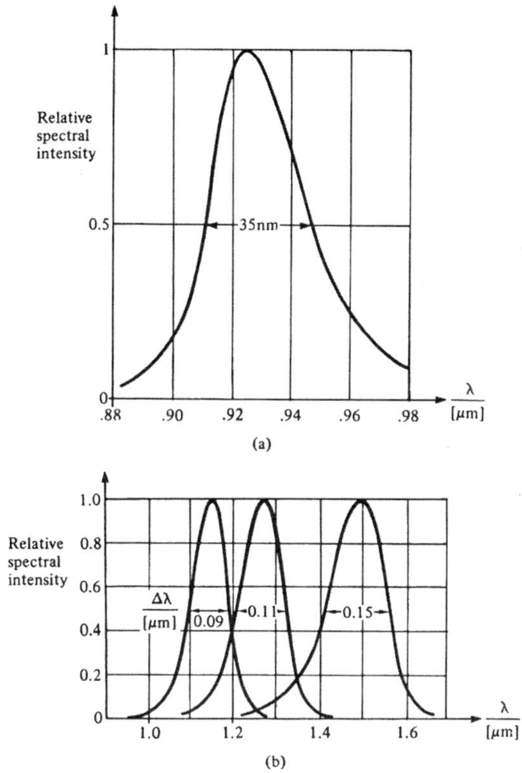

FIGURE 4.19. Examples of measured LED emission spectra: (a) typical spectrum of a si-doped GaAs duiode; (b) spectra from a set of InGaAsP LEDs with three different active layer compositions. From Wada, O., (1982), "Performance and Reliability of High Radiance InGaAsP/InP DH LEDs Operating in the 1.15–15 μm Wavelength Region," IEEE, *J. Quantum Electronics, QE* **18**, 368–374.

In an *edge-emitting* LED, or ELED for short, the light-generating layer is sandwiched between layers of material with wider energy bandgaps and lower index of refraction. This *double heterojunction* configuration (Saleh and Teich, 1991), Fig. 4.22, serves to confine both the current flow and the light, making the device more efficient and capable of coupling more optical power into single mode optical fibers than SLEDs. The adjacent layers of lower index of refraction combined with the *stripe-geometry* of the light-generating layer to form an effective *optical waveguide* that confines the light via the process of *total internal reflection* at the boundaries to this region, Fig. 4.22. A more detailed discussion of *light guiding* within optical fibers will be provided in Chapter 5.

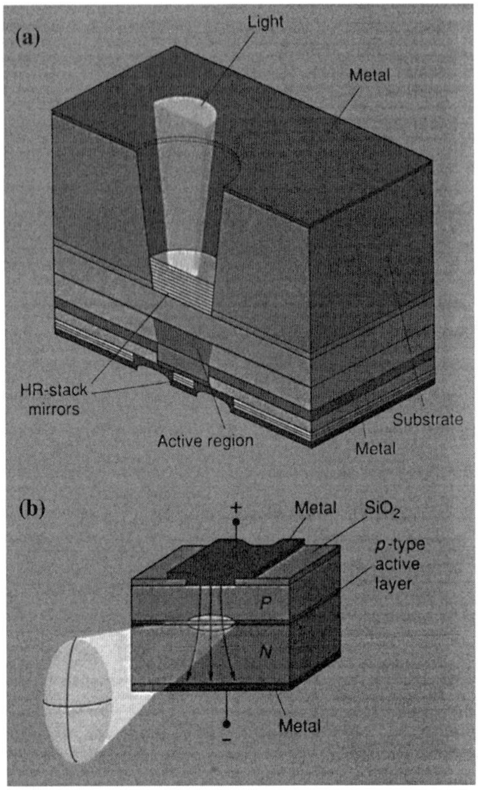

FIGURE 4.20. Schematic illustrations of (a) surface and (b) edge emitting light-emitting diodes. From Lerner, E.J., (1998), "Diode Lasers Light Up Disks, Communications, and Printers," *Laser World*, **34**(1). © Penwell, used with permission.

Individual ELEDs are cleaved along crystal planes from the original high-quality substrate disc leaving very smooth, flat and parallel end faces on the ELEDs. These end faces constitute mirrors with respectable reflectivities (~30%) due to the very high refractive index of these semiconductor materials ($n = 3.6$ for GaAs and $n = 3.5$ for InP). To avoid laser action, the length of an ELED is made relatively short compared to that of a laser diode, about 100 μm compared to 300 μm, and the reflectivity of the output end face is suppressed through the application of an anti-reflection coating. An etched slot is often used to restrict the length of the device subject to the injection current, as in

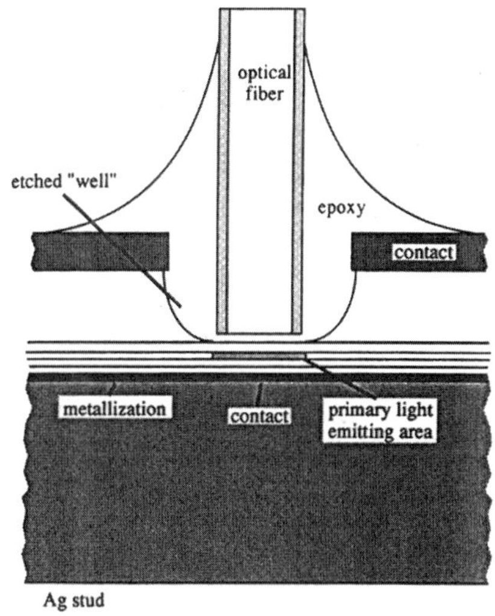

FIGURE 4.21. Design of a high-radiance surface-emitting Burrus diode. From Yurek, A.M., and Dandridge A., *"Optical Sources"* Reprinted with permission from *"Optical Fiber Sensors, Principles and Components"*, 1988 Culshaw and Dakin (Editors). Artech House, Inc., Norwood, MA, USA, www.artechhouse.com.

Fig. 4.23. This allows the device to be long enough to handle, while the light-generating zone is too short to lase even at the highest pumping rate.

Optical confinement and guidance of the light in the plane of the junction is provided by the adjacent layers of low index of refraction. This helps to constrain the angular spread of the output beam in the vertical direction (perpendicular to the junction plane) to about 30°. By contrast, the angular spread of the beam in the junction plane can be as large as 120°, see Fig. 4.23. This low vertical angular spread combined with the small emissive region of the endface makes the ELED well suited for efficient lens coupling to single-mode optical fibers.

A *superluminescent diode* (SLD) represents a more powerful broadband source than an ELED as it is based on *amplified spontaneous emission* (ASE). This device can be thought of as a *frustrated laser*, for the injection current is boosted to the point where a population inversion is created and stimulated emission intensifies the spontaneous emission arising in its double heterojunction. However, laser action is prevented by severe suppression of the reflection from both endfaces. This is accomplished either through the application of an antireflection coating, or by forming

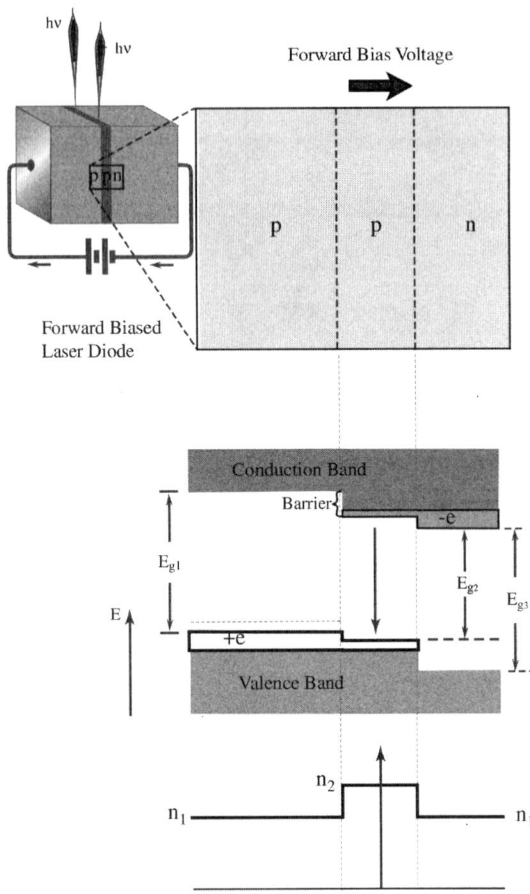

FIGURE 4.22. Potential configuration and bandgaps for an edge-emitting double heterojunction diode.

the light-guiding stripe at a small angle to the end facets. Care is required in coupling optical fibers to SLDs as they are quite susceptible to back reflections.

The output spectrum of early versions of SLDs were plagued with severe ripple due to residual reflections establishing cavity resonances. Recently, techniques such as the introduction of tapered absorption regions (Superlum) and well-designed optical connections have all but eliminated this problem, (Fig. 4.24). Both the spectral linewidth and the angular emission pattern of SLD's are narrower than conventional LEDs,

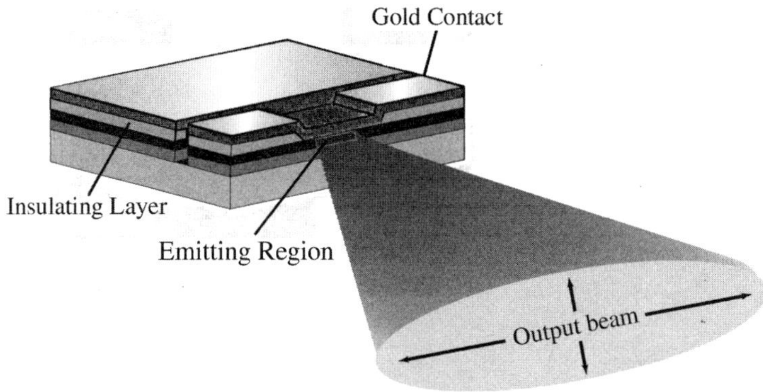

FIGURE 4.23. Schematic representation of an edge-emitting LED based on an etched slot design.

as a result of the gain process. The angular narrowing of the output from a SLD allows more effective coupling of its optical power into a single-mode optical fiber. In addition, the power output of SLDs can be 4 or 5 times that from an ELED. Indeed, as indicated in Fig. 4.24, an output optical power of 8 mW is readily achieved with such a device.

4.6 SEMICONDUCTOR LASER DIODES

Lasers made from semiconductor materials are similar to LEDs in that they are also *diode junction devices* that are either *surface-* or *edge-emitting*. The vital difference between LEDs and LDs is that the light emitted from

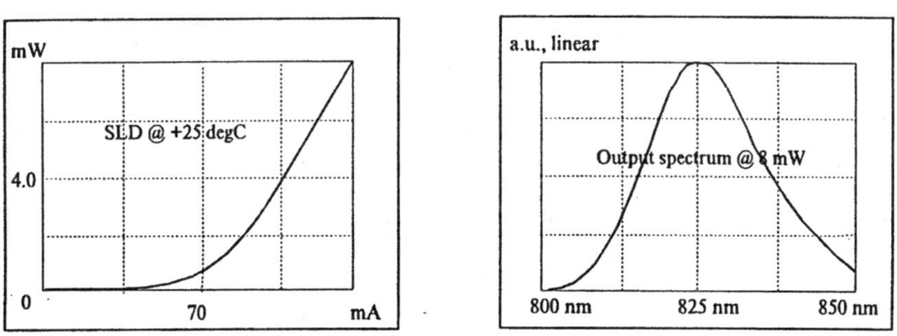

FIGURE 4.24. Light current characteristics and the optical spectrum of a superluminescent diode model SLD-360A. Permitted by Superlum Ltd.

the former arises principally from *spontaneous emission*, while in the case of the latter *stimulated emission* produces the bulk of the radiant power. Virtually all lasers used to date for fiber optic structural monitoring are edge emitting and so we shall restrict our discussion to this type of LD. It should be noted in passing, however, that *vertical cavity surface-emitting lasers*, (VCSELs) underwent an explosive development in the early 1990s and offer some interesting characteristics that may prove eventually useful for fiber optic structural monitoring (Jewell *et al.*, 1992a, b). In comparison with other types of lasers, LDs are extremely small, rugged, efficient and readily integrated into microelectronic systems because of their low voltage and current requirements. Furthermore, their high optical power permits many sensors to be simultaneously interrogated, which brings down the cost per channel, see Chapter 13.

In order to derive the equation that describes the laser diode output power we draw upon the empirical relation between the *peak gain coefficient* (g_p), and the *injection carrier concentration* (Δn), (Saleh and Teich, 1991)

$$g_p = \zeta \left\{ \frac{\Delta n}{\Delta n_T} - 1 \right\}, \tag{4.40}$$

where ζ and Δn_T are parameters that satisfy the following limits:

When $\Delta n = 0$, $g_p = -\zeta$, where ζ represents the *absorption coefficient* of the semiconductor material in the *absence of carrier injection*.

When $\Delta n = \Delta n_T$, $g_p = 0$; thus, Δn_T represents the *transparency injected carrier concentration*, that is to say at the value of Δn for which emission just balances absorption.

This is made clearer in Fig. 4.25, where it is apparent that for values of Δn greater than the transparency injected carrier concentration the medium changes from an *absorber* of the laser wavelength to an *amplifier* of that radiation.

It is easy to see that if 100% of the injected current, J, goes into the generation of charge carriers within the gain region

$$J = \frac{ewdL\Delta n}{\tau}, \tag{4.41}$$

where τ is the *lifetime* (including radiative and nonradiative processes) and e the *charge* of the carriers, and w, L, d are the respective *width, length*, and *thickness* of the gain region Fig. 4.26. Equation (4.41) assumes all of the

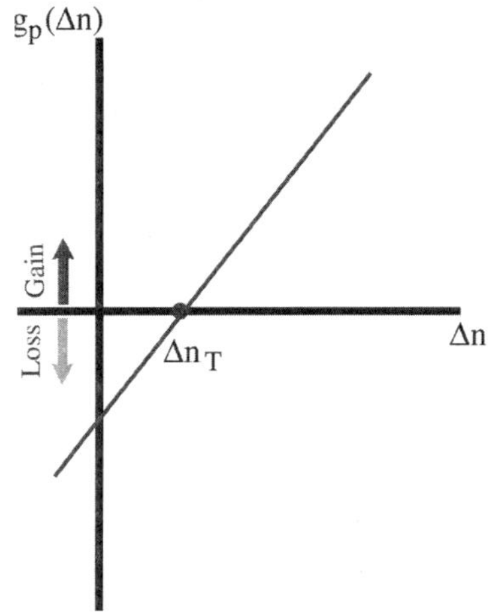

FIGURE 4.25. Linear dependence of the optical gain, $g_{\mathrm{p}}(\Delta n)$, for a laser diode with an excess of injected charge carrier density Δn over the tranparency number density, n_{T}.

injected current flows through the gain layer. It then follows that we can rewrite (4.40) in terms of the injected current

$$g_{\mathrm{P}} = \zeta\left\{\frac{J}{J_{\mathrm{T}}} - 1\right\}, \tag{4.42}$$

where J_{T}, is the corresponding transparency injected current. As mentioned earlier, the cavity mirrors for a laser diode arise from the

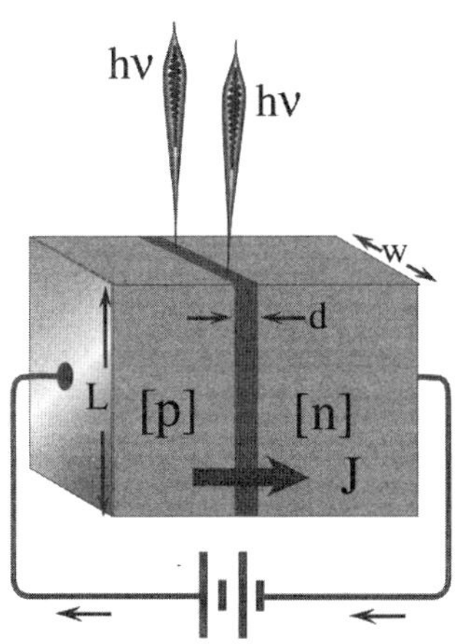

FIGURE 4.26. Schematic of basic *p–n* junction.

very high value for the index of refraction of these semiconductor materials and cleaving the crystal planes normal to the junction. In other cases, the end faces of the device are polished. The *intensity reflection coefficient* for normal incidence at the semiconductor–air interface is

$$R = \left\{\frac{n-1}{n+1}\right\}^2. \tag{4.43}$$

In the case of GaAs, $n = 3.6$ and $R = 32\%$.

We introduced earlier the cavity loss per pass, Eq. (4.31):

$$\gamma = \frac{1}{2}\log\left\{\frac{1}{r_1 r_2}\right\}.$$

We convert this to the *effective cavity loss per unit length* of the gain region, ζ_r, by dividing (4.31) by L to yield

$$\zeta_r = \frac{1}{2L}\log\left\{\frac{1}{r_1 r_2}\right\}. \tag{4.44}$$

Lateral losses of light can be accounted for by introducing a *confinement factor*, Γ, which represents the fraction of optical energy confined to the gain region ($\Gamma < 1$). In which case the *total loss per unit length* is

$$\zeta_t = \frac{1}{2\Gamma L}\log\left\{\frac{1}{r_1 r_2}\right\}. \tag{4.45}$$

Laser action occurs when the *gain per length* exceeds the *loss per length*, i.e., $g_p > \zeta_t$. The threshold condition is established by setting $g_p = \zeta_t$ in Eq. (4.42)

$$\zeta_t = \zeta\left\{\frac{J_c}{J_t} - 1\right\}. \tag{4.46}$$

From this equation we see that the *critical injection current*, J_c, takes the form:

$$J_c = J_T\left\{\frac{\zeta_t + \zeta}{\zeta}\right\}. \tag{4.47}$$

If we allow for recombination modes that do not generate light, then we must introduce the *internal quantum efficiency*, η_i

$$\eta_i = \tau/\tau_r, \tag{4.48}$$

where τ_r, is the *electron–hole radiative recombination lifetime*. Using this in Eq. (4.41) yields

$$J_T = \frac{ewdL\Delta n_T}{\eta_i \tau_r}, \tag{4.49}$$

When this is used in (4.47), the *critical injection current* can be evaluated.

We shall undertake a representative calculation (Saleh and Teich, 1991) to acquaint the reader with the order of magnitude of the numbers involved with a laser diode. We shall determine the *critical injection current* for an InGaAsP/InP double-heterostructure semiconductor laser diode with junction dimensions L = 200 μm, w = 10 μm, and $d = 0.1$ μm. The confinement of light is assumed to be perfect (i.e., $\Gamma = 1$), and the mirror loss coefficient: $\zeta_r = 118\ \text{cm}^{-1}$, assuming a surface reflectance of 31%. We also assume that: $\Delta n_T = 1.25 \times 10^{18}\ \text{cm}^{-3}$, $\zeta = 600\ \text{cm}^{-1}$, $\tau_r = 2.5$ ns, and $\eta_i = 0.5$ at $T = 300$ K.

The *transparency current density* based on these quantities $J_T = 32$ mA, and the corresponding threshold or critical current density $J_c = 38$ mA, which is low enough to permit continuous operation of the double-heterostructure laser diode at room temperature.

Above threshold, the *output laser power* (P_L), can be directly related to the difference between the actual injection current and its threshold value (Saleh and Teich, 1991)

$$P_L = \eta_L \{J - J_c\} \frac{1.24}{\lambda_L}, \tag{4.50}$$

where η_L is *laser efficiency* and is essentially a product of the *internal quantum efficiency* and the *emission efficiency*. The output laser power, P_L, will be in milliwatts, if the laser wavelength, λ_L, is in millimeters and the injection current is in milliamps. For the InGaAsP laser diode considered above and operating with an injection current of 45 mA, at a wavelength of 1.3 μm with a laser efficiency of 40%, the output laser power is 2.7 mW.

The *spectral distribution* of the laser output is primarily governed by three factors:

- The wavelength interval for which the gain coefficient, $g_p(\lambda)$, exceeds the loss coefficient, ζ_r
- The nature of the spectral broadening process, and in particular, if it is homogeneous
- The wavelength spacing of the Fabry–Perot cavity modes in comparison to the spectral width of the gain profile

A necessary condition for laser action is that: $g_p(\lambda) > \zeta_{r,}$, so this condition, in effect, defines the maximum wavelength interval for the laser radiation.

We term this the *net gain spectral width*, $\Delta\lambda_g$, (Fig. 4.27). The question of how many of the *Fabry–Perot cavity* modes lie within $\Delta\lambda_g$, depends on the wavelength separation of these longitudinal modes, $\Delta\lambda_{FSR}$. The wavelength separation between adjacent FPC modes is derived from Eq. (4.26) using the basic relation: $\Delta\lambda = -(\lambda^2/c)\Delta\nu$, and is referred to as the *free spectral range*

$$\Delta\lambda_{FSR} = \left|\frac{\lambda^2}{2nL}\right|. \tag{4.51}$$

If this free spectral range is much smaller than the net gain spectral width, the question then becomes which of the FPC modes will lase? The answer depends on the dominant form of spectral broadening for the laser transition. The very rapid intra-band relaxation processes that arise within semiconductor materials ensures that all of the FPC modes within the net gain spectral width probably lie within the *homogeneous linewidth*.

Under these circumstances, each laser wavelength can extract energy from all of the excited electrons so the wavelength with the greatest power grows the fastest via stimulated emission and leaves less and less excited energy (in the form of electron–hole pairs) for other wavelengths, see Fig. 4.10. Clearly in this mode competition for the given inverted population energy, the mode closest to the maximum of the gain profile has the

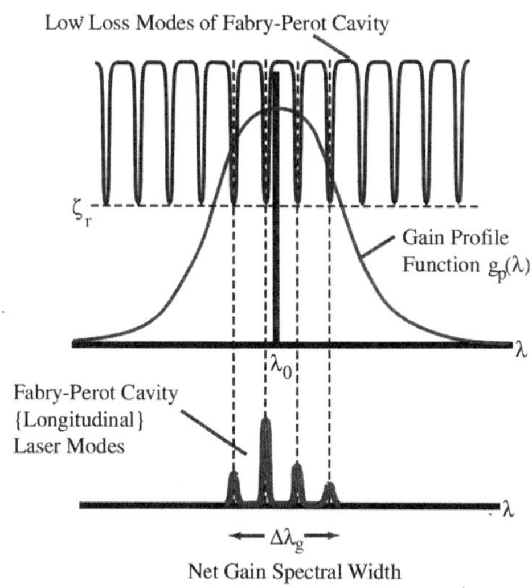

FIGURE 4.27. Gain and loss curves for a laser diode revealing the laser modes and the net gain spectral width.

advantage and this mode will grow at the expense of the other FPC modes. In principle this should ensure laser oscillation on only one FPC mode, however, another process called spatial hole burning often permits several *longitudinal* (FPC) modes to lase simultaneously, as seen for a InGaAsP laser operating at 1.3 μm, (Fig. 4.28).

Spatial hole burning refers to the depletion of the inverted population at the positions along the cavity where the **E**-field of the standing wave for a given frequency is highest. The inverted population, however, is much less depleted at other positions where the **E**-field is minimum. It is thus possible, under the right conditions, for another frequency with its standing wave peak **E**-field overlapping the position of minimum **E**-field of the first frequency to extract energy from the residue inverted population (Fig. 4.29), (Saleh and Teich, 1991). The linewidth of these *longitudinal modes* depends on the design of the laser, especially its architecture. They can range from a few gigahertz to 30 MHz, (Note: 10 MHz $\sim 8 \times 10^{-5}$ nm, at 1.55 μm).

The form of the spatial output from a semiconductor laser diode is determined by the architecture of the gain region. If the ratio of the gain layer thickness to the wavelength of light (d/λ) is somewhat less than unity, the laser can only sustain a single mode in the transverse direction perpendicular to the junction plane. Since the width of the gain region w is considerably larger than the wavelength of light λ the laser waveguide will support several transverse modes in the plane of the junction. Clearly, the smaller the width of the gain region the fewer the number of these *lateral modes* (Fig. 4.30). In general the number of lateral modes is limited because the higher order modes suffer greater loss.

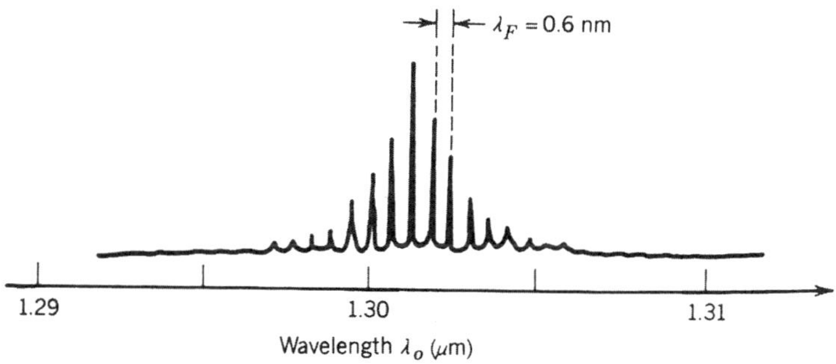

FIGURE 4.28. Spectral distribution of a 1.3-μm InGaAsP index guided buried-heterostructure laser. From Saleh, B.E.A., and Teich, M.C. "*Fundamentals of Photonics*," © 1991, Wiley Interscience. Reprinted by permission of John Wiley and Sons, Inc.

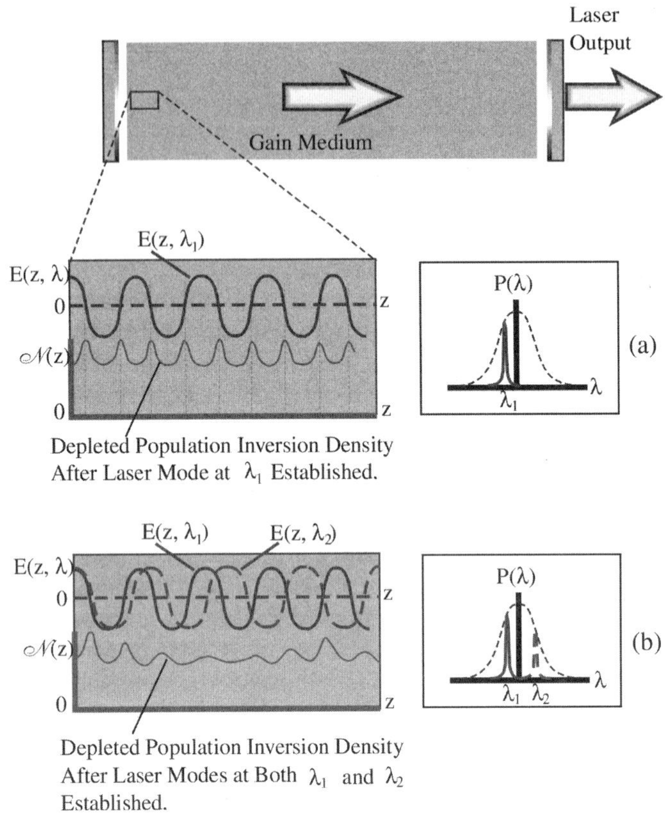

FIGURE 4.29. Spatial hole burning in a laser. (a) Highest gain Fabry–Perot mode at λ_1 lases first and depletes the inverted population at its nodal planes. (b) Next highest gain longitudinal mode that can draw energy effectively from the residual population inversion distribution commences laser action at λ_2.

The popular *buried-heterostructure* laser diode, BH–LD, (Fig. 4.31), is classified as an *index guided* laser diode, as lower index material on either side of the *gain region* confines the light in the plane of the junction, while the lower index of refraction layers adjacent to the gain region provide confinement in the direction perpendicular to the junction plane. The strong optical and current confinement designed into this BH–LD guarantee's a low threshold injection current and the small width of the gain region ensures that it operates on a single lateral mode. The lower loss associated with the lowest-order spatial modes ensures that the BH–LDs output is the *fundamental transverse mode*. The near-Gaussian form of this

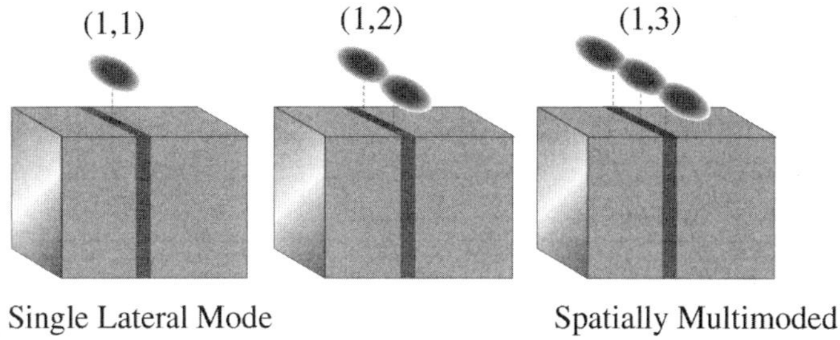

FIGURE 4.30. Schematic illustration of spatial distribution of laser output intensity for laser spatial modes $(l, m) = (1, 1), (1, 2)$ and $(1, 3)$ associated with three different widths of the gain region.

mode makes it attractive for efficiently coupling the laser output power into a single-mode optical fiber. For some applications not requiring the same degree of lateral confinement of both current and light, *ridge waveguide* lased diodes are less expensive (Tamir, 1990).

If the width of the gain region is w and the thickness is d, the *far-field angular divergence* will be $\approx \lambda / w$ (radians) in the plane parallel to the junction, and $\approx \lambda / d$ in the plane perpendicular to the junction, (Fig. 4.32). For a BH–LD operating at 1.5 μm, with $w = 3.0$ μm, and $d = 1.0$ μm, the far-field angular divergence of the output beam is 0.50 radians (29°) in the

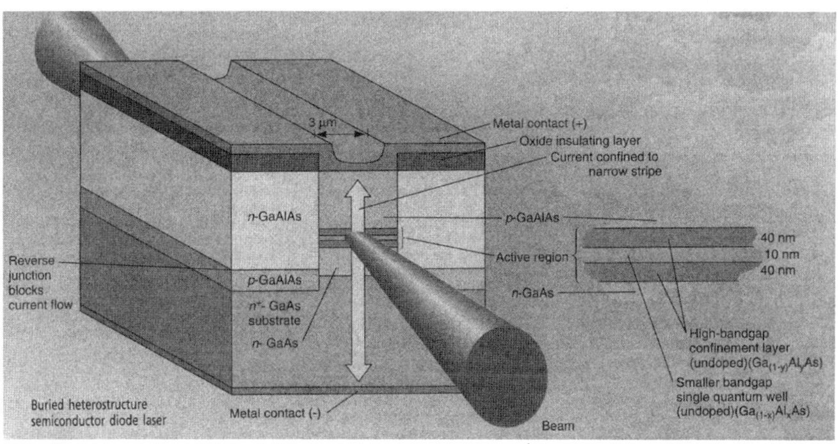

FIGURE 4.31. Composition and design of a representative buried-heterojunction semiconductor laser diode based on a GaA/As quantum-well gain region. From Hecht J., "Diode-laser performance rises as structures sink" *Laser Focus World,* **28**(5) May, 1999, p.127 ©Penwell, used with permission.

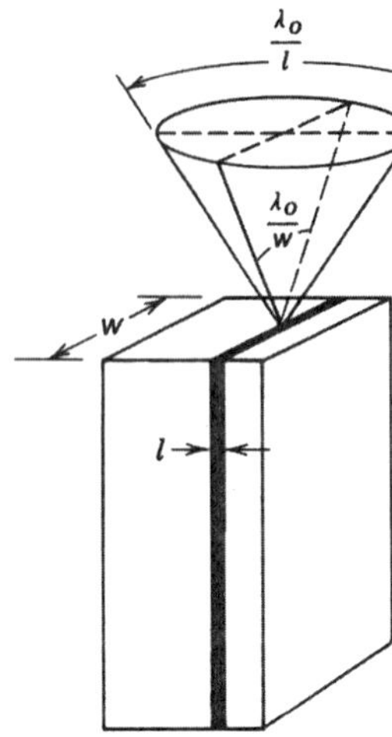

FIGURE 4.32. Angular distributrion of the intensity output from a laser diode. From Saleh, B.E.A. and Teich, M.C. *"Fundamentals of Photonics,"* © 1991, Wiley Interscience. Reprinted by permission of John Wiley and Sons, Inc.

plane parallel to the junction, and 1.5 radians in the plane perpendicular to the junction.

4.7 NARROWBAND (DBR AND DFB) LASER DIODES

For many applications it is highly desirable that the laser also operates on a single FPC, longitudinal mode, that is to say with a very narrow linewidth. As implied earlier, it is possible to make the laser diode operate on a single longitudinal mode by reducing its cavity length, L, to the point where the free spectral range of the FPC modes, see Eq. (4.51), is greater than the net gain spectral width, i.e.

$$\Delta\lambda_{\mathrm{FSR}} > \Delta\lambda_{\mathrm{g}}.$$

Unfortunately, this approach would lead to low output power and a high threshold current (Miller and Kaminow, 1988). One of the best ways to achieve very narrow linewidth laser operation is to replace one, or both, cleaved endface broadband mirrors with frequency-selective reflectors. There are two different laser diode architectures that are popular today based on this approach. Both have been shown to produce laser bandwidths in the few to tens of megahertz range, which is adequate for most sensing purposes.

The first uses one or two *distributed Bragg reflectors* (DBR) external to the gain region and is termed a *DBR laser diode*. The most popular design uses one grating to replace one of the semiconductor cleaved end face broadband mirrors, (Fig. 4.33). This grating is a periodic (corrugated) structure formed in a section of the waveguide that is just outside the gain region. As shown in Section 3.10, a narrow band of radiation, centered at a wavelength

$$\lambda = 2n_e\Lambda, \tag{4.52}$$

is back-reflected from the grating, where n_e is the *effective* index of refraction of the light-guiding region and Λ the period of the *grating corrugation* of the laser waveguide. A representative Bragg grating reflection spectrum is depicted in Fig. 4.34. This narrow Bragg grating reflection spectrum now defines the *effective gain curve* and since the *Bragg grating bandwidth*, $\Delta\lambda_B \ll \Delta\lambda_g$, it is possible to ensure that only one FP cavity mode falls within $\Delta\lambda_B$ without requiring the length of the laser cavity to be too short for reasonable output power, Fig. 4.35. Unfortunately, absorption of the laser radiation would arise in the grating section, as it is made of the same material as the gain region but is not pumped. To avoid this loss, the bandgap in the passive grating section, Eq. (4.37), has to be made wider than in the gain region; otherwise, the laser threshold current would be too high.

One of the most important features of a DBR laser is that its laser wavelength can be tuned electronically over an appreciable wavelength

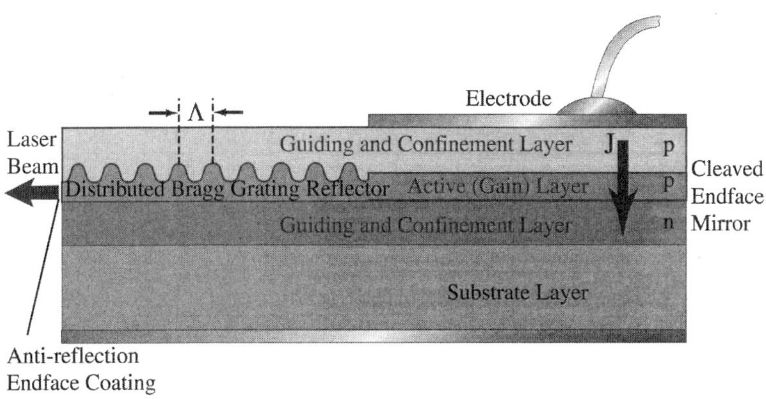

FIGURE 4.33. Schematic representation of a distributed Bragg grating reflector (DBR) laser diode.

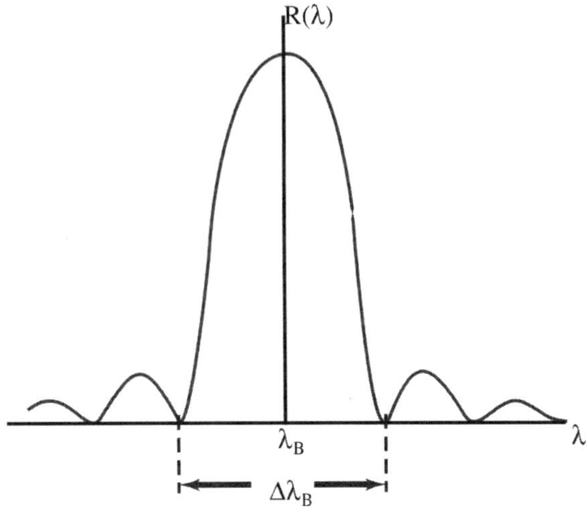

FIGURE 4.34. Representative Bragg grating reflection spectrum for DBR laser diode.

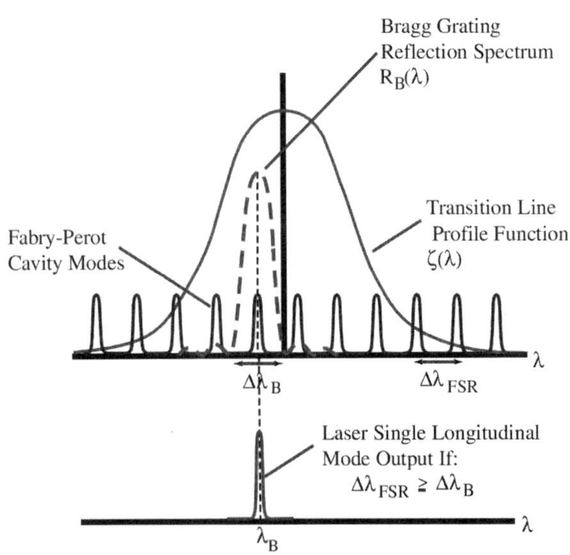

FIGURE 4.35. Single-mode laser operation with a DBR laser diode.

interval (~10 nm). To achieve such tuning the laser is built in three sections: a Bragg grating reflector, a phase control section, and a gain section, each driven by its own current: J_B; J_p; and J_g, respectively, Fig. 4.36, (Tamir, 1990). Injecting carriers into the grating section changes n_e and consequently the centre wavelength of the Bragg grating reflector shifts to shorter wavelength. Changing the injection current in the gain region shifts the gain peak, while injecting carriers into the phase control section helps attain continuous tuning by locking onto a single FPC mode. *Continuous tuning*, free of *mode hops*, has been obtained over a wavelength interval of at least 3 nm, (Ohn *et al.*, 1997). Recently, an advanced form of *sampled grating* DBR laser diode has been tuned over 60 nm, (Jayaraman *et al.*, 1994). Unfortunately, the complexity of fabricating this kind of DBR laser makes it generally more expensive than most others.

The second type of narrowband laser diode that uses a frequency-selective reflector in place of the broadband endface mirrors is termed a *distributed feedback* (DFB) laser. In this *DFB laser diode* a corrugated Bragg grating is formed along the entire length of the gain region providing distributed reflections and avoiding the establishment of Fabry-Perot cavity modes, (Fig. 4.37). The grating is created in a layer that is adjacent to the gain layer, but has good optical coupling to the gain region. The primary use for DFB lasers has been as a very narrowband, stable frequency source. However, these lasers can also be tuned by varying their temperature through changing their injection current. A shift of 0.1 nm per °C in the laser wavelength can be attained, (Ikegami, 1989), (Fig. 4.38).

Continuous tuning over a wavelength interval of 10 nm has recently been achieved by ramping the injection current, and thereby changing the temperature of operation, of a special *gain-coupled DFB laser diode*, supplied to the author's group by Nortel, (Li *et al.*, 1993). What makes this achievement of particular value from the structural monitoring

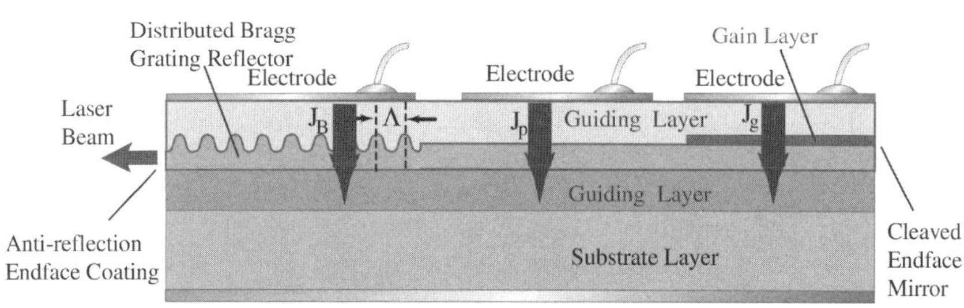

FIGURE 4.36. Schematic representation of a three-section DBR laser diode used for spectral tuning of its narrowband output.

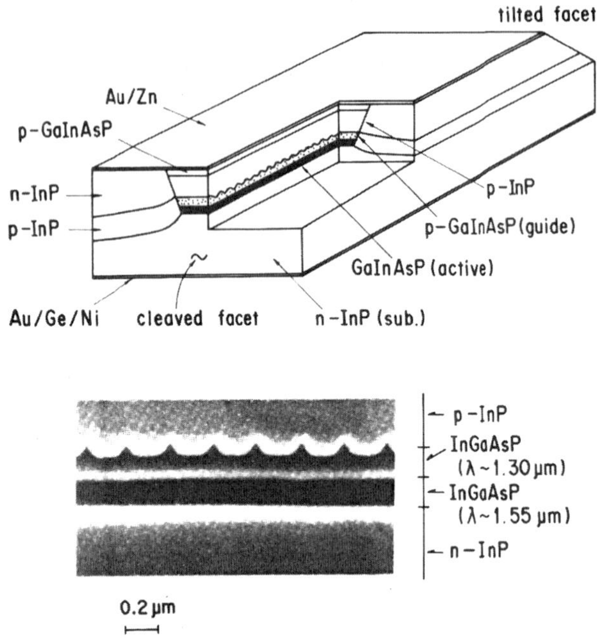

FIGURE 4.37. Cross-sectional view and composition of a DFB InGaAsP-laser designed with tilted cleaved facets. The Bragg grating corrugations in the gain region are revealed in the scanning electron micrograph. From Ikegami, T., "Longitudinal Mode Control in Laser Diodes", *"Optoelectronic Technology and Lightwave Communications Systems"* C. Lin (Ed). Van Nostrand Reinhold, 1989, p.277.

perspective is that the ~10 nm shift to longer wavelength was accomplished within a 1 ms time frame. Indeed, we have demonstrated that real-time dynamic monitoring with a frequency response of at least 100 Hz is readily attainable; see Chapter 13. It is important to point out, however, that the laser output power changes considerably during this period of rapid tuning. Fortunately, when using fiber Bragg grating sensors, the sensing information is spectrally encoded and so source power fluctuations do not cause a problem, providing the photodetector signal-to-noise ratio is above some threshold.

External tuning of semiconductor lasers has been undertaken using bulk optics and gratings. These systems can continuously tune the laser over a wavelength interval of 60 nm, or more. They are expensive and rather large, making them impractical for use in portable fiber optic sensor demodulation systems. One exception to this, however, is the external tuning of a laser diode by a fiber optic Bragg grating. If the tuning is

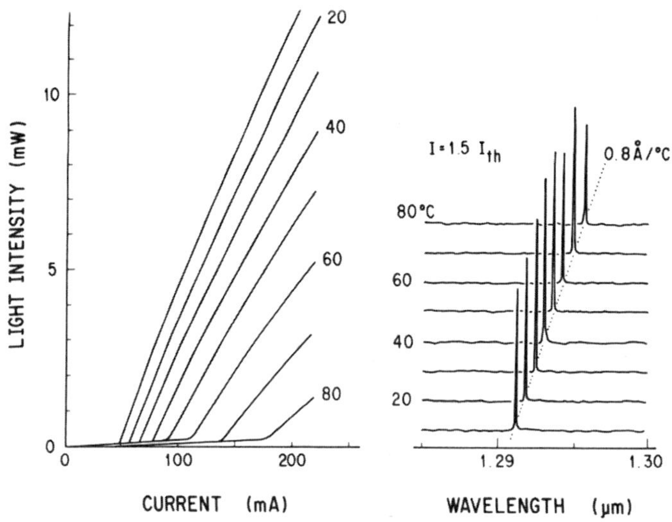

FIGURE 4.38. Lasing properties of a well-behaved distributed feedback laser diode. From Ikegami, T., "Longitudinal Mode Control in Laser Diodes", *"Optoelectronic Technology and Lightwave Communications Systems"* C. Lin (Ed). Van Nostrand Reinhold, 1989, p.276.

achieved by imposing a tensile strain on the FBG its range and lifetime are limited. Ball *et al.*, (1994) have shown that single-frequency tuning over 32 nm can be achieved by *compressive tuning* of a FBG.

4.8 JUNCTION PHOTODETECTORS

There are a number of very different photodetectors, ranging from photomultipliers to pyroelectric detectors (Wilson and Hawkes, 1989), so the choice for any sensing system has to be made based on the performance and characteristics desired. Some of the most important considerations include the following:

- Small size, low power, rugged, and inexpensive
- Spectral response, or the wavelength region of interest
- Responsivity, or detected current per light power received (mA/mW)
- Signal-to-noise detection limit
- Quantum efficiency

The first set of requirements is in many respects the most important for when combined with the selection of wavelength (1.3 or 1.55 μm) semiconductor photodiodes appear to be best suited for fiber optic structural monitoring. We shall therefore restrict our attention to this class of photodetectors.

The formation of a *p–n* junction in a semiconductor material leads to the creation of a depletion region. As indicated in Fig. 4.17, the depletion region is formed by the diffusion across the *p–n* junction of electrons from *n*-type material, and holes from the *p*-type material. This exposes un-neutralized (positive) donor ions in the boundary of the *n*-type material and un-neutralized (negative) acceptor ions in the boundary of the *p*-type material and it establishes a strong electric field across the depletion region. If an incident photon generates an *electron–hole pair* within this depletion region, the internal electric field will force the electron and hole to separate as they drift in opposite directions. This *absorption* of a photon and the subsequent formation of the electron-hole pair can be detected in two rather different ways, indicated as (a) or (b) in Fig. 4.39.

If the photodiode is left on *open-circuit*, an externally measurable potential will develop between the *p* and *n* regions, (Fig. 4.39a). This *photovoltaic* mode of operation is the same as used in the generation of electricity by *solar cells*. The advantages of this method of light detection are very low dark current and subsequent high signal-to-noise ratio for low levels of illumination. Unfortunately, this approach involves a nonlinear interaction that leads to a relatively slow time response.

$$V_{ext} = V_d - i_{ext}R_s$$

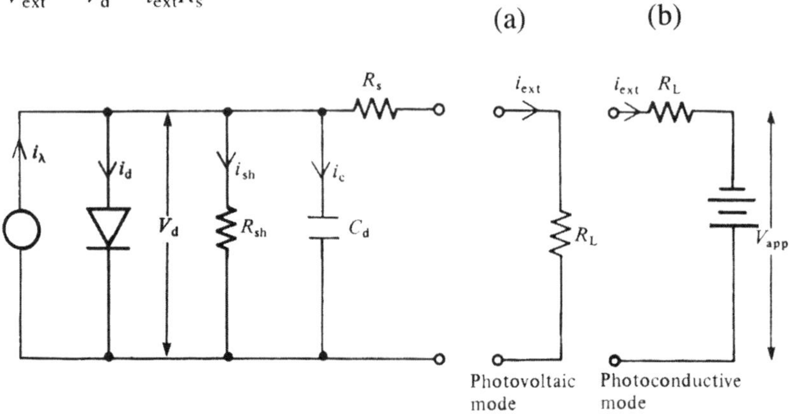

FIGURE 4.39. Basic circuit representation of the two detection modes of operation from a photodiode: (a) photovoltaic and (b) photoconductive. From Wilson, J., and Hawkes, J.F.B., (1989), *"Optoelectronics: An Introduction"*, Prentice Hall Inc., p. 281

The second mode of operation involves applying a reverse-bias voltage across the junction causing an external current to flow between the *p* and *n* regions, (Fig. 4.39b). This *photoconductive* method of detection has the advantage of linear response over many orders of magnitude of light exposure, but is susceptible to dark current, which is the creation of electron–hole pairs due to thermal agitation in the absence of light. Nevertheless, this mode of detection has a high sensitivity and can respond to optical energies of picowatts (pW) per mm^2, (Lerner, 1996). The difference between these two modes of operation of the photodiode can be seen by reference to the voltage–current curves presented as Fig. 4.40.

Photodiodes used in fiber optic sensing applications are generally operated in the strongly *reverse-bias mode* for the following reasons (Saleh and Teich, 1991):

- The presence of a strong **E**-field in the junction increases the drift velocity of the charge carriers, and decreases their chance of

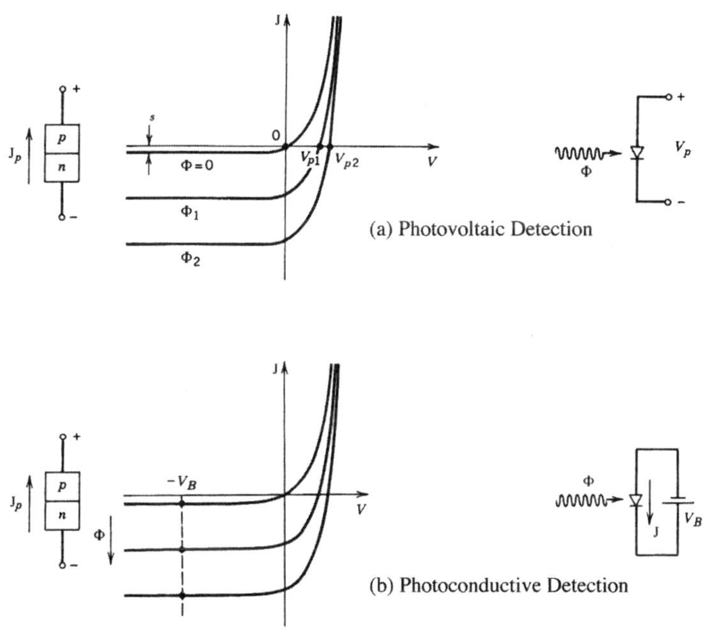

FIGURE 4.40. (a) Photovoltaic operation of a photodiode. (b) Photoconductive reverse-biased operatrion of a photodiode without a load resistor. The operating point lies on the dashed line. From Saleh, B.E.A. and Teich, M.C. "*Fundamentals of Photonics*" © 1991. Wiley Interscience, p.659.

recombination.

- The strong reverse-bias voltage increases the width of the depletion layer, thereby reducing the junction capacitance and improving the response time.
- The increased width of the depletion layer leads to a larger photosensitive volume and an improvement in the photon capture efficiency.

We shall use the term *photodetector* to designate photodiode detector in the remainder of this book. Figure 4.41 illustrates the *reverse-bias operation* of a *p–n* junction photodetector. Although photons can be absorbed in either the depletion or diffusion regions, it is only in the depletion region that the electron–hole pair is quickly separated by the electric field (note that the reverse-bias field reinforces the inherent field) so they contribute to the current. The width of the depletion region increases with lower doping concentrations for a given applied reverse bias voltage. Silicon (Si)-based photodetectors can be used for visible wavelengths and are optimum for

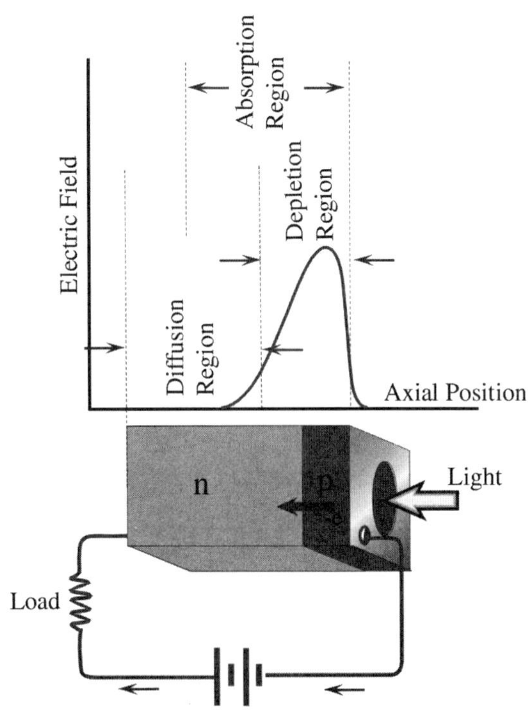

FIGURE 4.41. Reverse-bias operation of a *p n*-photodiode detector showing the depletion and diffusion regions.

0.8 to 0.9 μm, although their cutoff is at about 1.1 μm, (Tamari, 1997). In general they provide high performance with low dark current. Germanium (Ge) photodetectors can be used from 0.5 μm through to 1.8 μm, with an optimum response from 1.1 to 1.6 μm. Unfortunately, their performance is poorer than Silicon photodetectors, so more costly InGaAs-photodetectors are often used for 1.3 and 1.55 μm radiation.

The *photogenerated current*

$$J_\lambda = \eta \frac{P_\lambda e \lambda}{hc}, \tag{4.53}$$

can be seen to be proportional to the *optical power* P_λ (at wavelength, λ) incident upon the photodetector receiver area; the *free electron charge, e*; and the photodetector *quantum efficiency,* η. It is inversely proportional to the *photon energy* hc/λ. The *responsitivity,* of a photodetector,

$$\mathcal{R} \equiv \frac{J_\lambda}{P_\lambda} = \eta \frac{e\lambda}{hc}, \tag{4.54}$$

is in units of A/W or sometimes mA/mW.

4.9 PIN AND AVALANCHE PHOTODIODE DETECTORS

In order to permit operation at longer wavelengths, where the penetration depth is greater, a wider depletion region is required. To achieve this, the *n*-type material is doped so lightly that it can be considered *intrinsic,* and to make a low resistance contact a highly doped n^+-type layer is added. This creates a *p–i–n* (or PIN) *sandwich structure,* as seen in Fig. 4.42, that ensures all absorption takes place in the *depletion layer* (Senior, 1985). The high resistivity of the intrinsic region ensures that only a few volts of reverse bias is required to extend the depletion region right through to the n^+ region. Absorption considerations dictate that the *p*-type layer should be thin to allow short wavelength radiation to penetrate the depletion region, and the depletion region should be thick to absorb the long wavelengths.

Two examples of small-area ternary InGaAs/InP homojunction *PIN-photodiode detectors* are shown in Fig. 4.43, and the corresponding Responsivity curve for the *back-illuminated photodetector* is presented in Fig. 4.44. The values of the photodetector *quantum efficiency,* η, can be derived from this figure by the intersection of the responsivity curve with the set of η curves. The substrate for both devices is InP giving a long wavelength cut-

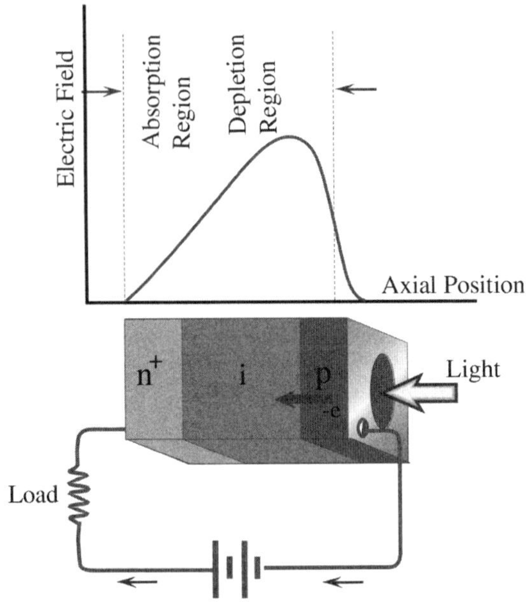

FIGURE 4.42. PIN photodiode detector revealing combined absorption and depletion regions.

off of 1.65 μm. Note that the transparent nature of InP, at the wavelengths of interest, allows both front- and back-illuminated architectures.

When a photodetector operating in the *photoconductive mode* is used at high frequencies a *load impedance* of 50 Ω is usually required. Under these conditions the voltage signal from the photodetector, which is the product of the current from the photodiode times the load resistance, is very small and considerable amplification is needed. The lack of internal amplification means that the photodetector performance is limited by the noise produced in the required external amplification circuitry.

Internal amplification can be produced in a photodiode if a strong reverse-bias voltage (100 to 400 V) is applied to the device. Under these circumstances, photogenerated electron–hole pairs are separated and each of the constituents accelerated by the large **E**-field to the point where they produce additional electron-hole pairs via *impact ionization*. In Fig. 4.45, we show a schematic of an *avalanche photodiode* (APD) and illustrate how one photogenerated electron–hole pair can create several new electron-hole pairs under the action of a strong reverse-bias field. Carrier gain, or multiplication, of 10 to 1000 is possible depending on the magnitude of the reverse-bias voltage applied to the device.

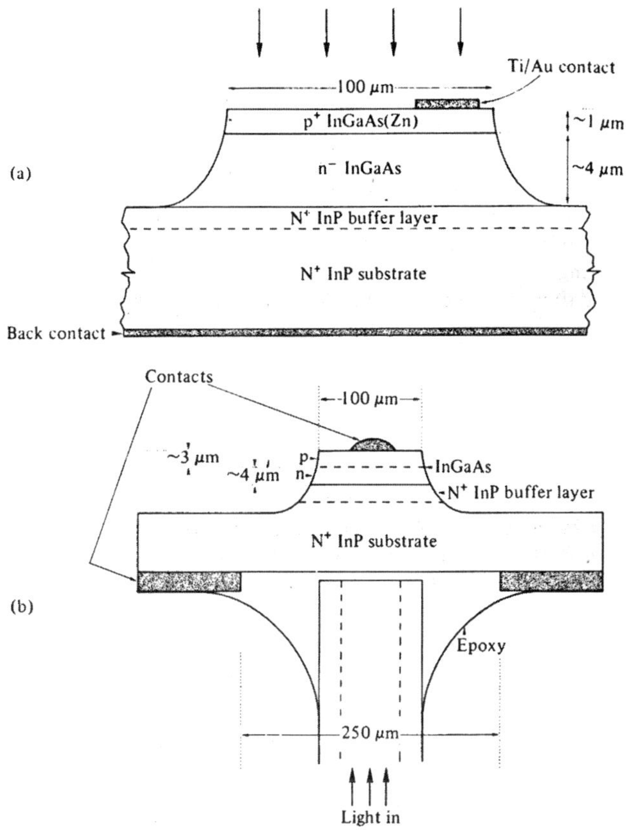

FIGURE 4.43. Schematic illustration of a small area InGaAs homojunction photodiodes: (a) front illuminated, and (b) back illuminated. From Gowar, J. (1984). *"Optical Communications Systems,"* Prentice Hall Inc., p. 366.

Although early designs of avalanche photodiodes were found to suffer from excessive dark currents at the high reverse-bias voltage required to attain sufficient carrier gain, later versions achieved improved performance (Senior, 1985). It is also desirable to fabricate APDs in a way that permits only one of the carriers, electrons or holes, to produce the bulk of the gain through impact ionization, (Saleh and Teich, 1991). If this is not done a circulating gain loop can be established that reduces the device bandwidth, adds gain noise, and can lead to *avalanche breakdown.*

An example of an InGaAsP/InP heterostructure APD with high electric field (gain) region in the InP layer, is depicted as Fig. 4.46. For *field-site structural sensing* the high sensitivity of the gain to fluctuations in

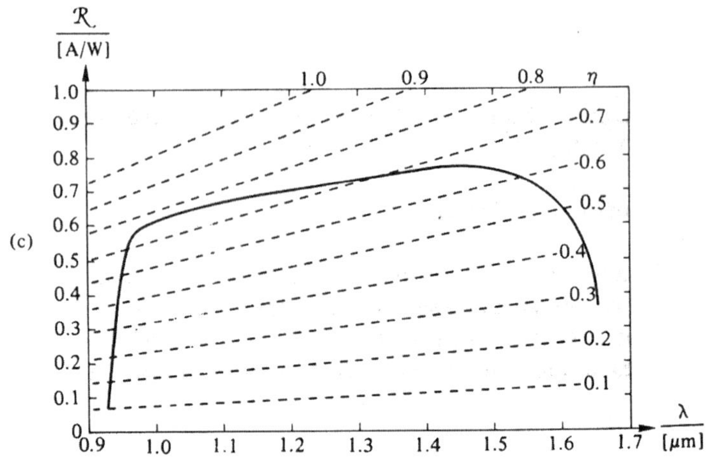

FIGURE 4.44. Responsivity of InGaAs back-illuminated homojunction photodiode shown in Fig. 4.43b. From Gowar, J. (1984) "Optical Communications Systems" Prentice Hall Inc., p. 366

voltage or temperature necessitates very stable voltage drivers and some form of *temperature compensation*, such as a balanced pair of APDs, only one of which is exposed to the light signal, the second serving as a reference (Wilson and Hawkes, 1989). In other cases thermoelectric devices are used to stabilize the temperature. The *photogenerated current* in an APD can be expressed in the form

$$J_\lambda = \eta \frac{P_\lambda G_c e\lambda}{hc}, \tag{4.55}$$

where, we have introduced G_c to account for the *carrier gain*.

4.10 CHARGE-COUPLED DETECTOR ARRAYS

The technology of image capture underwent a radical transformation with the invention of the *charge-coupled device* (CCD) in the early 1970s. One- and two-dimensional arrays of these tiny optoelectronic elements led to the development of relatively inexpensive: spectrum analyzers, copying machines, camcorders, and facsimile machines. The strength of CCDs lies

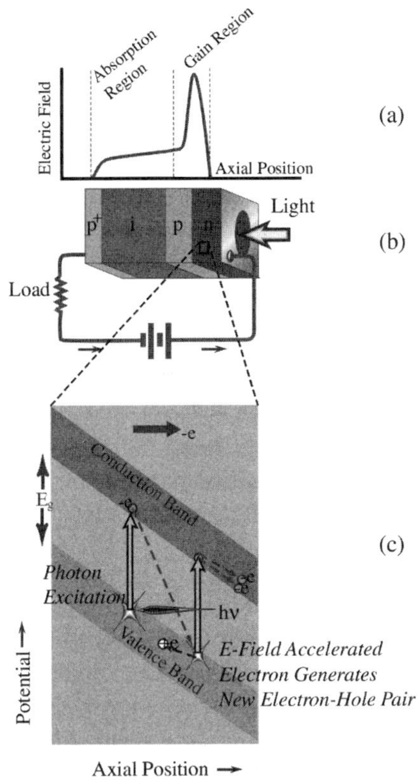

FIGURE 4.45. Avalance photodiode showing (a) the electric field distribution, (b) the configuration, and (c) the ionization process leading to carrier multiplication and amplification of the current.

in their high sensitivity, wide spectral response (X-rays to near-infrared), rugged design, and low cost. A linear CCD array is particularly useful for reading spectra from a *dispersive element*, or extracting the cross correlation from a *low coherence interferometer*, (Fig. 4.47a,b) respectively.

A *CCD array* is built up out of picture elements, or *pixels* for short, each comprising a *metal–oxide–silicon* (MOS) capacitor. Each device comprises a thin metal gate electrode on top of an insulating silicon dioxide layer that in turn lies over a *p*-type silicon substrate (Fig. 4.48, Lerner, 1996a). When a suitable reverse-bias voltage is applied to the device (the *gate electrode* is positive relative to the *p*-type layer) a depletion

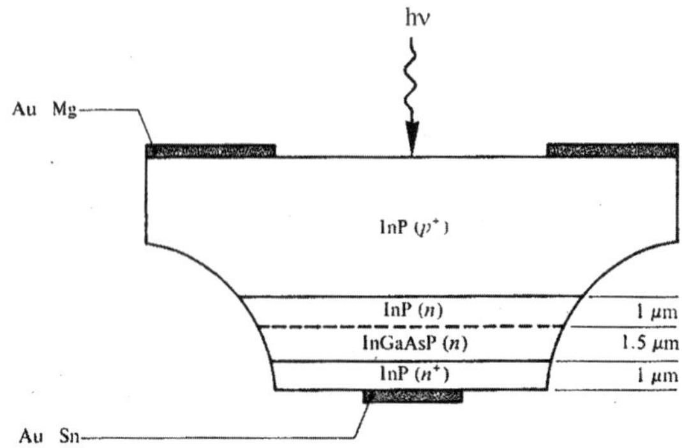

FIGURE 4.46. Cross-section of an InGaAsP/InP heterostructure avalanche photodiode with high electric field (multiplication) region in the InP layer. From Senior, J.M. (1985). "*Optical Fiber Communications, Principles and Practice,*" Prentice Hall Inc., p. 343.

region is created in the positively doped silicon as holes drift away from the gate electrode under the action of the **E**-field. Any photogenerated electron-hole pairs, arising from exposure to light separate, and the electrons tend to be trapped in the *potential well* created by the depletion region. The charge accumulated is proportional to the light energy falling on the device during the time the gate voltage is held positive.

Since it is impractical to wire each pixel, the array is read by repetitively cycling the voltage to any given pixel and its neighbors in a manner that effectively sweeps the charge from one pixel to the next (Wilson and Hawkes, 1989). The neighboring pixel is shielded from the light so when its gate electrode is switched to the same voltage as the exposed pixel it shares the charge developed by the incident radiation. If the voltage to the first pixel is then switched off, the electrons drift to the *light-shielded pixel* and the charge has effectively been transferred (Fig. 4.49). As the process is repeated the charge moves sequentially into one capacitor after another until it is eventually drawn off at the edge of the array. This *line address* leads to smearing of the image (Lerner, 1996). To avoid this image degradation, *interline transfer* is employed (Fig. 4.50). In this approach each column's capacitors shift their charge during readout to an adjacent, optically shielded column.

A number of new types of detector array have recently emerged that have some important advantages over CCD technology. One of the most interesting is termed the *active-pixel sensor*, (APS), (Fossum, 1993). In APS each pixel contains its own selection and readout transistors. The

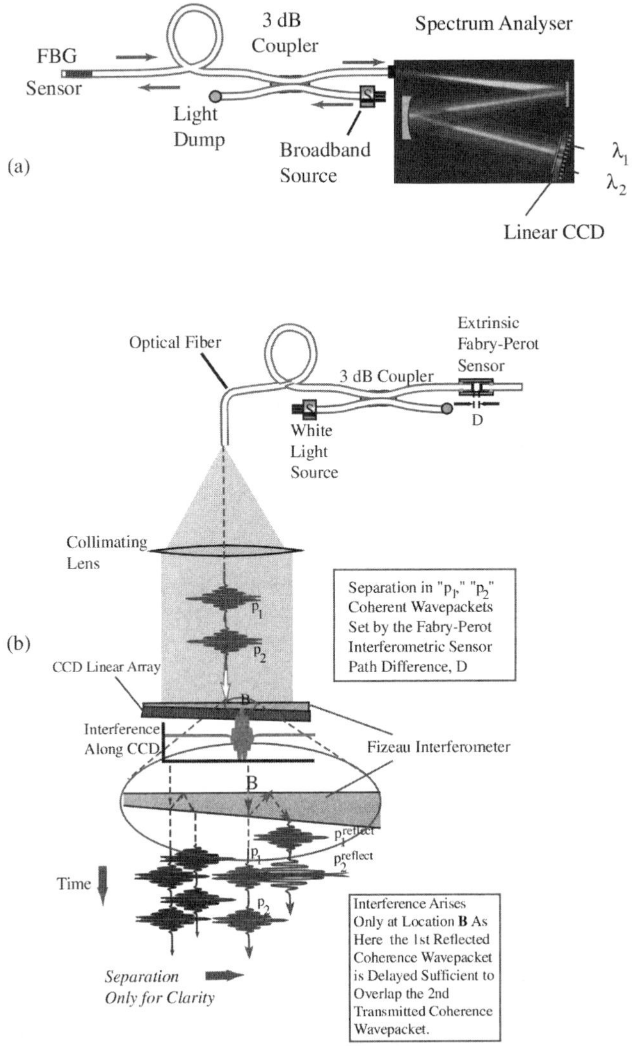

FIGURE 4.47. Use of a linear CCD array: (a) in spectral analysis of a fiber optic Bragg grating sensor, and (b) as a light cross-correlator for demodulating a low-coherence Fabry–Perot interferometric sensor.

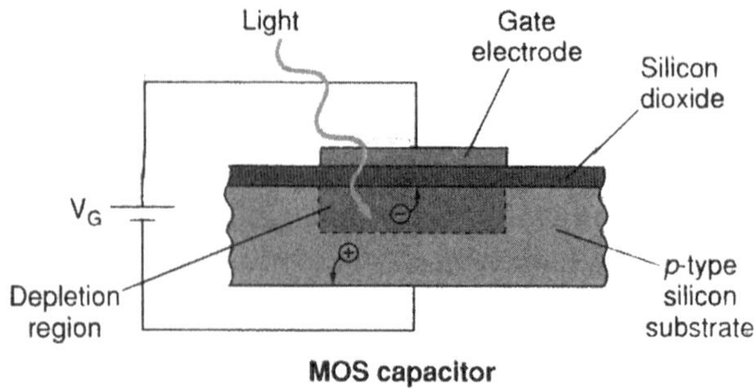

FIGURE 4.48. An applied voltage creates a depletion region in positively doped silicon by causing the holes to drift out of the region. In such a single CCE pixel, electrons created by light are trapped by the positive potential of the depletion zone. From Lerner, E.J., "Charge-coupled devices capture image information," *Laser Focus World*, **32**(8), 1996, 104, © Penwell, used with permission.

signal readout then takes place over wires rather than by shifting charge, which gives APS the advantage of *fast random access* and *on-chip electronics*.

4.11 PHOTODETECTOR SIGNAL-TO-NOISE

In general, a *photodetector* (PD), converts optical power incident upon its receiver surface into current that is measured electronically. The accuracy of this measurement and the lower limit of optical signal that can be detected are determined by a variety of *noise-generation processes*:

- *Photon noise*. This fundamental source of noise results from quantum fluctuations in the arrival of photons.
- *Photoelectron noise*. This is a consequence of the randomness in the photogeneration of an electron–hole pair when the quantum efficiency, $\eta < 1$.
- *Gain noise*. This arises due to the randomness in the internal gain within an APD.
- *Receiver circuit noise*. This is a result of current noise in the various electronic components.

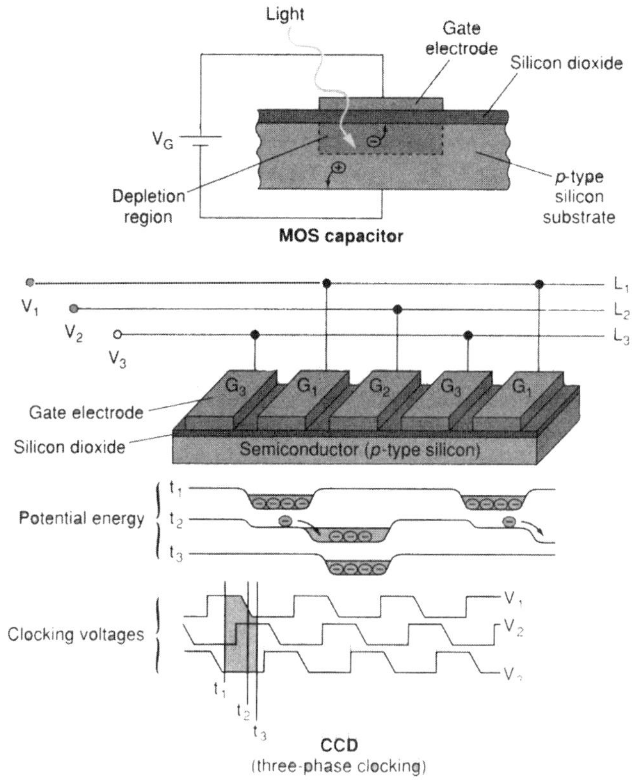

FIGURE 4.49. Light absorbed by the substrate of a *p*-type metal oxide silicon (MOS) capacitor (top) creates mobile electrons that become trapped in the depletion region under the gate. The sequence of gate voltages applied to the CCD (bottom) move electrons across the array of MOS capacitors. Note the potential energies and voltages at times t_1, t_2, and t_+3. Gates G_1, G_2, and G_3 define a single point. From Lerner, E.J., "Detectors span spectrum of sensing applications," *Laser Focus World*, **32**(8), 1996, © Penwell, used with permission.

The mean number of photons incident upon the receiver surface of a photodetector in the *response time* of the PD, τ_d, can be expressed in the form

$$\langle n \rangle = \frac{\lambda P_\lambda \tau_d}{hc}, \tag{4.56}$$

where P_λ represents the *optical power* incident upon the PD receiver surface in that time interval. The random fluctuations in the arrival rate of signal photons give rise to a significant source of noise. This is

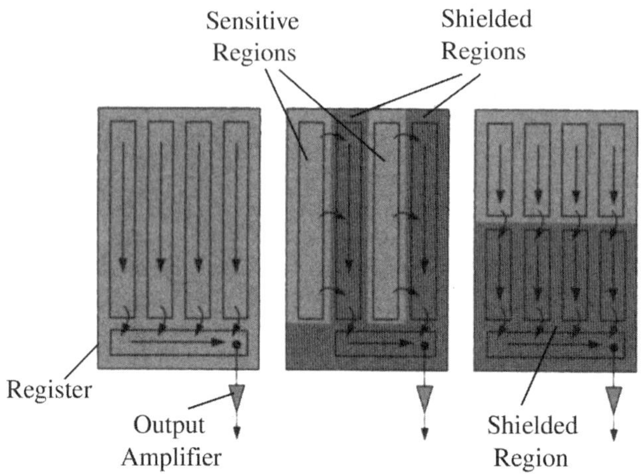

FIGURE 4.50. The in-line-address scheme for CCD readout (left) involves shifting the charge contents of the columns sequentially to the output register where they are transferred to the output amplifier. To prevent image smearing, the interline transfer process (middle) shifts the charge to shielded regions before transferring it to the output register. An entirely separate shilede region isolates the charge in the frame-transfer readout scheme (right). From Lerner, E.J., "Charge-coupled devices capture image information," *Laser Focus World*, **32**(8), 1996, 108, © Penwell, used with permission.

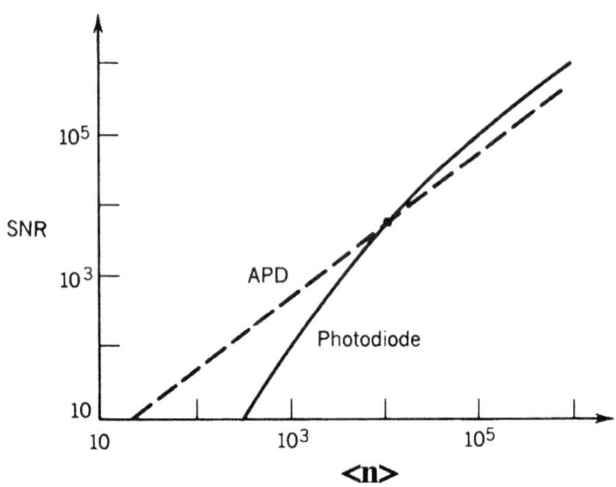

FIGURE 4.51. Variation of photocurrent SNR plotted against the mean number of photons received in the photodetector response time for both a photodiode and an APD with a mean electron gain of 100 and an excess noise factor of 2. From Saleh, B.E.A. and Teich, M.C. *"Fundamentals of Photonics"* 1991. © Wiley Interscience. Reprinted by permission of John Wiley and Sons, Inc.

described by the variance, or standard deviation, in the number of photons arriving in that period, σ_n^2. *Poisson statistics* has been found to apply to this random arrival of photons (Saleh and Teich, 1991), in which case we can relate the variance directly to the mean; i.e.

$$\sigma_n^2 = \langle n \rangle. \tag{4.57}$$

In general, the *signal-to-noise ratio*

$$\text{SNR} = \text{mean}^2/\text{variance},$$

can be introduced for any detection parameter. The *photon arrival signal-to-noise ratio* is

$$\text{SNR}_n \equiv \frac{\langle n \rangle^2}{\sigma_n^2}. \tag{4.58}$$

A more important performance parameter for a photodetection system is the *photocurrent signal-to-noise ratio*

$$\text{SNR}_J \equiv \frac{\langle J \rangle^2}{\sigma_{\mathrm{J}}^2}. \tag{4.59}$$

Here, the mean value of the photogenerated current can be expressed in terms of the mean number of *photogenerated electrons*, $\langle n_{\mathrm{e}} \rangle$, in the *PD response time* i.e.

$$\langle J \rangle = \frac{e}{\tau_{\mathrm{d}}} \langle n_{\mathrm{e}} \rangle = \frac{e}{\tau_{\mathrm{d}}} \eta \langle n \rangle = \frac{e\eta\lambda P_\lambda}{hc}, \tag{4.60}$$

where η, the *quantum efficiency*, is defined as the ratio of the mean number of photogenerated electrons to the mean number of incident photons of wavelength λ. The *variance* in the *photogenerated current*, σ_{J}^2, is proportional to the *variance* in the *mean number of photogenerated electrons*, σ_{e}^2, i.e.

$$\sigma_{\mathrm{J}}^2 = \frac{e^2}{\tau_{\mathrm{d}}^2} \sigma_{\mathrm{e}}^2, \tag{4.61}$$

Since *Poisson statistics* also applies to the photogenerated electrons, we can write

$$\sigma_{\mathrm{e}}^2 = \langle n_{\mathrm{e}} \rangle, \tag{4.62}$$

which leads to the relation

$$\sigma_{\mathrm{J}}^2 = \frac{e^2}{\tau_{\mathrm{d}}^2} \eta \langle n \rangle = e^2 \eta 2B \frac{\lambda P_\lambda}{hc}, \tag{4.63}$$

where we have introduced the *photodetection bandwidth*, *B*, in terms of the PD response time, i.e.

$$B \equiv \frac{1}{2\tau_{\mathrm{d}}}. \tag{4.64}$$

In this case we can rewrite *photocurrent signal-to-noise ratio*

$$\mathrm{SNR}_{\mathrm{J}} = \eta \frac{\lambda P_{\lambda}}{2Bhc}. \tag{4.65}$$

In the case of an avalanche photodetector with constant electron gain, G_{e}, it is simple to show that the same signal-to-noise ratio applies (Saleh and Teich, 1991), as both the mean photogenerated current, and the root-mean-square value of the photogenerated current variance are equally affected by the electron gain, G_{e}. In the more realistic case of an APD subject to some degree of randomness in the electron gain we introduce the *mean electron gain*: $\langle G_{\mathrm{e}} \rangle$, the *variance* in the *electron gain* $\langle G_{\mathrm{e}}^2 \rangle$, and the *excess noise factor,*

$$F \equiv \frac{\langle G_{\mathrm{e}}^2 \rangle}{\langle G_{\mathrm{e}} \rangle^2}. \tag{4.66}$$

Under these circumstances the mean electron current out of the device is

$$\langle J \rangle = \frac{e}{\tau_{\mathrm{d}}} \langle G_{\mathrm{e}} \rangle \eta \langle n \rangle = e \langle G_{\mathrm{e}} \rangle \eta \frac{\lambda P_{\lambda}}{hc}. \tag{4.67}$$

The variance in the photogenerated current under these conditions is

$$\sigma_{\mathrm{J}}^2 = \frac{e^2}{\tau_{\mathrm{d}}^2} \sigma_{\mathrm{e}}^2 \langle G_{\mathrm{e}}^2 \rangle = e^2 \eta \langle G_{\mathrm{e}}^2 \rangle 2B \frac{\lambda P_{\lambda}}{hc}, \tag{4.68}$$

and the corresponding *photocurrent signal-to-noise ratio*

$$\mathrm{SNR}_{\mathrm{J}}^{\langle G_e \rangle} = \frac{\eta \lambda P_{\lambda}}{2BhcF}. \tag{4.69}$$

We see that this expression is the same as Eq. (4.65) (the signal-to-noise ratio in the absence of electron gain) divided by the excess noise factor, *F*. This *effective gain noise* is associated with the randomness in the location at which ionization occurs and feedback gain linked to the presence of two carriers of opposite charge. In general, $2 < F < 100$, but it has a minimum value, $F = 2$, when only one carrier experiences gain. In Fig. 4.51, the variation in photocurrent SNR is plotted against the *mean number of photons* (4.56) received in the photodetector response time for both a photodiode and an APD with a *mean* electron gain of 100, and an excess noise factor of 2. The circuit noise parameter (Saleh and Teich, 1991), which is a measure of the quality of the optical receiver, is the same for both cases. Reference to Fig. 4.51 indicates that for small optical powers {where circuit noise is dominant} the APD yields a higher SNR than the photodiode. However, the SNR for the APD drops below that of photodiode at high optical powers.

CHAPTER 5

Fiber Optic Technology

5.1 INTRODUCTION

In this chapter we shall describe optical fibers and examine their radiation modes and transmission characteristics. We shall then explore their physical properties, strength, and the issue of packaging (connecting) them to sources and detectors. We will next describe a number of fiber optic devices that are of interest in structural sensing, especially fiber Bragg gratings. Finally, we shall discuss multiplexing, or the merging of a number of optical channels of information into one, and indicate its relevance to structural monitoring.

5.2 OPTICAL FIBERS

Optical fibers are made from fused silica, are about the diameter of a human hair, and transmit light over large distances with very little loss (Fig. 5.1). They can also be made to be sensitive to their state and environment and are therefore well suited as sensors. Optical fibers comprise two essential components: a *core* region surrounded by an annular *cladding* region. The core of the optical fiber serves to guide light along the length of the optical fiber. The cladding region has a slightly lower index of refraction than the core and physically supports the core region. Its primary function is to ensure that very little light is lost as it propagates along the core of the optical fiber. Light can travel for tens of kilometers in some types of optical fiber, and this important property lies at the heart of the fiber optic telecommunications industry. A

FIGURE 5.1. A fiber optic bundle.

representative optical fiber used by this industry is Corning SMF-28 with a core diameter, *d*, of 5 μm and a cladding diameter *D* of 125 μm. Optical fibers are manufactured in kilometer lengths for the telecommunication industry and operate over a range of wavelengths. However, the 1.55-μm wavelength is the standard for minimal transmission loss; see Section 5.5.

Optical fibers are invariably coated to protect them from moisture in the environment, which would otherwise greatly reduce their life. Indeed, unprotected optical fibers can disintegrate in days when exposed to moisture and subject to loads. When they are hermetically sealed they have been shown by the telecommunication industry to suffer no failure for periods of greater than 20 years even under modest load. This protective layer often takes the form of a thick *acrylate coating*. The overall diameter of the coated optical fiber is about 250 μm (Fig. 5.2). For certain applications, such as embedding within FRP composite material, a *thin polyimide coating* is sometimes preferred. For those applications where the

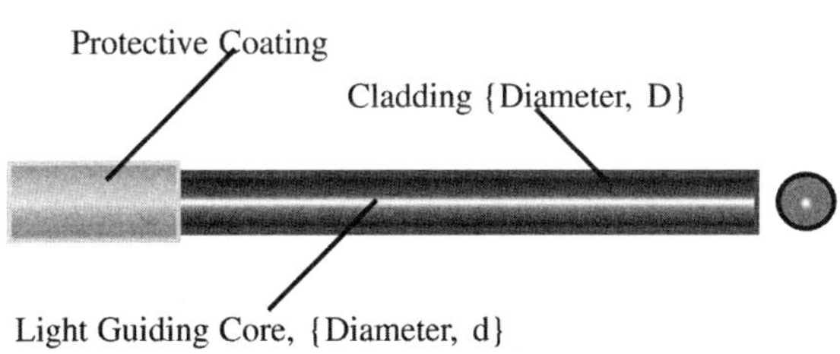

FIGURE 5.2. Primary constituents of an optical fiber.

optical fibers are likely to be subject to harsh physical environment, a tough protective jacket is required.

Optical fibers are generally divided into two kinds: single mode and multimode. In *single mode optical fibers* the light is confined to a very small core (5 to 10 μm diameter, depending on the wavelength of the light for which it is optimally designed) and is transmitted as a plane wave perpendicular to the propagation direction. This reduces temporal spreading of short light pulses and permits a *wide transmission bandwidth.* Single mode fibers are thus of prime importance to the telecommunications industry and are also essential for many forms of fiber optic sensor.

Multimode optical fibers can carry much higher power than their single-mode counterparts as their core diameters can be as large as 100 μm for the same outer diameter of the fiber (Fig. 5.3). This type of optical fiber is commonly used in the medical field and for industrial processing where it delivers high-power laser light to a target for cutting or welding. Although, a variety of sensors (especially chemical sensors based on fluorescence) use multimode optical fibers, all interferometric-based sensors, including Bragg gratings, require single mode optical fibers.

Most optical fibers have a radial variation in the form of a step in their index of refraction, as indicated in Fig. 5.4. This allows their light-propagating behavior to be very consistent and modeled quite accurately. Light is only guided along the core of such a *step index optical fiber* if it enters one of the endfaces within an angular cone defined by the *acceptance angle,* θ_A, see Fig. 5.5. This angle is related, through *Snell's law* of refraction at the air/fiber interface to the *critical angle,* θ_c, for internal reflection at the core/cladding boundary, see Fig. 5.6, by the expression

$$n_0 \sin \theta_A = n_1 \sin(\pi/2 - \theta_c), \qquad (5.1)$$

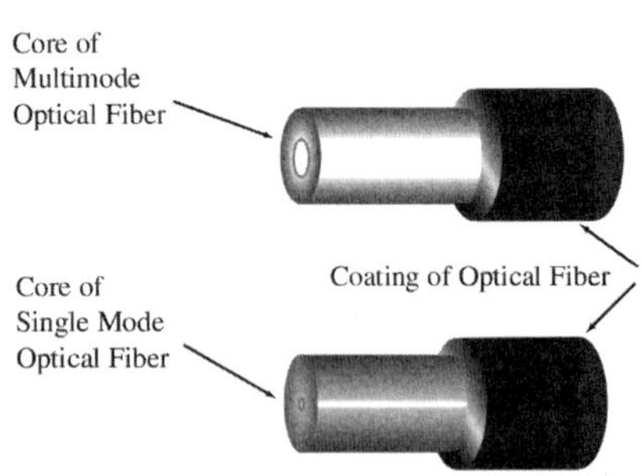

FIGURE 5.3. Single- and multimode optical fiber.

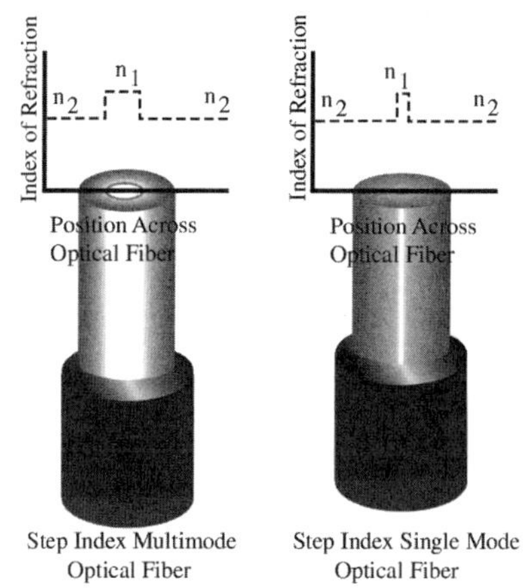

FIGURE 5.4. Radial profiles for the index of refraction for single and multimode step index fibers.

where n_1 and n_0 are the respective core and air indices of refraction. Combining (5.1) with Snells law of refraction at the core/cladding boundary

$$n_1 \sin \theta_c = n_2,$$

allows us to write

$$\sin \theta_A = \sqrt{[n_1^2 - n_2^2]}, \tag{5.2}$$

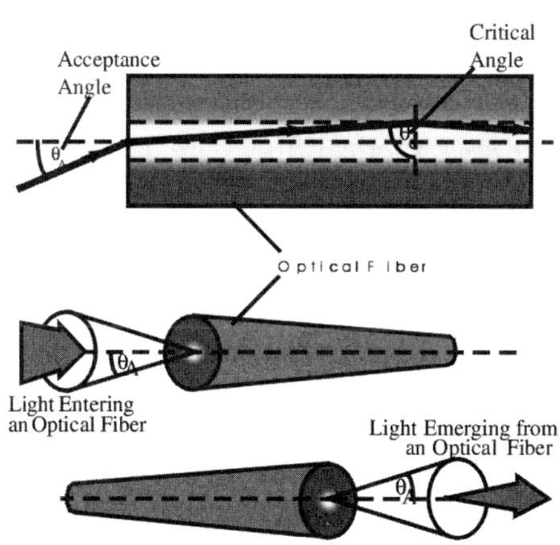

FIGURE 5.5. Core guiding of light within the acceptance angle of an optical fiber.

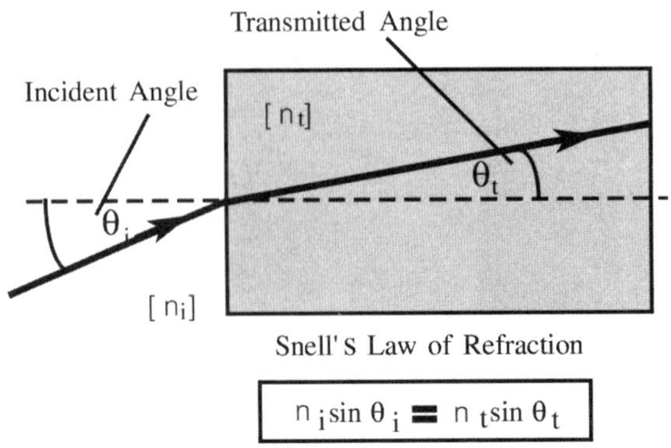

FIGURE 5.6. Schematic illustration of Snell's law of refraction.

where n_2 is the cladding index of refraction and the index of refraction for air is taken as unity. The *numerical aperture* (NA) of an optical fiber is defined by the expression

$$NA = \sqrt{[n_1^2 - n_2^2]}. \tag{5.3}$$

As can be seen by reference to Eqs. (5.2) and (5.3), the larger the NA (i.e., the difference in the core and cladding index of refraction), the larger the *acceptance angle* and the greater the amount of light accepted and guided by the optical fiber. Representative values for NA range from 0.1 to 0.2. The most common material used for telecommunication, standard optical fibers is *fused silica* glass (SiO_2) of exceptional high purity.

Multimode optical fibers have been made with graded index cores, and recently more complex geometrical configurations have been developed that involve inner and outer claddings (Hecht, 1996) designed to trap pump light for a *fiber laser.* Optical fibers are also made with a variety of designs for making them either *polarization-preserving* or *polarizing* in nature. The former are used when it is important to maintain a given state of polarization as light propagates along an optical fiber (Fig. 5.7), while the latter is used where it is necessary to eliminate one component of polarization (Wilson and Hawkes, 1989).

In general, *polarization-maintaining optical fibers* use either an asymmetry in the core geometry or a residual strain field across the core to create a strong birefringence in the optical fiber. For example, the formation of an elliptic, as opposed to a circular, core serves this purpose, see Fig. 5.7. Alternatively, forming the optical fiber with the *bow-tie* elements results in the creation of a residual strain on cooling of the optical fiber due to the

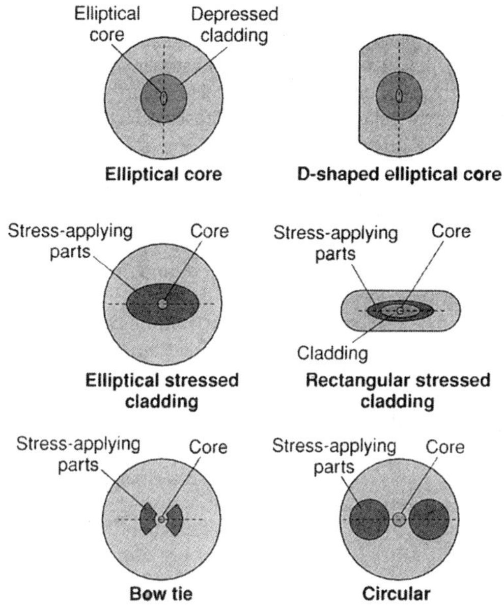

FIGURE 5.7. A number of different polarization-maintaining optical fiber designs.

difference in the thermal expansion properties of the bow-tie elements relative to the remainder of the cladding. The establishment of a strong *intrinsic birefringence* permits linearly polarized light, aligned with either the fast (low index) or the slow (high index) axis of the optical fiber to propagate unchanged along the optical fiber (Wilson and Hawkes, 1989). Although the use of polarization-maintaining optical fiber can avoid stress-induced changes to the state of polarization of light traveling along the optical fiber, and prevent the sensing system from being vulnerable to such effects, it is a costly solution and other techniques are usually employed.

5.3 OPTICAL FIBER GUIDED WAVE MODES

We treat the propagation of electromagnetic radiation along an optical fiber by making some reasonable assumptions. The optical fiber is assumed to have axial symmetry and comprises a small light-guiding core that is centered on the fiber axis. The radius of the core region is a, and its index of refraction n_1. Surrounding the core is a cladding region, which has a slightly lower index of refraction, n_2, and a radius that extends to infinity. We assume that the optical fiber *fractional refractive index difference*

$$\Delta \equiv \frac{n_1 - n_2}{n_1} \ll 1. \tag{5.4}$$

This small index difference is produced by doping the core region of fused silica glass fibers with trace materials such as germanium. The refractive index of the core can range from 1.44 to 1.46 depending on wavelength and Δ typically lies between 0.001 and 0.02.

To describe the propagation of light along an optical fiber we assume monochromatic, single-wavelength, radiation that satisfies the *electromagnetic wave equation* EWE, see Eq. (3.9). Although we shall treat only the electric field component of the radiation an identical analysis is possible for the magnetic field component. Since all optical fibers of interest have a circular cross-section, we will use a cylindrical coordinate system in this discussion. In such a coordinate system the electric field vector

$$\mathbf{E} = (E_r,\ E_\phi,\ E_z), \tag{5.5}$$

is seen to comprise three components. Since each of them individually satisfy the EWE, we shall use E, for the sake of brevity, to represent the generic **E**-field component. The EWE in a cylindrical coordinate system takes the form

$$\frac{\partial^2 E}{\partial r^2} + \frac{1}{r}\frac{\partial E}{\partial r} + \frac{1}{r^2}\frac{\partial^2 E}{\partial \phi^2} + \frac{\partial^2 E}{\partial z^2} = \frac{n^2}{c^2}\frac{\partial^2 E}{\partial t^2}. \tag{5.6}$$

If we assume that the index of refraction has no axial or azimuthal dependence, i.e.

$$n = n(r), \tag{5.7}$$

then we can assume a *monochromatic plane wave* solution for the **E**-field of the form

$$E(r, \phi, z, t) = E(r, \phi)e^{i(\beta z - \omega t)}. \tag{5.8}$$

This light wave is seen to propagate in the z-direction. Substitution of this solution into the EWE (5.6) yields the *Helmholtz* stationary form of the EWE

$$\frac{\partial^2 E(r, \phi)}{\partial r^2} + \frac{1}{r}\frac{\partial E(r, \phi)}{\partial r} + \frac{1}{r^2}\frac{\partial^2 E(r, \phi)}{\partial \phi^2} + [n^2k^2 - \beta^2]E(r, \phi) = 0, \tag{5.9}$$

where k and β are the free-space and fiber optic *axial propagation constants*, respectively.

The form of this equation is similar to that of a harmonic oscillator

$$\frac{d^2x}{dt^2} + A^2x = 0, \tag{5.10}$$

where x is the *displacement* from the *equilibrium position* and t is the *time* from some reference state. Now it is well known that two very different kinds of solution satisfy this form of equation depending on the sign of A^2. If $A^2 > 0$ oscillatory solutions apply; while if $A^2 < 0$ exponential decay, or growth solutions hold.

A similar situation is found with Eq. (5.9). That is to say, if $n^2k^2 - \beta^2 > 0$, *oscillatory solutions* hold, while if $n^2k^2 - \beta^2 < 0$, *decaying solutions* apply. Note that growing solutions require an energy input. *Guided modes* of radiation require oscillatory solutions within the core of the optical fiber and decaying solutions in the cladding (Neumann, 1988). We thus introduce the square of the *transverse propagation constant*

$$p^2 \equiv n_1^2k^2 - \beta^2, \tag{5.11}$$

and the square of the *transverse decay constant*

$$\gamma^2 \equiv \beta^2 - n_2^2k^2, \tag{5.12}$$

and assume that both p and γ are real.

The independence of n on the *azimuthal angle*, ϕ, as seen from Eq. (5.7), permits us to assume separable solutions of the form

$$E(r, \phi) = E(r)E(\phi). \tag{5.13}$$

Substituting (5.13) into (5.9) gives

$$\frac{r^2}{E(r)}\left[\frac{d^2E(r)}{dr^2} + \frac{1}{r}\frac{dE(r)}{dr} + \{n^2k^2 - \beta^2\}E(r)\right] = -\frac{1}{E(\phi)}\frac{d^2E(\phi)}{d\phi^2}. \tag{5.14}$$

It is apparent that the left side of (5.14) is only a function of r, while the right side is only a function of ϕ. The only way the two sides of this equation can always be equal is if both sides are equal to the same constant. Let us introduce ℓ^2 to be that constant, in which case we can write

$$-\frac{1}{E(\phi)}\frac{d^2E(\phi)}{d\phi^2} = \ell^2. \tag{5.15}$$

The solution of Eq. (5.15) then takes the form

$$E(\phi) = E_0 \cos\{\ell\phi\}, \tag{5.16}$$

with the *azimuthal eigenvalue* of the **E**-field, $\ell = 0, 1, 2, \ldots,$ an integer. This is required to ensure that $E(\phi)$ is a single valued function, i.e.,

$$E(\phi + 2\pi) = E(\phi).$$

Under these circumstances (5.14) can be rewritten in the form

$$\frac{d^2E(r)}{dr^2} + \frac{1}{r}\frac{dE(r)}{dr} + \left[n^2k^2 - \beta^2 - \frac{\ell^2}{r^2}\right]E(r) = 0. \tag{5.17}$$

For *guided modes* it is necessary that $p^2 > 0$, within the core ($r < a$), and so we can write (5.17) in the form

$$\frac{d^2E_1(r)}{dr^2} + \frac{1}{r}\frac{dE_1(r)}{dr} + \left[p^2 - \frac{\ell^2}{r^2}\right]E_1(r) = 0. \tag{5.18}$$

At the same time it is necessary that $\gamma^2 > 0$, in the cladding ($r > a$), and here we can write (5.17) in the form:

$$\frac{d^2E_2(r)}{dr^2} + \frac{1}{r}\frac{dE_2(r)}{dr} - \left[\gamma^2 + \frac{\ell^2}{r^2}\right]E_2(r) = 0, \tag{5.19}$$

where E_1 and E_2 correspond to the **E**-field in the core and cladding regions, respectively.

The core solutions take the form of *Bessel functions of the first kind* of order ℓ

$$E_1(r) = AJ_\ell(pr). \tag{5.20}$$

These are oscillatory types of solution with a decreasing amplitude of oscillation, (Fig. 5.8a) that are the cylindrical analogue of cosine functions in a Cartesian coordinate system. A is the constant of integration.

The cladding solutions take the form of *modified Bessel functions of the second kind* of order ℓ

$$E_2(r) = BK_\ell(\gamma r). \tag{5.21}$$

These are purely decaying types of solution (Fig. 5.8b) that are the cylindrical analogue of negative exponential functions in a Cartesian coordinate system. B is the constant of integration.

Continuity of the **E**-field across the core/cladding interface (at $r = a$) requires that

$$AJ_\ell(pa) = BK_\ell(\gamma a). \tag{5.22}$$

This permits us to express B in terms of A, $J_\ell(\gamma a)$, and $K_\ell(\gamma a)$. By redefining A and taking account of (5.13) and (5.16), we can express the **E**-field solutions in the optical fiber in the form

$$E(r, \phi) = A\frac{J_\ell(pr)}{J_\ell(pa)}\cos\{\ell\phi\}, \qquad \{r < a\}, \tag{5.23}$$

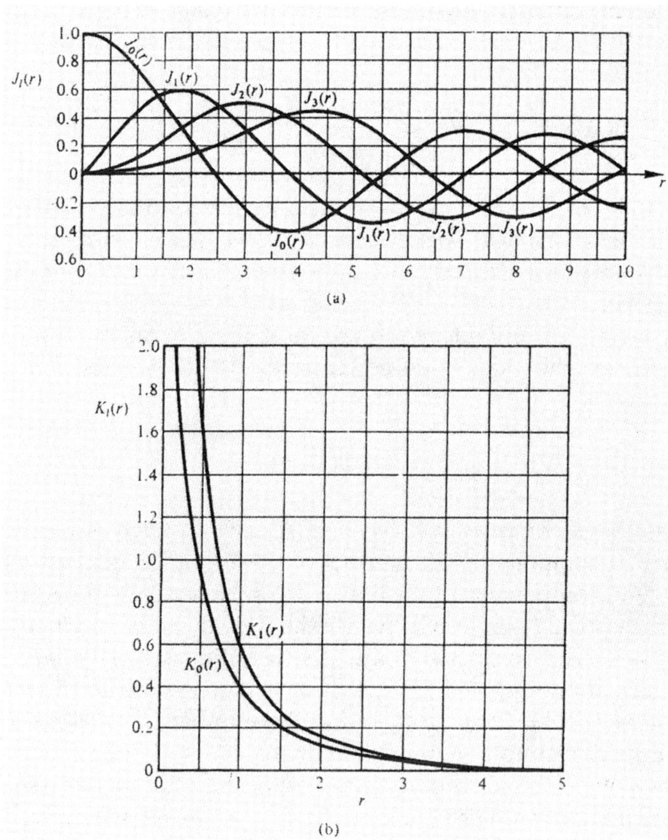

FIGURE 5.8. (a) Bessel functions of the first kind of order ℓ the first few orders. (b) Modified Bessel functions of the second kind of order ℓ for the first two orders. From Senior, J.M., (1985), "*Optical Fiber Communications, Principles and Practice*" Prentice Hall Inc., p. 37.

and

$$E(r, \phi) = A\frac{K_\ell(\gamma r)}{K_\ell(\gamma a)}\cos\{\ell\phi\}, \qquad \{r > a\}. \tag{5.24}$$

Continuity of the **B**-field across the core/cladding interface is equivalent, according to Maxwell's equations, to requiring continuity of the radial derivative of the **E**-field across this boundary (Neumann, 1988). When this is executed in conjunction with the recurrence relations between the Bessel functions the *characteristic equation* for a step index optical fiber is obtained

$$p\frac{J_{\ell-1}(pa)}{J_\ell(pa)} = -\gamma\frac{K_{\ell-1}(\gamma a)}{K_\ell(\gamma a)}. \tag{5.25}$$

Since both p and γ are dependent on β, see Eqs. (5.11) and (5.12), it is apparent that (5.25) is a *transcendental equation* which can be solved for β, the *axial propagation constant*, and ℓ, the *azimuthal eigenvalue* of the **E**-field (Cheo, 1985).

In general, for any integer choice for ℓ, there exists m solutions for β. That is to say, the eigenvalue equation (5.25) has *m-roots*. We can thus write

$$\beta = \beta(\ell,\ m). \tag{5.26}$$

In the *weakly guided wave approximation* ($\Delta \ll 1$) we have assumed that the longitudinal components (E_z and B_z) of the **E**- and **B**-field are neglected and the field components of the lightwave are essentially transverse to the axis of the optical fiber. In which case the **E**-field can be resolved into two linearly polarized components and the resulting set of *linearly polarized (LP)* modes are designated $LP_{\ell,m}$, where, ℓ corresponds to the number of *azimuthal nodal* ($E = 0$) *surfaces*, and m the number of *radial nodal surfaces* for the **E**-field. Examples of the radial intensity distribution (normalized to the same power) for some of the simplest of these LP-modes (Ghatak and Thyagarajan, 1989) are presented in Fig. 5.9, while a schematic representation of the **E**-field modal patterns are displayed in Fig. 5.10. In the case of a single-mode optical fiber, we shall see shortly that only the LP_{01}-mode is relevant. The radial **E**-field distribution for this mode is very close to a Gaussian distribution (see Fig. 5.11) (Neumann, 1988).

5.4 CUTOFF WAVELENGTH AND SINGLE-MODE FIBER

As stated earlier the conditions for *guided modes* of the optical fiber are: (i) oscillatory solution of the **E**-field within the core ($\beta < n_1k$) and (ii) decaying solutions in the cladding ($\beta > n_2k$). According to (5.11) and (5.12), this condition necessitates that both $p > 0$ and $\gamma > 0$. If $p > 0$, but $\gamma \leq 0$, then a *radiation mode* of the optical fiber exists and the wave's power is radiated outside the fiber so that light traveling along in this mode is quickly dissipated. Thus, $\gamma = 0$ represents the *limiting condition for guided modes*.

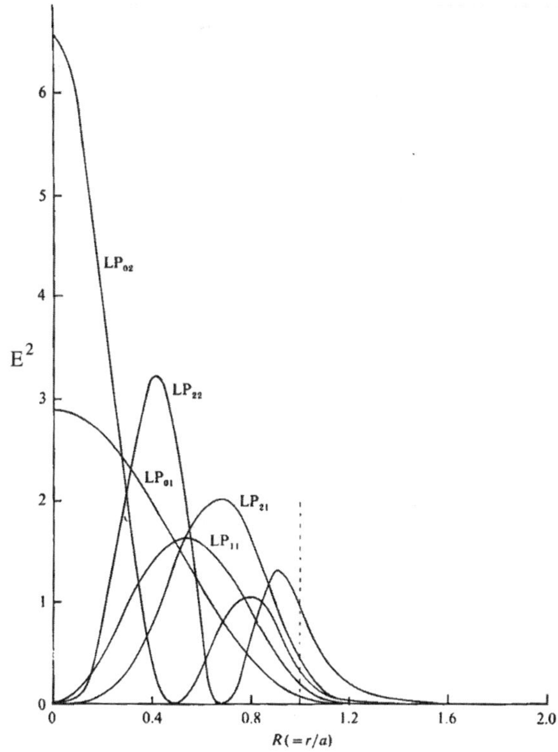

FIGURE 5.9. Radial intensity distributions (normalized to the same power) for a step index optical fiber with $V = 8$. Note that the higher order modes have a greater fraction of power in the cladding. From Ghatak, A.K. and Thyagarajan K., (1989), "*Optical Electronics*" Cambridge University Press, p. 382.

Under these circumstances the *characteristic equation* (5.25) leads to the condition

$$J_{\ell-1}(pa) = 0. \tag{5.27}$$

At this point it is convenient to introduce the *V-parameter*, which is defined by the relation

$$V^2 \equiv a^2(p^2 + \gamma^2). \tag{5.28}$$

Drawing upon the definitions of p and γ, we see this is equivalent to writing

$$V = ka\sqrt{n_1^2 - n_2^2}. \tag{5.29}$$

If $\gamma = 0$, it follows that

$$V = pa, \tag{5.30}$$

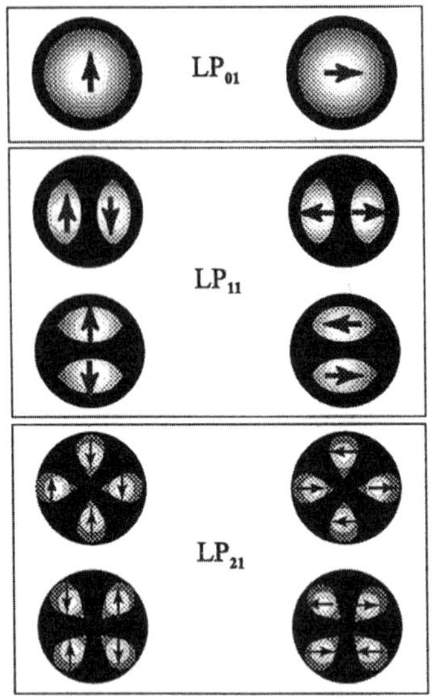

FIGURE 5.10. Schematic representation of the modal **E**-field distributions for a number of low order spatial modes of a step index optical fiber. The arrows indicate the direction of the **E**-field.

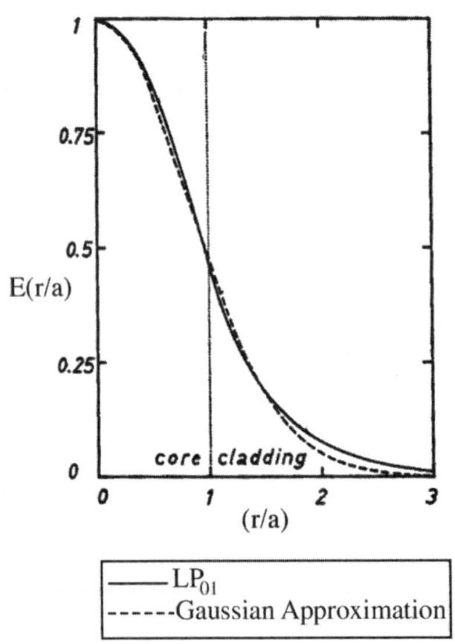

FIGURE 5.11. Comparison between the radial **E**-field distribution of the LP_{01} mode of optical fiber and a Gaussian distribution.

and the condition that light is no longer guided along the optical fiber, (5.27), takes the form

$$J_{\ell-1}(V) = 0. \tag{5.31}$$

Let us now examine some examples of this *cutoff condition* as it is applied to a number of specific radiation modes.

First we consider the case of radiation modes with $\ell = 0$, that is to say; LP_{01}, LP_{02}, $LP_{03}, \ldots$ Each of these modes is described by the expression

$$E(r, \phi) = \text{constant } J_0(pr),$$

as $\cos\{(\ell = 0)\phi\} = 1$. What distinguishes one of these modes from another is the number of radial nodal surfaces (i.e., the *m-value*), see Fig. 5.12. The *cutoff equation* (5.31) in this instance takes the form

$$J_{-1}(V) = 0. \tag{5.32}$$

At this point we draw upon a general relation for Bessel functions of the first kind, (Ghatak and Thyagarajan, 1989)

$$J_{-\ell}(V) = (-1)^{\ell} J_{\ell}(V), \tag{5.33}$$

which permits us to rewrite (5.32) in the form

$$J_1(V) = 0. \tag{5.34}$$

Reference to Fig. 5.8a reveals that the smallest value of V that satisfies this condition is $V = 0$. This means that there is no cutoff for the first mode, LP_{01}. The higher order solutions for this cutoff equation are $V\{LP_{02}\} = 3.832$, $V\{LP_{03}\} = 7.06, \ldots$ (Ghatak and Thyagarajan, 1989).

We saw earlier (5.29) that $V = ka\sqrt{n_1^2 - n_2^2}$. This can be used to determine the *cutoff wavelength* for any given mode. In essence

$$\lambda_{\text{cutoff}}^{LP(\ell,m)} = \frac{2\pi a}{V(LP_{\ell,m})}\sqrt{n_1^2 - n_2^2}, \tag{5.35}$$

where $V\{LP_{\ell,m}\}$ represents the value of V that satisfies the cutoff equation (5.31).

For radiation modes with $\ell = 1$, the *cutoff equation* takes the form

$$J_0(V) = 0. \tag{5.36}$$

The lowest value of V that satisfies this condition is $V = 2.405$, which arises for $m = 1$, (Ghatak and Thyagarajan, 1989). Thus, the LP_{11} mode cutoff wavelength

$$\lambda_{\text{cutoff}}^{LP_{11}} = \frac{2\pi a}{2.405}\sqrt{n_1^2 - n_2^2}. \tag{5.37}$$

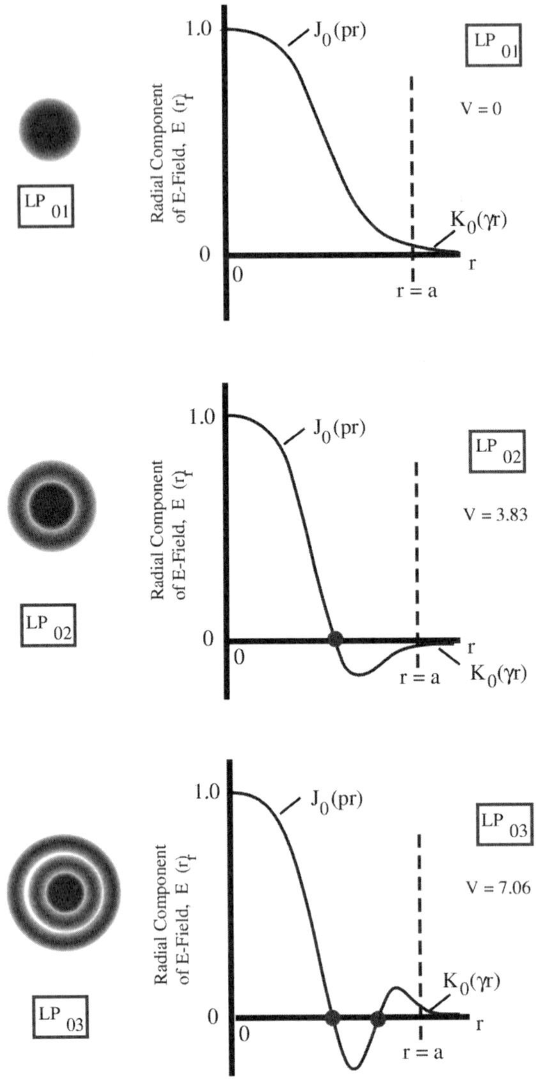

FIGURE 5.12. E-field radial profiles for the LP_{01}, LP_{02}, and LP_{03} modes of an optical fiber. The radial positions of the corresponding nodal surfaces are shown with the large dots.

By way of illustration we assume an optical fiber with $a = 5$ μm; $n_1 = 1.45$; and $n_2 = 1.448$. Then according to (5.37) the LP_{11} cutoff wavelength is

$$\lambda_{\text{cutoff}}(LP_{11}) = 2\pi \times 5[\mu\text{m}] \times (1.45^2 - 1.448^2)^{0.5}/2.405 = 0.9945\ \mu\text{m}.$$

The *transcendental characteristic equation* (5.25) can also be solved to yield β as a function of V for any given mode. The resultant $\beta(V)$ curves for some of the lowest order modes are presented as Fig. 5.13 (Cheo, 1985). For a given optical fiber the properties are fixed and so these $\beta(V)$ curves can really be thought of as $\beta(\lambda)$ curves for the different modes (indeed, $V \propto 1/\lambda$). From the preceeding discussion and Fig. 5.13, it becomes clear that a given optical fiber behaves as a "single mode" optical fiber for light with a wavelength "longer" than the LP_{11} cutoff wavelength. As the wavelength of the light decreases below this value the optical fiber guides two modes, LP_{01} and LP_{11}; then for even shorter wavelengths, further higher order modes become guided, and eventually the optical fiber becomes truly a *multimode optical fiber.* Reference to Fig. 5.13 makes it clear that the definition of a *single-mode optical fiber* is one for which the core radius is small enough that the wavelength of interest is longer than the LP_{11} cutoff wavelength.

The proportion of light in the core, as opposed to the cladding, depends on the wavelength and the mode for a given optical fiber. At a wavelength somewhat greater than the single-mode cutoff wavelength, the LP_{01} field distribution is closely approximated by a *Gaussian profile* (Neumann, 1988) and a significant proportion of the light exists in the cladding in the form of an evanescent field. We shall see in the next section that this makes the optical fiber more vulnerable to bend losses (Cheo,

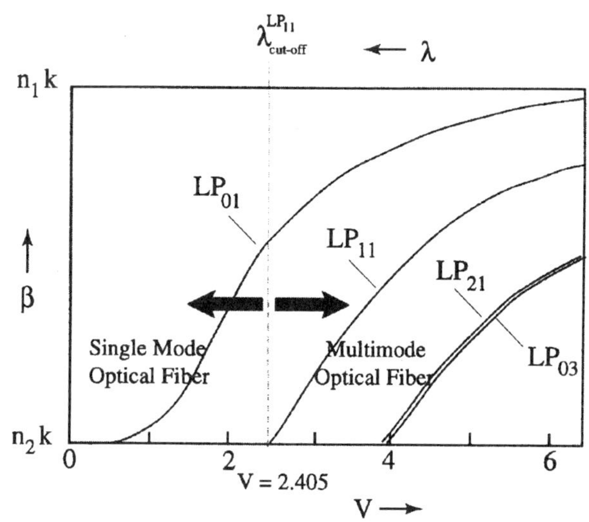

FIGURE 5.13. Variation of the propagation constants β with the V-parameter for several low-order modes of an optical fiber. The demarcation between single and multimode transmission of the optical fiber is clearly indicated.

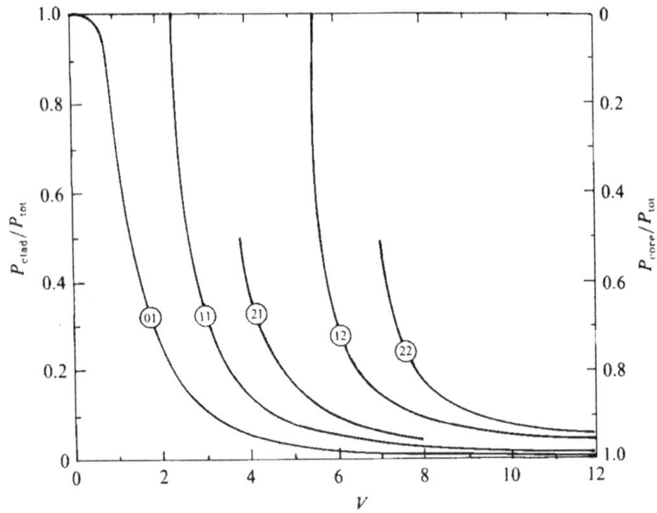

FIGURE 5.14. Variation of the optical power in the cladding versus that in the core, as a function of the *V*-parameter. From Ghatak, A.K., and Thyagarajan, K., (1989), "*Optical Electronics*". Cambridge University Press, p. 386.

1985). In Fig. 5.14, the ratio of the optical power in the cladding to the total optical power in the fiber is plotted as a function of V (note that $V \propto 1/\lambda$). Furthermore, since $P_{core} = P_{tot} - P_{clad}$, the ratio of the power in the core, to the total is also presented in Fig. 5.14. Clearly, it is not desirable to use a wavelength of light that is much longer than the single-mode cutoff wavelength, or a large fraction of the optical power is in the cladding, where it can be lost through a number of mechanisms, including bending. For sensing, where the lengths of optical fiber are usually modest, this is not normally a serious issue.

5.5 OPTICAL FIBER TRANSMISSION PROPERTIES

Anyone who has looked through a fairly thick plate of glass might well be surprised to learn that pure silica glass can have so little attenuation that light can travel though it for kilometers with very little loss. In the late 1960s it was discovered that most of the attenuation in glass could be attributed to absorption by trace metal impurities such as iron, copper, and manganese (Senior, 1985). The use of optical fibers for communications began to look attractive in the 1970s when attenuation was reduced below 20 $dBkm^{-1}$. This *attenuation* is defined to be 10 times the logarithmic (base 10) ratio of the *input/output optical power* for a *unit length* of optical fiber.

Thus the number of decibels (dB) of *attenuation* is

$$\alpha_A = 10 \log_{10}\{P_i/P_o\}, \tag{5.38}$$

where P_i and P_o represent the input and output powers, respectively.

Today the attenuation achieved in ultra *low-loss single mode optical fibers* is close to the theoretical limit of around 0.2 $dBkm^{-1}$ at a wavelength of 1.55 μm. For comparison purposes note that electrical signals in a copper conductor experience about 5 $dBkm^{-1}$. The attenuation of light in an optical fiber is a strong function of wavelength with two local minima, one at 1.3 μm, the other at 1.55 μm (see Fig. 5.15). The general trend is seen to be a decrease in attenuation with increasing wavelength to about 1.6 μm, when the attenuation suddenly increases quite dramatically with increasing wavelength. Rayleigh scattering tends to dominate attenuation at the shorter wavelengths, while infrared absorption leading to the excitation of lattice vibration kicks-in at the longer wavelengths.

Rayleigh elastic scattering arises from small optical inhomogeneities in the index of refraction of the glass that become frozen in during manufacture. This loss mechanism increases with temperature and decreases

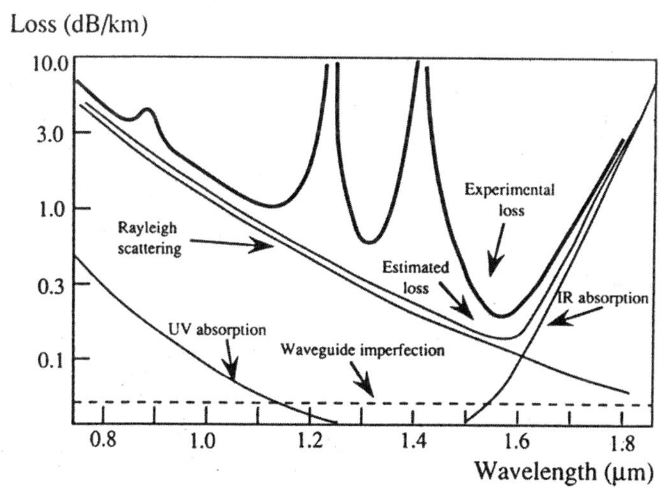

FIGURE 5.15. Attenuation of light within an optical fiber as a function of wavelength. Also indicated are some of the loss mechanisms. From Midwinter, J.E., and Guo, Y.L., (1992), "*Optoelectronics Lightwave Technology*". Wiley, New York, p. 9.

with (λ^{-4}) (Senior, 1985). There are additional nonlinear losses that become significant when high-power lasers are used to send very intense pulses along optical fibers. The two most important processes are *stimulated Brillouin scattering* and *stimulated Raman scattering*. The former has a much lower power threshold and to some extent sets the maximum power that can be transmitted along a single-mode optical fiber. It should be noted that these interactions do not simply attenuate the light energy; they actually convert the light to other wavelengths and are therefore termed *inelastic scattering processes* (Senior, 1985). As we shall see in Chapter 6, they can also be used for distributed sensing over large distances.

The spike in attenuation seen at about 1.4 μm arises from the presence of the hydroxyl (OH^-) group in silica (Senior, 1985) due to exposure of the optical fiber to moisture. The two windows of minimum loss at 1.3 and 1.55 μm are of considerable importance in telecommunications. Although in sensing the attenuation characteristics are not so important as the lengths used are relatively small, it is the low price of the optical fibers and the ancillary technology at the telecommunication wavelengths that make them of interest to the sensing community.

As we indicated in the previous section the radial **E**-field distribution for the LP_{01} mode within a single-mode optical fiber can be approximated by a *Gaussian profile* of the form

$$E_G(r) = E_0 e^{-r^2/w_G^2}, \tag{5.39}$$

where w_G represents the *mode field radius*, defined as the half-width between the $1/e$ points of the LP_{01} field distribution, and E_0 the on-axis value of the **E**-field. In reality, the **E**-field in the cladding is a little larger than given by this Gaussian profile (Neumann, 1988); Fig. 5.11. The intensity at the mode field radius for a Gaussian distribution is about $1/e^2 = 0.135$, so that we can treat the *mode field diameter* ($= 2w_G$) as containing most of the mode's power. The presence of a significant evanescent **E**-field in the cladding leads to additional types of light loss and attenuation. In particular, macro- and microbends can lead to significant losses of the light being guided within the optical fiber.

Macrobend losses (Marcuse, 1982) are associated with radiation of power from the cladding due to bending of the wavefront of the lightwave in the cladding (Neumann, 1988). This curving of the wavefront can be understood in terms of the inability of the wavefront to remain perpendicular to the local axis of the optical fiber without exceeding the speed of light in the medium at some finite radius (Fig. 5.16). Note in this figure the skew in the mode power profile that arises from this phenom-

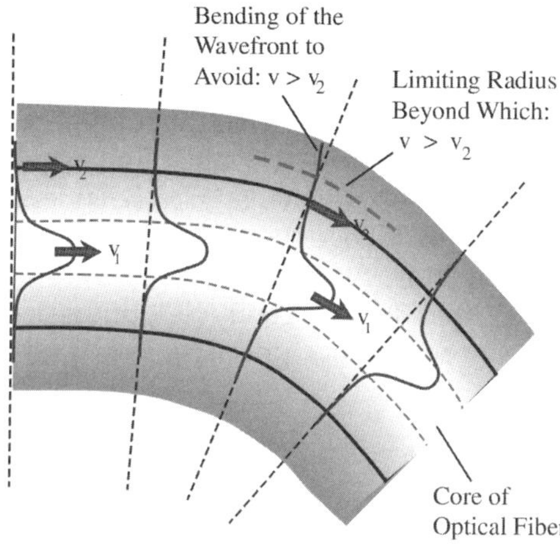

FIGURE 5.16. Schematic illustration of "macrobend" loss of light from an optical fiber. As lightwave propagates around the bend, the **E**-field in the outer section of the bend has to speed up compared to the part in the inner section. Since its speed cannot exceed the speed of light in the cladding, the wavefront becomes distorted and as a result part of the wave propagates out of the fiber.

enon and the change in the index of refraction due to bending stresses. Losses due to micro-deformations can be greatly reduced by decreasing the mode field radius so $w_G < a$. This tightening in the power confinement can be achieved by increasing Δ and decreasing the core diameter. Invariably, optical fibers become more prone to macrobend losses at longer wavelengths for, as we saw earlier (Fig. 5.14), the mode field spreads so that more power is in the cladding at longer wavelengths. Some optical fibers are made relatively (*bend-insensitive* by the incorporation of an annular *depressed cladding zone*) (Lin, 1989).

Macrobend loss can be represented by an *attenuation loss coefficient*:

$$\alpha_{mac} = Ce^{(-R/R_c)}, \tag{5.40}$$

where C is a function of the optical fiber *bend radius* of curvature R and the fiber V-number. R_c corresponds to a *critical bend radius of curvature* (Wilson and Hawkes, 1989), and is approximately given by

$$R_c \cong \frac{a}{n_1^2 - n_2^2}. \tag{5.41}$$

Clearly, the exponential nature of this relation ensures that macrobend losses will be small, if $R > R_c$, but they can become substantial when $R \approx R_c$. Representative ranges in the value of α_{mac} and R_c are 0.1 to 0.01 dBm^{-1} and 10 to 20 mm, respectively (Jeunhomme, 1983).

When an optical fiber is pressed against an irregular surface leading to a succession of small deformations, or *microbends*, another loss mechanism arises. *Microbend losses* (Gloge, 1975) are associated with *periodic small deformations* and result from coupling between the guided and radiation modes of the optical fiber. In a single-mode optical fiber, microbend loss can become significant for surface undulations with a *spatial frequency* Ω given by

$$2\pi\Omega = \beta - n_2 k, \tag{5.42}$$

where β and $n_2 k$ represent the respective *propagation constants* for the guided and radiation modes of the optical fiber (Miller and Kaminow, 1988). In the case of a multimode optical fiber, appreciable *mode coupling* will occur between modes with propagation constants β_1 and β_2, if the *spatial frequency* of the surface undulation matches the difference in these propagation constant (Wilson and Hawkes, 1989), i.e.,

$$2\pi\Omega = \beta_1 - \beta_2. \tag{5.43}$$

Micro-bend transducers (Fig. 5.17) based on this loss mechanism have been used with optical fibers to measure a wide range of parameters, including displacements, (Krohn, 1988) and pressure (Kersey, 1992; Dakin, 1995), see Chapter 6. The problem with any fiber optic sensor that encodes measurand information in changes of the intensity of light is that many other nonmeasurand factors influence the intensity of light. These include source power or temperature fluctuations, photodetector sensitivity varia-

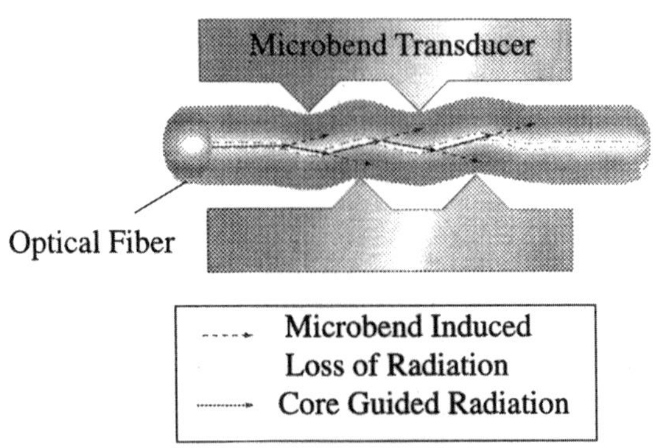

FIGURE 5.17. Basic design of a microbend sensor.

tions due to changes in voltage or temperature, degradation of detector window transmission or even the part of the detector surface exposed to light. In addition there are unknown and uncontrolled bends of the lead optical fiber and variations in connector loss.

5.6 OPTICAL FIBER STRENGTH AND FATIGUE LIFE

Optical fibers are drawn from a molten glass preform, then coated for their protection from the environment and surface damage (Senior, 1985). The optical fibers of primary interest for structural monitoring are made of fused silica and are perfectly elastic until their breaking point. They are brittle and do not yield as do metals when overstressed. As drawn, the *intrinsic strength* of commercial fibers when tested in short lengths ($\sim$1 *m*) is quite high: 5.5 GN m^{-2}, or using the more current terminology 5.5 GPa which is close to 800 k psi[1] (Kurkjian and Yuce, 1991) corresponding to a *breaking strain* of close to *10%*. Because the strength and useful life of an optical fiber depends critically on the: handling, environment and stress load to which it is subjected, a tough coating is normally used to protect optical fibers. A good example is the 60 μm thick acrylate coating used by Corning on its SMF-28 optical fibers. However, certain operations, such as connecting two optical fibers by splicing (discussed in the next section) or the manufacture of many types of optical fiber sensor, requires that the protective coating be stripped off.

Strength degradation of unprotected optical fibers is associated primarily with *microscopic crack growth* on the surface of the optical fiber due to moisture and temperature conditions and the application of short-term or long-term stress. A fiber is weakest at the deepest of these surface flaws. For example, a 0.7-μm-deep crack limits the fiber strength to approximately 100 kpsi, while a fiber with a 2.3-μm flaw will break at about 50 kpsi (Eccleston, 1991). In order to predict the functional life of an optical fiber, it is necessary to account for the deleterious effect of many small surface cracks. This requires some attention to fracture mechanics and a statistical treatment of the subject.

The *Griffith model* of *brittle failure* is used to describe the tensile stress concentration that develops ahead of a crack. According to this model, the stress field ahead of a crack tip in the surface of an optical fiber is described in terms of the *stress intensity factor* (Kapron and Yuce, 1991)

$$K_{\mathrm{I}}(t) = Y\sigma(t)\sqrt{c_{\mathrm{L}}(t)}, \tag{5.44}$$

[1] Note: 1 psi = 6894.76 Nm^{-2} (Senior, 1985)

where Y is a *geometrical shape factor* (for a semicircular crack, $Y = 1.24$), $\sigma(t)$ is the time varying *applied stress*, and c_L is the corresponding *crack length* (normal to the applied stress) (Fig. 5.18). Catastrophic micro-crack growth and fracture of the optical fiber arises if the *stress intensity factor* of the crack attains its critical, or maximum, value, K_ℓ^c, termed the *fracture toughness*. For silica glass, this value is 7.89×10^5 N m$^{-3/2}$ (Bogatyrjov *et al.*, 1991).

If an optical fiber is in a hostile environment, such as one with high humidity, then crack growth is accelerated via the process of *stress corrosion*. Under these circumstances the molecular bonds at the tip of the crack are attacked by water when they are under stress (Roberts *et al.*, 1991). Since the fracture stress and the time to failure due to fatigue depends not only upon the initial crack size and geometry but also the loading rate and environment, the notion of inert strength, S, of a crack is introduced. Thus, S, is the *fracture strength* of the optical fiber in an inert environment where no subcritical crack grows prior to fracture as described by fracture mechanics. We thus define the *inert strength*, S, of the crack through the equation

$$K_I^c = YS\sqrt{c_L}, \tag{5.45}$$

The inert strength always exceeds the tensile strength measured under normal laboratory conditions.

Under conditions of *stress corrosion*, the growth in the *crack velocity* is described by the relation:

$$\frac{dc_L}{dt} = AK_I^n(t). \tag{5.46}$$

In this equation A is a *material parameter*, and the dimensionless exponent, n, is the *stress corrosion factor*, or n-value (Kapron and Yuce, 1991). Substituting (5.44) for the *stress intensity factor* on the right side of (5.46) yields

$$\frac{dc_L}{dt} = AY^n\sigma^n(t)c_L^{n/2}, \tag{5.47}$$

This equation can be integrated directly, and eliminating the crack velocity in terms of the inert strength, given in Eq. (5.45), we arrive at an expression for the *initial* and *final inert strengths*, S_1 and S_2, respectively, in terms of the applied load history, i.e.,

$$S_2^{n-2} = S_1^{n-2} - \frac{1}{B}\int_{t_1}^{t_2} \sigma^n(t)dt, \tag{5.48}$$

where

$$\frac{1}{B} = \tfrac{1}{2}(n-2)AY^2\{K_{\mathrm{I}}^{\mathrm{c}}\}^{n-2}, \tag{5.49}$$

We see that the specific applied stress history [given by the second term on the right side of (5.48)] has a smaller effect on the final inert strength for an optical fiber with a large B-value. A *critical failure* condition arises when the final inert strength in (5.48) equals the particular value of the applied stress. Thus, the *instant of failure*, t_{f}, corresponds to t_2 as $\sigma(t_{\mathrm{f}}) = S_2$, in which case the general *failure life* equation takes the form:

$$\frac{1}{B}\int_0^{t_{\mathrm{f}}} \sigma^n(t)dt + \sigma^{n-2}(t_{\mathrm{f}}) = S^{n-2}. \tag{5.50}$$

Here the stress history begins at the *initial time* $t = 0$, and the subscript on the *initial inert strength* is dropped. This expression gives the lifetime for any stress history, if n, B, and S are known. If we assume that in general an optical fiber at failure is somewhat weaker than its original strength, we can neglect the second term on the left side of (5.50), and we have a simplified failure life equation of the form:

$$\int_0^{t_{\mathrm{f}}} \sigma^n(t)dt = BS^{n-2}. \tag{5.51}$$

Because a distribution of initial flaw sizes occurs on the surface of an optical fiber, there exists a distribution of initial strengths, and as a consequence a statistical methodology is required to treat the issues of *failure strength* and *fatigue life* of optical fibers. The distribution of inert strengths is found empirically to be described by the *Weibull statistical model*. This approach is based on a weakest failure model and claims that the *cumulative probability of failure*, $F(S, L)$, for an optical fiber with an initial inert strength S and a length L is given by the expression (Kapron and Yuce, 1991),

$$F(S, L) = 1 - P(S, L), \tag{5.52}$$

where

$$P(S, L) = e^{-L/L_0(S/S_0)^m}, \tag{5.53}$$

is the *cumulative survival probability*, with S_0 and L_0 representing the *Weibull strength* and length scaling parameters, and m is introduced as the *Weibull shape parameter*, or m-value. Note that S_0 is the inert strength corresponding to a survival probability of e^{-1} (36.8%), measured at a fiber length of L_0. The normalized slope of the Weibull logarithmic plot, or m-value, reflects the variance of the distribution with respect to strength, that

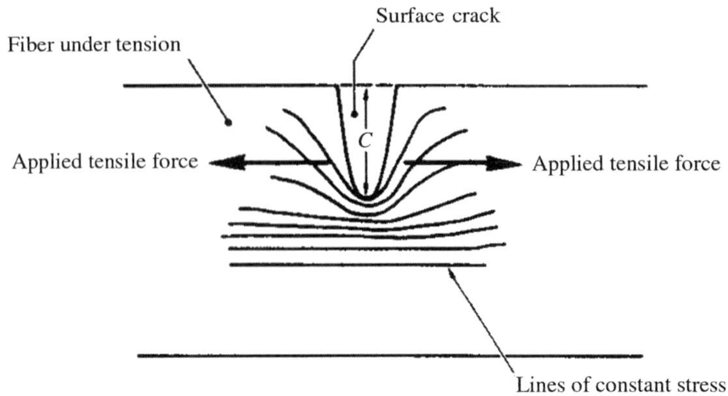

FIGURE 5.18. Schematic illustration of the stress intensification that arises at the tip of a crack within an optical fiber loaded in tension. From Senior, J.M., (1985), "*Optical Fiber Communications, Principles and Practice*". Prentice Hall Inc., p. 134.

is to say it is a measure of the spread in the *strength-at-failure* for the optical fibers. Small m-values are indicative of large range of surface flaw sizes and a commensurate large variation in failure strengths of the optical fibers. By contrast, large m-values are indicative of high surface quality and a close approach to the tensile strength of a flaw-free optical fiber under inert conditions, see the illustrative Weibull plot presented as Fig. 5.19 (Mulvihill, 1997).

In *static fatigue* the optical fiber is subject to constant stress, σ_a, until the weakest crack breaks at an observed failure time, t_f. From Eq. (5.51) the failure time for this loading situation is given by

$$t_f(\sigma_a) = B\frac{S^{n-2}}{\sigma_a^n}. \tag{5.54}$$

We see that the time to failure during static loading depends upon the initial strength, S, the factor, B, and the applied stress, σ_a. A plot of $\log t_f$ against $\log \sigma_a$ should lead to a best linear fit line with a slope of "$-n$" and a vertical intercept of $\log(Bs^{n-2})$. Note all logarithms cited, unless otherwise stated, are to be taken as *natural logarithms*, i.e., to base e.

In *dynamic fatigue* an optical fiber is subject to a constant applied stress rate, $\dot{\sigma}_a$, until the weakest crack breaks the optical fiber at an *observed failure stress*

$$\sigma_f = \dot{\sigma}_a t_f. \tag{5.55}$$

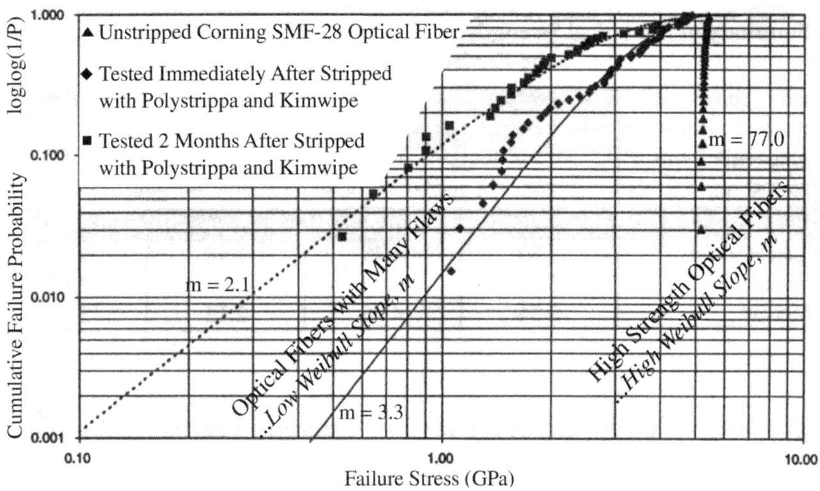

FIGURE 5.19. Plot of cumulative failure probability against failure stress for poor- and good-quality optical fibers. The optical fibers with many flaws have a low Weibull slope, while optical fibers with few flaws have high strength and large Weibull slope. From MSc Thesis, P. Mulvihill, (1997), "*Manufacturing Optical Fibre Bragg Grating Strain Sensors with an Excimer Laser for High-Strain Multiplexed, Embedded Applications,*" University of Toronto, Institute for Aerospace Studies.

For this case Eq. (5.51) takes the form:

$$\int_0^{t_f} \{\sigma_a(t)\}^n dt = \int_0^{t_f} \{\dot{\sigma}_a t\}^n dt = BS^{n-2}, \tag{5.56}$$

and if the stress rate is constant we can remove it from the integral in (5.56) and obtain an expression for the *dynamic failure time*

$$t_f = \left[\frac{(n+1)BS^{n-2}}{\dot{\sigma}_a^n}\right]^{1/n+1}. \tag{5.57}$$

This can be substituted back into (5.55) to yield an expression for the *dynamic failure stress*

$$\sigma_f = [(n+1)B\dot{\sigma}_a]^{1/(n+1)} S^{(n-2)/(n+1)}, \tag{5.58}$$

The magnitude of the *stress corrosion factor* {or susceptibility}, n, can be determined from the slope of the best linear fit line to the $\log \sigma_f$ against $\log \dot{\sigma}_a$ data, which is $1/(n+1)$.

Rewriting Eq. (5.58) in terms of S and then substituting for it in the expression (5.53) for the *cumulative survival probability*, we arrive at the revised Weibull distribution

$$P(\sigma_f) = e^{-[\sigma_f/\sigma_0]^{m_d}}, \tag{5.59}$$

where we introduce the *dynamic Weibull slope parameter*

$$m_d = m\frac{(n+1)}{(n-2)}, \tag{5.60}$$

and the *stress scaling parameter*

$$\sigma_0 = [(n+1)BS_0^{n-2}\dot{\sigma}_a]^{1/n+1}\left\{\frac{L_0}{L}\right\}^{1/m_d}. \tag{5.61}$$

Both, m_d and $\dot{\sigma}_0$ can be obtained from a plot of the failure probability, loglog$(1/P)$, against the log of the failure stress (Fig. 5.20a).

It is important to realize that the Weibull plot can have a *low strength knee* for certain optical fibers. Such is the case for optical fibers made via the *modified chemical vapour deposition* (MCVD) process by Corning and shown in Fig. 5.20b. Such a distribution indicates a greater probability for fracture of the optical fiber at low stress. For applications in which a constant low-grade stress may be anticipated Corning developed *outside vapor deposition* (OVD) process (Eccleston, 1991). The Weibull plot for optical fibers made from this preform shows "no" low-strength knee and the majority of fiber breaks are in the high-strength region, with few breaks at low strength, indicating few large flaws.

Although the pristine strength of an optical fiber is in excess of 4.8 GPa (~7% strain), the practical stress limit for long lengths is much lower due to the presence of many microscopic flaws. Manufacturers of optical fibers consequently *proof-test* them in order to guarantee a safe operating stress range and a predictable life. *Proof-test screening* applies a specified stress for a given period of time and causes the fiber to break at the site of any large flaw. Typically, optical fibers are proof-tested to 0.35 GPa, corresponding to about 0.5% strain (or 40 kpsi) for most applications, (Gartside *et al.*, 1988). The minimum time to failure, t_f, for an optical fiber that is proof-tested to a stress σ_p, and then subject to an applied constant tensile stress σ_a can be evaluated using Eq. (5.54), i.e.,

$$t_f = \frac{B}{\sigma_p^2}\left[\frac{\sigma_p}{\sigma_a}\right]^n. \tag{5.62}$$

This relation is used to produce the two curves of minimum time to failure versus the ratio of applied stress to proof-test stress (σ_a/σ_p) presented in

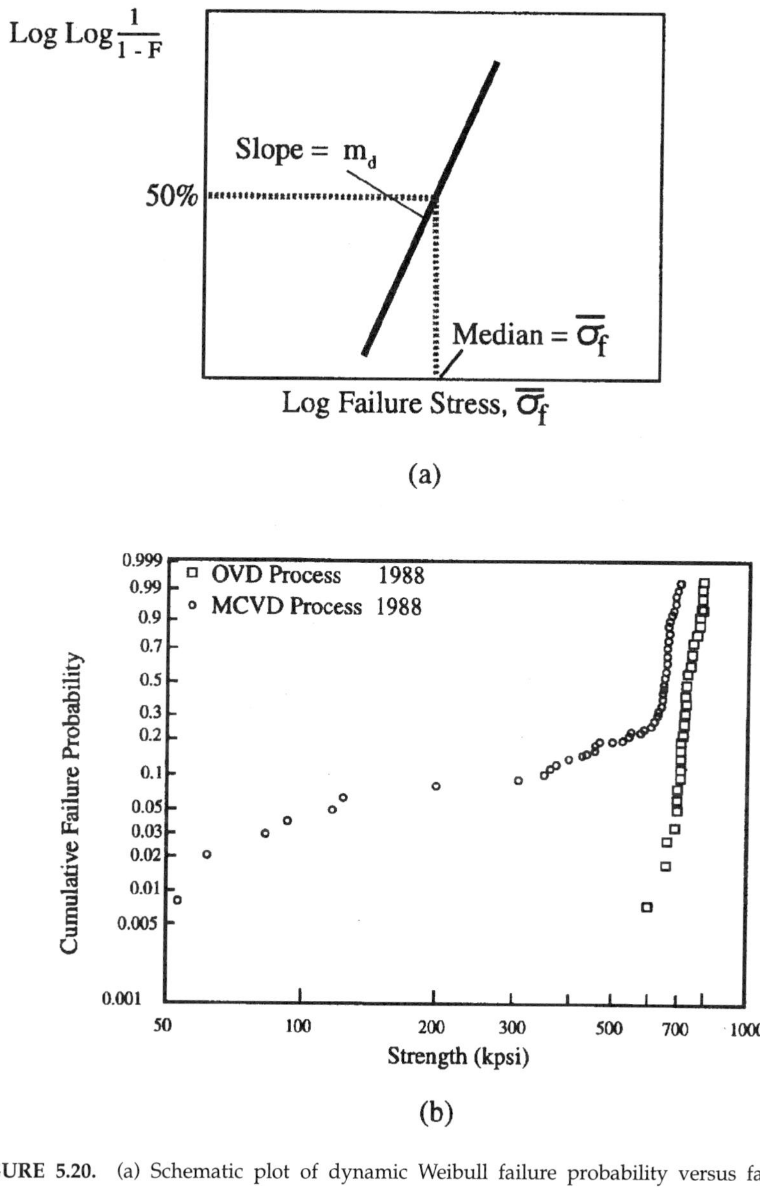

FIGURE 5.20. (a) Schematic plot of dynamic Weibull failure probability versus failure stress. (b) Low-strength "knee" associated with optical fibers manufactured by MCVD perform significantly worse than OCV-processed optical fibers.

Fig. 5.21 for 0.35 and 1.38 GPa proof-stress of fused silica optical fibers in a wet environment, (Gartside *et al*, 1988). These results clearly show that if it is desirable for the optical fiber to have a long working life, then its operating stress must be a small fraction of the proof stress. Put another way, if it is suspected that an optical fiber is likely to be subjected to certain stress, then it is necessary to proof-test the fiber to a stress of maybe *five times* that *operating stress* to guarantee a life in excess of 20 years.

In some applications bending the optical fiber maybe unavoidable. Since small bends can lead to high tensile loads over a part of the optical fiber cross-section, it maybe necessary to take account of this form of stress loading in any proof-test. This can be done by assuming a tensile stress equivalent to the maximum bending stress. However, since the bending stress is nonuniform and is concentrated in a short length of the fiber, such an approach should be quite conservative with regard to failure time predictions.

Considerable interest also exists in the strength of optical fiber Bragg gratings, as these constitute an important class of structural sensor and some mechanical tests indicate that the formation of these devices lead to a substantial weakening of the optical fiber (Feced *et al.*, 1997). Their results, presented as Fig. 5.22a, suggest that writing of FBGs with 193 nm laser radiation appears to degrade the strength of the optical fiber a little less than writing at 248 nm. However, the degradation with 193 nm is still

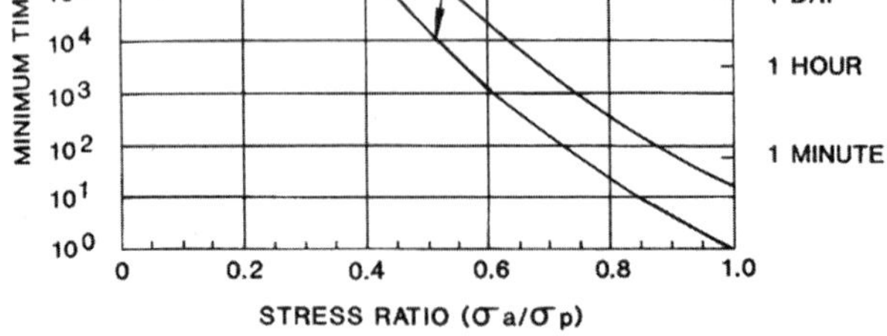

FIGURE 5.21. Minimum time to failure versus the ratio of applied stress to proof-test stress for 0.35 and 1.38 GPa proof-tested optical fibers in a wet environment. From Gartside III, C.H., Patel, P.P. and Santana M.R. "*Optical Fiber Cables*" (1998), "*Optical Fiber Telecommunications II*" Miller and Kaminow (Eds), Academic Press, p. 227, used with permission.

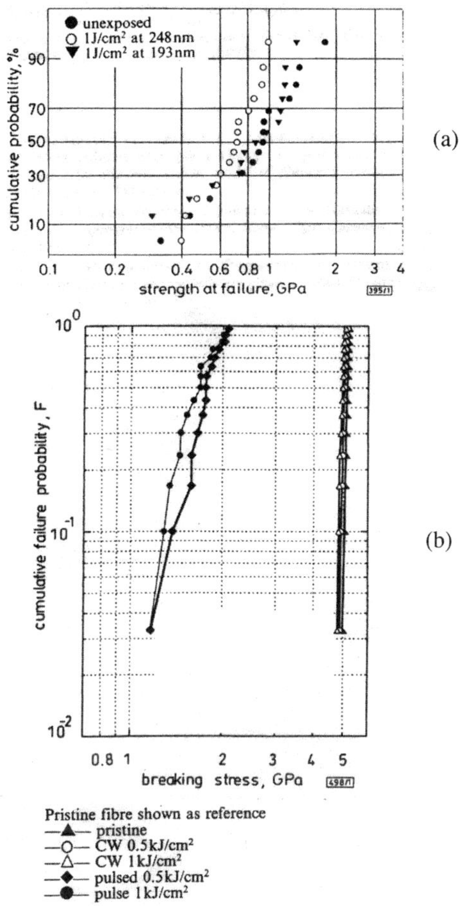

FIGURE 5.22. (a) Weibull plot of cumulative faijure probability against strength of failure for Coming SMF 28 optical fiber subject to different laser wavelengths. From Feced, R., Roe-Edwards, M.P., Kanellopoulos, S.E., Taylor, N.H., and Handerek, V.A., "Mechanic strength degradation of UV exposed optical fiber," *Electronics Letters*, **33**(2), 157–159, with permission from IEE Publishing. (b) Weibull plot comparing CW and pulsed irradiation influence on the strength of the optical fiber. From Varelas, D., Limberger, H.G., and Salathé, R.P., "Enhanced Mechanical Performance of Singlemode Optical Fibres Irradiated by a cw UV Laser" *Electronics Letters*, **33**, 704–705, with permission from IEE Publishing.

significant when compared to the pristine optical fiber. In contrast, Varelas *et al.* (1997) have demonstrated that producing FBGs with a continuous wave, frequency doubled Ar^+ ion laser at 240 nm causes minimal reduction in the strength of the optical fiber, see Fig. 5.22b. Unfortunately, this technique takes some time (minutes) to form the FBGs, whereas Askins *et al.* (1997) showed that KrF laser radiation at 248 nm could be used to write FBGs with *almost no degradation* in the strength of the optical fiber when this is undertaken on the optical fiber draw tower. The results of Askins *et al.* (1997) are presented later as Fig. 5.55. In this figure they directly compare their results with those of Feced *et al.* (1997).

5.7 FIBER OPTIC CONNECTORS, SPLICES, AND PIGTAILS

Optical fibers invariably have to be connected to something, whether it be a source, a photodetector or another optical fiber. Loss of light is of primary concern in regard to optical connections. Connections between two optical fibers are intended to be relatively permanent, or changeable, allowing disconnection and reconnection. Permanent connections are termed *splices* and can be mechanical or based on the fusion of the fibers. Temporary connections are made with optical connectors. Although optical splices incur 10 to 100 times the loss found in electrical splices, nevertheless when performed properly, the insertion loss is still negligible for most applications. We define the *insertion loss*, $\mathscr{K}$, in decibels by the relation:

$$\mathscr{K} = -10\log_{10}\{P_o/P_i\}, \tag{5.63}$$

where P_o/P_i is the ratio of output to input optical power for the join between the two optical fibers. Note this is negative as the definition involves the inverse ratio of optical powers used in the definition of attenuation, Eq. (5.38).

Simple *butt-join* is the predominant method of connection between two optical fibers. As the name implies, the two optical fibers are virtually butted up against each other in this form of connection. In general, low-loss connections require (i) fiber end preparation, (ii) fiber alignment, and (iii) alignment retention. The endface of each optical fiber must be smooth and perpendicular to the axis of the fiber. Examples of good and poor fiber endfaces are schematically illustrated in Fig. 5.23 photographs of actual cleaved optical fibers are presented in Fig. 5.24. Good optical coupling of light from one fiber to the next requires that the cores of the two optical fibers be aligned well, within a micron in both lateral dimensions, and that the endfaces be touching, or within a micron of each other. In some mechanical splices an *index-matching fluid* is introduced between the optical fibers to enhance the coupling of light. The index of refraction of this fluid is chosen to be between the indices of the two cores, so if two

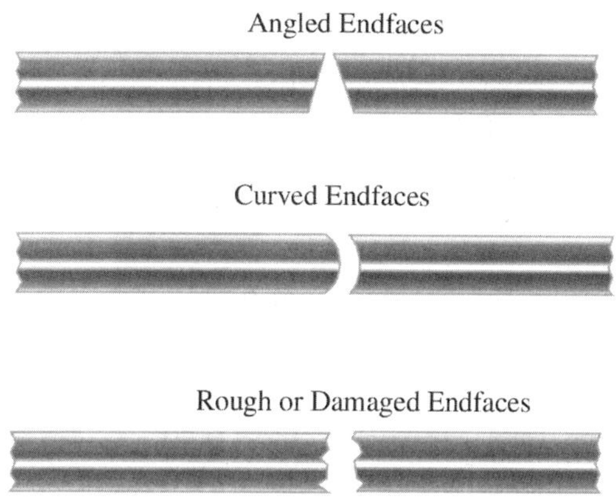

FIGURE 5.23. Schematic illustration of different types of optical fiber endface.

identical fibers are involved, the index of the fluid would ideally be the same as that of their optical fiber cores.

In the absence of any index-matching fluid, the magnitude of this *Fresnel insertion loss* incurred at a glass/air interface (Senior, 1985)

$$\mathcal{K}_{\mathrm{F}} = -10\log_{10}\{1 - R_F\}, \tag{5.64}$$

where we have introduced the *Fresnel reflection coefficient*: for normal incidence at the interface

$$R_{\mathrm{f}} = \left[\frac{n_{\mathrm{g}} - n_{\mathrm{a}}}{n_{\mathrm{g}} + n_{\mathrm{a}}}\right]^2, \tag{5.65}$$

with n_{g} and n_{a} representing the indices of refraction for the glass optical fiber core and the air, respectively. The loss at the join of two optical fibers is the sum of the loss from each interface. In the case of two optical fibers, each with a core index of refraction $n_{\mathrm{g}} = 1.5$, the loss of light at the join due to Fresnel reflection from the two surfaces would be 0.36 dB, assuming $n_{\mathrm{a}} = 1.0$.

The very small core and limited angular acceptance of single-mode optical fibers makes connecting them quite challenging, as misalignment can lead to much higher losses than arise from Fresnel reflections. The three primary types of misalignment: longitudinal; lateral offset; and angular, are schematically illustrated in Fig. 5.25. *Longitudinal misalignment* refers to the axial separation between the ends of the two optical fibers, (Fig. 5.25a), *lateral offset misalignment* appertains to the offset perpendicular to the core axes of the two fibers, and *angular misalignment*

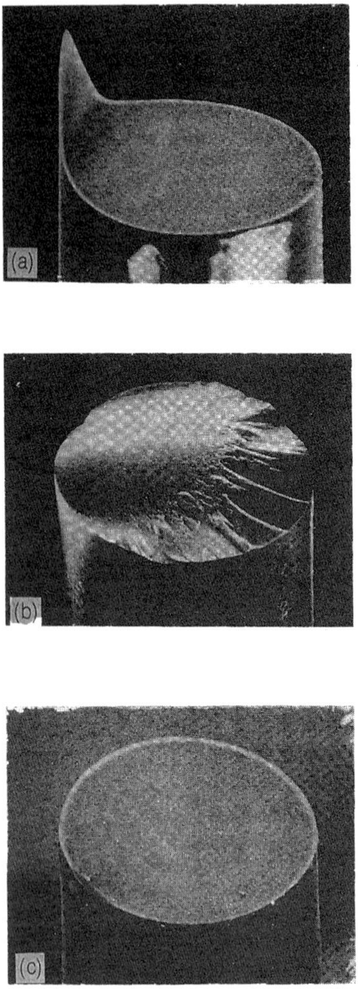

FIGURE 5.24. Three examples of prepared ends of an optical fiber. (a) Lip fault, (b) mist and hackle surface, and (c) smooth endface. From Miller, C.M., Mettler, S.C., and White, I.A., *"Optical Fiber Splices and Connectors: Theory and Methods,"* © 1986, Marcel Dekker Inc. p. 31.

accounts for the angle between the core axes. The magnitude of the loss from these three types of misalignment depends upon the fiber architecture, the core diameters and the optical power distribution between the different modes. Examples of the measured optical losses arising from these different forms of misalignment for a pair of single-mode optical

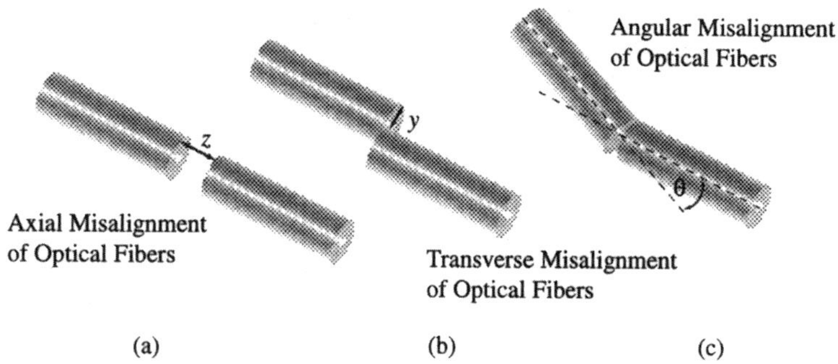

FIGURE 5.25. Three types of optical fiber misalignment: (a) axial or longitudinal, (b) transverse or lateral, and (c) angular.

fibers with a mode field diameter of 10 μm are presented in Fig. 5.26. It is clear that low-loss joins (< 0.5 dB) for single-mode optical fibers requires the lateral offset to be controlled to submicrometer accuracy (Young, 1989).

Fiber optic connectors must be capable of repeated connections and disconnections with negligible deterioration in alignment (Young, 1991).

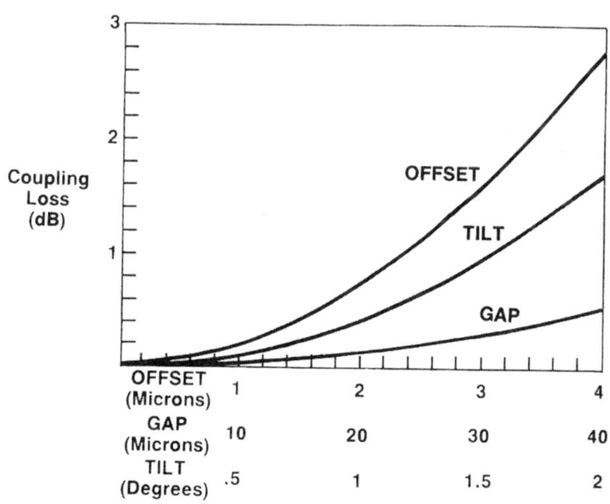

FIGURE 5.26. Coupling loss as a function of transverse (offset), angular (tilt), and axial (gap) misalignment for single-mode optical fibers having a mode field diameter of 10 μm. From Young, W.C., (1989), "Optical Fiber Connectors, Splices and Jointing Technology." In: *Optoelectronic Technology and Lightwave Communications Systems*, Lin, C., (Ed), Van Nostrand Reinhold, pp. 155–174.

They must protect the endfaces of the optical fibers from damage during repeated disconnections and reconnections, and must be insensitive to their environment, be it dust or moisture. They must also be robust to prevent varying cable tensile loads from increasing their insertion loss. One of the most common *butt-joined* connectors has the industrial designation FC for *Fiber Connector* and relies on a *ceramic ferrule* and a locking ring to ensure excellent optical alignment (Fig. 5.27, Miller *et al.*, 1986). The optical fibers fit accurately (< 1 μm) within the ferrules of each connector, and the two ferrules are mated by means of an accurate alignment sleeve. An example of such an arrangement using a *biconical alignment sleeve*. To eliminate Fresnel reflection losses, *fiber connector/angle physical contact* (FC/APC) are most often used. The insertion loss of this type of connector is usually better than −0.5 dB (Spillman and Lord, 1995).

More sophisticated connectors are available that employ *beam-expanding optics* to more efficiently couple light from one fiber to another. This type of connector relies on the near-Gaussian distribution of the LP_{01} mode from a single-mode optical fiber (Section 5.4). The endface of such an optical fiber can thus be treated as the *beam waist* of a Gaussian beam, (Fig. 5.28a) An appropriate lens that is suitably sited can form a well collimated output beam (Fig. 5.28b) that forms half of a *beam expanding connector* (Fig. 5.28c) Several types of lens can be used in this form of beam-expanding connector, including *graded-index* and *ball lenses.* This form of connector tends to be bulky and expensive and not usually necessary for most fiber optic structural monitoring systems.

Where a more permanent join of two optical fibers is required, or it is desirable to have an even smaller insertion loss (<0.2 dB), the fibers are spliced together either mechanically or by electric arc induced fusion. A *mechanical splice* is subject to the same kind of alignment requirements as connectorization but a lower insertion loss can be achieved because a continuous reference surface can be employed in this case for both fibers.

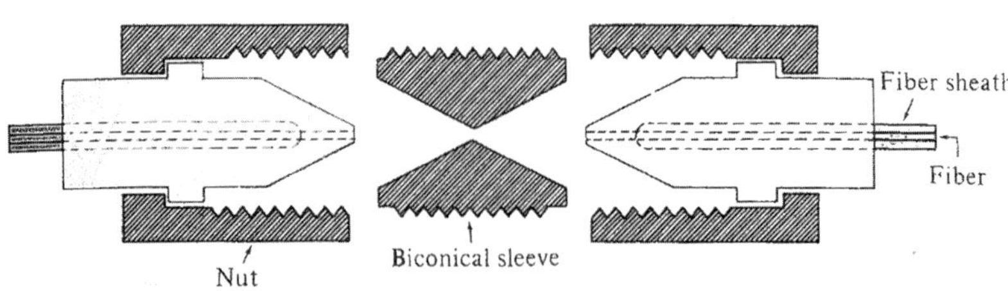

FIGURE 5.27. Cross-section of a biconical alignment sleeve used in fiber connectors. Also shown are the two guiding ceramic ferrules. From Wilson, J. and Hawkes, J.F.B., (1989), "*Optoelectronics, An Introduction,*" 2nd Edn, Prentice Hall Inc., p. 341.

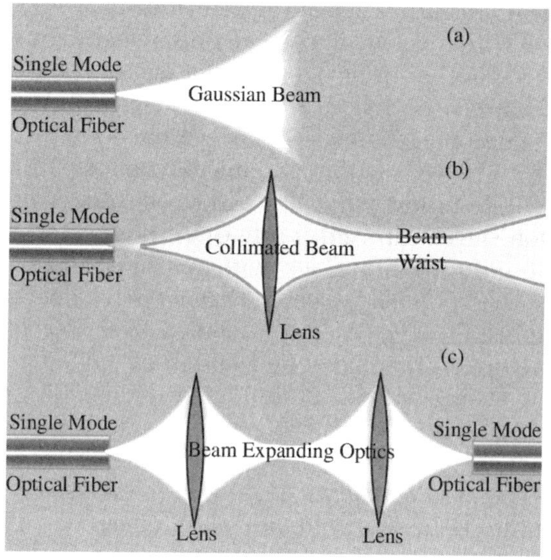

FIGURE 5.28. Three methods of treating the light output from the end of a single-mode optical fiber: (a) as a Gaussian beam, (b) with a single lens to form a collimated beam, and (c) with beam expanding optics designed to efficiently couple light between two optical fibers.

The two ends of the optical fibers are inserted into an accurate glass or ceramic capillary and a transparent adhesive, such as an epoxy resin, is injected through a transverse bore to give mechanical sealing and index matching of the splice (Senior, 1985). The ULTRA splice from Advanced Custom Applications Inc., of Belle Mead, New Jersey, is an inexpensive system (~$10 per splice) that uses a visible glass capillary alignment section that allows the user to inspect the splice. A tight fitting mechanical splice can also be achieved with a collapsed sleeve technique which is particularly useful for fibers of different diameters (Senior, 1985). A *fiber cleaver* is an instrument used to cut, or cleave, optical fibers for use with connectors or splices. An example is the Precision Cleaver S321 from Furukawa Electric, Japan, which produces good quality cleaves with smooth endfaces.

Fusion splicing is accomplished with a sophisticated machine that uses an electric arc between two fine electrodes to soften and fuse the ends of the two optical fibers while they are held in precise alignment with vacuum chucks. An example of a good and relatively inexpensive automated *fusion splicer* is the model S148S from *JDS Uniphase*, Ontario. This instrument permits the user to visually line up the fiber endfaces and view in real-time the splicing operation through small CCD video monitor. The sequence of operations are schematically illustrated in Fig. 5.29 and are made easy to use with this machine via an on-screen program. The fusion splicer includes a splice protection component that heat-fuses a protective sleeve over the completed splice as the final step in the fusing splicing operation. For single mode optical fibers we have seen that the lateral alignment needs to be well within one micrometer. As it happens a *self-alignment* process resulting from the action of surface tension forces during the fusion stage helps ensure that insertion losses of better than 0.1 dB are achieved. One drawback with fusion splicing is a possible reduction (as much as 50%) in the *tensile strength* of the fiber in the vicinity of the fusion splice (Young, 1989). A detailed overview of splices and connectors is provided in the book by Miller *et al.*, (1986).

Illuminating a single-mode optical fiber with light from a light-emitting diode, or laser diode is challenging as both of these semiconductor sources have very small emitting surfaces and radiate into large angles. Various lens arrangements have been employed to improve the coupling of light from the source to the optical fiber, (Fig. 5.30). Although *spherical microlenses* and *graded index lenses* were popular for a while, the recent development of *tapered* and *shaped* optical fibers has simplified the task to some extent. A scanning electron micrograph of the tapered end of an optical fiber fitted with a *microlens tip* is displayed in Fig. 5.31 (Young, 1989). More recently it has been found possible to produce this kind of lensed optical fiber in one piece by tapering the optical fiber either by

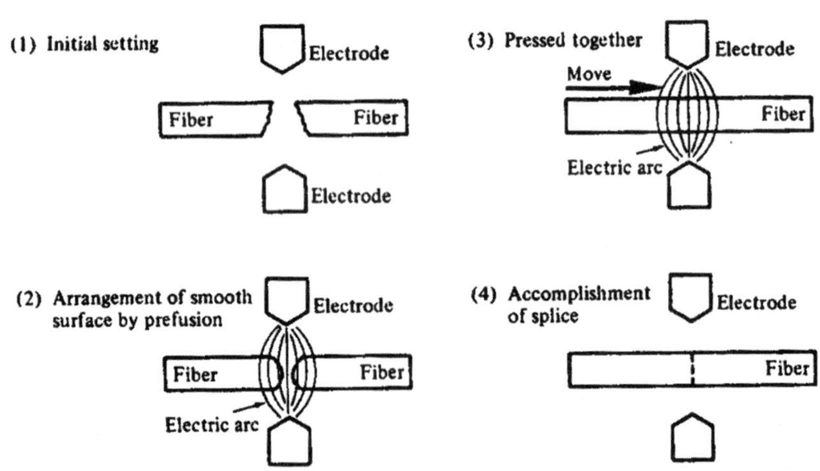

FIGURE 5.29. Sequence of operations involved in the fusion splicing of two optical fibers, Senior, J.M. (1985). "*Optical Fiber Communications, Principles and Practice*" Prentice Hall Inc., p. 158.

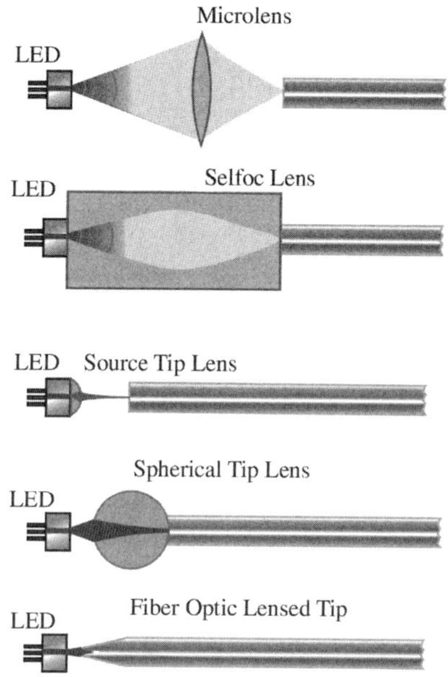

FIGURE 5.30. Schematic representation of several methods of using lenses to couple the light output from an LED into an optical fiber.

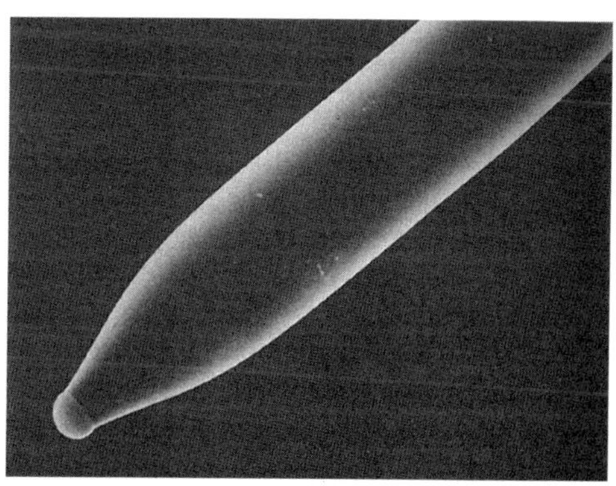

FIGURE 5.31. A scanning electron micrograph of the tapered and lensed tip of an optical fiber. From Young, W.C., (1989), "Optical Fiber Connectors, Splices and Jointing Technology." In: *Optoelectronic Technology and Lightwave Communications Systems*, Lin, C., (Ed), Van Nostrand Reinhold, pp. 155–174.

etching or polishing and then creating the round tip by momentary melting the end of the fiber. Light coupling efficiencies in excess of 30% have been achieved with such *fiber lenses*, (Edwards *et al.*, 1993), which is much better than the small amount (~8%) attained by butt-coupling of a single-mode optical fiber to a laser diode.

We can understand why *butt-coupling* is so ineffective even though the emitting area of the laser diode A_s is much smaller than the core area A_c of the optical fiber, (Fig. 5.32). Note that in this schematic we have exaggerated the separation between the optical fiber and the laser diode for the sake of clarity. For each point on the laser diode emitting surface, only light emitted within the *optical fiber acceptance cone* of angle θ_A will be guided along the core of the optical fiber. According to Eq. (5.2)

$$\theta_A = \sin^{-1}(\text{NA}), \tag{5.66}$$

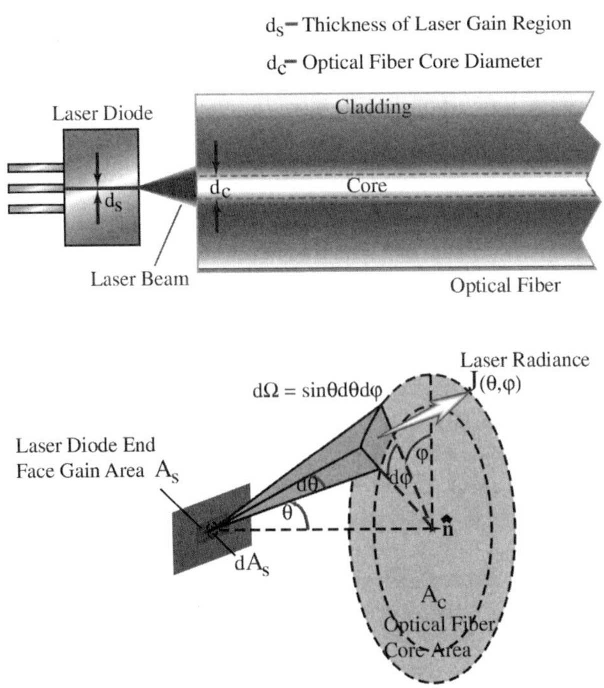

FIGURE 5.32. Schematic illustration of butt-coupling between a laser diode and an optical fiber. The separation is exaggerated for clarity. Also shown is the angular configuration used in the text calculation.

where NA is the *numerical aperture* of the optical fiber and we have assumed that the index of refraction for air is unity. If the source radiance as a function of angle is written $J(\theta)$, the total optical power coupled from the source into the core of the optical fiber is

$$P_c = A_s \int_\Omega J(\theta) d\Omega, \tag{5.67}$$

where $d\Omega$ is the solid angle subtended by the light rays in the angular interval θ and $(\theta + d\theta)$, and the integral is carried out for all of the light within the acceptance solid angle, Ω. Since $d\Omega = 2\pi \sin\theta d\theta$, (see lower part of Fig. 5.32), we can write (5.67) in the form

$$P_c = 2\pi A_s \int_0^{\theta_A} J(\theta) \sin\theta d\theta. \tag{5.68}$$

If we assume a *Lambertian angular distribution*, which is not too bad for an LED, then

$$J(\theta) = J(0)\cos\theta, \tag{5.69}$$

where $J(0)$ is the on-axis source radiance, in which case

$$P_c = 2\pi A_s J(0) \int_0^{\theta_A} \cos\theta \sin\theta d\theta. \tag{5.70}$$

Using (5.66) we can write

$$P_c = \pi A_s J(0)[\mathrm{NA}]^2, \tag{5.71}$$

and the *light coupling efficiency* η_c, defined as the ratio of the optical power coupled into the core of the optical fiber to the total power emitted by the source, is given by

$$\eta_c = \frac{\pi A_s J(0)[\mathrm{NA}]^2}{\pi A_s J(0)} = [\mathrm{NA}]^2. \tag{5.72}$$

A representative value for the *numerical aperture* of a single mode optical fiber is 0.15 (Senior, 1985) so that the coupling efficiency is unlikely to be much better than 2.25%.

Most semiconductor sources are now *pigtailed* to an optical fiber to ensure that good lateral, longitudinal, and angular alignment are maintained. In this method of *packaging* the optical fiber is affixed to some form of support structure that is rigidly attached to the mount for the semiconductor chip, Fig. 5.33. The optical fiber is either soldered (if metal coated) or bonded with an epoxy resin to this support structure. In the case of an epoxy bond it is necessary to avoid shifts of the optical fiber during the bonding process. In Fig. 5.33 we clearly see, at the greatest

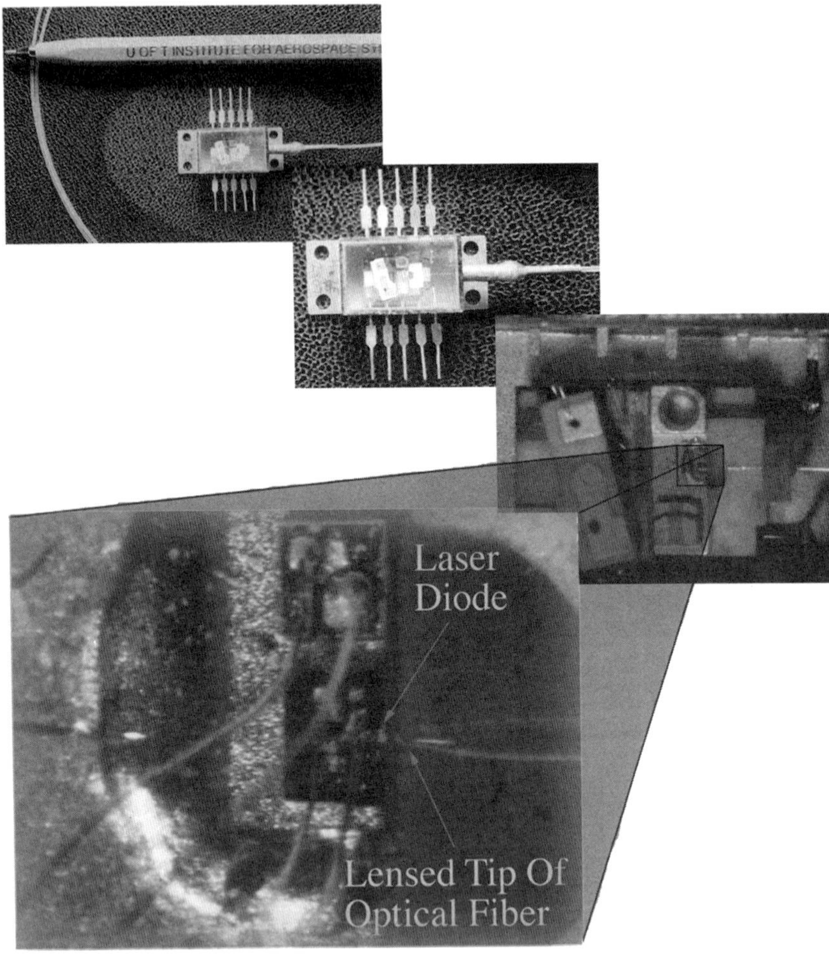

FIGURE 5.33. Set of photographs revealing, in a progressive manner, the finer details of packaging involved in a pigtailed laser diode. The most magnified image shows the lensed tip of the optical fiber and the laser diode.

magnification, the tapered and shaped optical fiber used in this pigtailed laser diode. Tapering the optical fiber effectively increases its numerical aperture so it more nearly matches that of the laser diode. The curved tip of the lensed optical fiber also reduces the amount of laser radiation reflected back into the laser diode from the end of the optical fiber. This avoids the degradation of the laser output characteristics that can arise when the laser's radiation is coupled directly back into the laser. *Hermetic encapsulation* is often employed to make the laser and its fiber optic pigtail robust.

5.8 OPTICAL ISOLATORS, COUPLERS, FILTERS, AND SPECTRAL ANALYZERS

As mentioned earlier, back reflections of laser radiation from within a fiber optic circuit can cause instability and degradation of the laser performance. Damage to lasers has even been observed if the reflections are high enough. For situations where it is imperative to prevent back reflections, or where it is important to have light travel in only one direction around an optical circuit, an *optical isolator* is used. This device is essentially a one-way light valve that most often uses the special property, namely its *non reciprocity,* of the *Faraday effect*. For other linear optical processes the path taken by light is completely reversed when the light travels in the opposite direction. Thus, if light is able to get from a source to a certain point where it is reflected, then the *principle of reciprocity* states in effect; "on reflection the light reverses its path and returns to the source."

5.8.1 Faraday isolators

A *Faraday optical isolator* comprises a Faraday rotator placed between two *plane polarizers,* as indicated in Fig. 5.34. The plane of transmission for the second polarizer is oriented at 45° to that of the first polarizer. When linearly polarized light passes through a *Faraday rotator,* its plane of polarization is rotated through an angle χ that depends upon the strength of the component of the magnetic field, B, in Tesla (T) in the direction of propagation within the device and the length of path traveled in this field, L, i.e.,

$$\chi = \mathscr{V} BL, \tag{5.73}$$

where $\mathscr{V}$ is known as the *Verdet constant*. The value of the Verdet constant for quartz (SiO_2) is 4.0 (rad m^{-1} T^{-1}) at 589.3 nm, while in the case of *terbium–aluminum–garnet* (TbAlG) the value is 331.6 (rad m^{-1} T^{-1}) at a wavelength of 500 nm. In an *optical isolator,* the product of path length and magnetic field strength is arranged so that on the outward journey a rotation angle, $\chi = 45°$, is imposed on the plane of polarization permitting it to pass through the second polarizer (Fig. 5.34). The dependence of this *Faraday rotation angle* on the direction of propagation is the key to the

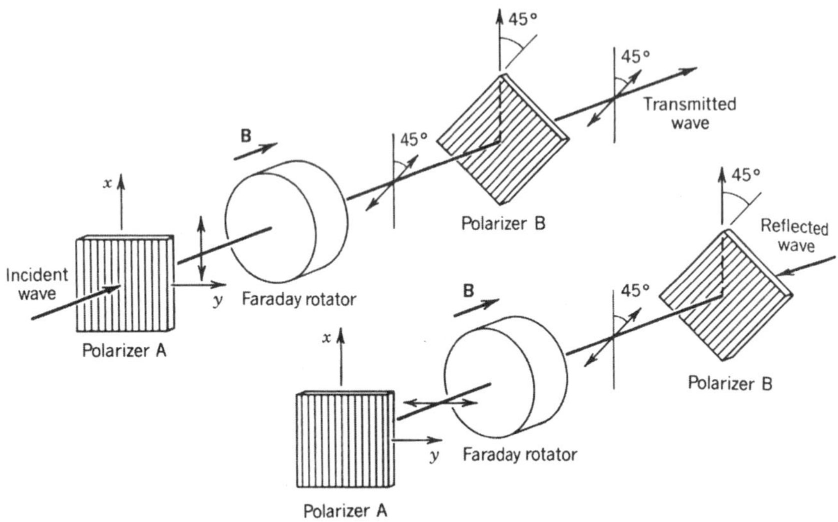

FIGURE 5.34. Schematic illustration of the principle involved in an optical isolator based on the Faraday rotation of light.

function of the optical isolator, for when light is reflected back through the device in the opposite direction, the rotation angle retains its sense of rotation and a further rotation of 45° is imposed on the plane of polarization, see lower part of Fig. 5.34, in which case the plane of polarization after a double pass through the device is now orthogonal to its original direction, and is thus blocked by the first polarizer. Optical isolator based on a terbium–gallium–garnet Faraday rotator have been shown to provide 90 dB attenuation for the reflected light (Saleh and Teich, 1991). Very small optical isolators that are pigtailed for ease of integration into a fiber optic system are available commercially (Fig. 5.35), and some laser diodes now come packaged with tiny optical isolators.

5.8.2 Direction couplers

For most fiber optic circuits it is necessary to split off some fraction of the light propagating along a given optical fiber and send it down another optical fiber. The device most often chosen for this purpose is called a *directional coupler.* A representative fiber optic sensing circuit is depicted in Fig. 5.36. We see in this figure that some fraction of the light from the source is sent by means of the *directional coupler* towards the reference

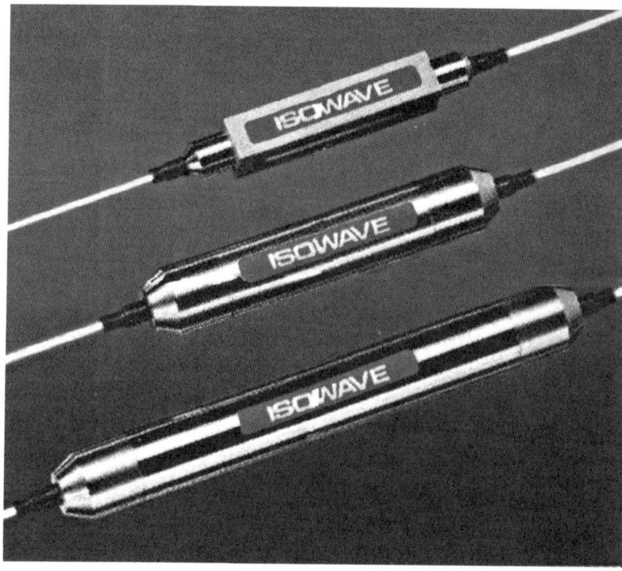

FIGURE 5.35. Three examples of optical isolators that are integrated into an optical fiber. In general the smaller versions have a lower rejection ratio between the forward and backward waves. Used with permission, Isowave, Inc.

photodetector, while the remainder is allowed to travel to the sensor. We also note that light reflected from the sensor is also divided on passing through the direction coupler in the reverse direction, with some fraction going to the sensing photodetector while the remainder goes back to the source, and the need for the optical isolator is apparent.

As we saw in Chapter 3, *evanescent coupling* is the mechanism by which light is transferred from one optical fiber to an adjacent optical fiber

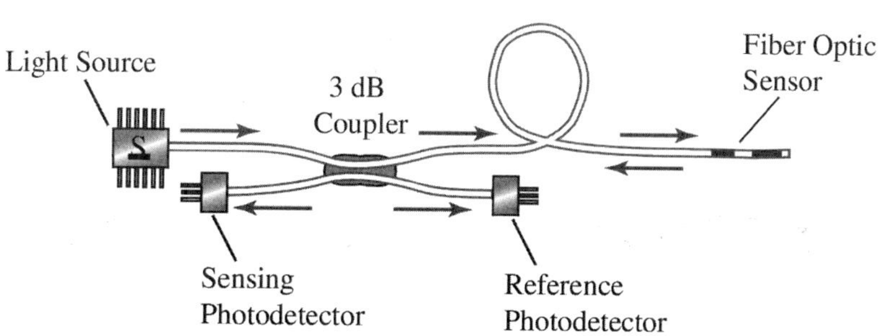

FIGURE 5.36. Generic fiber optic sensor arrangment indicating the central role played by the direction coupler.

within the *directional coupler.* The directional coupler *power transfer ratio,* $\mathscr{F}$, defined, in general, by $P_2(z)/P_1(0)$ takes the form:

$$\mathscr{F} = \left\{\frac{\pi}{2}\right\}^2 \mathrm{sinc}^2\left[\frac{z}{2L_{\mathrm{exc}}}\sqrt{1+\left\{\frac{\Delta\beta L_{\mathrm{exc}}}{\pi}\right\}^2}\right] \tag{5.74}$$

if we draw on Eqs. (3.91), (3.99), and (3.103) from Chapter 3. Note that:

$$\mathrm{sinc}(z) \equiv \sin(\pi z)/(\pi z).$$

The dependence of $\mathscr{F}$ on the product of the *spectral mismatch* between the propagation constants of the two optical fibers, $\Delta\beta$, and the power exchange length, L_{exc} (Saleh and Teich, 1991) is presented in Fig. 5.37. Clearly, the peak value of unity arises at the phase matched condition, i.e., $\Delta\beta = 0$. Also note from this figure that if $\Delta\beta L_{\mathrm{exc}} = \pi\sqrt{3}$, no light is transferred between the optical fibers.

This raises the prospect of an *optical switch* that can either send all of the light down one optical fiber, or by the application of an electrical signal, switch the light to the other optical fiber (Fig. 5.38). Such an *electro-optic switch* is based on an *integrated optic directional coupler* with optical waveguides formed in a material that can have its index of refraction changed by the application of a voltage to electrodes placed over the coupling region. A good example of a material with a high *electro-optic coefficient* is lithium niobate $\{LiNbO_3\}$.

The most common use for directional couplers, however, is to simply divide the light from one optical fiber equally between two optical fibers.

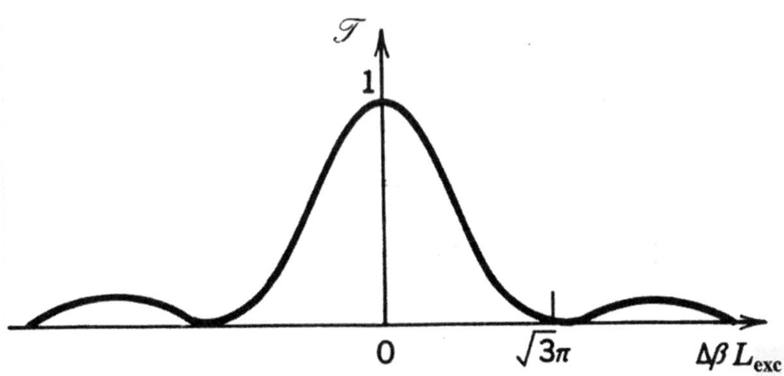

FIGURE 5.37. Variation of the power transfer ratio, $\mathscr{F}$, for a directional coupler with phase mismatch parameter $\Delta\beta L_{\mathrm{exc}}$. From Saleh, B.E.A., and Teich, M.C. "*Fundamentals of Photonics,*" © 1991 Wiley Interscience. Reprinted by permission of John Wiley and Sons, Inc.

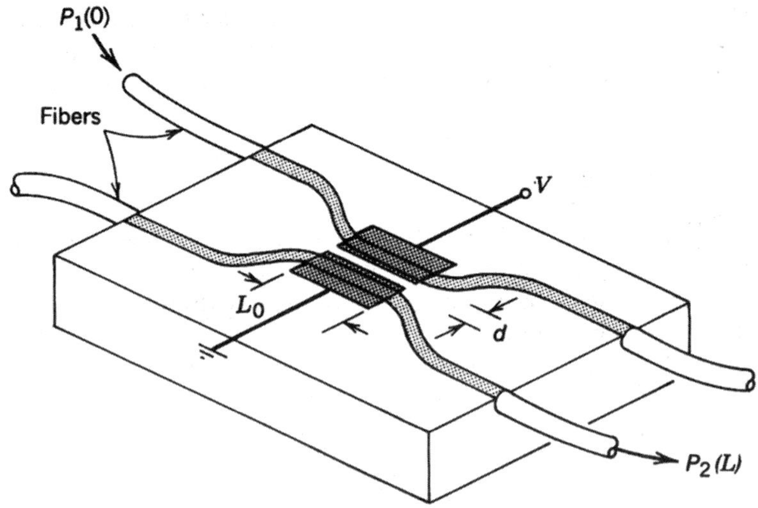

FIGURE 5.38. A schematic illustration of an integrated electro-optic directional coupler. Fom Saleh, B.E.A., and Teich, M.C. "*Fundamentals of Photonics,*" © 1991 Wiley Interscience. Reprinted by permission of John Wiley and Sons, Inc.

Such a 50/50, or 3 dB coupler, is produced by making the *coupling length* of the device

$$L_{3\mathrm{dB}} = L_{\mathrm{exc}}/2. \tag{5.75}$$

Under these circumstances the directional coupler operates at one of the crossover points in the power curves for the two coupled optical fibers, (see Fig. 5.39a). As discussed in Chapter 3, total power exchange occurs between the pair of optical fibers only if the phase matched condition applies. Under these circumstances Eqs. (3.100) and (3.101) describe the power exchange, and it is important to realize that since the coupling coefficient is wavelength dependent, the performance of a 3 dB coupler can vary by as much as several percent at wavelengths that are 50 nm different from the optimum for which the coupler was designed. Special single-mode wavelength flattened directional couplers can be manufactured (Alavie, private communication).

The most common form of directional coupler is based on a biconical taper that involves fusing two optical fibers after heating and pulling them. In this arrangement the cores are reduced in size and forced together. Since the coupling coefficient increases exponentially with

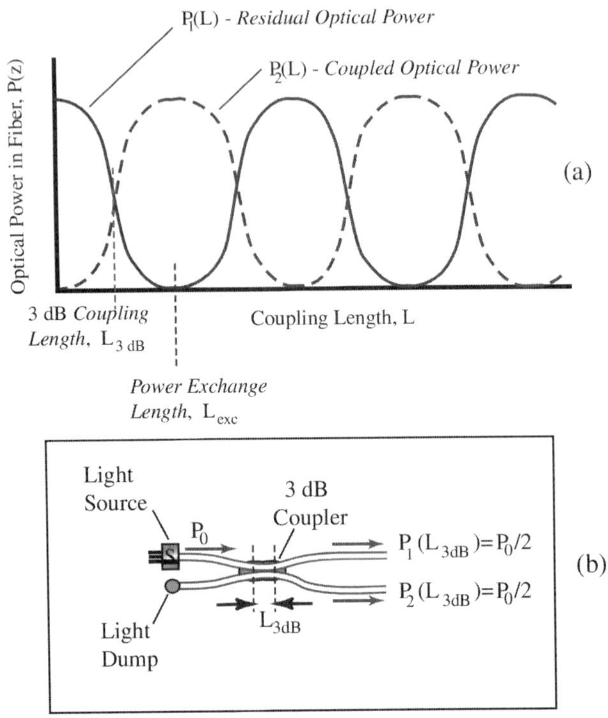

FIGURE 5.39. Schematic illustration of (a) the variation in power evanescently coupled between two optical fibers as a function of the coupling length, and (b) the equal division of light between the two output optical fibers when the coupling length of a directional coupler equals the 3-dB splitting value.

decreasing separation of the two cores, this type of coupler can be made with low *insertion loss* (5.63):

$$\mathscr{K}_{3dB} = -10\log_{10}\{P_{out}/P_{in}\} = -10\log_{10}\{(P_1 + P_2)/P_0\}, \tag{5.76}$$

The *coupling ratio* R_c for this type of device is defined as

$$R_c \equiv P_1/(P_1 + P_2), \tag{5.77}$$

where P_1, P_2, and P_0 represent the optical power in the two output optical fibers and the input optical fiber, respectively (see Fig. 5.39b). A representative insertion loss for a single mode 3 dB fused biconical tapered coupler is 0.05 dB. A lower insertion loss (~0.005 dB) can be achieved with more expensive *polished couplers* (Dakin and Culshaw, 1988). Although the coupling length for a fused biconical tapered coupler can be of the order of a millimetre the overall device with its protective sleeve can be several centmetres long.

5.8.3 Spectral filters

In a number of fiber optic sensing applications it is desirable to perform some form of *spectral filtering* or analysis. There are many bulk optic systems that can accomplish these tasks, but we will concentrate on those devices that lend themselves to compact practical instrumentation for structural monitoring. The simplest optical filters are segments of *rare-earth doped optical fiber* (Poole *et al.*, 1985) that serve as either *short-* or *long-pass* filters. In the former case all wavelengths longer than some chosen wavelength are strongly absorbed, while in the latter case all wavelengths shorter than a specified wavelength are absorbed (Fig. 5.40). In some filters reflection replaces absorption as the mechanism for preventing certain wavelengths from being transmitted. Such filters are termed *dichroic*. A *band-pass* filter is one that permits only a specified band of wavelengths to be transmitted, and these are defined in terms of the their peak transmission wavelength and their half-peak transmittance wavelength interval (Fig. 5.41a). The opposite of a band-pass filter is a *band-rejection filter*. These are also referred to as *notch filters*, as they involve a narrow band of absorption, or reflection, within a broad band of high

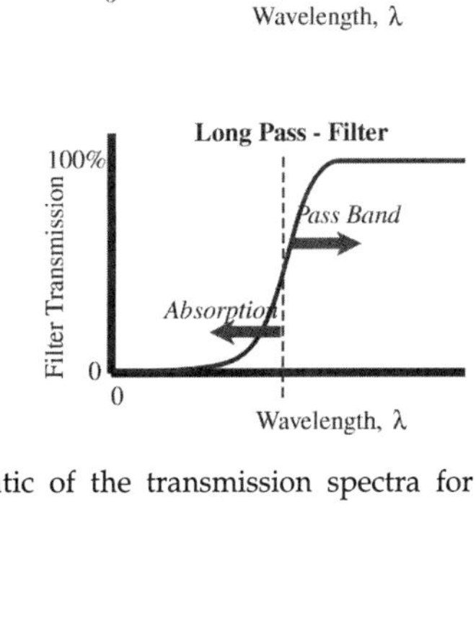

FIGURE 5.40. Schematic of the transmission spectra for band-pass and notch-rejection filters.

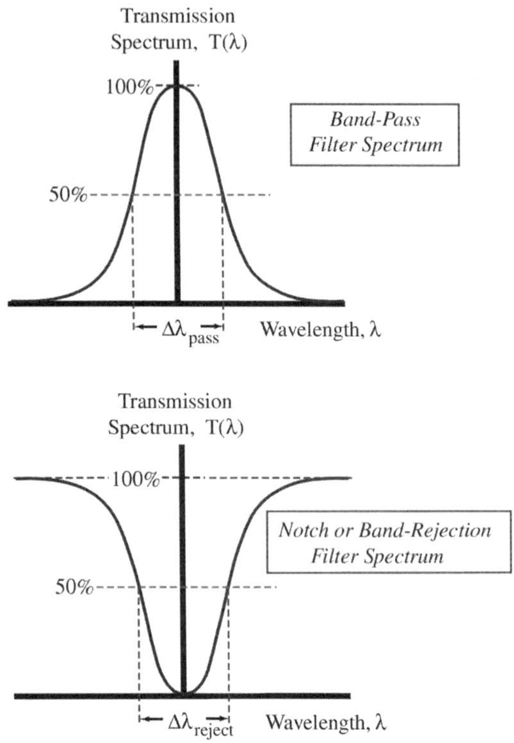

FIGURE 5.41. Schematic illustration of the transmission spectra for band-pass and notch-rejection filters.

transparency (Fig. 5.41b). An extensive list of various filters can be found in the book by Levi (1980).

5.8.4 Fabry–Perot etalon

As we saw in Section 3.7, the *Fabry–Perot interferometer* can be used as a *narrowband transmission filter* provided the spectral interval of interest is less than the wavelength interval between adjacent resonances. The frequency spacing between adjacent modes (Fig. 3.25), can be seen from Eq. (3.71) to be given by:

$$\Delta \nu_{m,m+1} = \frac{c}{2nL}. \tag{5.78}$$

The corresponding wavelength interval, termed the *free-spectral range* (4.51), can be expressed in the form

$$\Delta\lambda_{\mathrm{FSR}} = \frac{\lambda^2}{2nL}. \tag{5.79}$$

A Fabry–Perot interferometer with a solid spacer between the two mirror surfaces, as oppose to an air gap, is called an *etalon*. We see that in order to have an appreciable free-spectral range the cavity spacing, L, needs to be very small. For example, a fiber optic Fabry–Perot interferometer with an air-gap separation of the end mirrors $L = 50\ \mu\mathrm{m}$ would have a free-spectral range, $\Delta\lambda_{\mathrm{FSR}} = 20$ nm, at 1.55 μm. The *spectral width* of the Fabry–Perot transmission spike can be determined from the Airy function. From Eq. (3.69) the *cavity half-height half-width frequency*, $\Delta\nu_{\mathrm{c}}$, is ascertained by setting:

$$\mathscr{F}\sin^2\{2\pi n(\nu_{\mathrm{m}} + \Delta\nu_{\mathrm{c}})L/c\} = 1, \tag{5.80}$$

where ν_m is the frequency of the mth resonance which is used for filtering. From Eq. (3.71), we know

$$2\pi nL\nu_m/c = m\pi,$$

and since $\Delta\nu_{\mathrm{c}}/\nu_m \ll 1$, Eq. (5.80) can be approximately written in the form

$$\Delta\nu_{\mathrm{c}} = \frac{c}{2\pi nL\sqrt{\mathscr{F}}}, \tag{5.81}$$

with the definition of the *coefficient of finesse* (3.70), the *cavity frequency width* (for short) becomes

$$\Delta\nu_{\mathrm{c}} = \frac{c(1-R)}{4\pi nL\sqrt{R}}, \tag{5.82}$$

Using the same cavity spacing and wavelength ($L = 50\ \mu\mathrm{m}$ and $\lambda = 1.55\ \mu\mathrm{m}$) cited in the example given above and assuming $R = 0.9$ and $n = 1.0$

$$\Delta\nu_{\mathrm{c}} = 50\ \mathrm{GHz}.$$

The corresponding *wavelength half-height half-width*

$$\Delta\lambda_{\mathrm{c}} = 0.4\ \mathrm{nm},$$

is a measure of the wavelength resolution possible. Note that increasing the cavity length narrows the spectral width and therefore improves the resolution attainable with this type of filter. However, the price to pay is a corresponding decrease in the *free-spectral range* so the wavelength interval over which this filter can be used is reduced. Tuning of this type of

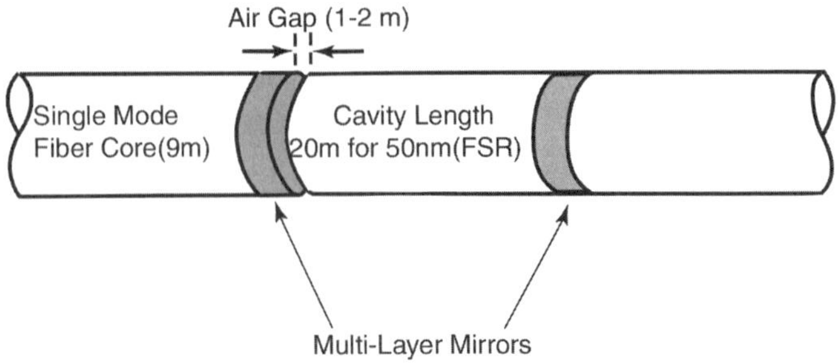

FIGURE 5.42. An all-fiber Fabry–Perot tunable filter.

narrowband filter is possible by controlling the *cavity spacing*. This is often accomplished by means of *piezoelectric elements* that change their length with applied voltage. The *all-fiber Fabry–Perot tunable filter* manufactured by Micron Optics Inc. Atlanta, US, is schematically illustrated in Fig. 5.42. This compact system has a bandwidth of 0.01 nm and uses less than 10 volts to sweep across its 50 nm free-spectral range.

5.8.5 Acousto-optic tunable filter

There is another form of tunable narrowband filter that is based on the acousto-optic effect and uses a standing acoustic wave to control the transmission of light of a specific wavelength (Xu *et al.*, 1993). In such an *acousto-optic tunable filter* (AOTF) a lithium niobate crystal, with both piezoelectric and electro-optic properties, permits only a narrow band of wavelengths to be transmitted through the device. The AOTF involves a planar optical waveguide that is collocated with a *surface acoustic waveguide* and a pair of orthogonal polarizers (Fig. 5.43). At resonance

$$\lambda_{res} = \Delta n_{pol} \Lambda_a, \tag{5.83}$$

and light of wavelength λ_{res} passing along the optical waveguide has its plane of polarization rotated from the orientation necessary to pass through the first polarizer to that required to pass through the second polarizer. In Eq. (5.83), Δn_{pol} represents the difference in the index of refraction between the two orthogonal states of polarization and Λ_s is the acoustic wavelength. Off-resonance wavelengths are not transmitted, as the standing wave in the surface acoustic waveguide fails to rotate the plane of polarization through 90°. The acoustic wavelength is controlled

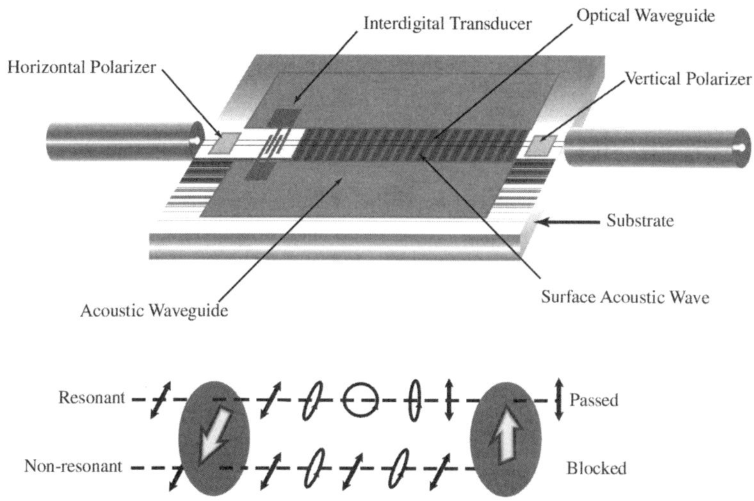

FIGURE 5.43. Schematic illustration of the configuration and principle of a narrowband acousto-optic tunable filter (AOTF).

by the frequency of the voltage applied to the interdigital electrodes of the acoustic wave transducer. Thus, this device can be operated in a *sweep mode* wherein the narrow band of transmitted wavelengths can be scanned over a wavelength interval determined by the change in the frequency of the voltage signal applied to the interdigital electrodes.

A major advantage of the AOTF is its wide tuning range, typically $>1\ \mu m$. Although the spectral resolution, which is in the range: 1 to 4 nm, (Xu *et al.*, 1993; Voltanthen *et al.*, 1996) of an AOTF is not as high as can be achieved with a Fabry–Perot filter, we see that its spectral range is very much larger (>10 times). By way of example, the AOTF supplied by Crystal Technology has a wavelength tuning range of 1.2 to 2.5 μm and a bandwidth of 4 nm. These devices can also be used to track single peak optical signals (Xu *et al.*, 1995) or multiplexed to simultaneously monitor several fiber Bragg grating sensors (Volanthen *et al.*, 1996). This involves opening several wavelength windows in the AOTF by the application of a number of different voltage frequencies at the same time. This will be important later when wavelength division multiplexing is considered as a method of handling large arrays of fiber optic sensors. As we shall see shortly, a fiber optic Bragg grating can also be used as a narrowband tunable filter. However, its wavelength tuning range is limited by the compressive strain that can be imposed on this device, and this is less than possible with an air-gap Fabry–Perot filter or an AOTF.

5.8.6 Wavemeters and optical spectrum analysers

Where it is desirable to measure the wavelength of a narrowband optical signal, or source, a *Burleigh Wavemeter* (Burleigh Instruments Inc., New York) can be used to high precision (Fig. 5.44). A wavelength measurement accuracy of better than 0.001 nm is possible in general, and in the case of a CW laser an accuracy of ± 1 in 10^7 can be achieved with the model WA-1500 if the bandwidth is less than 1 GHz. Where the spectral profile of an optical signal is required a spectrometer can be used as an alternative to a tunable filter. One of the most compact diffraction-grating-based spectrometers available today is produced by *Ocean Optics Incorporated*, Florida, and is shown in Fig. 5.45. This instrument is pigtailed to an optical fiber and offers 0.3-nm resolution over a 600-nm wavelength range.

Much more sophisticated instruments called *optical spectrum analyzers* (OSA) are also available but are generally very expensive and intended as laboratory test equipment. Examples are the Hewlett-Packard OSA [HP-71451A], and the Ando OSA [AQ-6315A]. These instruments have a

FIGURE 5.44. Photograph of a Burleigh wavemeter for measuring the wavelength of a narrowband optical signal. Used with permission, Burleigh Instruments, Inc., Fishers, N.Y., USA. "The WA-1500 Wavemeter® Optical Wavelength Meter".

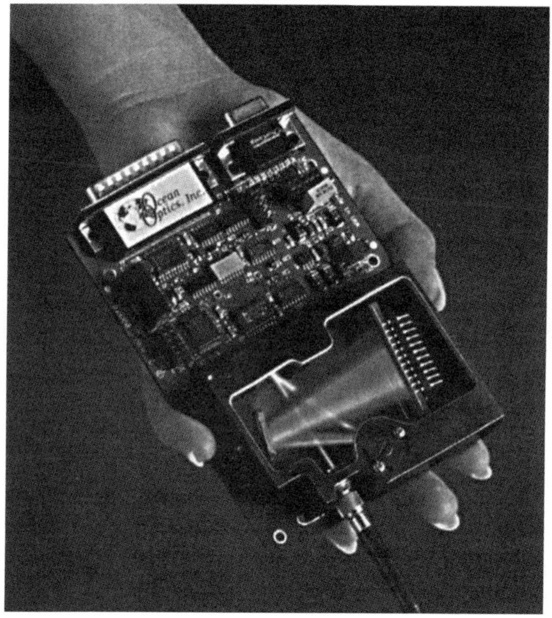

FIGURE 5.45. Illustration of a handheld miniature fiber optic spectrometer. Used with permission, Ocean Optics, Inc.

wavelength measurement accuracy of ±0.5 nm, a spectral range of 900 to 1700 nm, and a 0.1 nm wavelength resolution.

5.9 FIBER BRAGG GRATINGS

As discussed in Section 3.10, a *fiber optic Bragg grating* is formed when a periodic variation of the index of refraction is created along a section of an optical fiber by exposing the core of the optical fiber to an interference pattern of intense UV-laser light. The formation of permanent gratings in an optical fiber was first demonstrated by Hill *et al.* (1978). They excited a germania-doped optical fiber with intense argon-ion laser radiation at 488 nm and observed that after several minutes the intensity of reflected light increased until eventually almost all the light was reflected from the fiber. The growth in back-reflected light was explained in terms of a new nonlinear effect called *photosensitivity*, which permits the index of refraction in the core of the fiber to be increased by exposure to intense laser radiation. In this early experiment a *fiber Bragg grating* (FBG) was formed when a small amount of the laser light reflected back from the end of the optical fiber interferes with the exciting laser light to establish a standing wave pattern. "Photosensitivity" causes the index of refraction to be increased to a much greater extent at positions where constructive interference results in a maximum of laser intensity (Fig. 5.46). As the strength of the grating (proportional to the depth of its index modulations) increases, the intensity of the back-reflected light increases until it saturates near 100%.

Although photosensitivity appeared to be an ideal means for fabricating these early *Hill gratings* in optical fibers, their usefulness was extremely limited because they only reflected at wavelengths in the visible close to the wavelength of the writing light, were spread along the optical fiber with varying strength and took a long time to produce. These limitations were overcome about 10 years later in an experiment by Meltz *et al.* (1989) who recognized from the work of Lam and Garside (1981) that *photosensitivity* was a *two-photon process* that could be made much more efficient if it were a one-photon process corresponding to the

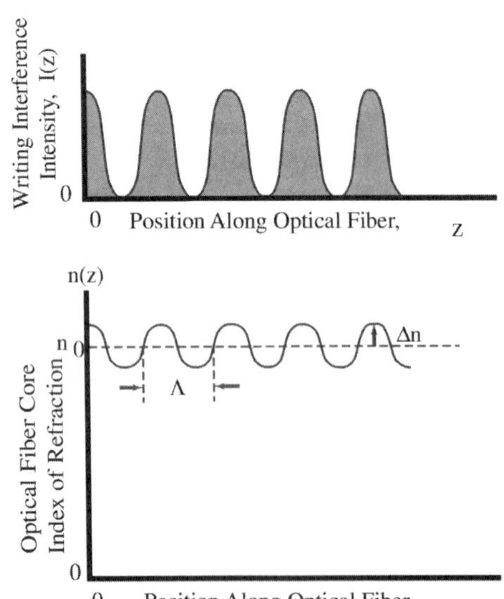

FIGURE 5.46. Schematic of the UV interference pattern used to write Bragg gratings into the core of suitably doped optical fibers. Lower diagram indicates resulting axial modulation in the core index of refraction.

germania oxygen vacancy defect band, at a wavelength of 245 nm (i.e., 5 eV) (Hill and Meltz, 1997).

In the experiment of Meltz *et al.* (1989), the fiber was irradiated from the side with two intersecting coherent ultraviolet laser beams of wavelength 244 nm which corresponds to one-half of the 488 nm, the wavelength of the blue argon laser line. Their *transverse holographic technique* worked because the fiber cladding is transparent to this ultraviolet light, whereas the fiber core is highly absorbing of this radiation. The fiber Bragg grating transverse holographic writing technique, illustrated in Fig. 5.47, has another principal advantage with regard to grating fabrication, the spatial period of the photo-induced grating depends on the angle between the two interfering coherent ultraviolet light beams. This permits Bragg gratings to be made that function at much longer wavelengths than the writing wavelength. This advance enormously increased the scope of applications for FBGs which includes fiber optic communications and sensing. A broad review of FBG has been provided by Russell and Archambault (1996).

A theoretical discussion of FBGs was presented in Section 3.10 and a representative spectrum displayed in Fig. 3.37. The periodic perturbation in the core index of refraction gives rise to successive coherent scattering for a narrow band of the incident light. The grating thus effectively acts as a *stop-band filter,* reflecting light with wavelengths close to the Bragg wavelength λ_B, and transmitting wavelengths sufficiently different from λ_B, (Fig. 5.48). The Bragg wavelength is given by Eq. (3.129): $\lambda_B = 2n_0\Lambda$, where n_0 is the mean core index of refraction and Λ is the *grating period* of the index perturbation. Each reflection from a peak in the index perturbation is in phase with the reflection from the next peak when the wavelength of the light corresponds to λ_B.

The FBG is an *intrinsic sensor,* as any change in the local strain or temperature modifies the index of refraction and the *grating period* and consequently changes the Bragg wavelength. This issue will be discussed in detail in Chapters 6, 7, 9, and 12. Concerns about the possible life of these FBG sensors were partly alleviated when it was discovered that the photoinduced changes in the core index of refraction can be made relatively permanent (lifetimes of 25 years are predicted) if the FBG is annealed for a few hours at a temperature of 50°C above the FBGs likely maximum operating temperature (Douay *et al.*, 1993; Erdogan *et al.*, 1994).

A variety of different continuous wave and pulsed lasers with wavelengths ranging from the visible to the vacuum ultraviolet have been used to write gratings in optical fibers. In practice, *krypton fluoride* (KrF) and *argon fluoride* (ArF), excimer lasers that generate (~10 ns) pulses at wavelengths of 248 and 193 nm, respectively, are used most frequently

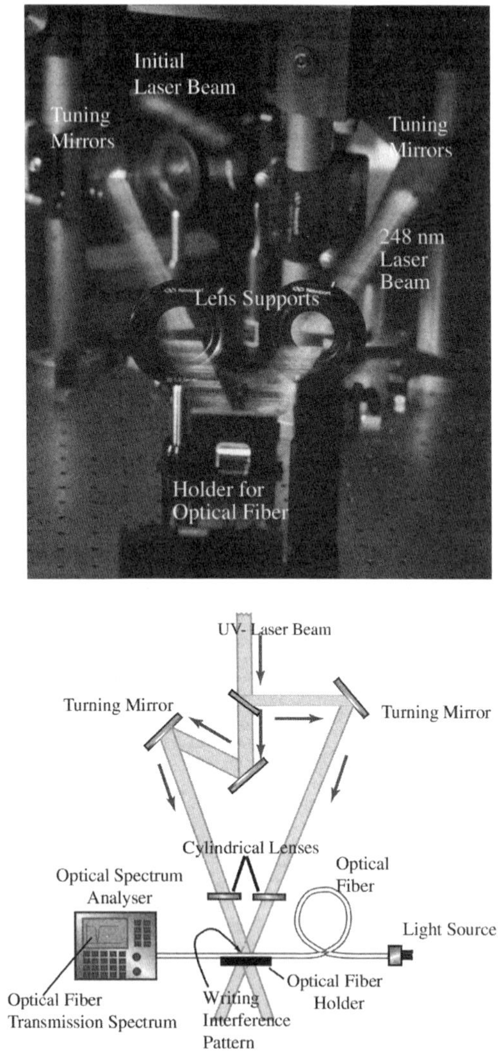

FIGURE 5.47. Photograph of the transverse holographic optical arrangement used to write Bragg gratings into the core of an optical fiber. The lower diagram displays a schematic of the optical arrangement.

to produce FBGs. The exposure required to produce a FBG is typically a few minutes with laser intensities of 100 to 1000 mJ cm^{-2} and pulse rates of 50 to 75 s^{-1}. Under these conditions the change in the core index of refraction, Δn, is between 10^{-5} and 10^{-3} in germanium-doped single-mode optical fiber.

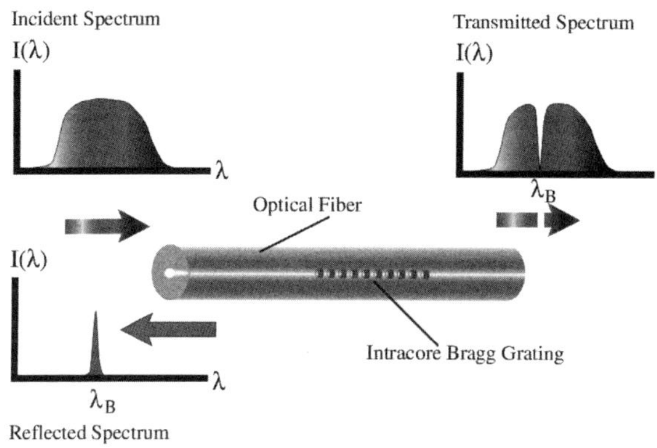

FIGURE 5.48. Schematic illustration of a fiber Bragg grating and its influence on broadband light propagating along the optical fiber.

Techniques such as *hydrogen loading* (Lemaire *et al.*, 1993) can be used to enhance the optical fiber's photosensitivity prior to laser irradiation. Changes in Δn as high as 10^{-2} have been achieved by this means. *Hydrogen loading* usually requires that the relevant section of an optical fiber be placed in a high-pressure vessel and left for many weeks exposed to hydrogen gas at several atmospheres. Some of the hydrogen diffuses into the optical fiber and makes the core more susceptible to the UV laser radiation. Excitation by laser radiation with an intensity greater than 1000 mJ cm^{-2} leads to the onset of a different, nonlinear interaction that enables a single excimer laser pulse to photoinduce large index changes in a small localized region near the core/cladding boundary. In this case the change in the index of refraction is sufficient to damage the optical fiber and is observable with a phase contrast microscope (Hill and Meltz, 1997).

One of the most exciting promises held out for FBGs was that they would one day be produced during the manufacture of the optical fiber itself. This would ensure high strength, as the fiber coating would not have to be removed and the naked fiber handled as is the normal practice in off-line fabrication of FBGs. Automated production on a fiber optic draw tower should also lead to low cost and more consistent properties for the FBGs. A special line-narrowed KrF laser was used by Dong *et al.*, (1993) in the first single excimer laser pulse writing of gratings during the draw tower production of an optical fiber. Askins (1994) and Friebele *et al.*, (1995), demonstrated that an array of FBGs could be written at different wavelengths using a computer-controlled agile transverse holographic writing system.

An example of the *reflectance spectrum* for a series of FBGs with 5-nm wavelength separation produced in this manner is presented as Fig. 5.49.

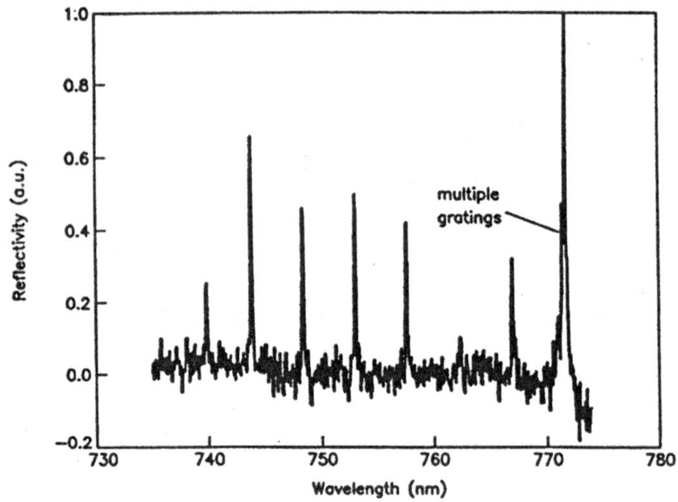

FIGURE 5.49. Reflection spectrum from a series of fiber Bragg gratings formed along an optical fiber by the transverse holographic writing technique. From Askins, C.G., Putnam, M.A., Williams, G.M., and Friebele, E.J., "Stepped-Wavelength Optical-Fiber Bragg Grating Arrays Fabricated in Line on a Draw Tower" *Optics Letters*, **19**, 147–149.

The quality of these gratings was not very good and their peak reflectivity was around 3% or worse. Recently, the use of an improved system, including a *phase conjugate mirror* (PCM) to make the laser more coherent have led to draw tower on-line production of FBGs with reflectivities of 26% and much cleaner spectral profiles (Putman *et al.*, 1995). This is indicated in Fig. 5.50, where the variation in *FBG reflectivity* with the length of the grating produced with and without the PCM is displayed.

The transverse holographic method of writing fiber Bragg gratings has largely been superseded by the *phase mask technique* (Hill *et al.*, 1993). The *phase mask* is a thin slab of silica glass into which is etched (using photolithographic techniques) a one-dimensional square-wave *periodic surface relief structure*. Since this material is transparent to the ultraviolet laser radiation the primary effect of the phase mask is to diffract the light into the 0, +1, and −1 diffraction orders (see Fig. 5.51). Careful control of the depth of the corrugations in the phase mask suppresses the zero-order diffraction, leaving the ±1 *order* diffracted beams to interfere and produce the periodic pattern of intense laser radiation needed to photoimprint a Bragg grating within the core of an optical fiber. If the period of the phase mask is Λ_{mask}, the period of the *photoimprinted index grating* is $\Lambda_{mask}/2$. Note that the grating period is independent of the wavelength of ultraviolet light irradiating the phase mask. Although the usual practice is to

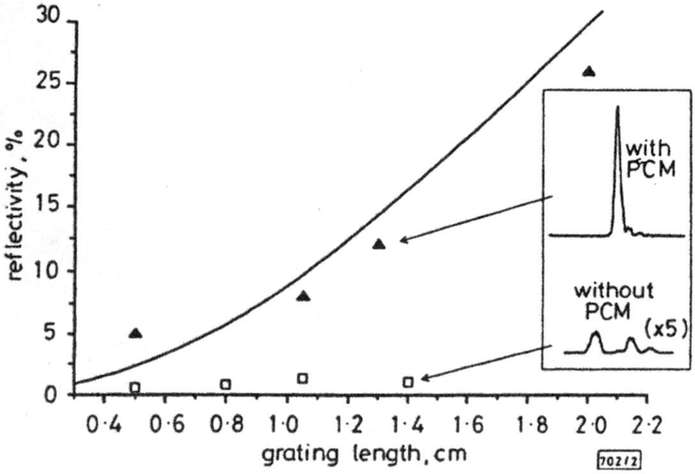

FIGURE 5.50. Variation in FBG reflectivity as a function of grating length with and without a phase conjugate mirror (PCM) used to control the coherence properties of the writing laser beam. From Putnam, M.A., Askins, C.G., Williams, G.M., Friebele, E.J., Baskansky, M., and Reintjies, J., (1995), "Single Pulse Fabrication of Fiber Bragg Gratings Using a Phased-Conjugated KrF Excimer Laser," *SPIE*, **2444**, 403–40.

bring the optical fiber almost into contact with the phase mask, Othonos and Lee (1995) demonstrated that improvements in the *spatial coherence* of the laser writing beam relaxed the need for such close contact, Fig. 5.52.

The *phase mask technique* greatly simplifies the manufacture of FBGs through easier alignment, reduced stability requirements on the photo-

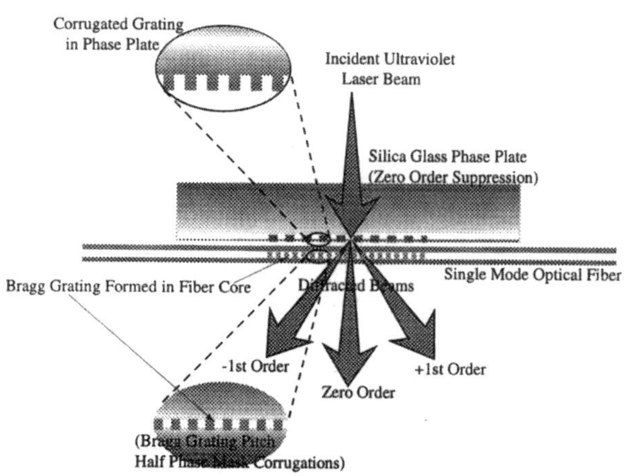

FIGURE 5.51. Schematic of the phase mask technique for producing fiber Bragg gratings.

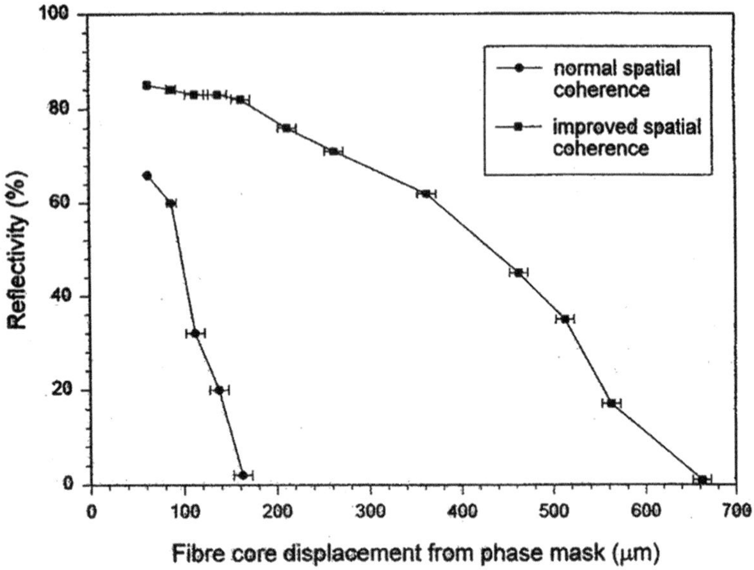

FIGURE 5.52. Variation in the FBG reflectivity with displacement between the core of the optical fiber and the phase masl, with and witout improved spatial coherence of the writing UV-laser beam. From Othonos, A. and Lee, X., "Novel and Improved Methods of Writing Bragg Gratings with Phase Masks" *IEEE Photon Technology Letters*, **7**, 1183–1185, © 1995, IEEE.

imprinting apparatus, and lower coherence demands on the laser beam. It also permits the use of a cheaper ultraviolet excimer laser source and tends to consistently yield *high performance gratings*. Furthermore, several gratings have been manufactured at once in a single exposure by irradiating a number of parallel optical fibers through the same phase mask. The prospect of manufacturing high-performance gratings at low cost is critical to the large-scale implementation of this technology for structural monitoring. Phase masks are now commercially available, and by way of illustration, a selection of the specifications for phase masks that can be supplied from QPS Technology Inc., Dorval, is provided in Table 5.1.

The phase mask technique suffers one disadvantage compared to the holographic writing technique and that is that a separate phase mask is required for each FBG with a different Bragg wavelength. However, wavelength tuning of about 2 nm is possible by applying tension to the fiber during the photo-imprinting process, (Meltz, 1997) and some degree of tuning in regard to the inscribed Bragg grating center wavelength is possible when the writing beam has extended coherence (Othonos and Lee, 1995; Fig. 5.53).

Table 5.1. Selection of QPS Technology Inc. Phase Mask Specifications

Parameter	Minimum	Typical	Maximum
Bragg wavelength range	780 nm	–	1650 nm
Mask period	525 nm	–	1150 nm
Zero order	–	2%	3%
First order	38%	40%	–
Grating length	5 mm	–	125 mm
Grating width	3 mm	–	10 mm

The phase mask technique not only yields high-performance devices but is very flexible in that it can be used to fabricate gratings with controlled spectral response characteristics. For instance, the typical spectral response of a finite-length grating with a uniform index modulation along the fiber length has secondary maxima on both sides of the main reflection peak (see Fig. 3.36). In certain applications this type of response is not desirable. These secondary maxima can be suppressed by a procedure called *apodization*. *Apodized fiber gratings* have been fabricated using *non-uniform diffraction efficiency phase masks* (Albert *et al.*, 1995), see

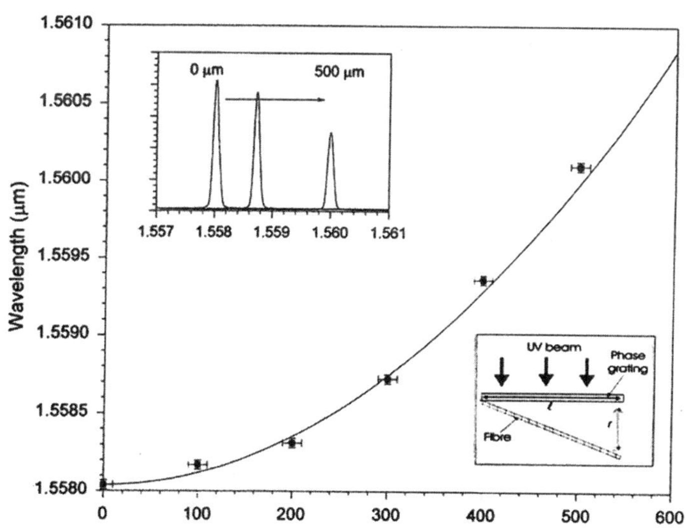

FIGURE 5.53. Wavelength variation in FBG produced with a single phase mask as a function of displacement between the fiber core and the phase mask. From Othonos, A. and Lee, X., "Novel and Improved Methods of Writing Bragg Gratings with Phase Masks" *IEEE Photon Technology Letters*, **7**, 1183–1185, © 1995, IEEE.

Fig. 5.54. Gratings have been produced with side-lobe suppression of 20 dB relative to FBG made with uniform phase masks (Malo *et al.*, 1995).

The phase mask technique can also be used to fabricate *chirped FBG,* or *aperiodic gratings* (Loh *et al.*, 1995). A *chirped grating* is one in which the index modulation depth, or period, becomes a function of position along the grating. Normally these are taken as constant so the grating is uniform. *Chirping* of a grating is sometimes undertaken in order to broaden its spectral response. We shall see in Chapter 12 that strain-induced chirping of an FBG leads to the prospect of truly *distributed strain sensing* (Measures *et al.*, 1994; Huang *et al.*, 1995; Ohn *et al.*, 1997). This type of application requires long fiber gratings. Several techniques using phase masks have been demonstrated for making such *extended FBG* (Martin and Ouellette, 1994; Asseh *et al.*, 1997). It is also noteworthy that whereas in a uniform FBG light at the Bragg wavelength can be thought to be reflected from the center of the FBG, in a chirped FBG the *depth of penetration* of light becomes a function of wavelength. This has important ramifications in both the telecommunications and sensing fields, and leads to a number of valuable applications (see Chapter 12).

As discussed earlier in Section 5.6, there is concern over the possible weakening of the optical fiber by the formation of FBG sensors. Recent tests, however, have shown that the strength of optical fibers unaffected by on-line single pulse laser writing of FBGs during the manufacture of

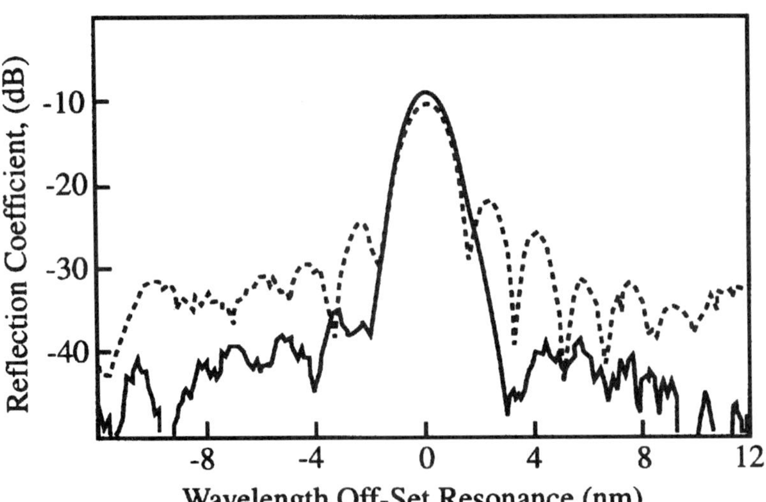

FIGURE 5.54. Reflection spectra for FBG photoimprinted with a uniform difffracting phase mask (broken curve) and a Gaussian profile of diffraction efficiency (full curve). Spectral resolution of measurmeent is 0.1 nm and 3 dB bndwidth of both gratings is 1.5 nm.

the optical fiber (Askins *et al.*, 1997, Fig. 5.55). This confirms one of the reasons for producing FBG on the *draw tower.* As an alternative to producing the FBG during the manufacture of the optical fiber and before the protective coating is applied, Starodubov *et al.* (1997) and Espindola *et al.* (1997) have demonstrated that it is possible to inscribe a grating in the core of a polymer-coated optical fiber by using laser radiation at 334 nm and 257 nm, respectively. These techniques avoid stripping and recoating the optical fiber, offering considerable time savings and greatly reducing the risk of damaging and weakening the optical fiber. The only problem with this approach is that it appears to require boron co-doped germanosilicate optical fibers, which are more expensive then regular telecommunications optical fibers.

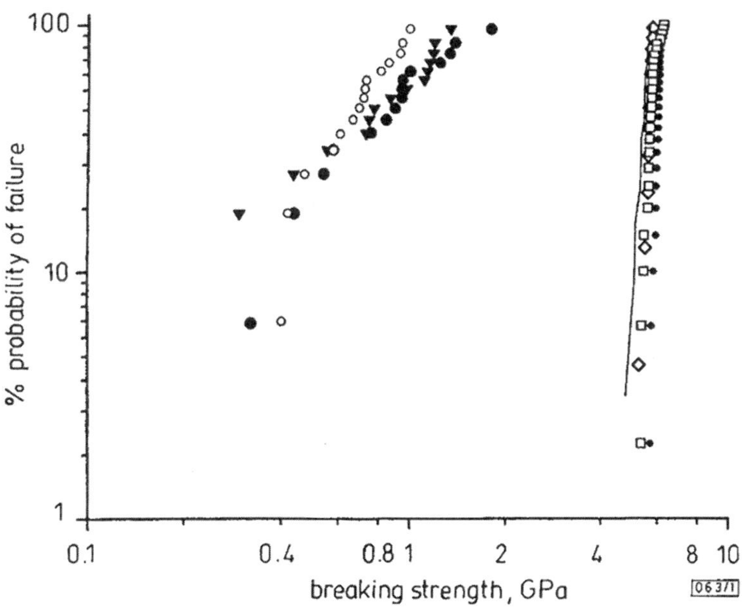

FIGURE 5.55. Weibull plots of fiber bvraking strengths. The high strength results on the right come from optical fibers that were written during the fiber draw process and immediately thereafter coated. The low strength results on the left correspond to optical fibers that were stripped and lft uncoated for 24 hours. Note the line on the right corresponds to unirradiated fibers from the draw tower, while the filled circles on the left correspond to unirradiated fibers that were stripped, like those that were irradiated. Askins, C.G., Putnam, M.A., Patrick, H.J., and Friebele, E.J., "Fiber Strength Unaffected by on-line writing of single-pulse Bragg Gratings," *Electronics Letters*, **33**(15), 1997, 1333–1334, with permission from IEE Publishing.

5.10 MULTIPLEXING AND DEMULTIPLEXING

We shall see later that fiber optic sensors are well suited for serial multiplexing, as the optical fiber serves as both the sensor and the information conduit through which the sensing data passes. Thus a string of fiber optic sensors can function along a single optical fiber. *Multiplexing* (MUX) in general refers to the merging of several information channels into one, while *demultiplexing* (DEMUX) is the opposite and refers to the separation of merged information channels (Fig. 5.56). There are two kinds of multiplexing: serial and parallel. In *serial multiplexing* a string of sensors are positioned along a single optical fiber, while in *parallel multiplexing* an array of optical fibers is excited (or read) with a single source (or photodetector) (Fig. 5.57). Combination of both serial and parallel multiplexing is also possible with fiber optic sensors.

There are a number of multiplexing/demultiplexing schemes and many of them are reviewed by Kersey (1995, 1997a,b). We shall limit our attention to the most common used in fiber optic structural monitoring:

Time division multiplexing (TDM): Short light pulses are used to identify the relevant sensor by means of time-of-flight measurements. For practical structures this requires very short pulses, which leads to low signal strength and very high-speed electronics.

Wavelength division multiplexing (WDM): Each sensor operates at a different wavelength. The relevant wavelength encoding can be performed by the light source, spectral filters, couplers, resonant cavities, or dispersive elements. Demultiplexing is often achieved by means of a spectral analyzer.

Coherence division multiplexing (CDM): A reference interferometer is used to path-match one of a number of sensing interferometers.

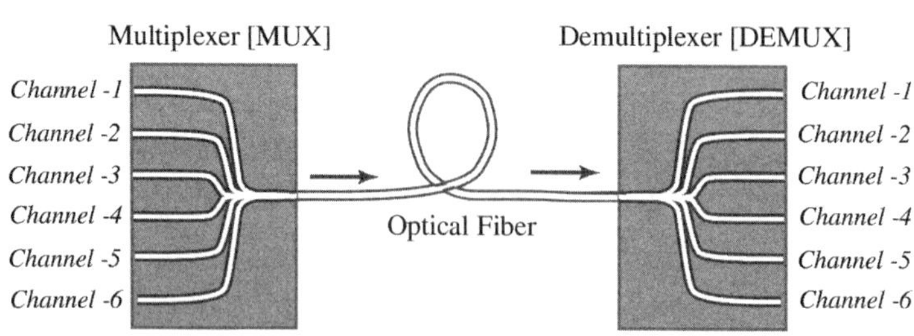

FIGURE 5.56. Schematic illustration of multiplexing and demultiplexing.

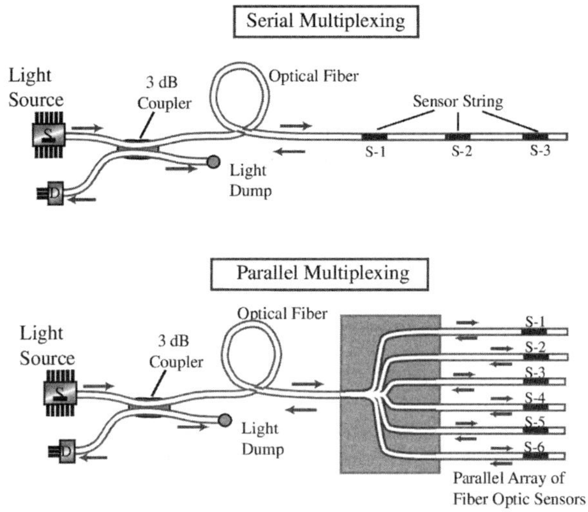

FIGURE 5.57. Serial and parallel multiplexing fiber optic sensors.

This elegant approach is restricted by the coherence length of the source and can require sophisticated signal processing.

Frequency division multiplexing (FDM): Each interferometric sensor operates with a signal that is modulated at a different frequency. Demultiplexing of the various signals is achieved by means of appropriate bandpass filters to ensure that each detector only receives signals from a specific sensor.

Spatial division multiplexing (SDM): Each sensor, in turn, is connected physically to the data link and a detector by means of a mechanical or electro-optical scanning or switching mechanism.

In TDM the sensors are positioned at different distances along the optical fiber so that an interrogating short laser pulse produces a series of distinct short signals at the photodetector (Fig. 5.58). The duration of the laser pulse has to be much less than the optical round-trip time between the sensors, T_{delay}, i.e.,

$$\tau_{\text{L}} \ll T_{\text{delay}} = 2nL_{\text{sep}}/c, \tag{5.84}$$

where τ_{L} is the laser pulse duration, L_{sep} is the separation between adjacent sensors, n is the optical fiber core index of refraction, and c is the free-space velocity of light. To gain an appreciation of the parameters involved in a practical situation, let us assume that it is desirable to have a

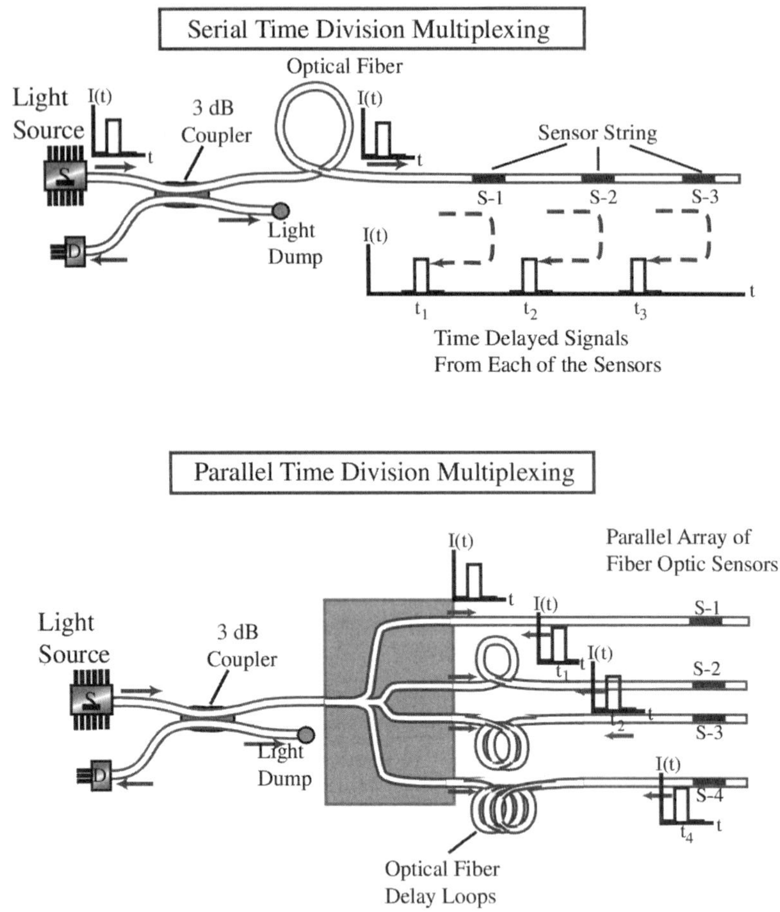

FIGURE 5.58. Schematic illustration of serial and parallel time division multiplexed fiber optic sensing systems.

spatial resolution of 10 cm. This means that at least two sensors should be within 10 cm of each other and the round-trip time would, according to (5.84), be close to 1 ns assuming $n = 1.5$. The duration of the laser pulse would thus have to be subnanosecond and high-speed electronics would be required to process such short-duration signals.

Repetitive pulsing of the light source allows each sensor to be addressed by simple *time-selective gating* of the photodetector. The use of *partial reflectors* along a length of optical fiber permits in-line serial multi-

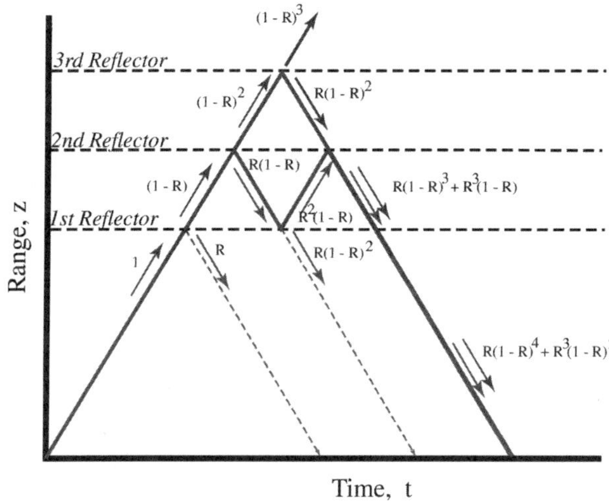

FIGURE 5.59. Optical range and time diagram revealing the relative strengths of optical signals reflected from one to three reflectors, each with reflectivity coefficient R.

plexing of many interferometers. One of the first *serially multiplexed fiber optic strain sensors* was based on the Fabry–Perot interferometer (Lee and Taylor, 1988). The introduction of recirculating optical fiber loops provides a self-referencing capability (Spillman and Lord, 1987).

Inevitably, a series of reflective sensors along an optical fiber gives rise to multiple pulse reflections and optical crosstalk. That is to say, the optical signal arriving at the photodetector as a result of a single reflection from the third sensor is mixed in with an optical signal that arises from reflections at the second, and first sensors. This is illustrated in the *range* and *time* (RAT) diagram presented as Fig. 5.59. The magnitude of this *multiple reflection crosstalk*, C_{MR}, is directly related to the reflectivity of the sensors. Indeed, the ratio of the two optical signals that arrive together at the photodetector after suffering three reflections at the second, first, and second sensors, respectively, and a single reflection from the third sensor is given by

$$C_{MR} = \frac{R^3(1-R)^2}{R(1-R)^4} = \frac{R^2}{(1-R)^2}, \tag{5.85}$$

where R is the *reflectivity* of each sensor and we assume that the fraction of light transmitted through any sensor is $(1 - R)$. For the example of three reflective sensors the multiple reflection crosstalk: $C_{MR} = 1.0$ for $R = 50\%$,

$C_{MR} = 0.0625$ for $R = 20\%$, and $C_{MR} = 0.00277$ for $R = 5\%$. It is evident that the lower the reflectivity, the smaller the crosstalk.

The fabrication of low-reflectivity splices with consistent reflectance proved to be quite a challenge. The development of fiber optic *extrinsic air-gap Fabry–Perot interferometers* (Murphy *et al.*, 1991) made this task a little easier, but it was the fiber Bragg grating that has proven to be the ideal sensor for serial multiplexing (Kersey, 1997). Another advantage of FBGs is that their reflectivity can be easily varied so that those furthest from the source and photodetectors have higher reflectivities. This permits the optimized design of a system involving a large number of serially multiplexed reflective sensors by satisfying both power budget and crosstalk considerations and ensuring that all sensors have a reasonable value for their signal-to-noise ratio.

Wavelength division multiplexing has emerged as a powerful method of handling fiber optic sensor arrays. In WDM each sensor in a serial or parallel arrangement operates within a distinct spectral interval that does not overlap with that of any other sensor. Two simplified schematics of this method of multiplexing for linear (serial) and area (parallel) arrays of sensors are presented in Fig. 5.60. Although in the figure we show an array of different wavelength sources, in general the range of wavelengths needed for this approach is provided by either a *single broadband source* or a *narrowband laser* that can be tuned over the desired spectral interval. Clearly WDM is best suited to sensors that wavelength encode the measurand information.

Implementation of WDM requires technology that has recently become readily available because of the growing importance of WDM to the fiber optic telecommunications industry. This includes, broadband LEDs, tunable DFB and DBR laser diodes, reviewed in Chapter 4. Also required are wavelength-selective couplers, splitters, and recombiners. We shall now discuss the underlying principle of such devices by considering the operation of a *wavelength-selective coupler.*

The wavelength dependence of the *coupling coefficient* in a fiber optic directional coupler stems from the stronger *evanescent coupling* that arises at longer wavelengths (Digonnet and Shaw, 1983). This wavelength dependence of the coupling coefficient leads to the kind of power transfer variation with wavelength depicted in Fig. 5.61. A *wavelength-selective coupler* is a form of fiber optic directional coupler in which the power coupling ratios for light incident at the two wavelengths λ_1 and λ_2 are

$$\mathscr{F}(\lambda_1) = \mathscr{F}_0 \sin^2\{C(\lambda_1)L_{WDM}\}, \tag{5.86}$$

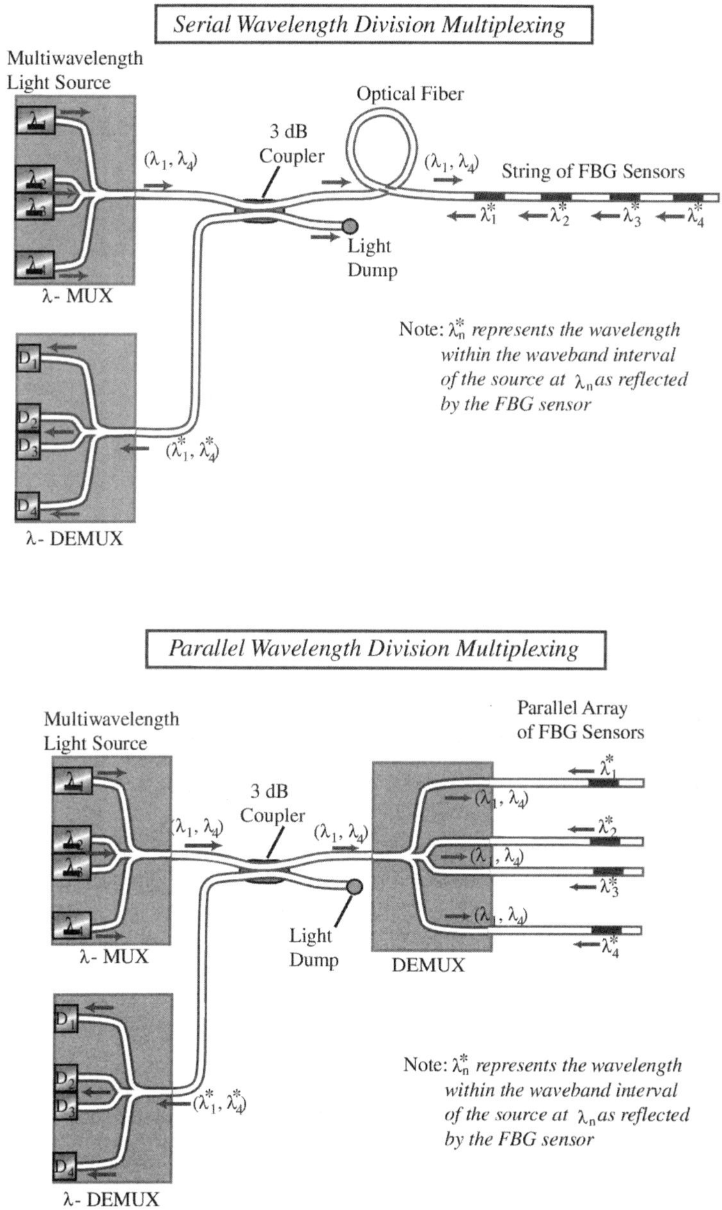

FIGURE 5.60. Schematic illustration of serial and parallel wavelength division multiplexing as applied to multiple FBG sensing arrays.

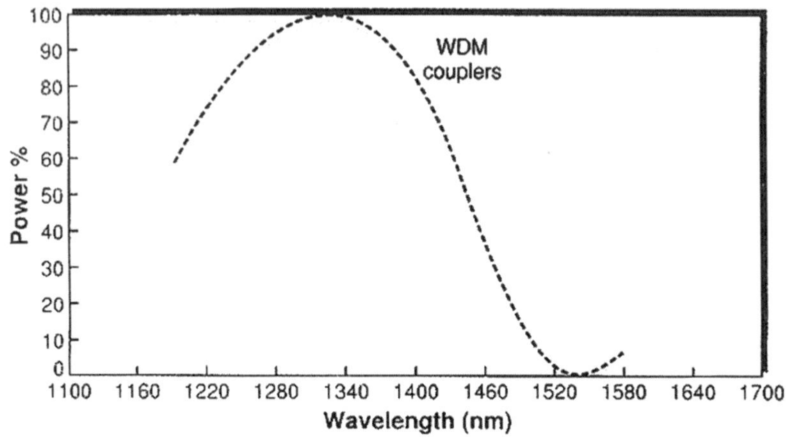

FIGURE 5.61. Example of the performance of a wavelength division multiplex (1300 nm/1500 nm) based on a biconical tapered wavelength selective coupler. Bhagavatula V.A., Holmes G.T., Nolan D.A., Jansen R., McCourt M., Fig. 3, p. 158, LFW 25(10) Oct. 1997, "Planar Technology enhances coupler performance"© Penwell. Used by permission.

and

$$\mathscr{F}(\lambda_2) = \mathscr{F}_0 \sin^2\{C(\lambda_2)L_{\rm WDM}\}, \tag{5.87}$$

respectively, where $C(\lambda_1)$ and $C(\lambda_2)$ are the coupling coefficients at the two wavelengths λ_1 and λ_2, and $L_{\rm WDM}$ is the interaction length.

As a result, the power transfer curves at each wavelength will be very different Fig. 5.62, and it is possible to tailor the directional coupler to transmit light at λ_2 and totally couple the light at λ_1 to the other optical fiber (see the inset to Fig. 5.62). This is accomplished by arranging that

$$C(\lambda_1)L_{\rm WDM} = \pi/2, \tag{5.88}$$

and

$$C(\lambda_2)L_{\rm WDM} = \pi, \tag{5.89}$$

in Eq. (5.74), and assuming identical optical fibers so $\Delta\beta = 0$. Note that the *power exchange length* $L_{\rm exc}$ in (5.74) is now wavelength dependent, and in general we should write

$$L_{\rm exc}^{\lambda_j} = \frac{\pi/2}{C(\lambda_j)}. \tag{5.90}$$

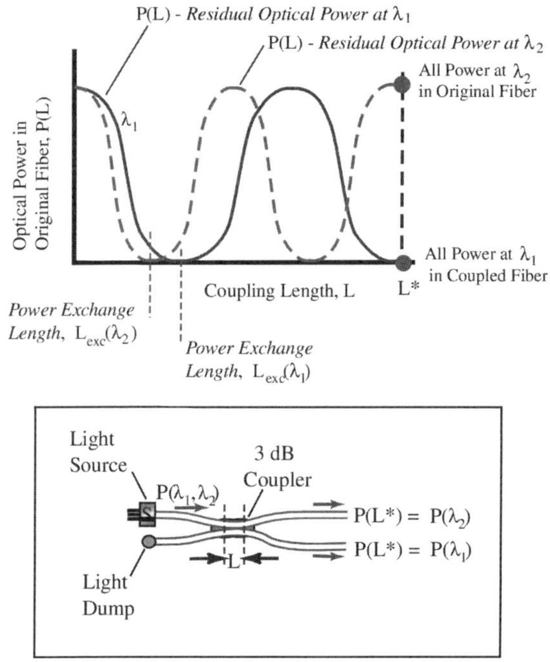

FIGURE 5.62. Principle behind a wavelength division multiplexer based on selective-wavelength evanescent coupling between two optical fibers.

The *minimum interaction length* for the coupler to act as a wavelength division multiplexer for the two wavelengths λ_1 and λ_2, is given by the expression

$$L_{\mathrm{WDM}} = \frac{\pi/2}{|C(\lambda_1) - C(\lambda_2)|}. \tag{5.91}$$

The minimum wavelength spacing for the separation of two signals by means of wavelength division multiplexing in a coupler (Digonnet and Shaw, 1983) is

$$\Delta\lambda = \frac{\pi}{2L_{\mathrm{WDM}}}\left[\frac{dC}{d\lambda}\right]^{-1}. \tag{5.92}$$

where $dC/d\lambda$ represents the wavelength derivative of the coupling coefficient at the wavelength of interest.

In all-fiber couplers, the temperature dependence of both the thermal expansion coefficient and that of the index of refraction makes the coupler increasingly sensitive to changes in temperature, the longer the interaction length. A separation of 0.25 nm was achieved in the 600 nm range in a device that was 40 cm long (Dakin and Culshaw, 1988).

CHAPTER 6

Fiber Optic Structural Sensors and Their Merits

6.1 MERITS OF FIBER OPTIC STRUCTURAL SENSORS

Why use fiber optic technology for structural monitoring? This is a reasonable question that can be expected from many engineers faced with this new technology. It must be addressed before we proceed. The bottom line is that fiber optic sensors permit measurements that are either impractical or uneconomic with conventional measurement technology, such as resistive foil strain gauges. *Fiber optic sensors* (FOS) have a number of advantages over their electrical-based counterparts, and certain FOS possess additional merits that we shall consider later. Fiber optic sensors are extremely small in diameter, very light, sensitive to strain and temperature changes, resistant to corrosion and fatigue, and capable of wide bandwidth operation. They are dielectric in nature, so they are compatible with FRP composite materials or concrete; safe to handle; unlikely to initiate fires or explosions, immune to electrical interference, and do not represent electrical pathways within a host structure.

The fact that no protection is required against lightning and various forms of electromagnetic interference figures prominently in their selection for large civil structures and aircraft. Even in the absence of a direct lightning strike, during a severe storm magnetic coupling can generate large voltages within electrical sensing systems based on wires that are distributed over large structures. These voltages can then burn out various components and fragile circuits used in the electronic systems.

The dual role for the optical fiber of sensor and pathway for the sensing signals makes the sensing architecture much simpler for FOS than conventional sensing technology, especially when arrays of many sensors are needed. We touched on the topic of multiplexing in Chapter 2 and Chapter 5. Multiplexing of fiber optic sensors also forms the basis of Chapter 11. An example of the unique monitoring applications made possible by this advantage is "smart" FRP reinforcement that could

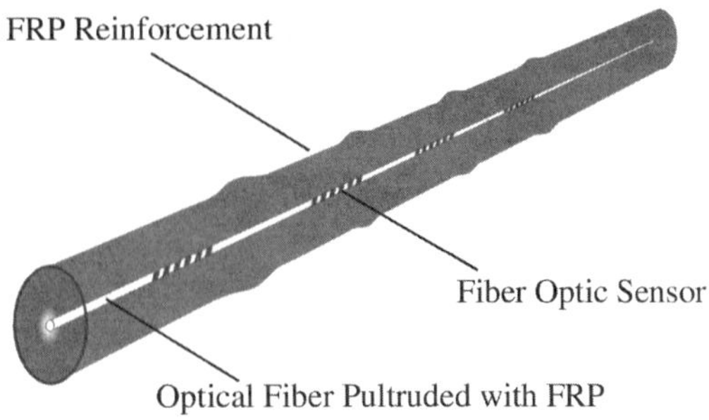

FIGURE 6.1. A "smart FRP reinforcement" intrumented with a string of fiber optic sensors pultruded with the FRP rod.

be used by the construction industry to replace steel. The fiber-like form, small diameter, and dielectric properties of the optical fiber permit it to be pultruded with the FRP reinforcement. This provides mechanical and environmental protection to the optical fiber and bestows on the pultruded FRP a small degree of "smartness" in the form of a resident monitoring system (Fig. 6.1).

Smart FRP reinforcement represents an ideal application for the fiber Bragg grating sensors, as the optical fiber can be manufactured with strings of such sensors. Indeed, initial results have demonstrated the viability of pultruding FBG sensors (see Section 8.6). It would be quite impossible to produce the electrical equivalent of such a built-in quasi-distributed strain sensing system, for the multiple electrical leads required by a set of resistive foil strain gauges would represent a severe mechanical perturbation to the reinforcement and probably would not survive the pultrusion process.

In terms of structural monitoring, the extraordinary versatility of certain fiber optic sensors is unrivaled by conventional measurement technology. FBGs can be used as short- or long-gauge-length sensors, the former with spatial resolution of better than 1 mm, the latter with gauge lengths that can extend well beyond 1 m. As we have already indicated in Chapter 5, it is possible to have both serial and parallel multiplex arrays of such sensors and, as we shall see in Chapter 12, truly distributed measurements are also practical with FBGs. The extreme sensitivity possible with fiber optic strain sensors has been demonstrated with *nanostrain resolution* of fiber Bragg gratings (Kersey *et al.*, 1992), while the almost unlimited bandwidth potential has been indicated by the megahertz frequency response of interferometric fiber optic sensors (Liu

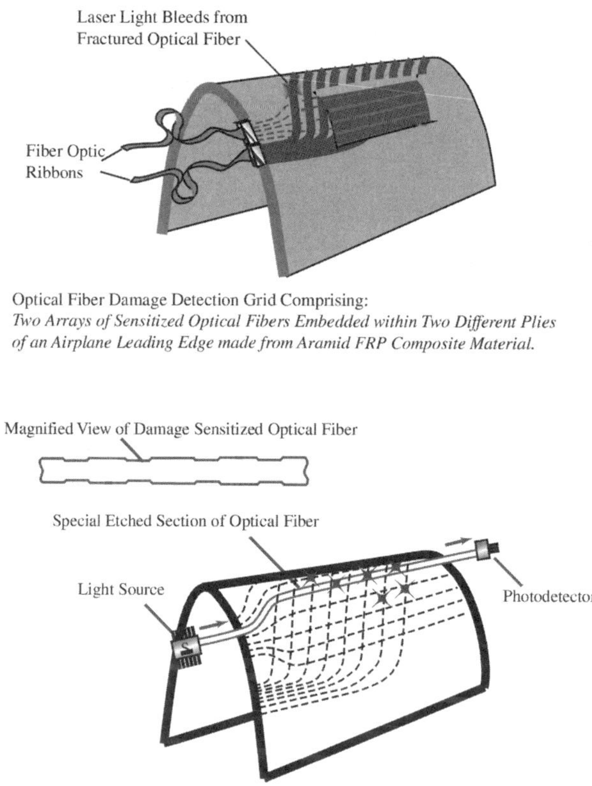

(a)

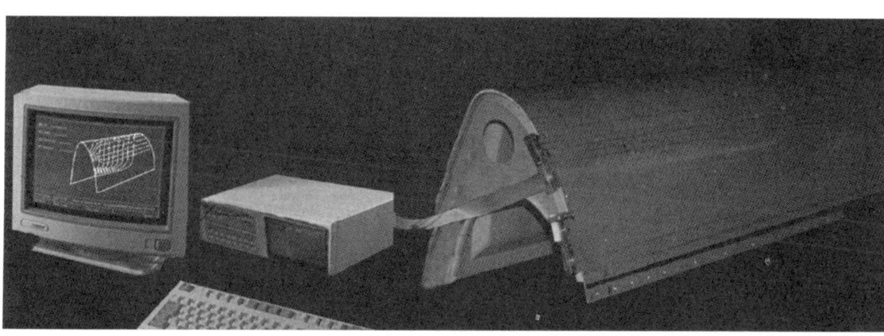

(b)

FIGURE 6.4. (a) Schematic and (b) view of a fiber optic damage detection system based on an embedded 2D grid of damage-sensitized optical fibers within an FRP leading edge of an airplane wing.

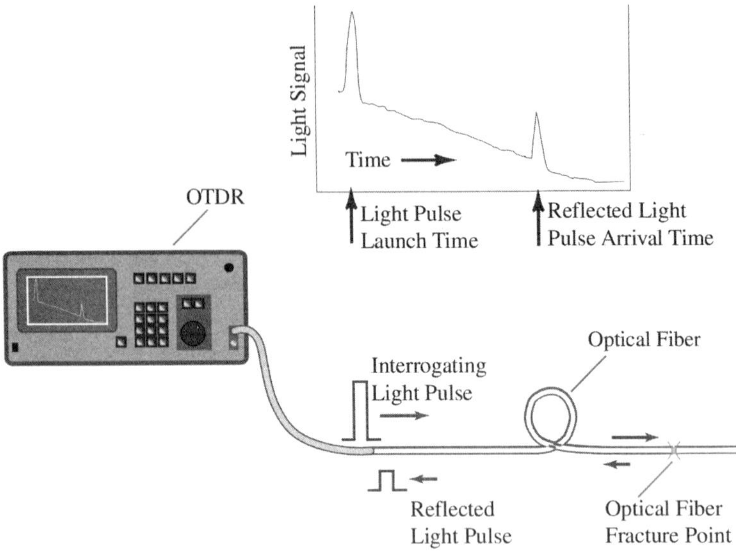

FIGURE 6.5. Principle of optical time domain reflectometry, OTDR.

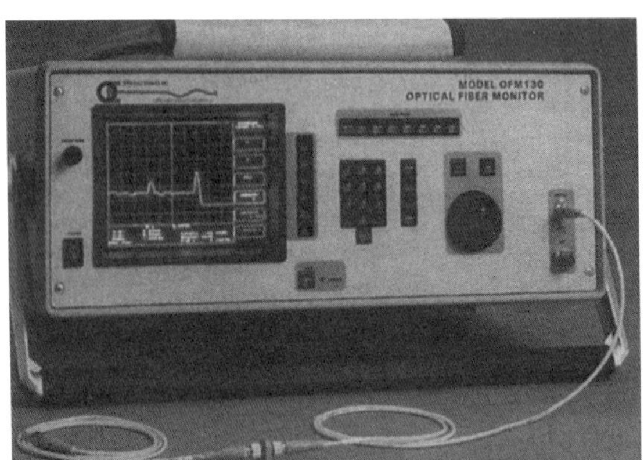

FIGURE 6.6. Photograph of a commercial OTDR with a spatial resolution of about 1 cm. The two signals seen on the screen correspond to Fresnel reflections from the connector and the end of the optical fiber. Used with permission, Opto-Electronics Inc.

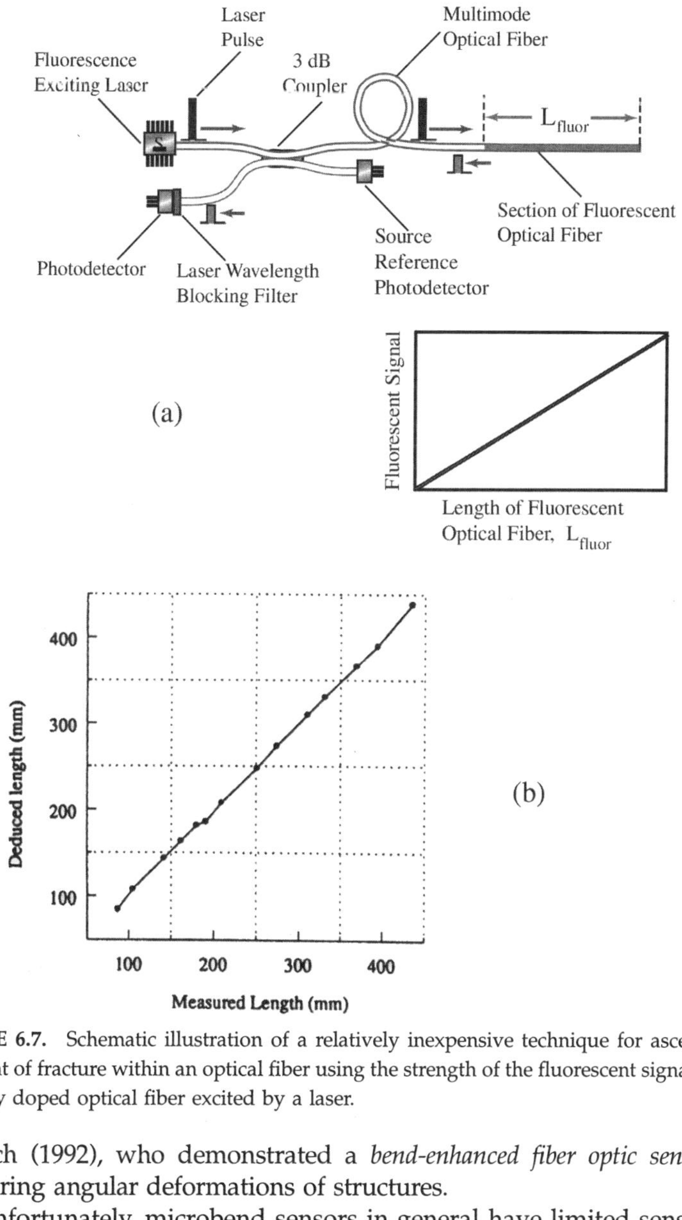

FIGURE 6.7. Schematic illustration of a relatively inexpensive technique for ascertaining the point of fracture within an optical fiber using the strength of the fluorescent signal from a specially doped optical fiber excited by a laser.

Danisch (1992), who demonstrated a *bend-enhanced fiber optic sensor* for measuring angular deformations of structures.

Unfortunately, microbend sensors in general have limited sensitivity, measurement range, and accuracy. Furthermore, many phenomena that are unrelated to strain can cause changes in the output of the source, or attenuation of light in the optical fiber, and the need for a reference optical

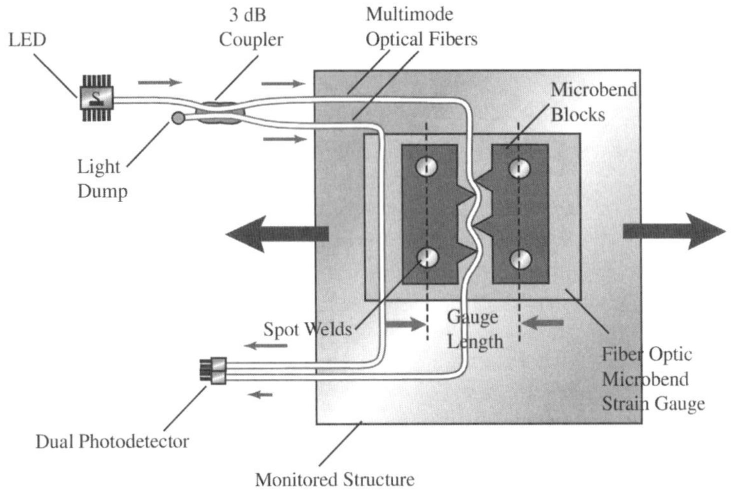

FIGURE 6.8. Basic configuration of a microbend fiber optic strain gauge.

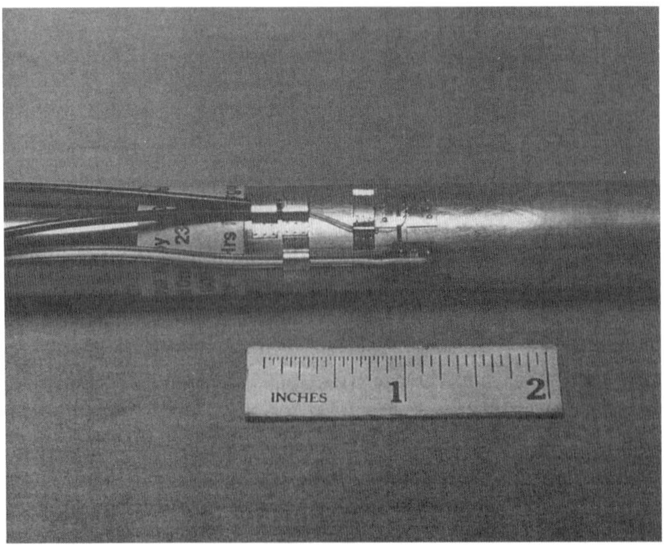

FIGURE 6.9. Photograph of a microbend fiber optic strain gauge mounted above a resistive foil strain gauge on a metal tube. From Berthold, J.W., (1997). "Sensors in Industrial Systems". Reprinted with permission from *Optical Fiber Sensors, Applications Analysis, and Future Trends*, vol. 4, Dakin, J. and Culshaw, B. (Eds), Artech House, Inc., Norwood, MA, USA. www.artechhouse.com.

fiber complicates the physical arrangement. The footprint of a microbend sensor is comparable to that of a conventional strain gauge, as seen in Fig. 6.9. Strain and displacement measurements can be made using OTDR to evaluate the time of flight between the reflections of light from two or more intrinsic mirrors within a multimode optical fiber (Zimmermann *et al.*, 1989). Unfortunately, the spatial resolution of this system is in the range of meters and the strain resolution is limited to hundreds of microstrain. Intensiometric fiber optic sensors for determining temperature are based on a number of approaches, including fluorescence decay, differential absorption, Brillouin and Raman scattering, and blackbody emission (Measures, 1989; Kersey, 1997b).

6.4 INTERFEROMETRIC FIBER OPTIC SENSORS

Interferometric sensors respond to measurand induced changes in the phase of a lightwave propagating along a single-mode optical fiber and represent a large class of extremely sensitive devices (Jackson and Jones, 1986). The *Michelson fiber optic interferometer* is one of the simplest to understand and easiest to build for laboratory experiments. It is schematically illustrated in Fig. 6.10. The principle of operation of the bulk optic version of the interferometer was explained in Chapter 3 (see Fig. 3.22). The essential difference between the bulk optic and fiber optic versions is that in the latter the lightwave is confined to travel within the core of a single-mode optical fiber instead of being in the form of a collimated light beam. Light from a highly coherent (laser) source is divided by the 3-dB coupler and sent down two optical fibers, as shown in Fig. 6.10. The mirrored tips of the sensing and reference optical fibers of the Michelson interferometer reflect the light back to the coupler, where it again is divided, and part of this optical signal is sent to a photodetector.

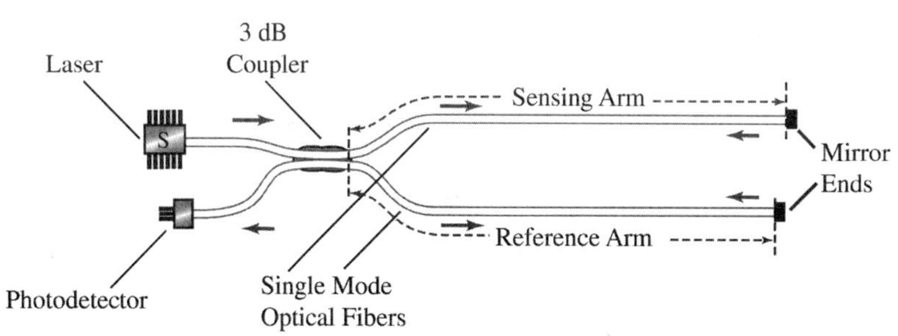

FIGURE 6.10. Schematic illustration of Michelson interferometric fiber optic sensor.

Although the two lightwaves that constitute this signal have a common origin and start out with the same phase at the coupler, after traveling their respective paths along the sensing and reference optical fibers and being reflected, there exists (when they return to the coupler) a phase difference $\Delta\phi$, which is directly proportional to the difference in these optical paths, i.e.,

$$\Delta\phi = 2k(n_s L_s - n_r L_r), \tag{6.1}$$

where k is the free-space propagation constant, n_s and n_r are the respective core indices of refraction, and L_s and L_r are the respective physical lengths (measured from the coupler to each of the mirrored tips) of the sensing and reference optical fibers. The factor of 2 in Eq. (6.1) is required as each path is traversed twice.

Unfortunately, the nonlocalized nature of this sensor and its need for a separate reference optical fiber make it unsuitable for many structural monitoring applications because of its greater intrusiveness, poor common mode rejection, and required phase preservation at the structure's interface. The latter two effects are likely to represent particularly serious problems for practical structures that are invariably subject to high levels of mechanical vibration. The Michelson fiber optic sensor is also vulnerable to stress-induced birefringence effects along the entire length of the optical fibers (Liu *et al.*, 1990). Nevertheless, some of the early vindications of the prospects for strain sensing with fiber optics, particularly in the case of high-frequency acoustic wave detection, were undertaken with this type of sensor (Liu *et al.*, 1989, 1990).

Some of these problems have been overcome by using a balanced polarization fiber optic Michelson interferometer (Furstenau *et al.*, 1993). Another method of overcoming some of these problems involves integrating the entire fiber optic sensor, including the coupler, within some form of measuring rod. This will be discussed further in Chapter 10, where it will be revealed that a practical long gauge structural sensing system, based on a fiber optic Michelson interferometer has been developed by Inaudi *et al.* (1997b) for a number of construction industry applications. This system has been commercialized by Smartec, Switzerland.

As described in Chapter 3, the Fabry–Perot interferometer comprises a cavity defined by two mirrors that are parallel to each other and perpendicular to the optical axis of the system (see Fig. 3.24). The fiber optic Fabry–Perot sensor exists in two forms: the *intrinsic* version depicted in Fig. 6.11a (Lee and Taylor, 1988; Hogg *et al.*, 1991a,b), or the *extrinsic* version presented as Fig. 6.11b (Murphy *et al.*, 1991b,c; Kruschwitz *et al.*, 1992). The former, a *fiber Fabry–Perot* (FFP) sensor, is built by creating a

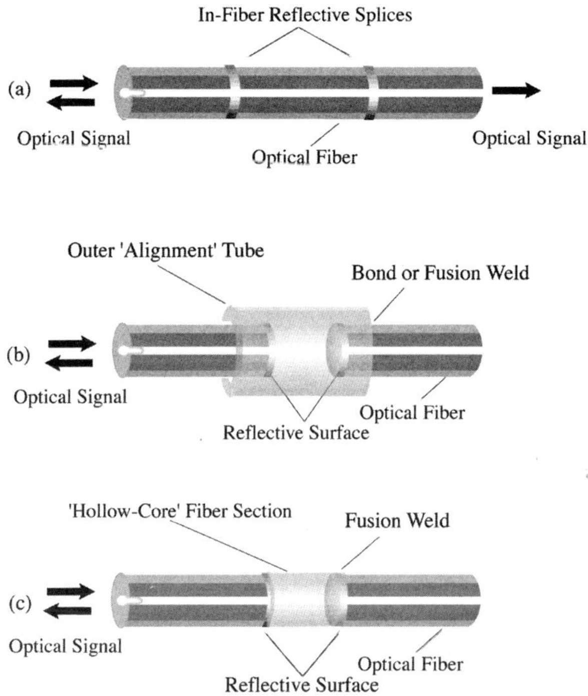

FIGURE 6.11. Three types of fiber optic Fabry–Perot sensor: (a) intrinsic fiber Fabry–Perot (FFP); (b) extrinsic Fabry–Perot interferometer (EFPI); and (c) in-line fiber etalon (ILFE).

cavity defined by two reflective fusion splices (or one fusion splice and a mirrored tip) and has the advantage of being the least perturbative for embedding within FRP materials. It is also capable of being used in the form of a strain rosette (Fig. 1.6; Valis *et al.*, 1992) as the cavity constitutes the localized sensing region; see Fig. 6.12.

A fracture procedure has been developed by LeBlanc (1998) to create an almost zero-cost, intrinsic Fabry–Perot displacement sensor after an optical fiber was embedded within its host structure. *Fresnel reflections* from the two ends of the fractured optical fiber serve as the mirrored surfaces of the cavity. By measuring the size of the resulting Fabry–Perot cavity (i.e., the "air gap" in the optical fiber). LeBlanc has been able to make some valuable micromechanical measurements of the role played by an optical fiber coating in transferring strain to the sensor.

In the *extrinsic (air-gap) Fabry–Perot interferometer* (EFPI) sensor, the cavity is formed between two mirror-tipped optical fibers that are supported within a microcapillary alignment tube. As a consequence of

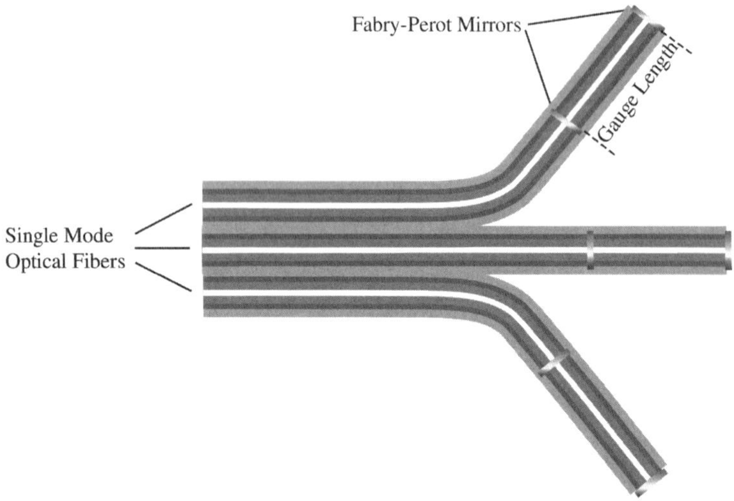

FIGURE 6.12. Fiber optic strain rosette using three intrinsic Fabry–Perot sensors.

the separation between gauge length and cavity length in the EFPI, the sensor can be built with a gauge length that is greater than the cavity length. This allows the EFPI to work with lasers of modest coherence length. The EFPI has the advantage of no transverse coupling and can therefore evaluate more directly the axial component of strain in a host material (Sirkis and Haslach, 1991). Although it is easier to manufacture than the intrinsic version involving mirrored fusion splices, the exact determination of its gauge length requires careful calibration. A third variant, referred to as an *in-line fiber etalon* (ILFE), uses a small length of hollow-core optical fiber that is welded to the two mirror-tipped optical fibers (Fig. 6.11c) to form a sensor without the physical discontinuities of the EFPI (Marcuse, 1989; Sirkis *et al.*, 1994a).

In some ways the *Fabry–Perot sensor* can be thought of as a form of Michelson interferometer wherein the reference and sensing optical fibers are one and the same, up to the first mirror that constitutes the start of the sensing region. This bestows several advantages on the Fabry–Perot sensor: it involves a single optical fiber, it has common mode rejection, and it can be a single-ended sensor. A high *finesse Fabry–Perot sensor* (one with high-reflectivity mirrors, i.e., $R > 50\%$) behaves as a *multipath interferometer* and reflects light back along the optical fiber only if its frequency closely coincides with one of the *cavity mode frequencies*,

$$\nu_{\mathrm{m}} = \frac{mc}{2nL}, \tag{6.2}$$

introduced earlier as Eq. (3.71). This multipeaked reflection spectrum was portrayed in Fig. 3.25. A strain- or temperature-induced change in the optical path length, nL, between the two mirrors of the Fabry–Perot sensor leads to a corresponding shift in the cavity mode frequencies. The extent of this shift is limited to the free-spectral range of the Fabry–Perot interferometer, given by Eq. (5.79), and this restricts the dynamic range of the sensor.

Early versions of the fiber optic intrinsic Fabry–Perot sensor were fabricated using single layers of dielectric, TiO_2, sputtered on the tips of the optical fibers prior to fusion splicing (Lee and Taylor, 1989). These mirrors had a low reflectivity ($<10\%$), so these low-finesse Fabry–Perot sensors behaved more like two-path interferometers with a sinusoidal phase–strain relationship, akin to Eq. (3.47). Subsequent work focused on the use of multilayer TiO_2/SiO_2 films deposited on the fiber tips. Fusion splicing of seven-layer coated fibers to uncoated fibers produced intrinsic mirrors with reflectances greater than 85%, and Fabry–Perot interferometers fabricated with these mirrors produced finesses of >20 (Lee and Taylor, 1992). The initial Fabry–Perot demodulation schemes were designed to extract only the relative measurand information from sensors (Hogg *et al.*, 1991a; Murphy *et al.*, 1991c).

Later approaches based on an analysis of the full reflectance spectrum of the sensor has permitted the optical path difference to be uniquely determined, bestowing interrupt immunity on the Fabry–Perot sensor (Bhatia *et al.*, 1994; Fig. 6.13a). The EFPI cavity air-gap spacing, L, can be determined from the wavelength separation of any pair of adjacent transmission modes, say λ_1 and λ_2 (see Fig. 6.13b). If we draw upon the free-spectral range for the EFPI, the relation

$$L = \frac{\lambda_1 \lambda_2}{2(\lambda_2 - \lambda_1)}, \tag{6.3}$$

is derived as Eq. (9.20) in Chapter 9.

In general, the circular symmetry of optical fibers used to make fiber optic interferometric sensors, discussed earlier, may not be adequate to stabilize the two orthogonal *linear polarization eigenmodes*. This can lead to *polarization fading* and the possibility of cross-coupling between the polarization modes. These effects represent a source of noise in interferometric measurements based on small changes (on the order of a few fringes) (Sheem *et al.*, 1982). Although a number of methods for alleviating this problem have been proposed (Kersey and Dandridge, 1988), one of the most effective involves the use of polarization-maintaining single-mode optical fibers.

The high birefringence designed into this kind of optical fiber stabilizes the *state of polarization* (SOP) because the two polarization eigen-

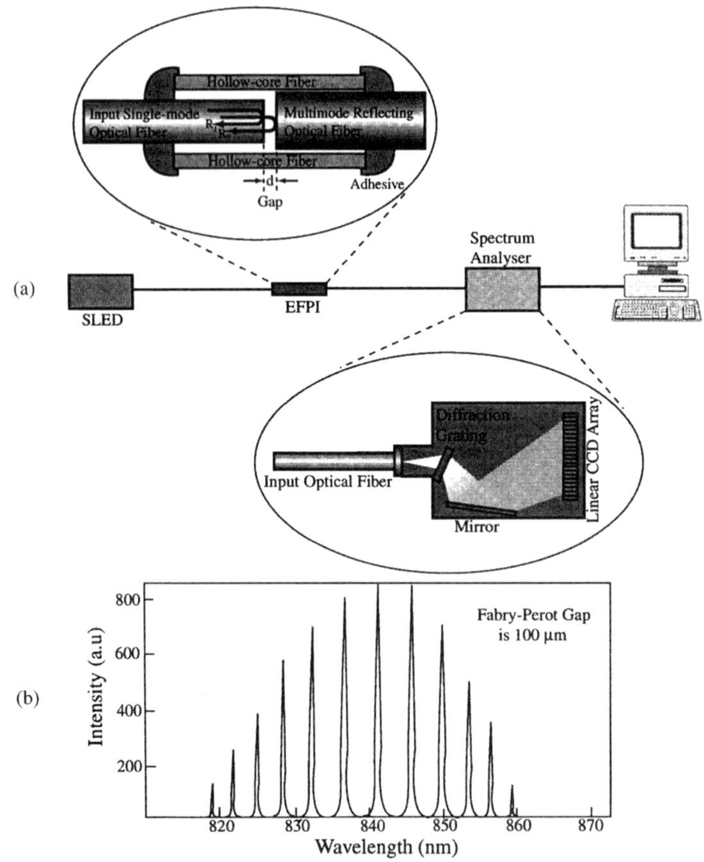

FIGURE 6.13. (a) Schematic of an EFPI sensing system using a spectrum analyzer to make the system interrupt immune. (b) The multispiked spectrum from the EFPI. From Lewotsky, K., "Self Calibrated Fiber Optic Sensor Measures Strain" *Laser Focus World*, Dec 1999, pp. 33–35, © Pen Well, used by permission.

modes are forced to have an appreciable difference in their respective propagation constants (i.e., indices of refraction), and this limits the exchange of energy between the two eigenmodes. This high birefringence is created either through geometrical asymmetry, such as an elliptic core, or thermal stress loading of the fiber core (see Sections 3.3 and 5.2) (Gambling and Poole, 1988). This, however, can be expensive, as special couplers that are polarization insensitive have to be employed in addition to the much more expensive polarization-maintaining optical fiber.

Another method of avoiding polarization fading is to employ a high-frequency polarization scrambler. This can be a relatively low-cost solution in the case of a network of sensors; however, this gain is at the expense of the interferometric fringe visibility, which declines to 0.5.

Many of these problems have been avoided by Belleville and Duplain (1993) who used low-coherence, *white-light interferometry* with a multimode optical fiber EFPI to produce a short-gauge strain sensor. The strain resolution achieved with this sensor was 0.25 $\mu\varepsilon$ when the strain range was set to 1000 $\mu\varepsilon$. The underlying principles of this kind of sensor were discussed in the context of a bulk optic system in Section 3.8. A schematic of this form of fiber optic low-coherence EFPI is presented as Fig. 6.14.

In essence, a *Fizeau interferometer* is used to determine the optical path length of the EFPI sensor. This is accomplished by identifying the position along the Fizeau interferometer where a narrow interference pattern arises (see discussion accompanying Fig. 3.29). Although this figure describes the use of a *Fizeau interferometer* in terms of bulk optic low-coherence interferometry, the principle of its operation is the same for the EFPI. In this instance the location on the Fizeau interferometer where interference arise and the transmission is a maximum, corresponds to where the optical path length of the Fizeau interferometer matches that of the EFPI. Instrumentation based on this concept has been developed into a commercial system by Fiso Technologies, Quebec, and is available through Roctest. Rao *et al.* (1994) demonstrated how a similar dual-wavelength low-coherence EFPI sensor could be used for temperature compensated pressure measurements.

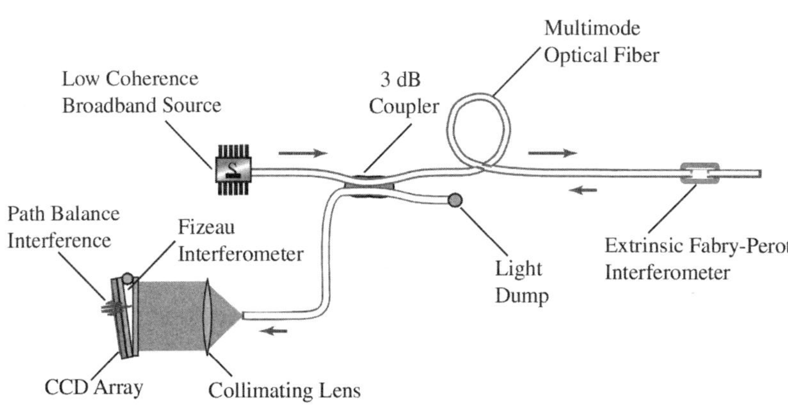

FIGURE 6.14. Low-coherence EFPI sensing system using a Fizeau interferometer and a CCD linear array as a cross-correlator.

6.5 POLARIMETRIC AND MODALMETRIC FIBER OPTIC SENSORS

The *polarimetric sensor* measures some external factor, such as host strain or temperature, by its influence on the two *polarization eigenmodes* of a single-mode optical fiber. In some sense it is similar to an interferometric sensor, with these two eigenmodes taking on the role of the sensing and reference arms of an interferometer. The difference in sensitivity to the measurand for each of the polarization eigenmodes makes it possible to infer the extent of this physical factor from the change in the *state of polarization* of the lightwave as it propagates along the optical fiber.

To this end, linearly polarized light is launched into a highly birefringent optical fiber with its plane of polarization at 45° to the two polarization eigenaxes. As the lightwave propagates along the optical fiber, its SOP changes because of the difference in the phase velocity of the two polarization components in the birefringent fiber (see Fig. 6.15). If no external influences are at work, then the intensity of light transmitted through a linear polarizer aligned with one of the polarization eigenaxes will not change with time. If the optical fiber is now subject to strain or a temperature change, then the SOP of the light propagating along the optical fiber will be modified by the resulting change in the birefringence. This change in the SOP is determined by monitoring the variation in the intensity of light transmitted through a linear polarizer (Krohn, 1988). Bock (1989) used a fiber optic polarization sensor to directly measure hydrostatic pressure.

Localization of this type of sensor is achieved by rotating the polarization eigenaxes through 45° in the "sensing section" relative to their orientation in the lead-in/out sections of optical fiber. A single-ended version of this form of polarization sensor is schematically illustrated in Fig. 6.16. This type of localization was first introduced by Varnham *et al.*

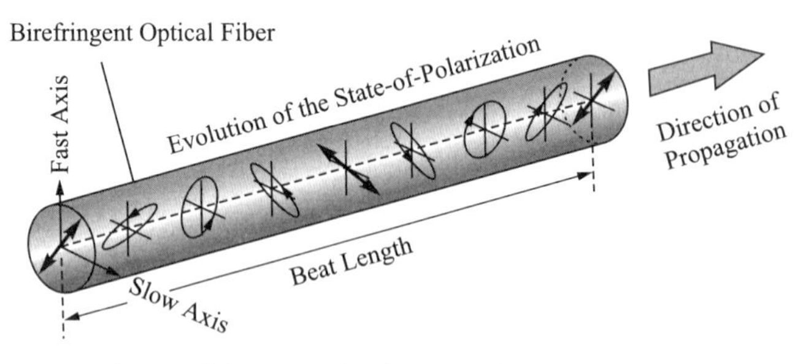

FIGURE 6.15. Evolution of the state of polarization for light initially linearly polarized at 45° to the fast and slow eigenaxes of a highly birefringent optical fiber.

(1983) in the case of a *polarization temperature sensor.* The concept was later used for strain sensing (Hogg *et al.*, 1990).

When linearly polarized light is launched into one of the two polarization eigenaxes of the lead-in optical fiber (i.e., the **E**-field of the light is aligned with one of the polarization eigenaxes), its SOP does not change until it enters the sensing section of the optical fiber. Here the 45° rotation of the eigenaxes means that both eigenmodes are equally excited, and as a result of the difference in sensitivity of the index of refraction for each axis to the measurand, the two components of the original wave now travel at different velocities, changing the SOP of the lightwave. After reflection, the lightwave reenters the lead-in section of the optical fiber, now called the lead-out optical fiber, and only the change in the SOP experienced in the sensing section varies the intensity of light transmitted through the linear polarizer placed after the coupler (Bock and Wosinski, 1990). The need for polarization-maintaining optical fiber and polarization-insensitive couplers and the long gauge lengths needed for good strain resolution (Measures, 1993) have considerably limited the use of this type of sensor.

The most well-developed form of localized *modalmetric sensor* is called an *elliptic-core two-mode sensor* and involves changes in the *transverse spatial mode distribution* of light within an optical fiber (Blake *et al.*, 1987a,b; Kim *et al.*, 1987; Murphy *et al.*, 1989; Lu and Blaha, 1991). In this sensor the *lowest-order transverse mode*, LP_{01}, propagates along the lead-in optical fiber to the sensing region. Here a section of elliptic-core optical fiber that permits two *transverse* LP_{01} and the *even* LP_{11} modes to propagate, at the wavelength for which the lead-in fiber is single mode, has been fusion spliced with its core slightly offset so that both spatial modes are excited (Fig. 6.17). A second section of single-mode fiber can be fusion spliced to the end of the sensing section, or the end of the two-mode sensing section can be mirrored. Either way, the spatial variation in the *transverse light*

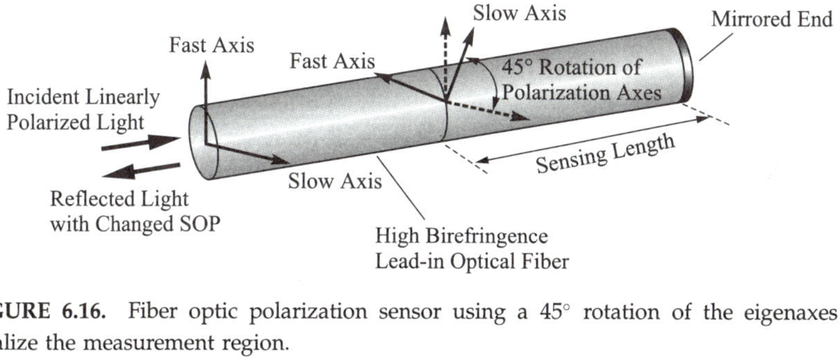

FIGURE 6.16. Fiber optic polarization sensor using a 45° rotation of the eigenaxes to localize the measurement region.

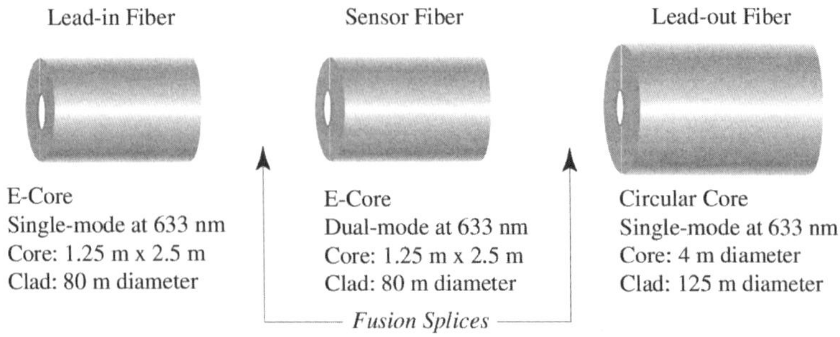

FIGURE 6.17. Exploded view of a two-mode fiber optic sensor. The localized sensing section comprises a segment of optical fiber that permits the first two spatial modes: LP_{01} and LP_{11}, to propagate through it. This section is offset and fusion spliced to the lead-in and lead-out sections of optical fiber.

distribution is converted into a nonlinear (cosine) relation between the signal and the strain in the host structure, typical of an interferometer. The difficulty of manufacture and the long gauge length required for reasonable strain resolution have diminished interest in this sensor, and it will not be considered further.

6.6 SPECTROMETRIC FIBER OPTIC SENSORS

6.6.1 Raman and Brillouin backscatter

In this section we show that Raman and Brillouin (inelastic) scattering processes can be used to determine the temperature profile along an optical fiber. Distributed fiber optic temperature sensing based on *Raman scattering*, first demonstrated by Dakin *et al.* (1985) and Hartog *et al.* (1985) was developed into commercial instruments by several companies during the late 1980s. The Raman scattering process produces components in a broad band about the exciting (pump) wavelength comprising *Stokes* (lower photon energy) and *anti-Stokes* (higher photon energy) emissions, as illustrated in Fig. 6.18a. The peak intensity in the Raman spectrum is at a wavenumber (inverse wavelength) shift of about $\pm 40\ \text{cm}^{-1}$, or a Raman frequency shift of $\sim 13\,\text{THz}$, and the ratio of intensities r_R at the anti-Stokes and Stokes wavelengths λ_a and λ_s, respectively, is given by

$$r_R = \left[\frac{\lambda_s}{\lambda_a}\right]^4 e^{-(hc\Delta v/kT)}, \tag{6.4}$$

where h is *Planck's constant*, c is the *free-space velocity* of light, k is *Boltzmann's constant*, Δv is the *frequency shift* of the Stokes (or anti-

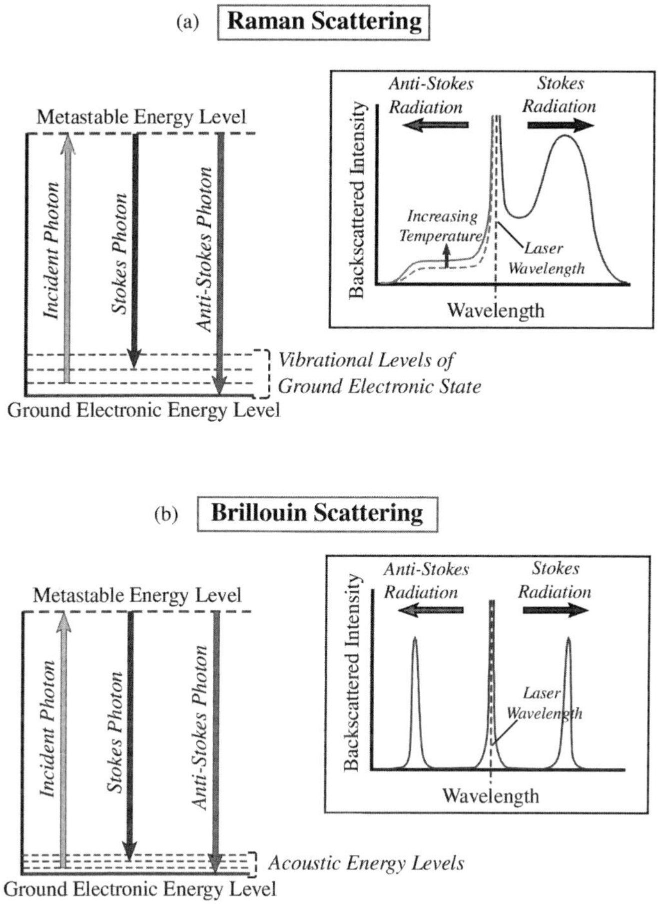

FIGURE 6.18. Energy level diagrams and spectra for Raman and Brillouin scattering.

Stokes) line from the optical laser (pump) frequency, and T is the temperature in kelvins. This ratio has a magnitude of $\sim$0.15 at room temperature and a temperature dependence of $\sim$0.8%/°C in the range 0 to 100°C (Dakin, 1995).

Detection of the Raman backscatter typically involves the use of a modified OTDR technique in which the intensity ratio of the *Stokes to anti-Stokes* backscattered light is determined as a function of time, and therefore space. Systems capable of operation over several kilometers of fiber length with $\sim$1°C resolution and $\sim$3- to 10-m spatial resolutions have been demonstrated. The major drawback of this technique is the low Raman scattering coefficient, necessitating the use of high input powers from the interrogating lasers and long signal averaging times. This technique, however, is particularly useful in the case of long lengths of optical fiber, up to 10 km, and the technique has been one of the best examples of fiber optic sensor commercial success (York Sensors, UK).

Brillouin scattering in optical fibers arises due to the interaction of light with *phonons* (quantized acoustic waves). In essence, the light is scattered from variations in the index of refraction associated with acoustic waves (Horiguchi *et al.*, 1992a). Light scattered off these phonons is frequency shifted by an amount determined by the acoustic velocity of the phonons, which in turn is dependent on the density of the glass and thus the material temperature. The *Brillouin frequency shift*, ν_B, is very much smaller than experienced in Raman scattering and is of the order of 11 GHz, or 0.5 μm at 1.5 μm radiation in optical fibers (Fig. 6.18b). The bandwidth of Brillouin scattered light is much narrower than Raman scattered light and is around 0.001 nm, or 100 MHz, because of the thermal distribution of the phonons.

The Brillouin frequency shift in an optical fiber is given by

$$\nu_B = 2nV_a/\lambda, \tag{6.5}$$

where V_a is the velocity of sound in glass, n is the effective index of refraction of the optical fiber, and λ is the free-space wavelength of operation (Horiguchi *et al.*, 1992a,b). For silica glass at 1.3 μm, the Brillouin frequency shift is approximately 13 GHz, with a bandwidth of 20 to 30 MHz. The frequency shift is linearly dependent on both the temperature and strain in the fiber (Bao *et al.*, 1993b). The measured fractional frequency shifts have been shown to have coefficients of $4.6 \times 10^{-6}\ \mu\varepsilon^{-1}$ for strain and $9.4 \times 10^{-5}\ ^\circ\mathrm{C}^{-1}$ for temperature. At 1.32 μm, the strain and temperature coefficients are $58\ \mathrm{kHz}\ \mu\varepsilon^{-1}$ and $1.2\ \mathrm{MHz}\ \mathrm{C}^{-1}$ for silica-based single-mode communication optical fiber. The spontaneous Brillouin scattering efficiency is approximately 20 dB weaker than that of Rayleigh scattering, but stimulated Brillouin scattering allows stronger signals to be produced (Horiguchi *et al.*, 1997).

The first approach at measuring temperature with this form of inelastic scattering process was termed *Brillouin optical time domain analysis* (BOTDA). The experimental arrangement used for this purpose by Kurashima (1990) is shown schematically in Fig. 6.19. Here a short laser (pump) pulse is launched into one end of the optical fiber, while a *continuous wave* (CW) probe light beam excites the other end. The probe optical frequency is offset from that of the pump by the nominal Brillouin frequency $\pm\nu_B$. The CW probe light experiences gain (loss) at those spatial locations along the fiber where the frequency offset is matched to the peak Brillouin gain (loss). Gain arises if the frequency of the CW light is lower than that of the pump frequency by ν_B; alternatively, if it is higher than that of the pump frequency by ν_B, then loss will occur.

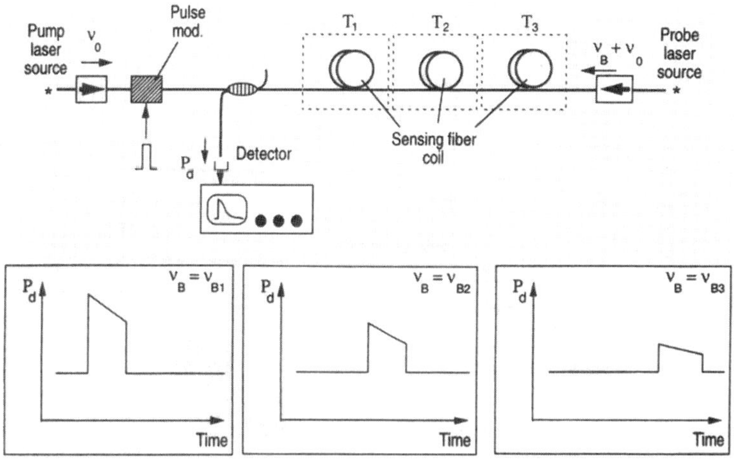

FIGURE 6.19. Experimental arrangement for a Brillouin optical time domain analysis (BOTDA) system. From Kersey, A.D., (1997), "Optical Fiber Sensors" In: *Optical Measurement Techniques and Applications*, P. K. Rastogi, (Ed.), Artech House, Norwood, 1997. www.artechhouse.com.

The time dependence of the detected CW light thus provides a profile of the gain experienced by the probe light as the pump pulse passes along the optical fiber (Fig. 6.19). If this measurement is made for a wide range of frequency offsets, a full picture of the Brillouin frequency for each fiber location (within the spatial resolution determined by the pump pulse-width) is obtained, thus allowing the temperature distribution along the optical fiber to be assessed. A novel signal processing approach by Nikles *et al.* (1994), using a single pump laser and an electro-optic modulator, permits Brillouin sensing in a single-ended configuration. An example of the temperature distribution measured with this system is presented as Fig. 6.20.

Since the Brillouin frequency shift varies with applied strain as well as temperature change (see Fig. 6.21), simultaneous measurements of strain and temperature along an optical fiber are possible (Bao *et al.*, 1994). These authors configured the sensing fiber in such a way that one half responded to the strain and temperature, while the other half was only influenced by temperature. We shall discuss this further in Chapter 7. Although the temperature and spatial resolutions are on the same order as those achieved with Raman scattering, that is $\sim$3- to 10-m spatial resolution and $\pm 1^{\circ}$C temperature resolution. Advances made in the past few years have allowed a spatial resolution of 1 m to be achieved (Horiguchi *et al.*, 1997).

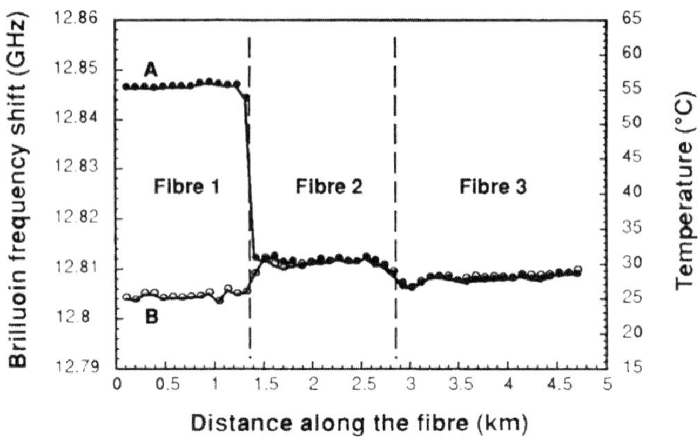

FIGURE 6.20. Measurement of the temperature distribution along two optical fibers "A" and "B" undertaken with BOTDA. From Nikles, M., Thevenaz L. and Robert, P., (1994), "Simple Distributed Temperature Sensor Based on Brillouin Gain Spectrum Analysis" 10th Int. Conf. Optical Fibre Sensors, Glasgow, Scotland, *SPIE* **2360**, 138–141.

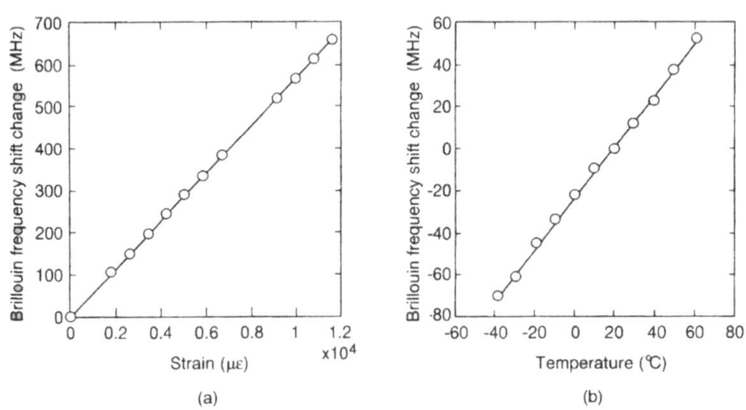

FIGURE 6.21. Linear variation of Brillouin frequency shift with imposed strain and temperature change. From Horiguchi, T., Rogers, A., Michie, W.C., Stewart, G. and Culshaw, B., "Distributed Sensors: Recent Developments". Reprinted with permission from "*Optical Fiber Sensors, Applications, Analysis and Future Trends,*" Vol. 4, Dakin, J. and Culshaw, B. (Eds), 1997. Artech House, Inc., Norwood, MA, USA. www.artechhouse.com.

Fellay *et al.* (1997), has suggested that the ultimate spatial resolution for strain measurements based on a stimulated Brillouin scattering is about 1 m and is limited by the spectral broadening associated with pulse narrowing; see Section 12.9. However, Brown *et al.* (1998) claim to have achieved a spatial resolution of 50 cm and a strain resolution of $40\,\mu\varepsilon^{-1}$. Although this claim is not properly substantiated, if true it would represent a significant breakthrough and would give reason to ponder the role of this technique in structural monitoring of fairly large structures.

However, the opto-electronics required for Brillouin systems are quite complex, requiring long coherence length, very stable lasers, high-speed modulators, detectors, and frequency discriminators. Furthermore, it takes several minutes to complete a measurement due to the averaging required by the very small optical signals. An extensive overview of distributed sensing techniques has been provided by Horiguchi *et al.* (1997). Ando Corporation, Japan, has a commercial fiber optic strain/loss analyzer based on Brillouin scattering and designated model AQ8602. This instrument is quite expensive but can provide modest spatial and strain resolution.

6.6.2 Fiber optic Bragg grating sensors

The fiber optic intracore Bragg grating sensor relies on the narrow-band reflection from a region of periodic variation in the core index of refraction of a single mode optical fiber (see Section 3.10) and the in-depth review of fiber Bragg gratings provided by Russell and Archambault (1997). In this FBG sensor, the *Bragg wavelength* λ_B (center wavelength of the reflected signal) is linearly dependent upon the product of the grating period, Λ, and the mean core index of refraction, n_0 [see Eq. (3.129)], i.e.,

$$\lambda_B = 2n_0\Lambda. \tag{6.6}$$

Changes in strain or temperature to which the optical fiber is subjected linearly shift this Bragg wavelength (Fig. 6.22), leading to a wavelength-encoded optical measurement (Morey *et al.*, 1989) that is self-referencing. This type of sensor is thus insensitive to power fluctuations in the source, variations in the photodetector response, or changes in the loss associated with connectors or macrobends of the optical fiber. As discussed in Section 5.10, wavelength encoding also provides a convenient and simple method of multiplexing. WDM is possible if the spectral width of the light source greatly exceeds the likely measurand-induced wavelength excursion range of an individual FBG. Extensive overviews of FBG sensing have been provided by Kersey *et al.* (1997b) and Hill and Meltz (1997).

As we shall show in Chapter 7, the FBG measured "strain response" or *strain-gauge factor* is

$$G_\varepsilon \equiv \frac{1}{\lambda_B}\left[\frac{\partial\lambda_B}{\partial\varepsilon}\right]_T = 0.78 \times 10^{-6}\ \mu\varepsilon^{-1}, \tag{6.7}$$

which gives rise to an FBG strain-wavelength sensitivity of roughly 1 nm per 1000 $\mu\varepsilon$ at a wavelength of 1.3 μm. In silica optical fibers, the thermal

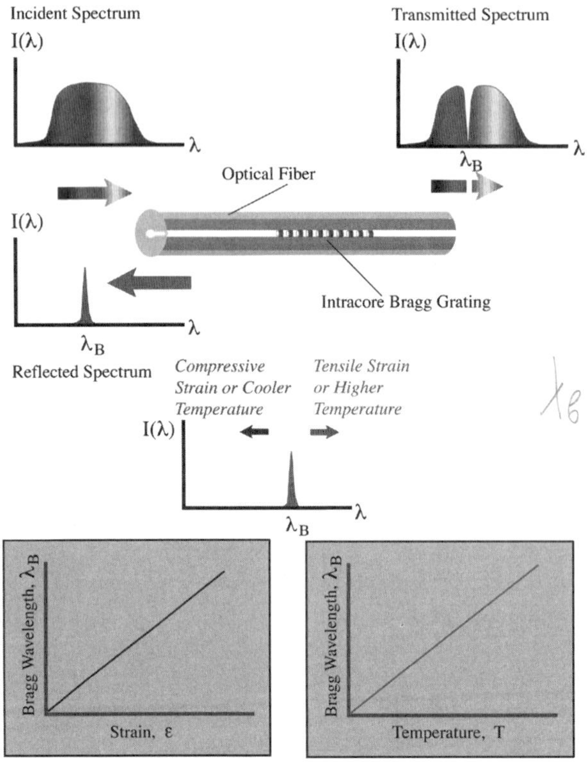

FIGURE 6.22. Basic fiber Bragg grating sensor and the linear variation of its center (Bragg) wavelength with strain and temperature.

response is dominated by the *thermo-optic effect*, that is to say, the change of the index of refraction with temperature. The corresponding FBG "thermal response," or *temperature-gauge factor*, is

$$G_T \equiv \frac{1}{\lambda_B}\left[\frac{\partial \lambda_B}{\partial T}\right]_{\varepsilon} = 6.67 \times 10^{-6}\ \mathrm{K}^{-1}. \tag{6.8}$$

A wavelength resolution of $\sim$1 pm (0.001 nm) is required at a wavelength of 1.3 μm to resolve a temperature change of 0.1 K, or a strain of 1 με. At a wavelength of 1.55 μm, this *wavelength resolution* scales to approximately 1.21 pm. The values of these gauge factors are only provided as a guide, for they are dependent not only on the wavelength of operation, but also on the specific optical fiber used. Thus, calibration of the FBG sensors, as with all sensors, is required for high-precision work. This information is normally provided by the supplier of the FBG or the sensing system.

There are many different approaches to evaluating the spectral shift of the Bragg wavelength for a FBG sensor (Kersey *et al.*, 1997b). One of the simplest involves the use of a spectral filter to assess the Bragg wavelength reflected by a FBG sensor. This *passive spectral ratiometric* approach uses the variation in the transmission of the spectral filter as a function of wavelength to relate the strength of the detected signal to the absolute wavelengh of the light reflected from the FBG, as indicated in Fig. 6.23 (Melle *et al.*, 1992). This demodulation technique has subsequently been engineered into an effective series of commercial fiber optic structural sensing systems by ElectroPhotonics Corporation, Ontario, Canada.

The preceding discussion is based on the use of uniform FBGs, wherein the only parameter of importance is the Bragg wavelength. One of the most significant features that distinguish FBGs from all other fiber optic sensors, and indeed all conventional strain or temperature sensors, is the prospect of intragrating sensing, that is to say, the ability to undertake measurements within the FBG. This unique property permits truly distributed sensing with chirped FBGs on a practical scale (i.e., mm to cm), unlike Raman and Brillouin scattering, which are limited to very much larger scale lengths (>1 m).

A *chirped FBG* is one in which the grating period, the mean index, or both are functions of position along the optical fiber. Figure 6.24 depicts how a FBG can be viewed as an assembly of many small *subgrating*

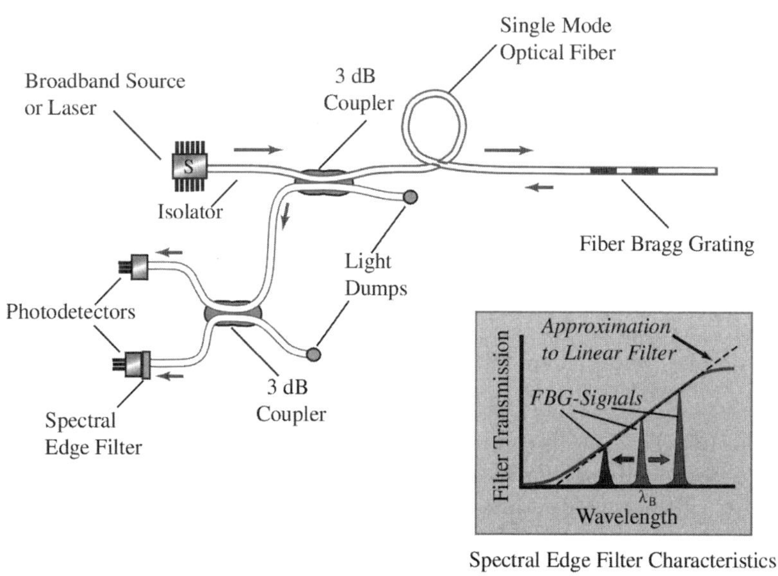

FIGURE 6.23. Passive spectral ratiometric demodulation of a fiber Bragg grating sensor.

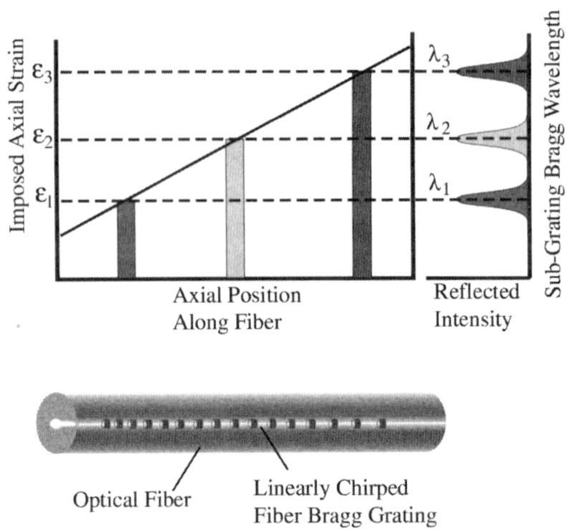

FIGURE 6.24. Strain-induced chirp of an FBG and subsequent spread in its reflectivity spectrum.

elements, each having its own reflection spectrum centered at a wavelength determined by the "local" mean strain. One of the simplest methods of chirping an FBG is to subject it to a linear strain gradient (see Fig. 6.25). This figure clearly reveals the progressive broadening of the reflection spectrum of the FBG associated with its increasing chirp. This *stress-induced chirping* was accomplished by bonding a uniform FBG to a suitably tapered cantilever beam that was increasingly deflected (LeBlanc *et al.*, 1994). The ramifications of chirped FBGs with regard to sensing, and in particular *distributed strain sensing*, will be discussed in detail in Chapter 12.

6.7 SELECTION OF A FIBER OPTIC STRUCTURAL SENSOR

In order to assess the potential suitability of any fiber optic sensor for undertaking localized, as oppose to distributed, structural measurements, the following criteria can be used as a guide (Turner *et al.*, 1990). An ideal fiber optic structural sensor should be:

1. Localized, so it can operate remotely with insensitive leads.
2. Well-behaved with reproducible response.
3. All-fiber for operational stability.
4. Able to provide a linear response.
5. A single optical fiber for minimal perturbation and common-mode rejection.
6. Single-ended for ease of installation and connection.

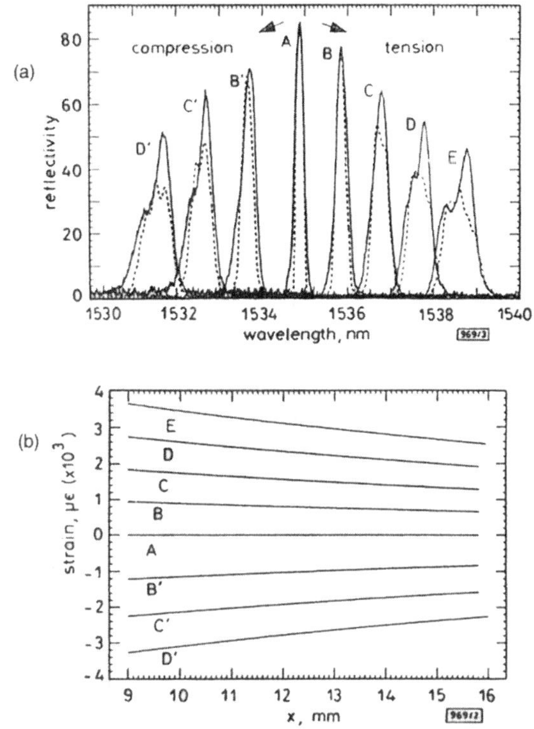

FIGURE 6.25. (a) Experimentally measured (full profile) and theoretically predicted (broken profile) change in the reflection spectra of an FBG subject to a range of linear strain gradients. (b) Corresponding strain gradients imposed along the FBG.

7. Sufficiently sensitive with adequate measurement range.
8. Insensitive to phase interruption at the structural interface.
9. Nonperturbative to the structure and robust for installation.
10. Interrupt-immune and capable of absolute measurement.
11. Amenable to multiplexing and forming a sensor network.
12. Capable of automated manufacture for low cost and consistency.

There are a number of fiber optic sensors that comply with criteria (2), (3), (5), (6), (8), and (9) and might be feasible for structural measurements. These include the polarimetric, two-mode, Fabry–Perot and Bragg grating sensors (Kersey, 1997b). A further "down selection" is made based on how well the sensors comply with the remaining criteria. This can be gauged by reference to Table 6.1.

Localization of the sensing domain for an optical fiber ensures that there is no sensitivity to the measurand of interest anywhere along the optical fiber, except in the designated region. This avoids *lead sensitivity*, a

Table 6.1 Comparison of Potential Fiber Optic Structural Sensors

Criteria	Fabry–Perot	Bragg Grating	Two-Mode	Polarimeter
1. Localized	y	y	(a)	(b)
7. Sensitivity	y	y	(c)	(c)
11. Multiplexing	(d)	y	(d)	(d)
12. Auto-Manufact.	(e)	y	(e)	(e)

Compliance is indicated by "y" for yes.
[a] The sensing length is two-mode while the lead optical fibers are single-mode.
[b] Sensing length starts where polarization eigenaxes rotated by 45° and ends at mirror.
[c] Sensitivity is much less than Fabry–Perot or grating, so spatial resolution quite poor.
[d] Multiplexing is possible using costly time-of-flight measurements.
[e] Require bonding or fusion splicing, assembling, and handling.

problem that plagued early fiber optic strain sensors based on Mach–Zehnder and Michelson interferometers and to this day is a major weakness of intensiometric sensors. The importance of this "localization" concept was recognized in the case of strain measurements early in the development of this technology (Valis *et al.*, 1989a). An appreciation for this lead-fiber sensitivity problem can be acquired by considering a *Michelson fiber optic interferometer*. This is a simple and low-cost sensor as shown in Fig. 6.10. The sensing arm of this interferometer is essentially the segment of optical fiber that runs from the 3 dB coupler to its mirrored tip. Any change in the *optical length* (defined as the product of core index of refraction and physical length of the optical fiber) of this sensing arm relative to the optical length of the reference arm of the interferometer gives rise to interference fringes in the photodetector. The problem is that changes in strain or temperature anywhere along this arm of the interferometer will be misinterpreted as changes in the region of interest, the embedded portion of the optical fiber, in this illustration. Furthermore, random modulation in the phase of the lightwave crossing any connector used in this length of optical fiber will also provide misinformation.

It is apparent from the foregoing discussion that both the Fabry–Perot and the fiber Bragg grating sensors seem to best meet the criteria for structural sensing, and it is not surprising to find that short-gauge-length structural sensing systems based on them are now commercially available (see Chapter 9). In the long term, the FBG sensors will probably dominate because of their greater versatility, multiplexing capability, lower cost, and consistent manufacture (once they are made during the fabrication of the optical fiber on the draw tower).

CHAPTER 7

Fiber Optic Strain and Temperature Sensitivity

7.1 INTRODUCTION

In this chapter we shall assume that the primary measurement of interest is strain. Since any strain measurement is inevitably corrupted by changes of temperature, it is necessary in most practical situations to simultaneously measure the temperature of the structure being monitored. By this means *thermal expansion effects* and the influence of temperature on the index of refraction of the optical fiber can be taken into account and the pure mechanical strain determined. *Temperature compensation* therefore constitutes an important part of this chapter. Another assumption, often implicitly made, in the discussion of strain measurements with optical fiber sensors is that the strain field is purely axial, that is to say, the strain is applied only in the direction of the optical fiber. We shall explore the validity of this assumption and consider the influence of *transverse strain* on any measurement. Such concerns are, of course, not unique to fiber optic sensing and have been confronted by the conventional resistive foil gauge community for many years (Sirkis, 1993c).

In reality it is the strain in the host structure that is relevant, but this may not correspond to the observed strain in the optical fiber. The chain of inferences permitting the host strain to be derived from the electrical signal on the computer is presented as Fig. 7.1. In this chapter, however, we shall focus on the transduction mechanisms that modify the lightwave propagating along the optical fiber in such a manner that the local values of strain and temperature can be evaluated from the observed changes in the lightwave. During calibration of the sensing system, a known strain and temperature is imposed on the host structure. This strain field is transferred to the optical fiber via some form of bondline. The resulting change in the fiber optic sensor measurement parameter (be it phase or wavelength) is converted from an optical signal in the optical fiber to an electrical signal by the *demodulation unit*. The electrical signal is then

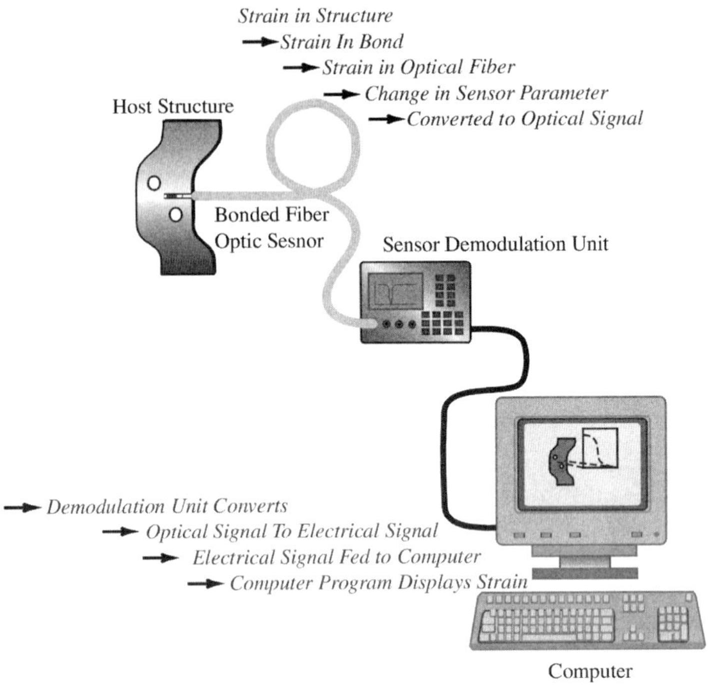

FIGURE 7.1. Schematic illustration of the chain of inferences involved in fiber optic strain sensing.

interpreted as a strain measurement by the computer. The calibration experiment permits the system parameters to be characterized so that when an unknown strain is imposed on the host structure, the electrical signal can be used to determine the strain in the host structure (Valis and Measures, 1992).

The earliest demonstration of a fiber optic strain gauge was undertaken by Butter and Hocker, (1978). It is interesting to note that this happened to be the same year the fiber optic Bragg grating was invented. Ironically, it took nearly a decade before a connection was made between these two landmark papers. Although many of the subsequent fiber optic sensor designs were shown to be capable of responding to transient strains, for example, Claus and Cantrell (1980), it was not until "localization" of the sensor was recognized as important (Valis *et al.*, 1989), that meaningful fiber optic strain measurements were undertaken. The invention of the Fabry–Perot interferometric sensor (Lee and Taylor, 1988) and the fiber Bragg grating sensor (Morey *et al.*, 1989) subsequently paved the way to true fiber optic strain gauges that could be made smaller than resistive foil strain gauges; see Fig. 1.6.

7.2 OPTOTHERMOMECHANICAL EQUATIONS

The fundamental parameter that lies at the heart of most fiber optic structural sensors is the optical path length of the sensing section. Under uniform conditions, the *sensor optical path length* (SOPL),

$$\zeta_{\mathrm{L}} = nL, \tag{7.1}$$

is just the product of the *core index of refraction*, n, and the gauge length of the optical fiber, L. For the fiber Bragg grating, it is the period of index modulation that constitutes the appropriate length parameter for (7.1), as opposed to the length of the grating (i.e., $\zeta_{\mathrm{L}}(\mathrm{FBG}) = n\Lambda$). In general, the SOPL is a function of the applied stress and temperature (σ, T), and can be expressed in the form

$$\zeta_{\mathrm{L}} = \zeta_{\mathrm{L}}(\sigma, T). \tag{7.2}$$

The incremental change in the SOPL with respect to some reference state of stress and temperature, (σ_0, T_0)

$$\Delta\zeta_{\mathrm{L}} = \zeta_{\mathrm{L}}(\sigma, T) - \zeta_{\mathrm{L}}(\sigma_0, T_0). \tag{7.3}$$

A Taylor expansion of Eq. (7.2), retaining only the linear terms, yields

$$\Delta\zeta_{\mathrm{L}} = \left[\frac{\partial\zeta_{\mathrm{L}}}{\partial\sigma}\right]_T \Delta\sigma + \left[\frac{\partial\zeta_{\mathrm{L}}}{\partial T}\right]_\sigma \Delta T, \tag{7.4}$$

where $\Delta\sigma$ and ΔT are the respective incremental changes in the local stress and temperature from the reference state, and $\left[\partial\zeta_{\mathrm{L}}/\partial\sigma\right]_T$ and $\left[\partial\zeta_{\mathrm{L}}/\partial T\right]_\sigma$ are the derivatives of ζ_{L} with respect to σ and T, respectively, at the reference values of (σ_0, T_0). Thus, $\left[\zeta_{\mathrm{L}}/\partial T\right]_\sigma$ is the derivatives of ζ_{L} with respect to T, at $\sigma = \sigma_0$, and $T = T_0$, assuming σ is constant.

In light of the relation for the SOPL given by Eq. (7.1), the variation in ζ_L due to incremental changes with respect to σ and T can be further expanded (Measures, 1992a):

$$\Delta\zeta_{\mathrm{L}} = \left\{n\left[\frac{\partial L}{\partial\sigma}\right]_T + L\left[\frac{\partial n}{\partial\sigma}\right]_T\right\}\Delta\sigma + \left\{n\left[\frac{\partial L}{\partial T}\right]_\sigma + L\left[\frac{\partial n}{\partial T}\right]_\sigma\right\}\Delta T. \tag{7.5}$$

This can be rewritten in a slightly different form

$$\Delta\zeta_L = nL\left[\left\{\left[\frac{\partial\varepsilon}{\partial\sigma}\right]_T + \frac{1}{n}\left[\frac{\partial n}{\partial\varepsilon}\right]_T\left[\frac{\partial\varepsilon}{\partial\sigma}\right]_T\right\}\Delta\sigma + \left\{\left[\frac{\partial\varepsilon}{\partial T}\right]_\sigma + \frac{1}{n}\left[\frac{\partial n}{\partial T}\right]_\sigma\right\}\Delta T\right]. \quad (7.6)$$

If we introduce *Young's modulus*, Y_F, and the *coefficient of thermal expansion*, α_F, for the optical fiber, then (7.6) takes the form

$$\Delta\zeta_L = nL\left[\left\{1 + \frac{1}{n}\left[\frac{\partial n}{\partial\varepsilon}\right]_T\right\}\frac{\Delta\sigma}{Y_F} + \left\{\alpha_F + \frac{1}{n}\left[\frac{\partial n}{\partial T}\right]_\sigma\right\}\Delta T\right], \quad (7.7)$$

where the second term in the first curly bracket constitutes the *strain-optic effect* and the second term in the second curly bracket constitutes the *thermo-optic effect*.

In the case of a fiber Bragg grating sensor, Eq. (7.7) takes the same form, except the length of the FBG L on the right side of Eq. (7.7) is replaced with Λ, the period of index modulation for the grating.

The relevant SOPL for a fiber optic *polarimetric sensor*, takes the form

$$\zeta_L \equiv (n_f - n_s)L, \quad (7.8)$$

where n_f and n_s, respectively, represent the core indices of refraction for the lightwave **E**-field components aligned with the fast and slow axes of the optical fiber. In this case the appropriate form of (7.7) is

$$\begin{aligned}\Delta\zeta_L = L\Bigg[&\left\{(n_f - n_s) + \left[\frac{\partial n_f}{\partial\varepsilon}\right]_T - \left[\frac{\partial n_s}{\partial\varepsilon}\right]_T\right\}\varepsilon_1 \\ &+ \left\{\alpha_F(n_f - n_s) + \left[\frac{\partial n_f}{\partial T}\right]_\sigma - \left[\frac{\partial n_s}{\partial T}\right]_\sigma\right\}\Delta T\Bigg]. \quad (7.9)\end{aligned}$$

In the case of the *elliptic-core two-mode* fiber optic sensor, the relevant SOPL,

$$\zeta_L = (n_i^{01} - n_i^{11})L, \quad (7.10)$$

where n_i^{01} and n_i^{11} respectively, represents the *effective core indices of refraction* for the LP_{01} and LP_{11}^{even} *transverse modes* of the optical fiber that are linearly polarized in the *i*-direction, in which case the appropriate form of (7.7) is

$$\begin{aligned}\Delta\zeta_L = L\Bigg[&\left\{n_i^{01} - n_i^{11}) + \left[\frac{\partial n_i^{01}}{\partial\varepsilon}\right]_T - \left[\frac{\partial n_i^{11}}{\partial\varepsilon}\right]_T\right\}\varepsilon_1 \\ &+ \left\{\alpha_{F(n_i^{01} - n_i^{11})} + \left[\frac{\partial n_i^{01}}{\partial T}\right]_\sigma - \left[\frac{\partial n_i^{11}}{\partial T}\right]_\sigma\right\}\Delta T\Bigg]. \quad (7.11)\end{aligned}$$

7.3 STRAIN AND TEMPERATURE SENSITIVITY AND GAUGE FACTORS

If we introduce the *strain* and *temperature sensitivities*, S_ε and S_T, respectively, defined by the relations

$$S_\varepsilon \equiv \left\{1 + \frac{1}{n}\left[\frac{\partial n}{\partial \varepsilon}\right]_T\right\} \tag{7.12}$$

and

$$S_T \equiv \left\{\alpha_F + \frac{1}{n}\left[\frac{\partial n}{\partial T}\right]_\sigma\right\}, \tag{7.13}$$

then Eq. (7.7) can be written

$$\frac{\Delta\zeta_L}{\zeta_L} = S_\varepsilon \Delta\varepsilon + S_T \Delta T, \tag{7.14}$$

where $\Delta\varepsilon$ and ΔT represent the change in strain and temperature from the reference values, i.e.,

$$\Delta\varepsilon = \varepsilon - \varepsilon_0 \tag{7.15}$$

$$\Delta T = T - T_0. \tag{7.16}$$

Representative values for the fiber optic strain and temperature sensitivities at a wavelength of 1550 nm are $S_\varepsilon = 0.8 \times 10^{-6}\ \mu\varepsilon^{-1}$; and $S_T = 6.0 \times 10^{-6}\ {}^\circ C^{-1}$, respectively.

A useful relation at this point is

$$\left[\frac{\partial n}{\partial \varepsilon}\right]_T \Delta\varepsilon = \Delta n, \tag{7.17}$$

where Δn is the incremental change in the fiber core index of refraction resulting from the imposition of the mechanical strain, $\Delta\varepsilon$. According to the *strain-optic theory*, the incremental change in the index of refraction for light linearly polarized in the i-direction, Δn_i, in the presence of a strain field ε_j with $j = 1$ to 6 ($j = 1$ to 3 represents the three *principal strains* and $j = 4$ to 6, the three *shear strains*), is described by the relation (Nye, 1954)

$$\Delta n_i = -\frac{n_i^3}{2} p_{ij}\varepsilon_j, \tag{7.18}$$

where p_{ij} are the coefficients of the *strain-optic* (or photoelastic) *tensor*. In Eq. (7.18) we are using the *summed index convention*, that is to say, $p_{ij}\varepsilon_j$ means the sum of all such terms over the repeated j-index. In this notation the index value of 1 will be taken to represent the optical fiber direction,

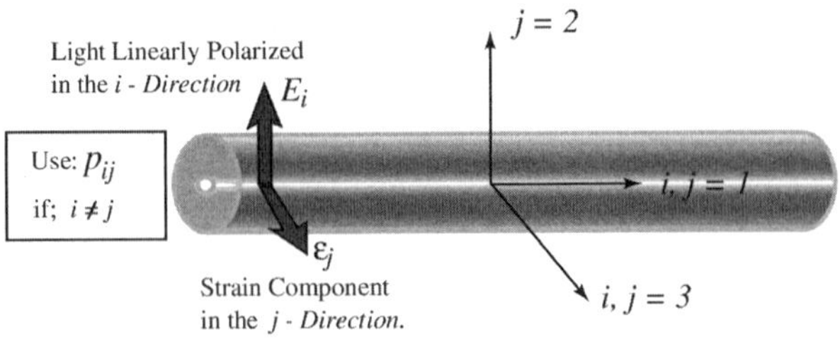

FIGURE 7.2. Set of orthogonal directions used in connection with fiber optic strain sensing.

and 2 and 3 will represent the two orthogonal directions in the plane that is transverse to the fiber direction; see Fig. 7.2. Clearly, the p_{ij} strain-optic coefficient ($i \neq j$) appertains to the change in the index of refraction experienced by the **E**-field of the lightwave arising from the presence of a strain in the material in a direction orthogonal to the **E**-field, while the p_{ii} strain-optic coefficient describes the situation where the **E**-field and the strain direction are colinear.

Using (7.12) with (7.14) and (7.18), we arrive at the fundamental *optothermomechanical equation* for fiber optic interferometric, or Bragg grating, sensors used to determine the influence of strain and temperature changes on the property of the lightwave in the core of an optical fiber

$$\left[\frac{\Delta\zeta_{\mathrm{L}}}{\zeta_{\mathrm{L}}}\right]_i = \varepsilon_1 - \varepsilon_{10} - \frac{n_i^2}{2} p_{ij}\varepsilon_j + S_T \Delta T. \tag{7.19}$$

Equation (7.19) describes the normalized incremental change in the SOPL for light linearly polarized in the *i*-direction. In many situations the reference strain in the direction of the fiber, ε_{10}, can be taken to be zero, and there is no change in temperature, so $\Delta T = 0$, in which case (7.19) takes on the form that is appropriate for *isothermal mechanical engineering*:

$$\left[\frac{\Delta\zeta_L}{\zeta_{\mathrm{L}}}\right]_i = \varepsilon_1 - \frac{n_i^2}{2} p_{ij}\varepsilon_j. \tag{7.20}$$

In the case of an isotropic, homogeneous, medium the strain-optic tensor comprises only two independent components (the strain-optic coefficients p_{11} and p_{12}) and takes the form

$$p_{ij} = \begin{bmatrix} p_{11} & p_{12} & p_{12} & 0 & 0 & 0 \\ p_{12} & p_{11} & p_{12} & 0 & 0 & 0 \\ p_{12} & p_{12} & p_{11} & 0 & 0 & 0 \\ 0 & 0 & 0 & p_{44} & 0 & 0 \\ 0 & 0 & 0 & 0 & p_{44} & 0 \\ 0 & 0 & 0 & 0 & 0 & p_{44} \end{bmatrix}, \tag{7.21}$$

where

$$p_{44} \equiv (p_{11} - p_{12})/2. \tag{7.22}$$

If the strain field is assumed to be *purely axial*, that is to say, solely in the direction of the optical fiber, then we may write

$$\varepsilon_i = \begin{bmatrix} \varepsilon_1 \\ -\nu\varepsilon_1 \\ -\nu\varepsilon_1 \\ 0 \\ 0 \\ 0 \end{bmatrix}, \tag{7.23}$$

where ν is the *Poisson ratio* for the optical fiber. This is essentially the simplification made by Butter and Hocker (1978) in their early paper on the subject. They further assumed that the strain in the optical fiber matches the component of strain in the host that is in the direction of the optical fiber, and that zero host strain in the 2- and 3-transverse directions is coupled to the optical fiber. The assumptions made by Butter and Hocker have continued to be made by people using fiber optic sensors for strain measurements. In practice, however, the assumptions are only valid for surface-mounted sensors or for embedded sensors subject to axis-symmetric, generalized plane strain loading (Sirkis, 1993b). The subject of strain coupling between the sensor and the host structure will be discussed in more detail in Chapter 8.

With the assumptions of Butter and Hocker, substitution of (7.21) and (7.23) into (7.20) yields the basic fiber optic strain gauge relation for the case of axial strain in an isotropic, homogeneous optical fiber with no temperature change,

$$\frac{\Delta\zeta_L}{\zeta_L} = G_\varepsilon \varepsilon_1, \tag{7.24}$$

where G_ε is introduced as the fiber optic axial strain gauge factor, or *strain-gauge factor* for short, and is given by

$$G_\varepsilon \equiv 1 - \frac{n_0^2}{2} P_\varepsilon, \tag{7.25}$$

where P_ε is termed the *effective strain-optic* (or photoelastic) coefficient

$$P_\varepsilon \equiv p_{12} - \nu(p_{11} + p_{12}). \tag{7.26}$$

Note that in the case of axial strain, i.e., strain only in the direction of light propagation, one might be tempted to believe there should not be a term involving p_{11}, since the **E**-field in the lightwave is orthogonal to the applied strain direction. However, a little thought reveals that the Poisson effect is responsible for such a term. It is interesting to note that the form of Eq. (7.24) is very similar to that used for conventional resistive foil strain gauges, i.e.,

$$\frac{\Delta R}{R} = G_\varepsilon^R \varepsilon_1, \tag{7.27}$$

where G_ε^R represents *the electrical strain-gauge factor* and has a value of 2, compared to the fiber optic strain-gauge factor value of about 0.8 (see Fig. 7.3).

In the case of a fiber optic *polarimetric strain sensor* subject only to axial strain, Eq. (7.9) can be used with arguments similar to those just presented to yield an equation identical in form to (7.24), except that

$$G_\varepsilon \equiv 1 - \frac{\eta^2(n_f, n_s)}{2} P_\varepsilon. \tag{7.28}$$

In a similar manner we use (7.11), for the elliptic-core two-mode fiber optic strain sensor, to produce again an equation identical to (7.24) with

$$G_\varepsilon \equiv 1 - \frac{\eta^2(n_i^{01}, n_i^{11})}{2} P_\varepsilon. \tag{7.29}$$

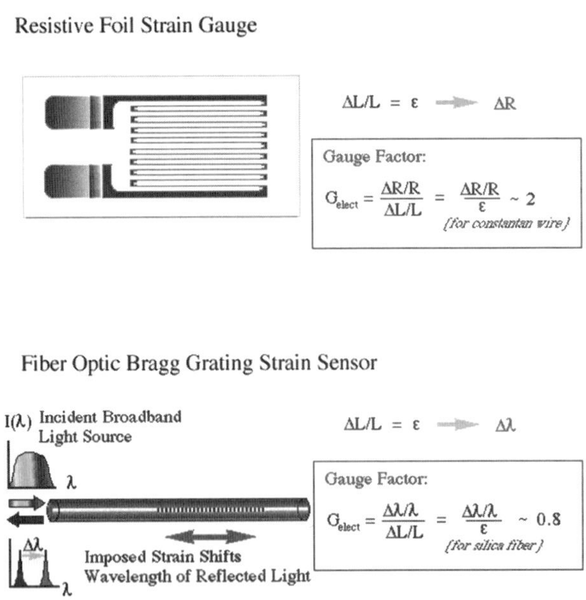

FIGURE 7.3. Comparison between the resistive foil strain gauge factor and its optical analogue, the fiber optic Bragg grating strain gauge factor.

In both (7.28) and (7.29), Measures (1995) introduced the *effective squared index*,

$$\eta^2(n_a, n_b) \equiv n_a^2 + n_a n_b + n_b^2, \tag{7.30}$$

where n_a and n_b represent n_f and n_s, respectively, in the case of the polarimetric sensor and n_i^{01} and n_i^{11}, respectively for the elliptic-core two-mode fiber optic strain sensor. In both cases P_ε is defined by (7.26). Sirkis *et al.* (1994) developed a *phase-strain model* for polarimetric sensors that provides a nonzero phase during uniaxial tension and includes contributions from transverse strain components. This analysis is applicable to a number of different types of optical fiber.

It is noteworthy that Eq. (7.24) is perfectly general for all SOPL-based sensors, provided the appropriate strain-gauge factor G_ε is employed. In fact, all sensors based on changes to their optical path length divide into those in which the normalized phase change is directly proportional to the normalized change in the SOPL, or in the case of fiber Bragg gratings, the normalized change of wavelength is proportional to the normalized change in the SOPL. We can thus write

$$\frac{\Delta\phi}{\phi} = \frac{\Delta\zeta_L}{\zeta_L} = \frac{\Delta\lambda_B}{\lambda_B}. \tag{7.31}$$

It is of interest to consider the *scale length* dependence of the incremental change in phase, $\Delta\phi$, for phase measurement based sensors, or the incremental change in wavelength, $\Delta\lambda_B$, for FBG sensors. In the former case, the incremental change in phase, $\Delta\phi$, is proportional to the gauge-length of the sensor through the *normalization factor* ϕ, where $\phi = nkL$ for interferometric sensors, $\phi = (n_f - n_s)kL$ for polarimetric sensors, and $\phi = (n_i^{01} - n_i^{11})kL$ for elliptic-core two-mode fiber optic sensors. In the case of FBG sensors, the *normalization factor*, $\lambda_B = 2n\Lambda$, depends on the grating period and is independent of the grating length.

For the more general case of the optical fiber subject to a strain field defined by three principal strains $\{\varepsilon_1, \varepsilon_2, \varepsilon_3\}$, we shall assume that the fiber axis is aligned with ε_1. For light with **E**-field components in the 2- and 3-directions, respectively (see Fig. 7.4), we can write the normalized incremental changes in the SOPL for each of these lightwave components,

$$\left[\frac{\Delta\zeta_L}{\zeta_L}\right]_2 = \varepsilon_1 - \frac{n_2^2}{2}\{\varepsilon_2 p_{11} + p_{12}(\varepsilon_1 + \varepsilon_3)\} \tag{7.32}$$

and

$$\left[\frac{\Delta\zeta_L}{\zeta_L}\right]_3 = \varepsilon_1 - \frac{n_3^2}{2}\{\varepsilon_3 p_{11} + p_{12}(\varepsilon_1 + \varepsilon_2)\}, \tag{7.33}$$

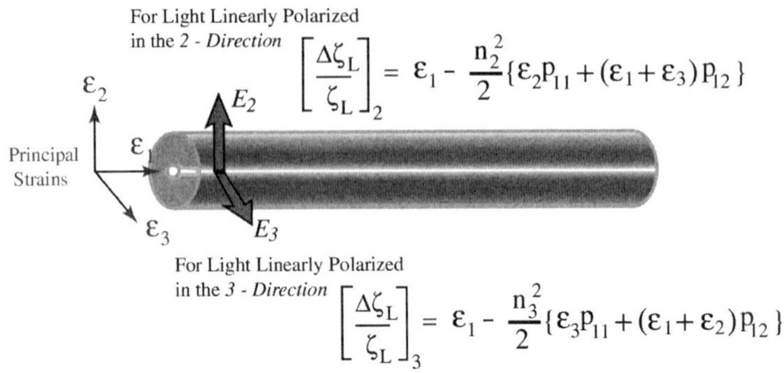

FIGURE 7.4. Schematic of the normalized change in the sensor optical path length (SOPL) as a function of the strain, the strain optic coefficients, and the relevant index of refraction for a fiber optic sensor.

where we have assumed that the optical fiber is *birefringent*, so that $n_2 \neq n_3$. If this were not so, then we would use just one value for the index of refraction. If we assume it to be n_0, and further assume that the two transverse strain components are equal, then the two equations would take on the same form,

$$\left[\frac{\Delta\zeta_L}{\zeta_L}\right]_{i=2\text{ or }3} = \varepsilon_1 - \frac{n_0^2}{2}\{\varepsilon_t p_{11} + p_{12}(\varepsilon_1 + \varepsilon_t)\}, \tag{7.34}$$

where ε_t is the *transverse component of strain*. If the photodetection system and the optical components, such as couplers, are insensitive to the state of polarization of the light, it is realistic to take an average of (7.32) and (7.33) so we can write

$$\frac{\Delta\zeta_L}{\zeta_L} = \varepsilon_1 - \frac{n_0^2}{2}\left\{\varepsilon_1 p_{12} + \frac{1}{2}(p_{11} + p_{12})(\varepsilon_2 + \varepsilon_3)\right\}. \tag{7.35}$$

It should be noted that resistive foil strain gauges are also sensitive to transverse strain and a correction term is required in (7.27) to account for this effect (Sirkis, 1993c).

The foregoing treatment has relied on an implicit assumption that the sensor length was small compared to any spatial variation in strain or temperature. That is to say, we have restricted our interest to effectively short gauge-length sensors. The more general form of Eq. (7.20) that is free of such assumptions can be expressed in the form of an integral,

$$\left[\frac{\Delta\zeta_L}{\zeta_L}\right]_i = \frac{1}{L}\int_{z=0}^{z=L}\left\{\varepsilon_1 - \frac{n_i^2}{2}p_{ij}\varepsilon_j\right\}dz, \tag{7.36}$$

where z defines the position along the fiber. Equation (7.36) has the same form as that proposed by Sirkis and Haslach (1990) for fiber optic interferometric sensors. The difference is that their equation uses the normalized change in phase, $\Delta\phi/\phi$, on the left side of (7.36).

The equations we have developed to this point are independent of the type of sensor just as long as the underlying measurement parameter is the SOPL. This is certainly true for all forms of fiber optic interferometric and fiber Bragg grating sensors, since as indicated in Eq. (7.31),

$$\frac{\Delta\phi}{\phi} = \frac{\Delta\zeta_L}{\zeta_L} = \frac{\Delta\lambda_B}{\lambda_B}.$$

Thus, for short gauge-length interferometric sensors, we can write the general form of the equation,

$$\frac{\Delta\phi}{\phi} = \varepsilon_1 - \frac{n_i^2}{2}p_{ij}\varepsilon_j, \tag{7.37}$$

and for fiber Bragg grating sensors,

$$\frac{\Delta\lambda_B}{\lambda_B} = \varepsilon_1 - \frac{n_i^2}{2}p_{ij}\varepsilon_j, \tag{7.38}$$

where λ_B is the Bragg wavelength (see Chapter 3, Eq. (3.129)]. Indeed, all of the above relevant [Eqs. (7.14), (7.20), (7.24), and (7.31) through (7.35)], can be used for either kind of sensor by substituting the appropriate ratio on the left side, i.e., $\Delta\phi/\phi$, or $\Delta\lambda_B/\lambda_B$. As mentioned earlier, it should be remembered that in the case of the FBG sensor the SOPL does not involve the length of the grating, L, but rather the modulation period for the core index of refraction, Λ.

In summary, the *basic structural sensing equation* (7.14), for fiber optic interferometric sensors, takes the form

$$\frac{\Delta\phi}{\phi} = S_\varepsilon\Delta\varepsilon + S_T\Delta T, \tag{7.39}$$

while for *fiber Bragg grating sensors* it takes the similar form

$$\frac{\Delta\lambda_B}{\lambda_B} = S_\varepsilon\Delta\varepsilon + S_T\Delta T. \tag{7.40}$$

In the case of *axial strain*, this pair of equations simplify to the form

$$\frac{\Delta\phi}{\phi} = G_\varepsilon \Delta\varepsilon + S_T \Delta T, \tag{7.41}$$

$$\frac{\Delta\lambda_B}{\lambda_B} = G_\varepsilon \Delta\varepsilon + S_T \Delta T, \tag{7.42}$$

where G_ε is the strain-gauge factor as defined in (7.25), and $\Delta\varepsilon$ and ΔT are the respective changes in strain and temperature from the reference values and are defined in Eq. (7.15) and (7.16). S_T is the *temperature sensitivity* and was defined by (7.13). This commonality in the equations describing different fiber optic interferometers and fiber Bragg gratings was well understood by Sirkis in his two papers published in 1993 [Sirkis, 1993].

Bertholds and Dandliker (1988) measured the properties of optical fibers with pure silica core and B_2O_3 doped cladding and found that $n_0 = 1.458$, $p_{11} = 0.113$, and $p_{12} = 0.252$, at a wavelength of 632.8 nm. These values of the strain-optic coefficients are about 11% and 3% respectively, lower than the values for bulk silica [Borrelli and Miller, 1968]. Although these coefficients were measured at a wavelength somewhat different from the bulk of the work reported in this book, the very weak dependence of these coefficients on wavelength have led most people to use these values at other wavelengths.

In most sensing work the strain-gauge factor G_ε is directly measured in calibration experiments and has been found to lie between 0.7 and 0.8 (Morey *et al.*, 1994; Alavie *et al.*, 1995, Russel and Archambault, 1996). If we take the representative value of $G_\varepsilon = 0.78$, the *incremental phase change* for an interferometric fiber optic sensor operating at $\lambda = 1550$ nm with a gauge-length L can be written

$$\Delta\phi = \phi G_\varepsilon \varepsilon = nkL\, G_\varepsilon \varepsilon, \tag{7.43}$$

Note that although $G_\varepsilon = 0.78$ is correct when we work with strain, many authors choose to set $G_\varepsilon = 0.78 \times 10^{-6}\ \mu\varepsilon^{-1}$ when they work in units of *microstrain*, $\mu\varepsilon$.

Using the preceding values,

$$\Delta\phi = \{1.458 \times 2\pi \times 0.78/1550 \times 10^{-9}\} L\varepsilon = 4.61[\text{rad}\ \mu\varepsilon^{-1}\ \text{m}^{-1}] L\varepsilon,$$

so the phase-strain sensitivity is

$$[S_\varepsilon]_\phi = 4.61[\text{rad}\ \mu\varepsilon^{-1}\ \text{m}^{-1}] = 2.64[\text{degree}\ \mu\varepsilon^{-1}\ \text{cm}^{-1}].$$

In the case of a Fabry–Perot interferometric sensor, L, is replaced with $2L$, because of the double pass through the device, and so the corresponding phase-strain sensitivity is

$$[S_\varepsilon]_\phi = 5.28[\text{degree } \mu\varepsilon^{-1} \text{ cm}^{-1}].$$

If the sensor gauge-length $L = 1\,\text{cm}$, then

$$\Delta\phi = 0.0461[\text{rad } \mu\varepsilon^{-1}]\varepsilon.$$

Thus, an imposed strain of 1000 $\mu\varepsilon$ would lead to a phase change $\Delta\phi = 46.1$ radians, or (dividing by 2π) roughly 7.3 fringes. Note that if $\lambda = 633\,\text{nm}$, then

$$\Delta\phi = 0.0461 \times 1.55/0.633 = 0.11[\text{rad } \mu\varepsilon^{-1}]\varepsilon,$$

a value close to the measurements of Farahi *et al.* (1990).

In the case of an FBG sensor, the change in the Bragg wavelength is

$$\Delta\lambda_B = \lambda_B G_\varepsilon \varepsilon, \tag{7.44}$$

so if the grating was operated at about 1550 nm, and again G_ε is set to 0.78, then

$$\Delta\lambda_B = 1550 \times 10^{-9} \times 0.78\varepsilon = 1.21[\text{pm } \mu\varepsilon^{-1}]\varepsilon,$$

Thus, an imposed strain of 1000 $\mu\varepsilon$ would lead to a 1.21 nm shift in the Bragg wavelength of an FBG sensor. As indicated earlier the phase measurements, (7.43), require the length of the sensor, while the length of a FBG does not enter into the calculation of the shift in wavelength, (7.44). Alavie *et al.* (1995) performed a definitive set of measurements on 24 FBGs in the 1532 to 1554 nm range and observed a strain-gauge factor, $G_\varepsilon \approx 0.80$, and a temperature sensitivity S_T, $\approx 6 \times 10^{-6}\ ^\circ\text{C}^{-1}$ for a temperature range of -20 to 80°C.

7.4 TRANSVERSE STRAINS AND THEIR MEASUREMENT

Since the strain field within a structure is described by a second-rank symmetric tensor and is characterized by six independent variables (Przemieniecki, 1985), the determination of strain requires six independent measurements. Measures *et al.* (1988) demonstrated the importance of evaluating several strain components with fiber optic sensors. For those host structures where there could be substantial transverse strain components, but it is only necessary to determine the axial strain component, either the EFPI or ILFE sensors can be used, as they have no transverse strain sensitivity (Sirkis, 1993b). By contrast, the polarimetric sensor is better suited to measure transverse strain than axial strain.

For those situations where a complex strain field needs to be measured, two approaches have emerged to tackle this challenging task. The first involves multifiber sensor rosettes (Measures *et al.*, 1988; Valis *et al.*, 1992; Carman *et al.*, 1992; Liu and Measures, 1993), while the second utilizes multi-parameter fiber optic sensors (Sirkis and Lo, 1994; Lawrence *et al.*, 1996; Udd *et al.*, 1998). Whereas the three fiber optic strain sensor rosettes of Valis *et al.* (1992) and Carman *et al.* (1992) were designed for the more limited, but useful, in-plane strain measurements, the two-parameter, three fiber optic sensor configuration proposed by Liu and Measures (1993) could be used to measure all six strain components. Most of the early fiber optic strain rosettes were based on Fabry–Perot interferometric sensors, as these devices have highly localized short gauge lengths, a necessity in a complex strain field. An example of an FFP strain rosette and its measured strain field is presented as Fig. 7.5, (Valis *et al.*, 1992). Also shown in this figure is the result from a conventional resistive foil strain rosette. Figure 1.6 shows a picture of these two very different types of strain rosette affixed to an aluminum cantilever beam.

One early multiparameter sensor used a combination of a polarimetric and a Michelson interferometric fiber optic sensor to simultaneously measure the axial and transverse strain components in a single optical fiber bonded to a vibrating cantilever beam (Sirkis and Lo, 1994). The experimental setup using two (2 × 2) couplers is displayed as Fig. 7.6. An example of the two intensity traces recorded from the pair of photodetectors is presented as Fig. 7.7a and the demodulated *strain time histories* for the two strain components are shown in Fig. 7.7b. *Hanning windows* in the Fourier frequency domain were used to separate the polarimetric and Michelson signals. For comparison purposes the strain history recorded by two suitable foil strain gauges is also shown in Fig. 7.7b. The combination of an intrinsic Fabry–Perot and in-line fiber etalon sensors have also been used to simultaneously measure both axial and transverse strain components within a structure (Singh and Sirkis, 1991).

Normally, fiber Bragg gratings are written into low-birefringent single-mode optical fiber, and as discussed in Chapter 3, there is only a single peak in the reflection spectrum; its center wavelength is given by Eq. (3.129). The application of a substantial *diametric load* has been shown, however, to broaden and at a load of about 40 N, bifurcate the spectrum, as evident in Fig. 7.8, (Wagreich, 1996). The intensity of the light is split fairly evenly between the two peaks, showing negligible loss even under high loads. When the Bragg wavelengths for each of the peaks are plotted

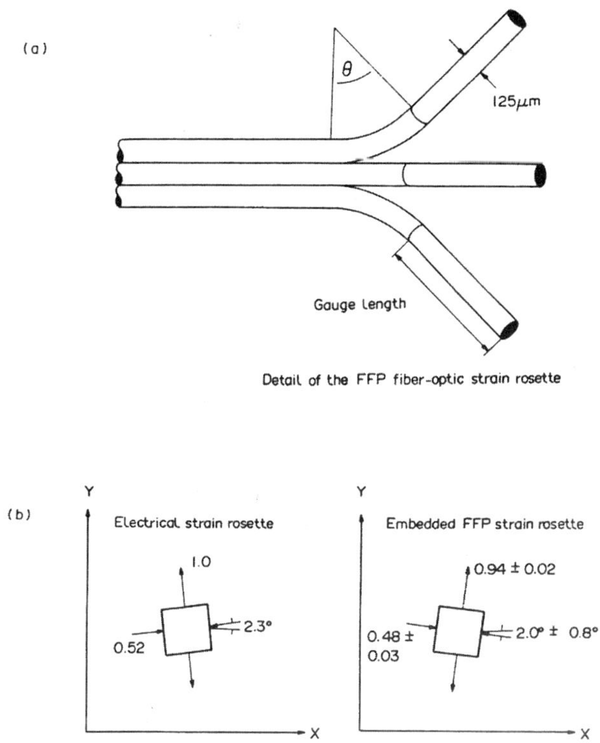

FIGURE 7.5. (a) A fiber optic strain rosette involving three fiber Fabry–Perot interferometric sensors. (b) Comparison of the principal strain components in a test FRP specimen as measuered by a resistive foil strain rosette and an embedded fiber Fabry–Perot strain rosette.

as a function of the applied load, linear curves are seen (Fig. 7.9). From the slopes of these lines we estimate that

$$\Delta\lambda_x/\Delta F = 0.00326 \text{ nm N}^{-1} \quad \text{and} \quad \Delta\lambda_y/\Delta F = -0.00051 \text{ nm N}^{-1}.$$

Nonlinearities in the variation of wavelength with applied compressive diametric load were observed by Bjerkan *et al.* (1997) for loads in excess of 0.5 N mm^{-1}. This was attributed, in this instance, to degradation of the coating material, which was absent in some of the other experiments. The presence of a *polyimide coating* led to the more severe deviation from linearity.

If a Bragg grating is formed within a highly birefringent optical fiber, such as 3M polarization-maintaining optical fiber (3M-PS-6621; Fig. 7.10), the reflection spectrum comprises two peaks at slightly different wave-

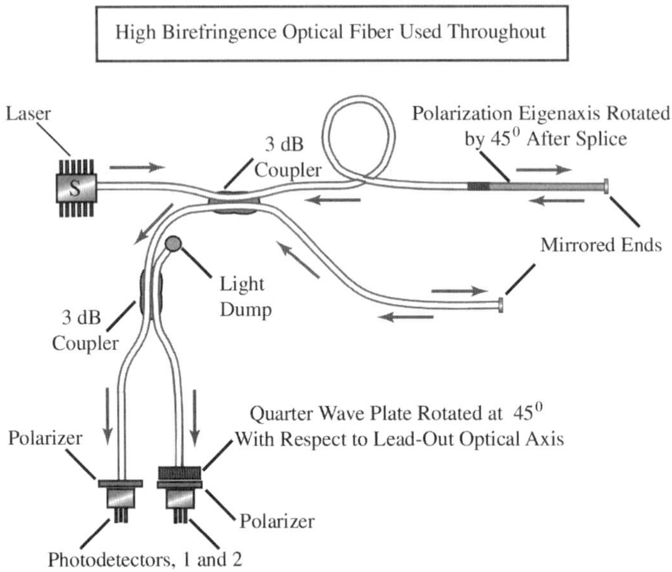

FIGURE 7.6. Schematic illustration of a combined fiber optic polarimetric and Michelson interferometric sensor.

lengths centered at around 1300 nm (Fig. 7.11). This can be understood in terms of the difference in the core index of refraction for slow and fast components, n_2 and n_3, respectively, of the lightwave, i.e., from Eq. (3.129), the Bragg wavelength for the unstrained, slow wave is

$$\lambda_{2,0} = 2n_2\Lambda, \tag{7.45}$$

while in the case of the unstrained, fast wave it is

$$\lambda_{3,0} = 2n_3\Lambda. \tag{7.46}$$

These were observed to be 1303.160 and 1302.705 nm, respectively (Fig. 7.11).

Combining Eqs. (7.31), (7.32), and (7.36) for the case of an optical fiber subject to a plane strain state (implying $\varepsilon_1 = 0$), we arrive at a pair of equations for the shifts in the Bragg wavelengths for the slow and fast waves, which can be expressed in the form of a matrix equation,

$$\begin{bmatrix} \Delta\lambda_2 \\ \Delta\lambda_3 \end{bmatrix} = \mathbf{K}\begin{bmatrix} \varepsilon_2 \\ \varepsilon_3 \end{bmatrix}, \tag{7.47}$$

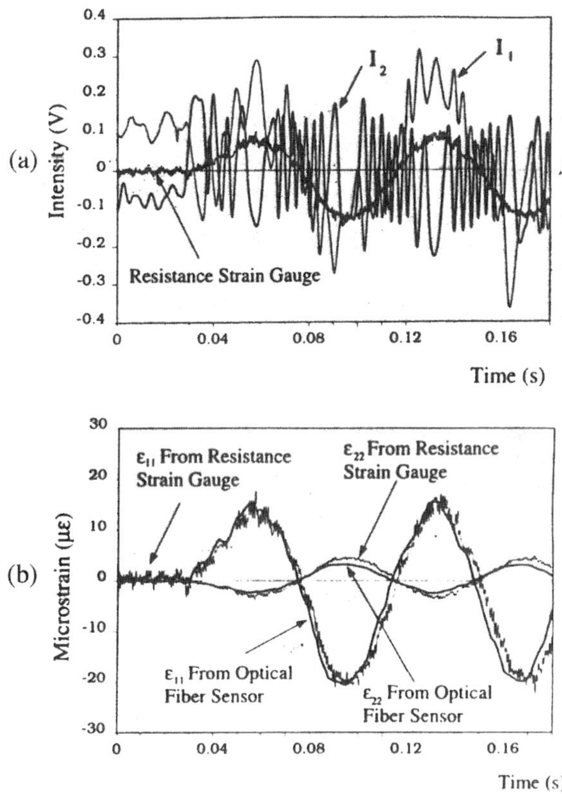

FIGURE 7.7. (a) Intensity traces from the two photodetectors used in the sensor configuration shown in Fig. 7.6, and the strain variation as measured by a resistive foil strain gauge. (b) Demodulated temporal strain variation for axial and transverse strain as measured by the polarimetric and Michelson sensors and two resistive foil strain gauges. From Sirkis J.S. and Lo Y.L., "Simultaneous Measurement of Two Strain Components Using 3×3 and 2×2 Coupler-Based Passive Demodulation of Optical Fiber Sensors," *J Lightwave Technology,* **12**(12), 2153–2161, © 1994, IEEE.

where the appropriate matrix is

$$\mathbf{K} = -\frac{1}{2}\begin{bmatrix} \lambda_{2,0}n_2^2p_{11} & \lambda_{2,0}n_2^2p_{12} \\ \lambda_{3,0}n_3^2p_{12} & \lambda_{3,0}n_3^2p_{11} \end{bmatrix} \tag{7.48}$$

Lawrence *et al.* (1997b) used the nominal value for the core index of refraction, $n_0 = 1.46$, and the Bertholds and Dandliker (1988), values of $p_{11} = 0.113$ and $p_{12} = 0.252$, for the strain-optic coefficients, to arrive at the

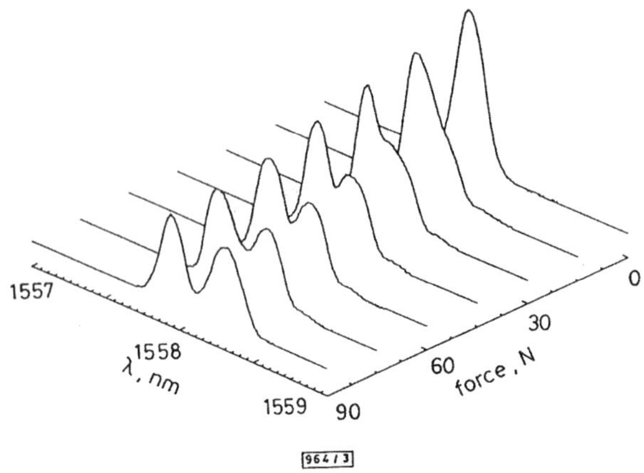

FIGURE 7.8. Experimentally observed broadening and eventual bifurcation of a fiber Bragg grating sensor subject to an increasing diametric load. From Wagreivh, R.B. "Effects of diametric load on fiber Bragg gratings fabricated in low birefringent fiber," *Electronics Letters*, **32**(13), 1223–1224, with permission from IEE Publishing.

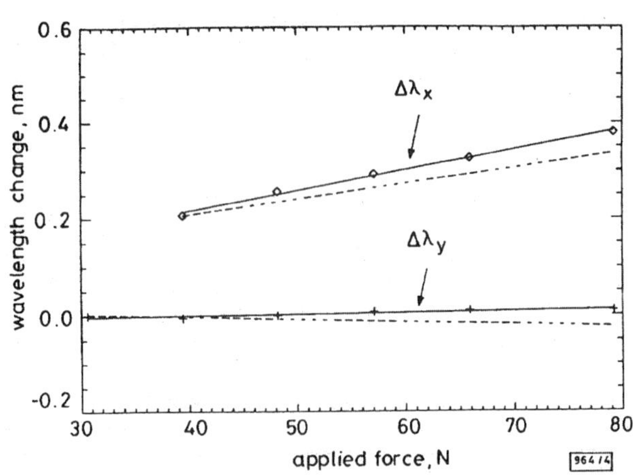

FIGURE 7.9. Variation of the center wavelengths with applied diametric load for a bifurcated fiber Bragg grating reflection spectrum. From Wagreivh, R.B. "Effects of diametric load on fiber Bragg gratings fabricated in low birefringent fiber," *Electronics Letters*, **32**(13), 1223–1224, with permission from IEE Publishing.

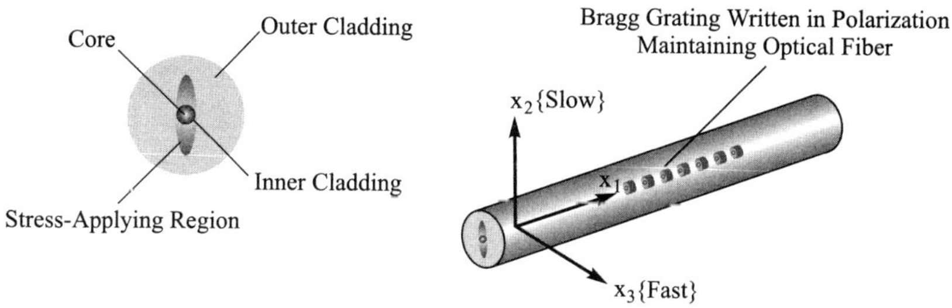

FIGURE 7.10. Schematic of a highly birefringent, polarization-maintaining optical fiber. Also shown is the cross-section of a commercial polarization-maintaining optical fiber.

components of the **K**-matrix and check if it is well-conditioned. They determined that

$$\mathbf{K} = -\begin{bmatrix} 156.89 & 349.88 \\ 350.00 & 156.95 \end{bmatrix}, \tag{7.49}$$

which is marginally acceptable with a condition number of 2.6.

An important observation of Lawrence *et al.* (1997) was that the response of the sensor (Fig. 7.12), strongly depends upon the orientation

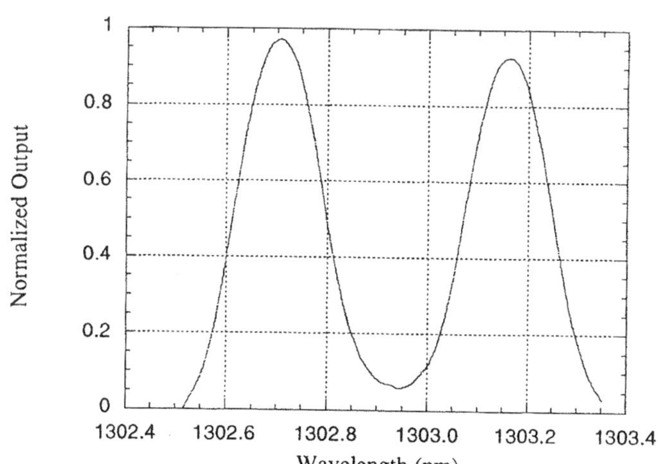

FIGURE 7.11. Bifurcated reflection spectrum of a Bragg grating formed in a 3M polarization maintaining optical fiber. From Lawrence, C.M., Nelson D.V., and Udd, E., (1997b), "Measurement of Transverse Stains with Fiber Bragg Gratings" *SPIE* **3042**, 218–228.

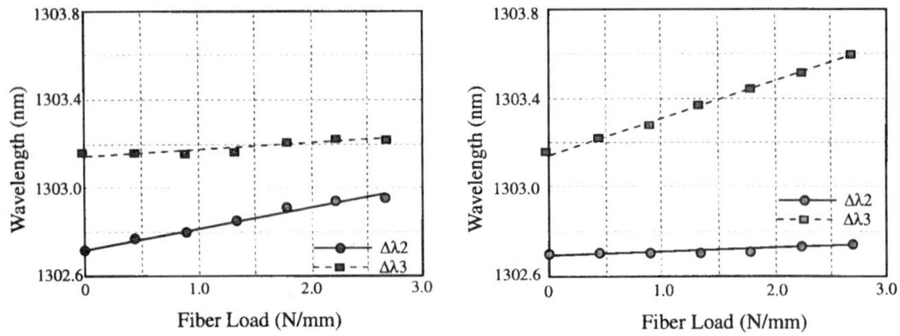

FIGURE 7.12. Variation in the center wavelength of the bifurcated FBG reflection spectrum with load for two different orientations of the polarization axes of the optical fiber relative to the direction of the applied transverse load (see inset to Fig. 7.13).

of the optical fiber relative to the direction of the applied transverse load (Fig. 7.13). This clearly indicates that it will be highly desirable to control the orientation of the optical fiber relative to the load direction. This may require a specially shaped optical fiber to prevent random orientation of the polarization axes as the fiber is embedded within its host material.

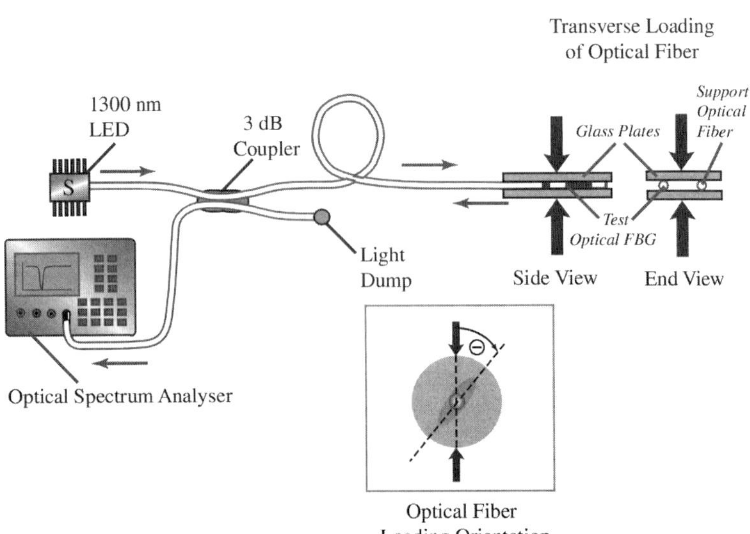

FIGURE 7.13. Experimental arrangement used to derive the results displayed in Fig. 7.12.

Calibration experiments revealed that the **K**-matrix derived above for the values of low birefringent optical fibers is not adequate to describe the results. A more accurate calibration factor was derived based on a finite-element model of the optical fiber. This work suggests that the two transverse components of strain can be measured using the shifts in the birefringent Bragg wavelengths and inverting Eq. (7.47).

The ability to write several Bragg gratings in the same section of optical fiber (Othonos *et al.*, 1994) opens the possibility of expanding the two transverse strain measurement of Lawrence *et al.* (1997a) to three strain and one temperature measurement. This has been demonstrated by Udd *et al.* (1998) using two collocated FBGs that are written with very different center wavelengths (1300 and 1550 nm) in polarization-maintaining optical fiber. This sensor provides four wavelengths with which to monitor the temperature and three strain components (Fig. 7.14). It should be noted that the polarization-preserving optical fiber need be used only in the vicinity of the FBGs since the sensing information on all four effective fiber gratings is wavelength encoded. Figure 7.15 shows a schematic of the spectral response of a single grating in polarization maintaining optical fiber to both axial and transverse strains. It is apparent that axial loading shifts the entire spectrum, while transverse loading increases the separation in the peaks associated with the two polarization axes of a single grating.

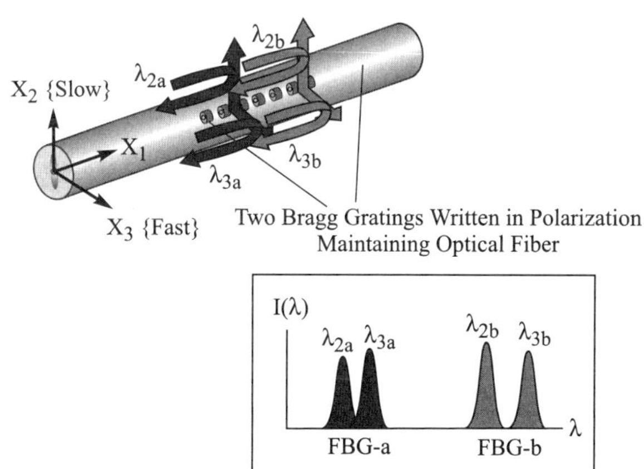

FIGURE 7.14. Schematic of two FBGs written in polarization-maintaining optical fiber, showing that each of the wavelength components are reflected from different points along the gratings and that the resulting reflection spectrum for this dual grating comprises four peaks.

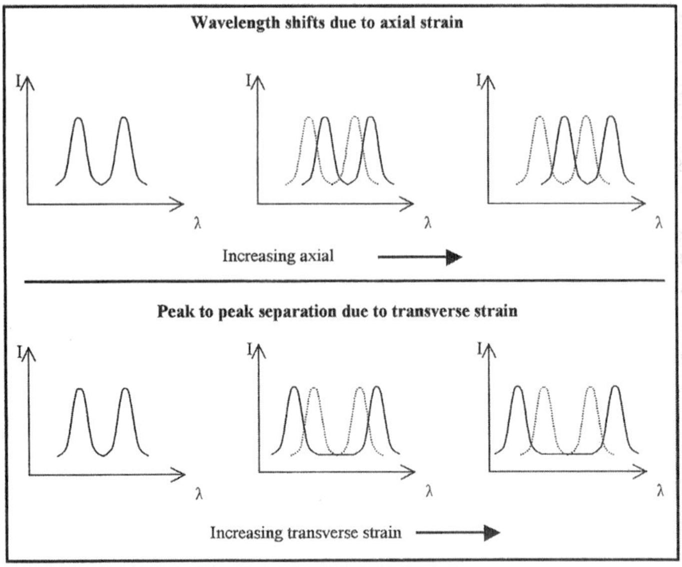

FIGURE 7.15. The wavelength shift in the dual peaked reflection spectrum for a grating written in polarization-maintaining optical fiber and subject to increasing axial strain is portrayed in the upper diagram, while the increasing separation of the peaks resulting from a greater transverse strain is indicated in the lower diagram. From Schulz, W.L., Udd, E., Seim, J.M., Martin Laylor H. and Mcgill, G.E., (1998), "Single and Multiaxis Fiber Grating Based Strain Sensors for Civil Structure Applications", *SPIE*, **3489**, 71–78.

A schematic illustration of the experimental arrangement used by Udd *et al.* (1998) is depicted in Fig. 7.16, and two traces from the optical spectrum analyzer, revealing the change in the reflection spectrum of the dual gratings with load, is presented in Fig. 7.17. The two peaks centered at 1550 nm are on the left, while the two peaks centered at 1300 nm are on the right. It is quite apparent from these spectra that transverse loading causes an overall redshift of the peaks and an increase in the wavelength separation of the peaks from each grating. The representative response of the sensors to increasing load at different temperatures is displayed as Fig. 7.18 (Udd *et al.*, 1998). These tests were performed with bare optical fibers, as the presence of coatings has been shown to produce a nonlinear response (Bjerkan *et al.*, 1997).

Although the experiments of Udd *et al.* (1998) are encouraging, a significant amount of research and development will be required to turn this into a practical sensing system that can routinely measure the temperature and three strain components within a host structure. For

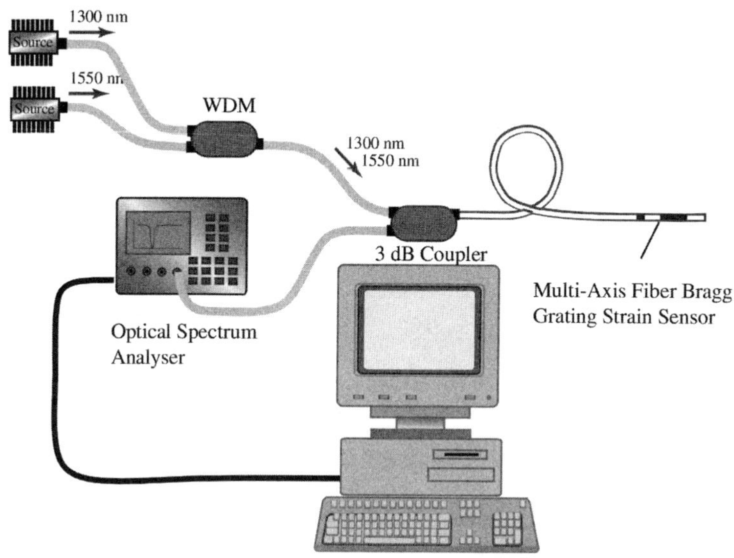

FIGURE 7.16. Experimental arrangement used to study the reflection spectrum from dual FBGs written in polarization-maintaining optical fiber. From Udd, E., Schulz, W., Seim, J., Mcgill, G. and Laylor, H.M. "Distributed Multiaxis Fiber Grating Strain Sensor Applications for Bridges". In: *Fiber Optic Sensors for Construction Materials and Bridges*, edited by F. Ansari, Proceedings of the Intl. Workshop on Fiber Optic Sensors for Construction Materials and Bridges, pp.168–180, May 3–6, 1998, used with permission from Technomic Publishing Company.

starters, the strain-optic coefficients and the effective core indices of refraction for highly birefringent (polarization-maintaining) optical fibers are unknown, so careful calibration experiments will be required. At high loads gradient effects can confuse the results and the issue of cross sensitivity has also to be considered.

7.5 THERMAL APPARENT STRAIN

It is apparent from Eq. (7.14) that any deviation in temperature from the reference value gives rise to a change in the SOPL that is indistinguishable from the change associated with the application of a purely mechanical force to the optical fiber. This thermal apparent strain problem is well known to the strain measuring community, and conventional strain gauges have to be temperature compensated if there is any likelihood of

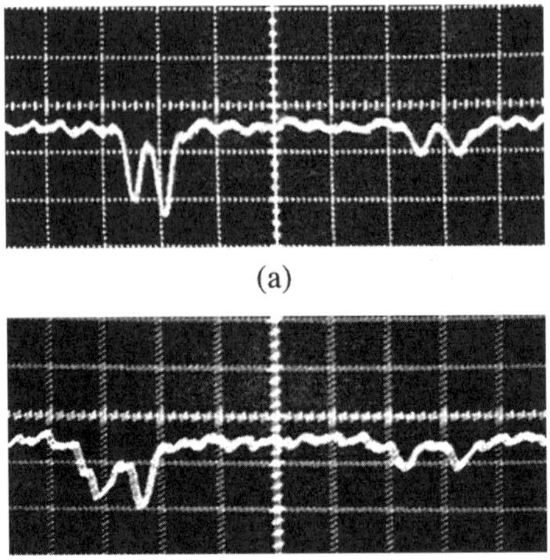

(a)

(b)

FIGURE 7.17. Two traces revealing the double bifurcated reflection spectrum of a dual FBG formed in polarization-maintaining optical fiber. The redshift of the spectrum and the increasing separation of the peaks with increasing transverse load can be seen as we compared (a) with (b). From Udd, E., Lawrence, C. and Nelson, D., (1997), "Development of a three axis strain and temperature fiber optic grating sensor," *SPIE*, **3042**, 229–236.

the temperature changing during the course of measurements. Some form of temperature compensation is obviously necessary for fiber optic strain sensing systems, and this will be treated in the next section.

To better understand thermal apparent strain, we consider an optical fiber that is bonded to a host structure at the reference temperature, T_0. If the structure is free of any externally applied mechanical load, but subsequently experiences a small change in temperature, ΔT, an incremental stress will develop in the optical fiber due to the mismatch in the thermal expansion coefficients between the optical fiber and the host structure. This *thermally induced axial* stress can be expressed in the form

$$\Delta\sigma = Y_{\mathrm{F}}(\alpha_{\mathrm{H}} - \alpha_{\mathrm{F}})\Delta T, \tag{7.50}$$

where Y_{F} is Young's modulus for the optical fiber and α_{H} and α_{F}, are the thermal expansion coefficients for the host and fiber, respectively. Substituting (7.50) for the incremental stress in Eq. (7.7) enables us to write

$$\frac{\Delta\zeta_{\mathrm{L}}}{\zeta_{\mathrm{L}}} = \left\{1 + \frac{1}{n}\left[\frac{\partial n}{\partial \varepsilon}\right]_T\right\}(\alpha_H - \alpha_{\mathrm{F}})\Delta T + \left\{\alpha_{\mathrm{F}} + \frac{1}{n}\left[\frac{\partial n}{\partial T}\right]_\sigma\right\}\Delta T, \tag{7.51}$$

(a)

(b)

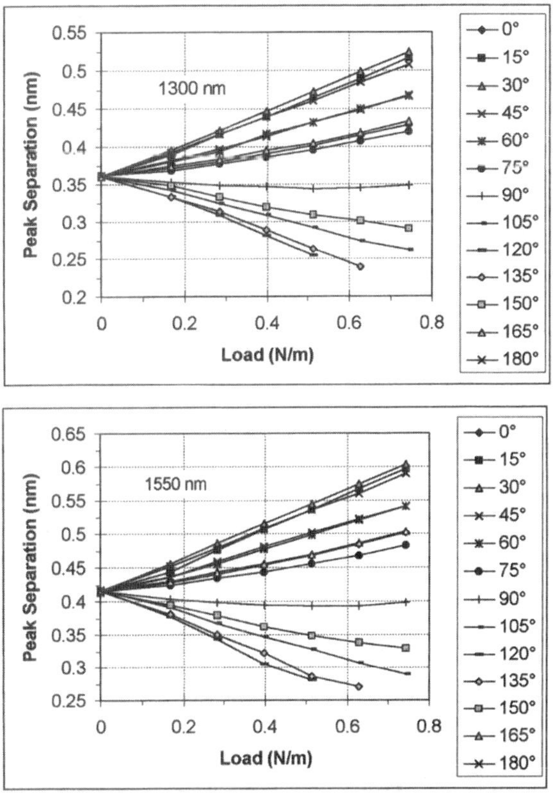

FIGURE 7.18. Increase in peak separation with load for different temperatures of the reflection spectrum for the FBG (a) at 1300 nm and (b) at 1550 nm. From Udd, E., Schulz, W., Seim, J., Mcgill, G. and Laylor, H.M. "Distributed Multiaxis Fiber Grating Strain Sensor Applications for Bridges". In: *Fiber Optic Sensors for Construction Materials and Bridges,* edited by F. Ansari, Proceedings of the Intl. Workshop on Fiber Optic Sensors for Construction Materials and Bridges, pp.168–180, May 3–6, 1998, used with permission from Technomic Publishing Company.

for the normalized incremental change in the SOPL resulting from this change of temperature. Using Eqs. (7.12) and (7.13) in (7.51) yields

$$\frac{\Delta\zeta_L}{\zeta_L} = [S_\varepsilon(\alpha_H - \alpha_F) + S_T]\Delta T, \tag{7.52}$$

and expressing this normalized incremental change in the SOPL in terms of an effective strain change, we can write

$$\frac{\Delta\zeta_L}{\zeta_L} = S_\varepsilon \Delta\varepsilon_T, \tag{7.53}$$

where we introduced the *apparent thermal strain*

$$\Delta\varepsilon_T \equiv \left[\alpha_H - \alpha_F + \frac{S_T}{S_\varepsilon}\right]\Delta T. \tag{7.54}$$

If we assume that a silica optical fiber is bonded to an aluminum plate, then $\alpha_H = 23.8\ \mu\varepsilon\ ^\circ C^{-1}$ and $\alpha_F = 0.5\ \mu\varepsilon\ ^\circ C^{-1}$. Now, the values of S_ε and S_T depend upon the type of optical fiber, the wavelength of light, and the temperature change. As indicated earlier, representative values at a wavelength of 1550 nm are

$$S_\varepsilon = 0.8 \times 10^{-6}\ \mu\varepsilon^{-1} \quad \text{and} \quad S_T = 6.0 \times 10^{-6}\ ^\circ C^{-1},$$

in which case,

$$\Delta\varepsilon_T = [23.8 - 0.5 + 7.5]\mu\varepsilon\ ^\circ C^{-1}\Delta T^\circ C = 30.8\Delta T\ \mu\varepsilon.$$

Thus, 10°C increase in temperature will lead to a thermal apparent strain of 308 με. As indicated in this example, thermal apparent strain can be a serious source of error in strain measurements, unless the host temperature is constant. It is also clear in the preceding example that the mismatch in the thermal expansion coefficients dominates this source of error. Most FRP materials have a much lower thermal expansion coefficient than aluminum (for example, Mitsubishi Leadline TM, $\alpha_H = 0.68\ \mu\varepsilon\ ^\circ C$), so the extent of the error arising from temperature fluctuations would not be so severe. Nevertheless, even if the thermal expansion coefficients matched, the fact that the temperature sensitivity S_T is close to an order of magnitude larger than the strain sensitivity S_ε means thermal apparent strain could still be a problem.

7.6 TEMPERATURE COMPENSATION FOR FIBER OPTIC SENSORS

One of the most common methods of dealing with thermal apparent strain is to independently measure the temperature as well as the strain. The strain measurement can then be corrected for thermal apparent strain. When we make allowance for stress-induced strain in (7.52), we arrive at the more complete fiber optic sensor equation

$$\frac{\Delta\zeta_L}{\zeta_L} = S_\varepsilon\{\Delta\varepsilon + (\alpha_H - \alpha_F)\Delta T\} + S_T\Delta T. \tag{7.55}$$

This can be written in a more compact form,

$$\frac{\Delta\zeta_L}{\zeta_L} = S_\varepsilon \Delta\varepsilon + S_\alpha \Delta T, \tag{7.56}$$

if we introduce the new *apparent temperature sensitivity coefficient*

$$S_\alpha \equiv S_T + S_\varepsilon(\alpha_H - \alpha_F). \tag{7.57}$$

If an independent measurement of temperature is available, either from a conventional device or a fiber optic sensor that has been designed and positioned so that it only responds to temperature, then the change in temperature is used in Eq. (7.56) and the only unknown is then the strain.

An alternative strategy is to simultaneously make two measurements, each of which depends on both the strain and temperature. We can then determine the two unknowns (strain and temperature) from the two measurements. For the sake of brevity we shall introduce the ith measurement of the normalized change in the SOPL

$$M_i \equiv \left[\frac{\Delta\zeta_L}{\zeta_L}\right]_i. \tag{7.58}$$

In this case, if two such measurements of the SOPL are made at, say, two different wavelengths λ_1 and λ_2, we can express this in a matrix form,

$$\begin{bmatrix} M_1 \\ M_2 \end{bmatrix} = \begin{bmatrix} {}^1S_\varepsilon & {}^1S_\alpha \\ {}^2S_\varepsilon & {}^2S_\alpha \end{bmatrix} \begin{bmatrix} \Delta\varepsilon \\ \Delta T \end{bmatrix}, \tag{7.59}$$

where ${}^{i=1,2}S_\varepsilon$ represents the strain sensitivities at wavelengths λ_1 and λ_2, respectively, and ${}^{i=1,2}S_\alpha$ represents the apparent temperature sensitivities at wavelengths λ_1 and λ_2, respectively. Inverting this matrix equation permits us to evaluate $\Delta\varepsilon$ and ΔT, i.e.,

$$\begin{bmatrix} \Delta\varepsilon \\ \Delta T \end{bmatrix} = \frac{1}{\Delta S_{\varepsilon,\alpha}} \begin{bmatrix} {}^2S_\alpha & -{}^1S_\alpha \\ -{}^2S_\varepsilon & {}^1S_\varepsilon \end{bmatrix} \begin{bmatrix} M_1 \\ M_2 \end{bmatrix}, \tag{7.60}$$

where the determinant,

$$\Delta S_{\varepsilon,\alpha} \equiv {}^1S_\varepsilon {}^2S_\alpha - {}^2S_\varepsilon {}^1S_\alpha, \tag{7.61}$$

must be *nonzero*. Indeed, the greater the difference in the strain and apparent temperature sensitivities at the two measurements (i.e., at the two wavelengths in the example), the more accurate this inversion process. Small inaccuracies in the two measurements, δM_1 and δM_2, can be translated into significant errors in the solution of (7.60), if the S-matrix in (7.59) is ill-conditioned (Vengsarkar *et al.*, 1990). To see this, we note that the errors in the strain and temperature, $\delta\varepsilon$ and δT, respectively, are related to the corresponding inaccuracies in the SOPL measurements. δM_1

and δM_2, through the pair of equations that can be derived from (7.60), [Michie *et al.*, 1996]

$$|\delta\varepsilon| = \frac{{}^2S_\alpha|\delta M_1| + {}^1S_\alpha|\delta M_2|}{|\Delta S_{\varepsilon,\alpha}|} \tag{7.62}$$

and

$$|\delta T| = \frac{{}^2S_\varepsilon|\delta M_1| + {}^1S_\varepsilon|\delta M_2|}{|\Delta S_{\varepsilon,\alpha}|}. \tag{7.63}$$

This clearly reveals the importance of undertaking measurements with sensitivities that are sufficiently different that the determinant (7.61) is not small, for the smaller it is, the larger the error in the measurements. A comparative error analysis of a number of different types of fiber optic sensor was undertaken by Michie *et al.* (1996a). They conclude that although FBGs have many advantages, they require very precise measurement of wavelength changes to accurately recover strain and temperature information.

Equation (7.59) is in a general enough form that it can apply to any kind of dual measurement technique for strain and temperature. For example, it can be used if the "1" designated a FBG measurement and the "2" an EFPI measurement, as indeed has been suggested by Liu *et al.* (1997a). Jones (1997) provides a fairly extensive overview of the different dual measurement techniques for this purpose. A few of the more important representative examples of the different techniques that have been proposed to undertake simultaneous strain and temperature measurements are discussed further in the next section.

7.6.1 FBG and FP combination

Liu *et al.* (1997d), reported on a novel design of EFPI that incorporates an FBG within the device (Fig. 7.19). The FBG is used as a strain-free temperature sensor and the EFPI acts as the strain sensor. The EFPI was constructed using a single-mode input fiber and a multimode reflecting fiber. The cleaved ends of the optical fibers were housed within a precision-bore capillary tube and secured in position via two fusion joints. It should be noted that the gauge length of the EFPI is defined as the distance between the two fusion welds, which can be quite different from the gap between the two optical fiber endfaces, which serve as a Fabry–Perot cavity. Since the reflectivity of the optical fiber/air interface is about 4%, the EFPI acts as a low-finesse two-beam interferometer. The

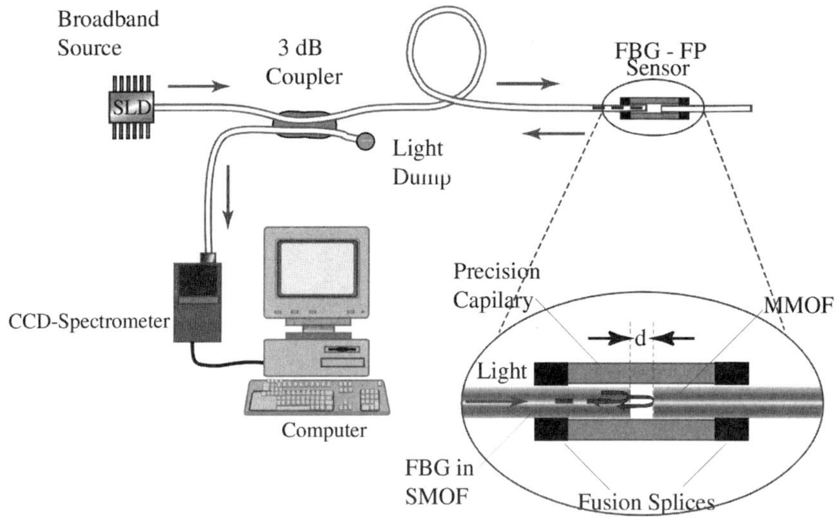

FIGURE 7.19. Spectral demodulation system for a combined FBG and EFPI sensor in which the FBG is designed to be strain-free and used to measure the temperature alone, while the EFPI responds to both strain and temperature. From Liu T., Fernando G.F., Zhang L., Bennion I., Rao Y.J. and Jackson D.A. (1997b) "Simultaneous Strain and Temperature Measurements Using a Combined Fiber Bragg Grating/Extrinsic Fabry–Perot Sensor" OSA 12th Int. Conf. on Optical Fiber Sensors, Williamsburg, 20–23.

sensor was illuminated by a super luminescent diode with a broad peak wavelength at 860 nm.

The spectral intensity of light reflected from the EFPI (see Section 3.6) is

$$I(\lambda) = AI_0(\lambda)\left[1 + \tilde{v}\cos\left\{\frac{4\pi d}{\lambda}\right\}\right], \tag{7.64}$$

where A is a constant, $I_0(\lambda)$ is the *spectral intensity* incident on the EFPI cavity, λ is the *free-space wavelength*, $\tilde{v}$ is the *fringe visibility*, and d is the EFPI *cavity length*. Since the EFPI is used here to determine the strain imposed on the host structure, we need to recognize that the cavity length is

$$d = \frac{\lambda_1\lambda_2}{2(\lambda_1 - \lambda_2)}, \tag{7.65}$$

where, in this instance, λ_1 and λ_2, are any two wavelengths that are separated by one fringe of the interference pattern described by (7.64). An

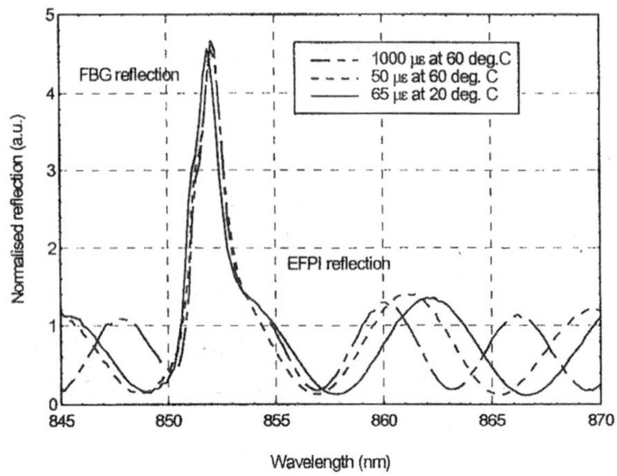

FIGURE 7.20. Sinusoidally modulated Bragg reflection spetrum of the FBG–EFPI hybrid sensor of Fig. 7.19. From Liu T., Fernando G.F., Zhang L., Bennion I., Rao Y.J. and Jackson D.A. (1997b) "Simultaneous Strain and Temperature Measurements Using a Combined Fiber Bragg Grating/Extrinsic Fabry–Perot Sensor" OSA 12th Int. Conf. on Optical Fiber Sensors, Williamsburg, 20–23.

example of the reflection spectrum from this hybrid sensor is presented as Fig. 7.20. It is seen to comprise two components: a narrow FBG peak and an EFPI oscillation.

When the sensor is embedded within, or surface mounted to, FRP composites, the cavity length can be influenced by a number of factors: the magnitude of the applied load and its orientation relative to the sensor direction; the adjacent ply direction; the nature of the bond; the change in temperature; and the thermal expansion coefficient of the composite material. Liu *et al.* (1997a) used a matrix equation similar to (7.59) to describe the EFPI cavity length change Δd and the FBG normalized change in the Bragg wavelength $\Delta\lambda_B/\lambda_B$. Inversion of their equation gave the matrix equation for the change in the strain and temperature,

$$\begin{bmatrix} \Delta\varepsilon \\ \Delta T \end{bmatrix} = \begin{bmatrix} 62.68 & -82.12 \\ 0.3257 & 149.6 \end{bmatrix} \begin{bmatrix} d - 42.5 \\ \lambda - 851.75 \end{bmatrix}, \tag{7.66}$$

where the reference cavity length was 42.5 μm, and the reference Bragg wavelength was 851.75 nm. Figure 7.21 compares the results obtained from this hybrid sensor with those from a conventional strain gauge and a thermocouple, for 63 sets of test data randomly selected across the ranges 30 to 60°C, and 0 to 1200 με.

(a)

(b)

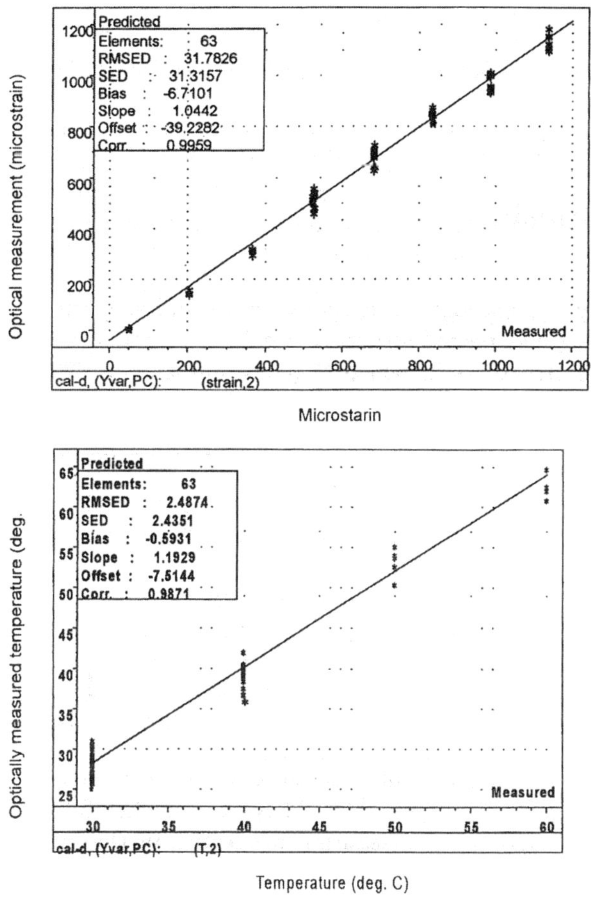

FIGURE 7.21. Comparison of strain and temperature measurement recorded from the FBG–EFPI sensor shown in Fig. 7.19 with those from a conventional foil strain gauge (a) and thermocouple (b). From Liu T., Fernando G.F., Zhang L., Bennion I., Rao Y.J. and Jackson D.A. (1997b) "Simultaneous Strain and Temperature Measurements Using a Combined Fiber Bragg Grating/Extrinsic Fabry–Perot Sensor" OSA 12th Int. Conf. on Optical Fiber Sensors, Williamsburg, 20–23.

Jin and Sirkis (1997) have similarly combined a FBG with an ILFE sensor. In this instance, however, both the FBG and the ILFE respond to strain and temperature, but with very different sensitivities. An erbium-doped fiber amplifier was used as a light source at 1534 nm and a path-matched Fabry–Perot cavity was used to demodulate the ILFE by low-coherence interferometry. A second FP cavity was used to demodulate the

FBG by converting the strain- and temperature-induced shifts in the Bragg wavelength to an optical phase change. They demonstrated their strain/temperature sensor in both surface bonded and embedded configurations and observed agreement with conventional strain and temperature gauges to within 5 $\mu\varepsilon$ and 0.5°C, respectively, for a sensor gauge length of 1 cm.

Henderson *et al.* (1997) have demonstrated a concept first proposed by Morey *et al.* (1991) of using pairs of FBGs (with 0.2 nm linewidth) to form a string of wavelength-multiplexed fiber Fabry–Perot interferometers. The reflectivities of the first and second grating of each FFP were chosen to be about 38% and 100%, respectively, in order to achieve good fringe visibility. Using a broadband superfluorescent fiber source (with a center wavelength at 1545 nm and 40 nm linewidth), the dynamic strain acting on any FFP sensor is read by a processing Mach–Zehnder interferometer as a change in the SOPL. Temperature changes are ascertained from a shift in the reflected wavelength, which is measured by an optical spectrum analyzer. The experimental arrangement is depicted schematically in Fig. 7.22.

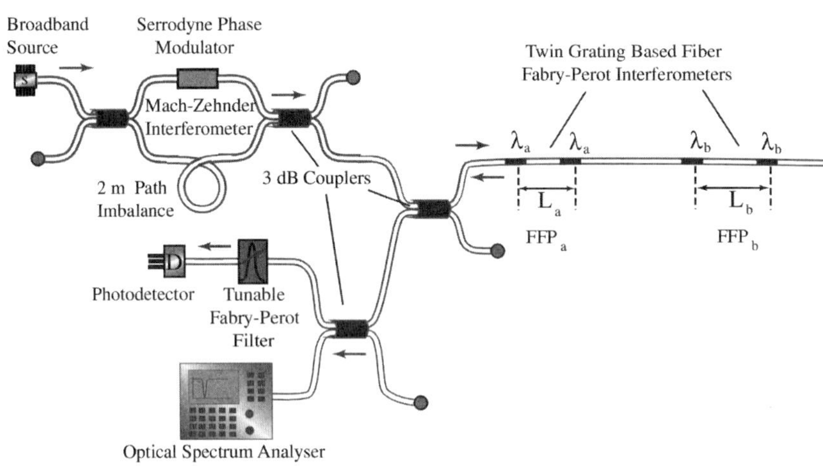

FIGURE 7.22. Schematic illustration of an experimental sensing system involving a serial multiplexed pair of fiber Fabry–Perot sensors formed by pairs of identical FBGs. A tunable Fabry–Perot filter is used to select which fiber Fabry–Perot sensor is to be read, and the Serrodyne phase modulator allows low coherence interferometric demodulation of each sensor. From Henderson P.J., Rao Y.J., Jackson D.A., Zhang L. and Bennion I. (1997), "Simultaneous Dynamic-Strain and Temperature Monitoring Using a Wavelength Multiplexed Fibre-Fabry–Perot Array with Low-Coherence Interrogation," OSA 12th Int. Conf. on Optical Fiber Sensors, Williamsburg.

The effective coherence length of the light reflected from any pair of FBGs is about 1.2 cm (i.e., $\sim$1545 nm^2/0.2 nm) and a tunable Fabry–Perot filter, placed ahead of the detector, is used for selecting the specific FFP sensor to be monitored from the wavelength of its FBGs. The imbalance in the Mach–Zehnder interferometer (MZI), placed after the source, serves to bring the signals reflected from the FBGs forming any FFP into coherence, permitting interference to be observed at the photodetector. This can be better understood in terms of our coherence wavepacket (CWP) concept, introduced in Chapter 3, and shown in terms of this configuration in Fig. 7.23. The imbalanced MZI generates a pair of CWPs, S_1 and S_2. Interference can arise when the second CWP reflected from the first FBG (S_{2a}) of FFP_a overlaps, in space, with the first CWP reflected from the second FBG(S^*_{1a}) of FFP_a. A strain resolution of about 0.1 nε Hz^{-05} has been shown feasible with a 1 kHz detection bandwidth, while the average temperature sensitivity was about 10.6 pm °C^{-1}.

Ferreira *et al.* (1998) also used an imbalanced MZI to read an EFPI, in this instance configured to measure displacement, and an FBG to correct for temperature changes; see Fig. 7.24. They used an SLD at 833.5 nm,

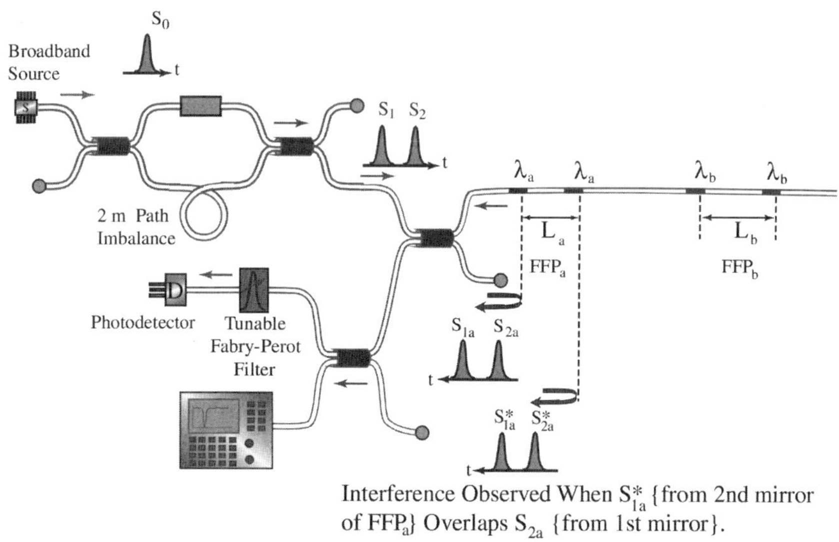

FIGURE 7.23. Coherent wavepacket explanation of the experimental technique illustrated in Fig. 7.22. From Henderson P.J., Rao Y.J., Jackson D.A., Zhang L. and Bennion I. (1997), "Simultaneous Dynamic-Strain and Temperature Monitoring Using a Wavelength Multiplexed Fibre-Fabry–Perot Array with Low-Coherence Interrogation," OSA 12th Int. Conf. on Optical Fiber Sensors, Williamsburg.

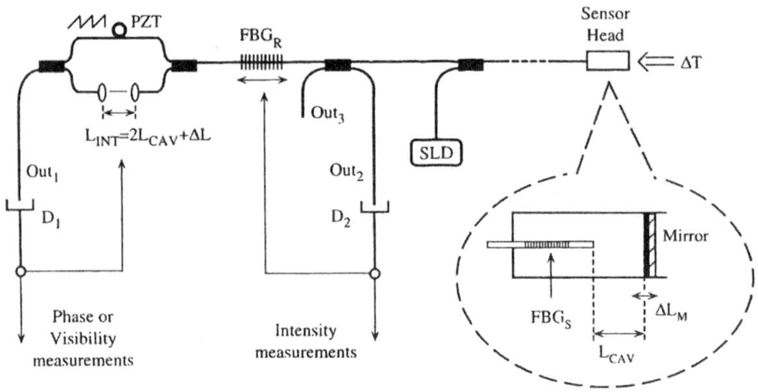

FIGURE 7.24. A hybrid FBG–EFPI sensing system that uses an imbalanced Mach–Zehnder interferometer to interrogate the EFPI and a matched reference FBG to read the sensing FBG. From Ferreira, L.A., Lobo Ribeiro, A.B., Santos, J. L., Farahi, F. "Simultaneous displacement and temperature sensing using a white-light interrogated low-finesse cavity in-line with a fibre Bragg grating" *Smart Materials and Structures* – Special Issue on Fibre Optic Structural Sensing, 7, 1998, 189–198. Used with permission from the Institute of Physics Publishing Ltd.

having a 25 nm linewidth, and positioned the MZI ahead of the photodetector. They also used an actively tuned FBG to read the temperature compensating FBG in the EFPI by maximizing the spectral overlap between the two FBGs. The system was tested by simultaneously changing the temperature of the sensor by 16.2°C and displacing the mirror by 20.3 μm. These results show that the sensing system is capable of discriminating temperature changes from displacements of the mirror. The sensor dynamic displacement sensitivity was observed to be 0.04 nm $Hz^{-0.5}$, while its static displacement sensitivity was estimated to be 6.1 nm $Hz^{-0.5}$.

7.6.2 Combinations of FBGs

In the case of FBG strain gauges, a number of techniques have been proposed for the simultaneous measurement of strain and temperature. One of the simplest involves bonding one half of a FBG to its host structure, leaving the other half free of mechanical strain (Fig. 7.25a). The section of the FBG that is bonded experiences both strain and temperature changes while the free section only experiences the temperature change (Lo and Sirkis, 1997). An example of the resulting *split spectrum* from a FBG that is partially bonded to a cantilever beam is

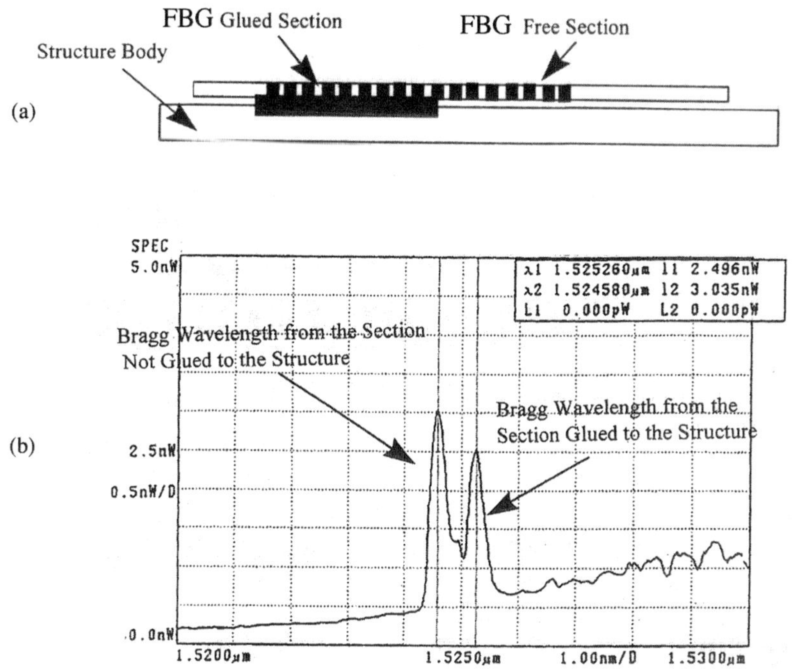

FIGURE 7.25. (a) A temperature-compensated FBG strain sensor in which half of the FBG is strain-free and responds only to the local temperature. (b) Double-peaked reflection spectrum from the FBG that is partially bonded to host structure as indicated in (a). From Lo, Y.L. and Sirkis, J.S., (1997), "Simple Method to Measure Temperature and Axial Strain Simultaneously Using One In-Fiber Bragg-Grating Sensor", *SPIE Smart Sensing, Processing and Instrumentation*, **3042**, 237–246.

shown in Fig. 7.25b. The authors show that the minimum detectable axial strain depends strongly on the linewidth of the FBG and that 10 με, with a 2 με error, might be achieved with a 0.02 nm FBG. They also suggest that a cascade of two FBGs with slightly different periods might give better results. Unfortunately, this simple technique is only viable for surface-mounted FBG sensors. However, the use of a capillary tube to strain decouple one-half of a FBG (or a second grating) may permit this approach to also be used when the sensor is embedded within a host material.

We have seen in Section 7.4 that application of unequal transverse stress to a FBG will cause the reflection peak to bifurcate (Wagreich, 1996). Dunphy *et al.* (1995b) recognized that the wavelength separation of the subsequent two spectral peaks had a different sensitivity to strain and

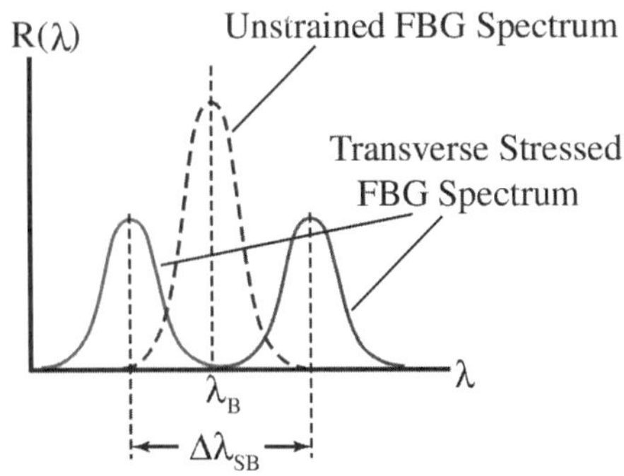

FIGURE 7.26. Splitting of the reflection spectrum from an FBG sensor subject to transverse stress.

temperature than that of the average wavelength of the reflection spectrum (Fig. 7.26). The form of the matrix equation (7.59) that is suitable for this situation can be written

$$\begin{bmatrix} \Delta\lambda_{SB} \\ \lambda_B \end{bmatrix} = \begin{bmatrix} S_\varepsilon^* & S_\alpha^* \\ S_\varepsilon & S_\alpha \end{bmatrix} \begin{bmatrix} \Delta\varepsilon \\ \Delta T \end{bmatrix}, \tag{7.67}$$

where $\Delta\lambda_{SB}$ and λ_B represent, respectively, the spectral separation of the two peaks and the average wavelength of the FBG reflection spectrum at the strain and temperature of interest S_ε^* and S_α^* represent the strain and apparent temperature sensitivities of the FBG spectral separation, respectively, while S_ε and S_α represent the strain and apparent temperature sensitivities of the FBG average wavelength, respectively. The components of the sensitivity (**S**) matrix of (7.67) as determined by Dunphy *et al.* (1995) have the values

$S_\varepsilon^* \approx 0;$ $\quad$ $S_\alpha^* = -1.44 \times 10^{-3}\ \text{nm}\ {}^\circ\text{C}^{-1}$

(between 20 and 75°)

$S_\varepsilon = 1.17 \times 10^{-3}\ \text{nm}\ \mu\varepsilon^{-1};$ $\quad$ $S_\alpha = 9.82 \times 10^{-3}\ \text{nm}\ {}^\circ\text{C}^{-1}.$

It is interesting to note that the separation of the peaks in the bifurcated FBG spectrum were observed to be invariant to axial strain out to at least 4000 με.

This elegant approach permits a single grating to provide two kinds of signal that can be used for the temperature compensation of strain measurements. It should be noted that the transverse stress induced birefringence is only required at the location of the FBG. To create the appropriate transverse stress distribution, the optical fiber has to be embedded within a FRP composite layup at some angle to the adjacent ply (reinforcing fiber) direction, Fig. 7.27. On curing of the composite material, the residue stresses associated with the *resin eye* that forms around the angled optical fiber are responsible for the formation of the unequal transverse stresses. The greater the angle between the optical fiber and the adjacent ply direction, the more severe the stresses and the larger the induced birefringence and spectral splitting. Figure 7.28a, displays the strongly bifructated spectrum of a FBG that is embedded perpendicular to the local reinforcing fibers of the FRP host material (see Fig. 7.28b). As elegant as this technique appears, it suffers one major drawback. Embedding optical fibers orthogonal to the adjacent plies has been shown to perturb the FRP material and can appreciably weaken the structure, unless certain precautions are undertaken (see Chapter 8).

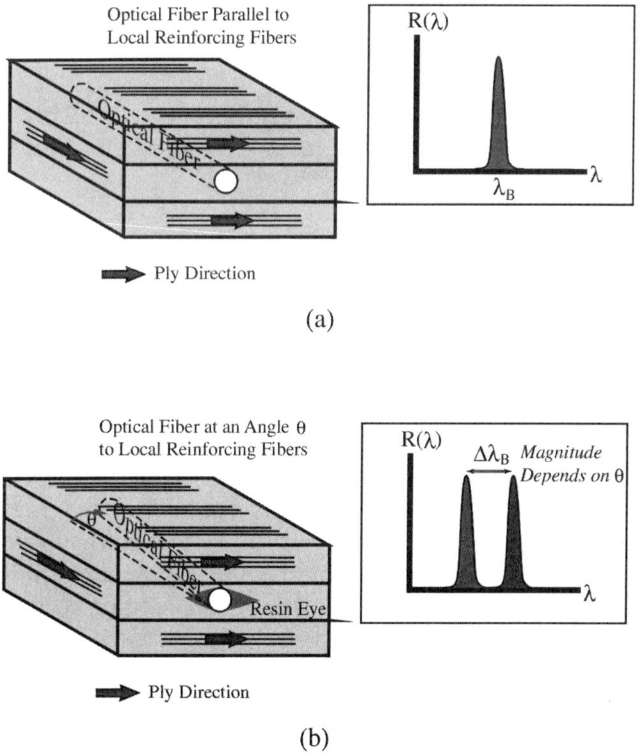

FIGURE 7.27. Schematic illustration of a technique from inducing bifurcation in the FBG reflection spectrum by residual strain arising when the FBG is embedded at some angle to the local ply direction of the FRP host structure.

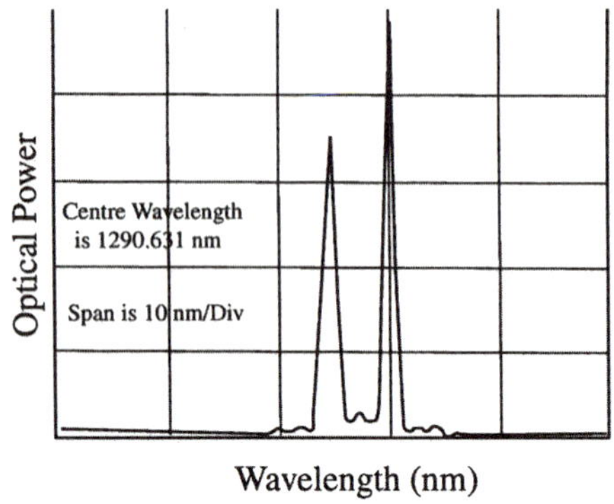

(a)

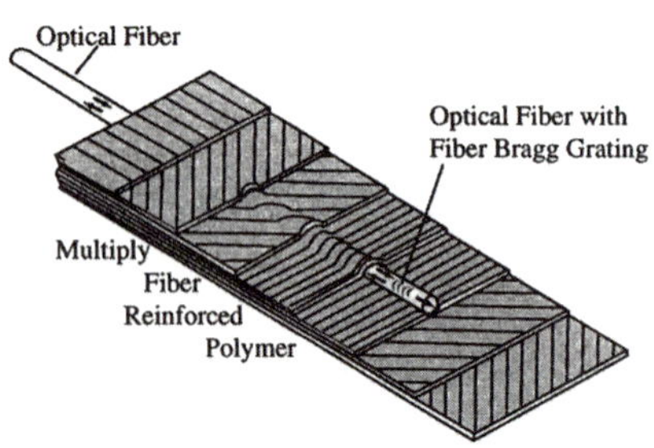

(b)

FIGURE 7.28. (a) Experimental evidence of bifurcastion of the FBG spectrum caused by embedding the sensor within a FRP composite material at right angles to the local ply direction. (b) Corresponding configuration of FBG sensor and the FRP layup.

One way to overcome this drawback is to inscribe the FBG within highly birefringent optical fiber rather than use the FRP host residue stress to generate the birefringence. Udd *et al.* (1998) proposed that FBG formed in polarization maintaining optical fibers could be used for simultaneously monitoring strain and temperature. However, they caution that this task is difficult with standard polarization-maintaining optical fibers because the **S**-matrix tends to be ill-conditioned, leading to large errors. Sudo *et al.* (1997) have shown that this technique is very promising if the FBG is written in the PANDA type of polarization-maintaining optical fiber (an optical fiber with "circular" thermal stress rods on either side of the core, illustrated in Fig. 5.8). They observed a difference of 0.48 nm in the wavelengths of the two peaks corresponding to the fast and slow axes (Fig. 7.29). Some of their results are presented in Fig. 7.30. A series of tests

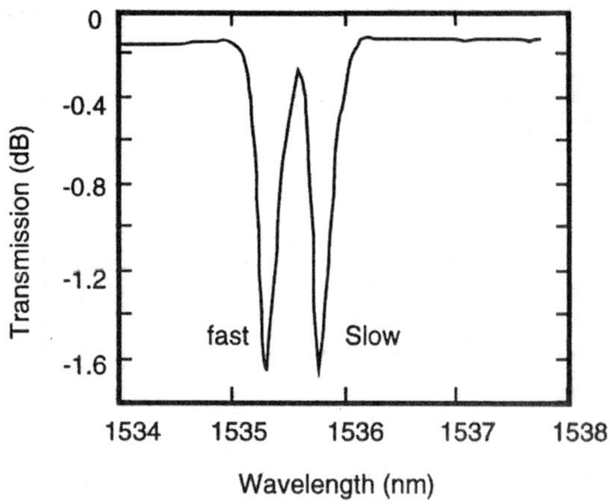

FIGURE 7.29. Bifurcation in the transmission spectrum of a FBG written in PANDA polarization-maintaining optical fiber. From Sudo M, Nakai M., Himeno K., Suzaki S., Wada A. and Yamauchi R., (1997), "Simultaneous Measurement of Temperature and Strain using PANDA Fiber Grating", OSA 12th International Conference on Optical Fiber Sensors, Williamsburg, 170–173.

indicated that this sensor was capable of simultaneously measuring strain and temperature with errors of $\pm 20\ \mu\varepsilon$ and $\pm 2°C$, respectively.

From Section 7.4, we envisage that there may be a problem with using FBG spectral peak separation for temperature compensation where significant changes in transverse loading might arise. This is because it may not be easy to distinguish changes of temperature from transverse loading, as both affect the wavelength separation of the FBG peaks. Although the complexity and cost of employing polarization-maintaining optical fiber are cited as disincentives for using any technique that involves this type of optical fiber, in this instance only the sensor need be made of this fiber, as the measurement is actually spectral in nature. On the down side, however, such an approach would require two fusion splices on either side of the sensor.

A fusion splice is required right in the center of a twin grating sensor proposed by James *et al.* (1996) (Fig. 7.31), and even though they demonstrated maximum errors of $\pm 17\ \mu\varepsilon$ and $\pm 1°C$ over a measurement range of 2500 $\mu\varepsilon$ and 120°C, the presence of this fusion splice and the difference in cladding diameters of the two FBGs poses several problems. The first is that strain transfer at the physical discontinuity may lead to errors in measurement; the second is that the bondline at the discontinuity may have a short life because of high stress concentrations; and the third is that

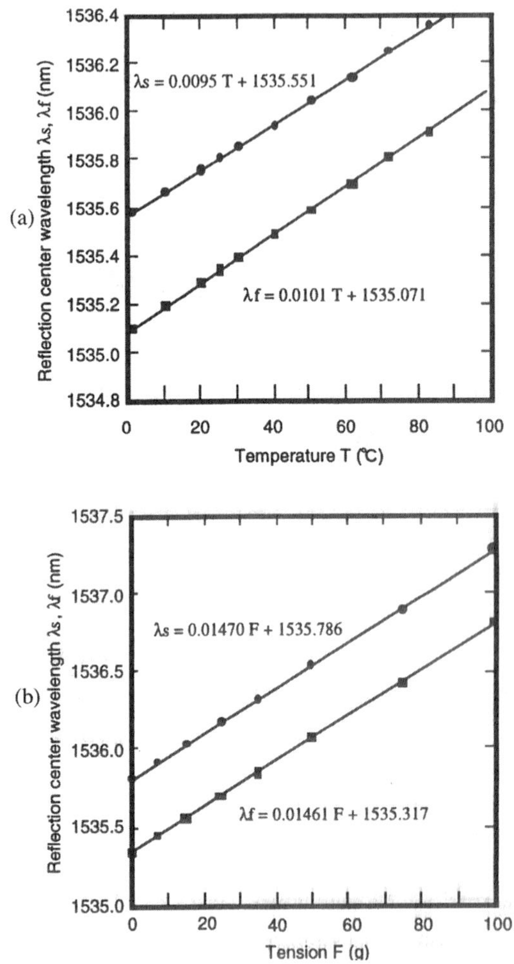

FIGURE 7.30. (a) Variation of the center wavelengths with temperature (a) and strain (b) for the double-peaked reflection spectrum of an FBG written in PANDA optical fiber. From Sudo M., Nakai M., Himeno K., Suzaki S., Wada A. and Yamauchi R., 1997, "Simultaneous Measurement of Temperature and Strain using PANDA Fiber Grating", OSA 12th International Conference on Optical Fiber Sensors, Williamsburg, 170–173.

the life of the sensor maybe compromised by the fusion splice if the sensor is subject to long-term stress loading.

One of the most expedient methods of simultaneously measuring strain and temperature is to use the wavelength dependence of the strain-optic and thermo-optic effects. Xu *et al.* (1994a) have demonstrated that it is possible to use two ELED sources operating at 850 and 1300 nm to

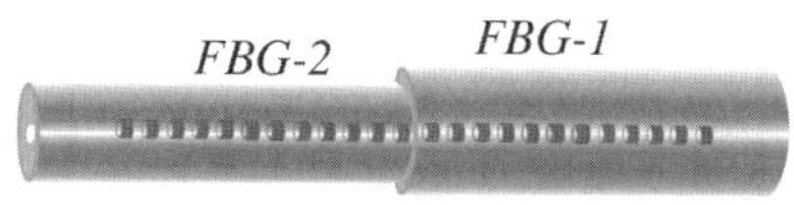

FIGURE 7.31. Schematic illustration of a sensor based on a twin fiber Bragg grating with different cladding diameter optical fibers.

illuminate two collocated FBGs, with nominal Bragg wavelengths of ~848 and 1298 nm (see Fig. 7.32). The subsequent variations in the reflected wavelengths with applied strain and temperature are presented in Fig. 7.33. The appropriate matrix equation for this case takes the form

$$\begin{bmatrix} \Delta\lambda_{B1} \\ \Delta\lambda_{B2} \end{bmatrix} = \begin{bmatrix} K_{\varepsilon,1} & K_{T,1} \\ K_{\varepsilon,2} & K_{T,2} \end{bmatrix} \begin{bmatrix} \Delta\varepsilon \\ \Delta T \end{bmatrix}, \tag{7.68}$$

where 1 and 2 refer to the two wavelengths, 1300 and 850 nm, respectively, and the measured wavelength strain, $K_{\varepsilon,j}(j = 1 \text{ or } 2)$ and wavelength temperature, $K_{T,j}(j = 1 \text{ or } 2)$ elements of this **K**-matrix are

$$\begin{aligned} K_{\varepsilon,1} &= 0.96 \pm 0.0065 \text{ pm } \mu\varepsilon^{-1} \\ K_{\varepsilon,2} &= 0.59 \pm 0.0034 \text{ pm } \mu\varepsilon^{-1} \\ K_{T,1} &= 8.72 \pm 0.077 \text{ pm } ^{\circ}\text{C}^{-1} \\ K_{T,2} &= 6.30 \pm 0.037 \text{ pm } ^{\circ}\text{C}^{-1}. \end{aligned}$$

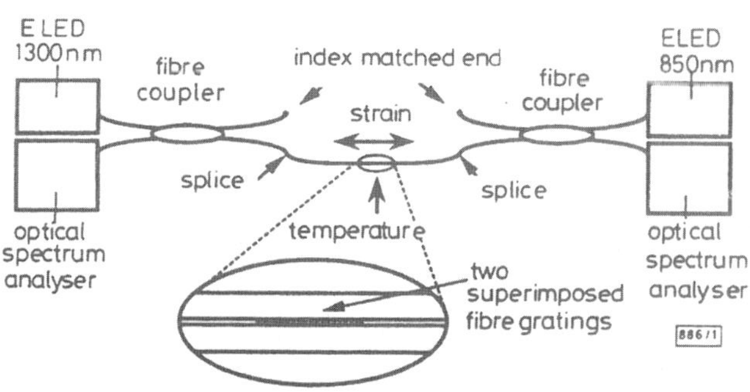

FIGURE 7.32. A fiber optic temperature-compensated strain measuring approach using two collocated FBGs operating at two very different wavelengths, 850 and 1300 nm. From Xu, M-G, Archambault, J.-L., and Dakin, J.P., (1994), "Structural Bending Sensor Using Fiber Gratings," *SPIE* **2292**, 407–413; Fiber Optic and Laser Sensors XII, De Paula, R.P. (Ed).

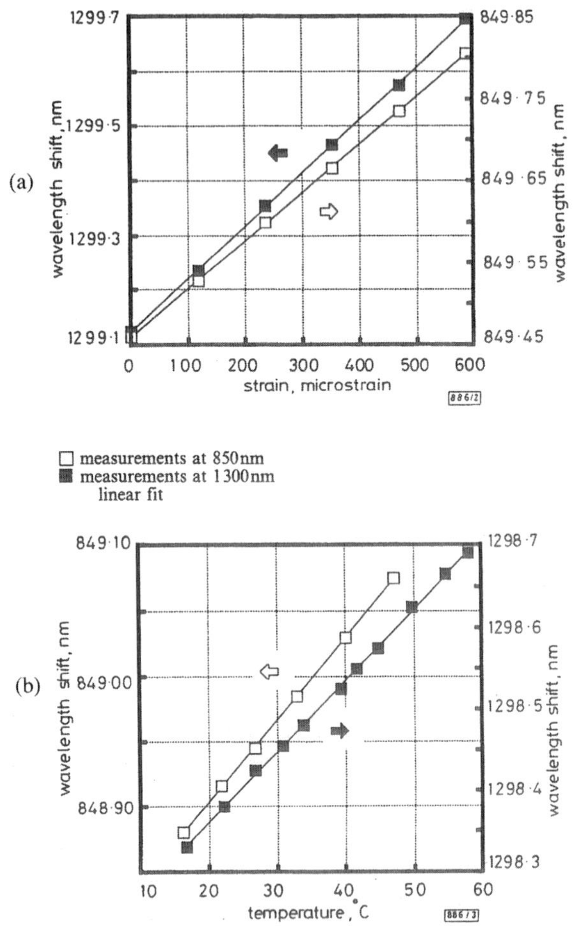

FIGURE 7.33. Wavelength variation with strain (a) and temperature (b) for the dual collocated FBGs used in the system illustrated in Fig. 7.32. From Xu *et al.* "Discrimination between strain and temperature effects using dual-wavelength fibre grating sensors", *Electronics Letters* **30**(13) 1085–1087. Used with permission from IEE Publishing Ltd.

The inverse matrix is used to determine the strain and temperature with typical errors of 10 με and 5°C over the measurement ranges: 600 με and 50°C. The disadvantages of this approach are the need to fabricate two FBGs and use two light sources, and the wide wavelength range (if the optical fiber is single-mode at 850 nm, it will not be optimized at 1300 nm).

7.6.3 Long-period gratings

For those application where spatial resolution requirements are not too severe (i.e., several centimeters), combinations of FBG with so-called *long-period gratings* (LPGs) have been shown to provide a means of simultaneously measuring strain and temperature. A long-period grating arises when the core index of refraction of a photosensitive optical fiber is forced to suffer an axial modulation with a spatial period that is resonant between the core guided mode and one of the cladding modes (Vensarkar *et al.*, 1996). Under phase-matched conditions, light from the fundamental guided mode is coupled to the forward-propagating cladding modes at distinct wavelengths given by the relation

$$\lambda_m = (n_{\text{eff}} - n_{\text{m}})\Lambda, \tag{7.69}$$

where λ_m represents the center wavelength of the notch at which light is coupled into the azimuthally symmetric cladding mode of order m, while n_{eff} and n_m are the effective indices of the guided mode and the azimuthally symmetric cladding mode of order m, respectively. Lastly, Λ is the spatial period of the index modulation.

Like FBGs, long-period gratings are fabricated by exposing a suitable single-mode optical fiber to an axial variation of intense UV (245 nm) laser radiation. Their properties are quite different, however. For example, the period of an LPG tends to be in excess of a hundred microns, as opposed to about half a micron for FBGs, and their minimum length is at least several centimeters. The most important difference, however, is that LPGs do not reflect a narrow band of light, as do FBGs; instead, their transmission spectrum involves a number of spectral notches, each corresponding to a band of guided mode radiation being coupled into a different cladding mode (Bhatia *et al.*, 1997a); see Fig. 7.34. When the LPG is subjected to a change in strain or temperature, each of these spectral notches experience a wavelength shift.

Let $\Delta\lambda_1$ and $\Delta\lambda_2$ represent the cumulative shifts of the two unperturbed notches of an LPG centered at λ_1 and λ_2 ($\lambda_1 > \lambda_2$), respectively, due to changes in strain and temperature. Then the relevant form of matrix equation (7.59) for this situation can be written

$$\begin{bmatrix} \Delta\lambda_1 \\ \Delta\lambda_2 \end{bmatrix} = \begin{bmatrix} K_{\varepsilon,1} & K_{T,1} \\ K_{\varepsilon,2} & K_{T,2} \end{bmatrix} \begin{bmatrix} \Delta\varepsilon \\ \Delta T \end{bmatrix}, \tag{7.70}$$

where $K_{\varepsilon,j}$ and $K_{T,j}$ ($j = 1$ or 2) are the effective wavelength-strain and wavelength-temperature elements of the **K**-matrix for the LPG.

Bhatia *et al.* (1997a) fabricated LPGs in *hydrogen-loaded* Corning SMF-28 optical fiber with a period $\Lambda = 280\ \mu\text{m}$. For unperturbed spectral

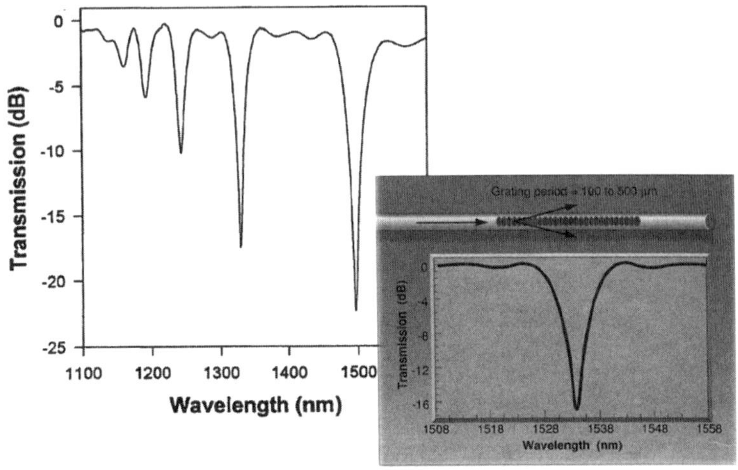

FIGURE 7.34. Multiple notched transmission spectrum for a long-period grating written in a single-mode optical fiber. Inset illustrates a forward scattering wave from long-period grating and one of the transmission notches.

notches centered at $\lambda_1 = 1607.9$ nm and $\lambda_2 = 1332.9$ nm the measured wavelength-strain and wavelength-temperature coefficients of the LPG **K**-matrix are

$$\begin{aligned} K_{\varepsilon,1} &= 1.68 \text{ pm } \mu\varepsilon^{-1} \\ K_{\varepsilon,2} &= 0.28 \text{ pm } \mu\varepsilon^{-1} \\ K_{T,1} &= 70.6 \text{ pm } ^{\circ}\text{C}^{-1} \\ K_{T,2} &= 45.2 \text{ pm } ^{\circ}\text{C}^{-1}. \end{aligned}$$

The shift in the resonance bands (transmission notches) at 1607.9 nm (circles) and 1332.9 nm (squares) as a function of the change in temperature are present in Fig. 7.35. Unfortunately, LPGs are prone to nonlinearities and cross-sensitivities, an example of which is shown in Fig. 7.36, [Bhatia *et al.*, 1997a]. In this instance the wavelength-strain coefficient for the 1607.9 nm notch is seen to be quite dependent on temperature. By taking account of this nonlinearity, a minimum detectable strain and temperature change of about 30 με and 0.6°C, respectively, were reported by the authors. Figure 7.37 demonstrates that their strain measurements are fairly independent of temperature. As indicated in Section 7.7, Bhatia *et al.* (1997b) were able to make a relatively temperature-insensitive strain

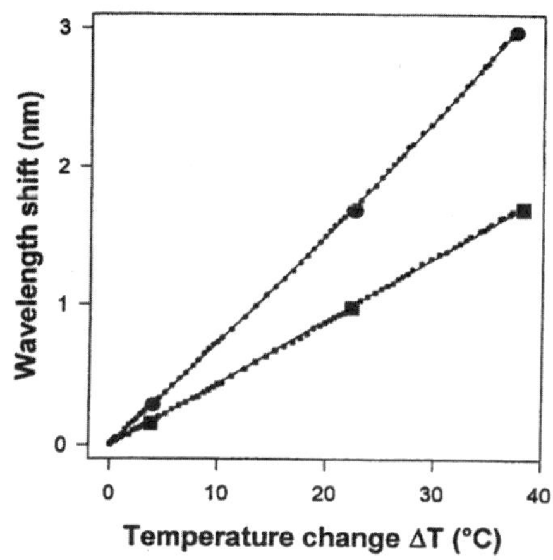

FIGURE 7.35. Variation in the center wavelength of two transmission notches (at 1607 and 1332.9 nm) with temperature for a long-period grating. From Bhatia *et al.*, (1997a), "Simultaneous Strain and Temperature Sensing Using Long-Period Gratings," *SPIE*, **3042**.

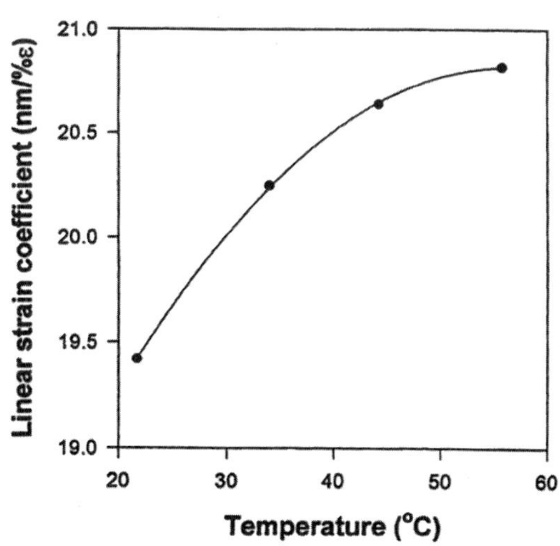

FIGURE 7.36. Wavelength-strain coefficient variation with temperature for the 1607.9 nm transmission notch of an LPG. From Bhatia *et al.*, (1997a), "Simultaneous Strain and Temperature Sensing Using Long-Period Gratings," *SPIE*, **3042**.

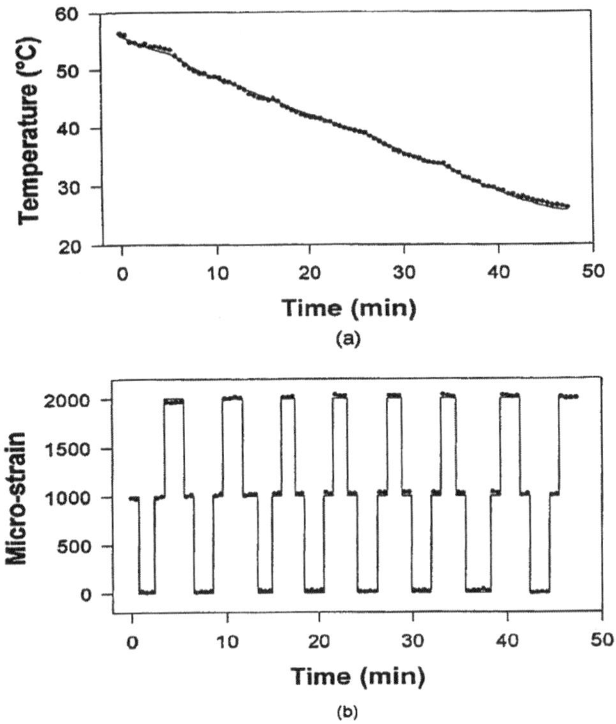

FIGURE 7.37. (a) Temperature variation with time for a LPG. (b) Strain measurement for a repetitive strain loading and unloading cycle during the time period when the temperature varies according to (a). From Bhatia *et al.*, (1997a), "Simultaneous Strain and Temperature Sensing Using Long-Period Gratings," *SPIE*, **3042**.

sensor from a modified form of LPG by reducing the temperature sensitivity by a factor of 50.

Patrick *et al.* (1996b) used the significant difference between wavelength shifts for strain and temperature for an LPG and an FBG to ascertain the strain and temperature. In this instance the appropriate form of the matrix equation is

$$\begin{bmatrix} \Delta\lambda_{\text{LPG}} \\ \Delta\lambda_{\text{FBG}} \end{bmatrix} = \begin{bmatrix} K_{\varepsilon,\text{LPG}} & K_{T,\text{LPG}} \\ K_{\varepsilon,\text{FBG}} & K_{T,\text{FBG}} \end{bmatrix} \begin{bmatrix} \Delta\varepsilon \\ \Delta T \end{bmatrix}, \qquad (7.71)$$

where $K_{\varepsilon,j}$ and $K_{T,j}$ are the effective wavelength-strain and wavelength-temperature elements of the appropriate **K**-matrix, but j, in this instance, does not refer to different wavelengths as in the previous cases; instead it refers to different devices (LPG or FBG).

The measured elements of this form of the **K**-matrix are

$$\begin{aligned}K_{\varepsilon,\mathrm{LPG}} &= 0.5\ \mathrm{pm}\ \mu\varepsilon^{-1}\\ K_{\varepsilon,\mathrm{FBG}} &= 1.0\ \mathrm{pm}\ \mu\varepsilon^{-1}\\ K_{T,\mathrm{LPG}} &= 60.0\ \mathrm{pm}\ {}^{\circ}\mathrm{C}^{-1}\\ K_{T,\mathrm{FBG}} &= 9.0\ \mathrm{pm}\ {}^{\circ}\mathrm{C}^{-1}.\end{aligned}$$

As seen by reference to Fig. 7.38, the sensor arrangement is rather involved, using two FBGs to essentially sample the transmitted spectral notch of the LPG. The FBG wavelengths were at $\lambda_{\mathrm{B1}} = 1293$ nm and $\lambda_{\mathrm{B2}} = 1321$ nm, respectively, with an LPG wavelength $\lambda_{\mathrm{LPG}} = 1306$ nm. In this sensing system the ratio of the signals reflected from the two FBGs, R_1 and R_2, respectively, determines the shift in the wavelength of the LPG notch, while the Bragg wavelengths themselves reflect the response of the FBGs to the strain and temperature. In essence, a change of strain produces a small decrease in R_1 and a small increase in R_2, because the shift in λ_{LPG} lags the shift in λ_{B1} and λ_{B2}. In contrast, a change in temperature creates a large increase in R_1 and a large decrease in R_2, because the shift in λ_{LPG} is much larger than the shifts in λ_{B1} and λ_{B2}. This is schematically illustrated in Fig. 7.38.

The sensing system was calibrated by simultaneously measuring the shift in one of the FBG wavelengths and the change in the *normalized reflectance ratio*

$$F(R_1, R_2) \equiv \frac{\sqrt{R_1} - \sqrt{R_2}}{\sqrt{R_1} + \sqrt{R_2}}. \tag{7.72}$$

The variation of $F(R_1, R_2)$ and λ_{B2} with strain, at a temperature held constant at 38°C, is presented as Fig. 7.39a, while their variation as a function of temperature at an applied strain of 590 με presented as Fig. 7.39b. Once the sensing system is calibrated any change in applied strain and temperature can be determined simultaneously by using the known calibration. This sensing system was shown to be capable of simultaneously measuring strain and temperature with resolutions of ±9 με and ±1.5°C, respectively, over the range from 290 to 1270 με and 25 to 50°C.

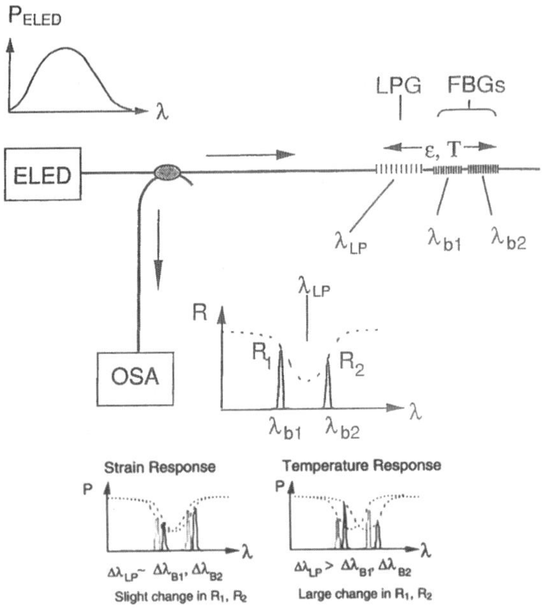

(a)

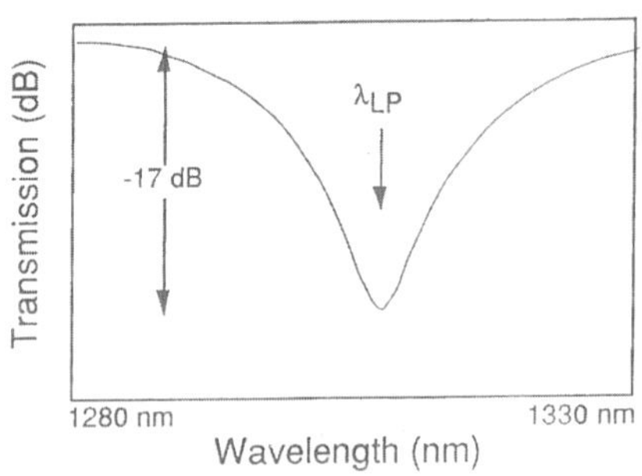

(b)

FIGURE 7.38. (a) An experimental arrangement for undertaking temperature-compensated strain measurements using a combination of an LPG and a pair of FBGs. Also shown are illustrations of how the two FBGs effectively sample the spectral shift of the LPG transmission spectrum. When combined with its own wavelength shifts, the system can distinguish strain from temperature variations. (b) Transmission notch of the LPG used in the experimental arrangement illustrated in (a). From Patrick, H.J., Williams, G.M., Kersey, A.D., Pedrazzani, J.R. and Vengsarkar, A.M., "Hybrid fiber Bragg grating/long period fiber grating sensor for strain/temperature discrimination," *IEEE Phot. Tech. Lett.* **8**, 1223, © 1996 IEEE; *and* (b) From Kersey, A.D., Davis, M.A., Patrick, H.J., LeBlanc, M.M., Koo, K.P., Askins, C.G., Putnam, M.A. and Friebele, E.J., "Fiber Grating Sensors" *J Lightwave Technology*, **15**, 1442–1463, © 1997, IEEE.

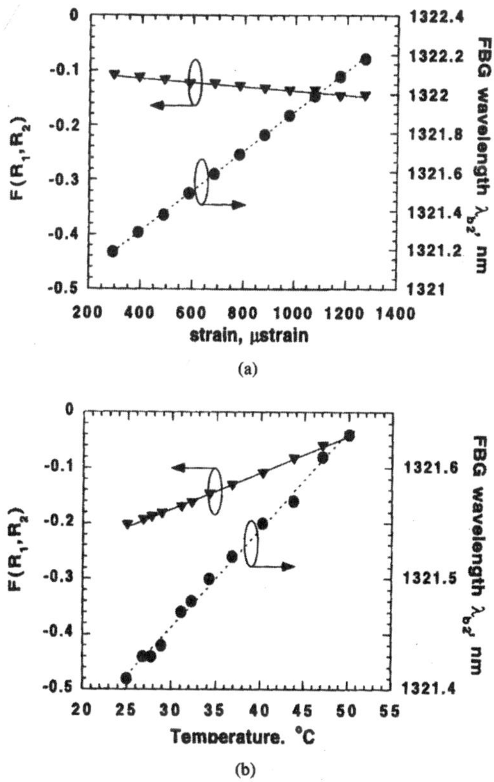

FIGURE 7.39. (a) Variation with strain of the reflectance ratio [see text, Eq. (7.72)] of both FBGs and one FBG center wavelength, for the experimental system illustrated in Fig. 7.38a. (b) Variation with temperature of reflectance ration [see text, Eq. (7.72)] of both FBGs and one FBG center wavelength, for the experimental system illustrated in Fig. 7.38a. From Patrick H.J., Williams, G.M., Kersey, A.D., Pedrazzani, J.R. and Vensarkar, A.M., "Hybrid Fiber Bragg Grating/Long Period Fiber Grating Sensor for Strain/Temperature Discrimination," *Photon Tech Lett*, 8, 1223–1225, © 1996, IEEE.

7.6.4 Use of Brillouin scattering

As indicated in Section 6.6, light propagating along an optical fiber loses a small fraction of its power through a process called Brillouin scattering. This loss mechanism involves the scattering of photons of light from *phonons* (quantized packets of acoustic waves) in the optical fiber. The extent of this scattering can be greatly amplified through a three-wave interaction referred to as *stimulated Brillouin scattering* (SBS). In SBS a pulse of *pump* (laser) *radiation* at frequency ν_{pump} transfers power to a *counter*

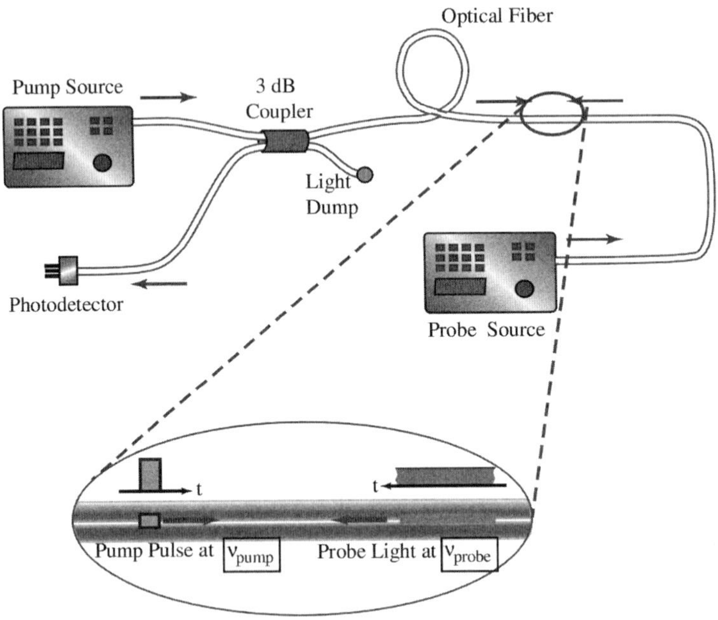

FIGURE 7.40. Schematic illustration of two-laser excitation of stimulated Brillouin scattering within an optical fiber. Expanded view of optical fiber illustrates the passage of the short pump pulse through the continuous wave probe light.

propagating continuous-wave beam of *probe radiation* at frequency ν_{probe} (Fig. 7.40), provided the following condition applies

$$\nu_{\text{pump}} - \nu_{\text{probe}} = \nu_{\text{B}}. \tag{7.73}$$

Here, ν_{B}, is the Brillouin frequency shift and is given by Eq. (6.5)

$$\nu_{\text{B}} = \frac{2nV_{\text{a}}}{\lambda_{\text{p}}}.$$

V_{a} is the acoustic wave velocity, n is the fiber core index of refraction, and λ_{p} is the wavelength of the pump pulse. It is no surprise to find that the Brillouin frequency shift is also a function of the strain and temperature state of the optical fiber. Davis and Kersey (1996a) have used this knowledge to propose that SBS can be used in serial multiplexed arrays of FBGs to provide quasi-distributed strain and temperature profiles along an optical fiber. A particular attraction of this concept is that the reflections from the FBGs occur at wavelengths well separated from those involving Brillouin scattering, so the strain and temperature sensitivities of the optical fiber are quite different. The appropriate form of matrix equation (7.59) for this pair of measurements can be written

$$\begin{bmatrix} \Delta\nu_{\text{SBS}} \\ \Delta\lambda_{\text{FBG}} \end{bmatrix} = \begin{bmatrix} K_{\varepsilon,\text{SBS}} & K_{T,\text{SBS}} \\ K_{\varepsilon,\text{FBG}} & K_{T,\text{FBG}} \end{bmatrix} \begin{bmatrix} \Delta\varepsilon \\ \Delta T \end{bmatrix}, \tag{7.74}$$

where $K_{T,\text{SBS}}$ and $K_{\varepsilon,\text{SBS}}$ represent the Brillouin frequency shift coefficients with respect to temperature and strain respectively, and $K_{T,\text{FBG}}$ and $K_{\varepsilon,\text{FBG}}$ are the corresponding FBG wavelength shift coefficients. As before, inversion of this **K**-matrix equation (7.74) yields the changes in the temperature and strain to be determined.

To test this concept, two single-mode Nd:YAG lasers operating at 1319 nm were used as the pump and probe signals. The FBG was interrogated at 1550 nm by the broadband source of an optical spectrum analyzer (see Fig. 7.41). The FBG and 50 m of optical fiber were placed in an oven. The duration of the pulsed pump laser was set to 100 ns, providing a spatial resolution for the Brillouin frequency shift measurement of 10 m. The frequency difference between the two lasers was measured by means of an interferometer and an example of the results attained presented as Fig. 7.42. Here the frequency spectra of the probe

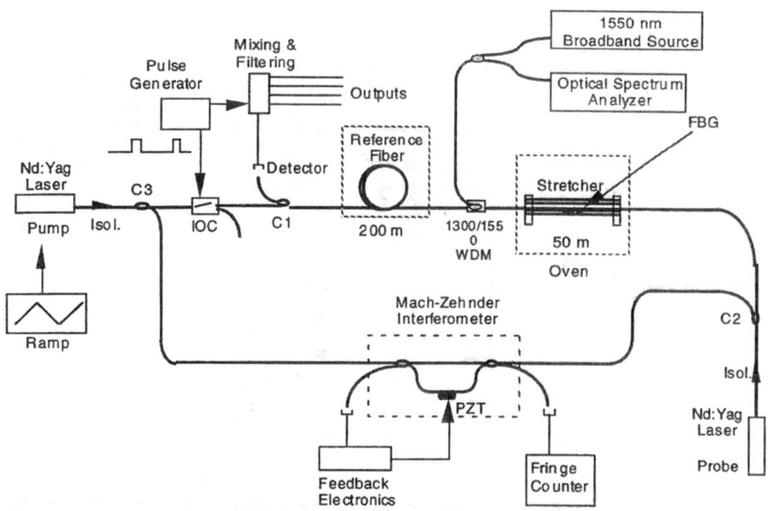

FIGURE 7.41. Experimental arrangement used to undertake temperature-compensated strain measurements based on a combination of an FBG sensor and stimulated Brillouin scattering. The latter is excited by two Nd:YAG lasers, the former by a 1550 nm LED. From Davis, M.A. and Kersey, A.D., (1996b), "Simultaneous Measurement of Temperature and Strain Using Fiber Bragg Gratings and Brillouin Scattering" *SPIE* **2838**, 114–123.

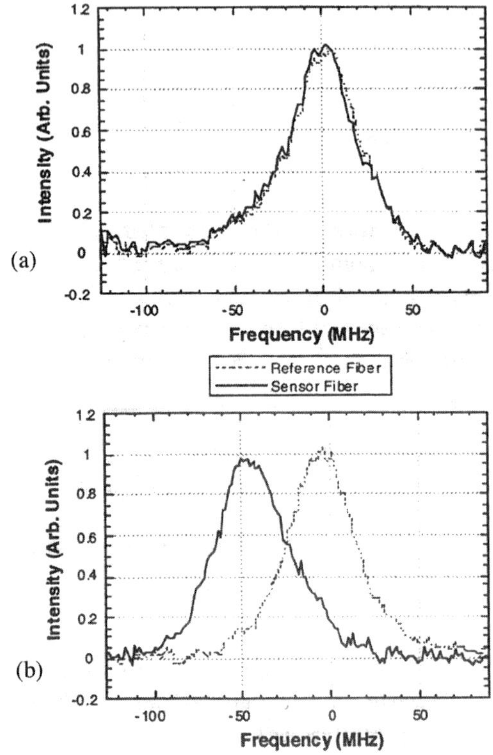

FIGURE 7.42. Frequency difference between the Nd:YAG lasers of the experimental setup illustrated in Fig. 7.41, corresponding to peak SBS for the reference and sensing segments of optical fiber (a) when both are at 23.8°C and (b) when the sensing length of optical fiber is heated to 60.5°C. From Davis, M.A. and Kersey, A.D., (1996b), "Simultaneous Measurement of Temperature and Strain Using Fiber Bragg Gratings and Brillouin Scattering" *SPIE* **2838**, 114–123.

signal, after passage through both the reference and sensing lengths of optical fiber, are seen to coincide when both are at 23.8°C. These results are attained by sweeping the pump laser frequency. However, as the sensing length temperature is raised, in the oven, to 60.5°C, about a 50-MHz frequency shift is observed between the spectra.

The variation in the Brillouin frequency shift and the FBG wavelength shift with temperature is presented as Fig. 7.43. The Brillouin frequency shift is seen to be about 13.9 times more sensitive to temperature changes than the FBG wavelength shift. In Fig. 7.44, the strain responsitivities are compared for the Brillouin and FBG techniques and a ratio of about 5.6 is

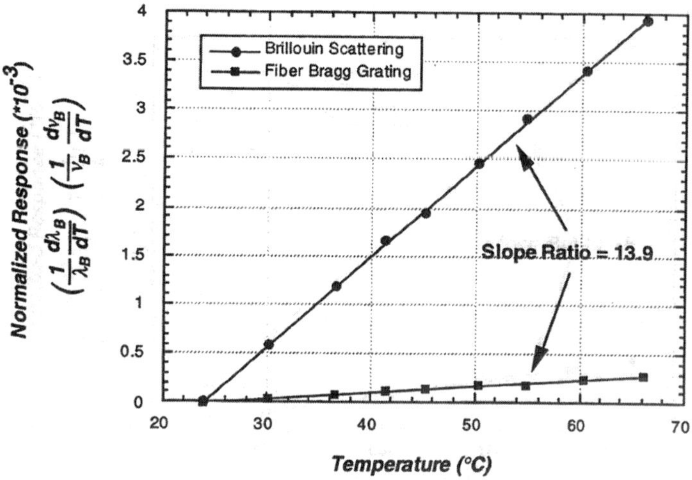

FIGURE 7.43. Variation in both FBG center wavelength shift and Brillouin frequency shift with temperature. From Davis, M.A. and Kersey, A.D., (1996b), "Simultaneous Measurement of Temperature and Strain Using Fiber Bragg Gratings and Brillouin Scattering" *SPIE* **2838**, 114–123.

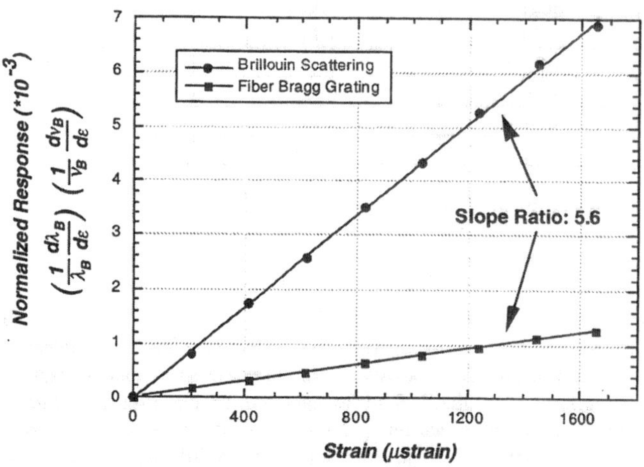

FIGURE 7.44. Variation in both FBG center wavelength shift and Brillouin frequency shift with strain. From Davis, M.A. and Kersey, A.D., (1996b), "Simultaneous Measurement of Temperature and Strain Using Fiber Bragg Gratings and Brillouin Scattering" *SPIE* **2838**, 114–123.

Table 7.1 Temperature and Strain Coefficient for FBG and Brillouin Scattering

	Temperature	Strain
FBG	(10.2 ± 0.1) pm °C^{-1}	(1.181 ± 0.005) pm με^{-1}
Brillouin	(1.21 ± 0.01) MHz °C^{-1}	(55.1 ± 0.4) KHz με^{-1}

observed. The respective strain and temperature responsitivities for the two techniques are presented in Table 7.1 (Davis and Kersey, 1996a).

The difference in the responsitivities ensures that the determinant of the **K**-matrix is nonzero, so the matrix can be inverted and the two techniques used to simultaneously evaluate the strain and temperature. Figure 7.45 displays a test of the system, where the strain was varied in a random fashion, while the temperature was approximately linearly increased. Although this approach shows great promise, at present the complexity, size and cost of the system do not lend themselves to the kind of compact, portable, and inexpensive system required for broad implementation.

7.7 TEMPERATURE-INDEPENDENT STRAIN SENSORS

For applications where the sensor is not bonded to, or embedded within, a host structure, or there is little difference between the thermal expansion coefficient of the host material and that of the fiber optic sensor, a number of temperature-insensitive fiber optic strain sensors have emerged.

The principle used in the novel temperature-independent strain sensor of Xu *et al.* (1995a) involves the observation that strain applied to a chirped FBG changes the bandwidth of the grating and therefore its overall reflectivity (Fig. 7.46). Change of temperature, however, does not influence the bandwidth of the chirped grating (Huang *et al.*, 1994). *Differential etching* of the grating region of the optical fiber creates a taper (Putnam *et al.*, 1995b) which leads to chirping of the FBG on the application of tension. The experimental arrangement used to test this sensor is depicted in Fig. 7.47. An ELED (1550 nm, bandwidth 96 nm) illuminates the chirped FBG and a 3-dB coupler directs a fraction of the reflected light from the FBG sensor to a sensing photodetector. The coupler also sends a fraction of the original light from the source to a reference photodetector. The length of the FBG was 20 mm, and the 25 mm taper varies the cladding diameter from 125 to 78 μm. The system output, defined as the ratio of the signals from the two photodetectors, is shown

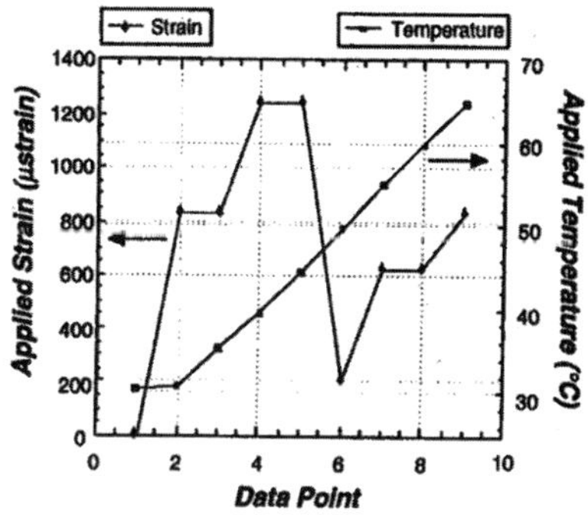

a)

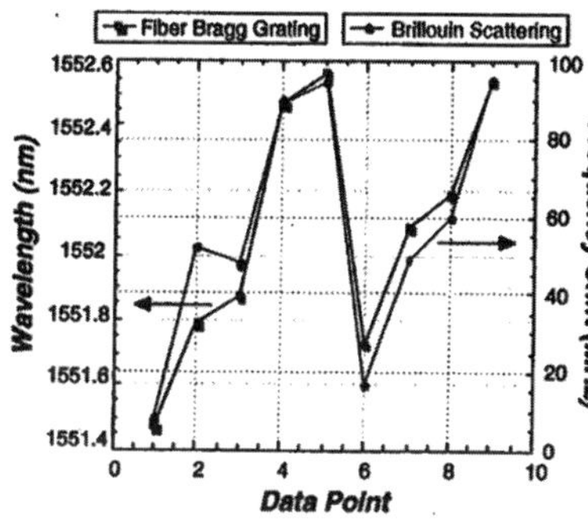

b)

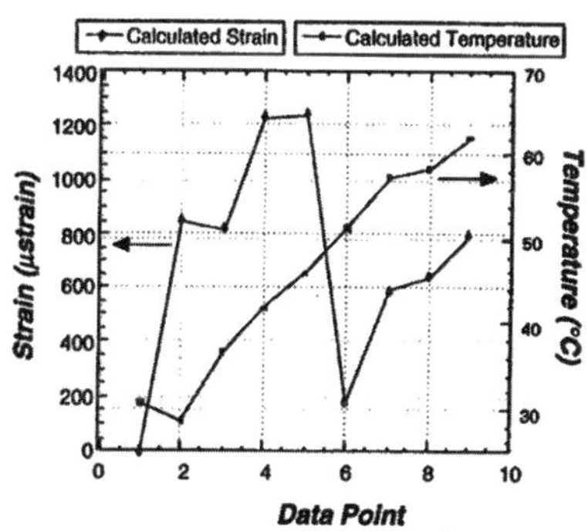

c)

FIGURE 7.45. (a) Strain and temperature variations imposed on the FBG sensor of the system illustrated in Fig. 7.41. (b) FBG wavelength shifts and Brillouin frequency shifts corresponding to the strain and temperature variations of (a). (c) Calculated strain and temperature variations derived from (b). From Davis, M.A. and Kersey, A.D., (1996b), "Simultaneous Measurement of Temperature and Strain Using Fiber Bragg Gratings and Brillouin Scattering" *SPIE* **2838**, 114–123.

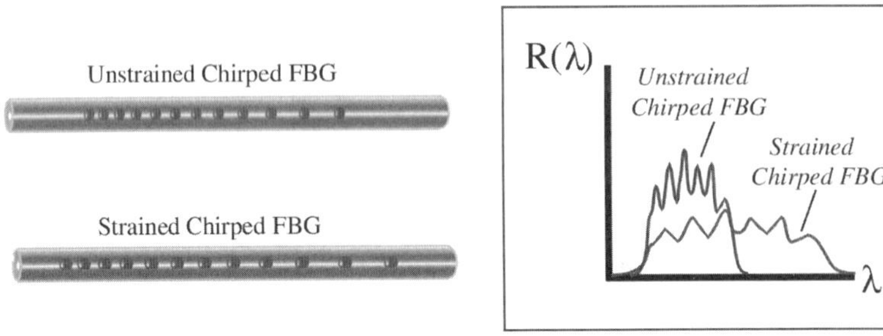

FIGURE 7.46. Schematic illustration of the change in the reflection spectrum of a chirped FBG when subject to strain.

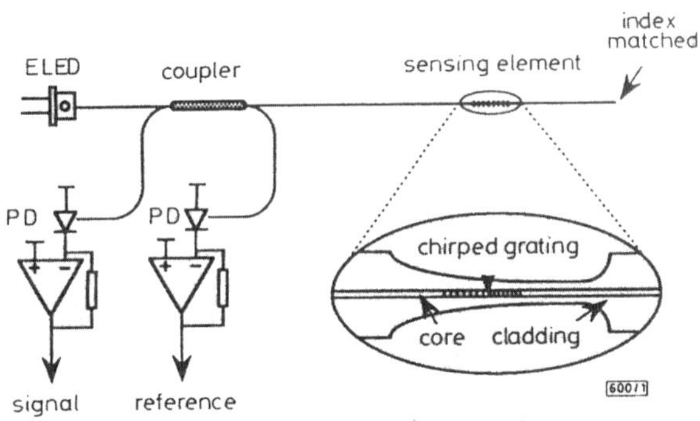

FIGURE 7.47. An experimental arrangement designed to test the concept of a temperature-independent fiber optic strain sensor based on the use of chirped FBG. From Xu *et al.*, "Temperature-Independent Strain Sensor Using a Chirped Bragg Grating in a Tapered Optical Fiber" *Electronics Letters* **31**(10), 823–825, used with permission, IEE Publishing Ltd.

in Fig. 7.48a to vary with strain but be independent of temperature, while its linear dependence on strain is revealed in Fig. 7.48b. The strain resolution was observed to be 4.4 με, corresponding to 1% of the full measurement range. *Prechirping* of the grating, by forming it under tension, permits a wider strain range to be accommodated. Unfortunately, this sensor, loses one of the virtues of FBGs, wavelength encoding of the sensing data; consequently, it is vulnerable to microbend lead loss.

Xu *et al.* (1994c) have demonstrated a thermally compensated bending gauge based on an opposing pair of surface mounted FBG sensors. This concept is very simple and is illustrated in Fig. 7.49. It has some similarity to the approach adopted by Butter and Hocker (1978) to eliminate temperature sensitivity from their strain measurements. In essence, Xu

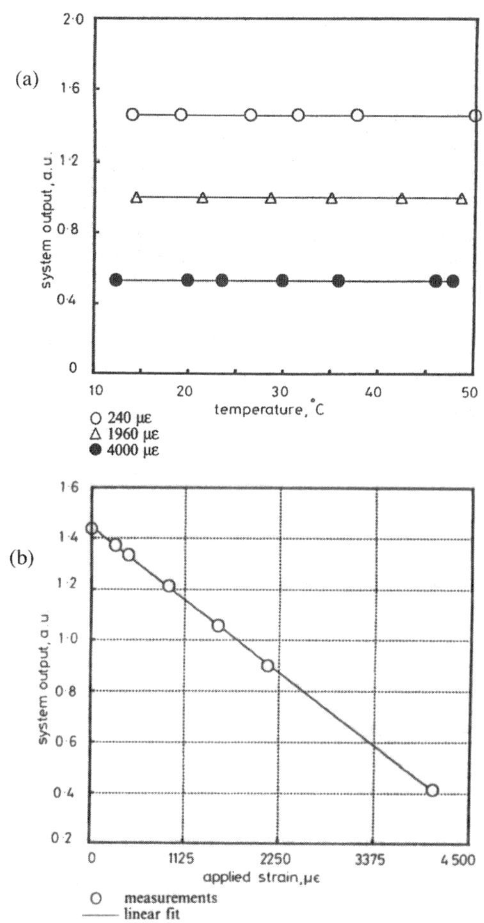

FIGURE 7.48. (a) Radio of the sensing and reference photodetector signals (of the system shown in Fig. 7.47) as a function of temperature for three values of strain applied to the chirped FBG sensor. (b) Variation of the ratio with applied strain. From Xu *et al.*, "Temperature-Independent Strain Sensor Using a Chirped Bragg Grating in a Tapered Optical Fiber", *Electronics Letters* **31**(10), 823–825, used with permission, IEE Publishing Ltd.

et al. (1997), proposed mounted two FBG sensors on the opposing sides of the structure to undergo bending. In their experiment a cantilever beam was used within an oven, and the two FBG sensors were positioned equally distant from the bending neural axis of the beam so one experienced tensile strain, ε, and the other an equal and opposite compressive

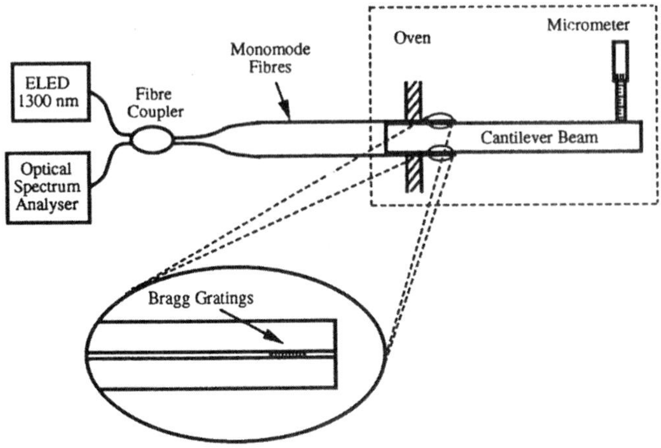

FIGURE 7.49. A temperature-independent bend sensor involving a pair of FBG sensors placed on opposing sides of the bend neutral axis. From Xu, M.-G., Archambault, J.-L., Reekie, L., Daking, J.P., "Structural Bending Sensor Using Fiber Gratings," *Proc. SPIE*, **2292**, 407–413: Fiber Optic and Laser Sensors XII, DePaula, R.P. (Ed.).

strain, $-\varepsilon$. Assuming the FBGs are identical and exposed to the same change of temperature, ΔT, the difference in their Bragg wavelengths,

$$\lambda_B(+\varepsilon, T) - \lambda_B(-\varepsilon, T) = S_\varepsilon \varepsilon + S_T \Delta T - S_\varepsilon\{-\varepsilon\} - S_T \Delta T = 2S_\varepsilon \varepsilon,$$

is clearly proportional to the strain and is independent of the change in temperature. This is confirmed in Fig. 7.50, where the strain readings are seen to be unaffected by a change in temperature of 28°C.

Special long-period gratings, produced by careful control of the core index of refraction radial profile, have been shown to make temperature-insensititive strain gauges (Bhatia *et al.*, 1997b). The wavelength–temperature and wavelength–strain coefficients for an LPG with a period $\Lambda = 166\ \mu m$, and length $L = 3.65$ cm were determined to be -2.5 pm $°C^{-1}$ and -2.7 pm $\mu\varepsilon^{-1}$, respectively; see Fig. 7.51. To test the *thermal cross-sensitivity* of this LPG, it was heated from 25 to 125°C for three settings of strain. The results are presented as Fig. 7.52 and show the relative insensitivity of this strain gauge to temperature fluctuations. Although LPG sensors are fabricated with relatively inexpensive amplitude masks (Vensarkar *et al.*, 1996), they only work in transmission. This rules out the highly desirable single-ended sensor architecture, unless the

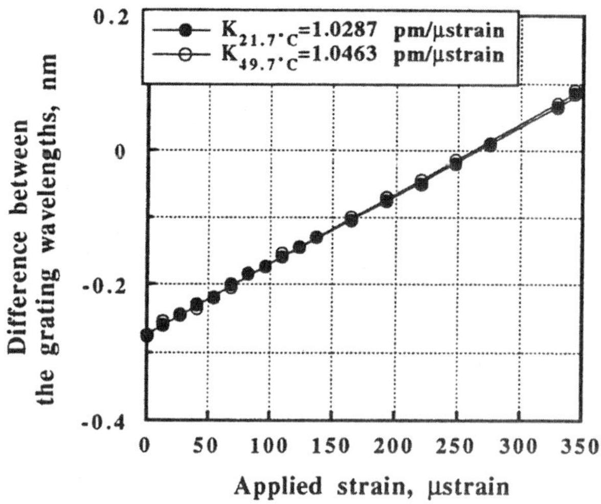

FIGURE 7.50. Experimental measured difference in FBG center wavelengths (for setup shown in Fig. 7.49) as a function of applied strain for two temperatures. From Xu, M.-G., Archambault, J.-L., Reekie, L., Daking, J.P., "Structural Bending Sensor Using Fiber Gratings," *Proc. SPIE*, **2292**, 407–413: Fiber Optic and Laser Sensors XII, DePaula, R.P. (Ed.).

endface of the optical fiber is mirrored. If the limited spatial resolution of these sensors is not a problem, this approach is quite encouraging. However, the multinotched transmission spectrum of LPGs makes it difficult to use wavelength division multiplexing on a serial array of such sensors.

In addition to the foregoing approaches to develop temperature-insensitive strain sensors, there has also been some attempt to built fiber optic strain sensors that are designed to "cancel" thermal apparent strain. Belleville *et al.* (1995) demonstrated that the residual thermal drift of their low coherence extrinsic Fabry–Perot strain sensor could be limited to 1 $\mu\varepsilon\ {}^{\circ}C^{-1}$. This was accomplished by replacing the reflective fiber in their sensor with a metal fiber having a thermal expansion coefficient that is as close to that of the host structure as possible (Fig. 7.53). Unfortunately, such sensors are very host-material-specific, which makes them expensive, as they have to be tailor-made to the host. Furthermore, this type of strain sensor would not work with FRP materials, for the value of the host average thermal expansion coefficient depends on the particular configuration, or layup, of the FRP.

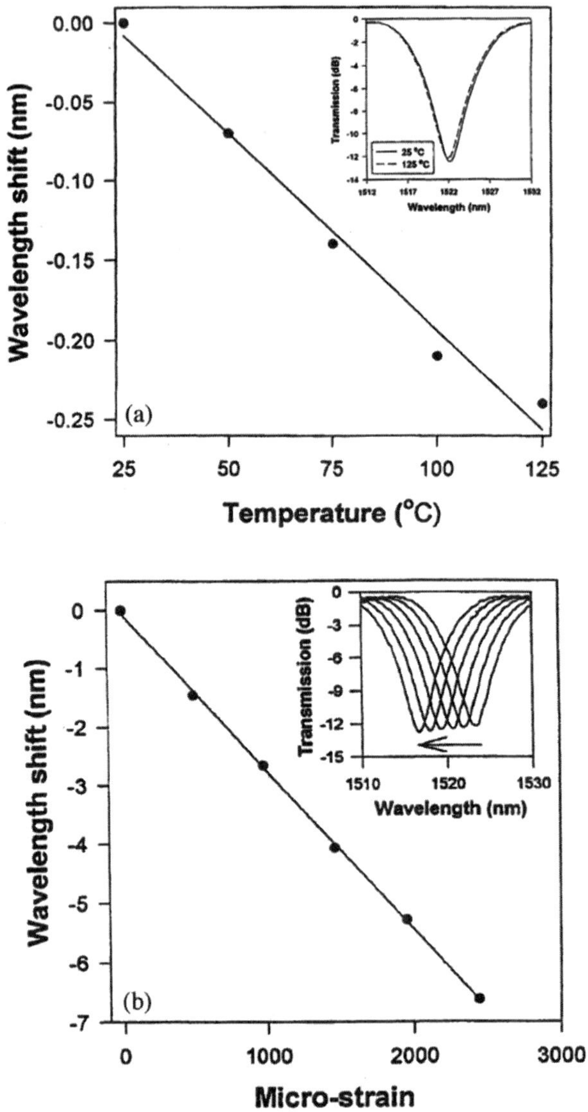

FIGURE 7.51. (a) Small variation in the center wavelength of one of the transmission notches of a specially designed LPG with temperature. Inset shows shift in the spectrum of the transmission notch for 100°C change in temperature. (b) Large variation in the center wavelength of one of the transmission notches of a specially designed LPG with strain. Inset shows the corresponding shift of the transmission notch spectrum. From Bhatia *et al.*, (1997b), "Temperature-Insensitive Long-Period Gratings for Strain and Refractive Index Sensing," *SPIE*, **3042**, 25.

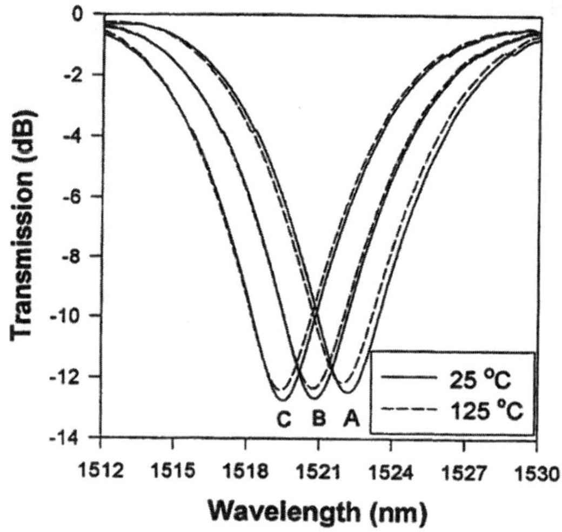

FIGURE 7.52. Transmission notch spectra of a special temperature-insensitive LPG for the two temperatures shown (25 and 125°C) and three strain values [A (0 $\mu\epsilon$), B(500 $\mu\epsilon$), and C (100 $\mu\epsilon$)]. From Bhatia *et al.*, (1997b), "Temperature-Insensitive Long-Period Gratings for Strain and Refractive Index Sensing," *SPIE*, **3042**, 25.

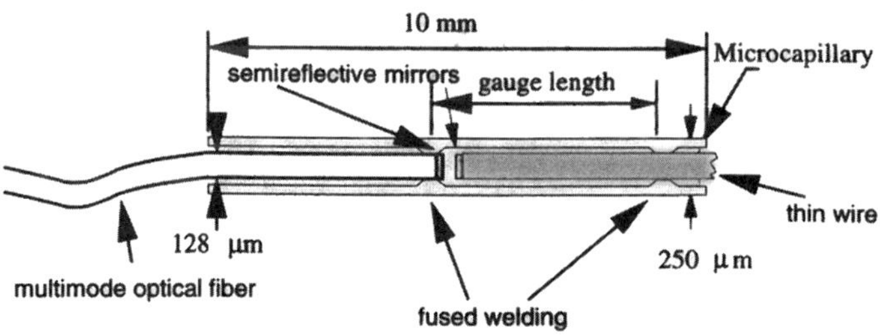

FIGURE 7.53. Schmatic of a temperature-insensitive EFPI based on using a thin wire with a thermal expansion coefficient designed to match that of the host structure. From Belleville, C. and Duplain, G. (1993), "White Light Interferometric Mutimode Fiber-Optic Strain Sensor" *Optics Letters* **18**, 78–80.

7.8 STRAIN–TEMPERATURE CROSS-SENSITIVITY

An important underlying assumption that has been made in all of the above work is that the strain-optic and thermo-optic coefficients are constant and independent of the applied strain or change of temperature. If this assumption were not valid, nonlinear terms would have to be introduced and these would make interpretation of sensor signals much more difficult. This nonlinearity arises because Eq. (7.4) was based on neglecting all but the linear terms in the original Taylor expansion. For large values of $\Delta\varepsilon$ and ΔT, this may no longer be acceptable, and we need to take account of higher-order terms in this expansion. The next term likely to be of interest is the *strain and temperature cross-sensitivity term* given by the product of the strain and temperature *cross-sensitivity coefficient* $C_{\varepsilon,T}$ and the incremental changes $\Delta\varepsilon$ and ΔT, i.e.,

$$\frac{\Delta\zeta_L}{\zeta_L} = S_\varepsilon \Delta\varepsilon + S_T \Delta T + C_{\varepsilon,T} \Delta\varepsilon \Delta T. \quad (7.75)$$

Farahi *et al.* (1990) and Vengsarkar *et al.* (1990) have shown that for free optical fibers, this cross-term is negligible unless large strain and temperature excursions arise. However, they also observed that this cross-sensitivity coefficient is proportional to the length of the optical fiber, so that long gauge-length sensors are more likely to be influenced by large strain and temperature excursions than short gauge sensors. Farahi *et al.* (1990) have determined that $C_{\varepsilon,T} \sim 10^{-8}$ rad°C^{-1} $\mu\varepsilon^{-1}$ m^{-1}. A more complete treatment of second-order sensitivities has been undertaken for bow-tie optical fibers at 633 and 670 nm by Ma *et al.* (1997). Farahi (1993) has shown that in terms of polarization-maintaining optical fiber, elliptic-core fiber is more vulnerable to cross-sensitivity than bow-tie optical fiber.

CHAPTER 8

Sensor Installation and Material Integration Issues

8.1 INTRODUCTION

To function reliably and accurately, a fiber optic structural sensor has to be installed properly, whether on the surface of its host structure, or embedded within it. In general this type of sensor has to be strain coupled to the host via a bondline so this is a crucial part of any correctly functioning structural sensor. Between the optical fiber and the host structure there often lies a protective coating. The properties of this coating can have a major influence on the life and functionality of the sensor, as well as the affect the optical fiber has on the host structure. Indeed, careful consideration must be given to the perturbative nature of the inclusion created by an embedded optical fiber within a FRP composite host material. This perturbation can give rise to a region of *stress concentration* around the embedded optical fiber that can affect the interpretation of the sensor's strain measurement and even compromise the host's strength and life. Where optical fibers are to be incorporated within pultruded structures, the effects of the pultrusion process on embedded sensors and their functionality also has to be investigated. In this chapter we examine these and other issues that have to be addressed if fiber optic structural sensors are to find broad application and work correctly over an extended lifetime with minimal influence on the host's strength and durability.

8.2 INSTALLATION OF FIBER OPTIC STRUCTURAL SENSORS

In any installation of fiber optic sensors it is important to avoid light loss through bends with a radius of curvature less than about 3 cm, or high

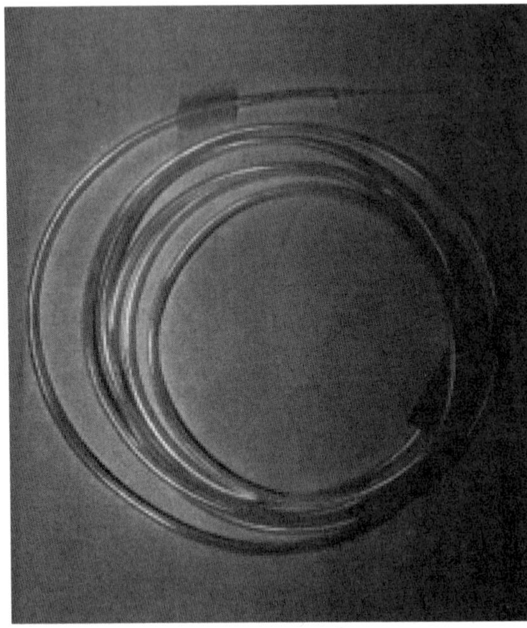

FIGURE 8.1. Optical fiber protected within Penflex^{TM} crush-resistant tubing.

pressure points. The optical fiber can be protected to some extent by housing it within a crush resistant tubing, such as Penflex^{TM} (Fig. 8.1). We shall often use the term *fiber optic cable* to mean a *jacketed* and therefore well protected optical fiber. Melloni *et al.* (1997) have developed a pultruded flat, fiber glass ribbon that contains *8*-optical fibers. This *fiber optic ribbon cable* could find broad application where a number of fiber optic sensors are required at one location and the interrogation system is located some distance away. Indeed, it represents the optical analogue of the flat wire ribbon which greatly improved complex wiring patterns within todays electronic systems.

Connectors and ports should be cleaned with acetone or ethanol, whenever there is a reconnection. There are several ways of mounting fiber optic structural sensors on a structure. The procedure used depends on whether the installation is on: concrete, metal or FRP composite material. The first stage of any installation is surface preparation. Since strain transfer invariably has to go through the bondline between the sensor and its host, it is crucial that this bond is strong and complete as possible.

8.2.1 Surface installation on metal structures

The surface preparation technique for bonding a fiber optic structural sensor to a metal structure, such as steel, is essentially the same as used for conventional strain gauges and starts with a thorough cleaning of the surface. After selecting the location for the sensor it is extremely important to remove any loosely bound material, such as rust or old paint, then clean the surface with acetone or ethanol to remove any oil, grease or other contaminants. It is possible to purchase fiber optic strain sensors that have already been mounted onto small metal plates that can be directly welded to steel (Fig. 8.2). These are quite robust strain gauges and include FBG sensors from *ElectroPhotonics Corporation* and EFPI sensors from *Roctest*.

Where a fiber optic strain sensor is to be mounted directly on a metal surface draw two short pencil lines along the desired bondline spaced

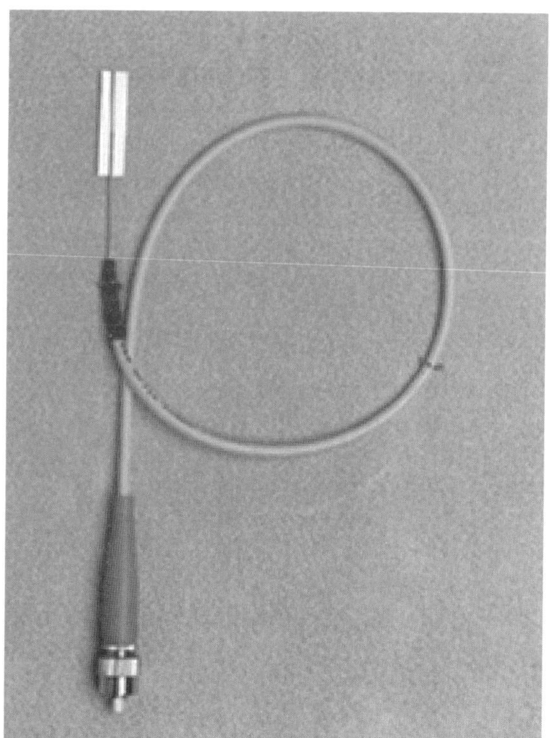

FIGURE 8.2. Weldable fiber optic Bragg grating strain sensor. Used with permission, ElectroPhotonics Solutions Corporation.

about *5 cm* apart, *A* and *A* in Fig. 8.3a. Using two pieces of flashbreaker tape, *B* and *B*, create a *2 mm* wide channel along the bondline (Fig. 8.3b). This tape will control the depth of the epoxy adhesive and can be removed later leaving no residue. A syringe cartridge, without the needlepoint, is used to distribute the epoxy in a series of small beads every *4* to *6 mm* along the projected bondline (Fig. 8.3c). Smooth the beads into a continuous line to a uniform depth of about *1 mm* over the channel between points *A* and *A*, and remove any excess epoxy from the flash-breaker tape (Fig. 8.3d). The fiber optic sensor should then be laid into the epoxy, making sure it is aligned with the points *A* and *A*, and that the

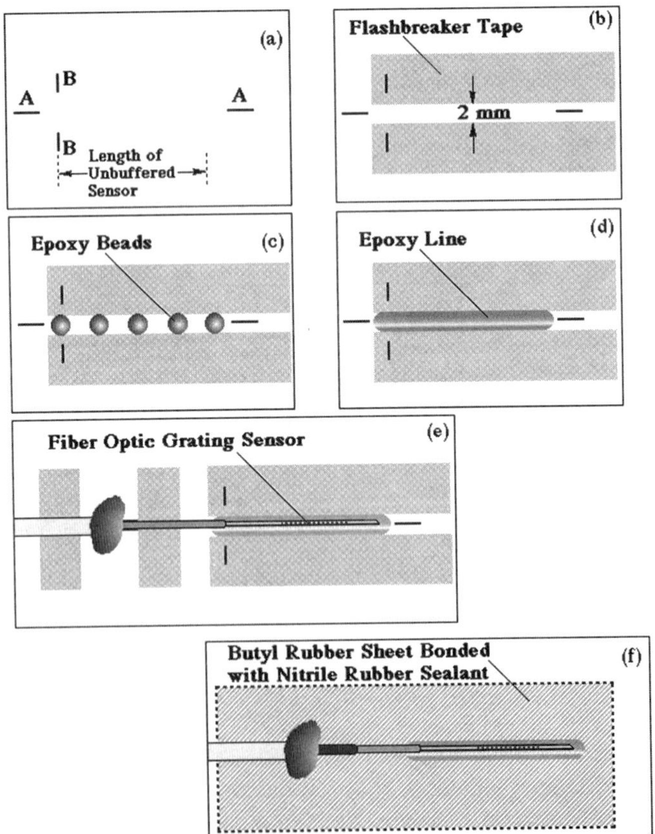

FIGURE 8.3. Schematic illustration the six steps described in the text for surface installation of a fiber optic strain sensor. "Installation, Use and Repair of Fiber Optic Sensors" 1998. Used with permission, ISIS Canada.

coated and jacketed sections are affixed with small sections of the flash-breaker tape (Fig. 8.3e). Note that the bare section of the optical fiber containing the *sensor should never be touched*, only the coated {or jacketed} section should be handled. Finally the sensor is protected from any pulling action applied to the cable, by one or more epoxy beads added to secure the jacketed portion of the optical fiber to the metal. Once the epoxy gels remove the tape and allow the epoxy to fully cure. If the sensor is likely to be exposed to a hostile environment a further protection can be afforded in the form of a butyl rubber sheet that is bonded over the sensor with nitrile rubber sealant (Fig. 8.3f).

8.2.2 Installation on or within concrete structures

In the case of *existing concrete structures* it is best to use some form of protective wrap rather than try and install the optical fibers directly to the concrete. This overwrap could, for example, take the form of a small coupon of carbon FRP, such as Mitsubishi Replark Prepreg material, or a complete FRP wrap used to rehabilitate and strengthen the concrete (Fig. 8.4). In the case of laboratory experiments it is possible to directly mount the fiber optic sensors on the surface of a concrete structure (Fig. 8.5), however, they are vulnerable unless protected. The surface of the concrete should be ground smooth, then washed and thoroughly dried. A procedure similar to that described above (and illustrated in Fig. 8.3) should then be utilized. If the sensor is embedded within an FRP laminate, an epoxy primer glue should first be applied to the surface until it is no longer absorbed. After 24 hours a second coat is applied and allowed to dry. The laminate with the fiber optic sensor is then bonded to the concrete with epoxy glue. It is important to avoid the formation of bubbles and ensure that a uniform layer of the epoxy bonds the entire laminate to the concrete. The optical fiber should be aligned parallel with the adjacent ply direction in the laminate to avoid the formation of a resin eye and a bump in the wrap. Leave adequate optical fiber cable emerging from the laminate for connectorization at a site not easily reached by vandals. The fiber optic cabling should also be covered, if possible, with some form of material wrap to avoid drawing attention to it.

In the case of a rebar, to be installed in a new concrete structure, ideally a small area sufficient to accommodate the sensor is milled flat, and the procedure just outlined undertaken. An example of a fiber optic structural sensor mounted on a shear reinforcement used in the Taylor Bridge (Maaskant *et al.*, 1998), is shown in Fig. 8.6. The strong alkalinity of concrete {pH of ~ 12} can corrode unprotected optical fibers, so addi-

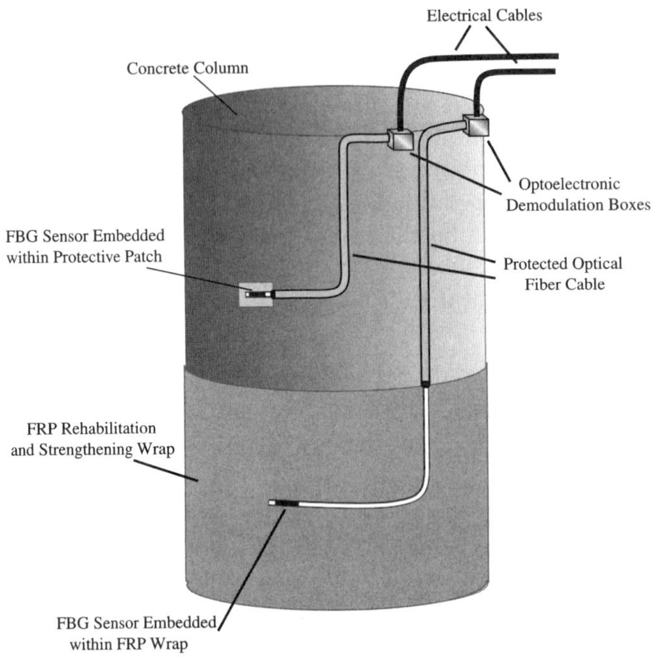

FIGURE 8.4. Illustration of two methods of installing fiber optic strain sensors on existing concrete structures. (1) The sensor is embedded within a protective coupon of FRP that is bonded to the structure. (2) The sensor is embedded within a rehabilitation and strengthening FRP wrap.

tional protection of the sensor is necessary in this environment. This can be provided by the application of a *nitrile rubber sealant* and *butyl rubber sheets*, see Fig. 8.3f. This form of protection is especially well suited to protect the sensors installed on steel rebars from the powerful vibration mixers used in the construction of large concrete structures. In most cases an additional covering of tape, for moisture and alkaline protection, is used and serves as an obvious indicator of the presence of the fiber optic sensor (Fig. 8.7). If possible it is desirable to build the instrumented section of rebar in a controlled environment then ship it the construction site where it is used with the regular rebars.

Carbon FRP material is especially well suited to replace steel as prestressing tendons within various types of concrete structure. As indicated in Chapters 2 and 9, a number of bridges have recently been constructed with two types of CFPR prestressing tendons in place of steel in some of the deck-support girders. One type is in the form of a seven stranded rope, the other involves a pultruded rod, see Fig. 2.4a,b,

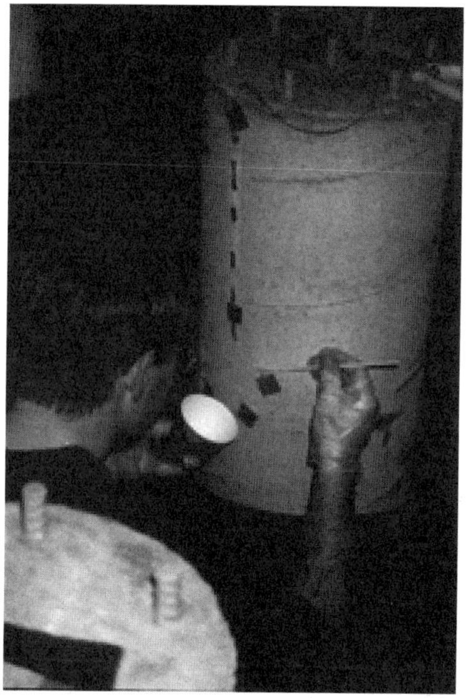

FIGURE 8.5. Surface mounting of a fiber optic Bragg grating sensor on the outside of a concrete column.

respectively. Mounting the optical fiber between strands of the CFRP rope is quite a time-consuming operation and requires the local removal of the wrapping used to protect the rope and provide extra bond strength with the concrete. The pultruded CFRP rod has ridges to enhance bonding to concrete. These ridges must be ground down to make a smooth surface for mounting the sensor. However, care is needed not to cut too deep as this could destroy some of the carbon fibers and this could weaken the tendon. It is desirable, if possible, to glue the fiber optic sensor to the tendon only once the tendon has been put under its maximum tensile load. The sensor will then be in a mild state of compression for the remainder of its life and this is much more desirable than having it in a perpetual high state of tension, see section 5.6. As described earlier, the tendon surfaces have to be thoroughly cleaned prior to bonding the sensors to them and a procedure similar to that described for a metal surface should be followed, as far as possible.

The fiber optic cables are guided along reinforcing, or prestressing, elements to their desired point of egress from the concrete structure

FIGURE 8.6. Fiber optic Bragg grating sensor mounted on the cleaned and smoothed surface of a steel rebar to be installed within a concrete structure.

(Fig. 8.8). At this location a *metal housing* is placed before any concrete is poured and the excess length of optical fiber is stored and protected within this housing during fabrication of the concrete structure, see Fig. 8.9 and in addition Figs 9.39 and 9.40. The optical fiber cables stored in these housings are also shielded from damage during any required transportation of such instrumented concrete structures.

Where direct strain measurements from within the concrete are desired the optical fiber sensor can be mounted inside a specially designed *metal tube with two end flanges* for better adhesion to the concrete (Fig. 8.10). Both *ElectroPhotonics Corporation* and *Roctest* have products of this nature. This type of sensor is often set in concrete in one of two ways: it can be cast directly into the wet mix, or encapsulated into concrete briquettes which are cast into the wet mix. The use of these *instrumented briquettes* is particularly valuable where there are large aggregates that would otherwise produce misleading strain readings. As we shall see in Chapter 10, a company named *Smartec*, of Switzerland has expanded on the early work of Inaudi *et al.* (1996) and developed *long gauge* fiber optic deformation

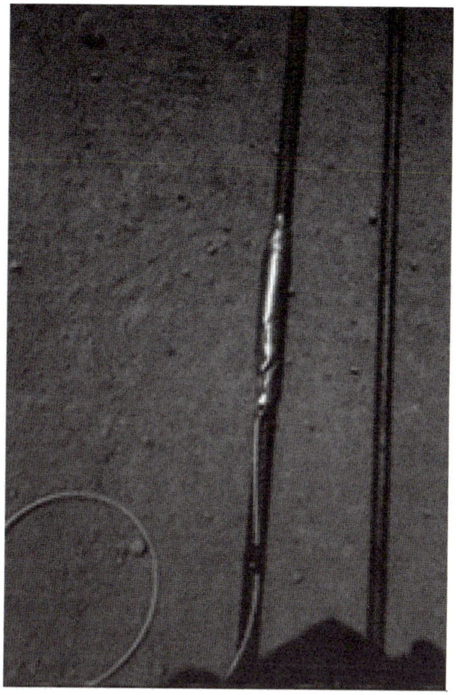

FIGURE 8.7. Steel rebar instrumented with a surface bonded fiber optic sensor and ready for installation within a concrete structure. Clearly seen is the tape covering the protective butyl rubber sheet and the tough optical fiber cable lead to the sensor.

sensors that are mounted within metal pipes for complete protection. These instrumented pipes can then be attached to other reinforcements prior to concrete pouring.

Since the environment at a construction site is as different from a fiber optics laboratory as night is from day, it is very desirable to do as much preparation as possible in a clean and well lit location before moving to the actual construction site. It is important that people working along side the fiber optic sensor installers be as enlightened as is feasible so they contribute to the solution of problems rather than their creation. More detailed information on installation procedures is available from ISIS-Canada in the form of a handbook titled: "Installation, Use and Repair of Fiber Optic Sensors". *Prefabrication* of the *fiber optic structural monitoring* system and *remote communications* will probably be the way to integrate this technology into structures in a way that makes it relatively transparent to construction site workers. A section of the CFRP shear reinforcement used in the Taylor Bridge, see Chapter 9, is shown in Fig. 8.11, being

FIGURE 8.8. Several fiber optic instrumented rebars are seen within the mesh of steel reinforcement for a concrete structure. The optical fiber cables are routed along this mesh to the planned egress points for the concrete structure.

FIGURE 8.9. Example of the special metal housings provided at the egress point for the fiber optic sensor cables used on the Confederation Bridge. These housing protect the optical fiber cables during construction (and if necessary, transportation) of the concrete structure. Used with permission, ISIS Canada.

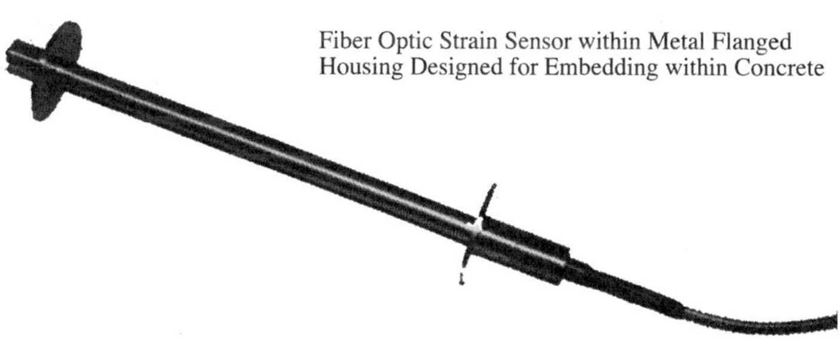

FIGURE 8.10. Specially designed fiber optic Bragg grating strain sensor for direct embedding within a concrete structure during fabrication. Used with permission, ElectroPhotonics Solutions Corporation.

carried by Professor Urs Meier, Director of EMPA and one of the first to encourage the use of this type of material for Civil Engineering. Eventually, structures like this that are instrumented with fiber optic structural sensors should be almost indistinguishable from uninstrumented structures as far as the construction site workers are concerned.

FIGURE 8.11. A section of CFRP shear reinforcement being carried by Professor Urs Meier, demonstrating its light weight in comparison to its steel counterpart. Used with permission, ISIS Canada.

8.2.3 Installation within FRP laminated structures

To install fiber optic structural sensors within *FRP laminated structures* a set of procedures needs to be adopted which minimizes their impact on the performance of the host structure and its fabrication process. First the direction and number of optical fibers has to be established with the location in terms of the ply lay-up scheme. The exact siting of the sensor section of the optical fiber also has to be determined in terms of the measurement requirements. The *ingress/egress points* need to also be established in terms of manufacturing constraints and possibly siting of the host structure in terms of other structural elements.

If the optical fiber is to exit its FRP host through an edge, the usual trimming cannot be undertaken after the fabrication of the structure. If trimming is necessary, the optical fiber cannot egress through the edge and instead must be fed through slits created in the successive FRP prepeg layers until the optical fiber finally emerges from the surface of the host structure as an optical fiber pigtail. In the cured FRP composite structure, the point of exit for the *optical fiber pigtail* represents a stress point and some form of *strain relief* must be provided. Silicone can be used for this purpose.

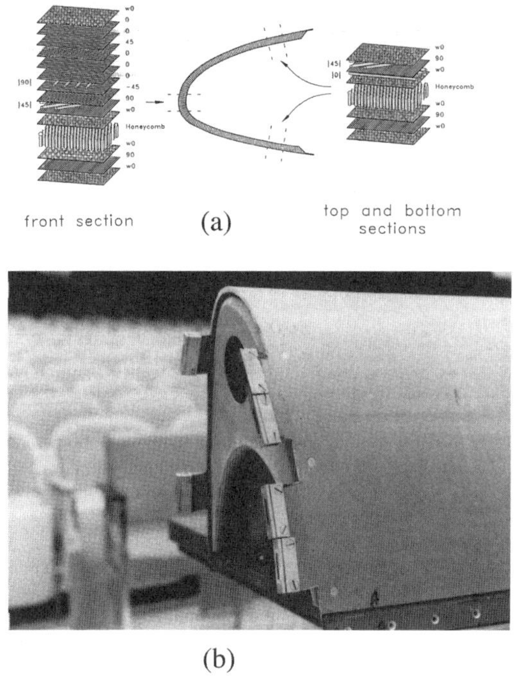

FIGURE 8.12. (a) Location and orientation of embedded optical fiber damage sensors within the aramid fiber/epoxy layup used in the leading edge of a DASH-S aircraft wing. (b) Fabricated FRP aircraft wing leading edge revealing the optical congectors used with the embedded optical fiber damage sensors. Reprinted from LeBlanc M. and Measures R.M., (1992), "Impact Damage Assessment in Composite Materials with Embedded Fiber Optic Sensors," *J Composite Engineering* **2**, 573–596, used with permission from Elsevier Science.

The aramid fiber/epoxy leading edge of an aircraft wing, shown as Fig. 6.4b, is an example of a prepreg FRP structure that was instrumented with a large *2D*-array of embedded optical fibers (Glossop *et al.*, 1990). In this instance, the optical fibers were embedded within several layers of this structure, as shown in Fig. 8.12a, and emerged through the edge where they were collected and interfaced to the light source and photodetector array by means of the polished connectors shown in Fig. 8.12b (LeBlanc and Measures, 1992).

Most FRP laminated structures are created from a number of prepreg layers and use some type of form or mold to define their shape during the cure process as their matrix material often becomes liquid-like at some time during the cure-cycle. The optical fiber is cleaned in the area to be embedded and placed on the appropriate prepreg layer in such a manner that its path through the mold ensures its safety and location during the cure. If the fiber is inadequately secured and its location poorly sealed, matrix material will be lost and the embedded position of the fiber may shift relative to the exit location, causing at worst fiber breakage, or at best incorrect positioning.

8.3 FIBER OPTIC SENSOR INTEGRATION WITHIN FRP MATERIALS

This section and those that follow deal with fiber optic structural sensors that are embedded within FRP composite materials and address a number of key questions:

(i) do embedded optical fiber sensors degrade the performance and/or life of the host structure,
(ii) can the far-field host strain be accurately determined from a measurement of the strain within the optical fiber,
(iii) how effective is the strain transfer from host to sensor, especially when steep strain profiles are being measured.

As we shall see the properties of the optical fiber coating plays a key role in these matters. Other significant considerations that need to be addressed include: ingress and egress of optical fibers with respect to their host structure, and optical connectivity in the case of pultruded optical fiber sensors.

The first question is really on two levels. Do embedded optical fiber sensors cause: (i) *microscopic stress concentrations* and or damage which will eventually shorten the life of the host, especially when fatigue loaded, and (ii) *macroscopic damage* that will lead to a reduction in the structure's load carrying capacity. In trying to appreciate what influence the presence of an embedded optical fiber has on its host structure, we turn to *continuum mechanics*. Elastic theory predicts that a small hole in a plate loaded under tension, leads the stress in the immediate vicinity of the hole to be intensified by as much as a factor of three (Fig. 8.13). This so called *stress concentration* can lead to degradation of the material.

We saw in Fig. 2.28, that optical fibers embedded within FRP composite material parallel to the adjacent reinforcing fibers represent a cylindrical inclusion within the host material. If there is a mismatch in the stiffness at the interface between the two materials a *stress concentration* can result. The issue of what does an embedded fiber optic strain sensor actually measure is particularly significant in light of this observation, for in some sense the presence of the optical fiber sensor is perturbing the very parameter it is there to measure. Sirkis and Dasgupta (1995) used the

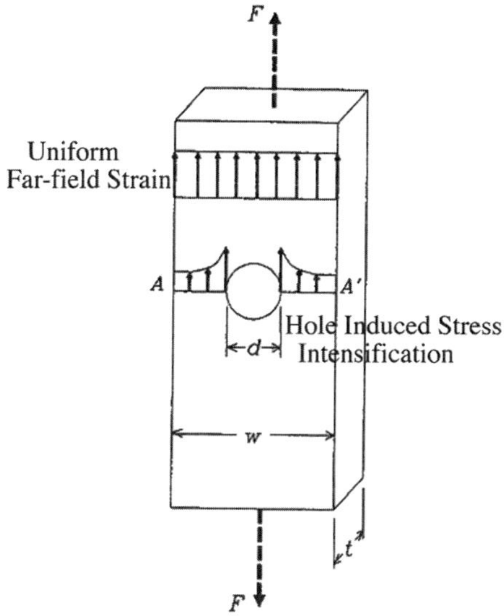

FIGURE 8.13. Schematic of stress intensification assoiciated with the presence of a hole in a tensile-loaded plate. From Sirkis J.S. and Dasgupta A., "Optical Fiber/Composite Interaction Mechanics" In *"Fiber Optic Smart Structures"* Eric J. Udd, (Ed.), © 1995 Wiley Interscience. Reprinted by permission of John Wiley and Sons, Inc.

term *obtrusivity* to indicate the extent of the perturbation in the average value of state variables {stress or strain} due to the presence of the embedded optical fiber. *Obtrusivity* not only affects the calibration of the sensor but can compromise the host structure if its stress concentration is excessive.

The first documented question about potential mechanical implications of placing optical fibers within load-bearing structures were raised by Udd *et al.* (1987) who were concerned about potential degradation in the strength of laminated composite materials with embedded optical fibers. Czarnek *et al.* (1989) responded to this concern by performing the first *Moiré interferometric study* of the strain fields created in the vicinity of optical fibers embedded in graphite/epoxy laminated structures. This research also drew attention to the existence of *lenticular resin-rich* regions that formed around optical fibers embedded at some *angle* to the *indigenous reinforcing fibers.* These *resin eyes,* as they are sometimes referred to in the literature, are illustrated in Fig. 8.14, which also provides an indication of the *notation* used to describe both the stacking sequence for the plies of the FRP and the orientation of an embedded optical fiber. Czarnek *et al.* (1989) also investigated the size and shape of these resin-rich zones and

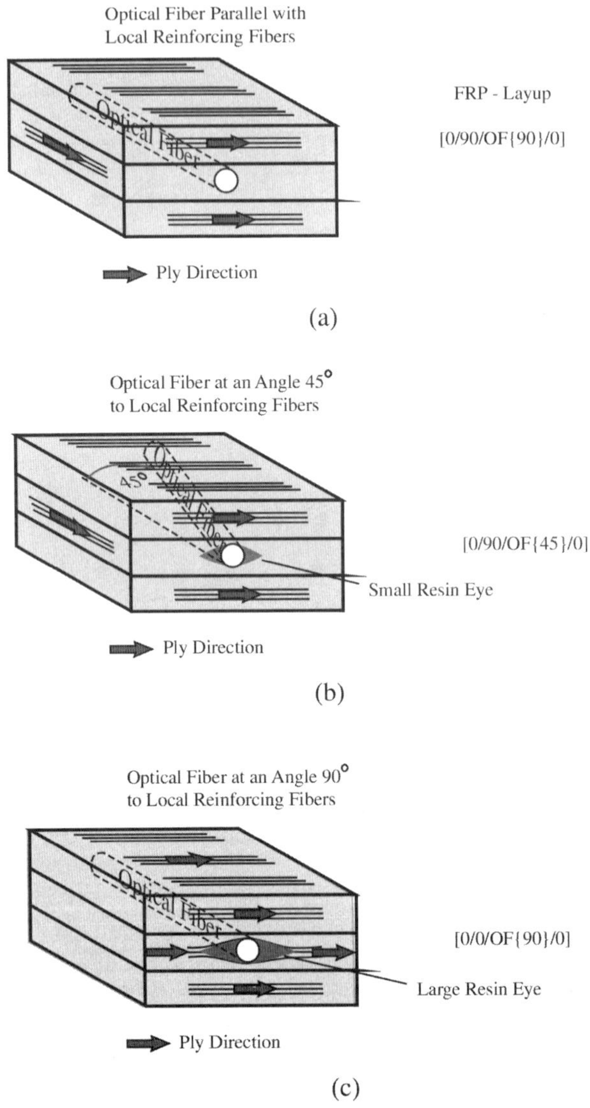

FIGURE 8.14. (a) Optical fiber embedded parallel to local ply direction. (b) Optical fiber embedded at 45° to local ply direction creates a modest resin eye. (c) Optical fiber embedded perpendicular to the local ply direction leads to the formation of a large resin eye.

how the stacking sequence of adjacent laminates contributed significantly to the induced strain concentrations. They measured the vertical displacement fringe patterns observed using *Moiré interferometry* applied to $[0_2/90_2/OF\{90\}/90_2/0_2]$ and $[90_2/0_2/OF\{90\}/0_2/90_2]$ uniaxial tension specimens and observed that the optical fiber causes a much larger perturbation {i.e., more fringes} in the local strain field when the optical fiber is at *90°* to the reinforcing fibers, than when aligned with them. Note the subscript indicates the number of plies at the specified orientation. Thus 90_2 means two plies at *90°*.

Even without interpreting the Moiré fringe patterns, it was quite apparent that the optical fiber has a pronounced effect on the local strain field. Similar studies were conducted by Salehi *et al.* (1989) who reproduced some of the original tests of Czarnek *et al.* (1989) and added specimens of greater thickness. One interesting fringe pattern recorded by Salehi *et al.* (1989), revealed the presence of a transverse matrix crack which has propagated directly through the optical fiber/host composite interface of a $[90_4/0_4/OF\{90\}/0_4/90_4]$ graphite-epoxy laminate. This type of matrix crack is symptomatic of the low transverse strength of most laminated composites and is not generally indicative of an optical fiber/laminated composite system performance. Indeed, several optical fiber size-effect studies have suggested that *strain perturbations* due to embedded optical fibers become insignificant for diameter of the order of *100 μm* or less, see section 8.5. These experiments were confirmed by the theoretical prediction of Case and Carman (1996).

8.3.1 Obtrusivity of embedded fiber optic structural sensors

To understand the way the *obtrusivity* of the embedded fiber optic strain sensor affects its calibration, we go back to the *basic opto-thermo-mechanical* equation (7.35) derived for unpolarized light. In the absence of any deviation from the reference temperature this equation for the normalized change in the *sensor optical path length* can be written:

$$\frac{\Delta\zeta_L}{\zeta_L} = \varepsilon_1^F - \frac{n_0^2}{2}[\varepsilon_1^F p_{12} + \tfrac{1}{2}\{p_{11} + p_{12}\}(\varepsilon_2^F + \varepsilon_3^F)], \tag{8.1}$$

where, ε_1^F is the *axial strain*, and ε_2^F, ε_3^F are the *transverse components of strain* in the optical fiber. To this point we have made the tacit assumption that the *strain in the optical fiber is the same as that in the host*; i.e.,

$$(\varepsilon_1^F, \varepsilon_2^F, \varepsilon_3^F) = (\varepsilon_1^H, \varepsilon_2^H, \varepsilon_3^H), \tag{8.2}$$

In reality this is *rarely* the case for as we have seen above the presence of the optical fiber can give rise to a stress concentration that makes the local value of the host strain quite different from its far-field {unperturbed} value. Sirkis and Dasgupta (1995) reviewed this subject and stated that a number of stress analysis techniques yield the same important findings: "the state of strain in the optical fiber core is quite different from the far-field value in the host". In order to relate the change in the SOPL, to the change in the *host far-field strain* the micromechanic interaction between the optical fiber and its host has to be taken into consideration. This interaction is governed by the relevant constitutive properties and the condition of the interface between the two materials since all displacements and stresses are transferred from the host to the optical fiber via their interface (Davidson *et al.*, 1990).

There has been a considerable amount written on this subject (Sirkis and Haslach, 1991; Sirkis and Dasgupta, 1992; Sirkis and Mathews, 1993; Kim *et al.*, 1993). Most papers treat the optical fiber as an *elastic cylindrical inclusion* within the host and use classical displacement formulations combined with far-field boundary conditions determined by the load applied to the host structure. Sirkis and Mathews (1993), demonstrated the extent to which *obtrusivity* can influence the *calibration* of an embedded fiber optic strain sensor by experimentally measuring the phase sensitivity of a Mach-Zehnder interferometer as a function of the ratio of the fiber to host stiffness under compressive loading. The *phase sensitivity,* in this instance, was defined:

$$S_\phi \equiv \frac{\Delta\phi}{L\varepsilon_1}, \tag{8.3}$$

where, $\Delta\phi$, represents the change in phase of the interferometer for a given length, L and axial strain, ε_1. The axial strain in the fiber is taken to be the same as that in the host in this experiment as a state of *generalized plane strain* in the *1-direction* is assumed to exist. The results of these experiments together with the predictions of a theoretical analysis by Sirkis and Haslach (1991), are presented as Fig. 8.15.

The familiar Butter and Hocker (1978) model is independent of the sensor/host stiffness ratio since it assumes that there is *zero coupling of host transverse strain* to the optical fiber and that the *axial strain in the fiber matches that of the host*. Consequently, on the graph displayed as Fig. 8.15, the *Butter and Hocker model* is represented by the horizontal line. It is noteworthy that the results presented in this figure suggest that the simple Butter and Hocker's model can be used when the *optical fiber stiffness is 10-times greater than that of the host material*, and the optical fiber locally reinforces the host material. Fortunately, this is often case when the optical

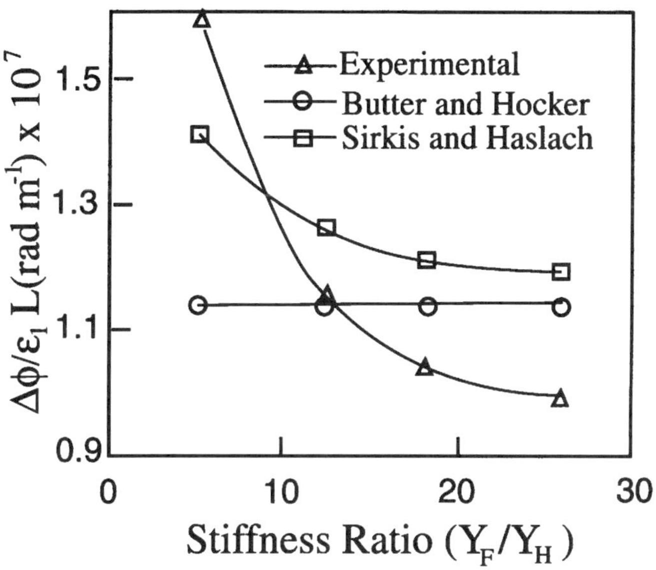

FIGURE 8.15. Variation of the interferometric phase sensitivity with the fiber to host stiffness ratio for experimental data (△), Butter and Hocker model (○), Sirkis and Haslack model (□).

fiber is embedded within a FRP composite material collinear with the adjacent reinforcing fibers.

Although these results are derived for a specific fiber optic strain sensor, they have broader application and are equally valid for *intrinsic Fabry–Perot*, and *fiber Bragg grating* sensors. However, fiber optic interferometric sensors based on air-gap cavities, such as the EFPI and ILFE, are *insensitive to transverse strain* and are consequently less effected by local stress concentrations. Indeed, their response is much more in keeping with the Butter and Hocker model and described by equation (7.41).

8.4 THE INFLUENCE OF FIBER OPTIC COATINGS

To this point we have neglected the influence of *optical fiber coatings* on the sensor/host strain transfer and the stress concentration associated with the embedded optical fiber. Coatings have long been used by the telecommunication industry to protect the glass fibers from surface abrasion and exposure to moisture as these can lead to the development of *surface microcracks* and premature fracture. The two most popular coatings are acrylate and polyimide, the latter is applied in a much thinner layer and is more high temperature resistant than the former. (Nath *et al.*, 1997) observed that certain *thermoset polyimide coatings* survived processing temperatures of *300°C* compared to the *100°C* failure of acrylate coatings.

Pak (1992) and Sirkis and Dasgupta (1993), have reviewed the subject of embedded optical fibers and modeled them as concentric cylinders within an infinitely large transversely isotropic host material (Fig. 8.16). The coating of the optical fiber is treated as the outer annular region with its own properties. Fig. 8.17a depicts how the shear strain induced in the optical fiber is affected by the thickness and shear modulus of the *optical fiber coating* under longitudinal shear loading conditions. In particular, the thinner the coating, the greater the shear strain in the optical fiber when the shear modulus of the coating is less than that of the host matrix. Conversely, for a coating that is stiffer than the surrounding host matrix, the strain transfer increases with the coating thickness (Pak, 1992). In the case of uniaxial tension perpendicular to the axis of the optical fiber, maximum strain transfer occurs when there is no coating (Fig. 8.17b), Sirkis and Dasgupta (1993).

Dasgupta and Sirkis (1992) suggested that the optical fiber coating provides an additional degree of freedom in the design process and presented evidence that supports the existence of mechanically "optimal" coatings for the embedded optical fiber sensor. The optimization criteria were chosen both from damage mechanics {*obtrusive*} rationale as well as

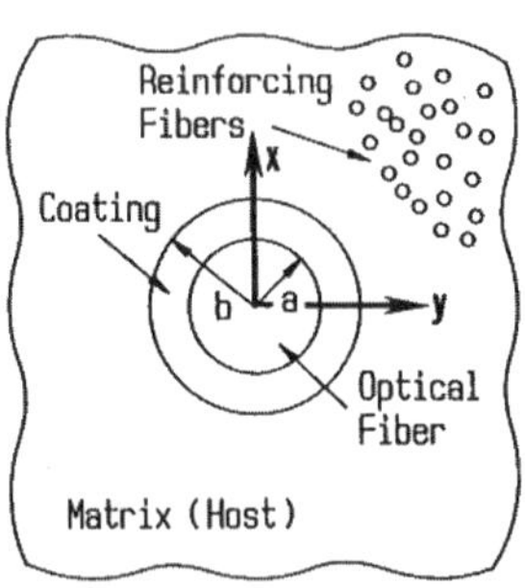

FIGURE 8.16. Schematic of the model used to examine the influence of coatings on strain transfer from a host structure to its embedded optical fiber. From Sirkis, J.S. and Dasgupta, A., "The Role of Local Internaction in Fiber Optic Smart Structures," *J. Intell. Mater. Syst. & Structures*, **4**, 260–271, Technomic Publishing Co. Inc., © 1993.

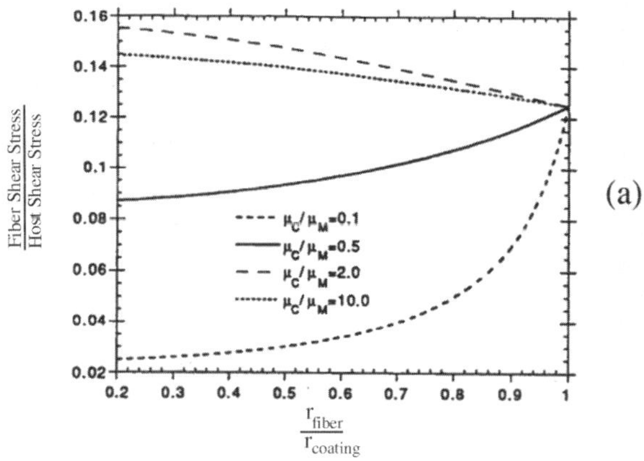

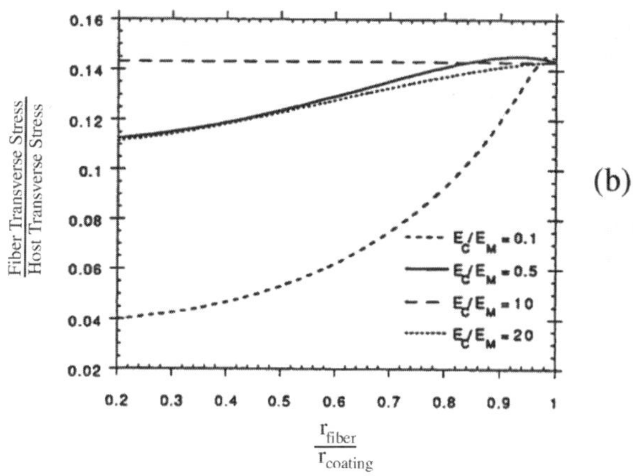

FIGURE 8.17. (a) Variation in the fiber to host shear stress ratio as a function of the ratio of the fiber to coating (plus fiber) radii for several ratios of the coating to host shear moduli. (b) Variation in the fiber to host transverse stress ratio as a function of the ratio of the fiber to coating radii for several values of the coating to host material ratio of Young's moduli. (a) From Pak Y.E., (1992), "Longitudinal Shear Transfer in Fiber Optic Sensors", *Smart Materials and Structures J*, **1**, 57–62, used with permission, Institute of Physics Publishing Ltd. (b) From Sirkis J.S. and Dasgupta A., "Optical Fiber/Composite Interaction Mechanics" In *Fiber Optic Smart Structures*, Udd E., (Ed.), with permission from Technomic Publishing Co. Inc., © 1995.

from the perspective of improving the accuracy of the strain sensor. As implied earlier, regions of high tensile stress around the optical fiber sensor can promote nucleation of fatigue cracks and propagation of pre-existing flaws introduced by *thermal expansion mismatches* during the manufacturing process. Dasgupta and Sirkis (1992) argued that coatings that produce a zero, or fully, compressive stress state in, or near, the embedded optical fiber are desirable, and restricted their investigation to coated optical fibers embedded parallel to the reinforcing fibers and loaded in uniaxial tension.

As an example of their calculations (Fig. 8.18), shows contours of optimal ratios of *coating radius to fiber radius*, r_C/r_F, in the *coating design space* for a graphite-epoxy host with longitudinal Poisson's ratio of 0.3. *Design space* is defined by the *coating Young's modulus* {normalized by steel Young's modulus} plotted against *coating Poisson's ratio*. The family of curves shown in Fig. 8.18), provides the appropriate combination of coating thickness and elastic constants to eliminate the transverse stresses in the host composite material for uniaxial tension loading.

Carman and Reifsnider (1992) advanced the *optimal coating concept* for structurally embedded optical fiber sensors by building on the earlier work of Carman *et al.* (1991a, 1991b). They produced a set of *optimal coating parameter curves* for minimizing the stress concentration in the host under transverse loading, see Fig. 8.19. The vertical axis represents the ratio of the *coating's Young modulus*, Y_c, to the *host's transverse modulus*, Y_H^t. The horizontal axis represents the ratio of the *coating radius*, r_c to that of

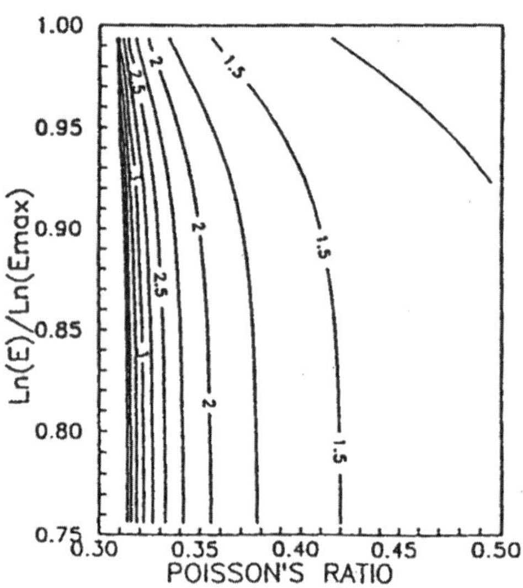

FIGURE 8.18. Variation of the optical fiber coating Young's modulus (normalized by the value for steel) against the coating's Poisson ratio for various values of the ratio of the coating radius to that of the optical fiber. From Sirkis J.S. and Dasgupta A., "Optical Fiber/Composite Interaction Mechanics". In: *Fiber Optic Smart Structures* Udd, Eric J., © 1995 Wiley Interscience. Reprinted by permission of John Wiley and Sons, Inc.

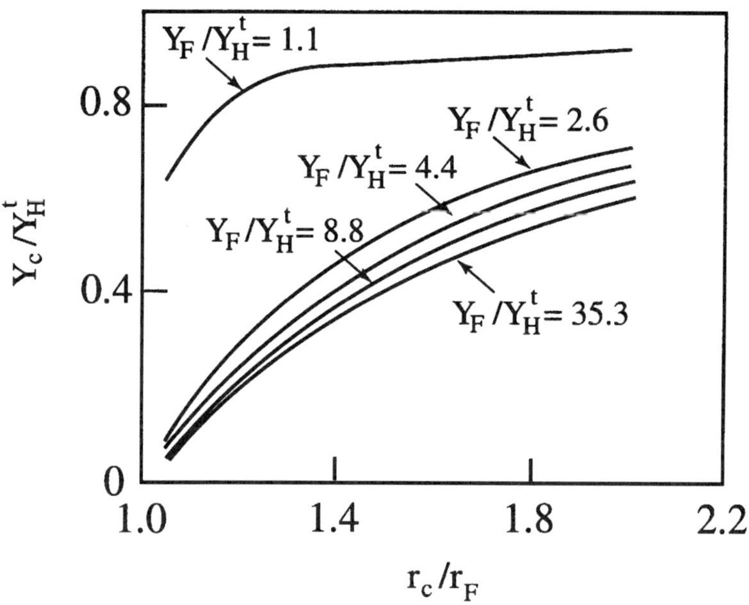

FIGURE 8.19. Variation of the optical fiber coating Young's modulus, normalized by the host Young's modulus, with the ratio of the coating to the optical fiber radius for several values of optical fiber to host Young's modulus.

the *optical fiber radius*, r_F. Each curve corresponds to a different ratio of the optical fiber's Youngs Modulus, Y_F, to that of the host's transverse modulus, Y_H^t. As this ratio becomes large the curves tend to merge into an asymptotic curve that is insensitive to the value of this ratio.

Fig. 8.20 shows the calculated variation for the *maximum principal stress* in the matrix material as a function of *angular position* around the embedded *fiber/host interface* for three different ratios of *coating to matrix stiffness* when the host is subjected to *transverse tensile loading*. The dashed line represents an optimum because the coating stiffness leads to an overall *minimum* in maximum principal stress. Carman *et al.* (1991a, 1991b), Sirkis and Dasgupta (1990), Dasgupta and Sirkis (1992) and Carman and Averill (1992) agree that it is not possible to simultaneously satisfy all desirable design criteria due to the competing nature of the in-plane stress components.

Roberts and Davidson (1991) concluded that besides *stress concentration* in the host, *weak optical fiber coating materials, weak optical fiber/coating* or *coating/host interfaces*, can lead to the *formation of cracks* at low stress

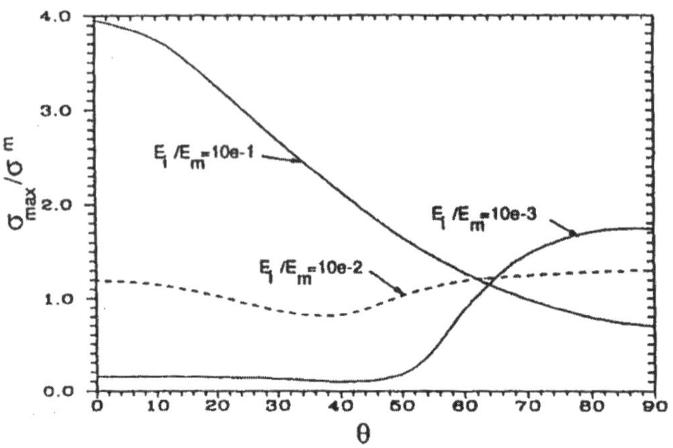

FIGURE 8.20. Variation of the maximum principal stress in the host matrix material as a function of the angular orientation about the fiber/host interface for three ratios of the coating to matrix stiffness (Young's modulus). From Carman, G. P. and Averill, R.C., (1992), "Analytical Modeling of Micromechanical Stress Variations in Continuous Fiber-Reinforced Composites", *Proc IUMTAM Conf.*, 27–61, © Springer Verlag GmbH.

levels. Such *cracks* would quickly lead to total *transverse cracking* of unidirectional FRP composite material host structures. They undertook a detailed *micro-fractographic analysis* of the region near the embedded optical fiber. Fig. 8.21 illustrates two different types of initial fracture observed for the acrylate and polyimide coated optical fibers. The formation of such cracks can affect the transverse strength of the host material, especially in the case of unidirectional FRP composites. However, this kind of FRP lay-up is known to be weak in the transverse direction and is rarely used in practice.

Pak *et al.* (1991 and 1992) also considered the design of *optical fiber coatings* for *transverse tensile* and *longitudinal shear loading* for an isotropic host structure. They showed that for these conditions, a *thinner coating enhances strain transfer* from the host to the coating when the coating is softer than the matrix. On the other hand, when the coating is stiffer than the matrix, *thicker coatings* provides better *shear strain transfer.* There is now experimental evidence to confirm the theoretical prediction that coating choice can significantly alter the performance of the embedded optical fiber sensing system, either by reducing *obtrusivity*, or *enhancing sensor performance* (Waite *et al.*, 1988, Dasgupta and Sirkis, 1992; Pak, 1992; Carman *et al.*, 1991a, 1991b, 1993; Singh *et al.*, 1991, and Roberts and Davidson, 1992).

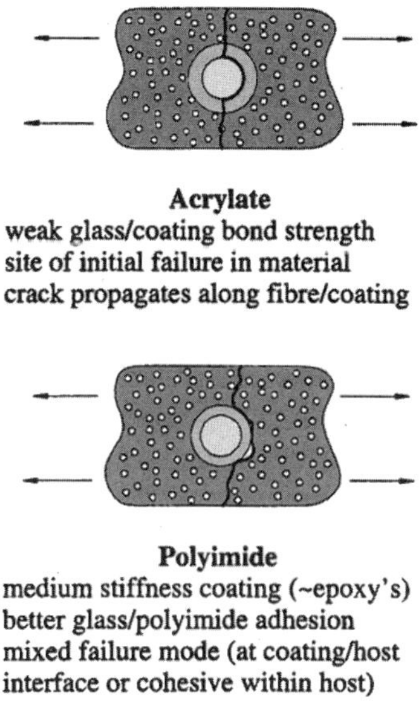

FIGURE 8.21. Schematic illustration of two types of initial fracture observed for acrylate and polyimide coatings of optical fibers. From Roberts, S.S.J. and Davidson, R., (1991), Mechanical Properties of Composite Materials Containing Embedded Fiber Optic Sensors, *Fiber Optic Smart Structures and Skins IV, SPIE* **1588**, 326–341.

Although there is no consensus yet on which coatings have the best overall characteristics, *acrylate coatings* appear to be losing favour due to their thermal instability at typical *composite curing temperatures*. This is dramatically illustrated in Fig. 8.22b, which shows a magnified picture of an *epoxy-acrylate coated 125 μm* optical fiber after curing of its $[0/90]_s$ CFRP host laminate. It is quite apparent that the acrylate coating had melted during the cure cycle and that a *resin-pocket* had formed as the optical fiber's orientation is orthogonal to the adjacent ply direction. By contrast, Fig. 8.22a shows an uncoated optical fiber embedded parallel to the local ply direction after cure of its [90/0]s CFRP host laminate. Clearly in the latter case no resin-eye is formed and the presence of the optical fiber has minimal influence on its host material.

8.4.1 Sensor/host bond and strain transfer

For a bonded (or embedded) sensor in a zone of relatively uniform axial host strain, the stress distribution along the fiber is illustrated in Fig. 8.23a.

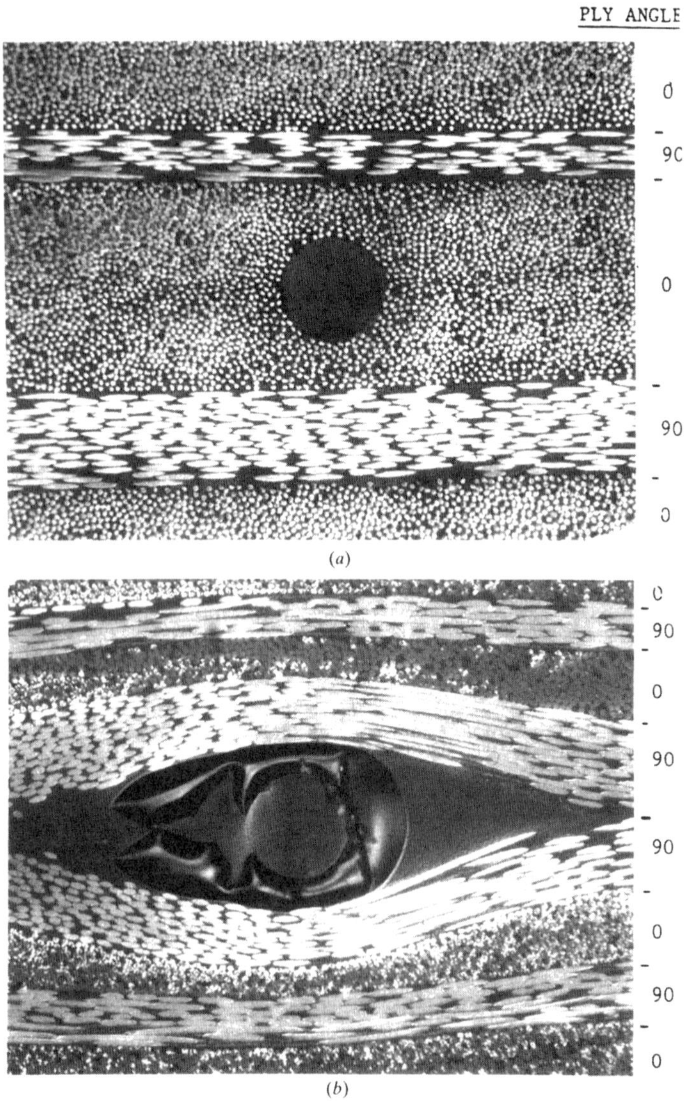

FIGURE 8.22. Microfractographs of embedded optical fibers. (a) An uncoated optical fiber embedded parallel to The local reinforcing carbon fibers. (b) An acrylate-coated optical fiber embedded orthogonal to the local reinforcing carbon fibers. Reprinted from Davidson, R., Bowen, D.H., and Roberts, S.J., "Composite Materials Monitoring through Embedded Fiber Optics," *International Journal of Optoelectronics* 5, 397–403, © 1990. Reproduced with permission of Taylor & Francis Publishing Inc., http://www.routledge-ny.com

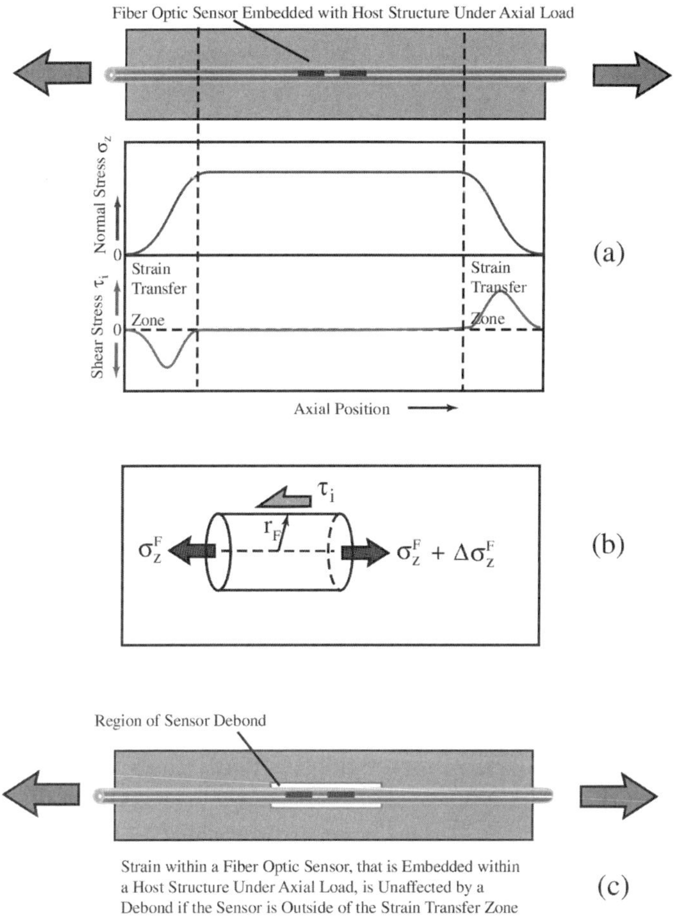

FIGURE 8.23. Schematic showing that debonding of an embedded optical fiber is not likely to disrupt strain measurements if the sensor is positioned beyond the strain transfer region.

High shear stresses only arise near the ends of the bonded region and over a development length of the order of a few millimeters in accordance with the relative stiffness and geometry of the adhesive and the fiber. This can be understood from a balance of interfacial *shear stress*, $\tau_i(z)$, to the *axial stress*, $\sigma_z^F(z)$, in the optical fiber, i.e.,

$$\pi r_F^2 \sigma_z^F(z) = 2\pi r_F \tau_i \Delta z + \pi r_F^2 \{\sigma_z^F(z) + \Delta\sigma_z^F(z)\}, \tag{8.4}$$

see Fig. 8.23b. In which case we can write:

$$\tau_i = -\frac{r_F}{2}\frac{d\sigma_z^F(z)}{dz}, \tag{8.5}$$

and as we might suspect the *shear stress* at the boundary of the optical fiber is directly proportional to the *gradient of the axial stress* within the optical fiber. The effects of *creep*, or *bond degradation*, are limited to this zone and will not affect the sensor's response providing the sensing region lies outside of the strain transfer zone, see Fig. 8.23c. A suitable choice of adequate bond-length can therefore protect the sensor from vagaries of *bond mechanics* (Maaskant *et al.*, 1997).

The *bond* between the *sensor and its host* is thus a key element in *strain measurements*. If slippage occurs between the host and the coating, or the coating and the optical fiber, the strain recorded will differ from what it would be for perfect strain coupling. DeFrancia *et al.* (1989 and 1991) studied the *adhesion* between *optical fibers and a FRP host*, but their pull-out tests were based on optical fibers embedded in a heat resin, rather than a FRP composite material. Waite and Sage (1988) suggested that although the bond between acrylate coating and the host material was adequate to transfer strains prior to failure, it was not sufficient to transfer the strain from the host to the optical fiber close to the ultimate strength of the host material, and slippage occurred.

Nath *et al.* (1991) observed that certain *polyimide coatings* bonded strongly to both the optical fiber and the host material, while Roberts and Davidson (1991) reported that the bond between the acrylate coating and its optical fiber was poorer than between the acrylate and the host. Indeed, as can be seen in the highly magnified view of a small section of embedded optical fiber, presented as Fig. 8.24 the *acrylate coating* remains bonded to the host but has clearly *separated from the optical fiber*. The lateral striations seen in the acrylate coating tend to indicate coating failure in a stick-slip manner as the crack front propagated from the sample edge to the optical fiber inclusion. In the comprehensive tests undertaken by Roberts and Davidson (1991) the presence of this medium *acrylate optical fiber*, with a *125 μm* cladding diameter and a coating diameter of *240 μm*, was to *reduce the FRP composite strength* by *13%* {i.e., *57.0 MPa* compared to *65.5 MPa*, or *1%* strain, in the absence of an embedded optical fiber}.

Paul *et al.* (1993) showed that the *bond strength* between *polyimide-coated optical fibers* and a *graphite-epoxy host* was sufficient to transfer the high strains associated with a notch. Since some manufacturers use a *silane coupling agent* to strengthen the adhesion between the optical fiber and its *polyimide coating* it cannot be assumed that all polyimide coated

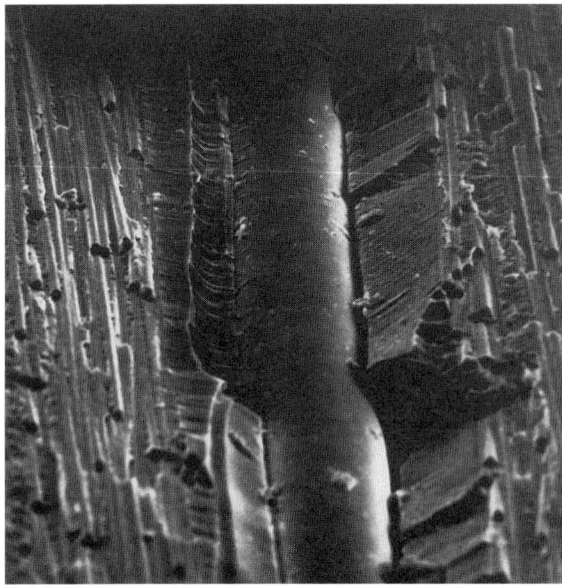

FIGURE 8.24. Electron micrograph of an optical fiber embedded within an FRP composite laminate showing that the aciylate coating remains bonded to the host matrx even though it has become separated from the optical fiber. From Roberts S.S. and Davidson R., (1991), "Mechanical Properties of Composites Materials Containing Embedded Fiber Optic Sensors", *Fiber Optic Smart Structures and Skins IV, SPIE*, **1588**, 326–341.

optical fibers will perform in the same manner. Carman and Sendeckyj (1995), have reviewed the mechanics of embedded optical sensors and suggest that where ever large stresses are likely to be encountered it would be prudent to *quantify the shear strength* of the *interface* between the optical fiber and its coating and between the coating and the specific host material. This is especially true for fiber optic sensing systems intended for an extended life.

Fiber optic cables usually include some form of plastic jacket to provide protection to the optical fiber. Mendez *et al.* (1993) demonstrated with pull-out tests that bonding between this optical fiber jacket and cement is extremely poor. On the other hand, Nanni *et al.* (1991) showed that bonding between cement and the acrylate optical fiber coating is sufficiently strong for good strain transfer. Escobar *et al.* (1992) observed that teltzelsilicon optical fiber coating was more resistant to damage during embedding in concrete, and transferred strain better than a soft acrylate coating. The properties of three common coatings are shown in Table 8.1.

Table 8.1 Properties of Optical Fiber Coatings (Habel *et al.*, 1994)

Coating	Elastic Modulus (MPa)	Tensile Strength (MPa)	Static Fatigue Resistance	Chemical Durability
Polyimide	2400	130.0	yes	poor
Acrylate	700	26.0	no	good
Silicone	2	5.5	no	—

The chemical durability of the three *optical fiber polymer coatings* have been studied (Habel *et al.*, 1994, and Habel and Hillemeier 1995) by soaking them in *alkaline solutions* and subsequently embedding them in concrete. The polyimide coating was affected more strongly by the basic environment than was the acrylate coating and even the fluorine-thermoelastic showed evidence of softening after embedment in concrete for 15 days. Over the last few years much experience has been gained in regard to integrating fiber optic sensors within large concrete structures and certain procedures have been found to be quite effective permitting the successful marriage of optical fibers with such structures as witnessed by the substantial number of structures recently instrumented with this technology (Merzbacher *et al.*, 1996), see Chapters 9 and 10.

8.5 INFLUENCE OF EMBEDDED OPTICAL FIBERS ON THE HOST STRUCTURE

In Chapter 2 a convincing case was made for integrating fiber optic structural monitoring systems into many different kinds of structure. Before engineers, responsible for designing and building structures that range from bridges to aircraft, widely implement the use of resident optical fiber sensing technology it will be necessary for them to know the consequences of such an action. For the most part they will have to know if the presence of this technology will compromise the performance or life of the structure it is designed to monitor. Specifically, it will be crucial to know how *embedded optical fiber sensors* affect the: *stiffness, strength and fatigue characteristics of their host structure*. A useful overview of this subject has been provided by Carman and Sendeckyj (1995).

It is fairly obvious that optical fiber sensors attached to the surface of most structures, with the *exception of aerodynamic surfaces*, will have no significant adverse affect. It is also true that optical fibers embedded within concrete structures or bonded to reinforcing elements placed

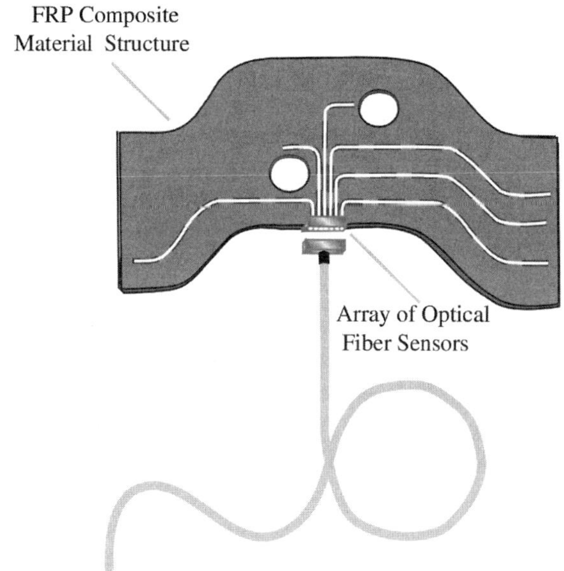

FIGURE 8.25. Schematic illustration of how parallel multiplexing leads to the bunching of optical fibers at the egress site.

within such structures will have a negligible influence. The issue is primarily of concern when we consider embedding optical fibers sensors within FRP composite materials. Since the number of combinations of *sensor configuration* and possible *FRP arrangement* is very large, it is necessary that we understand the influence of embedded fiber optic sensors and have reliable models for predicting their effect on the properties of their host structure.

Considerable progress has been made towards this goal in the past decade. In general the number of optical fibers within any lamina is likely to be very small, indeed one in most situations. The exception to this would be where a number of optical fibers are brought together at some common ingress/egress point, or in the future, to an optoelectronic microchip that collects the sensing data from a parallel multiplexed array of sensors (Fig. 8.25).

8.5.1 Optical fiber resin-pockets

As we have seen *interlaminar lenticular resin-rich pockets* form around optical fibers embedded within FRP composite host structures at an orientation other than parallel to the adjacent indigenous reinforcing

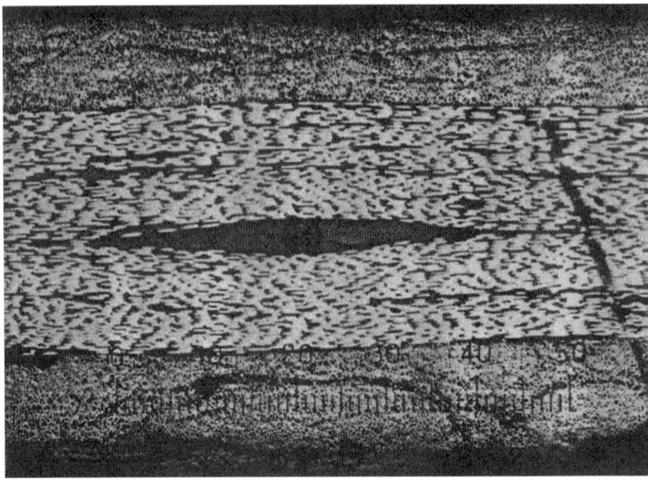

FIGURE 8.26. Photomicrograph of a resin eye (or resin pocket) formed around an optical fiber that is embedded orthogonal to the local ply direction. From Dasgupta, A., Wan, Y. and Sirkis, J.S., (1992), "Prediction of resin pocket geometry for stress analysis of optical fibers embedded in laminated composites." *Smart Materials & Structures*, **1**, 101–107, used with permission, Institute of Physics Publishing Ltd.

fibers. As we have just indicated the influence such *resin-pockets* have on the ultimate strength and life of the host structure (Davidson *et al.*, 1990, and Dasgupta *et al.*, 1992), is of more concern than the effect *obtrusivity* has on the calibration of embedded sensors. Fig. 8.26 displays another example of a *resin-pocket* when the optical fiber is embedded perpendicular to the adjacent reinforcing fibers. As suggested in Fig. 8.14, the size of such *resin-pockets* decrease the smaller the angle between the optical fiber and the reinforcing fibers. The resin pocket is also smaller, the smaller the thickness of the laminate. Large resin-pockets act as *interlaminar discontinuities* and represent a *potential reliability hazard* to both sensor and the host under applied static, cyclic, or shock loading.

Finite element analysis has been a popular method of investigating the *stress concentration* around an embedded optical fiber. It has also been used to help establish a relationship between the *optical fiber strain* and the *far-field host strain* for complex geometries. Investigations of the effect of the *resin-pocket geometry* on the calculated strain concentrations under tensile loading of certain laminate sequences reveals that the strain concentration can be altered by nearly a factor of two, depending on the resin-pocket geometry (Salehi *et al.*, 1989, and Tay *et al.*, 1990). The geometry of the resin pocket in these studies was extracted from observa-

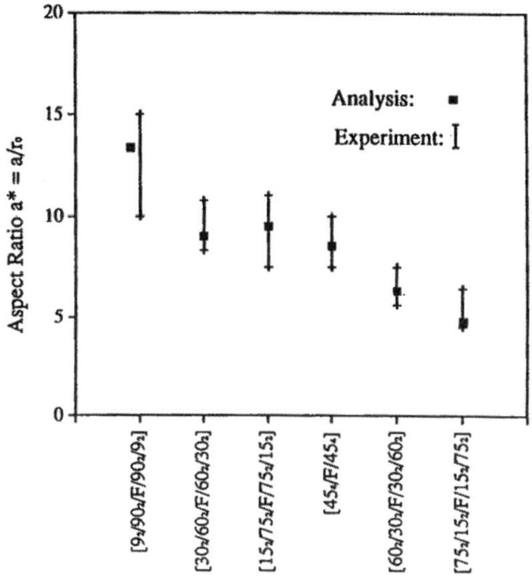

FIGURE 8.27. Resin pocket aspect ratio plotted against the FRP layup which is directly related to the angle between the optical fiber and the local ply direction) for both a theoretical model and experiments. From Dasgupta, A., Wan, Y. and Sirkis, J. S., (1992), "Prediction of resin pocket geometry for stress analysis of optical fibers embedded in laminated composites." *Smart Materials & Structures,* **1**, 101–107, used with permission, Institute of Physics Publishing Ltd.

tions of enlarged micrographs. Close agreement has been obtained between computational models and the observed resin-pocket geometries in graphite-epoxy specimens, as seen in Fig. 8.27. This figure shows the agreement between predicted and measured *resin-pocket aspect ratios* for six different laminate stacking sequences. The uncertainty in the experimental values stems from the error in measuring the tip of the resin-pockets.

There is some indication from the research undertaken to date (Jensen *et al.*, 1992 and Holl and Boyd 1993), that the presence of embedded optical fibers has little effect {*<5%*} on the *stiffness* of CFRP provided: the *optical fiber volume fraction* is low, and the diameter of the optical fiber is less than *250 μm*. Indeed, the study by Holl and Boyd, showed that *polyimide coated optical fibers* as large as *250 μm* in diameter had no measurable affect on either the *tensile* or *compressive stiffness* of both unidirectional and quasi-isotropic CFRP laminates. In these specimens five optical fibers were embedded orthogonal to the load direction and the reinforcing fibers.

8.5.2 Effect on tensile strength

A number of research papers (Measures *et al.*, 1989, Roberts and Davidson 1991, Carman *et al.*, 1993, Holl and Boyd 1993 and Paul and Schoeppner, 1993) have reported on studies of the influence of embedded optical fiber: *spacing*, *size* and *orientation* on the *tensile strength* of FRP composite material host structures. The results of this work show that no measurable degradation is evident for optical fibers embedded parallel to the host reinforcing fibers. As clearly seen in Fig. 8.28, no anomaly that would cause a stress concentration exists around the optical fiber for this configuration. When the optical fibers are embedded perpendicular to the reinforcing fibers, a small undulation is formed in the structural fibers as they bend around the resin pocket formed around the optical fiber. The stress increase associated with this local distortion is insignificant for small optical fiber diameters {<200 μm} so their presence does not appreciably change the *tensile strength* of the host structure.

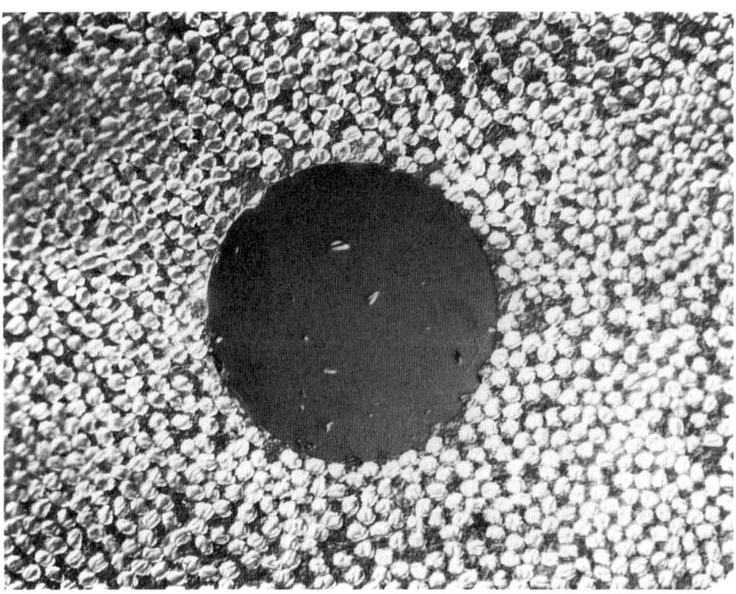

FIGURE 8.28. Highly magnified view of an uncoated optical fiber embedded parallel to the local reinforcing fibers of the host FRP material. From Roberts S.J. and Davidson R., (1992), "Short Term Fatigue Behavior of Composite Materials Containing Embedded Fiber Optic Sensors and Actuators" 10th Euopean Conference Smart Structures and Materials, Glasgow, *SPIE* **1777**, 255–262.

8.5.3 Effect on transverse strength

The *transverse strength* of unidirectional FRP composites, containing optical fibers embedded parallel to the reinforcement, depends on the properties and diameter of the optical fiber coating. Little effect has been observed in *static loading* if the diameter is less than *100 μm* (Carmen and Sendeckyj, 1995). As discussed earlier, Roberts and Davison (1991) revealed that weak coatings, or weak coating/host interfaces, or weak coating/optical fiber interfaces can lead to *transverse crack formation* under transverse loading, see Fig. 8.21. Carman and Reifsnider (1992) studied analytically the stress variation in the vicinity of a *coated optical fiber* embedded in FRP composite material subject to *transverse strains*. They suggested that *a coating with the appropriate compliance may avoid any premature failure* of the composite material. For a graphite/epoxy laminate the optimum coating modulus is typical of a *polyimide coating* {i.e., *3.45 GPa*}.

8.5.4 Effect on compressive strength

Although the presence of embedded optical fibers appears to have a negligible effect on the *tensile strength* of FRP composite host structures, the same cannot be said with regard to the *compressive strength*. In particular, Yarcho (1989) and Jensen *et al.* (1992a and 1992b) studied the effects of *250 μm* acrylate-coated optical fibers on the *compression strength* of various CFRP laminates. When the optical fibers were embedded parallel to the reinforcing fibers, no degradation in the *compression strength* was observed.

In contrast, quite significant decreases {up to *70%*} in the *compression strength* were measured when the optical fibers were *embedded perpendicular* to the adjacent reinforcing fibers. There is reason, however, to suspect the results of experiments involving the use of acrylate coated optical fibers in FRP composites, such as graphite/bismaleimide (Jensen *et al.*, 1992) requiring a relatively high curing temperature for the acrylate is known to melt, distort and possibly create voids, see Fig. 8.22.

Roberts and Davidson (1991), and Carmen *et al.* (1992) also found that large-diameter optical fibers degrade the *compression strength* of a FRP composite material if not embedded collinear with the structural fibers. The influence of small optical fibers and embedment angle on *compression strength* has not yet been fully addressed. Clearly, there is a need for an adequate model to predict the effect of embedded optical fibers on compression strength.

To this end, Case and Carmen (1994) developed a *compression strength model* based on the shape of the *lenticular region* constructed by Dasgupta *et al.* (1990) and the classic compression model described in Tsai and Hahn (1980). With this model they could predict the *degradation in compression strength* as a function of: *optical fiber size, laminate stiffness properties, laminate lay-up, and processing parameters.* They demonstrated that degradation of the compression strength occurs for optical fibers embedded at angles larger that *15 degrees* with respect to the loading direction. They also showed that as the stiffness of the system increased, the compression strength increased. This model, while not exact, provides an analytical foundation for the evaluation of the effect that various parameters have on *compression strength.*

8.5.5 Effect on shear strength

Yarcho (1989) measured a decrease of 7% in the *interlaminar shear strength* of an unidirectional FRP composite laminate with embedded *250 μm,* acrylate-coated optical fibers. Subsequently, Roberts and Davidson (1991) performed an *interlaminar shear test* using a three-point bend specimen on graphite-epoxy composites containing *100 μm* polyimide—and acrylate-coated fibers. Their results were similar to those obtained by Paul and Schoeppner (1993) and suggested *no degradation in the shear strength* due to the presence of the embedded optical fibers.

8.5.6 Effect on fatigue

There has been little attention paid to the influence of embedded optical fibers on the *fatigue life* of FRP composites. Roberts and Davidson (1992) cycled FRP composite materials containing embedded optical fibers at various tensile load levels below *80%* of ultimate load for less than *20,000* cycles to determine the effect of embedded fiber optic sensors on low-cycle fatigue life. The results of embedding a *140 μm* polyimide-coated optical fiber in a *cross-ply laminate* parallel to the *90°* reinforcement fibers at the [0/90] interface led them to conclude that optical fibers do not initiate premature fatigue degradation in this cross-ply configuration.

The work of Melvin *et al.* (1992) suggested that polyimide-coated optical fibers had a less detrimental effect on the *fatigue life* of CFRP laminates than did acrylate coated fibers, while extensive experiments of Loken (1990) indicated that no degradation in the fatigue life of unidirection CFRP laminates was produced by embedded optical fibers with a diameter of less than *150 μm*.

8.5.7 Effect on fracture toughness

Blagojevich *et al.* (1990), investigated the effect of embedded optical fibers on the *interlaminar fracture toughness* of Kevlar/epoxy laminates. A double cantilever beam test was chosen to investigate the tendency for a *Mode I delamination* to propagate into the laminate. They found that the presence of an embedded optical fiber did not decrease the *critical energy release rate* but rather increased it by about *15%* for a unidirectional lamina. In essence the optical fiber acted as a *stiffener* to the local resin matrix with respect to the propagation of cracks.

8.5.8 End-Effects of optical fibers terminating within host

Single ended fiber optic sensors often involve the *termination of the optical fiber* within the host structure. In the case of a FRP composite host, the termination of the optical fiber inside the material leads to the formation of a *resin-cone at the tip of the optical fiber.* This increases the risk of *interfacial failure* and raises the local stress concentration in the host material. It is therefore a less desirable configuration from a material and sensor performance point of view. Indeed, interfacial fracture in the form of *debonding* and *material yielding* has been observed, by LeBlanc (1998). However, the use of a coating with a Young's modulus softer than that of the optical fiber reduces stress intensification in the host, and choosing a *coating with the same modulus of elasticity* as that of the host's resin matrix, all but *eliminates this stress intensification.*

LeBlanc (1998) has studied the problem of optical fiber termination within FRP composite material and found:

1. As with other considerations the optical fiber should be embedded between two collinear plies and in the direction of the adjacent reinforcing fibers.
2. A coating with the appropriate combination of dimension and elastic modulus should be used so as to reduce or eliminate the stress concentration effects under transverse loading. Carman and Reifsnider (1992) specify for a given host, the combination of Young's modulus and radius that minimizes transverse stress concentration in the host.

3. The use of a coating that satisfies Carman and Reifsnider's requirements will also reduce the stress intensification in the host caused by termination of the optical fiber.

If the stiffness of the coating is the same as that of the host's resin matrix the effects of the sharp *stiffness discontinuity* at the end of the optical fiber are *eliminated*. The use of a *softer coating* {and, for a specific coating modulus, the *thicker* the coating} will avoid debonding of the sensor from its host. Unfortunately, it also leads to a *longer strain transfer length*. With increasing host strain both the *interfacial shear stress* as well as the *stress energy* present within the material increases. This means that at some point either *yielding* of the coating material, or *debonding*, takes place. This is referred to as *failure onset strain*. LeBlanc (1998) developed a computer model that calculates, for a given set of material properties, which kind of strain failure will occur first. In this model yielding and debonding are treated separately, and whichever is the smaller indicates which mode of failure occurs first.

LeBlanc (1998) also undertook careful experiments to test his computational work and demonstrated how to create Fabry–Perot sensors by designed fracture of the optical fiber. He showed that measurements of the resulting gap provided information on the yielding and debonding processes, and that FBGs can also be used to evaluate if debonding or yielding has occurred, see section 12.10.3. His results indicate that *single-ended embedded optical fiber sensors* can be safely used as long as the possibility of interfacial and end-zone damage is recognized. *Polyimide*, which is considered suitable for embedded sensors in general appears to also be *satisfactory for the end-zone problems*. Mechanical considerations, however, indicate that the thickness of the *polyimide coating* should be much *thicker* than is currently used: *60 μm* as compare to the *10* to *15 μm*. Epoxy is another material with mechanical properties similar to that of polyimide. The presence of *radial compression* in the optical fiber is very important in order to avoid debonding in the first place, and permit *efficient strain transfer* from the host to the optical fiber even in the presence of debonding via interfacial friction. Polyimide coatings often have *residual stresses* that are helpful in that regard.

8.5.9 Minimization of adverse effects

From our above discussion we can concur with Jensen and Sirkis (1995), who concluded that *minimization* of any *adverse influence* of embedded optical fibers on the structural integrity of the host material requires that:

(i) The optical fiber diameter {including the coating} be about 100 μm or less, and certainly no more than one-ply thickness.
(ii) The optical fiber coating should be more compliant than the surrounding host matrix. This will avoid debonding of the sensor from the host, but will increase the strain transfer length.
(iii) The optical fibers should be embedded parallel to the reinforcing structural fibers (i.e., parallel to the adjacent ply directions).
(iv) If possible the optical fibers should also be parallel to the loading direction.

8.6 PULTRUDED FIBER OPTIC STRUCTURAL SENSORS

Structures made from fiber reinforced polymer composite materials are produced by a number of different manufacturing techniques, including: *prepreg lay-up; pultrusion,* and *filament winding*. In the previous section we focused exclusively on FRP lay-up based structures. The literature is rather sparse on the integration of optical fiber structural sensors with FRP composite material structures manufactured via pultrusion, (Friebele *et al.*, 1994 and 1995). As we saw in Chapter 2, *FRP rebar and prestressing tendons* are finding increasing application in Civil Engineering and are viewed as a potential replacement for steel in many new concrete structures. This advance is going hand in hand with the development of a fiber optic structural sensor that are being used to monitor the long term performance of these new materials, see Chapters 9 and 11 (Measures *et al.*, 1995 and Nellen *et al.*, 1997). To date, most of the fiber optic sensors have been bonded to the outside of these FRP structures. For the technology to find broad implementation it will be necessary to integrate the sensing system with the structural elements during manufacture. Since most of the CFRP rods are made by *pultrusion* it is obvious that fiber optic sensors integrated with structural elements via the pultrusion process represents a significant advance in the technology of structural monitoring.

Kalamkarov *et al.* (1997), have demonstrated that *polyimide-coated optical fibers*, with *fiber Bragg grating sensors* included, *survived pultrusion* and bonded well to the host material. In contrast, *acrylate coated optical fiber* were not able to withstand the high curing temperatures and exhibited *severe debonding*. In this study *9.5 mm* diameter rods were produced by a customized pultrusion machine that used an electrically heated die shaper operating up to *150* °C. The polyimide-coated optical fiber had an overall diameter of *155* μ*m*, while the acrylate-coated optical fibers had an outer diameter of *250* μ*m*. Fig. 8.29, displays two SEM micro-graphs of the two

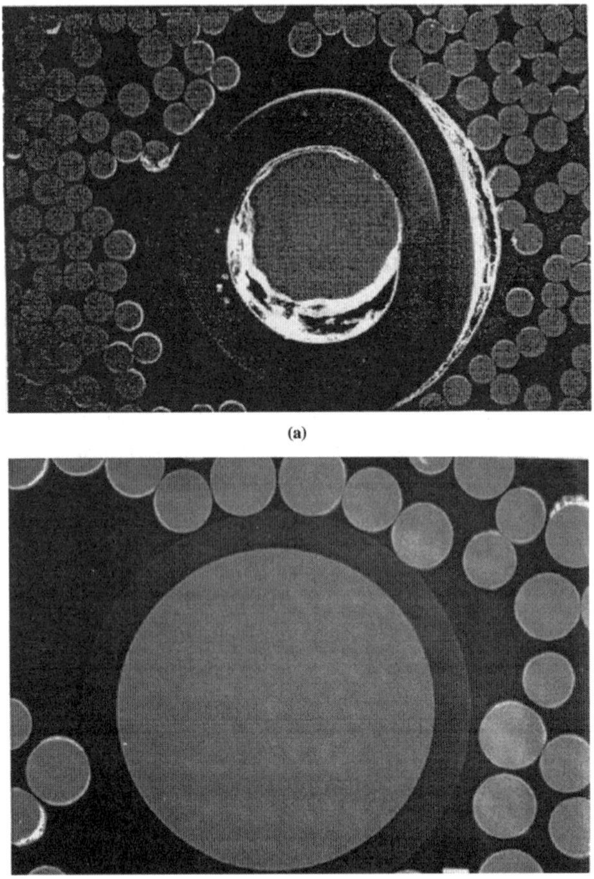

(a)

(b)

FIGURE 8.29. Two scanning electron micrographs of pultruded optical fibers: (a) acrylate-coated optical fiber, and (b) polyimide-coated optical fiber. From Kalamkarov, A.L., MacDonald, D.O., and Westhaver, P.A.D., (1997), "Pultrusion of Smart FRP Composites," Symposium on Smart Structures and Materials, Conference on Smart Sensing, Processing and Instrumentation, March 1997, San Diego, California, *SPIE Proceedings*, **3042**, 400–409.

types of pultruded optical fiber. Debonding of the acrylate-coated optical fiber from its host material is clearly visible in Fig. 8.29a, while the excellent interface between the polyimide-coated optical fiber and its host material is quite apparent in Fig. 8.29b.

The embedment of a single optical fiber parallel to the reinforcing fibers had no apparent detrimental influence on the *tensile properties* of the pultruded FRP rods. A slight deterioration of the *shear strength* {~ *10%*}

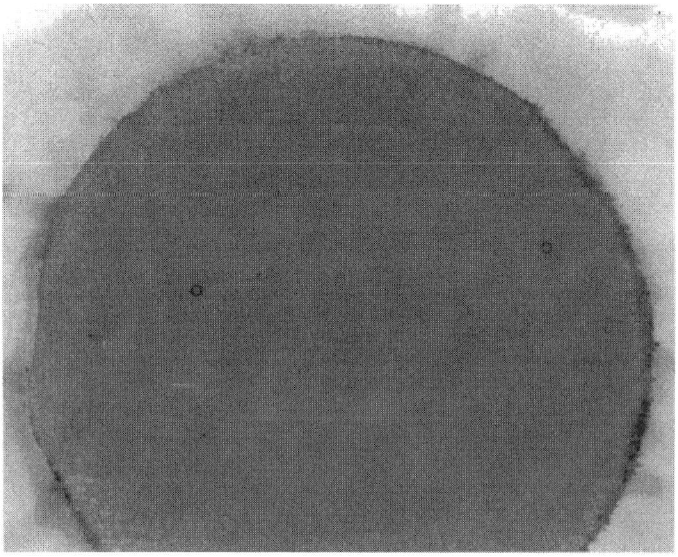

FIGURE 8.30. Magnified view of the end of a pultruded glass-fiber-reinforced polymer rod containing two embedded optical fibers. From Kalamkarov, A.L., MacDonald, D.O., and Westhaver, P.A.D., (1997), "Pultrusion of Smart FRP Composites," Symposium on Smart Structures and Materials, Conference on Smart Sensing, Processing and Instrumentation, March 1997, San Diego, California, *SPIE Proceedings*, **3042**, 400–409.

was, however, observed in the case of *glass FRP rods*, as opposed to *carbon fiber rods*, where the presence of an optical fiber had negligible influence on the shear strength. The inclusion of two optical fibers into the pultruded rod had a more significant effect on the shear strength of the glass FRP rod. A cross-section of this rod showing the two optical fibers is presented as Fig. 8.30. Variations in the radial positioning of the optical fiber was determined to be less than *0.5 mm* over a *10 m* test section. The ability to *accurately position* the optical fiber in any arbitrary section of the pultruded rod will be seen to be important from the perspective of optical connectivity.

8.7 FIBER OPTIC STRUCTURAL SENSOR CONNECTORIZATION

For surface mounted fiber optic structural sensors, including those embedded within concrete structures, a *tough optical fiber cable* protects the inner optical fiber up to the site of bonding to the host structure, see section 8.2. For most FRP composite structures instrumented with a *resident fiber optic structural sensing system* it will be necessary to avoid having *optical fiber pigtails*. The most obvious reason for this is that within the FRP host structure the optical fibers are well protected from physical abuse and environmental hazards and need no more than a coating. Furthermore, as we have seen above, efficient strain transfer and small diameter are requisites for the embedded sensors and these also dictate nothing more than a coating. Beyond the shelter provided by the host structure, however, the optical fiber requires substantial protection in the form of a tough jacket. Clearly, at the surface of the host structure there has to be a transition from one to the other.

Several approaches have been proposed for this *fiber optic/host interface* (Spillman and Lord 1995). If the host structure in made of FRP and comprises many plies, the *optical interface* design can be fairly straight forward and based on standard components, Fig. 8.31. In the case of a FRP host structure made by *lay-up fabrication* the optical fiber is mounted within a *connector ferrule* prior to embedding within the composite lay-up. The optical fiber is then cleaved and polished down to the end surface of the ferrule. The pigtailed ferrule is then integrated into the FRP lay-up and the structure cured making provision to shield the tip of the ferrule from the resin which would otherwise mar a clean optical connection. It is obvious that this type of *optical interface* is only acceptable if the relevant edge of the host structure is subject to only minor stress loading and is free of other structural encumbrances. An example of such an optical fiber sensor instrumented CFRP test specimen that is connectorized in this manner (Spillman and Lord 1995), is presented as Fig. 8.32.

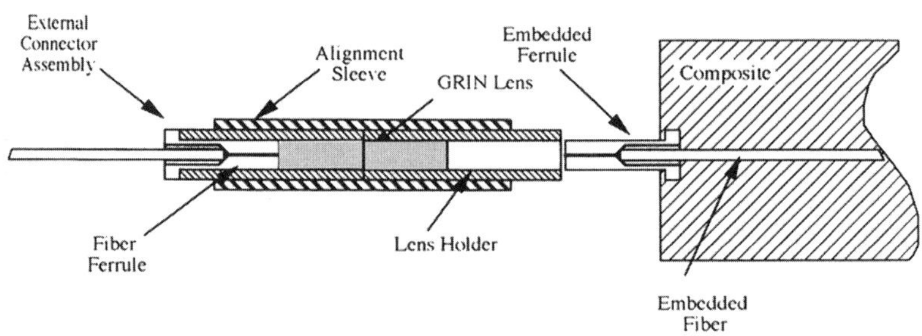

FIGURE 8.31. Schematic of an optical fiber connector interface to an optical fiber embedded within a FRP composite material structure. From Spillman, W.B., Jr. and Lord, J.R., "Methods of Fiber Optic Ingress/Egress for Smart Structures" In: *"Fiber Optic Smart Structures"* Ed. Eric J. Udd, © 1995 Wiley Interscience. Reprinted by permission of John Wiley and Sons, Inc.

FIGURE 8.32. Optical fiber connector interface for a CFRP test specimen containing an embedded optical fiber. From Spillman, W.B., Jr. and Lord, J.R., "Methods of Fiber Optic Ingress/Egress for Smart Structures" In: *"Fiber Optic Smart Structures"* Ed. Eric J. Udd, © 1995 Wiley Interscience. Reprinted by permission of John Wiley and Sons, Inc.

If the host is in the form of a thin FRP wrap, as used for the rehabilitation and strengthening of reinforced concrete structures, then a more gradual transition is required. Examples of this approach can be found in Chapters 9 and 10. For *pultruded FRP host structures* the endface is cut from a *continuous feedstock* leaving any embedded optical fiber flush with this endface, see for example, Fig. 8.30. Making a reliable and robust optical connection to such optical fibers represents a severe challenge, especially when it has to be done inexpensively. If the positioning of the optical fiber can be controlled with the kind of precision suggested by Kalamkarov *et al.* (1997), i.e., *0.5 mm*, then it might be possible to develop some form of *smart-endcap* to accomplish this task in an automated manner.

This *smart-endcap* would be mounted over the endface of the pultruded structure and automatically align either, an optical cable, or an optoelectronic module, with the optical fiber embedded within the pultruded structure. In the case of interferometric or Bragg grating sensors the embedded optical fiber is single moded, which means that the core diameter is of the order of *10 μm*. Good coupling between two such single mode optical fibers requires an alignment precision of better than *3 μm*. The other approach using an optoelectronic module might not require this degree of alignment precision (Fig. 8.33).

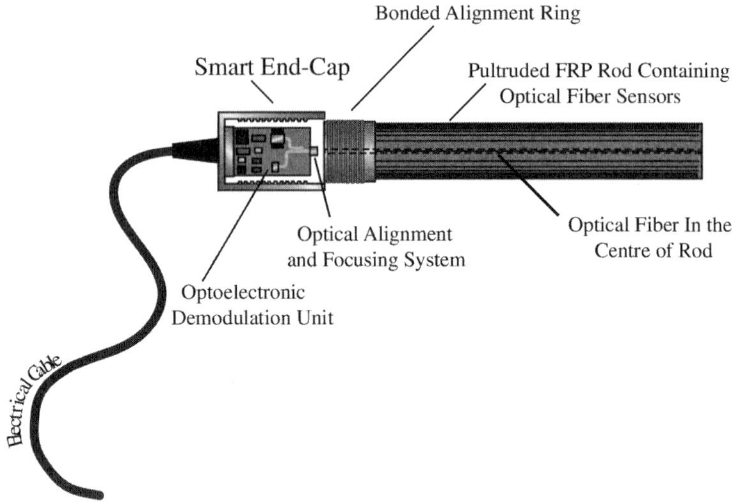

FIGURE 8.33. An illustration of the "smart end-cap" concept for a pultruded FRP rod instrumented with an optical fiber sensor array.

The fiber to fiber *smart-endcap* would be simpler, less expensive and smaller in size, but would require an electrical cable, in addition to the optical cable, to provide power to the alignment mechanism within the endcap.

The optoelectronic module approach would not need two different kinds of lead as some of the electrical power would be used for the actuation mechanism. Since only the tip of an optical fiber would have to be translated a distance of maybe only *0.5 mm* (Kalamkarov *et al.*, 1997), the actuation mechanism could be very small and require very little power. Tiny piezoelectric elements might be adequate for the task. Indeed, the actuator drive power would presumably be controlled by the level of signal received from the photodetector in order to achieve optimum light coupling.

Alternative and simpler strategies may sometimes be possible. For example it may be feasible to pultrude small pockets containing a length of optical fiber that is used as a pigtail once the pultruded rod is cut at the site of this pocket. Such an approach would pretty well be limited to applications where the lengths of pultruded material is known ahead of its manufacture.

CHAPTER 9

Short Gauge Sensor and Applications

9.1 INTRODUCTION

For many kinds of structural monitoring applications, a strain sensor with a gauge length in the range: 2 mm to 2 cm is most often desired. For other applications, for example, the direct measurement of strains within a concrete structure, somewhat larger gauge lengths (2 cm to 5 cm) are more suitable for averaging the strain variation associated with the inhomogeneity scale length of this type of medium. In this chapter we will use the range 2 mm to 5 cm to define a *short gauge length* fiber optic strain sensor. In Chapter 10 we will focus on fiber optic sensors with gauge lengths that are greater than 5 cm. Part of the rationale for dividing the material in this book according to gauge length is that conventional strain measuring technology is to some extent also distinguished by gauge length. Resistive foil strain gauges generally are used when spatial resolution of a few centimeters is required, while vibrating wire gauges or extensiometers are, for the most part, used for gauge lengths in excess of 5 cm.

Currently, there is a rich diversity of short gauge length fiber optic structural sensors. A brief overview was provided in Chapter 6, and useful reviews of fiber optic sensing technology are also available in the literature (Kersey *et al.*, 1997b; Kersey, 1997b; Meltz, 1996; Measures, 1995). In this chapter we will focus on those fiber optic short gauge structural sensors with the greatest potential for becoming the structural monitoring technology of the twenty-first century. Fiber Bragg grating sensors certainly look most promising from this perspective, but some useful groundwork has been accomplished to date with sensors based on the Fabry–Perot interferometer.

This chapter and the subsequent chapters have been written with the notion that it is as important to appreciate what has been accomplished by way of measurement applications as it is to understand the current state-of-the-art of the technology. This is particularly true for those wishing to see whether this technology has emerged from the laboratory, and if so, whether it can be of value to them? Furthermore, the details of the *sensor interrogation* (demodulation) technology will probably evolve with improvement in optoelectronics, so the fundamentals of the sensing systems are emphasized in this and the subsequent chapters. It is expected that 10 years from now fiber optic Bragg grating sensors will not be that different from those available today, especially in the way they are used for measurements. However, the sophistication of the optoelectronics used to read the sensor will probably have advanced dramatically. To further assist the reader who wishes to understand just what technology is available today, a number of applications using commercial fiber optic structural monitoring systems are included.

Thus, as indicated earlier, this chapter focuses on the use of single, short gauge-length sensors, while Chapter 10 will cover long gauge-length sensors. Although Chapter 11 will deal with multiplexing of sensors, since some of the most important applications to date have used parallel multiplexed arrays of single sensors this topic will be included in this chapter, particularly in terms of applications. The emphasis of Chapter 11 will be serial multiplexing of sensors. In Chapter 12, we explore truly distributed sensing via Bragg intragrating measurements.

The applications reviewed in this chapter span a number of disciplines and industries. The most important are the construction and aerospace industries. However, applications for ships, trains, pressure vessels, pipelines, and mines are also explored. Although a broad number of applications will be mentioned so the reader can appreciate the breadth of applications. A few of the more significant applications will be described in some detail in order that an insight into what is possible can be better gauged.

Since the greatest number of major applications of fiber optic structural monitoring, certainly in the construction industry, involve the use of fiber Bragg grating sensors, the first part of Chapter 9 will focus on this technology and applications of FBG sensors. The latter part will look into Fabry–Perot sensor technology and its applications. Thus, in Section 9.2 an overview of the various techniques used for demodulating FBG sensors will be presented. The nature of the FBG is such that quite a wide variety of techniques for reading them have evolved. These include techniques based on spectral and interferometric analysis. This section will lay the groundwork for understanding the specific technology used in each of the applications discussed in Section 9.3. The primary demodulation systems developed for short-cavity Fabry–Perot sensors will be reviewed in Section 9.4 and some of the more interesting applications of this technology will form the basis of Section 9.5. Although an attempt will be made to make these descriptions as self-contained as possible, inevitably they will draw upon the knowledge base laid out in the earlier chapters, particularly Chapters 3, 4, and 5.

9.2 FIBER BRAGG GRATING SENSOR DEMODULATION

As we have seen in Chapters 5 through 7, the FBG sensor encodes changes in the strain and temperature, to which it is subjected, to shifts of its center (Bragg) wavelength; see Fig. 6.22. One of the simplest techniques for interrogating FBG sensors is based on spectral filtering and involves a *passive ratiometric* approach. Several variants on this theme will be described. Another widely adopted technique is based on placing a tunable narrowband filter ahead of the photodetection system. Ultrahigh sensitivity has been demonstrated with interferometric approaches, especially when combined with laser sensors, a novel twist in which the sensor directly controls the wavelength of the exciting laser. Lastly, an approach based on a tunable laser has been demonstrated by members of my laboratory that promises to permit all of the different sensing modalities for which FBGs can be used to be undertaken with the same system.

9.2.1 Passive spectral ratiometric demodulation

The earliest spectral method of determining the center wavelength of a fiber Bragg grating sensor involved the use of either a spectrometer or a monochromator. Such instruments were unwieldy, expensive, and quite impractical for field use at the time. As such, they represented somewhat of impediment to extensive application of the FBG sensor technology.

A major breakthrough came with the recognition that the Bragg wavelength could be monitored by means of a simple spectral filter (Measures *et al.*, 1992b). A schematic illustration of this measurement concept was presented as Fig. 6.23. Light from a broadband source is sent to the FBG sensor via a 3 dB coupler, which subsequently permits half of the light reflected from the FBG sensor to pass to a second 3 dB coupler where it is divided and sent to two photodetectors. A spectral filter is placed ahead of one of the photodetectors so that the magnitude of the light signal recorded by this device depends on the wavelength of the optical signal. The second photodetector serves as a reference and compensates for intensity variations arising from source power fluctuations, connector alignment variations, and macrobend losses in the optical fiber.

Ideally, the transmission efficiency of the spectral filter should be a linear function of wavelength, as indicated in Fig. 6.23. If, the spectral filter has indeed a linear transmission function, $F(\lambda)$, then we can write

$$F(\lambda) = A(\lambda - \lambda_0), \tag{9.1}$$

where A is the slope of the filter function, and λ_0 the wavelength for which $F(\lambda) = 0$, Melle *et al.* (1992a) assumed that the narrowband light signal back-reflected from the FBG had a Gaussian form with a spectral width $\Delta\lambda$ and a center wavelength λ_B. They showed that the ratio of the filtered to reference (unfiltered) light signals

$$\frac{I_F}{I_R} = A\left\{\lambda_B - \lambda_0 + \frac{\Delta\lambda}{\sqrt{\pi}}\right\}, \tag{9.2}$$

is linearly proportional to the Bragg wavelength, λ_B.

The experimental system used to validate this approach employed a *superluminescent diode* operating at about 830 nm to excite the FBG sensor. An example of the linear variation in the signal ratio with the strain imposed on the (832 nm) FBG that was surface bonded to an aluminum cantilever beam is presented in Fig. 9.1. The inset shows part of the transmission curve of the RG830 IR high passband filter employed in this experiment (Melle *et al.*, 1993b). Highlighted is the relatively linear segment of this nonlinear curve used in this work. Deviations from linearity can be handled by using an *electronic lookup table* within the processing system. This relates the photodetector signals to the Bragg wavelength and thereby to the strain, or temperature, experienced by the FBG sensor. An example of the dynamic response of this passive ratiometric *wavelength determination system* (WDS) is displayed in Fig. 9.2. Clearly, the steeper the slope of the transmission function A, the higher the resolution attainable. However, this comes at the expense of the range, as illustrated in Fig. 9.3.

In reality three factors complicate this passive ratiometric WDS: (i) temperature sensitivity of the filter's transmission spectrum; (ii) nonlinearity of filter function and (iii) polarization sensitivity of the second coupler combined with stress-induced birefringence in the FBG. Extensive engineering by ElectroPhotonics Corporation (EPC) overcame each of these problems and allowed them to produce commercial systems based on this concept. Figure 9.4 displays a photograph of a EPC four-channel *fiber optic grating strain indicator* (FOGSI) unit. The specification of

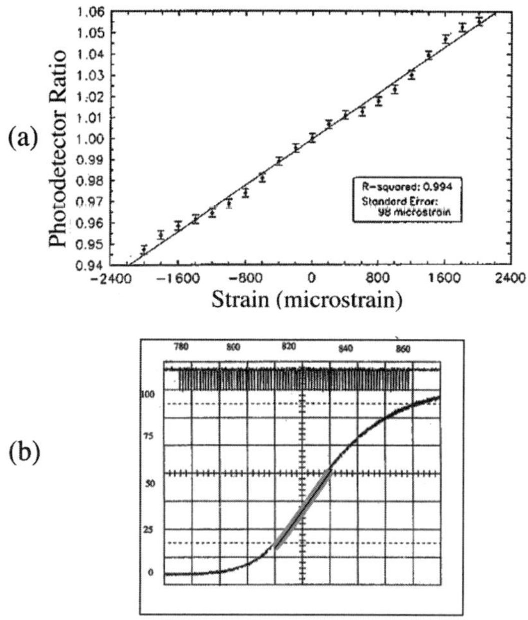

FIGURE 9.1. (a) Variation with strain of the signal ratio for a passive spectral ratiometric FBG demodulation system. (b) Expanded view of the spectral filter function used in the passive spectral ratiometric FBG demodulation system. The near linear portion is highlighted. From Melle, S.M., Liu, K., and Measures, R.M., (1993b), "Practical Fiber Optic Bragg Grating Strain Gauge System", *Applied Optics* **32**, 3601–3609.

the model FLS-3500 are as follows:

- Measurement range: 8500 $\mu\varepsilon$
- Strain resolution: 1 $\mu\varepsilon$
- Measurement precision: 5 $\mu\varepsilon$
- Single channel analog response: 1 kHz 3 dB

This system actually achieves four-channel status by rapidly switching a single optical channel detection system between four-optical fiber receiver channels. As will be shown later, in reference to the Taylor Bridge (Maaskant *et al.*, 1998), a special ruggedized system was built and demonstrated to function through the harsh winter environment of a bridge in a place such as Winnipeg. EPC also produces a true parallel eight-channel unit (EPC FLS-3100) with the same specification for applications where high frequency response is required for all channels. A photograph of a four-channel version of this demodulation system is shown in Fig. 9.5.

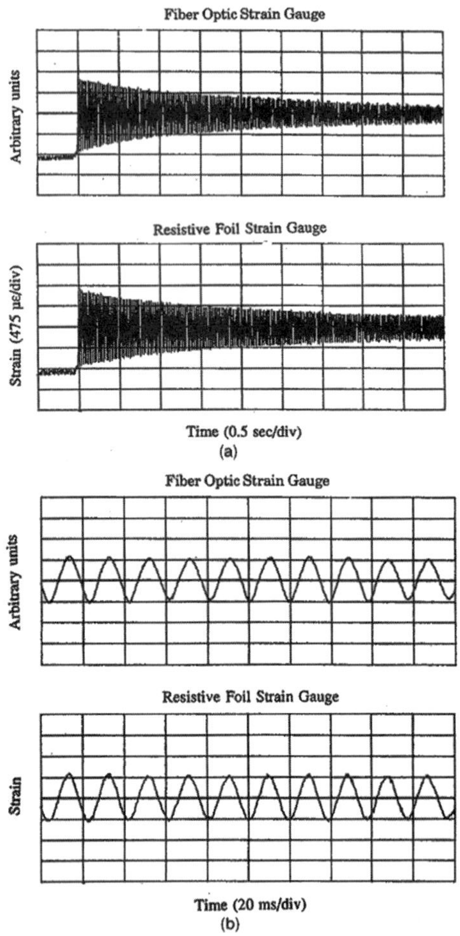

FIGURE 9.2. (a) Dynamic response of a FBG sensor and a resistive foil strain gauge that are bonded to a vibrating cantilever beam. (b) Magnified view of the temporal variation shown in (a). From Melle, S.M, Liu, K., and Measures, R.M., (1993b), "Practical Fiber Optic Bragg Grating Strain Gauge System," *Applied Optics* **32**, 3601–3609.

An alternative approach that is more efficient with the optical sensing signal, and somewhat more compact, was also proposed by Melle *et al.*, (1992a) and first demonstrated by Davis and Kersey (1994). To understand this advance we refer to Chapter 5. It was shown there that the coupling coefficient for a 2×2 coupler could have a strong wavelength dependence. This leads to a power transfer between optical fibers that is highly dependent on the wavelength of the incident signal (see Fig. 5.62). If the

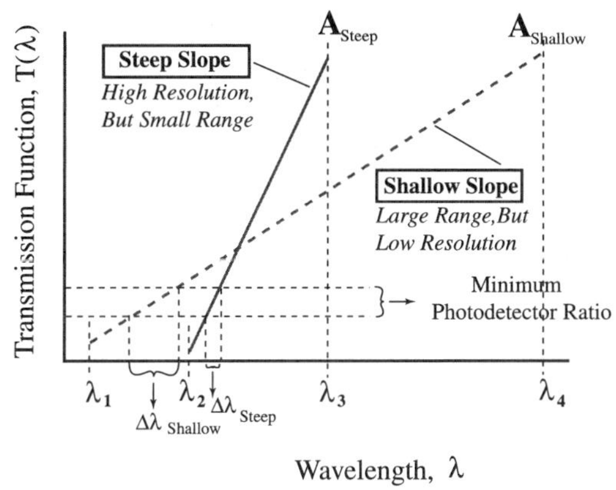

FIGURE 9.3. Trade-off between strain measurement range and resolution for the slope of the spectral filter function used in the FBG passive spctral ratiometric demodulation system. A steep slope provides high resolution, but small measurement range, while a shallow slope allows a large range, but low resolution.

second 3 dB coupler used in the experimental arrangement portrayed in Fig. 6.23 is replaced with such a *wavelength-dependent coupler,* then it essentially serves as both a signal divider and a double-sided spectral filter. In such a case, both photodetectors see optical signals that have in effect been processed by spectral filters however, their spectral slopes are of opposite sign.

The experimental configuration is schematically illustrated in Fig. 9.6, and the corresponding spectral transmission function of the Gould

FIGURE 9.4. Photograph of the ElectroPhotonics, four-channel fiber optic grating strain indicator unit, model FLS-3500. Used with permission, ElectroPhotonics Solutions Corporation.

FIGURE 9.5. Photograph of a truly parallel four-channel fiber optic grating strain indicator unit, model FLS-3100. Used with permission, ElectroPhotonics Solutions Corporation.

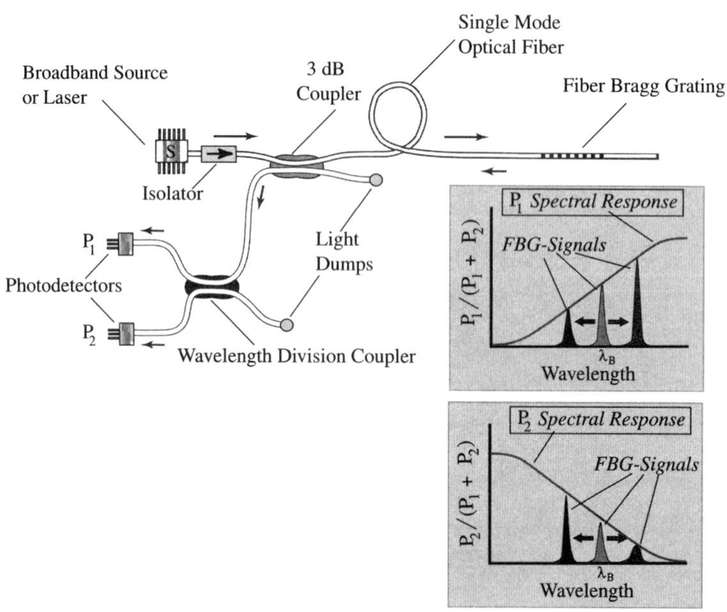

FIGURE 9.6. Passive spectral ratiometric demodulation system for FBG sensors using a wavelength-dependent coupler in place of a separate coupler and filter.

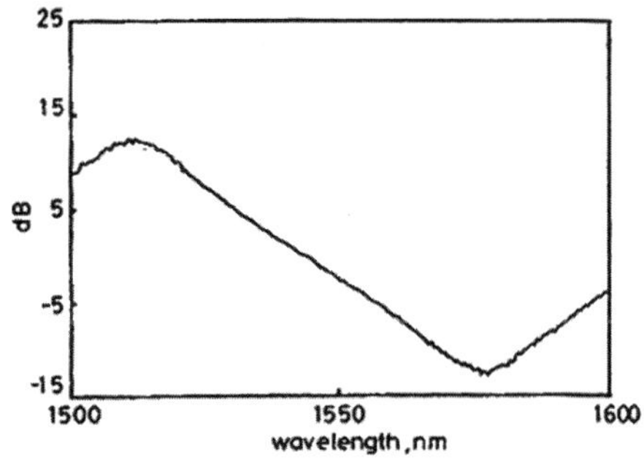

FIGURE 9.7. Spectral transmission function of the wavelength division coupler used in experimental system illustrated in Fig. 9.6. From Davis, M.A., and Kersey, A.D., (1994), "All-Fibre Bragg Grating Strain-Sensor Demodulation Technique Using a Wavelength Division Coupler", *Electronic Letters*, **30**, 75–77, used with permission, IEE Publishing.

wavelength division coupler is presented as Fig. 9.7. An erbium-doped fiber superfluorescent source delivered ~300 μW over a ~40 nm band centered at about 1540 nm. A linear response with strain can be obtained by taking the ratio of the difference to the sum of the two photodetector signals

$$R_{\mathrm{WDM}} = \frac{\mathrm{PD}_1 - \mathrm{PD}_2}{\mathrm{PD}_1 + \mathrm{PD}_2}, \tag{9.3}$$

A set of results from an FBG sensor subject to three-point bending strain is presented as Fig. 9.8 and indicates a strain resolution of ~±3 με, (Davis and Kersey, 1994). The dynamic strain of the system was evaluated by subjecting the metal strip to which the FBG sensor is attached to a 5-με amplitude vibration at 10.5 Hz. The resulting signal-to-noise measurement of 28.7 dB in a 0.15 Hz bandwidth, is shown as an inset to Fig. 9.8, and corresponds to a minimum detectable dynamic strain of ~0.5 με $\mathrm{Hz}^{-0.5}$. As we shall see later, this approach, as with all passive spectral ratiometric schemes, lends itself to serial multiplexing (Davis *et al.*, 1997b; see Fig. 9.60a. Unfortunately, the wavelength division coupler is fairly sensitive to temperature variations and changes in the state of polarization.

A third passive spectral ratiometric approach is based on the use of two long-period gratings, introduced in Chapter 7, to serve as opposing slope spectral filters (Patrick *et al.*, 1996a). The experimental arrangement is schematically illustrated in Fig. 9.9 and the corresponding spectral transmission curves of the two LPGs are displayed as an inset to this figure. The FBG was centered at 1276 nm at room temperature and zero strain and had a peak reflectivity of 80% and a linewidth of 0.2 nm. It was

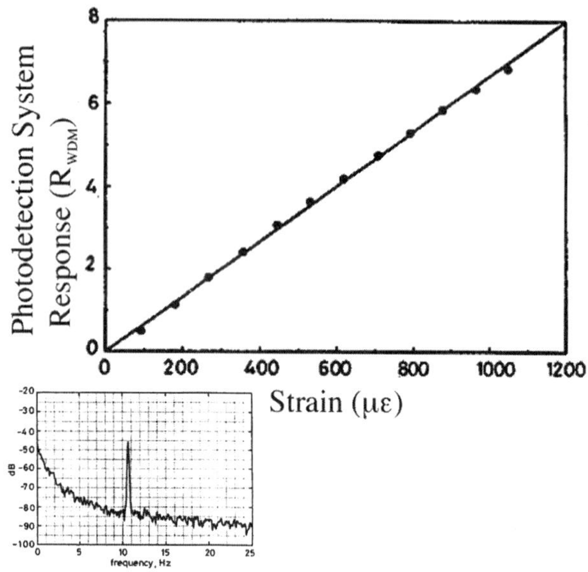

FIGURE 9.8. Photodetection system linear response with strain for experimental arrangement illustrated in Fig. 9.6. From Davis, M.A., and Kersey, A.D., (1994), "All-Fibre Bragg Grating Strain-Sensor Demodulation Technique Using a Wavelength Division Coupler," *Electronic Letters*, **30**, 75–77, used with permission, IEE Publishing.

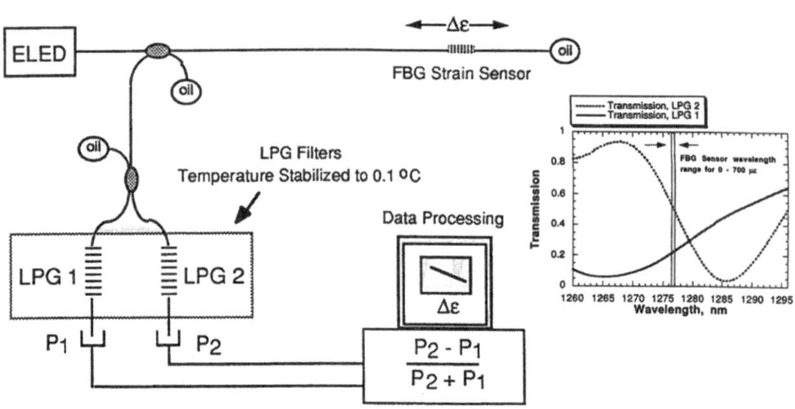

FIGURE 9.9. Schematic illustration of the passive spectral ratiometric FBG demodulation system baed on the use of two long-period gratings. The inset displays the spectral transmision functions ofr the two LPGs. From Patrick, H.J., Kersey, A.D., Pedrazzani J.R. and Vengsarkar, A.M. "Fiber Bragg Grating Sensor Demodulation System Using In-Fiber Long Period Grating Filters." In: *Distributed and Multiplexed Fiber Optic SensorsVI, Proc. SPIE* **2838**, (1996a) pp. 60–65.

excited by a 1.3-μm ELED. Examples of the sensing system response to quasi-static and dynamic loading are presented in Fig. 9.10. As with the WDM approach, the strain is obtained by taking the ratio of the difference to the sum of the two photodetector signals [Eq. (9.3)]. Although the short-term strain resolution was ~1 με in a 3.3-Hz bandwidth, drifts of around *90* με hindered absolute measurements. Once again, high sensitivity to temperature and polarization in both the coupler and the LPGs contributed to this instability.

9.2.2 Tunable narrowband filter demodulation

Tunable narrow-passband filters are central to a broad class of FBG sensor demodulation techniques. One of the most popular technique is based on the use of an optical fiber Fabry–Perot filter (Kersey *et al.*, 1993; Davis *et al.*, 1995; Kersey *et al.*, 1995) however, acousto-optic filters (Xu *et al.*, 1993; Dunphy *et al.*, 1993) and FBG-based filters (Jackson *et al.*, 1993; Davis and Kersey, 1995a) have also been successfully used for this purpose.

FFP-Filter. A FBG demodulation system based on a scanning *fiber Fabry–Perot* (FFP) filter is depicted in Fig. 9.11a, and was first demonstrated by Kersey *et al.* (1993). In this approach, light reflected from an array of Bragg grating sensors is passed through FFP filter with a *free-spectral range* (see Section 5.8) that is large enough to ensure that only one of the narrowband FFP wavelength resonances lies within the wavelength excursions of the sensing FBG. Use of piezoelectric stacks permits electrical control of FFP mirror spacing and therefore tuning of the FFP passband wavelength. In operation, the light reflected from the FBG sensor is returned via the coupler to the scanning fiber Fabry–Perot optical filter and subsequently to a photodetector. Several demodulation schemes have been used. The first involves applying a repetitive ramp voltage to the piezoelectric element so that FFP filter wavelength is repetitively scanned over the FBG spectral band. Simple peak detection is then employed to determine where in the voltage cycle (i.e., at what wavelength) the photodetector receives the maximum back-reflected light from the FBG.

A second, and more sophisticated approach involves forming the *effective derivative* of the received signal as the FFP wavelength is scanned over the FBG spectral band. This is accomplished by adding a small dither to the FFP filter ramp voltage so as to modulate the FFP filter bandpass wavelength (~0.01 nm). The photodetector signal is mixed electronically with the dither frequency and low-pass filtered so as to extract the component at the dither frequency. As the FFP filter sweeps over the

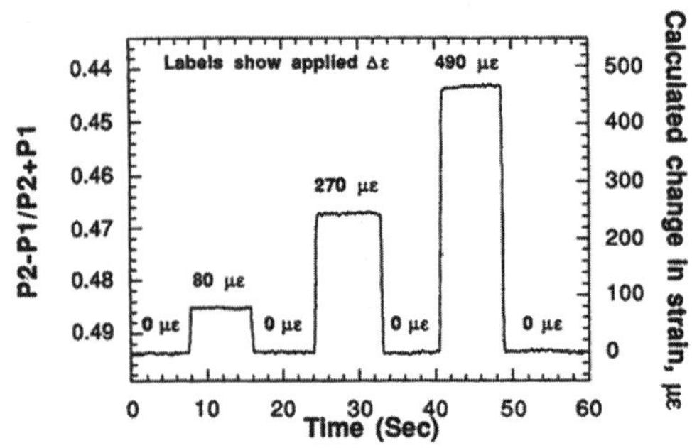

(a)

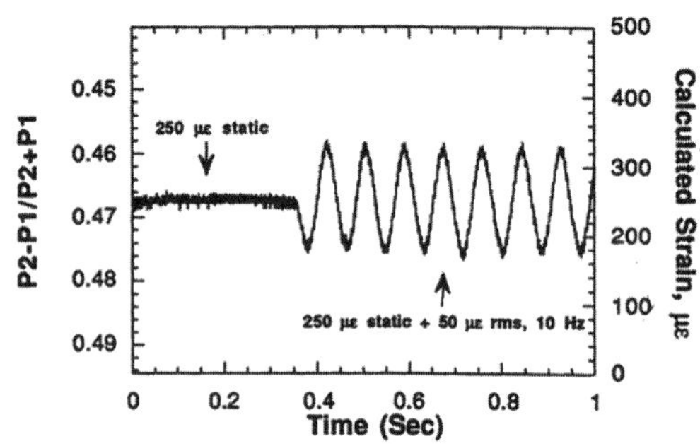

(b)

FIGURE 9.10. Temporal variations in strain as measured by the passive spectral ratiometric system illustrated in Fig. 9.9. From Patrick, H.J., Kersey, A.D., Pedrazzani, J.R. and Vengsarkar, A.M., (1996a), "Fiber Bragg Grating Sensor Demodulation System Using In-Fiber Long Period Grating Filters." In: *Distributed and Multiplexed Fiber Optic Sensors VI, Proc. SPIE* **2838**, pp. 60–65.

full wavelength range, the output of the mixer–filter arrangement provides the derivative of the photodetector signal see Fig. 9.11b. The zero-point crossing of this derivative signal corresponds to the peak of the FBG spectrum and consequently gives the Bragg wavelength.

The third approach can be used where it is desirable to track an individual FBG. In this instance the integrated low-pass filter output serves as an error-correcting signal for an electronic feedback loop (see Fig. 9.12), which effectively locks the center wavelength of the FFP filter to the selected Bragg wavelength. An external DC bias can be used to select which of the FBGs is tracked by this technique.

Typical characteristics for the FFP filter were a bandwidth of ~0.3 nm and a free-spectral range (FSR) of 50 nm. This allows up to 16 individual FBG sensors, spaced by ~3 nm, to be interrogated. For a FSR of 50 nm, generation of the scanning voltage for the FP filter is via a 16-bit digital-to-

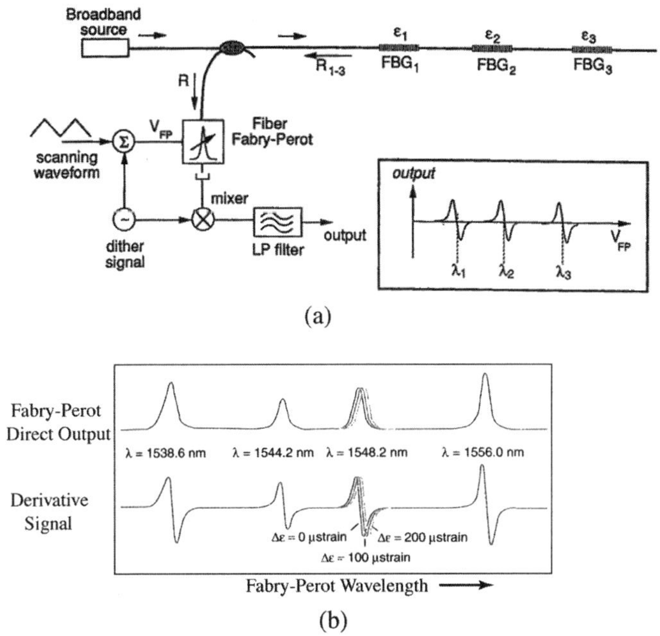

FIGURE 9.11. (a) A demodulation system for FBG sensors based on a scanning fiber Fabry Perot filter. (b) The derivative of the system response is attained by applying a dither to the scan of the Fabry Perot filter. From Kersey, A.D., Berkoff, T.A., and Morey W.W., (1993), "Multiplexed Fiber Bragg Grating Strain-Sensor System with a Fiber Fabry–Perot Wavelength Filter," *Optics Letters* **18**, 1370–1372.

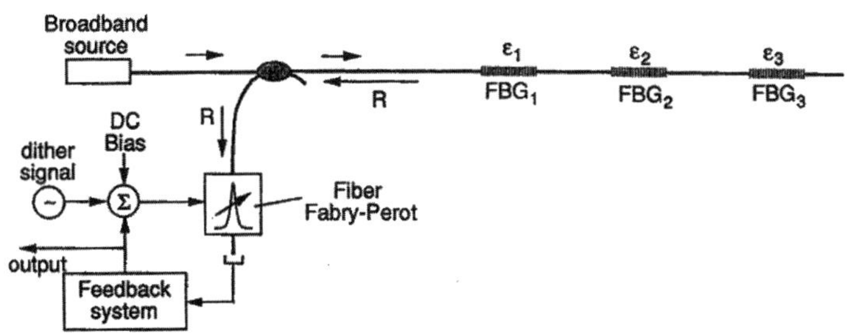

FIGURE 9.12. A closed-loop version of the FBG demodulation system, illustrated in Fig. 9.11 and also based on a tunable Fabry Perot filter. From Kersey, A.D., Berkoff, T.A., and Morey W.W., (1993), "Multiplexed Fiber Bragg Grating Strain-Sensor System with a Fiber Fabry–Perot Wavelength Filter," *Optics Letters* **18**, 1370–1372.

analog converter. This permits a minimum resolvable wavelength shift (least significant bit) of approximately 0.8 pm, or an equivalent strain resolution of 0.8 $\mu\varepsilon$. Currently available FFP filters can be scanned at rates $\geq$300 Hz, although scan rates to 1 kHz should be possible. A comparison of the strain monitored with an FBG sensor that is demodulated by a scanning FFP filter, and that of a conventional resistive foil strain gauge, is presented as Fig. 9.13. Strain resolutions on the order of $\pm$1 $\mu\varepsilon$ have been achieved, and up to 16 FBG sensors have been multiplexed on a single fiber.

The use of an optical switch allows this type of demodulator to address several parallel arrays of FBG sensors. A system for tracking 60 such sensors has been developed by Davis *et al.* (1996a) and is schematically illustrated in Fig. 9.14. Single-mode optical fiber switches driven under PC control are utilized to allow the measurement of strain along five-independent strings of 12 FBG sensors. The strings are illuminated using a single 1.3 μm ELED source ($\sim$150 μW power) through a 3-dB coupler and the switches. Processing for each of the 60-sensors in the array takes about 2.5 s (with 50 averages/sensor), which is more than an adequate sampling rate for static strain monitoring, for example, in civil engineering. The short-term resolution of the system was $\pm$1 $\mu\varepsilon$, and drift over a 30-min period was $< \pm$3 $\mu\varepsilon$, which was attributed to slight temperature fluctuations in the laboratory.

A commercial system based on the fiber Fabry–Perot tunable filter has been produced by Micron Optics Inc. This fiber Bragg grating interroga-

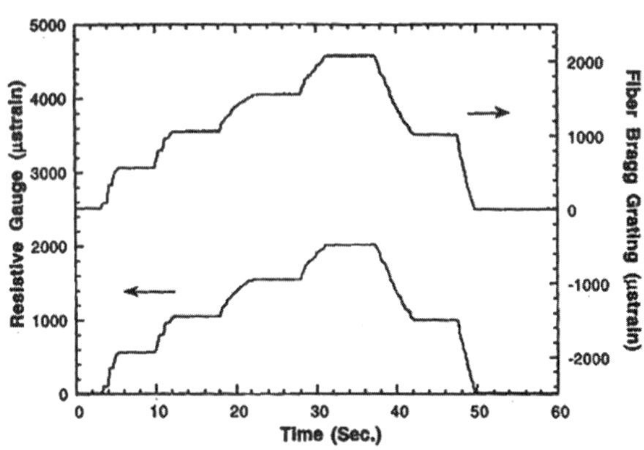

FIGURE 9.13. A comparison of strain measurements recorded by the fiber optic Bragg grating sensor and a resistive foil strain gauge. From Kersey, A.D., Berkoff, T.A., and Morey W.W., (1993), "Multiplexed Fiber Bragg Grating Strain-Sensor System with a Fiber Fabry–Perot Wavelength Filter," *Optics Letters* **18**, 1370–1372.

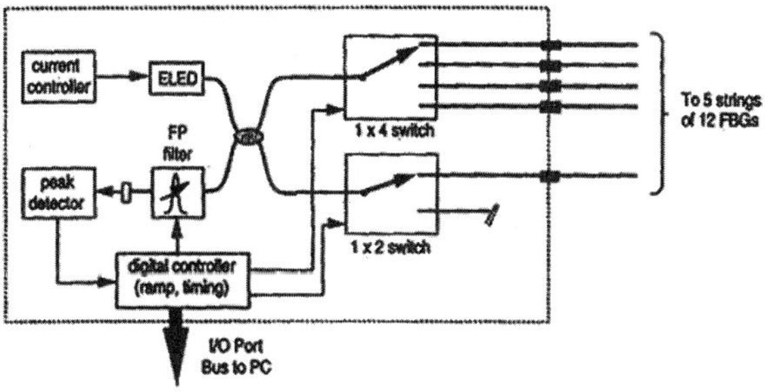

FIGURE 9.14. The experimental arrangement, based on a tunable FP filter and a fiber optic switch, used to demodulate an array of 60 FBG sensors. From Davis, M.A., Bellemore, D.G., Kersey, A.D., Putnam, M.A., Friebele, E.J., Idriss, R.L., and Kodinduma, M. (1996a), "High-Sensor Count Bragg Grating Instrumentation System for Large-Scale Structure Monitoring Applications," *SPIE* **2718**, 303–309.

tion system can simultaneously interrogate up to 31 FBG sensors arbitrary spaced along an optical fiber with a strain resolution of better than 1 με, a precision of 4 με, and a frequency response of 25 Hz. The total measurement range for all FBG sensors is, however (bandwidth) limited to 40 000 με, which means that for FBG sensors with a possible strain excursion of ±5000 με, only four sensors can in fact be interrogated with this system.

AOTF Filter. In Chapter 5, Section 5.8, the *acousto-optic tunable filter* (AOTF) was introduced as another form of tunable narrowband filter (see Fig. 5.43). At constant temperature the peak transmission wavelength of an AOTF is determined solely by the frequency of *radiofrequency* (RF)-drive signal applied to the device. Furthermore, several transmission bands can simultaneously be created by using the appropriate set of RF frequencies. Important merits of this technique include very wide spectral range, so many FBGs can be WDM, and random access to any specific FBG because of its frequency-agile capability (Xu *et al.*, 1993b). The original AOTF (Crystal Technology, MID-IR AOTF) had a wavelength tuning range of 1.2 μm to 2.5 μm and a spectral resolution (FWHM) of 4 nm. An AOTF can therefore be used for both dynamic and quasi-static, multiplexed FBG sensing systems.

A schematic of the system used by Geiger *et al.* (1995) is presented as Fig. 9.15. A broadband ELED at 1300 nm illuminates the FBG sensors and a AOTF with a 3.3 nm transmission band is positioned ahead of the photodetector. Tuning the transmission band of the AOTF is accomplished through a computer-driven RF voltage-controlled oscillator (VCO). The computer can lock the AOTF onto a particular FBG or display all of the back-reflected wavelengths from a set of FBGs by sweeping over the appropriate wavelength range. The wide bandwidth of the AOTF compared to that of the FBGs (3.3 nm c.f 0.2 nm) makes this approach impractical as the strain resolution would be inadequate. The peak wavelength of a FBG can be measured with acceptable precision, by dithering the RF frequency applied to the AOTF about the nominal value required to sit on the FBG peak wavelength (i.e., FSK technique) so that the resulting amplitude modulated (AM) signal from the photodetector goes to zero, see inset to Fig. 9.15.

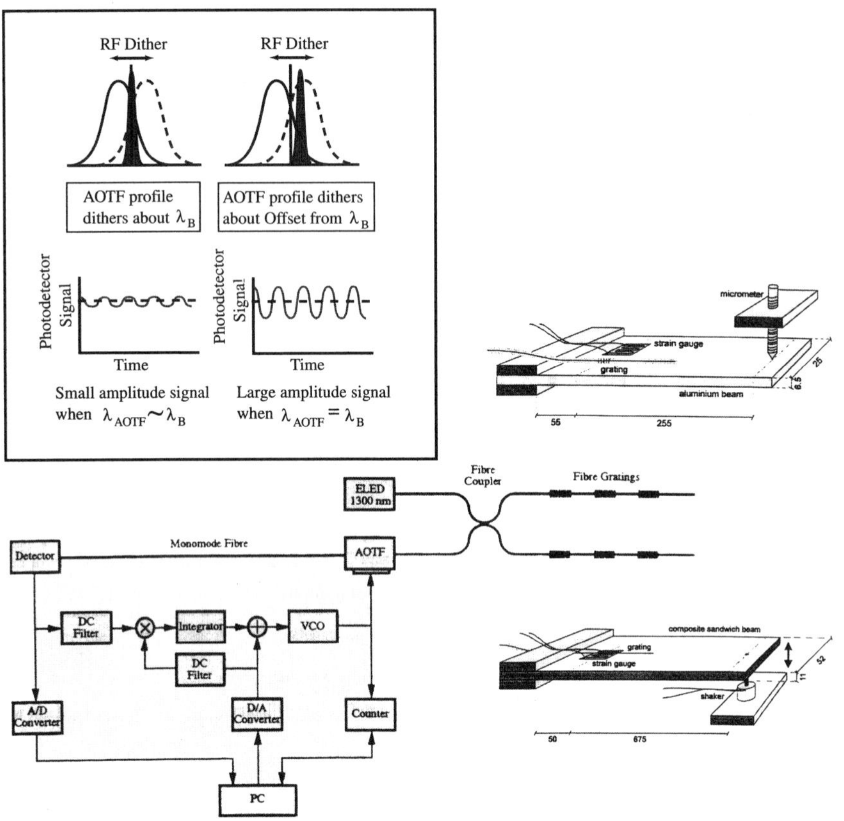

FIGURE 9.15. An acousto-optic tunable filter demodulation system for two serial arrays of FBG sensors. Also shown are two mechnical systems used to test the system, one based on micrometer deflection of a cantilever beam, the other a shaker for vibrating the beam. The inset indicates how an RF dither to the AOTF permits a precise measurement of the Bragg wavelength. From Geiger, H., Xu, M.G., and Dakin, J.P. (1995) "Multiplexed Measurements of Strain Using Short and Long Gauge Length Sensors", *SPIE*, **2507**, 25–34.

This type of interrogation system has been modeled and its performance analysed by Xu *et al.* (1996) and was tested on both quasi-static and dynamic loaded cantilever beams at low levels of strain, and with high levels of static tensile strain; see Fig. 9.16. The quasi-statically loaded aluminum cantilever beam was used to calibrate the system. The dynamically loaded cantilever beam was made of CFRP enclosing a honeycomb material. The tensile static test was taken to failure of the specimen. Figure 9.17 reveals the linear variation of the mean AOTF frequency against the applied strain. The static tensile tests demonstrated that the system operated right up to failure of the specimen at 4300 $\mu\varepsilon$ and the dynamic test demonstrated measurements at 5 Hz with a measurement accuracy of ± 2 $\mu\varepsilon$, (Geiger *et al.*, 1995). The AOTF *frequency–strain sensitivity* from the data presented in Fig. 9.17 is -96.7 Hz $\mu\varepsilon^{-1}$. Dunphy (1996)

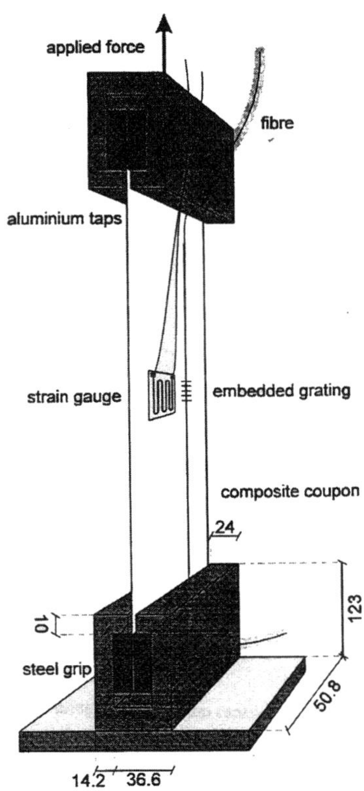

FIGURE 9.16. Test arrangement for FBG sensor demodulated by the AOTF system described in Fig. 9.15. From Geiger, H., Xu, M.G., and Dakin, J.P. (1995), "Multiplexed Measurements of Strain Using Short and Long Gauge Length Sensors", *SPIE*, **2507**, 25–34.

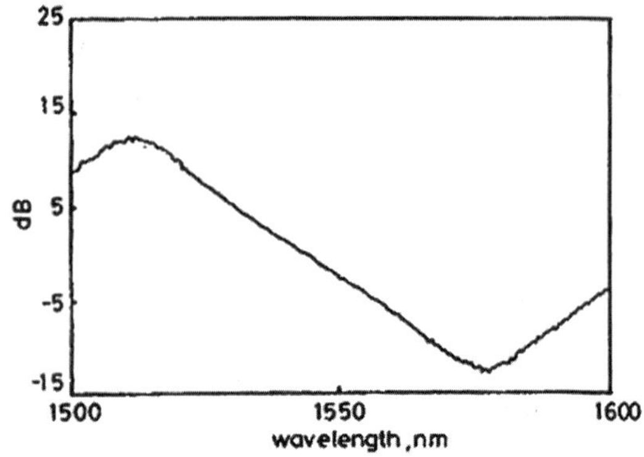

FIGURE 9.17. Linear variation of AOTF frequency aginst strain for the two FBG sensors used in the experimental setup illustrated in Fig. 9.15. From Geiger, H., Xu, M.G., and Dakin, J.P. (1995), "Multiplexed Measurements of Strain Using Short and Long Gauge Length Sensors," *SPIE*, **2507**, 25–34.

has reported on the development of an integrated optics AOTF system with a capacity to provide four-simultaneous, unsampled, real time signals of static and dynamic strain measurements from four-wavelength-multiplexed FBG sensors with a sensitivity of 2 nε $Hz^{-0.5}$.

Matched FBGs. The challenge of measuring the center wavelength for an FBG can also be answered by the old maxim "Use a thief to catch a thief." Jackson *et al.* (1993) were the first to show that an FBG sensor can be read by a *matching demodulation FBG*. In this system the *reading grating* was used in a reflectometric configuration and its wavelength controlled by means of a *piezoelectric stretcher* (Fig. 9.18). The system can be operated in either the scan mode, or the tracking mode. In the former, a voltage ramp is supplied to the piezoelectric stretcher that spectrally translates the receiving FBG reflection spectrum across the band of wavelengths likely to contain the sensing FBGs spectrum. When the photodetector signal reaches a maximum, the Bragg wavelengths coincide, and with the knowledge of the voltage applied to the receiving FBG at that point the strain being experienced by the sensing FBG can be determined. In the latter mode a *closed-loop servo* is used to lock the receiving FBGs wavelength to that of the sensing FBG and once again use the necessary piezoelectric voltage to evaluate the wavelength. These techniques can also be used in serial-multiplexed arrays of FBG sensors; see Chapter 11. A strain resolution of $\sim 4\,\mu\varepsilon$ was attained, but could be improved with narrower-bandwidth FBGs.

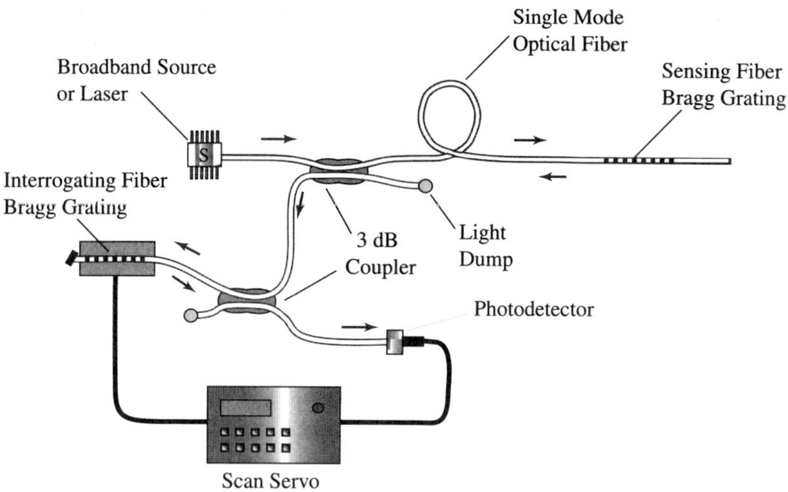

FIGURE 9.18. Schematic of the matched FBG technique for demodulating a sensing FBG.

The weakness of this scheme, the potential loss associated with two reflections in series and two couplers, was overcome by Davis and Kersey (1995). They took advantage of the efficient transmission characteristics of the receiving FBG. Both FBGs are produced with the same reference center wavelength. The sensing FBG is mounted on the structure to be monitored, while the interrogating FBG is mounted on a piezoelectric stretcher that is controlled by a feedback voltage signal obtained from the photodetector output signal. The experiment undertaken by Davis and Kersey (1995b) was aimed at demonstrating the multiplexing capability of this approach so a pair of sensing FBG (with Bragg wavelengths at 1551.8 nm and 1557.2 nm) were mounted on separate aluminum plates, each of which was subject to independent three-point bending. The corresponding matched pair of FBGs used for demodulation were mounted on individual piezoelectric stack stretchers.

The sensing FBGs were illuminated by a broadband superfluorescent erbium-doped fiber source that allowed serial wavelength multiplexing of a number of FBGs. It is worth noting that only one photodetector is needed in this arrangement. The signal from this photodetector was fed to two lock-in amplifiers, each referenced to the dither signals applied to the filter FBGs. The outputs of the two lock-ins were summed with the dither signals and fed back onto the interrograting FBGs. This feedback loop allowed the filter gratings to accurately track any strain induced wavelength shift experienced by the sensing FBGs with the lock-in output signals providing a voltage directly proportional to the imposed strain (Fig. 9.19). The inset to this figure shows the tracking voltage acquired when both tensile and compressive strain was applied to the two sensing FBGs. Independent operation of the two sensing FBGs is possible as different dither frequencies are applied to the two interrogating gratings. Although a favorable strain resolution was achieved, 10 nε Hz$^{-0.5}$, the strain range is limited by the piezoelectric stack stretchers. This approach also has an additional weakness in that any temperature-induced wavelength excursions must also be included in the response of the interrogat-

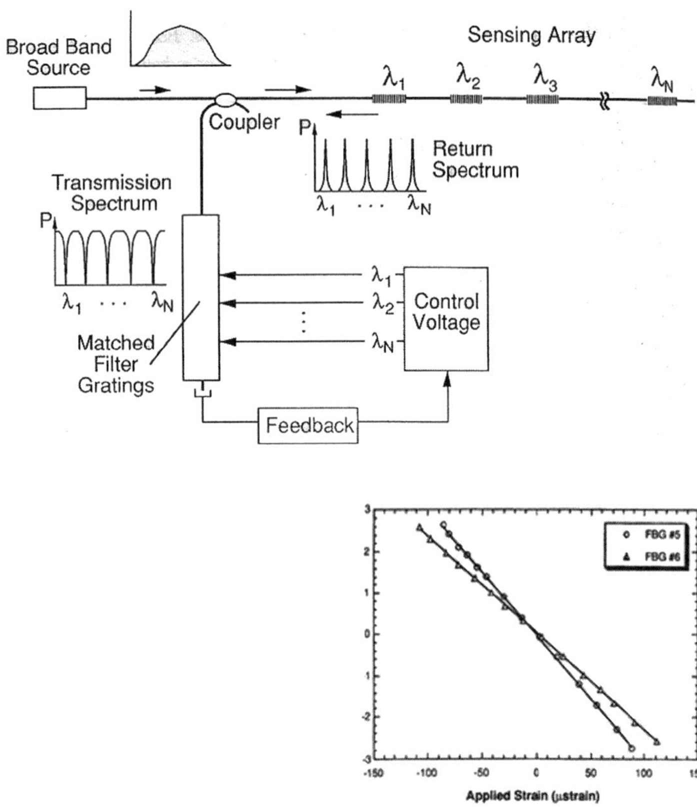

FIGURE 9.19. A demodulation system for FBG sensors based on closed-loop tracking with a matched set of interrogating FBGs. Inset shows linear variation with strain for the tracking voltage applied to piezoelectric elements controlling the interrogating FBG reflection spectrum. From Davis, M.A. and Kersey, A.D. (1995b), "Serially Configured Matched Filter Interrogaion Technique for Bragg Grating Arrays," *SPIE* **2444**, 295–312.

ing FBGs. It is also worth noting that matched FBGs have been used for referencing intensity-based displacement sensors (Cavaleiro *et al.*, 1995).

9.2.3 Interferometric-based demodulation

A high-resolution interferometric demodulation technique suitable for measuring small dynamic-strain-induced shifts in the reflection wavelength from an FBG sensor was first demonstrated by Kersey *et al.* (1992b). They employed an unbalanced Mach–Zehnder interferometer as an optical filter element with a transfer function of the form $[1 + \cos\phi(\lambda)]$ where the phase term, $\phi(\lambda)$, depends on the input wavelength, λ. Figure 9.20 schematically illustrates the experimental arrangement and indicates the general principle of this technique as applied to FBG sensors. Back-reflected light from the sensor is directed through the interferometer, where shifts in Bragg wavelength are converted into phase changes. According to Eq. (3.49), an optical path imbalance between the two arms of the Mach–Zehnder interferometer, $nd = n(d_1 - d_2)$, leads to a phase difference

$$\phi = \frac{2\pi nd}{\lambda} + \delta, \tag{9.4}$$

where λ is the wavelength of the light, d_1 and d_2 are the respective path lengths of the two arms, and δ is the reference phase factor. In such a case,

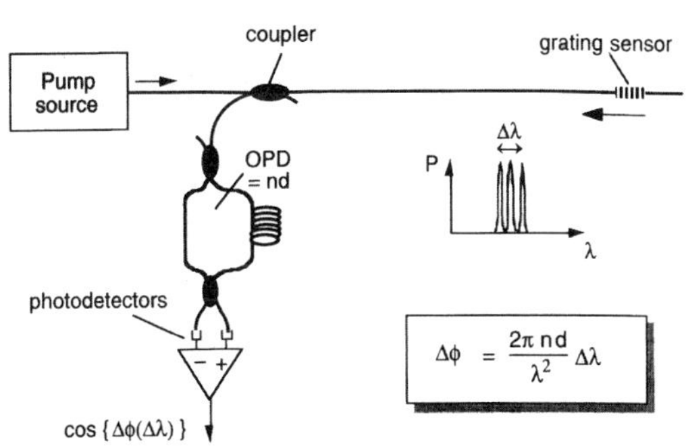

FIGURE 9.20. A Mach Zehnder interferometric system used for demodulatign FBG sensors is particularly good for dynamic strain measurements. From Kersey, A.D., Davis, M.A., Patrick, H.J., LeBlanc, M., Koo, K.P., Askins, C.G., Putnam, M.A. and Friebele, E.J., "Fiber Grating Sensors" *J. Lightwave Technology*, **15**, 1442–1463, © 1997, IEEE.

the phase change $\Delta\phi$ induced by a FBG wavelength shift $\Delta\lambda_B$ is given by the relation

$$\Delta\phi = -\frac{2\pi nd}{\lambda_B^2}\Delta\lambda_B. \quad (9.5)$$

For this approach to work the interferometer path difference must be kept less than the effective coherence length of the light reflected from the grating (i.e., ~1 cm for a 1-cm grating at 1.3 μm). Remember that the wider the spectral width of the light, the longer its coherence length [Eq. (3.58)] and that the spectral width of the FBG (which determines the bandwidth of the reflected light) depends on the length of the grating, [Eq. (3.138)].

Equation (9.5) can be used with the appropriate equation relating Bragg wavelength shift to strain to determine the strain from the phase change. In the case of axial strain at the reference temperature, Eq. (7.42) combined with (9.5) yields

$$\Delta\phi = -\frac{2\pi nd}{\lambda_B}G_\varepsilon(\varepsilon - \varepsilon_0), \quad (9.6)$$

Although the phase change can be determined with a single detector by fringe counting, a more accurate technique involves quadrature detection with a pair of detectors monitoring the intensity in each of the interferometer output arms, as indicated in Fig. 9.20. In this approach the interferometer is held in quadrature by a feedback voltage applied to a piezoelectric cylinder on which one arm of the optical fiber interferometer is wound. This piezoelectric fiber stretcher compensates for low-frequency phase drift fluctuations. Balanced photodetection was used to provide intensity noise rejection and a diode-pumped erbium-doped fiber superfluorescent source of ~300 μW output power and a 35-nm (1530–1565 nm) spectral width was used to excite the FBG sensor.

From Eq. (9.6) it is clear that a very high sensitivity to small changes in the Bragg wavelength can be achieved through an appropriate choice of the optical path imbalance for the interferometer. For example, with an interferometric optical path difference of 1 cm(~6.7 mm fiber length) the output wavelength-to-phase conversion factor is ~35 rad nm^{-1} at a wavelength of 1.3 μm (Kersey *et al.*, 1997b). Using the FBG strain sensitivity of approximation 1 pm $\mu\varepsilon^{-1}$(at 1.3 μm), the strain-to-phase response of the system is ~0.037 rad $\mu\varepsilon^{-1}$. Typically, with interferometric systems, phase measurements down to μrad $Hz^{-0.5}$ levels are possible, which results in a strain resolution of ~0.035 nε $Hz^{-0.5}$. Although this sensitivity has yet to be demonstrated, a strain resolution of 0.6 nε Hz$^{-0.5}$ has been demonstrated in the above experiments of Kersey *et al.* (1992b).

Although the interferometric technique just described is very sensitive to dynamic strains, drifts in the interferometer bias phase can prevent it from being suitable for quasi-static strain measurements. A technique to compensate for this drift using a reference wavelength to directly monitor the interferometer stability was reported by Kersey *et al.* (1993). The experimental arrangement is presented as Fig. 9.21 and uses a similar broadband source as the earlier work, but in this configuration the unbalanced interferometer is introduced immediately after the source rather than ahead of the photodetectors. The important difference, however, is the use of a second photodetector and a reference FBG to compensate for thermally induced drifts.

A repetitive ramp voltage applied to the piezoelectric fiber stretcher of the unbalanced Mach–Zehnder interferometer produced a phase modulation at an angular frequency, ω, of the light incident on both FBGs. The two photodetector signals were then bandpass filtered at the modulation frequency to yield two electrical signals of the form

$$S_s = A_s \cos[\omega t + \psi(t) + \phi(\lambda_s)], \tag{9.7}$$

and

$$S_R = A_R \cos[\omega t + \psi(t) + \phi(\lambda_R)], \tag{9.8}$$

where S_s and S_R represent the respective sensing and reference signals, and A_S and A_R are the respective sensing and reference photodetection

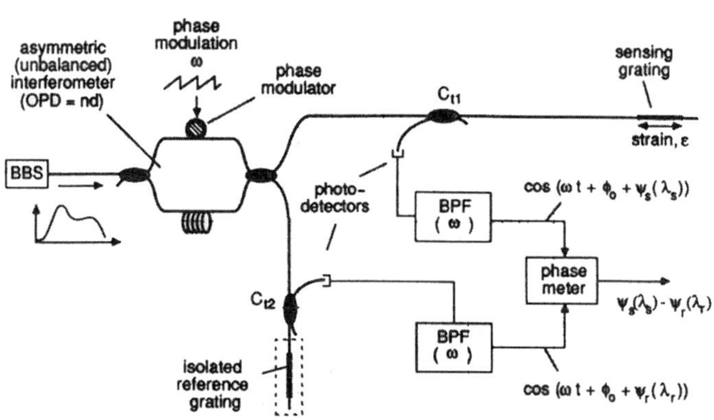

FIGURE 9.21. A drift-compensated Mach Zehnder interferometric FBG demodulation system. From Kersey, A.D., Berkoff, T.A. and Morey, W.W., (1993), "Mutiplexed Fiber Bragg Grating Strain Sensor-System with a Fiber Fabry-Perot Wavelength Filter". *Optics Letters* **18**, 1370–1372.

system constants, including reflected optical power and photodetector gain. These two phase-modulated signals are then fed to a *phase analyzer* that extracts the phase difference: $\phi(\lambda_s) - \phi(\lambda_R)$, which can be used to determine the strain imposed on the sensing FBG using the same procedure as outlined earlier. A strain resolution of $\sim 0.6\,\text{n}\,\varepsilon\,\text{Hz}^{-0.5}$ for low-frequency strain perturbations at 1 Hz was demonstrated with this system.

9.2.4 Laser sensor demodulation

All of the preceding demodulation techniques use broadband sources with narrowband FBGs. This is very inefficient of photons, and at best is wasteful of power; at worst the resulting poor signal-to-noise ratio limits the sensing system performance. A novel approach at improving the situation involves using the sensing FBG as one of the mirrors of a laser, (Measures *et al.*, 1992a). This concept can be understood by reference to the schematic presented as Fig. 9.22. The wavelength of operation for this laser is determined by the gain curve and the peak reflectivity of the FBG. If a high reflectivity FBG is efficiently coupled to the gain medium, the laser wavelength will coincide with the Bragg wavelength for the grating. Thus, as the state of strain, or temperature, of the FBG changes, the laser wavelength changes in a one-to-one relation. In essence the strain and temperature conditions are more than just wavelength encoded, they *are encoded into the laser wavelength*. Clearly this is an extraordinarily effective use of the laser output power.

Although the original proposal (Measures *et al.*, 1992a), was to have the sensing FBG tune the wavelength of a semiconductor laser diode, the first demonstration of this idea involved a sensing FBG controlling the wavelength of an erbium-doped fiber laser (Melle *et al.*, 1993a). The experimental arrangement is depicted in Fig. 9.23. A powerful laser

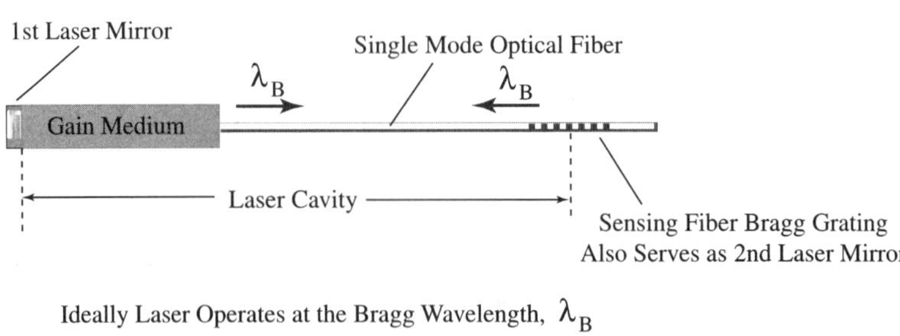

FIGURE 9.22. A simplified sketch revealing the basic configuration for a fiber Bragg grating tuned laser.

diode operating at 980 nm is used to pump a length of erbium-doped optical fiber by means of a 980/1550-wavelength division multiplexer. One end of the doped fiber is mirrored while the other goes to a FBG mounted on a cantilever beam. This pair of reflectors essentially define the fiber laser cavity. The wavelength of this fiber laser was measured by means of the passive spectral ratiometric technique (Melle *et al.*, 1992a,b).

The dramatic change in the spectral output of this fluorescent optical fiber, determined by an optical spectrum analyzer, with increasing pump laser power is clearly seen in Fig. 9.24. For low levels of pump laser power (< 2.2 mW) attenuation of light at the Bragg wavelength of the FBG can be seen as the notch in the fluorescent spectrum. As the pump power rises to 2.6 mW, this attenuation notch is replaced by a small peak corresponding to amplification of light at the Bragg wavelength. Further increases in pump power rapidly lead the emergence of a narrow strong fiber laser line at the Bragg wavelength. The linear variations in the fiber laser wavelength with strain and temperature are presented in Fig. 9.25, (Othonos *et al.*, 1993). A problem with this configuration is that the laser cavity length is large ($\geq$ 1 m) and so the laser FPC mode spacing is very small, and tens to hundreds of laser lines fall within the FBG reflectivity linewidth. This made continuous-wavelength tuning difficult. Nevertheless, this early work demonstrated that this approach could be used for both static and dynamic measurements with a strain resolution of $\sim$5 $\mu\varepsilon$ and a bandwidth of 13 kHz. Alavie *et al.* (1993) demonstrated that

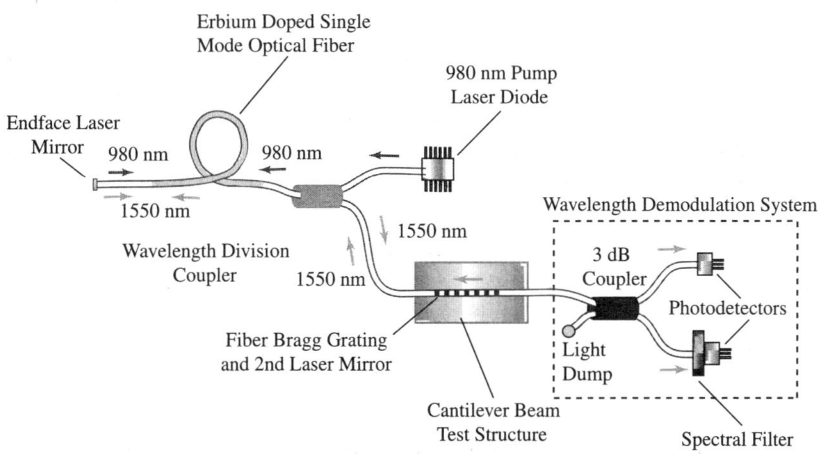

FIGURE 9.23. The experimental configuration for an FBG sensor tuned fiber laser. The wavelength of the erbium-doped fiber lasre is measured by the passive spectral ratiometric technique.

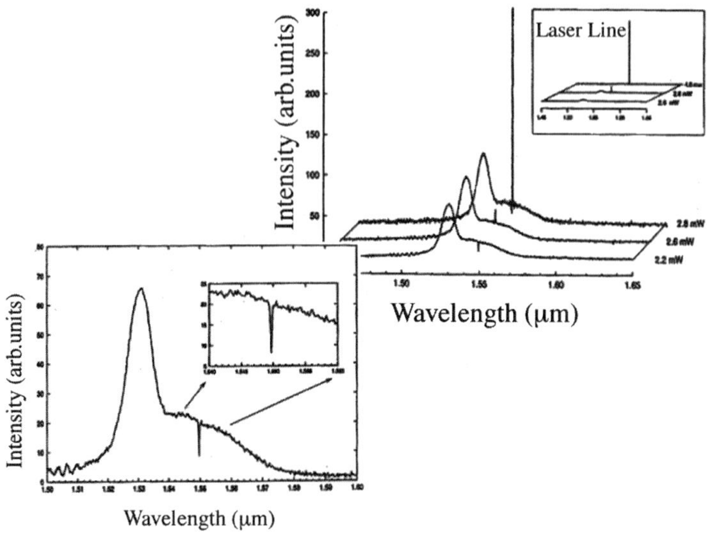

FIGURE 9.24. Spectral output of the FBG tuned fiber laser illustrated in Fig. 9.23. For low values of injection current in the 980 nm pump laser, the output is primary due to fluorescence and the absorption notch of the FBG can be seen in the spectrum. Once the threshold for laser action is exceeded, this spectral notch is replaced by the laser line.

several such fiber Bragg grating laser sensor (FBGLS) could be serially multiplexed.

Another form of FBGLS uses two fiber Bragg gratings to define the laser cavity (Kersey and Morey, 1993; Ball *et al.*, 1993), see Fig. 9.26. These lasers tend to be very short, as it is important to have both gratings in the same strain and temperature environment; otherwise their reflection spectra may not overlap. This bestows upon them the bonus that their mode spacing is large, so it is much easier to attain single mode operation and continuous wavelength tuning. The resulting laser line is much narrower than the passive FBG bandwidth and can fall to 50 kHz. When this is used in conjunction with the unbalanced Mach–Zehnder demodulation technique described earlier, very small dynamic strain perturbations can be detected. Kersey *et al.* (1997b) indicate that a dynamic strain resolution of better than a picostrain $Hz^{-0.5}$ might be attainable.

9.2.5 Tunable laser and quantum-well device demodulation

It has long been recognized that a tunable laser can be used to demodulate FBG sensors by scanning its wavelength over a spectral interval corre-

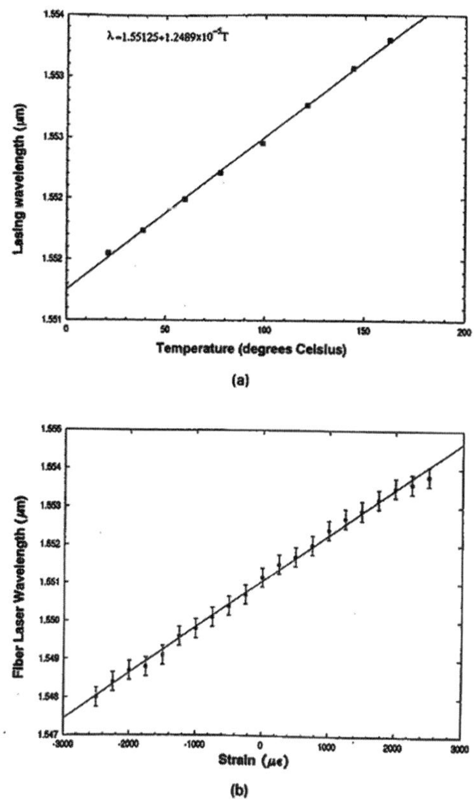

FIGURE 9.25. Linear variation of the fiber laser wavelength with strain and temperature changes applied to the tuning FBG. From Melle, S., Alavie, T., Karr, S., Coroy, T., Liu, K. and Measures, R.M., "A Bragg Grating-Tuned Fiber Laser Strain Sensor System", *IEEE Photon Technology Letters*, **5**, 263–266, © 1993, IEEE.

sponding to the expected dynamic range for the sensor (Measures, 1989). The high cost of suitable DFB or DRB laser diodes has deterred interest in this method of demodulation. Recent advances in this technology, combined with greatly reduced cost, have rekindled interest in this approach. Indeed, if one considers that the immense power available in such lasers enables one laser to demodulate a hundred or more multiplexed FBG sensors (see Section 5.10), the system cost per sensor starts to look quite reasonable.

With this consideration in mind, a group in the author's Fiber Optic Smart Structures Laboratory started to explore the prospect of using a tunable laser to demodulate FBG sensors. We have been able to show that

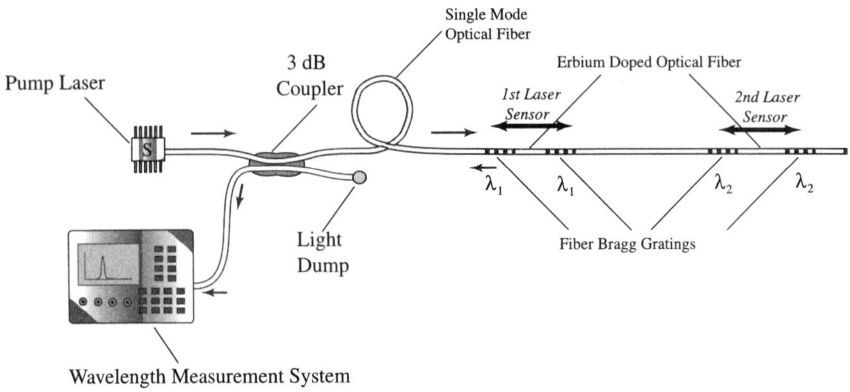

FIGURE 9.26. Schematic illustration of the twin FBG laser sensors based on erbium-doped optic fiber.

it will be possible to design a modular sensing architecture that permits one tunable laser to demodulate each of the different sensing modalities possible with FBGs (Measures *et al.*, 1998)—that is to say; short and long gauge-length sensors; serial and parallel arrays of multiplexed sensors; and truly distributed intra-grating sensors. At the heart of this *universal demodulation system* would be an extremely wavelength agile laser diode that can be scanned through 10 nm in about 1 ms.

A series of initial tests have been undertaken to confirm this concept (Measures *et al.*, 1998). A schematic illustration of the initial experimental arrangement we used to demonstrate this new approach is shown in Fig. 9.27. A special *gain-coupled distributed feedback* (GC-DFB) tunable laser (made available to us by Nortel, Ottawa) was used to interrogate a 1532 nm FBG sensor (Coroy *et al.*, 1998). The GC-DFB laser was temperature stabilized at 21°C, and provided smooth, continuous tuning from 1535.5 to 1544.5 nm with output signal powers varying between about 0.5 to 6.25 mW over this range. The FBG sensor was bonded to a cantilever beam that had been tapered to ensure a constant strain field along the length of the grating. A resistive foil strain gauge was bonded next to the fiber optic sensor to provide a reference strain measurement. The reflected signal from the FBG was collected by a photodetector. The photodetector signal and the laser injection current, which was used to tune the laser wavelength and evaluate it, were monitored using a modest computer with a data acquisition card.

The GC-DFB laser was repeatedly scanned over a wavelength range of approximately 2.5 nm, centered around the unstrained wavelength of

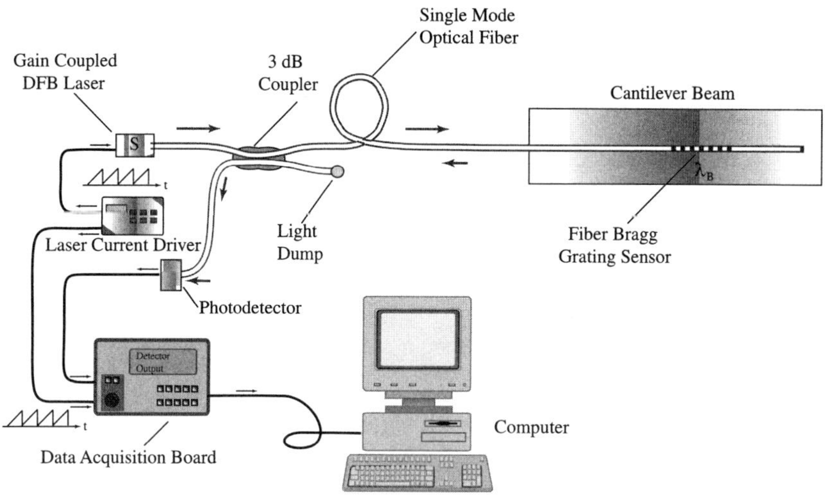

FIGURE 9.27. A demodulation system for FBG sensors based on the rapid scan of a gain-coupled DFB laser dode combined with a peak detection system.

the FBG, with a scan period of 10 ms. For this scan range and speed, the data acquisition system was able to resolve 3.2 points within the full width half maximum (FWHM) width of the FBG reflection peak, taking data at 10 000 samples per second. A standard peak detection technique was used to track the wavelength signal from the FBG. Figure 9.28 reveals the strain response of this FBG sensing system over a strain range of 2000 $\mu\varepsilon$. For each data point 60 successive measurements were taken. Since the sample period for these measurements was 10 ms, this corresponds to a bandwidth-normalized strain resolution of $\sim 0.076\ \mu\varepsilon\ \mathrm{Hz}^{-0.5}$.

In the same way electronics moved from systems based on individual components to those that involve integrated electronics, the use of the tunable DFB laser diode lends itself to the eventual development of an *integrated optoelectronic demodulation system* (Measures *et al.*, 1998). The last technique we shall consider also fits this promise and represents the first attempt at the kind of microminiaturization possible. The starting point for this advance was the trade-off required in designing the *passive spectral ratiometric demodulation system* just described. To achieve high strain or temperature resolution it is necessary to use a steep slope for the spectral edge filter transmission curve, while a reasonable wavelength (strain or temperature) range requires a much less steep slope; see Fig. 9.3. The answer to this dilemma is the kind of active spectral ratiometric demodulation system based on a quantum-well electroabsorption device first demonstrated by Coroy and Measures (1995).

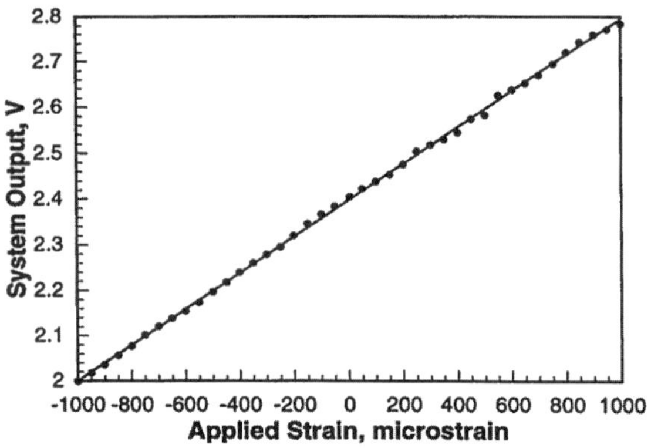

FIGURE 9.28. Linear variation with strain of the response of the tunable laser FBG sensor demodulation system illustrated in Fig. 9.27. From Coroy, T., Chappell, L.M., Guillermo, N.J., Huang, S.Y. and Measures, R.M., (1997), "Peak Detection Demodulation of a Bragg Fiber Optic Sensor Using a Gain-Coupled Distributed Feedback Tunable Laser", OSA 12th International Conference on Optical Fiber Sensors, Williamsburg, 210–212.

In this concept the spectral edge filter's transmission characteristics can be rapidly modified by the application of an appropriate voltage. To some extent this is similar to what is done with the narrowband filter approach described earlier and using a Fabry–Perot interferometer, an acousto-optic filter, or a piezoelectric controlled grating. In these solutions, however, the spectral filter is a very different device from the photodetector, and the resulting demodulation system is relatively expensive and bulky. We have pursued a more challenging solution involving integrated optoelectronics, as this offers the prospect of microminiaturization, low cost, and consistent performance that comes from automated microchip manufactured based systems.

The approach uses *quantum confined Stark-effect tuning* of the absorption edge of a *quantum-well electroabsorption filtering detector* (QWEFD). Although this is a fairly advanced concept, Fig. 9.29 is intended to help provide a basic understanding. We saw in Chapter 4 that both semiconductor lasers and photodetectors were made from specially prepared multiple-ultrathin layers of materials, such as indium–gallium–arsenide, or InGaAs for short. A QWEFD is a similar device, except that the transmission characteristics of the thin layer through which light travels are strongly influenced by the magnitude of the reverse bias voltage applied to the device. A simplified schematic of the QWEFD structure used by Coroy and Measures (1997) is presented as Fig. 9.29a. An indication of the variation in the relative response of this detector with wavelength, for a range of reverse bias voltages, is displayed as Fig. 9.29b. It is apparent from this diagram that if the reverse bias is, say, 7 V, the detector output when exposed to light at a wavelength of 1550 nm is

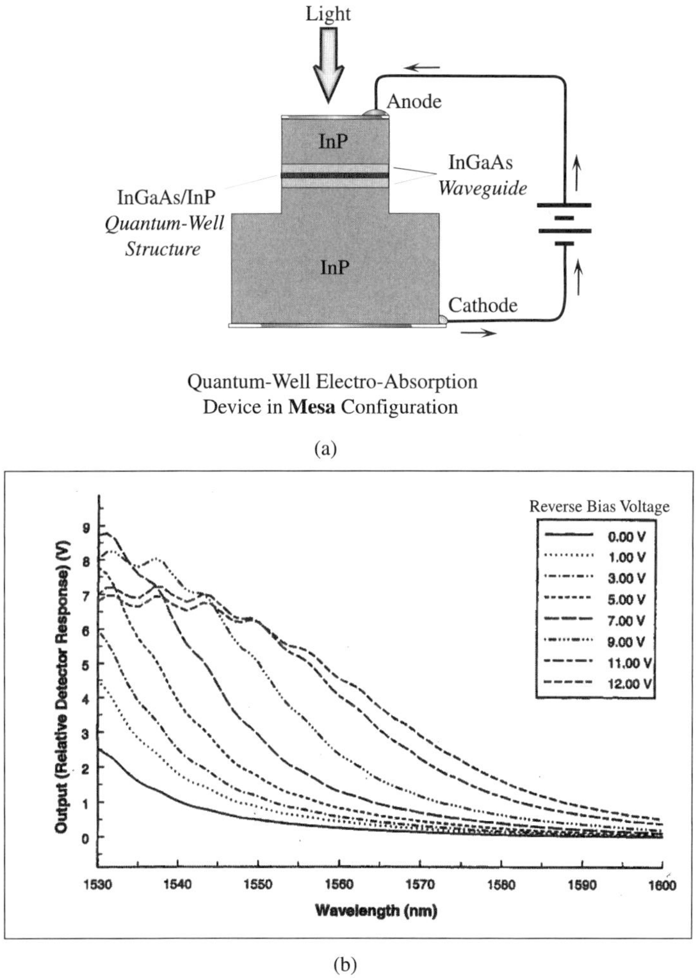

FIGURE 9.29. (a) Schematic diagram of the mesa quantum-well electro-absorption filtering detector (QWEFD) used for measuring the wavelength of a FBG sensor. (b) Open-loop spectral response for a QWEFD as a function of the applied reverse bias voltage. From Coroy, T. and Measures, R.M., (1995), "Active Wavelength Demodulation for Bragg Grating Strain Sensors Applications of Photonic Technology," Lampropoulos, Chrostowski and Measures, (eds), pp. 343–347, with permission from Plenum Press.

roughly half what it is for light at a wavelength of 1540 nm. Thus, if the reverse bias voltage is fixed, this device can work with a regular photodetector as a compact passive spectral ratiometric system.

Active wavelength demodulation (AWD), however, allows this device to be used in a more effective manner. In this mode of operation a ratiometric signal, V_{ratio}, is obtained from the filtering and reference detectors to provide intensity self-referencing. This signal is continuously compared with a reference signal, V_{ref}. If the system is initially in equilibrium then the reference signal, V_{ref}, is set equal to the ratiometric signal for light at wavelength λ_1, when the reverse bias voltage has some convenient value, say V_1 This corresponds to state-point (A) in the *system state diagram* (Fig. 9.30). If the wavelength of the light now shifts to a longer wavelength, say, λ_2, it is clear that $V_{ratio} = V_B$ (so the signal drops on the response curve from *state-point A* to state-point B, in Fig. 9.30), which is smaller than V_A $\{=V_{ref}\}$. The system responds by increasing the applied reverse bias voltage from V_1 to V_2, which shifts the spectral response curve to longer wavelengths until equilibrium is reestablished (state-point C in

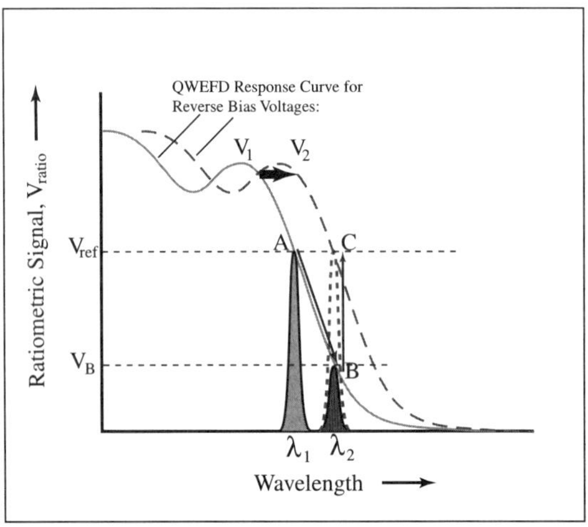

FIGURE 9.30. Schematic illustrtion of closed-loop tracking of the Bragg wavelength of a FBG sensor using the change in the reverse bias voltage (V_1 to V_2) for the QWEFD (Fig. 9.29) necessary to restore the ratiometric signal to the ratiometric signal to the reference value (B to C) after its drop (A to B) accompanying the shift in the Bragg wavelength. From Coroy, T. and Measures, R.M., (1998), "Demodulation Systems for Bragg Grating Fiber Optic Sensors Using a Quantum Well Electroabsorption Filtering Detector," *Smart Materials and Structures Journal*, 7, 265–271, with permission from the Institute of Physics Publishing Ltd.

Fig. 9.30). The change in the reverse bias voltage that is applied to the QWEFD ($\Delta V = V_1 - V_2$) is taken as the output signal from which the wavelength shift can be evaluated once the system is calibrated.

A preliminary experimental setup to test this concept with a mesa QWEFD is schematically illustrated in Fig. 9.31. In the mesa configuration the light passes through the device in a direction perpendicular to the plane containing the quantum wells. This makes coupling of light from an optical fiber to the device much easier than in the waveguide mode as the thickness of the quantum-well layers is much smaller than the core of the optical fiber (Fig. 9.32). A plot of the system output against strain imposed on the FBG sensor is presented as Fig. 9.33 (Coroy and Measures, 1998). A bandwidth normalized strain resolution of $\sim \pm 3\ \mu\varepsilon\ \mathrm{Hz}^{-0.5}$ was attained with FBG at 1553.5 nm using a superluminescent diode as the source. The application of antireflection coatings on the QWEFD window surfaces, to suppress the small Fabry–Perot modulation, seen on the response curves, could probably improve this resolution.

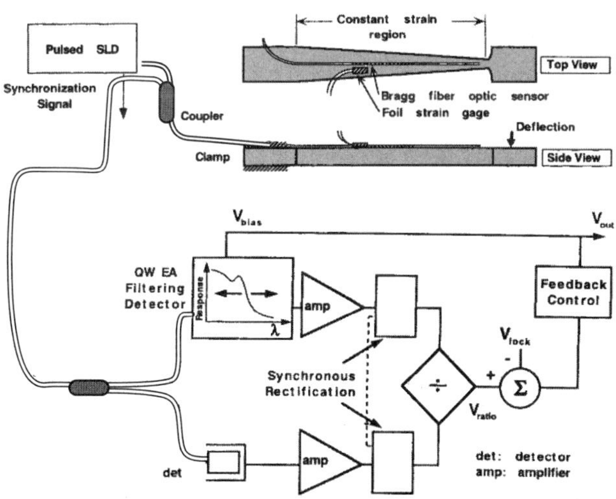

FIGURE 9.31. Experimental arrangement used to demonstrate the use of a mesa QWEFD (Fig. 9.29) for FBG sensor demodulation. From Coroy, T. and Measures, R.M., "Active wavelength demodulation of a Bragg grating fibre optic strain sensor using a quantum well electroabsorption filtering detector," *Electronics Letters* **32**(19) 1811–1812, with permission from IEE Publishing Ltd.

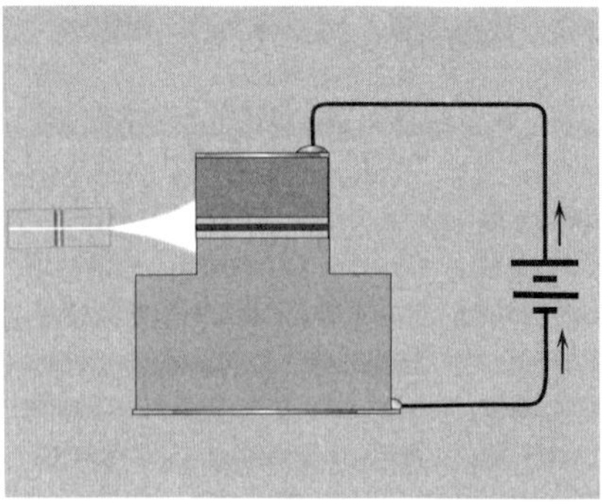

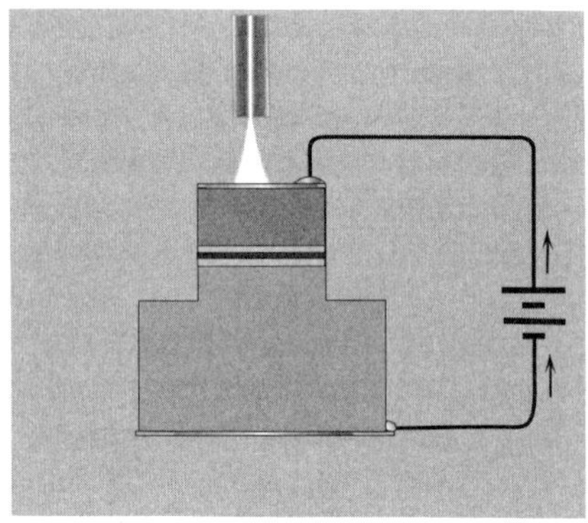

FIGURE 9.32. Schematic illustration of the "waveguide" and "mesa" configurations of the QWEFD described in Fig. 9.29.

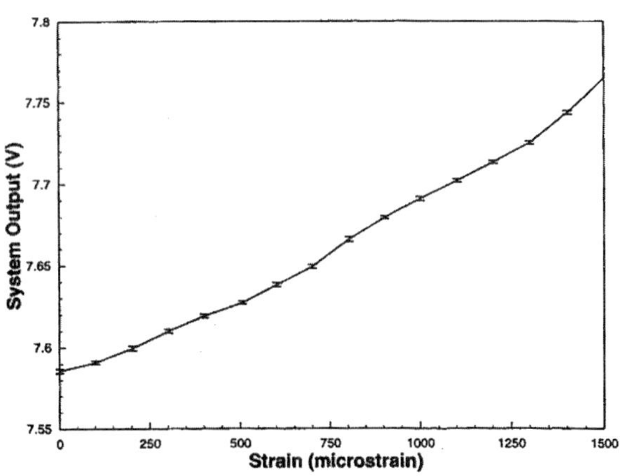

FIGURE 9.33. QWEFD system response against strain imposed on FBG sensor. From Coroy, T. and Measures, R.M., (1998), "Demodulation Systems for Bragg Grating Fiber Optic Sensors Using a Quantum Well Electroabsorption Filtering Detector," *Smart Materials and Structures Journal*, **7**, 265–271, with permission from the Institute of Physics Publishing Ltd.

9.3 FIBER BRAGG GRATING SENSOR APPLICATIONS

In 1993 the first major integration of a number of fiber optic Bragg grating structural sensors into the concrete girders of an innovative new bridge in the city of Calgary was undertaken by a team from the author's laboratory. This represented an important milestone, for although fiber optic strain sensors had been bonded to the surface of a variety of structures, this was the first time they had been structurally integrated into a large structure such as a bridge. This project also constitutes the first long term field-site testbed for fiber optic strain monitoring. To some extent, Fuhr *et al.* (1991) had led the way by showing it was possible to embed optical fibers within the concrete foundation of a building. Since we were only included in the Beddington Trail Bridge project a few months before the start of construction, a saying was coined in my laboratory: "We had to run as fast as we could to catch the bridge." I can assure the reader that indeed this team of young people worked night and day to make the schedule, which was quite literally cast in concrete.

As time has passed this project has turned out to be the first of many such undertakings, some by us, and some by a growing number of researchers and engineers keen to explore the bounds of this technology. In this next section we shall highlight a number of major demonstrations of fiber optic structural monitoring. Since most of these have been bridges, this is reflected in this section. To some extent, the division between Chapter 11 and this one is artificial, for as it will become apparent, many of the applications that we shall include in this chapter will involve some form of multiplexing, since any major installation invariably needs a large number of sensors. As indicated earlier, I have therefore tended to include parallel multiplexed sensor arrays in this chapter, while serial sensor arrays form the focus of Chapter 11.

9.3.1 Beddington Trail Bridge, Calgary

The primary task of the fiber Bragg grating sensor array installed in the *Beddington Trail Bridge* in Calgary, Canada, was to monitor and compare the use of three types of prestressing tendon: *conventional steel strand*, *Carbon Fiber Composite Cable* (CFCC), and *carbon fiber Leadline Rod*. CFCC was provided by Tokyo Rope Manufacturing Co. Ltd. Japan, and Leadline was made available by Mitsubishi Chemical, Japan; both were shown in Fig. 2.4. The three-lane, two span skew bridge, seen under construction, and complete in Fig. 9.34, comprised 26 precast, prestressed concrete

FIGURE 9.34. Beddington Trail Bridge in Calgary during construction and when open to the public in 1993.

girders, 6 of which used one or the other CFRP materials in place of steel for the prestressed tendons (Rizkalla and Tadros, 1994).

Long-term assessment of the tendon/concrete behavior and performance was the initial motivation for establishing the resident monitoring system in this bridge. In addition, there was the desire to see if monitoring of: traffic usage, extreme load events, and load history of such structures was also possible with this new kind of sensor technology. Such information is valuable for assessing bridge designs, comparing actual loading with loads assumed in the design and specified in the bridge design codes. It could also prove to be helpful for diagnostic and assessments with regard to bridge maintenance or retirement when structural damage has been sustained.

Since the tendons used in the Calgary bridge were to be prestressed to approximately 8 000 $\mu\varepsilon$, long life expectancy was ensured by bonding the FBG sensors to the tendons after application of the prestress so that subsequent destresssing and relaxation would place the sensors in compression, a favorable state for protection from stress-corrosion mechanisms.

Prior to construction of the bridge, a 1:33 scale test model of a bridge girder was built and tested at the University of Manitoba (Abdelrahman *et al.*, 1995). FBG sensors were installed alongside resistive foil strain gauges on the prestressing tendons and the beams subject to quasi-static and cyclic loading. The FBG sensors remained operational until failure of the beam, as indicated in Fig. 9.35. The ability of these sensors to sustain cyclic loading with very little creep was further demonstrated in a simulation of bridge loading (Alavie *et al.*, 1994c). In this experiment a FBG sensor was bonded to a prestressed $\pm 45^\circ$ carbon fiber/PEEK composite specimen and loaded 320 000 times between 0 and -2000 $\mu\varepsilon$. The FBG sensor spectrum revealed no degradation in the sensor response (Fig. 9.36) and the small degree of shift in the spectrum was commensurate with creep expected for the specimen due to microcracking.

One of the more daunting tasks in the instrumentation of concrete structures with fiber optic sensors is avoiding their destruction during placement and compaction. As described in Chapter 8, special cables have to be used that protect the optical fibers from both moisture and alkalinity and minimize pinching and bending. The lead optical fiber cables need to be carefully routed so as to be protected from the vibrating probes used to distribute the concrete through the array of reinforcement bars (Fig. 9.37). A cross-sectional view of the *bulb-tee precast prestressed concrete girder* used in the Beddington Trail Bridge is depicted in Fig. 9.38. The locations of the prestressing tendons and duct for post tensioning are indicated, as is the site for the FBG sensor installation.

The fiber leads were routed along the instrumented reinforcement to a recessed metal outlet box, which was attached to the inside of the wooden

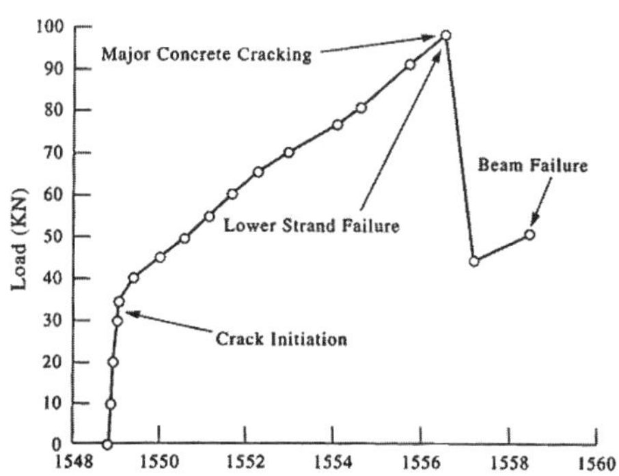

FIGURE 9.35. Variation of the Bragg wavelength for an FBG sensor attached to a concrete beam with the load applied to that beam in a failure test. Reprinted from Maaskant, R., Alavie, T., Measures, R.M., Tadross, G., Rizkalla, S.H. and Guha-Thakurta, A., (1997), "Fiber Optic Bragg Grating Sensors for Bridge Monitoring", *Journal of Cement and Concrete Composites*, **19**, p. 26, with permission from Elsevier Science.

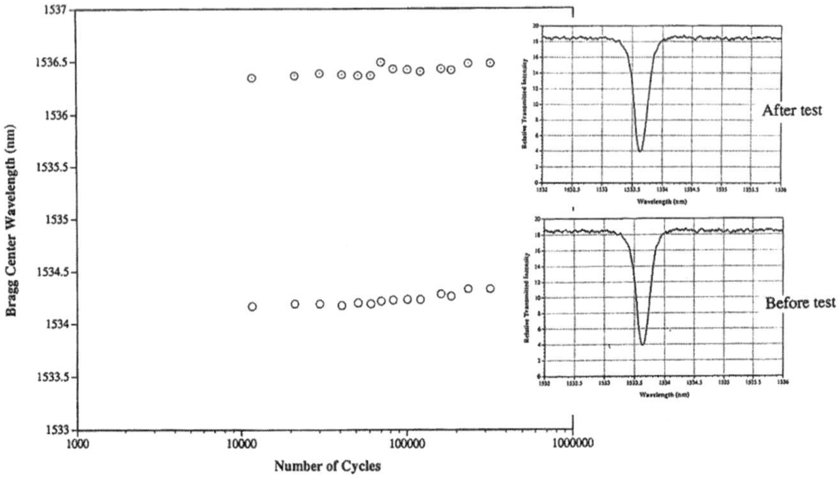

FIGURE 9.36. Variation in the Bragg wavelength of an FBG sensor attached to a carbon fiber/PEEK specimen as it is loaded through 320,000 cycles of 0 to − 2000 microstrain. The two small insets compare the FBG transmission spectrum before and after fatigue cyclic loading. From Alavie, A.T., Maaskant, R., Ohn, M.M., Rizkalla, S. and Measures, R.M., (1994a), "Application and Characterization of Intracore Grating Sensors in a CFRP Prestressed Concrete Girder," *SPIE* **2191**, 102–110.

form for the concrete (Fig. 9.39). To avoid embrittlement of the optical fibers, it is important to ensure that the outlet box is properly sealed prior to steam curing of the concrete. The fiber optic sensor array comprised 18 FBGs that were mounted at various locations on the three different kinds of prestressing tendons. The sensors within each girder were tested after fabrication of the girder (see Fig. 9.40) All of the fiber leads were then routed from the recessed outlet boxes in each of the girders to a central junction box placed at the top of one of the bridge abutments (Fig. 9.41); the other view in this figure reveals all of the optical fiber cables within the junction box. Each cable comes from a different FBG sensor embedded within one of the girders.

A number of the FBG sensors were located approximately 37.5% of the girder length from the abutment end in order to have maximum sensitivity to the bridge strain response to active loads (traffic). For those girders prestressed with CFRP, an additional sensor was placed near the stress transfer zone, about 1 m from the abutment, or pier, as indicated in Fig. 9.42. Static-strain measurements from the Beddington Trail Bridge FBG sensor array are presented as Fig. 9.43. Most of the original wavelength readings from the FBGs were obtained using an erbium-

FIGURE 9.37. Photograph showing the careful routing of an optical fiber cable in one of the Beddington Train Bridge girders prior to pouring of the concrete.

doped fiber laser as the source and a Burleigh Jr. Wavemeter (Alavie *et al.*, 1994a). The measurement precision of this configuration was ~40 $\mu\varepsilon$, and was limited primarily by the Wavemeter. The sensor measurements are designated by the tendon type (SS for steel strand; TR for Tokyo Rope CFCC; and LL for Mitsubishi Leadline Rod), girder identification number (#), and sensor location identified by the number to 1 to 4, as depicted in Fig. 9.42.

The measurements of September 1993 were taken immediately following posttensioning of the girders. These measurements represent the stress relaxation in the tendons from the combined effects of *destressing, concrete shrinkage, creep, dead loading of the bridge deck,* and the *posttensioning applied across the two spans.* Subsequent measurements began about 1 month after the bridge was opened, and the last recorded measurements took place in April 1995, some 18 months later.

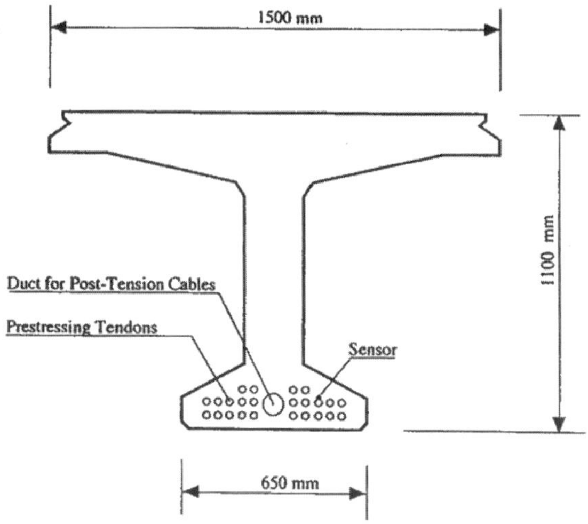

FIGURE 9.38. Diagram showing the placement of FBG sensors and prestressing tendons in one of the tee-bulb precast girders used in the Beddington Trail Bridge. Reprinted from Maaskant, R., Alavie, T., Measures, R.M., Tadross, G., Rizkalla, S.H. and Guha-Thakurta, A., (1997), "Fiber Optic Bragg Grating Sensors for Bridge Monitoring," *Journal of Cement and Concrete Composites*", **19**, p. 26, with permission from Elsevier Science.

FIGURE 9.39. One of the recessed metal outlet boxes used to house and protect the optical fiber leads during concrete pour for one of the Beddington Trail Bridge girders.

FIGURE 9.40. Testing of the embedded FBG sensor after fabrication of the hot girder for the Beddington Trail Bridge. Used with permission, ElectroPhotonics Solutions Corporation.

FIGURE 9.41. Two views of the junction box for the fiber optic sensor array of the Beddington Trail Bridge. Used with permission, ElectroPhotonics Solutions Corporation.

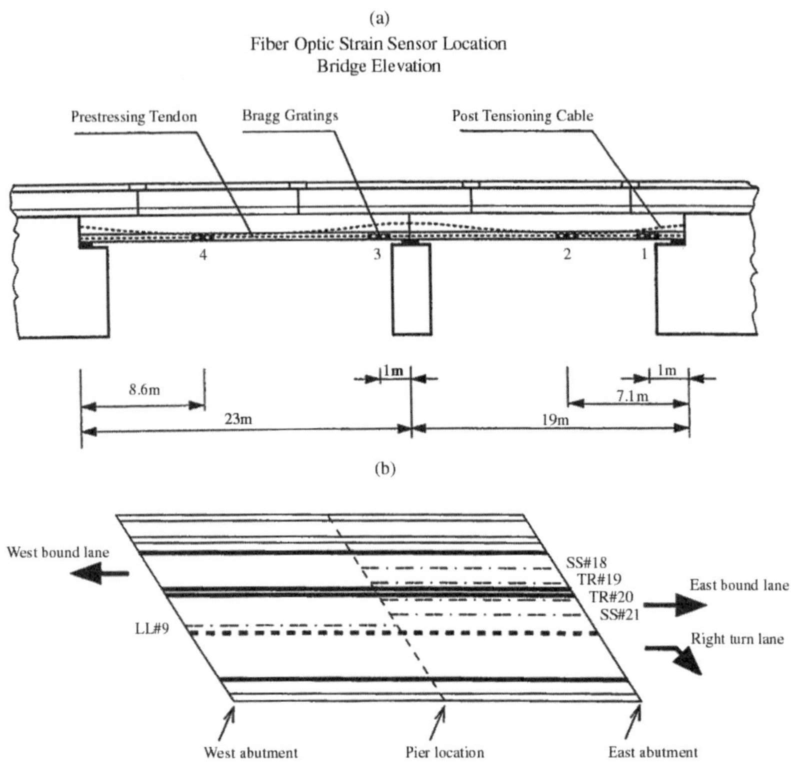

FIGURE 9.42. Schematic of FBG sensor placement in the Beddington Trail Bridge. Reprinted from Maaskant, R., Alavie, T., Measures, R.M., Tadross, G., Rizkalla, S.H. and Guha-Thakurta, A., (1997), "Fiber Optic Bragg Grating Sensors for Bridge Monitoring," *Journal of Cement and Concrete Composites*", **19**, p. 27, with permission from Elsevier Science.

The measurements presented in Fig. 9.43 have been corrected for thermal apparent strain, based on the prevailing temperature. Since the three prestressing materials possess widely different thermal expansivities, an attempt was made to normalize the data with respect to temperature so a meaningful comparison could be made regarding the permanent tendon relaxation. Since the steel reinforcement were thermally closely matched to the concrete, the girders were assumed to possess an overall expansivity equal to that of concrete, and the measurements normalized to 20°C. The resulting initial strain relaxation, and subsequent relief in the 18 months that the bridge was in-service are presented in Fig. 9.44. The total expected loss of prestress in the MICK steel tendons has been estimated as 380 MPa or 1900 $\mu\varepsilon$, (Maaskant *et al.*,

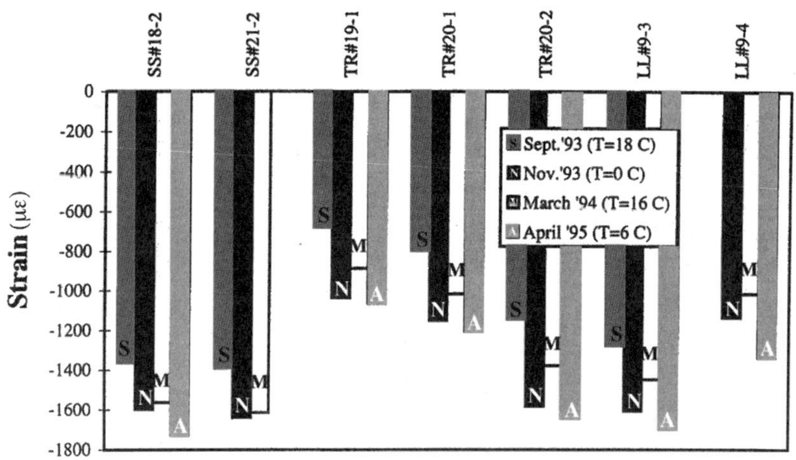

FIGURE 9.43. Thermally corrected strain relief for the steel (SS), Tokyo Rope (TR), and Leadline (LL) prestressing tendons used in the Beddington Trail Bridge and observed over an 18-month period by the embedded FBG sensors. Reprinted from Maaskant, R., Alavie, T., Measures, R.M., Tadross, G., Rizkalla, S.H. and Guha-Thakurta, A., (1997), "Fiber Optic Bragg Grating Sensors for Bridge Monitoring," *Journal of Cement and Concrete Composites*", **19**, p. 28, with permission from Elsevier Science.

1997). The lower Young's modulus for the CFRP tendons leads to 25% less stress loss than the steel tendons. Indeed, by and large, the *measurements show lower loss of strain for the CFRP tendons than for steel.*

One of the most useful observation to come from the work on the Calgary Bridge was the recognition that the embedded FBG sensors were

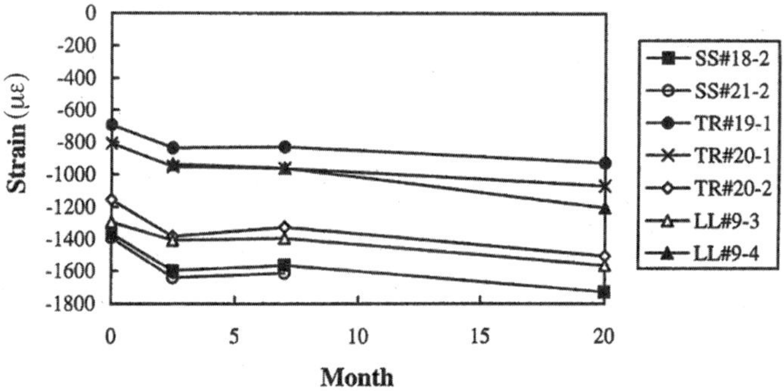

FIGURE 9.44. Temporal variation of the strain measurements presented in Fig. 9.43. Reprinted from Maaskant, R., Alavie, T., Measures, R.M., Tadross, G., Rizkalla, S.H. and Guha-Thakurta, A., (1997), "Fiber Optic Bragg Grating Sensors for Bridge Monitoring," *Journal of Cement and Concrete Composites*", **19**, p. 29, with permission from Elsevier Science.

not only capable of monitoring the long-term strain relief of the prestressing tendons; they were also able to *register the dynamic strain response of the bridge to live loads*, such as traffic. Figure 9.45 shows the response of an FBG sensor (LL#9-4) to dynamic loads imposed on the bridge by the passage of a 25-ton single-axle truck, in both the eastbound (a) and

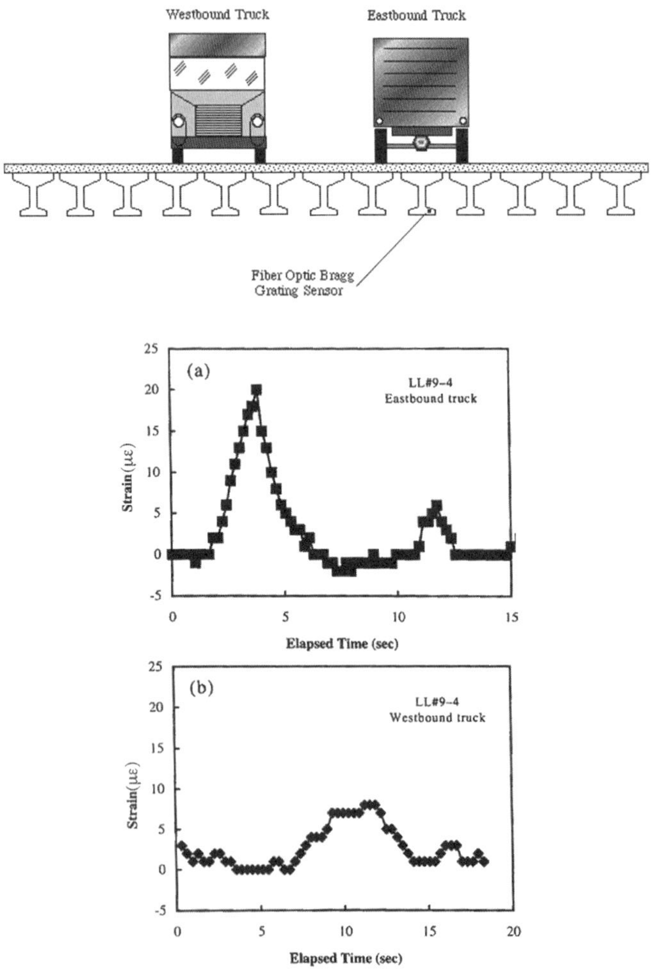

FIGURE 9.45. Dynamic response of one of the FBG sensors embedded within a girder of the Beddington Trail Bridge to the passage of a truck over the bridge. Reprinted from Maaskant, R., Alavie, T., Measures, R.M., Tadross, G., Rizkalla, S.H. and Guha-Thakurta, A., (1997), "Fiber Optic Bragg Grating Sensors for Bridge Monitoring," *Journal of Cement and Concrete Composites*, **19**, p. 29, with permission from Elsevier Science.

westbound (b) lanes at a slow speed (~20 kph). Note that sensor (LL#9-4) was located almost directly under the passage of the truck in the eastbound lane, and this accounts for the disparity in the signals observed in Figs. 9.45a and 9.45b. These measurements were undertaken with a commercial Fiber Optic Grating Strain Indicator from ElectroPhotonics Corporation, Ontario.

9.3.2 Taylor Bridge in Winnipeg

The Taylor Bridge at Headingley in Winnipeg (Figs. 2.5 and 9.46), is, to date, the largest span bridge in the world to use CFRP for both prestressing and reinforcement. Four of its girders and a portion of the concrete desk slab used this new material (Rizkalla *et al.*, 1997). The total length of the five-span bridge is 165 m, with each span consisting of eight 33-m I-shaped precast prestressed concrete girders, as shown in Fig. 9.47. Two different types of CFRP reinforcements were used in the bridge. Two girders were pretensioned with 15.2-mm diameter CFCC, produced by Tokyo Rope of Japan; at the other end of the bridge two further girders were pretensioned with 10-mm diameter (indented) Leadline bars

FIGURE 9.46. View of the innovative new Taylor Bridge in Winnipeg. Used with permission, ISIS Canada.

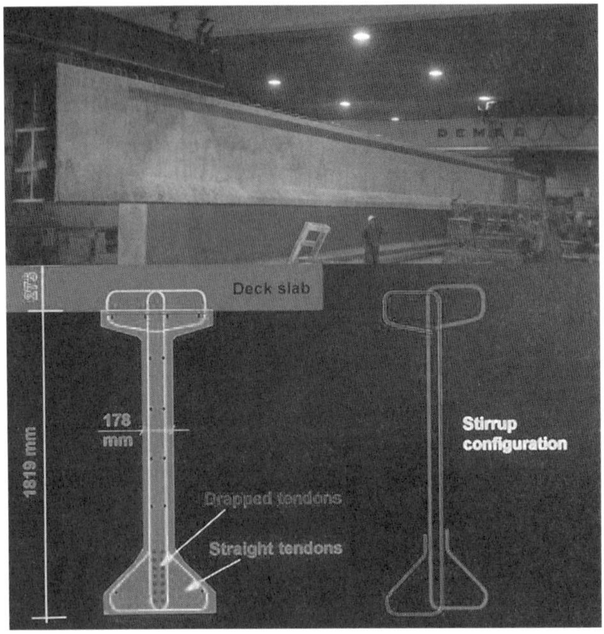

FIGURE 9.47. Photograph of one of the precast taylor Bridge I-beam girders, and (below) a schematic of the girder and its reinforcement design. Used with permission, ISIS Canada.

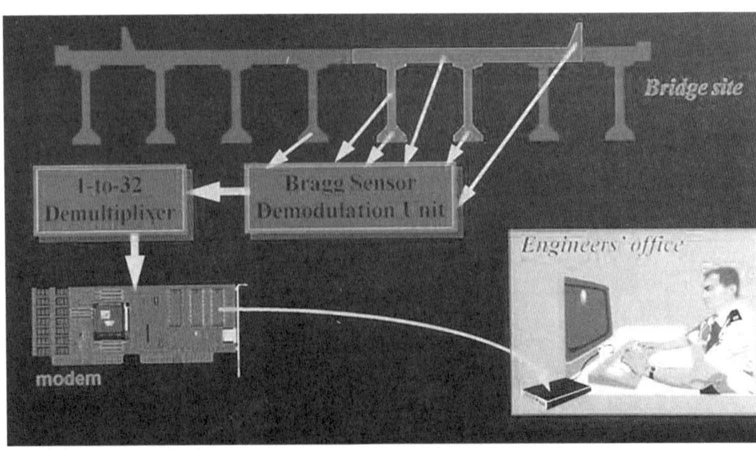

FIGURE 9.48. Schematic of the fiber optic sensing system incorporated into the Taylor Bridge. Used with permission, ISIS Canada.

produced by Mitsubishi Chemical Corporation of Japan. Both types of CFRP were also used to provide shear reinforcement in two of the CFRP prestressed girders. The bridge was opened on 28 October 1997 and is the first "smart bridge," which can be monitored remotely by engineers at a central monitoring station, using phone lines to communicate with the on-site demodulation system and computer that interrogates the imposing array of 66 fiber optic Bragg grating sensors integrated into the bridge (Fig. 9.48).

The bridge was instrumented with an array of fiber optic Bragg grating strain sensors and a number of conventional strain gauges. The 66 FBG sensors were distributed throughout the bridge to measure the strain in the: steel and CFRP prestressing tendons, steel and CFRP stirrups, concrete deck, and concrete barrier wall. A total of 17 thermocouples were also installed to provide temperature compensation for the strain sensing system. The installation of the sensors was undertaken by a team from ISIS—Canada and ElectroPhotonics Corporation (EPC). The scale of the girder can be gauged from Fig. 9.49, which shows some of the

FIGURE 9.49. Installation of FBG sensors into the CFRP shear reinforcement frame for one of the precast concrete girders for the Taylor Bridge. Used with permission, ISIS Canada.

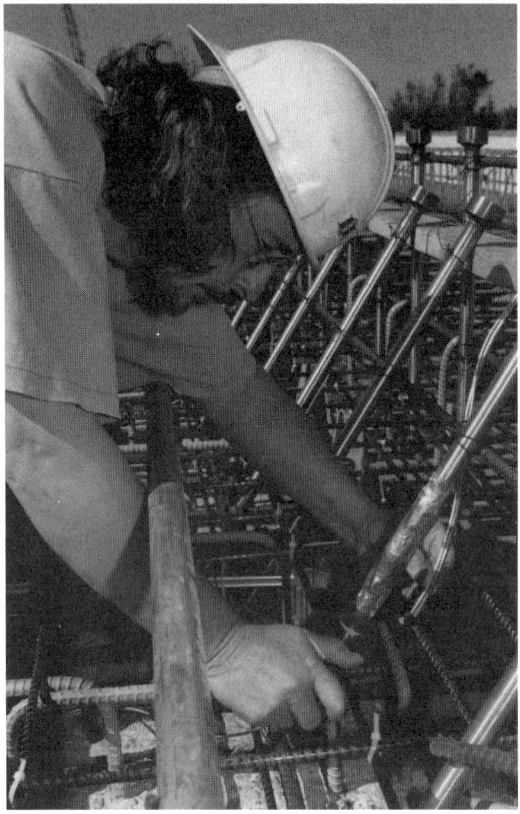

FIGURE 9.50. Installation of FBG sensors into the Taylor Bridge barrier-wall reinforcement frames. Used with permission, ISIS Canada.

fiber optic sensors being installed within one of the girders. Figure 9.50 shows a fiber optic sensor being installed within part of the bridge barrier wall that is reinforced with glass FRP.

The fiber optic sensing system was developed specially for this bridge by ElectroPhotonics Corporation, Ontario, and is designed to operate over the wide range of weather conditions expected at the bridge site (−40°C in the winter to +40°C in the summer). Demodulation of the FBG sensor array was undertaken by means of a ruggedized EPC-FLS3500R, equipped with a 32-channel multiplexing unit for mapping the strain from both ends of the bridge and tracking it over the life of the bridge. In addition, a fast-response EPC-FLS3100 single-channel demodulation unit

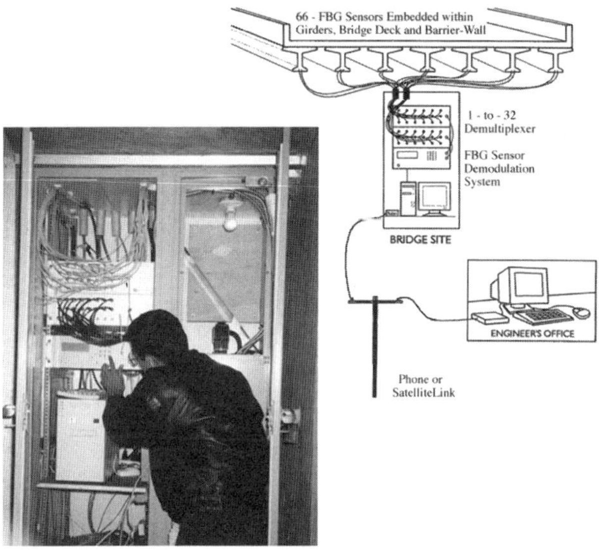

FIGURE 9.51. Schematic of the 66-FBG sensing system installed in the Taylor Bridge, its demodulation system, and the communication system linking it to the ISIS Central Office. Photograph shows the ElectroPhotonics Corporation demodulation and communication system installed at the bridge site. Used with permission, ElectroPhotonics Solutions Corporation.

was provided for monitoring the dynamic response of the bridge to traffic loads. An overview of the fiber optic sensing system is schematically illustrated in Fig. 9.51, with the inset showing the actual cables and demodulation system at the bridge site.

An example of the temporal strain variation recorded by one of the FBG sensors as the large truck shown in Fig. 2.5, passed four-times back and forth across the bridge is presented as Fig. 9.52. This sensor was located at the midpoint of one of the (CFRP Leadline) girders. It is evident from Fig. 9.52 that the magnitude of the bridge strain response to this truck was about 25 $\mu\varepsilon$ and that the resolution of the FBG sensing system is adequate to distinguish the impulse from each truck axle as it passes over the sensor. Indeed, even the direction the truck can be determined from the positioning of the two strain peaks; see arrows. The versatile range of FBG sensors that is available commercially with EPC's demodulation systems is shown in Fig. 9.53. Table 9.1 displays the respective specifications of these different type of sensors. The list includes FBG temperature sensors, weldable strain sensors, and a concrete-embeddable sensor.

TABLE 9.1. Specifications for Several Types of FBG Sensors

	Strain				Load	Temperature
Sensor type	Embeddable	Bondable	Weldable	Weldable high temperature	Bolt	Probe
Gauge length	150 mm	10 mm	25 mm	25 mm	>10 mm	—
Gauge material	Stainless steel	Fused silica glass	Stainless steel	Stainless steel	Bolt material	Metallic/nonmetallic
Cable length	1 m to 1 km	1 m to 1 km	1 m to 1 km	1 m to 1 km	1 m to 1 km	1 m to 1 km
Cable diameter	3 mm	1 to 3 mm	1 to 3 mm	1 to 3 mm	3 mm	1 mm to 3 mm
Connector type	FC, ST, SC	FC, ST, SC	FC, ST, SC	FC, ST, SC	FC, ST, SC	FC, ST, SC
Strain range	0.6%	1%	1%	1%	1%	—
Temperature range	−50°C to 80°C (−58°F to 176°F)	−50°C to 80°C (−58°F to 176°F)	−50°C to 80°C (−58°F to 176°F)	0°C to 325 °C (32°F to 617°F)	−50°C to 80°C (−58°F to 176°F)	−70°C to 350°C (−90°F to 660°F)

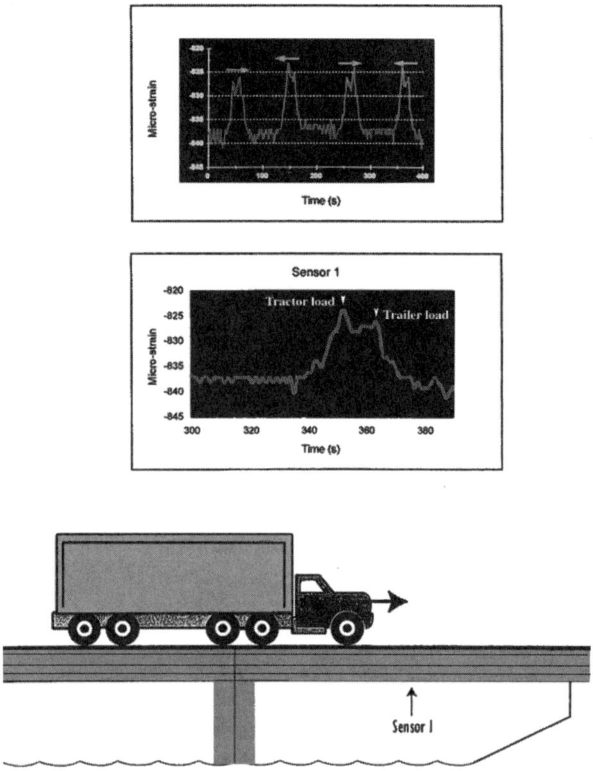

FIGURE 9.52. Temporal response of one FBG sensor embedded within the Taylor Bridge to several passes of a truck over the bridge. Used with permission, ElectroPhotonics Solutions Corporation.

9.3.3 The Confederation Bridge

The Confederation Bridge, or the PEI–fixed–link as it was originally called, is the world's longest bridge over ice-covered ocean and spans 12.9 km from Prince Edward Island to New Brunswick in Canada (Fig. 9.54). The fast and efficient construction of this bridge has made it an engineering masterpiece. The scale of production of the many elements that made up this bridge can be appreciated from the "bridge production and assembly yard" seen in Fig. 9.55. This is even more impressive when it is realized that some of the concrete box girders weighted 8 000 tons and were carried out to the bridge site on a massive catamaran that was specially modified for this task. There they were lowered onto huge concrete piers that were built to rise majestically out of the ocean, see Fig. 9.56.

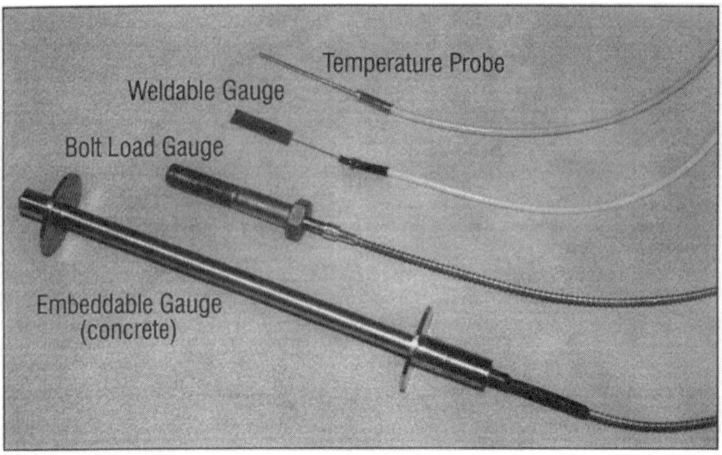

FIGURE 9.53. Array of different fiber optic Bragg grating sensors produced by Electro-Photonics Corporation, Ontario. Used with permission, ElectroPhotonics Solutions Corporation.

FIGURE 9.54. A view of the 12.9-km ocean-crossing Confederation Bridge, linking Prince Edward Island to New Brunswick in Canada.

FIGURE 9.55. An overview of the Confederation Bridge component construction and assembly yard.

The concrete box girder design of this bridge made the core hollow so that it serves as a utility corridor for electrical and telephone services. At some locations the concrete box girders are 14 m high and a double-decker bus could be driven through them. This is the part of the bridge that was instrumented with an array of FBG structural sensors, by a team from ISIS-Canada and ElectroPhotonics Corporation.

These sensors with their reels of optical fiber cable were attached to sections of steel reinforcement (Fig. 9.56), that were then tied to other steel reinforcements during installation of the sensors within the bridge structure (Fig. 9.57). A ruggedized EPC FLS3100 demodulation unit will allow multi-channel monitoring of both quasi-static and dynamic strains.

9.3.4 Structural health monitoring of bridges

Currently, engineers and inspectors faced with bridge evaluation have a difficult task, as their assessment often relies heavily on visual inspection. The decision to repair or replace the bridge is strongly influenced by their assessment. It is hoped that the development of integrated fiber optic structural sensing systems will prove to be a viable tool for bridge

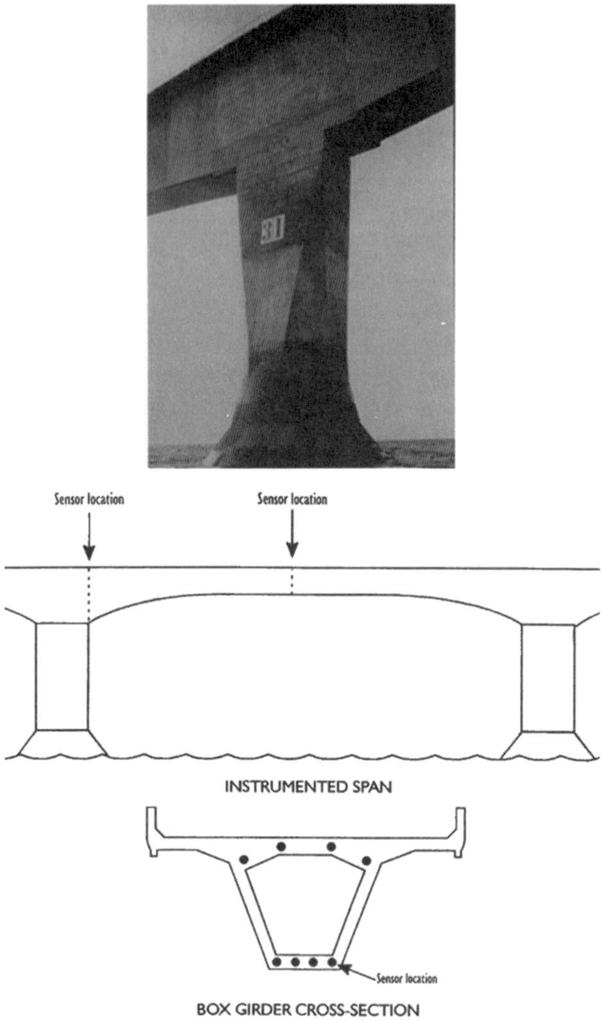

FIGURE 9.56. A view of Pier #31 of the Confederation Bridge showing the 14-m-high box girder. The lower part reveals a shcematic of the box girder cross-section and sensor location. Used with permission, ISIS Canada.

monitoring and maintenance. To accomplish this goal of structural health monitoring, the sensing system will have to do the following

1. Assess load history
2. Check performance against design assumptions

3. Flag abnormal conditions or behaviors
4. Evaluate the effectiveness of repairs and maintenance

Delamination in concrete results from corrosion of the embedded reinforcing steel. The corrosion products increase the volume of the reinforcing element and this results in cracking of the bonded concrete. Pressure may build to 4000 psi (27.6 MPa) as a result of the corrosion products. This delamination can lead to spalling of the concrete which in turn results in its removal by freeze and thaw cycles.

Kodindouma and Idriss (1996) used FBG sensors embedded within reinforced concrete beams to explore the feasibility of detecting the onset of failure at the steel–concrete interface. The FBG sensors were interrogated by means of a EPC FLS3000 from ElectroPhotonics Corporation. Conventional resistive foil strain gauges (RSG) were placed close to the fiber optic sensors. Several 3.05 m (10 ft) long test concrete beams with tension steel bars were specially manufactured. One test batch comprised

FIGURE 9.57. Three of the fiber optic instrumental steel rebars to be installed into the Confederation Bridge. Also shown are the reels of fiber optic cable used to connect the sensors to the demodulation system.

a control beam and three damaged beams with center delaminations of 0.61, 1.22, and 1.83 m, respectively. Two FBG sensors were used, one was embedded at the midspan of the control beam and the second was embedded at the delamination edge of the 1.22-m center delamination beam. Conventional RSG were placed close to the fiber optic sensors and others were used to map the strain field in the test beams. Figure 9.58 presents the strain in the tension steel for the set of beams. It is clear that the strain over the delamination in the control beam is much lower than in the damaged beams. The sharp increase in strain at the edge of the delamination is due to redirection of the tensile stress in the steel at the delamination edge of the 1.22-m center delamination beam. The strain is constant over the length of the delaminations as no bond to the concrete exists in that region.

As the beams were loaded, the strain was monitored and large discontinuities observed each time shear cracks opened and spread (see Fig. 9.59). In the control beam the FBG sensor shows two major discontinuities at 9 400 lb (41.8 kN) and 11 000 lb (49.4 kN) following the formation of two large and sudden shear cracks (Fig. 9.59a). It is also worth noting that the FBG sensor continued to provide strain measurements up to the point of failure of the beam at 15 800 lb. In the beam with the 1.22-m center delamination the FBG sensor displays a large discontinuity at about 9000 lb corresponding to the formation of a large shear crack that opens at the edge of the delamination (Fig. 9.59b). A similar pattern, though less pronounced, is observed for the RSG. The later onset of discontinuity

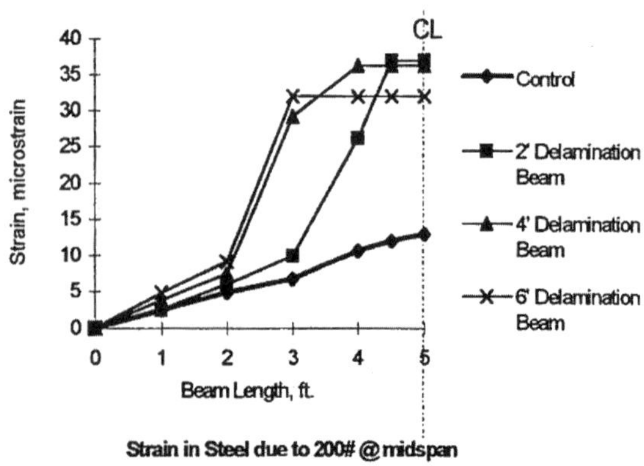

FIGURE 9.58. Variation of the strain distribution as measured with FBG snesors and foil strain gauges in a number of tests beams with different degrees of delamination. From Kodindouma, M.B. and Idriss, R.L. "An Integrated Sensing System for Highway Bridge Monitoring", *SPIE Smart Structures and Materials* 1996, Smart Systems for Bridges, Structures and Highways, 28–29 February 1996, San Diego.

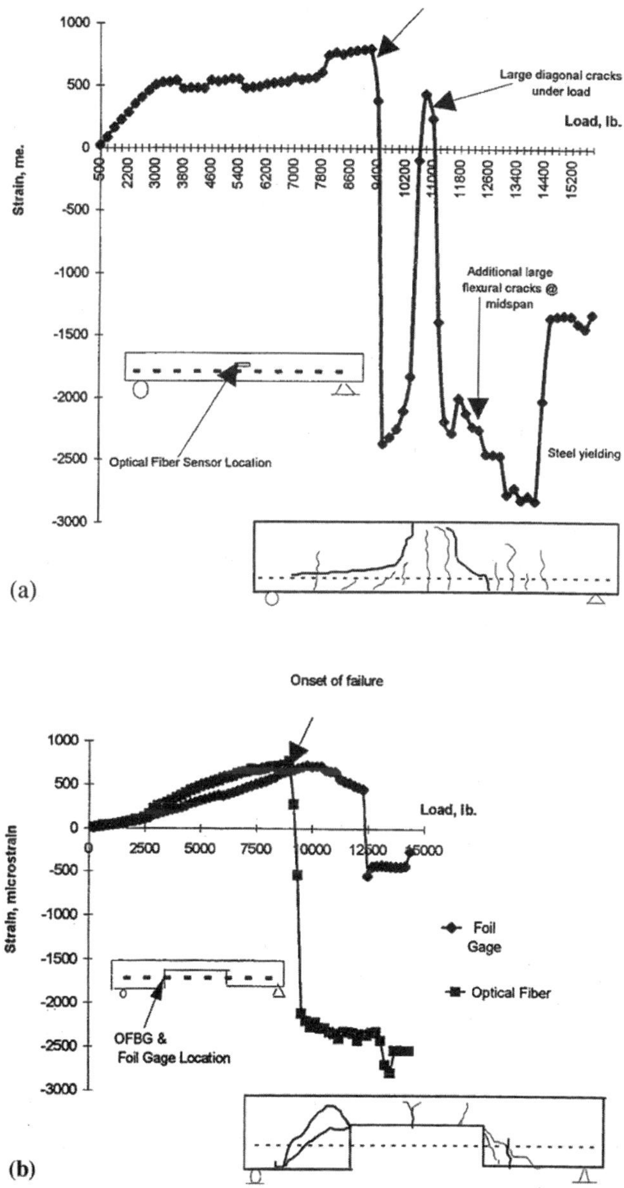

FIGURE 9.59. Strain variation with load for two test beams measured by FBG sensors and foil strain gauges. (a) Strain variation with load for the control beam. (b) Strain variation with load for damaged test beam. Insets show FBG sensor location and crack distribution in both specimens. From Kodindouma, M.B. and Idriss, R.L., (1996), "An Integrated Sensing System for Highway Bridge Monitoring" *SPIE* **2719**, 132–140.

recorded by RSG is probably due to the difference in location of the FBG and RSG.

The success of this early work led to a field-site test involving monitoring the dynamic strain response of the in-service I-10 interstate bridge in Las Cruces, New Mexico to traffic loading. Several FBG sensors were attached to the center support girder of the structure for this purpose (Davis *et al.*, 1997b). The spectral ratiometric demodulation technique (Measures *et al.*, 1992a), described in Section 9.2.1, was used with a broadband ELED source and a tunable Fabry–Perot bandpass filter to select which FBG is interrogated. A wavelength division multiplexer was used to split the Bragg grating reflected signal between the two detectors, and its wavelength dependence allowed it to also serve as a spectral filter, see Fig. 9.6. The WDM used in this work had a linear transmission response over 0.75 nm, which corresponded to about 750 με. A schematic of the experimental system is presented as Fig. 9.60a.

One of the two FBG sensors was attached to the lower flange of the girder, while the other was placed on the web of same girder. Both FBGs were on a single strand of optical fiber. A video camera recorded the traffic and was used to correlate the type of load to the strain detected by the FBG sensors. Figure 9.60b, reveals strain data recorded from the flange mounted FBG sensor over a 2-minute interval, during which five cars and two trucks passed over the bridge. As seen in the figures the cars generate about 6 με, while the truck response was typically an order of magnitude larger. An initial compression of about −20 με is observed just prior to the arrival of the trucks over the sensor location. This is thought to be due to the action of the truck on the bridge span before the instrumented span. As the truck passes over the FBG sensor, a tensile strain of between 60 and 80 με is registered.

A histogram of the vehicle-induced peak strains obtained from the flange-mounted FBG sensor over a 10-minute period is presented as Fig. 9.61. It can be seen that the majority of traffic crossing the bridge during this time interval was cars, which gave rise to strain in the 0 to 10 με range, followed by a much smaller number of vans in the 10 to 25 με range. A second grouping of peak strains is in the 40 to 65 με range, which is attributed to large trucks with light loads. Finally, the large fully loaded trucks are seen to fall in the 70 to 85 με range (Davis *et al.*, 1997b). The FBG sensor measurements undertaken in this project demonstrate that statistical data on traffic usage and loading of a bridge can be gathered by these sensors. This suggests that with proper data signal analysis, information on vehicle type and weight could be collected by this technology. In Chapter 11, new work undertaken by this group using networks of multiplexed FBG sensors will be described.

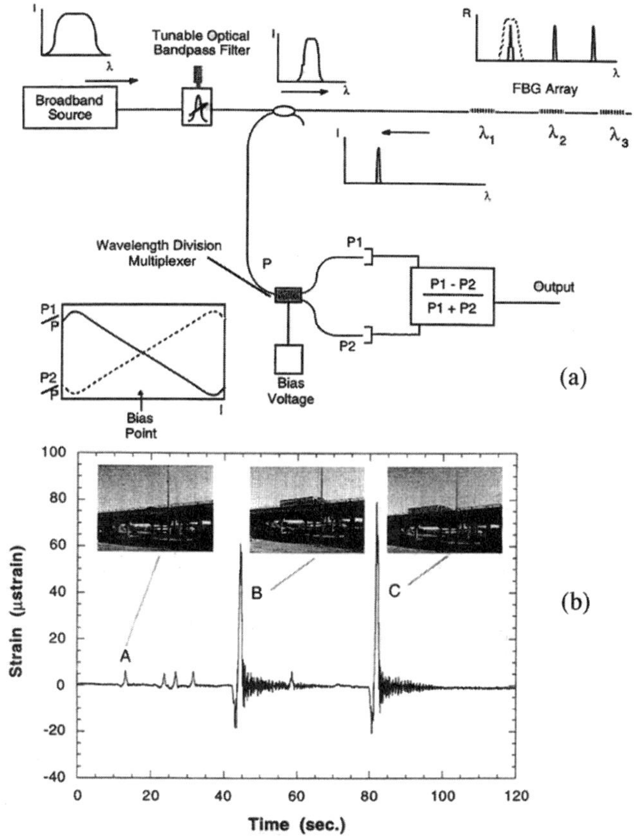

FIGURE 9.60. (a) Schematic illustration of the FBG sensor demodulation system used for monitoring the sensor array installed on the I-10 Bridge in Las Cruces. (b) Dynamic response of the sensors in response to the traffic shown in the video clip insets. From Davis, M.A., Kersey, A.D., Berkoff, T.A., Jones, R.T., Idriss R.L., Kodinduma, M. (1997b), "Dynamic Strain Monitoring of an In-Use Interstate Bridge Using Fiber Bragg Grating Sensors" *SPIE* **3043**, 87–107.

Meissner *et al.* (1997) installed FBG sensors into a prestressed concrete bridge that crosses the A4 motorway near Dresden, FRG. To protect the FBG from damage during construction of the bridge, a special sensor frame was designed of steel containing a groove within which two FBG sensors lie (see Fig. 9.62). This frame was positioned close to the tendon ducts (Fig. 9.63) and the changes in the Bragg wavelength for each sensor were monitored with an optical spectrum analyzer (HP70951) with a resolution bandwidth of 0.08 nm. All the sensors survived, and their

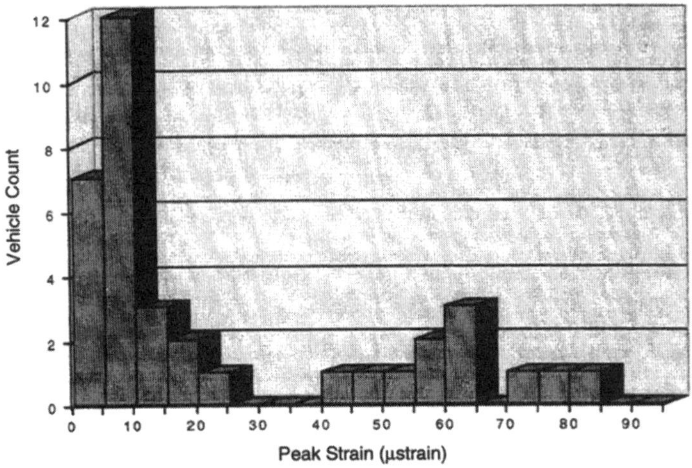

FIGURE 9.61. Histogram of vehicle-induced peak strains as measured during a 10-minute period by an FBG sensor mounted on the I-10 Bridge in Las Cruces. From Davis M.A., Kersey A.D., Berkoff T.A., Jones R.T., Idriss R.L., Kodinduma M. (1997b) "Dynamic Strain Monitoring of an In-Use Interstate Bridge Using Fiber Bragg Grating Sensors" *SPIE* **3043**, 87–107.

strains were very similar. Since they were placed after prestresing, their readings indicated negative strain (i.e., compression). A, FBG sensor demodulation system was then built using a broadband erbium doped fiber source (1520 to 1560 nm) and a long-period grating to serve as a broadband edge-filter, as described in Section 9.2.1. A schematic of the passive spectral ratiometric demodulation system is presented as Fig. 9.64. The change in the strain of one of the FBG sensors during the completion of the bridge is presented in Fig. 9.65. Note that the higher strain recorded 56 days after prestressing is probably due to weather, creep and shrinkage

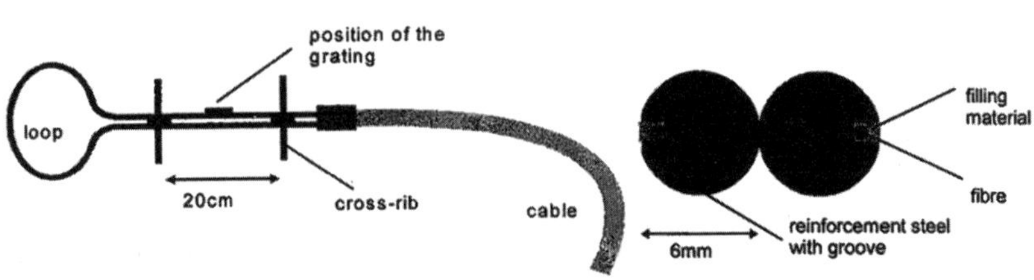

FIGURE 9.62. Special FBG sensor frame used for monitoring the traffice response of a bridge crossing the A4 motorway near Dresden. From Meissner, J., Nowak, W., Slowik, V., Klin, T., (1997), "Strain Monitoring at a Prestressed Concrete Bridge" OSA 12th Int Conf on Optical Fiber Sensors, Williamsburg, 408–411.

FIGURE 9.63. Photograph of the FBG sensor frame shown in Fig. 9.64, at its site location on the bridge. From Meissner, J., Nowak, W., Slowik, V., Klin, T., (1997), "Strain Monitoring at a Prestressed Concrete Bridge," OSA 12th Int Conf on Optical Fiber Sensors, Williamsburg, 408–411.

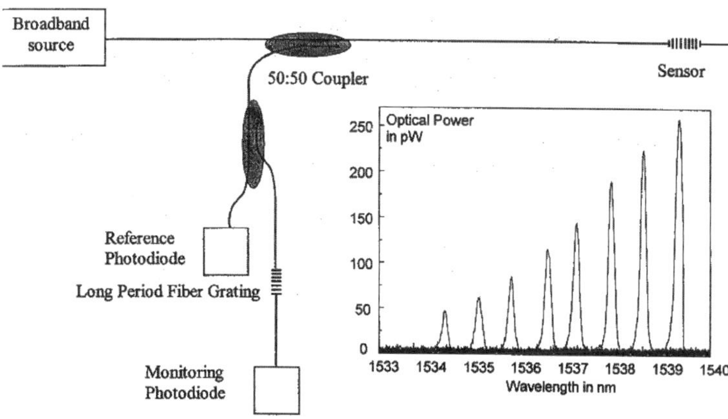

FIGURE 9.64. Demodulation system for the FBG sensor employing a long-period grating as the wavelength-dependent filter for the passive spectral ratiometric system. Inset reveals variation of FBG reflection spectrum with sensor load. From Meissner, J., Nowak, W., Slowik, V., Klin, T., (1997) "Strain Monitoring at a Prestressed Concrete Bridge," OSA 12th Int Conf on Optical Fiber Sensors, Williamsburg, 408–411.

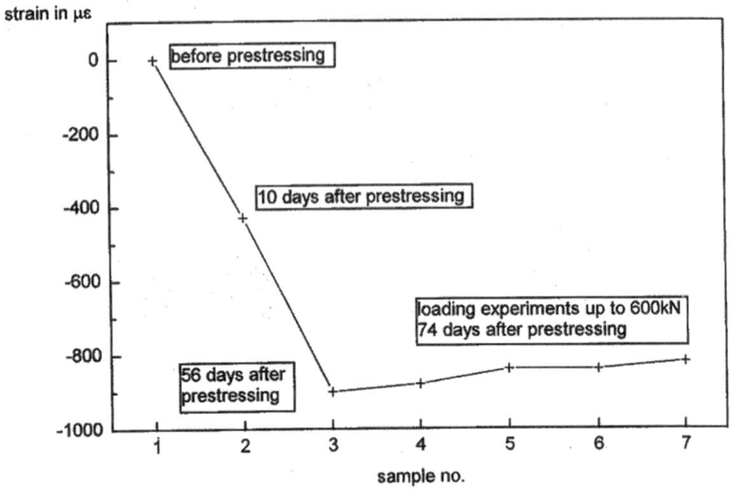

FIGURE 9.65. Variation in strain monitored by a FBG sensor during completion of the bridge near Dresden. From Meissner, J., Nowak, W., Slowik, V., Klin, T., (1997), "Strain Monitoring at a Prestressed Concrete Bridge," OSA 12th Int Conf on Optical Fiber Sensors, Williamsburg, 408–411.

of the concrete as it cooled. The bridge was loaded up to 600 kN, 74 days after prestressing.

9.3.5 Structural monitoring during a destructive bridge test

Major revisions in southern Norway required the removal of the Smedstua bridge. The Norwegian Public Roads Authority with two other institutions decided to instrument the bridge with a number of different types of fiber optic sensor and conventional monitoring systems, then load the bridge to failure (Storoy and Saether, 1997). The two-lane single-slab bridge had a span of 16.3 m and represented a common construction in Norway. Forty conventional strain gauges were attached to the upper and lower reinforcing bars of the bridge slab at the middle and at one of the pillars. Three FBG sensors were attached to reinforcing bars close to three of the conventional strain gauges. A 2.5-m-long fiber optic polarimetric sensor was compared with a manual extensiometer. Crack detection was performed with optical fibers attached to the concrete surface close to one of the pillars using an OTDR.

TABLE 9.2. Major Load States in Destructive Bridge Test

Load state	Load description
1	Zero load
2	Maximum traffic load (105 tons)
3	Test load limit (220 tons)
4	200 tons
5	400 tons
6	570 tons
7	700 tons
8	830 tons
9	One truck removed
10	Both trucks removed, bridge failure

The unloaded (dead load) was taken as the reference state. Various load states were imposed with trucks until the maximum allowed test load of 220 tons. To load the bridge to failure, a box located on the bridge was continuously filled with gravel. The various load states are described in Table 9.2. Since the temperature of the reinforcing bars did not vary by more than a few degrees during the test, no temperature compensation was undertaken, which led to an uncertainty of roughly 20 $\mu\varepsilon$.

One FBG sensor was interrogated by scanning the output wavelength of a semiconductor DFB laser (Fig. 9.66); the other two were made part of a fiber laser so that they controlled the laser wavelength (Fig. 9.67). Both techniques were described in Section 9.2. It should be noted that Storoy *et al.* (1997) claim that prior to mounting of the FBGs almost complete restoration of the original mechanical strength of the optical fiber was achieved by etching away a 1- to 2-μm layer with 10% hydrofluoric acid (results similar to those cited by Glossop *et al.*, 1990).

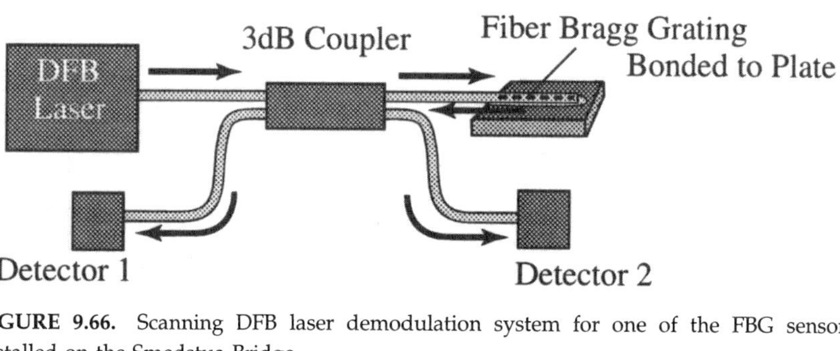

FIGURE 9.66. Scanning DFB laser demodulation system for one of the FBG sensors installed on the Smedstua Bridge.

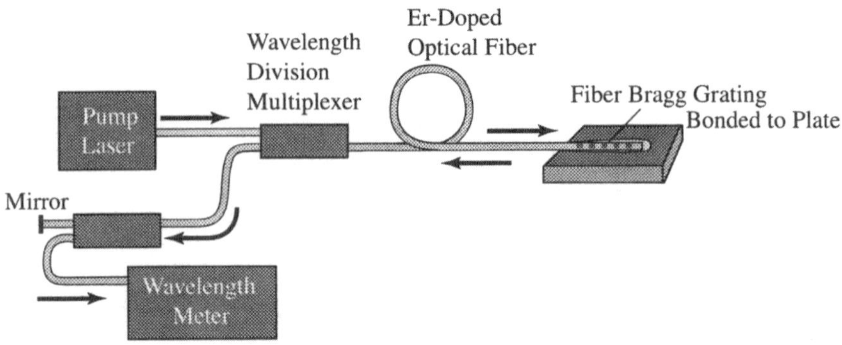

FIGURE 9.67. Fiber laser sensor demodulation scheme used for two of the three FBG sensors installed on the Smedstua Bridge.

Current turning of the DFB laser gives a dynamic range of 500 με, for the FBG sensor while temperature turning extends the range to ±2000 με. The resolution achieved was 1 με with a short-term stability of ± 5με. The measured strain at the different load states recorded with the FBG sensor demodulated in this manner are presented in Fig. 9.68. Comparison with the corresponding RSG shows a mean discrepancy of only 12 με; moreover, the RSG failed to provide readings beyond state 8 of Table 9.2, while the FBG continued to provide data to state 10.

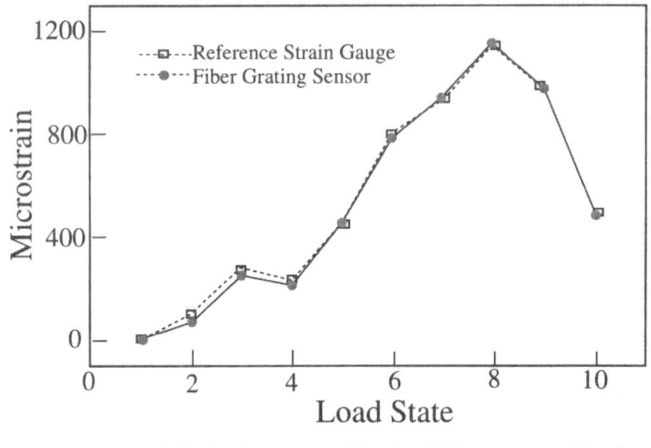

FIGURE 9.68. Variation of the strain measured by the FBG sensor read by the tunable DFB laser, and a reference foil strain gauge, during various load states of the Smedstua Bridge (see Table 9.2).

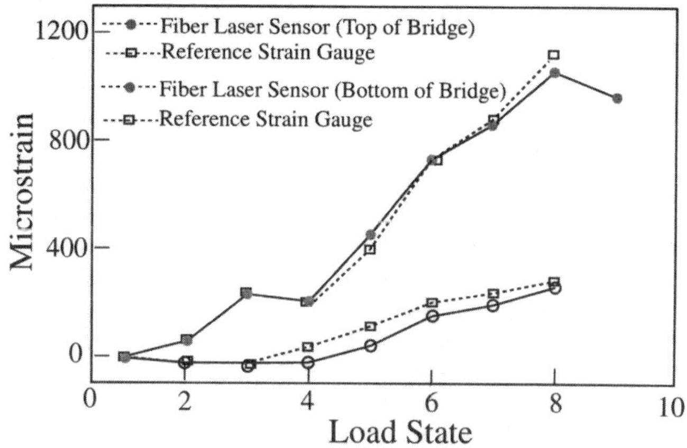

FIGURE 9.69. Variation of the strain measured by two FBG sensors and foil strain gauges during various load states of the Smedstua Bridge (see Table 9.2).

The fiber Bragg grating laser sensor system was used with the two remaining FBGs by mechanically switching between the one on the top of the bridge and the other under the bridge. This system also demonstrated a strain resolution of 1 $\mu\varepsilon$ but provided a short-term stability of ± 2 $\mu\varepsilon$. The FBG sensor measured strain at different loads is presented with the results from the corresponding RSGs in Fig. 9.69. A somewhat larger discrepancy is observed in these data between the FBG sensors and their conventional counterparts.

9.3.6 Fiber optic monitoring of a CFRP stay-cable bridge

A combination of FBG sensors and conventional strain gauges were used to instrument Storck's Bridge in Winterthur, Switzerland (Nellen *et al.*, 1997). This was the first stay-cable bridge in the world to use CFRP cables in place of steel in two of its many 35-m cables. This bridge has a length of 120 m crosses 18 railway tracks and was shown earlier as Fig. 2.6. Each CFRP cable was monitored by a sensor array comprising seven FBGs. Three of the seven FBGs (BG1, BG4 and BG7) were adhered to loaded 5-mm diameter wires at the circumference of the cable and spaced by 120° relative to the center of the cable (Fig. 9.70). This allows the average load and the load gradient to be measured.

The remaining four FBGs were used as dummies for temperature compensation and for monitoring long-term drifts. This set of FBGs (BG2,

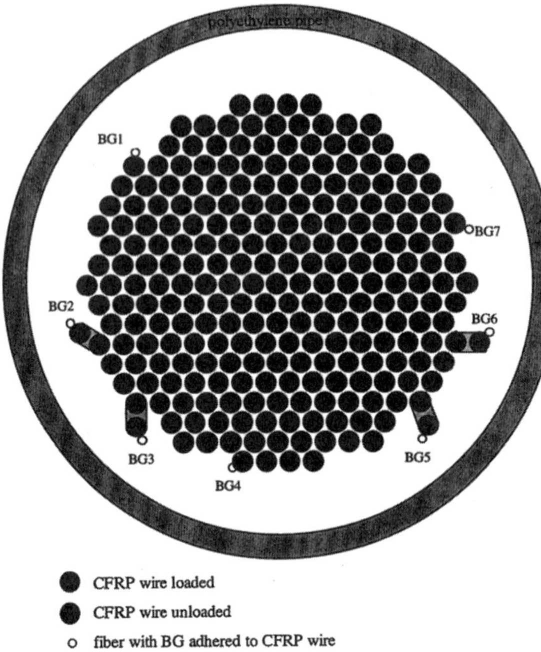

FIGURE 9.70. Cross-section of one of the two 35-m CFRP stay cables used on the Storck's Bridge, Switzerland. Also shown is the placement of seven FBG sensors. From Brönnimann, R., Nellen, P., Sennhauser, U., (1998), "Application and Reliability of a Fiber Optic Surveillance System for a Stay Cable Bridge," *Smart Materials and Structures Journal*, **7**, 229–236, with permission from the Institute of Physics Publishing Ltd.

BG3, BG5, and BG7) were adhered to 30-cm lengths of CFRP wire that were attached to the loaded wires in such a way that they were insensitive to the load on the cable (see Fig. 9.70). BG5 and BG6 were adhered in a prestrained state. All of the FBGs were adhered on the CFRP wires over a 200-mm length to minimize the influence of creep on the measurement. Both ends of the optical fiber containing the FBGs were accessible for redundancy against a break in the optical fiber. Decay of strain served as an indicator of creep or debonding of the fiber coating or the epoxy adhesive due to environmental conditions.

Each of the FBGs used in this work was written by a single laser pulse during fabrication of the optical fiber on a draw tower (Askins *et al.*, 1992). The experimental arrangement for interrogating the FBGs is schematically illustrated in Fig. 9.71, and involves a superluminescent diode source and a 0.46-m imaging spectrometer from Jobin Yvon equipped with a N-MOS 1024-channel detector head from Hamamatsu. A representative reflection spectrum from one of the FBG sensor arrays overlapped with the

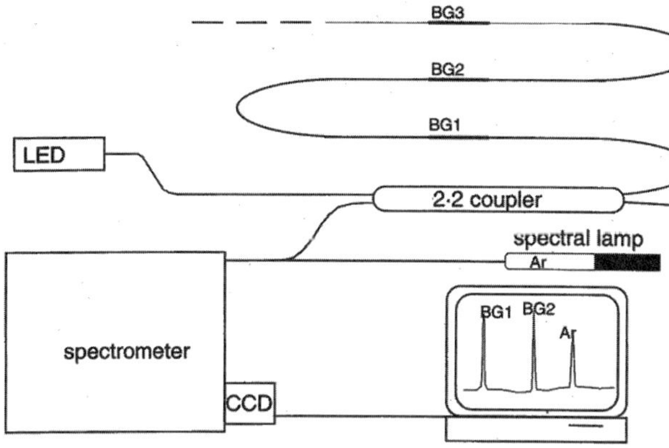

FIGURE 9.71. Experimental arrangement used for FBG sensor demodulation on the Storck's Bridge. From Brönnimann, R., Nellen, P., Sennhauser, U., (1998), "Application and Reliability of a Fiber Optic Surveillance System for a Stay Cable Bridge", *Smart Materials and Structures Journal*, **7**, 229–236, with permission from the Institute of Physics Publishing Ltd.

reference spectrum from an argon lamp (used for absolute wavelength calibration) is presented as Fig. 9.72. The optical fiber itself serves as the entrance slit for the spectrometer. The signal from the CCD is digitized and recorded by a computer.

The results of measurements made at different stages of construction of the bridge are presented in Fig. 9.73. The following key events occurred during the 6-month observation period:

1. The cables were equipped with FBG sensors
2. The cables were fixed on the bridge with no load
3. The first (reference) load was applied to the cables
4. The bridge was concreted and gained most of its deadweight
5. The load distribution on the cables was adjusted
6. The cable load was measured with a hydraulic press
7. The pavement was added 2 days prior to opening

The broken line in this figure corresponds to the mean value of measurements made with resistive foil strain gauges and shows good agreement with the FBG readings.

Since it was only practical to mount the FBGS before loading of the cables, the sensors are subject to a permanent tensile strain of from 1000 to 1500 $\mu\varepsilon$ *during their functional life*. The operating temperature range is expected to be −40 to 70°C with a relative humidity that ranges from 50 to 70%. In a series of experiments, polyimide-coated optical fibers were found to perform better under adverse environmental conditions than acrylate-coated fibers. Since the thermal expansion coefficient of CFRP in

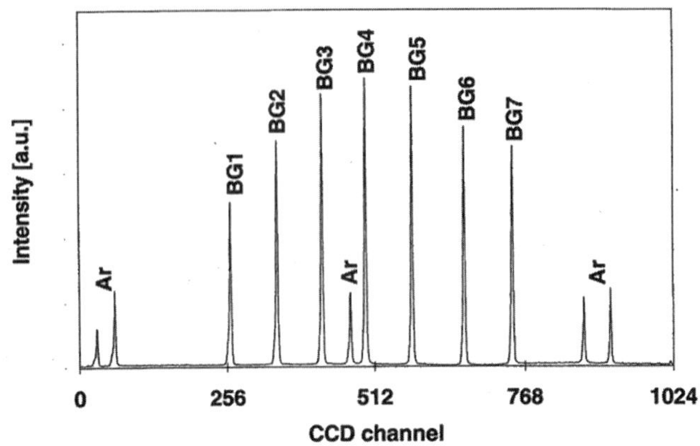

FIGURE 9.72. Representative reflection spectrum of the serial array of seven FBG sensors from one of the CFRP staty cables for the Storck's Bridge. Also evident are the argon reference spectral lines. From Brönnimann, R., Nellen, P., Sennhauser, U., (1998), "Application and Reliability of a Fiber Optic Surveillance System for a Stay Cable Bridge," *Smart Materials and Structures Journal*, 7, 229–236, with permission from the Institute of Physics Publishing Ltd.

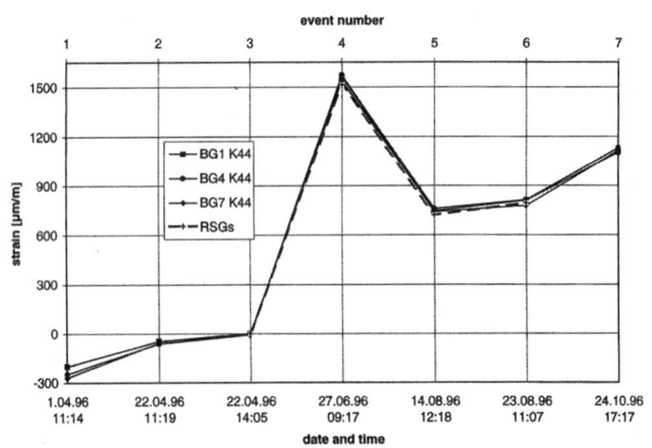

FIGURE 9.73. Variation in the strain as measured by several of the FBG sensors attached to one of the CFRP stay-cables of the Storck's Bridge during different phases of construction and loading of the bridge. From Brönnimann, R., Nellen, P., Sennhauser, U., (1998), "Application and Reliability of a Fiber Optic Surveillance System for a Stay Cable Bridge," *Smart Materials and Structures Journal*, 7, 229–236, with permission from the Institute of Physics Publishing Ltd.

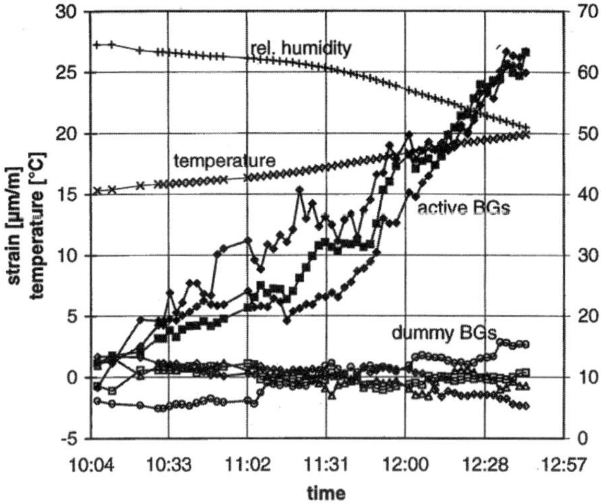

FIGURE 9.74. Variation in the strain experienced by one of the CFRP stay cables of the Storck's Bridge as measured by a number of sensors during a period of the day where there is a gradual rise in temperature. From Brönnimann, R., Nellen, P., Sennhauser, U., (1998), "Application and Reliability of a Fiber Optic Surveillance System for a Stay Cable Bridge," *Smart Materials and Structures Journal,* 7, 229–236, with permission from the Institute of Physics Publishing Ltd.

its longitudinal direction is almost negligible, the loads on the CFRP cables are expected to increase with ambient temperature as the steel cables expand. This is indeed observed (see Fig. 9.74).

The sensitivity of the fiber optic sensing system to cable loading is indicated in Fig. 9.75, where a hydraulic press was used to increase the load on CFRP cable K44 for a short period of time. Later loading of the two closest steel cables (K34 and K54) leads to a redistribution of load between the CFRP and steel cables resulting in a decrease of strain on K44. Note that a change of 100 $\mu\varepsilon$, in strain corresponds to a change of about 8 tons (78 kN) in the CFRP cable load. This work has demonstrated that the FBG sensors on the Storck's Bridge worked fine over the 10-month period after their installation, and their results were in good agreement with those of the resistive foil strain gauges (Nellen *et al.*, 1997).

9.3.7 Marine catamaran model measurements

The introduction of new materials often results from the need for cost-effective construction that can stand harsh environments. The use of stiff,

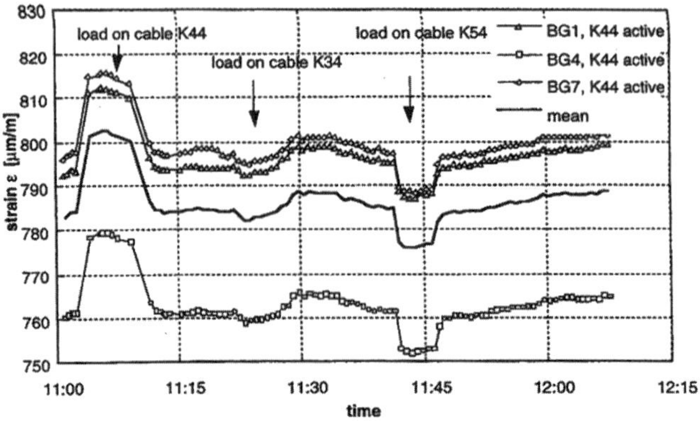

FIGURE 9.75. Variation in the strain measured by a number of the FBG sensors attached to one of the CFRP stay cables of the Storck's Bridge during hydraulic loading and unloading. From Nellen, P.M., Anderegg, P., Brönnimann, R., Sennshauser, U., (1997), "Application of Fiber Optical and Resistance Strain Gauges for Long-Term Surveillance of Civil Engineering Structures" *SPIE*, **3043**, San Diego, CA, March 5–6, 77–86. In: *Smart Structures and Materials* 1997, Conference on Smart Systems for Bridges, Structures and Highways. *SPIE*.

lightweight FRP composite materials is particularly attractive in the development of new, fast marine vehicles, such as catamarans. The complete characterization of the elastic behavior of such a structure requires many sensors to monitor bending moments, shear forces, and slamming forces at various positions on the model. When conventional strain-gauge technology is used for this purpose, huge amounts of wiring is required and installation is very labor-intensive. This is costly. Multiplexed fiber optic sensing arrays represent a much more elegant solution, with far less complexity and labor cost.

Hjelme *et al.* (1996, 1997) report on the application of FBG sensing to tests of a 1 : 20 scale model of a planned 82-m catamaran. A schematic illustration of this model catamaran is presented as Fig. 9.76. The twin 4.1-m hulls of this model were made of Divinycell foam and glass fiber polyester resin with a thickness of 30 mm. The FBG sensors were used for two kinds of measurement: (i) the *bending moment* at the middle of the catamaran model, and (ii) the *slamming loads* imposed on the wet deck.

Two FBG demodulation techniques were employed for these measurements. The first technique combines time division multiplexing (TDM) with a pulsed high-frequency modulation of a semiconductor DFB laser that is wavelength tuned at low frequency. This was used for the bending measurements. The second demodulation technique involves locking the wavelength of the DFB laser to the peak wavelength of the

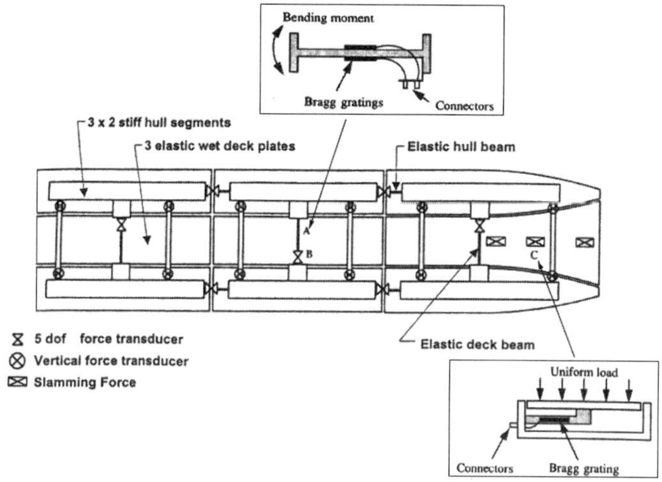

FIGURE 9.76. Schematic of a 4-m model of an 82-m twin-hull catamaran revealing the placement of the FBG sensors. Hjelme, D.R., Bjerkan, L., Neegard, S., Rambech J.S. and Aarsnes J.V. (1997), "Application of Bragg Grating Sensor in the Characterization of Scaled Marine Vehicle Models" *Applied Optics* **36**, 328–336.

Bragg grating. This approach offers higher resolution, at faster speed, but has a more limited strain range and is better suited for the slamming measurements.

The TDM approach was used to separate the signals from two FBGs (with the same wavelength, 1527 nm) that were mounted on the top and bottom of the stainless steel beam, which was positioned at the centre of the model, marked as "A" in Fig. 9.76 (Hjelme *et al.*, 1996). Demodulation of each FBG was achieved by scanning the wavelength of the DFB laser in combination with pulsed high-frequency modulation at 560 MHz (Fig. 9.77). A linear laser wavelength sweep of approximately 0.7 nm (corresponding to a sensor strain range of 630 $\mu\varepsilon$) was achieved at a rate of 700 Hz. The received signal is electronically processed to generate essentially the time derivative of the FBG response so that the Bragg wavelength can be determined from the zero-point crossing of this derivative signal (Fig. 9.78); see Section 9.2.

The virtue of this scheme of demodulation and demultiplexing is that standard telecommunication electronic and optoelectronic components are used. Optical delay lines (of length: 0, 25, 50 and 75 m) were used with a pulse width and repetition rate designed to ensure that the photodetector responded to the optical signal from only one FBG at a time (see Fig. 9.77). An example of measured bending moments recorded

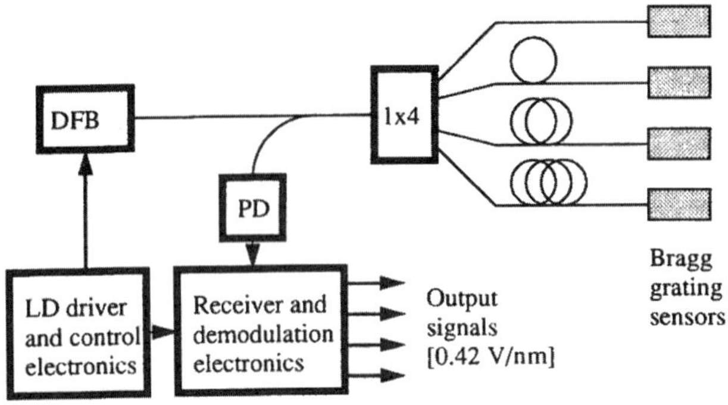

FIGURE 9.77. Time-division multiplexed demodulation system used with the FBG sensors installed on the model of the catamaran. Hjelme, D.R., Bjerkan, L., Neegard, S., Rambech J.S. and Aarsnes J.V. (1997), "Application of Bragg Grating Sensor in the Characterization of Scaled Marine Vehicle Models" *Applied Optics* **36**, 328–336.

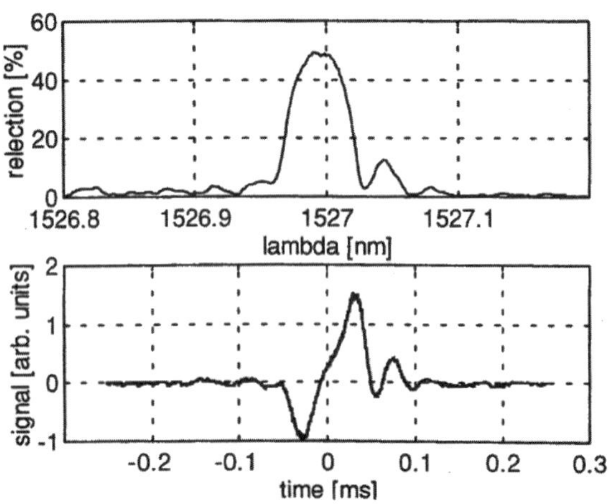

FIGURE 9.78. Reflection spectrum and corresponding derivative signal recorded from one of the FBG sensors by means of the DFB laser demodulation system of Fig. 9.80. From Hjelme, D.R., Bakke, B., Rambech J.S. and Neegard, S. (1996), "Multiplexed Fiber Optic Bragg Grating Strain Sensor System for use in Marine Vehicle Testing," SPIE International Symposium on Optical Science, Engineering and Instrumentation, Denver, USA, August 4–9, 1996.

by a FBG sensor, and RSG for a wave period of 1.6 s and a wave height of 3 cm is presented in Fig. 9.79. Unfortunately, the turning range of the DFB laser limits, somewhat, the dynamic range of the sensing system, and the use of optical delay lines can be impractical for certain situations.

The second measurement technique relies on laser locking the DFB laser output at 1548 nm while modulating it at 30 kHz. The detected signal reflected from the FBG is fed into a lock-in amplifier (Fig. 9.80). Since the harmonic output at the modulation frequency contains a term proportional to the derivative of the Bragg grating response, it will be zero at the peak wavelength for the FBG sensor. Strain changes imposed on the FBG

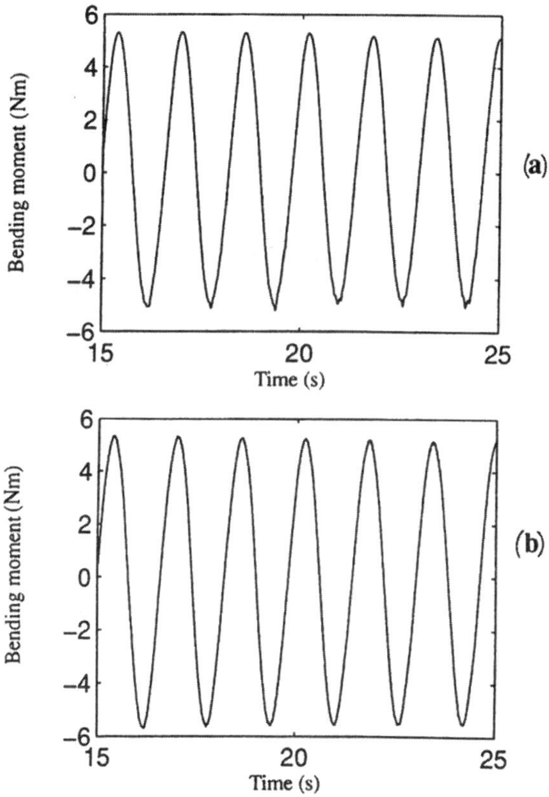

FIGURE 9.79. Bending moments determined as a function of time by an FBG sensor and a resistive foil strain gauge located at the centre of the catamaran model subject to 3-cm wave heights in a water tank test. From Hjelme, D.R., Bjerkan, L., Neegard, S., Rambech J.S. and Aarsnes J.V. (1997), "Application of Bragg Grating Sensor in the Characterization of Scaled Marine Vehicle Models" *Applied Optics* **36**, 328–336.

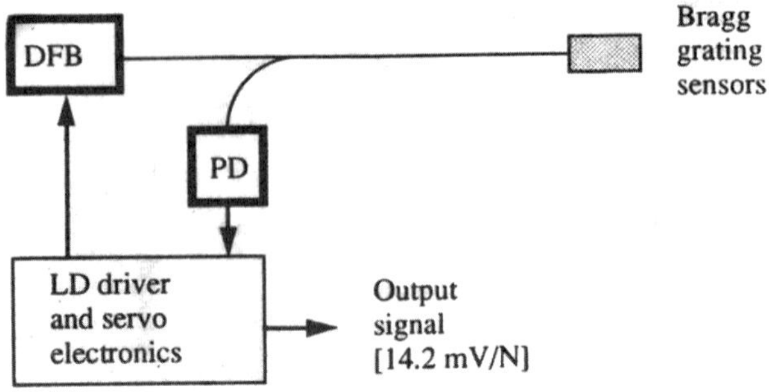

FIGURE 9.80. Schematic illustration of the second FBG sensor demodulation system for the catamaran model, based on locking the DFB laser wavelength to the back-reflected FBG signal. From Hjelme, D.R., Bjerkan, L., Neegard, S., Rambech J.S. and Aarsnes J.V. (1997), "Application of Bragg Grating Sensor in the Characterization of Scaled Marine Vehicle Models" *Applied Optics* **36**, 328–336.

sensor will lead to commensurate shifts in the wavelength location of this zero-point crossing. A feedback loop is created when the lock-in output signal is fed back to the input of the laser driver so that the laser wavelength tracks the Bragg peak wavelength. Changes in the strain field can be determined by monitoring the lock-in output signal. This system permitted a sampling frequency of 1000 Hz and was adequate for the slamming force measurements. The grating used for these measurements had a peak reflectivity of 70%, a FWHM of 230 pm, and a strain sensitivity of 1.082 pm $\mu\varepsilon^{-1}$. A polarizing fiber was used as the launch fiber to reduce errors caused by strain-induced birefringence. Figure 9.81 presents a sequence of wave impacts measured by both the FBG sensor and a resistive foil strain gauge. The larger noise level in the fiber optic sensor arose in its electronic circuitry. Figure 9.82 shows the power spectra corresponding to the time period shown in Fig. 9.81. The results verify that FBG sensors are adequate for measurement of both the bending moment of the marine vessel structure and its dynamic loading due to wave action (Hjelme *et al.*, 1997).

9.3.8 Structural monitoring of composite hull ship

For naval applications, ship hulls made of composite materials possess a number of advantages over conventional ships. Some of the most attractive features include low weight, design versatility, much reduced magnetic signature, and improved durability. The Royal Norwegian

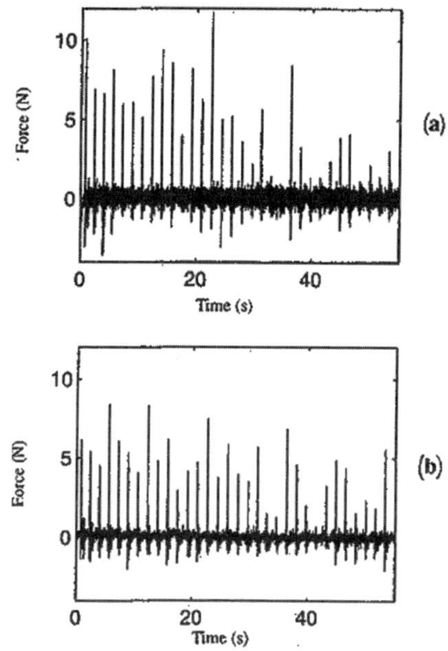

FIGURE 9.81. Comparison of wave impacts on the catamarn model as measured by both FBG sensors (a) and resistive foil strain gauges. From Hjelme, D.R., Bjerkan, L., Neegard, S., Rambech J.S. and Aarsnes J.V. (1997), "Application of Bragg Grating Sensor in the Characterization of Scaled Marine Vehicle Models" *Applied Optics* **36**, 328–336.

Navy (RNN) has developed a new class of Mine Counter Measures Vessel (MCMV) based on an advanced design concept that combines a composite dual hull with an air cushion catamaran. The air-cushion catamaran design allows the ship to ride higher on the sea, reducing drag and enabling operation at high speeds. Under moderate to high sea conditions, the area between the two hulls, or wet deck, experiences transient loading due to wave slamming. This loading can produce very high forces that can potentially damage the composite panels of the wet deck. Delamination of the GFRP from the foam separator can result in severe hidden damage and weakening, which can go undetected. Subsequent wave-slamming loads can result in potentially catastrophic damage to the structure. In 1996 an RNN mine-hunter ship of this class was instrumented with an array of fiber optic Bragg grating sensors designed to monitor the loading of critical hull and wet-deck locations during its sea trials (Fig. 9.83; Kersey *et al.*, 1997a).

The ship is close to 50 m in length and has a 13-m beam. The wet-deck area is approximately 50 m by 8 m and determined to be a critical location

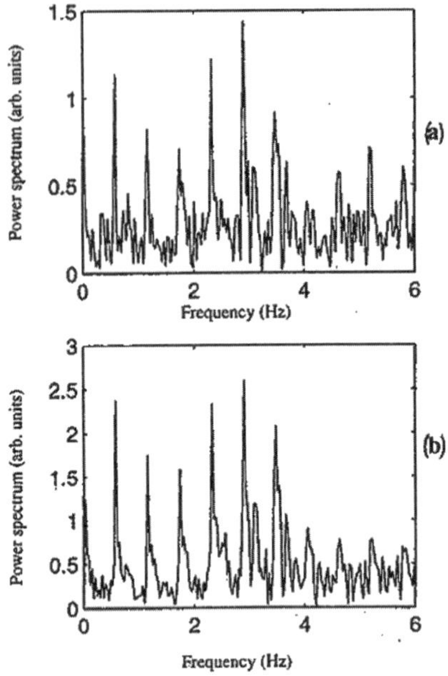

FIGURE 9.82. Power spectra derived from the catamaran model as measured by both FBG sensors (a) and resistive foil strain gauges. From Hjelme, D.R., Bjerkan, L., Neegard, S., Rambech J.S. and Aarsnes J.V. (1997), "Application of Bragg Grating Sensor in the Characterization of Scaled Marine Vehicle Models" *Applied Optics* **36**, 328–336.

FIGURE 9.83. Royal Norwegian Navy mine countermeasures vessel. From Kersey, A.D., Davis, M.A., Berkoff, T.A., Dandridge, A., Jones, R.T., Tsai, T., Codgell, G., Wang, G., Havsgaard, G.B., Pran, K., Knudsen, S., (1997a), "Transient Load Monitoring on a Composite Hull Ship Using Distributed Fiber Optic Bragg Grating Sensors," *SPIE* **3042**, 421–430.

for monitoring the loading from wave slamming. The goals of this Composite Hull Embedded Sensor System (CHESS) were as follows:

1. Determine the typical magnitude of strain transient generated by normal and high sea-state conditions
2. Evaluate the rise time of such transient loads
3. Demonstrate the use of FBG sensors for providing distributed sensing for such applications

Although by and large serial multiplexing is being treated in Chapter 11, this application follows on so naturally from the work on the model catamaran described earlier that it is included here. The MCMV was instrumented with three arrays of four FBGS that were bonded to the hull and wet areas of the ship, as indicated in Fig. 9.84. In each case the sensors were mounted both transversely (port to starboard) and longitudinally (stern to bow). Demodulation of the FBG sensors was accomplished using an unbalanced Mach–Zehnder interferometer, as described in Section 9.2.3 (Kersey *et al.*, 1992b), see Fig. 9.20. Wavelength-division multiplexing was employed to handle the serial arrays of FBGs (Fig. 9.85). Although more than 16 FBGs in a given optical fiber have been interrogated at low frequencies with this type of system, in this instance it was only capable of handling four FBGs the fast transient strains associated with this work.

During sea-state 2 (SS2) conditions, the strains were observed to range from 5 $\mu\varepsilon$ (rms) on the wet deck to 30 $\mu\varepsilon$ (rms) on the lower hull. Later the

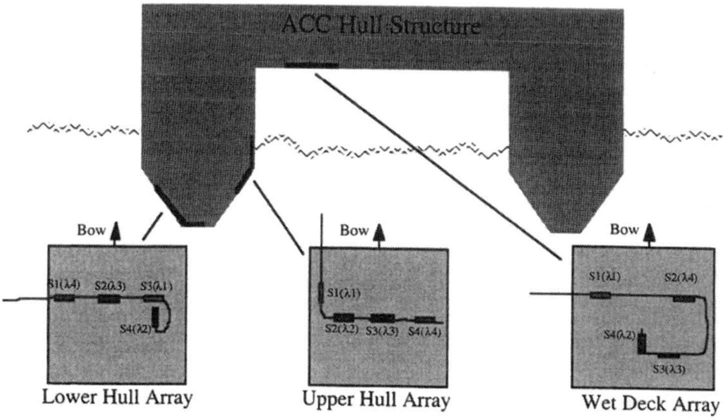

FIGURE 9.84. Distribution of three FBG sensor arrays on the twin-hulled Royal Norweigan Navy mine counter measures vessel. From Kersey, A.D., Davis, M.A., Berkoff, T.A., Dandridge, A., Jones, R.T., Tsai, T., Codgell, G., Wang, G., Havsgaard, G.B., Pran, K., Knudsen, S., (1997a), "Transient Load Monitoring on a Composite Hull Ship Using Distributed Fiber Optic Bragg Grating Sensors," *SPIE* **3042**, 421–430.

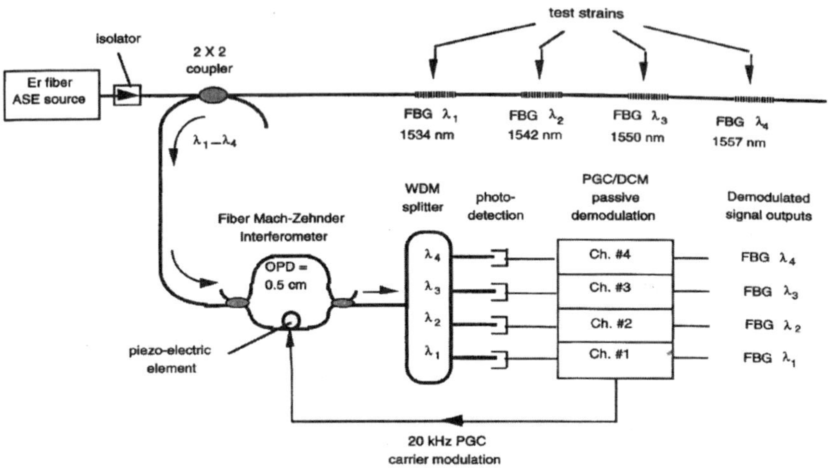

FIGURE 9.85. Demodulation system used for the FBG sensor arrays mounted on the Royal Norwegian Navy MCMV and based on a Mach Zehnder interferometer and wavelength demultiplexing. From Kersey, A.D., Davis, M.A., Berkoff, T.A., Dandridge, A., Jones, R.T., Tsai, T., Codgell, G., Wang, G., Havsgaard, G.B., Pran, K., Knudsen, S., (1997a), "Transient Load Monitoring on a Composite Hull Ship Using Distributed Fiber Optic Bragg Grating Sensors," *SPIE* **3042**, 421–430.

level of strains recorded increased dramatically as the ship engaged SS5 to SS6. Figure 9.86 displays strain variations recorded by the sensor array on the lower hull over a 2-h period. It is apparent that wave-induced loading of the hull can exceed several hundreds microstrains at times. The typical transients due to wave slamming were observed to have rise-times of less than 0.01 s suggesting that a 200-Hz sampling rate is required for monitoring such loading of the ship. The data recorded in Fig. 9.86, was reduced to the form displayed in Fig. 9.87 by taking the time average rms over a 6 s time window. This provides a measure of the wave-induced strain loading of the hull over a 2 h interval.

The FBG sensor arrangement on a bottom panel on one of the side hulls of the catamaran is shown in Fig. 9.88d. One grating (L) was mounted longitudinally and three gratings (T1 to T3) were mounted transversely. The reflection wavelengths of the FBGs were spaced approximately 7 nm apart in the 1550-nm region. The recorded data presented in Fig. 9.88, by Wang *et al.* (1997), shows a number of transient events including strain induced by a global long-ship vibration in the hull at 2 Hz (Fig. 9.88a). We see that the longitudinal strain, *L*, and the strain in the transverse direction T1, T2, and T3, are 180° out of phase. Transient local vibrations of the bottom panel of about 20 Hz are seen in Fig. 9.88b. The

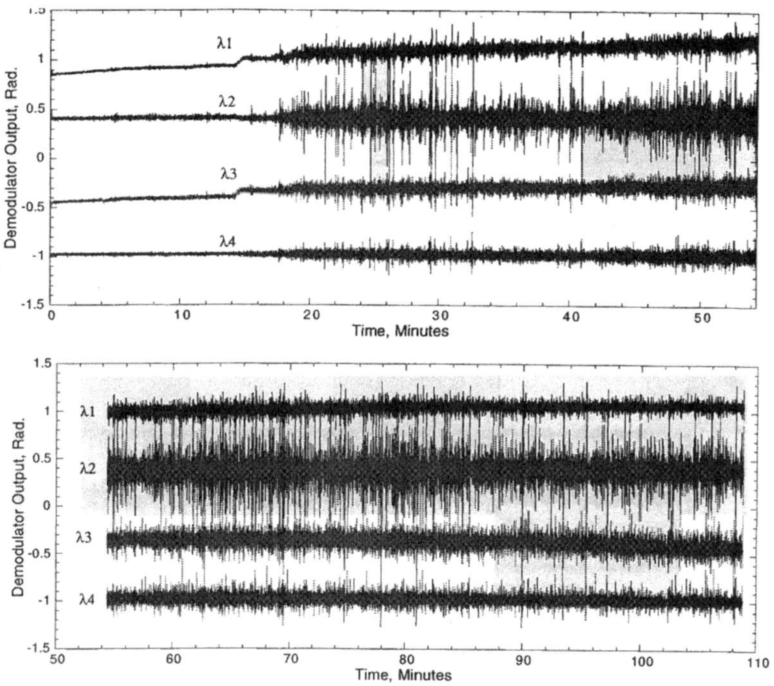

FIGURE 9.86. Strain variations recorded from FBG sensor arrays mounted on the lower hull of the Royal Norwegian Navy MCMV during a 2-hour period. From Kersey, A.D., Davis, M.A., Berkoff, T.A., Dandridge, A., Jones, R.T., Tsai, T., Codgell, G., Wang, G., Havsgaard, G.B., Pran, K., Knudsen, S., (1997a), "Transient Load Monitoring on a Composite Hull Ship Using Distributed Fiber Optic Bragg Grating Sensors," *SPIE* **3042**, 421–430.

strain in the transverse direction appears to dominate. Strain induced by the semiperiodic impact between the hull and the surface waves is shown in Fig. 9.88c. The low self-noise of the sensor system, about 5 nε $Hz^{-0.5}$, also made it possible to observe weak vibrations generated by machinery at about 200 Hz. Wang *et al.* (1997) demonstrated that wavelet transforms are well suited for this type of analysis and will probably be used in processing of the data collected from the sensor system.

9.3.9 Fiber optic crack monitoring of concrete structures

The presence of cracks in concrete is not unusual and may have little influence on the structure's functionality. Stability of existing cracks does, however, serve as an indicator of mechanical soundness, especially in the case of the new steel-free, *fiber-reinforced concrete* (FRC) decks being

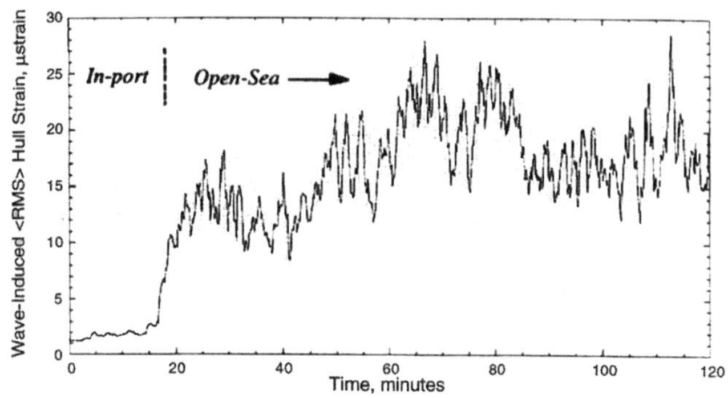

FIGURE 9.87. Temporal variation of the strain (time averaged over 6 s) recorded from one of the FBG sensors mounted on the Royal Norwegian Navy MCMV during a 2-hour run. From Kersey, A.D., Davis, M.A., Berkoff, T.A., Dandridge, A., Jones, R.T., Tsai, T., Codgell, G., Wang, G., Havsgaard, G.B., Pran, K., Knudsen, S., (1997a), "Transient Load Monitoring on a Composite Hull Ship Using Distributed Fiber Optic Bragg Grating Sensors," *SPIE* **3042**, 421–430.

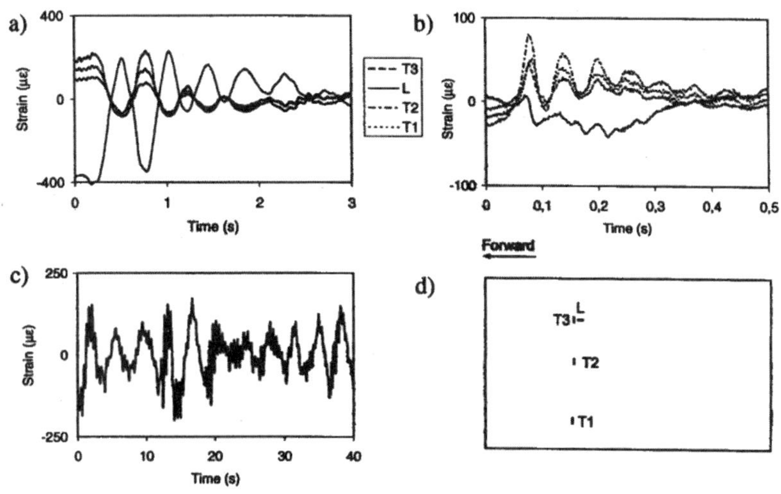

FIGURE 9.88. (a) Temporal variations in the strain measured for a number of FBG sensors mounted, as indicated, in the longitudinal and transverse directions relative to the axis of the MCMV (d), on the bottom panel of one of the side hulls. (b) High-frequency local straind transients due to vibrations of the bottom panel. (c) Strain in transverse direction. From Wang, G., Havsgaard, Urnes, E., Pran, K., and Knudsen, S., (1997), "Digital Demodulation and Signal Processing Applied to Fiber Bragg Grating Strain Sensor Arrays in Monitoring Transient Loading Effects on Ship Hulls." OSA, 12th Int Conf On Optical Fiber Sensors, Williamsburg, 612–615.

introduced in a number of bridges (Newhook and Mufti, 1996). The Chatham Bridge, Ontario is a newly restored, two-traffic-lane, steel girder bridge. The exterior 13.1-m spans were replaced with an innovative design consisting of an FRC deck slab that is completely devoid of steel.

A number of cracks have developed in this FRC deck ranging in width from 0.1 to 0.2 mm. These need to be monitored and a new *fiber optic crack* (FOC) monitoring system is under development for this purpose. A FBG sensor is used to measure the strain produced in a specially shaped metal strip by growth in the crack it straddles. The FOC gauge is bonded at both ends to the concrete structure, and has the FBG sensor attached to its external surface. The initial tests of this FOC gauge have used four RSGs in a full-bridge configuration to verify the readings from the fiber optic sensor. The calibration of the device was undertaken by imposing on the FOC gauge a prescribed displacement that was determined with 0.001-mm resolution (Hearn, 1998).

9.4 INTERFEROMETRIC SHORT-GAUGE STRUCTURAL SENSORS

In this section we will focus on demodulation techniques used for Fabry–Perot interferometric sensors. First we need to recognize that there are two very different forms of this type of sensor. The basic differences were indicated in Fig. 6.11. We will first discuss some of the demodulation approaches adopted for the "intrinsic" *fiber Fabry–Perot* (FFP) sensor, then we will progress to the demodulation approaches used for the more widely used *extrinsic Fabry–Perot interferometric* (EFPI) sensor, which has evolved into a practical short gauge-length structural sensing system that is commercially available.

9.4.1 Fiber Fabry–Perot interferometric sensor

As we saw in Section 3.6, the generic phase-response function, $R(\Delta\phi)$, for a two-beam interferometer can be expressed in the form

$$R(\Delta\phi) \equiv \frac{I}{I_0} = \frac{1}{2}\{1 + \tilde{\nu}\cos(\Delta\phi)\},$$

according to Eq. (3.61) (Jackson and Jones, 1989). In this equation I is the photodetector current from the sensor, I_0 is the photodetector current that arises directly from the source, and $\Delta\phi$ is the phase retardation introduced between the two paths of the interferometer. The *fringe visibility factor*, $\tilde{v}$, (or fringe contrast) is introduced to cover partial coherence situations.

The purpose of the *signal recovery* or demodulation system is to convert the complex interferometric output from a fiber optic sensor into an electrical signal proportional to the phase retardation. The problem with the fiber optic sensor response function indicated earlier is that it is sinusoidal, and therefore multivalued and nonlinear. Furthermore, the nature of the cosine function leads to sign ambiguity and signal fading. The purpose of any signal recovery scheme is to address as many of these shortcomings as possible. One of the simplest forms of signal recovery is "fringe counting." This is acceptable for large changes in $\Delta\phi\{\gg 2\}$ but fails to distinguish changes in direction of the measurand. A further problem of the phase-response function given by Eq. (3.61) is that it does not provide interrupt immunity (i.e., an absolute measurement). Signal recovery techniques should address all five of these issues:

1. Signal fading
2. Interrupt immunity
3. Sign ambiguity
4. Nonlinearity
5. Multivalued response

Signal recovery techniques can be broadly classified as: passive or active, and they can involve heterodyne or homodyne detection (Jackson *et al.*, 1982; Liu and Measures, 1992). In *heterodyne detection* the optical frequencies in the two interferometer arms are unequal, while in *homodyne detection* there is a common frequency in both arms. True heterodyne signal recovery is unlikely for structural sensors, as it is somewhat difficult to affect the frequency in one arm (path) of a single optical fiber sensor, such as a Fabry–Perot interferometer.

Switched Dual Wavelength Quadrature Technique. One of the most elegant methods of overcoming signal fading is to use a system that produces two outputs with a phase bias of $\pi/2$. Such *quadrature* (orthogonal) outputs ensure that one output has maximum sensitivity when the other experiences a minimum sensitivity. In the case of the single-fiber: polarimetric (Turner *et al.*, 1989), two-mode (Lu and Blaha, 1989), and Fabry–Perot (Hogg *et al.*, 1991a) sensors, quadature outputs can be achieved through an active process, such as switching or shifting the laser frequency by an amount corresponding to some multiple of a $\pi/2$ phase shift for the sensor. This type of technique is sometimes referred to as quadrature phase modulation. If the laser wavelength is switched between two values corresponding to a $\pi/2$ phase shift, quadrature phase signals are essentially extracted from the sensor. If these are sampled by orthogonal gating pulses, summed, and bandpass filtered, the resulting sinusoidal signal can be demodulated with a phase-sensitive detector (Kersey *et al.*, 1984a,b). The required *quadrature wavelength* shift

$$|\Delta\lambda_Q| = [2N + 1]\frac{\lambda^2}{4nL}, \tag{9.9}$$

where $N = 0, 1, 2 \ldots,$ is termed the *quadrature order.* For a small gauge-length ($L \leq 10$ mm) sensor, the quadrature wavelength *shift* falls well within the *tuning* range of a single longitudinal mode laser diode, i.e., $\Delta\lambda_Q = 0.02$ nm. By comparison, a quadrature wavelength shift of about 20 nm is required for a polarimetric sensor (with a birefringence *beat length* of 1.3 mm, and laser wavelength of 820 nm) and a gauge-length of a few centimeters (Turner *et al.*, 1989). This difference arises because it is the difference in the index of refraction between the fast and slow axes that is used in place of n in Eq. (9.9). This necessitates switching between two separate lasers and represents a problem in terms of phase control.

However, if the sensor gauge length is closer to 1 m as is possible, since the sensitivity of this type of sensor is much less than that of an interferometric sensor (see Section 7.3), the quadrature wavelength shift reduces to 0.7 nm and the sensor could be operated by switching a single laser between two wavelengths. Claus *et al.* (1993) used quadrature detection for their extrinsic Fabry–Perot interferometric sensors by using two such optical fiber interferometers in the same sensor head and making the gauge-length difference correspond to a $\pi/2$ phase difference at the same wavelength. This is not, however, a practical scheme for structural sensing, especially in the case of embedded sensors.

Pseuedo-Heterodyne Phase Detection. Since variation of the injection current in a semiconductor laser diode causes a corresponding wavelength change, the laser can be directly phase modulated (Jackson *et al.*, 1982). This approach is applicable to fiber Fabry–Perot interferometric fiber optic sensors with gauge lengths of several millimeters. In this *Serrodyne* or *chirped wavelength modulation* technique, the wavelength of the laser is swept over a given wavelength interval $\Delta\lambda_L$, and the corresponding change in phase $\Delta\phi$ is measured and used to determine the change in the cavity length of the Fabry–Perot sensor. From this phase change determination, the strain imposed on the sensor can be evaluated.

In Section 3.6 and earlier in this chapter, Eq. (9.4) we saw that when a lightwave is divided, then recombined after traveling a path difference L, the phase difference ϕ between the waves can be written in the form

$$\phi = \frac{2\pi nL}{\lambda} + \delta,$$

where n is the mean index of refraction along the path, λ is the free-space wavelength, and δ is the reference phase. The temporal change in this phase difference:

$$\Delta\phi(t) = -\frac{2\pi nL}{\lambda^2}\Delta\lambda(t), \tag{9.10}$$

is proportional to the temporal variation in the laser wavelength, $\Delta\lambda(t)$. If the wavelength of the laser is repeatedly *ramped* linearly over the interval $\Delta\lambda_L$, with a period T, then we can write

$$\Delta\lambda(t) = \begin{cases} \frac{\Delta\lambda_L}{T}(t - mT): & \text{for} \ldots mT < t < (m+1)T \\ 0: & \text{for} \ldots t = (m+1)T \end{cases} \tag{9.11}$$

where m is an integer.

This *sawtooth modulation* of the laser wavelength is illustrated in Fig. 9.89, and leads to a phase variation of the form

$$\Delta\phi(t) = \omega_{ch}t + \Delta\phi_0, \tag{9.12}$$

where we introduce the chirp frequency

$$\omega_{ch} \equiv -\frac{2\pi nL}{T\lambda^2}\Delta\lambda_L, \tag{9.13}$$

and the phase constant

$$\Delta\phi_0 \equiv m\frac{2\pi nL}{\lambda^2}\Delta\lambda_L. \tag{9.14}$$

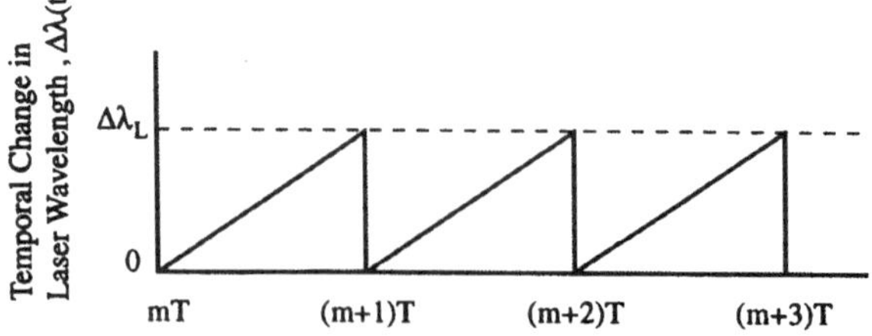

FIGURE 9.89. Schematic illustration of the sawtooth modulation in laser wavelength used to read a fiber Fabry Perot interferometric sensor.

This linear dependence on time for the phase change as expressed in (9.12), gives rise to a sinusoidal variation in intensity. However, the sudden drop in wavelength at the end of each ramp gives rise to a sinusoidal signal of much higher frequency (in essence, a spike). If a suitable bandpass filter is used to remove this *fly-back spike*, then to first approximation the signal is phase modulated with a carrier frequency, ω_{ch}, and we can express the self-normalized signal in the form

$$\frac{I}{I_0} = \frac{1}{2}\{1 + \tilde{\nu}\cos(\omega_{ch}t + \Delta\phi_0)\}. \tag{9.15}$$

Clearly, this *serrodyne*, or chirped wavelength, excitation leads to a sinusoidal optical signal onto which is impressed any phase modulation arising from changes in the strain field of the sensor.

If we assume $n = 1.5$, $\lambda = 1550$ nm, and a 1.0-nm wavelength tuning range $\Delta\lambda_L$ for the laser, then according to Eq. (9.10),

$$|\Delta\phi| = \frac{2\pi 1.5 \times 10^{-9}[\text{m}]L}{\{1.55 \times 10^{-6}\}^2[\text{m}^2]} = 3.925 \times 10^3\text{L.[rad.m}^{-1}] \tag{9.16}$$

For an FFP interferometric sensor we can take the cavity length $L = 10$ mm. In which case

$$|\Delta\phi| = 39.25 \text{ rad, or } 6.25 \text{ fringes,}$$

as 2(= 6.28 rad) corresponds to one fringe. We see that a 1-nm chirp of the laser wavelength yields a phase change of just over six fringes, which can easily be measured with good precision, enabling any small strain-induced changes in L to be accurately determined.

A schematic illustration of such a pseudo-heterodyne modulated Fabry–Perot sensor system is presented in Fig. 9.90 (Hogg *et al.*, 1991a,b). In this experiment the laser diode injection current was modulated by a linear sawtooth signal with an amplitude corresponding to one fringe. Figure 9.91, displays a representative laser diode driving current waveform and the resulting phase-generated optical carrier signal before and after filtering. After detection, the carrier signal is fed to a digital/analog hybrid phase tracker similar to one reported by Jackson (1981). This system used a carrier frequency of 31.25 kHz and had an effective strain-bandwidth figure of merit of about 2200 $\mu\varepsilon$ Hz cm^{-1} when operated at 850 nm with a gauge length of 1 cm. The theoretical strain resolution of this system is a small fraction of a microstrain and its time response is adequate to cover the normal loading of most practical structures. However, its limited temporal response can pose a serious problem in the event that the structure is exposed to any sudden high rate of loading, such as an impact. Under these conditions the strain sensing system can lose track of the strain.

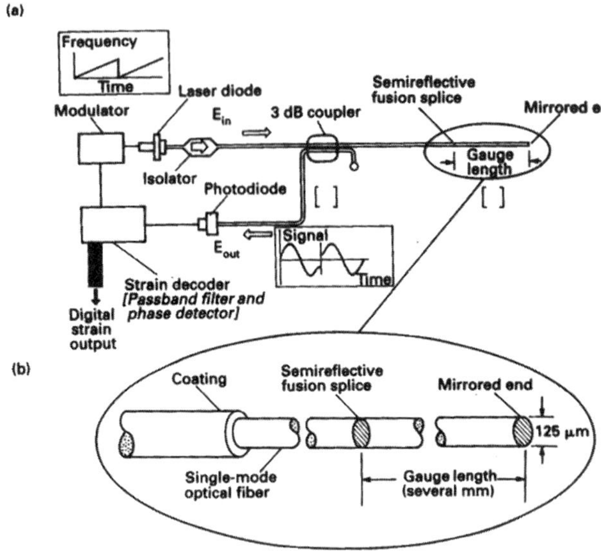

FIGURE 9.90. (a) A pseudo-heterodyne demodualtion system for a fiber Fabry–Perot sensor. (b) Expanded view of the intrinsic fiber FP sensor configuration. Reprinted from Measures, R.M., (1993), "Fiber Optic Sensing for Composite Smart Structures," *Composite Engineering*, **3**, 715–750, with permission from Elsevier Science.

An improved pseudo-heterodyne Fabry–Perot sensor demodulation system based on sinusoidal modulation of the wavelength of the laser diode and three-point phase sampling of the detected optical signal was subsequently developed (Mason *et al.*, 1992). This demodulation system provided a strain resolution of a few $\mu\varepsilon$ with a 2 cm gauge-length Fabry-Perot sensor, and could operate to a loading rate of 10^5 $\mu\varepsilon\ s^{-1}$ making it suitable for a broad range of strain sensing applications.

9.4.2 Extrinsic Fabry–Perot interferometric sensors

As we saw in Fig. 6.11b, the extrinsic Fabry–Perot interferometric (EFPI) sensor involves a small air-gap between two reflective surfaces, one of which is the lead-in optical fiber. EFPI sensors can be made using either single- or multimode optical fibers. Those based on multimode optical fibers have higher light-coupling efficiency, which makes optical connectivity easier, while single-mode EFPI sensors tend to provide a higher accuracy and greater strain range. This is a direct result of the larger

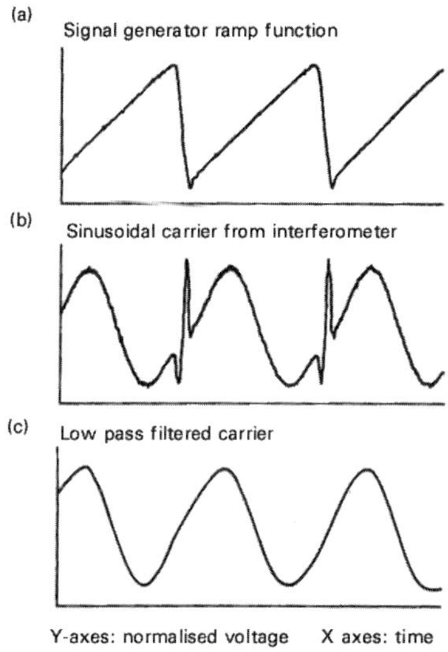

FIGURE 9.91. Signals associated with the demodulation system of Fig. 9.96. (a) Laser diode drive current waveform thzt modulates the output wavelength. (b) Sinusoidal carrier signal received from the fiber FP interferometric sensorl (c) Low-pass filtered signal from the photodetector. From Hogg, D., Janzen, D., Valis, T., Measures, R.M., (1991a), "Development of a Fiber Fabry–Perot Strain Gauge," *SPIE* **1588**, Fiber Optic Smart Structures and Skins IV, 300–307.

attenuation associated with the higher numerical aperture of the multimode optical fiber. The cavity length of multimode EFPI sensors lies in the range of 0 to 100 μm, while that of single mode-EFPI sensors lies in the range of 0 to 300 μm (Liu *et al.*, 1997b).

Claus *et al.* (1993) used the cleaved ends of a single-mode lead-in optical fiber and a multimode reflecting optical fiber to constitute a low-finesse Fabry–Perot cavity (Fig. 9.92a; Murphy *et al.*, 1994). The two optical fibers were held in place by means of epoxy bonds at the end of the silica microcapillary tube. A similar design was used by Belleville and Duplain (1993) and Liu *et al.* (1996), except that they used multimode optical fibers to form both sides of the Fabry–Perot cavity and they relied on fusion splices between the optical fibers and the precision-bore quartz microcapillary tube to define the gauge-length of the sensor. This ensured a more consistent gauge length than provided by the epoxy bond design, as the epoxy tended to be drawn by different amounts into the microcapillary tube, giving rise to a gauge-length ambiguity (Paul *et al.*, 1993). Furthermore, using the distance between the weld spots to define the

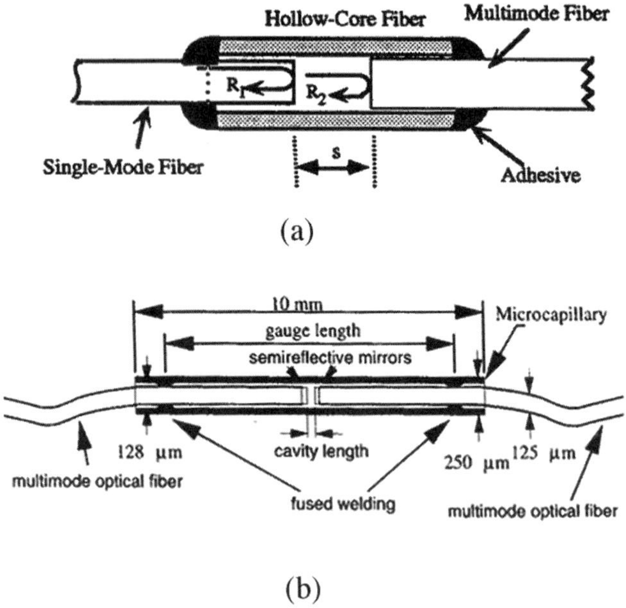

FIGURE 9.92. Schematic of two forms of extrinsic Fabry Perot interferometric (EFPI) sensor. (a) Basic design developed at Virginia Tech. (b) Basic design developed at the National Institute for Optics, From Belleville, C. and Duplain, G., (1993), "White Light Interferometric Multimode Fiber-Optic Strain Sensor," *Optics Letters* **18**, 78–80.

gauge length made it possible to control the gauge operating strain range, as this geometry behaves as a mechanical strain amplifier.

Belleville and Duplain (1993) further improved the design of this sensor with the use of dielectric mirror coatings on the ends of their optical fibers (Fig. 9.92b). This resulted in a higher-finesse Fabry–Perot interferometer and a better fringe contrast than possible with cleaved ends. In later versions, their cavity length was selected to lie between 10 and 20 μm (Morin *et al.*, 1996).

The use of such an air-gap cavity bestows upon the EFPI sensor an insensitivity to transverse strain and a reduced sensitivity to temperature changes compared to FBG and FFP sensors. Furthermore, this design lends itself to thermal compensation for certain types of applications (Belleville and Duplain, 1993). Since light crossing this air gap is unguided the length of the air gap, *L*, has to be small in order to avoid significant loss of light due to diffraction. This limits the strain range and also makes the pseudo-heterodyne FFP demodulation approach impractical.

To put it into perspective, the cavity length for a FFP lies typically between 3 and 20 mm (Lee and Taylor, 1988; Hogg *et al.*, 1991b), while the air-gap cavity length, for EFPIs are much shorter than those of FFPs and as indicated above rarely exceed 300 μm. If we take, $L = 100$ μm, as representative of the multimode EFPI sensors, then $\Delta\phi = 0.03925$ rad, or 0.0625 fringes. Thus, a 1-nm chirp of the laser wavelength in the case of an EFPI leads to a very small phase change, leading to poor resolution and accuracy, so the pseudo-heterodyne approach is not suitable for this type of fiber optic sensor. As we shall see in Chapter 10, this pseudo-heterodyne approach can be used for long gauge-length Fabry–Perot interferometric sensors.

9.4.3 Extrinsic Fabry–Perot interferometric sensor demodulation

One of the most serious disadvantages of using high-coherence interferometry to demodulate the short cavity of an extrinsic Fabry–Perot sensor is that only changes in the cavity length can be measured, not its absolute value. For many applications it is important to be able to measure the actual strain present as well as its variation with time. Two approaches have emerged that enable EFPI sensors to make absolute measurements of strain.

The first involves spectral demodulation and uses the wavelength separation of a pair of consecutive fringes to determine the EFPI cavity length (Claus *et al.*, 1993; Liu *et al.*, 1966). The experimental arrangement used by Liu *et al.* (1996) is schematically presented in Fig. 9.93. This is very similar to the arrangement used by Claus *et al.* (1993), except that instead of a scanning monochromator being placed just in front of the source, Claus *et al.* used an optical spectrum analyzer ahead of the photodetector.

A *low-finesse* Fabry–Perot has a phase response function that is well approximated by that of a two-beam interferometer (3.62) and can be written:

$$\frac{I}{I_0} = \frac{1}{2}\{1 + \tilde{v}\cos(\Delta\phi)\}, \qquad (9.17)$$

where in this instance the phase difference $\Delta\phi$ for lightwaves reflected from the front and rear surfaces of the Fabry–Perot interferometer can be expressed in the form

$$\Delta\phi = \frac{2\pi n\{2L\}}{\lambda}. \qquad (9.18)$$

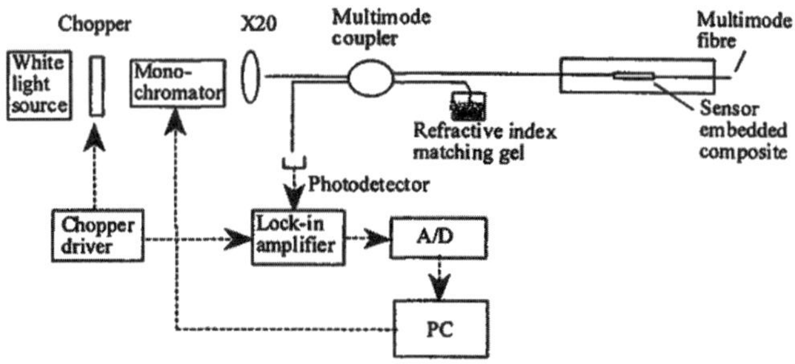

FIGURE 9.93. One form of spectral demodulation system used to undertake absolute measurements with an EFPI sensor. From Liu, T., Brooks, D., Martin, A., Badcock, R.., Fernando, G.F., (1996), "Design, Fabrication and Evaluation of an Optical Fiber Sensor for Tensile and Compressive Strain Measurements via the Use of White Light Interferometry," *SPIE* **3042**, 203–212.

In this instance L is the *Fabry–Perot cavity length*, or separation of the reflecting surfaces. If λ_1 and λ_2 are the wavelengths of two consecutive fringe maxima produced by scanning the source wavelength (see Fig. 9.94; Liu *et al.*, 1996), then we can write

$$\Delta\phi = \frac{2\pi n\{2L\}}{\lambda}, \tag{9.19}$$

This allows us to express the cavity length in terms of these two wavelengths,

$$L = \frac{\lambda_1 \lambda_2}{2\{\lambda_2 - \lambda_1\}}, \tag{9.20}$$

where we have set $n = 1$ for the air-gap cavity. Clearly, the larger the cavity length, the smaller the wavelength separation of the fringes in the sensor spectrum. This is experimentally demonstrated in Fig. 9.95, where the normalized sensor spectrum for two very different cavity lengths (3.4 and 44.6 μm) is presented (Liu *et al.*, 1997a).

In an optimization of the sensor geometry Liu *et al.* (1996) showed (Fig. 9.96), that the cavity length should be kept to a minimum in order to achieve the highest strain resolution for a given monochromator spectral resolution. However, because the length of the EFPI cavity ultimately decides the strain range, a compromise has to be made between the sensor strain sensitivity and the maximum strain range (tensile and compressive). The potential strain range was estimated to be ±10 000 με with an accuracy of approximately 30 με, and a temperature operating range of 38 to 180°C. The response time of the sensing system is limited by the time

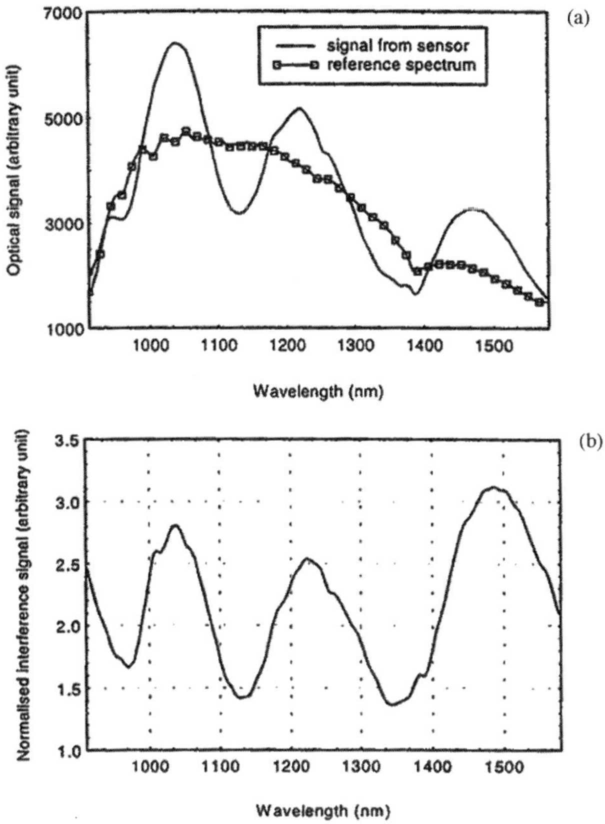

FIGURE 9.94. (a) Interferometric fringes produced by the EFPI sensor as the wavelength of the source increases (solid curve). As shown in Fig. 9.93 a monochromator is used to select a narrow band of light from the broadband source, the spectrum of which is also displayed (curve through squares). (b) Source-normalized interference signal from EFPI sensor. From Liu, T., Brooks, D., Martin, A., Badcock, R., Fernando, G.F., (1996), "Design, Fabrication and Evaluation of an Optical Fiber Sensor for Tensile and Compressive Strain Measurements via the Use of White Light Interferometry", *SPIE* **3042**, 203–212.

taken for one spectral scan, about 1 min (Liu *et al.*, 1997a). A faster time-response system uses a broadband source, a spectrometer, and a CCD photodetector array to capture the transmission spectrum of the EFPI: see Fig. 6.13 (Bhatia *et al.*, 1994). Here the spectra can be read within 0.1 ms.

Two EFPI sensors were serially multiplexed using a single SLED source and operating in the reflective mode with a reference EFPI (Bhatia *et al.*, 1994). Liu *et al.* (1997a) demonstrated that two EFPI sensors could be

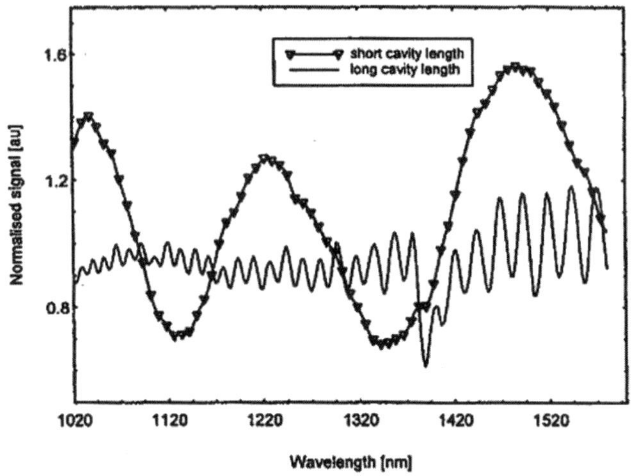

FIGURE 9.95. Normalized sensor spectra for a short and long cavity-length EFPI sensor. From Liu, T., Fernando, G.F., Rao, Y.J., Zhang, L., Bennion, I. and Jackson, D.A., (1997a), "Simultaneous Strain and Temperature Measurements in Composites Using a Multiplexed Fiber Bragg Grating Sensor and an Extrinsic Fabry-Perot Sensor," *SPIE* **3042**, 203–212.

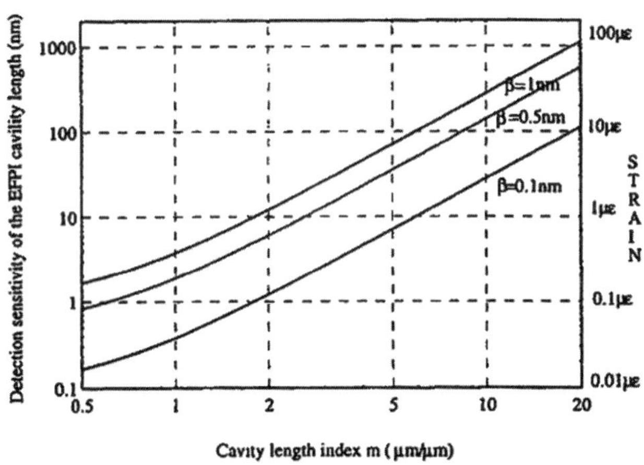

FIGURE 9.96. Variation in the detection sensitivity of an EFPI sensor as a function of cavity length index for three values of β. From Liu, T., Fernando, G.F., Rao, Y.J., Zhang, L., Bennion, I. and Jackson, D.A., (1997a), "Simultaneous Strain and Temperature Measurements in Composites Using a Multiplexed Fiber Bragg Grating Sensor and an Extrinsic Fabry-Perot Sensor," *SPIE* **3042**, 203–212.

parallel multiplexed by using two superluminescent sources sources and three 3-dB couplers as shown in Fig. 9.97. Each broadband source illuminates one EFPI, but their combined reflection spectrum is interrogated by an Ocean Optics, compact CCD spectrometer, displayed in Fig. 5.45. This instrument had a spectral resolution of 0.15 nm when used with a single-mode optical fiber, which in effect serves as its input slit. The Superlum SLD had a peak wavelength at 827.8 nm, while the Hammathu SLD had its peak at 855.7 nm. These spectra have very little overlap at wavelengths greater than 854 nm and less than 832 nm, as seen in Fig. 9.98a. The resulting spectral fringe pattern reflected from the two EFPI sensors is presented in Fig. 9.98b. In this figure the cavity length of sensor 1 was kept at 71 μm, while that of sensor 2 is changed from 18 to 100 μm. The accuracy of this system was observed to be 0.6 μm with a dynamic range of 500 or 300 μm.

The second approach uses low coherence interferometry combined with optical cross-correlation to evaluate the absolute strain experienced by an EFPI sensor (Belleville and Duplain, 1993; Murphy *et al.*, 1994). This approach was first demonstrated, however, on an FFP, by Zuliani *et al.* (1991). The broadband light, modulated and reflected from the EFPI sensor, is passed through a cylindrical lens and analyzed by a Fizeau interferometer, serving as an optical cross-correlator (see Fig. 9.99). A CCD linear array mounted on the Fizeau analyzer captures the light pattern transmitted through this wedge-shaped interferometer. The principle of operation was explained in Section 3.8 and illustrated in Fig. 3.29.

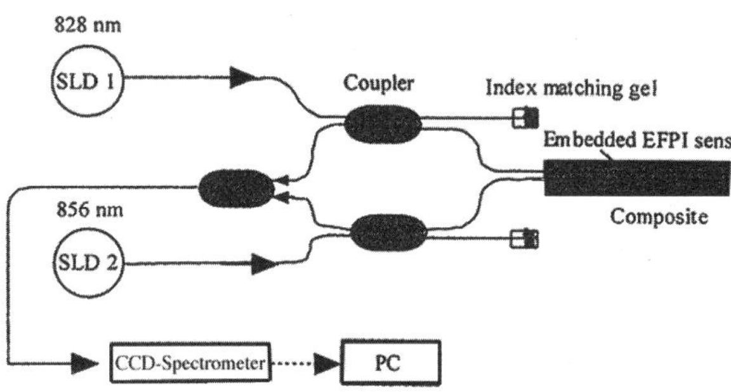

FIGURE 9.97. A demodulation system for parallel multiplexing two EFPI sensors using two SLD sources at slightly different wavelengths (828 and 856 nm) and a CCD spectrometer. From Liu, T., Fernando, G.F., Zhang, L., Bennion, I., Rao, Y.J., and Jackson, D.A., (1997b), "Simultaneous Strain and Temperature Measurements Using a Combined Fiber Bragg Grating/Extrinsic Fabry-Perot Sensor," OSA 12th International Conf. On Optical Fiber Sensors, Williamsburg, 20–23.

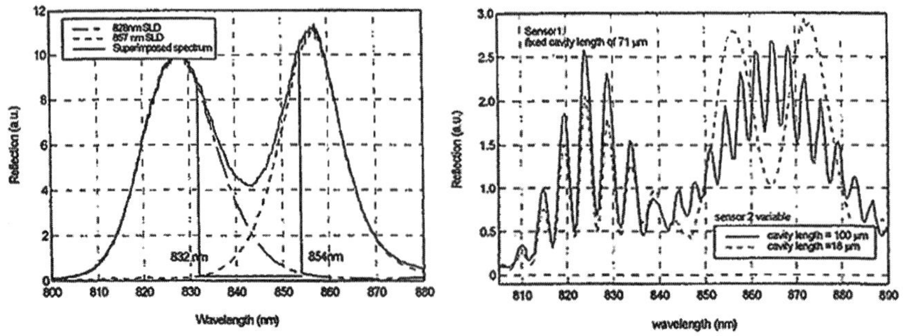

(a) (b)

FIGURE 9.98. (a) Spectra from two SLD sources used in the demodulation of two EFPI sensors (see Fig. 9.97). (b) Combined spectral interference pattern from the two EFPI sensors when each is illuminated by one SLD source. Note that the cavity length of one sensor was kept at 71 μm, while the other was changed from 18 (broken curve) to 100 μm (solid curve). From Liu, T., Fernando, G.F., Zhang, L., Bennion, I., Rao, Y.J. and Jackson, D.A., (1997b), "Simultaneous Strain and Temperature Measurements Using a Combined Fiber Bragg Grating/Extrinsic Fabry-Perot Sensor," OSA 12th Int. Conf. On Optical Fiber Sensors, Williamsburg, 20–23.

Figure 9.99 adapts the earlier figure to the sensing system at hand and once again uses the coherence wavepacket concept to try and explain how this low white-light interferometric demodulation system works.

A more complete discussion of the theory behind this approach is provided by Belleville and Duplain (1993). In essence, the wedge-shaped Fizeau interferometer makes an instantaneous, in-place correlation of the optical signal and transmits minimally at the exact location along the wedge where its spacing matches the EFPI cavity length. When the EFPI cavity length changes in response to strain, the position of the minimum in transmission shifts along the linear photodetector array accordingly. Knowledge of the slope of the wedge, the unstrained cavity length, and the gauge length is used to calculate the strain experienced by the EFPI sensor.

An example of the cross-correlation signal as read by the linear photodiode array is displayed as Fig. 9.100. The lower trace is the electrical signal as read directly from the CCD linear array, while the upper trace is filtered and amplified. The sensitivity of the system is estimated to be 0.025% of the strain range, which for a given slope of the Fizeau wedge is determined by the gauge length. A strain range of 1000 με is attained with a gauge length of 1 mm; while this can be increased to

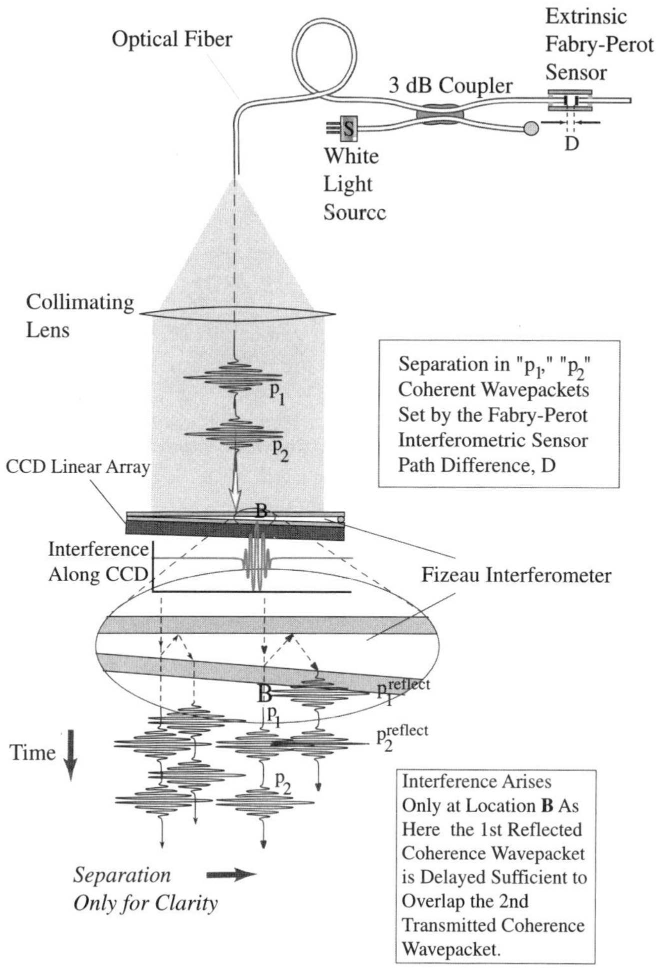

FIGURE 9.99. Schematic illustration of a low-coherence demodulation for an EFPI sensor using a Fizeau interferometer and a CCD linear array to constitute an optical cross-correlator. Cohrent wavepackets are used to help explain the system.

10 000 $\mu\varepsilon$ by adjusting the gauge length to 10 mm. A representative comparison of the dynamic response of a EFPI sensor and a conventional resistive foil strain gauge is presented in Fig. 9.101 (Belleville *et al.* 1995).

This approach has led to the development of a compact, fiber optic sensing system that is produced by Fiso Technologies, Quebec, and is available through Roctest Ltd.

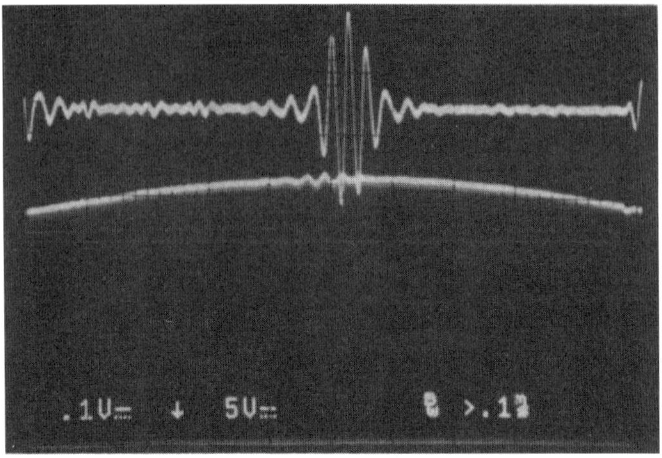

FIGURE 9.100. Lower trace reveals optical cross-correlator's output signal as read by the CCD linear array, while the upper trace displays the same signal after filtering and amplification. From Belleville, C. and Duplain, G., (1993), "White Light Interferometric Multimode Fiber-Optic Strain Sensor," *Optics Letters* **18**, 78–80.

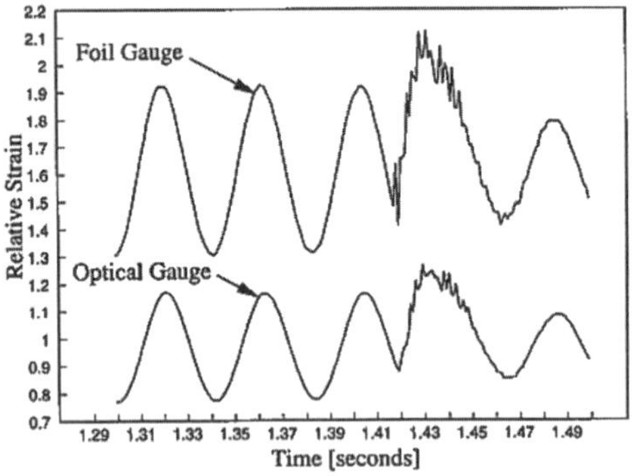

FIGURE 9.101. Comparison of the dynamic response of an EFPI sensor and a resistive foil strain gauge. From Belleville, C. and Duplain, G., (1993), "White Light Interferometric Multimode Fiber-Optic Strain Sensor," *Optics Letters* **18**, 78–80.

9.5 INTERFEROMETRIC SENSOR APPLICATIONS

9.5.1 Icebreaker propeller monitoring

Polar water navigation presents a formidable challenge to the propulsion systems of ships as large pieces of ice impinging on their propeller blades can result in stresses that exceed the yield strength of the blade material.

Damage to propellers is costly and can also spell disaster if a ship becomes disabled in a remote area. To avoid such situations, theoretical models of propeller/ice interaction must be validated against experimental data and the knowledge gained used to improve the design of ships. Today's sophisticated blade shape requires that the load be monitored at many locations in order to obtain an accurate picture of the stress and load distribution.

Resistance strain gauges have been found to have a very short useful life, typically a few months, in the harsh environment in which an icebreaker propeller operates, despite the best efforts to protect the sensors and associated cabling from abrasion, shock, vibration, and installation damage. Moisture and chemicals are also deleterious to foil gauges, since the minute changes in grid resistance are readily affected by electrolytes, resulting in offset drift and eventually complete gauge failure. Electromagnetic noise created by powerful electric motors, radars, and other shipboard equipment represents another challenge that must be addressed, a fact compounded by the very low signal levels attained with the long cable runs between the resistive foil gauges and the detection equipment on a large ship.

By contrast, fiber sensors are immune to electromagnetic and radiofrequency interference. Their all-glass construction also makes them impervious to chemical attack by electrolytes, thereby maintaining gauge integrity for much longer periods. These properties mean that shielding and grounding practices can be relaxed, while the small size of fiber optic sensors and their leads makes them less intrusive.

Morin *et al.* (1996) report the full-scale instrumentation of a Polar Class icebreaker propeller blade with 54 extrinsic Fabry–Perot interferometric fiber optic strain gauges and shaft-mounted electronics (Fig. 9.102). The Polar Class are the most powerful nonnuclear icebreakers so far constructed having a 75 000 HP engine and controllable-pitch propellers. The primary objective of this work was to obtain a detailed map of the location and distribution of the ice impact loads acting upon a propeller blade at full scale, and to accomplish this on a much larger and more powerful propeller than had been previously instrumented so that effects of scale on the measured ice loads could be explored. The variable-pitch propellers also make it possible to observe the influence of pitch variation upon ice impact loads.

As described earlier in Section 9.4.2, the EFPI sensors are fabricated by inserting two pieces of multimode optical fiber into the ends of a small quartz microcapillary tube (Morin *et al.*, 1996). The ends of the optical fibers are polished flat and thin-film coated to form semireflective mirrors. They are then positioned to leave a small 10 to 20 μm, air gap between the

FIGURE 9.102. Photograph of the Polar Class icebraker Polar Star instrumented with an array of 54 EFPI sensors. Inset shows the propeller blade monitored by this array of fiber optic sensors. From Troy, C.T., (1997), "Fiber Optic Smart Structures: A Technology Ahead of its Time is Finally Winning Acceptance," *Photonics Spectra*, May 97, pp. 112–118.

fiber tips, after which they are spot fused to wall of the micro-capillary tube. Since the gauges are operated in reflection and are therefore single-ended, one of the fibers is cut so that it barely extends from the microcapillary, and this end of the gauge is sealed off. The microcapillary and the fiber at the other end are bonded together so that any tension applied to the lead fiber is transmitted to the microcapillary and the glue bonding it to the substrate, rather than being resisted by the glass weld that fastens the fiber to the microcapillary. The gauge construction is illustrated in Fig. 9.103.

The conversion of the optical signal into an electronic representation of the strain information is accomplished by means of a Fizeau interferometer and *linear photodiode array* (LPA) combination as shown in Fig. 9.104 and discussed in Section 9.4.2. The wavelength-encoded signal reflected by the EFPI sensor, collimated by a cylindrical lens, is used to illuminate the whole width of the Fizeau/linear photodiode array assembly. The Fizeau interferometer performs an instantaneous, in-place corre-

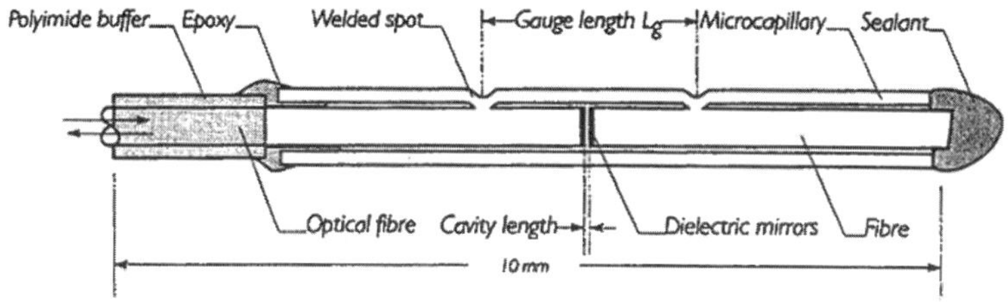

FIGURE 9.103. Schematic of the EFPI strain gauge design. From Morin, A., Caron, S., Van Neste, R., Edgecombe, M.H., (1996), "Field Monitoring of the Ice Load of an Icebreaker Propeller Blade Using Fiber Optic Strain Gauges," *SPIE* **2718**, 427–438.

lation of the optical signal. The data acquisition rate is set by the LPA scanning rate, while the resolution depends upon the number of pixels and upon the ability of the processing electronics to achieve subpixel interpolation. Table 9.3 summarizes the main system characteristics.

The design constraints imposed upon the instrumentation were very severe. For example, the EFPI gauge bonding had to withstand high vibration and shock without gauge debonding, ensuring that strain transfer remained effective. The data acquisition subsystem also represented a formidable challenge as the optoelectronic demodulation systems had to be mounted on the shaft; no suitable optical slip-ring technology was available. This meant that all the carefully aligned optics and the associated electronics would have to withstand the severe vibration and high *g*-forces caused by shaft rotation and ice impacts on the propeller.

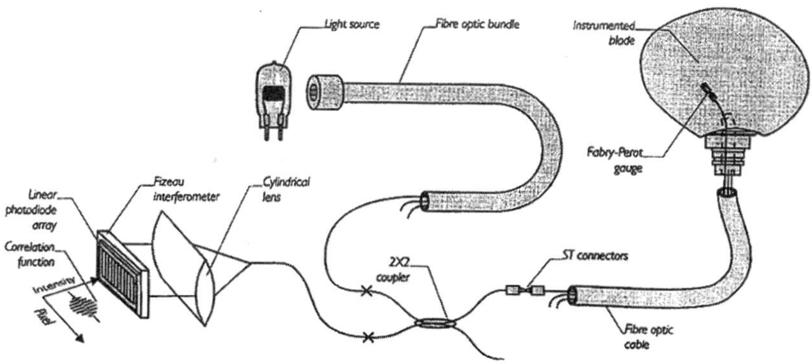

FIGURE 9.104. The layout of the optical part of the EFPI sensing system used aboard the U.S. *Polar Star.* From Morin, A., Caron, S., Van Neste, R., Edgecombe, M.H., (1996), "Field Monitoring of the Ice Load of an Icebreaker Propeller Blade Using Fiber Optic Strain Gauges," *SPIE* **2718**, 427–438.

TABLE 9.3. Fiso EFPI Sensor System Characteristics

Parameter	Value
Dynamic range	±4000 με
Resolution	2 με
Data acquisition rate	1000 Hz
Strain rate (mm/mm-s)	0.12 s^{-1}
Number of EFPI gauges	54 configured as 18 rosettes

The decision to place the demodulation systems on the shaft meant that electrical power would have to be transferred to the shaft while large amounts of digitized data would have to be telemetered away at high speeds.

Figure 9.105 illustrates the overall system architecture. EFPI gauges embedded within the blade were connected to the demodulation systems through individual fibers bundled in a fiber optic cable that was routed through the propeller shaft. The electronic and optical components required to illuminate the sensors and interpret the optical strain signals were installed in an annular housing mounted on the propeller shaft in the aft propulsion motor room. This electronics package, including the light source, 54 channels of optical signal conditioning, and a ruggedized industrial microcomputer to manage the on-shaft data acquisition, all rotated with the propeller shaft. A schematic and picture of the system are presented as Fig. 9.106.

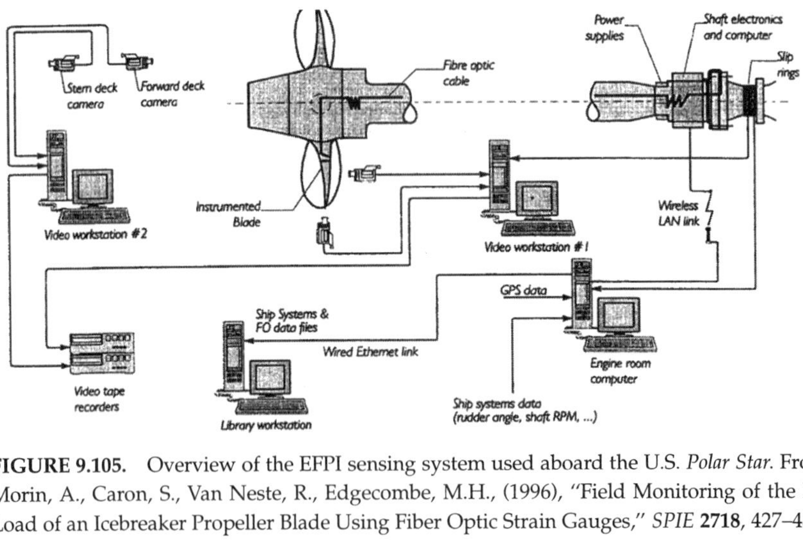

FIGURE 9.105. Overview of the EFPI sensing system used aboard the U.S. *Polar Star*. From Morin, A., Caron, S., Van Neste, R., Edgecombe, M.H., (1996), "Field Monitoring of the Ice Load of an Icebreaker Propeller Blade Using Fiber Optic Strain Gauges," *SPIE* **2718**, 427–438.

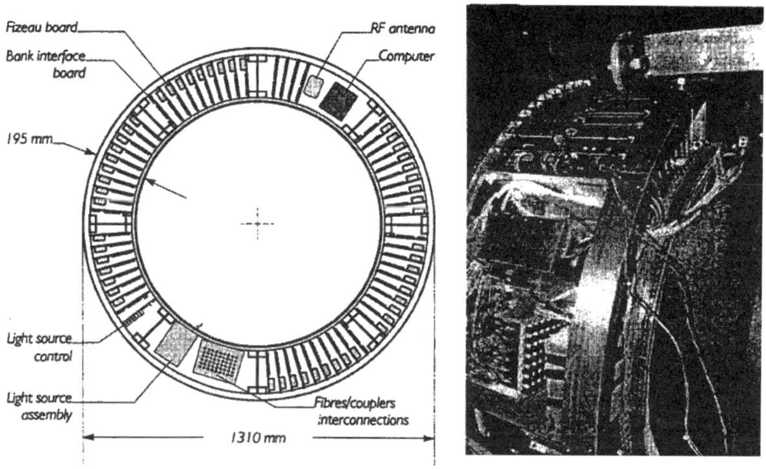

FIGURE 9.106. Schematic and photograph of circular arrangement of the 54 opto-electronic units mounted around the *Polar Star* propeller shaft. From Morin, A., Caron, S., Van Neste, R., Edgecombe, M.H., (1996), "Field Monitoring of the Ice Load of an Icebreaker Propeller Blade Using Fiber Optic Strain Gauges," *SPIE* **2718**, 427–438.

The ideal information to be provided by the instrumentation was identified during the project definition to include the following:

- Bending moment due to ice impact resisted by the blade at its root
- Spindle torque about the blade pitch axis resisted at its root
- Instantaneous location of the centroid of the ice impact load acting on the blade
- Direction of the ice impact load relative to the blade geometry
- Magnitude of the ice impact load
- Radial and circumferential distribution of the ice contact load acting on the blade

The propeller blade was instrumented for strain measurement at three circumferential locations at each of three radial displacements from the center of the propeller hub. At each of these nine locations a pair of three element rectangular rosettes were installed back to back, one on each face of the blade. The object of this arrangement was to provide strain measurements sensitive to blade loads acting on all parts of the blade and to provide a significant element of redundancy in the system's ability to measure the blade response to the modes of loading expected to be most important for blade strength design. By using three-element rosettes at each location, both the bending moment and shear about any axis contained in the plane of the blade can be determined. In addition, the principal strains and direction of the principal strain axes can be determined at each instrumented location providing important information to establish the location and extent of the ice contact load acting on the blade.

The performance of the fiber optic blade instrumentation system of the *Polar Star* during its first use under ice-breaking conditions substantially exceeded early expectations. In fact, even though the vessel encountered unusually severe ice-breaking conditions, the instrumentation functioned substantially as it was designed to during its initial deployment in ice and a large set of useful measurements was obtained. Of a total of 55 sensors installed in the blade, 53 were found to be functional. In general, the shaft optoelectronics package performed extremely well under levels of vibration significantly more severe than would normally be experienced on a mission of this type due to the unusually heavy ice conditions encountered.

With regard to the maximum strain rate, it was found that the instrumentation design was a good match with the extreme responses measured in this experiment. Because of the unusual thickness and strength of the ice cover encountered, a number of very high-amplitude blade impact loads were measured that exceeded the maximum loads that had been considered likely during planning for the deployment. In only a few of these extreme events did the strain rate in the blade approach the design capability, and then only for one or two milliseconds during the rising edge of the impact strain wave form.

The most significant difference between the performance of the fiber optic strain measurement system and those based upon the use of resistance foil strain gauges in similar applications were the much better signal-to-noise levels and the absence of long-term signal drift in the fiber optic system. Figure 9.107 compares the strain wave shapes measured by two fiber optic strain gauges with the same orientation but installed back to back on opposite sides of the blade during a 5-s period spanning the passage of an ice block through the propeller. As can be seen from the plot, the instrumented blade strikes this ice block seven times (once per revolution) before it passes aft of the propeller. The suction side of the blade experiences tensile strain and the pressure side compressive strain as the blade is bent aftward by each impact.

Figure 9.108 compares the strain wave shapes of the three component gauges of a single rosette located on the suction side of the blade on the blade spindle axis near the root during a half-second interval spanning impact between the blade and an ice block. The gauges with an identifier ending in *R* are oriented in the radial direction of the blade, while those ending in *T* are oriented 45° from this direction toward the trailing edge of

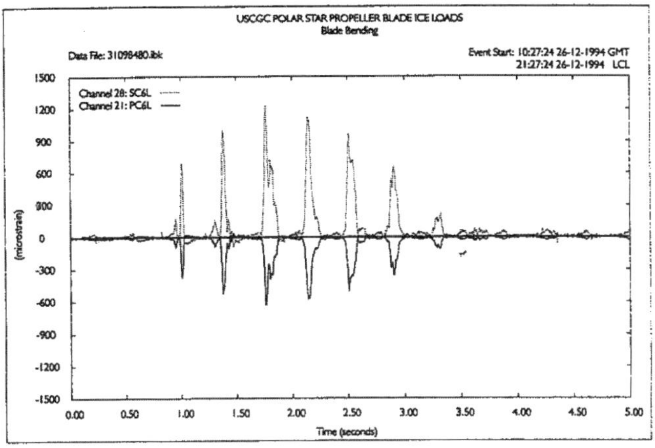

FIGURE 9.107. Comparison of the dynamic strain response of two of the fiber optic strain gauges mounted on opposite sites of the propeller blade to ice impacts. From Morin, A., Caron, S., Van Neste, R., Edgecombe, M.H., (1996), "Field Monitoring of the Ice Load of an Icebreaker Propeller Blade Using Fiber Optic Strain Gauges," *SPIE* **2718**, 427–438.

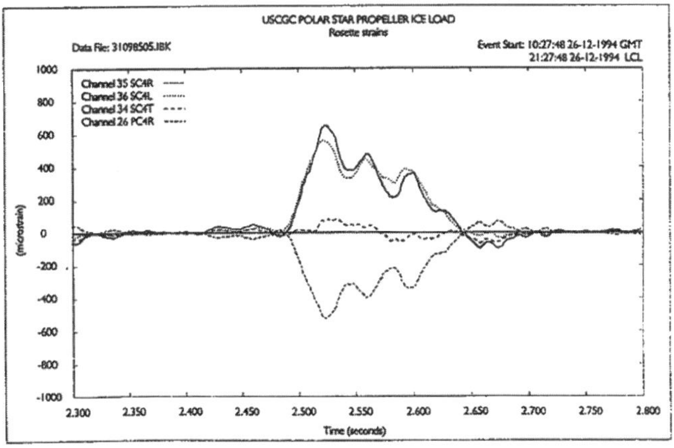

FIGURE 9.108. Expanded view of the responses from several fiber optic strain gauges to one ice impact revealing multiple strikes of the block of ice by the propeller as it rotates. From Morin, A., Caron, S., Van Neste, R., Edgecombe, M.H., (1996), "Field Monitoring of the Ice Load of an Icebreaker Propeller Blade Using Fiber Optic Strain Gauges," *SPIE* **2718**, 427–438.

the blade, and those ending in *L* are oriented at 45° toward the leading edge. As one would expect, the radial and leading edge direction strains are large while the strain in the direction of the trailing edge is small, corresponding to an impact load acting between the leading edge and the blade tip. In this detailed view of the strain wave shapes, the absence of noise artifacts is apparent. Again, for comparison, the response of a fourth gauge, the one back-to-back with the radial element of the rosette, is shown.

In conclusion, this experiment has undoubtedly demonstrated the reliability of both the Fabry–Perot fiber optic strain gauge and the associated optoelectronics, as the overall system response was almost unchanged from the original condition after 40 000 km of navigation. The long-term drift was found to be negligible for most of the fiber optic strain gauges and the measurement system proved to function correctly, even under extremely harsh conditions.

9.5.2 FRP composite cure monitoring

Fiber-reinforced composite materials are being used increasingly in many aerospace, civil, and industrial applications. Unfortunately, residual stresses can significantly degrade the strength of a composite material, resulting in cracking, delamination, reduced fracture toughness, and reduced fatigue strength. Additionally, these stresses can lead to warpage or springback, which presents difficulties in the assembly of composite structure, (Sarrazin *et al.*, 1995). Several causes of residual stress in composite materials have been reported in the literature. Differences in the coefficient of thermal expansion between the constituents of a composite can result in locked-in thermal stresses upon cooling from elevated processing temperatures. Residual microstresses are generated in unidirectional and cross-ply fiber-reinforced composites due to the different thermal deformations of the reinforcing fibers and the surrounding matrix. In cross-ply laminates, residual macrostresses result from the anisotropic thermal deformations of adjacent plies (Daniel and Liber, 1977). Additional residual stress in polymer-matrix composites may be caused by chemical shrinkage of the matrix during cure (White and Hahn, 1992) and moisture absorption.

Measuring residual stress in composites is difficult because the stresses of interest are often located below the surface of the material, requiring the use of destructive techniques or inconvenient measurement methods, such as X-ray diffraction, neutron diffraction, and photoelasticity. The residual strain in a ply, ε_R, is defined as the difference between the total strain, ε_{Tot}, in the ply developed during the cure and the sum of the chemical and thermal shrinkage strains, ε_C and ε_T, respectively, that the ply would experience if it were not constrained by adjacent plies, i.e.,

$$\varepsilon_R = \varepsilon_{Tot}, -\{\varepsilon_C + \varepsilon_T\}, \tag{9.21}$$

In general, the strains indicated in this equation will contain six components, corresponding to three normal components and three shear components.

The goal of the experiments undertaken by Lawrence *et al.* (1997c) was to determine the residual macrostrains and macrostresses in an AS4/3501-6 cross-ply composite laminate, with a $[0_5/90_5]_s$ construction. In order to determine the residual strain defined in Eq. (9.21), the total strain and the nonmechanical strains generated by chemical and thermal shrinkage need to be measured. The total strains can be measured from the laminate directly with embedded EFPI fiber optic sensors manufactured by Fiso Technologies, Quebec. The nonmechanical strains, however, cannot be determined directly from the laminate. Recall that the nonmechanical strains in a ply are the strains that would occur if the ply were free to expand or contract without the restraint provided by adjacent plies. In order to measure this strain, Lawrence *et al.* (1997c) used a unidirectional laminate $[0_{20}]$ of the same material and subjected it to an identical cure cycle. In theory, all of the strains measured by sensors embedded in this unidirectional laminate should be nonmechanical ($\varepsilon_C + \varepsilon_T$,) since the plies are not restrained from deformation by surrounding plies.

Figure 9.109 shows a schematic of the $[0_5/90_5]_s$ laminate indicating the location of the thermocouple and fiber optic sensors as well as a graph of the temperature measured by the thermocouple during the cure. In Fig. 9.110, the graph on the left presents the total strain ε_{Tot} in the 0^0 direction measured by EFPI-1 and the graph on the right shows the total strain in the 90° direction measured by EFPI-2. Fiber optic sensors EFPI-1 and EFPI-2 were embedded in different layers of the laminate so that both could be embedded parallel to reinforcing fibers. Both sensors indicated a gradual shrinkage of the laminate during the first two phases of the cure cycle, which can be attributed to chemical shrinkage. Also, both sensors indicated compressive strains as the composite was cooled to room temperature. Because of the symmetry of the laminate, we would expect the strains measured by the two sensors to be equal; however, we can see that the values differ. The final value of strain in the 0^0 direction, (−450 $\mu\varepsilon$) is less than the final value in the 90° direction, (−570 $\mu\varepsilon$). Possible explanations for this difference include errors in alignment and variation in adhesion of the sensors, errors in sensor calibration, or strain and temperature gradients in the panel.

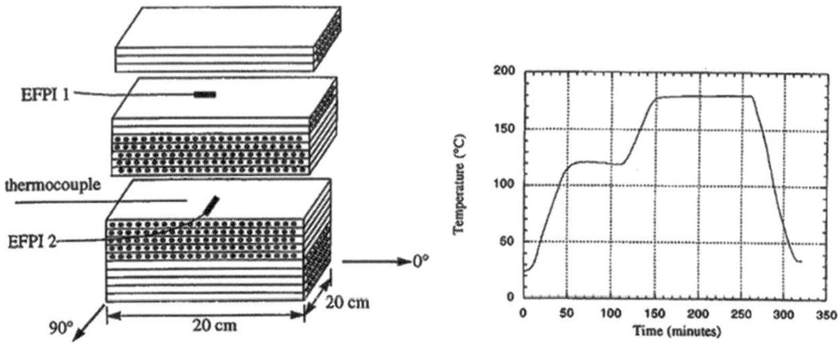

FIGURE 9.109. Temperal variation of temperature as measured by a thermocouple embedded with two EFPI sensors within a multilaminate FRP composite material specimen during the cure process. The schematic reveals sensor location and orientation. From Lawrence, C.M., Nelson, D.V., Bennett, T.E. and Springarn, J.R., (1997c), "Determination of Process-Induced Residual Stress in Composite Materials using Embedded Fiber Optic Sensors," *SPIE* **3042**, 154–167.

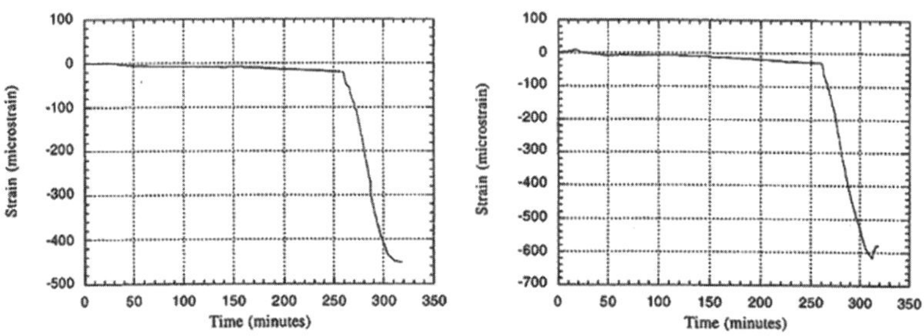

FIGURE 9.110. Total strain variations recorded by the two EFPI sensors during the cure of the multilaminated FRP composite structure shown in Fig. 9.117. From Lawrence, C.M., Nelson, D.V., Bennett, T.E. and Springarn, J.R., (1997c), "Determination of Process-Induced Residual Stress in Composite Materials using Embedded Fiber Optic Sensors," *SPIE* **3042**, 154–167.

In conclusion, the measurement of process-induced residual macro-stresses in composite materials was undertaken using embedded fiber optic strain sensors and thermocouples. This method offers the advantage of nondestructive determination of residual stress and can account for the effects of chemical and thermal shrinkage, as well as viscoelastic stress relaxation. Residual stresses are monitored in real time during cure, allowing this information to be used for process monitoring and control.

CHAPTER 10

Long Gauge-Length Fiber Optic Sensing

10.1 INTRODUCTION

For many structural monitoring applications, a *long gauge-length sensor* (defined here as length ≥ 5 cm) is required in order to accomplish the measurement task. This can be the result of inhomogeneity of the material creating local strain variations. Concrete is a good example, as it can sometimes have a granularity on the scale of 10 mm. This necessitates a measurement base of $\sim$50 mm when macroscopic measurements of the strain within such concrete are required (Inaudi, 1997b). In some sense strain can be thought of as a local parameter, relating the incremental change in length to some base length of a structure. *Deformation*, on the other hand, refers to shape variation of a structure and is more representative of a global change in that structure. The measurement base for monitoring deformation will generally be meters, or even hundreds of meters. Of course, deformation is just the integration of strain over the measurement base.

Fiber optic "long" gauge-length sensors can be divided into two kinds: *extensometers*, in which an optical fiber sensor is attached to a measurement device (often a rod or tube) at only two points (see Fig. 10.1a); and *structurally integrated fiber optic long gauge-length* (SIFOLG) sensors which are strain coupled (i.e., bonded) to the host structure, or an intermediary structure, all along the optical fiber between two selected points (see Fig. 10.1b). In the extensometer, the separation of these two points defines the gauge length of the sensor and the length of the actual sensing region is irrelevant provided it is smaller than this gauge length. In the case of the SIFOLG sensor, what is measured is the integrated strain, or the net displacement, between the two selected points.

The extensometer is more limited in terms of the types of structure it can monitor, for it is invariably straight. Furthermore, if the device is to measure compressive and tensile strain, the optical fiber is required to

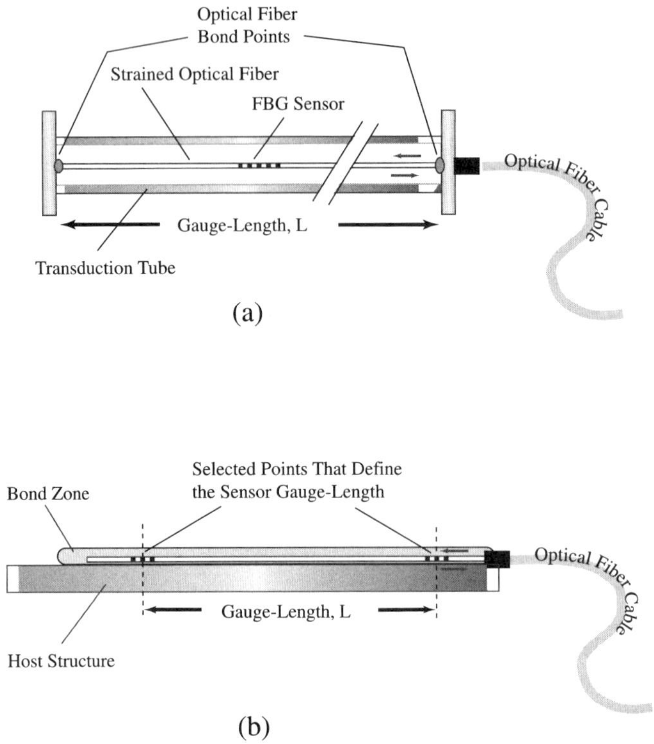

FIGURE 10.1. (a) A long gauge-length fiber optic strain sensor based on the use of a transduction tube and a tensile loaded optical fiber sensor, supported at only two points. (b) A long gauge-length fiber optic strain sensor that is bonded along its entire length so it responds to the strain integrated between the two points that define its gauge-length.

exist in a state of permanent tensile strain equal to the largest compressive strain it is ever likely to experience. As was pointed out in Section 5.6, this is not desirable if the fiber optic monitoring element is to have a long functional life. In contrast, the SIFOLG sensor can be bonded either directly to the host structure, or to an intermediary structure. The latter serves as a transduction device for the sensor. In both instances the sensor can respond to either compressive or tensile strains without the need for a tensile strain bias. The SIFOLG sensor can follow the shape of its host structure and be used to measure, for example, the hoop strain in an axially loaded column.

Deformation and displacement measurements are valuable in the case of structures that show dimensional stability. In this type of structure the load is well below the material failure limit and local strain measurements are of little interest. However, the functionality of the structure could be adversely affected by deformations that structurally have no significance. For example, the comfort of passengers traveling over a bridge in a bus might be dramatically affected by a deformation that actually reduces the strain in the bridge.

In space, weight considerations make it is quite challenging to build large structures that are rigid enough to maintain their desired shape. This is especially true for satellite antennas, which often are quite large and need to conform to a critically designed shape. The problem is exacerbated when there is a requirement to change the pointing direction, for movement of the structure can set up oscillations that then have to be monitored and damped out. *Active control* can be used for this purpose and to maintain the proper shape, but to function properly, it is highly dependent on adequate spatial deformation information.

Absolute strain measurements relative to the unstrained state are important for establishing the loading state of a structure. If a strain sensor is installed on an existing loaded structure, it can only measure the change in strain from that time on. This could be of little value in assessing how close the structure is to its failure limit, if the structure is heavily loaded. This could arise in the case of a bridge, for example, which if poorly designed could be under substantial strain just from its own weight. Unfortunately, there are a number of case histories of overloaded structures; see Chapter 2. To determine the absolute strain experienced by a structure, it is necessary to install the sensors while it is unstrained, or at least when its strain state is predictable. Long-term structural monitoring requires that the sensing system be free as possible of drifts in the measurement parameters. Dynamic measurements, on the other hand, require an adequate time response, and drifts are generally of less concern.

Deformation monitoring has in the past relied on a variety of techniques, such as triangulation, dial gauges, and invar wires, that can only work on the outside of the structure. One of the advantages offered by fiber optics is the ability to integrate the monitoring system into the structure, making it much more convenient and efficient to use: Just call up the instrument interface on your computer. It is also possible, in certain instances, to have a temporary measurement patch that can be applied to the structure as needed.

Inaudi *et al.* (1995) indicated that long gauge-length fiber optic sensors could play an important role in monitoring underground structures such as tunnels, foundations, piles, mines, and anchored walls, for unlike bridges these depend on environmental parameters that cannot always be taken into account in the design stage. Such structures are often instrumented by conventional monitoring systems and could therefore really benefit from the advantages provided by fiber optic sensors. Anchorage cables used to stabilize geological features, rock walls, slurry walls, or soil are also often monitored because ground movement can either reduce their tension (and efficiency) or increase it, inducing possible failure. The required strain resolution is ~10 $\mu\varepsilon$ over a gauge length (measurement base) of between 1 and 50 m.

The refurbishment of many bridges constructed 20, 30, and 40 years ago is urgently required because of degenerative processes such as steel corrosion and carbonation, use of poor materials, and increased loads and amount of traffic. A new method of strengthening and rehabilitating corroded concrete columns is to use FRP wraps. However, the steel reinforcements embedded within the concrete can continue to undergo corrosion, so monitoring of the hoop stresses in the wrap by a long gauge-length fiber optic sensor could indicate the degree of this ongoing degradation.

10.2 LONG GAUGE-LENGTH SENSORS AND THEIR DEMODULATION

It is clear that there are numerous applications for long gauge-length fiber optic strain sensors. In this section we shall review the basic forms of fiber optic long gauge-length sensors and the demodulation approaches used to read them. The methods of demodulation range from simple time-of-flight measurements of short light pulses to low-coherence interferometry. A number of more sophisticated techniques involving tunable lasers and chirped Bragg gratings have been demonstrated. The third section in this chapter will provide a number of examples of applications using this technology.

10.2.1 Fiber optic extensometers and their demodulation

As shown in Fig. 10.1a, an extensometer can use a short gauge-length fiber optic strain sensor to measure the displacement of two points on a transduction rod, or tube, and defined as the gaugelength of the sensor. The optical fiber containing the sensor is held under tension and bonded at only these two points. As indicated earlier, commercial versions of such extensometers, suitable for embodiment within concrete, are available from ElectroPhotonics and Roctest. Two illustrations of a similar device that has been developed at the Optical Measurement Laboratory of Cea-Leti, France (courtesy P. Ferdinand) are shown in Fig. 10.2. This form of extensometer uses a central metallic transduction rod to anchor the optical fiber and comes in 0.1 m and 0.2 m base lengths. Protection is provided by a waterproof tube that surrounds both the optical fiber and the transduction rod.

Demodulation of the sensor that constitutes the extensometer is exactly the same as previously described in Chapter 9, whichever kind of fiber optic sensor is used within the extensometer. In the case of the Cea-Leti device, an FBG is the sensor of choice (Dewynter-Marty *et al.*, 1997). The sensing system is schematically illustrated in Fig. 10.3. An SDL at 1.3 μm is used as a source and a high finesse scanning Fabry–Perot cavity is used to read its spectrum after transmission through a number of FBGs. The system has close to 9 pm spectral resolution, or approximately 10 με strain resolution. The 40-nm bandwidth of the source allows some degree of wavelength division multiplexing, which includes reference FBGs and a temperature FBG. The measurements attained with this instrumentation are discussed in Section 10.3.

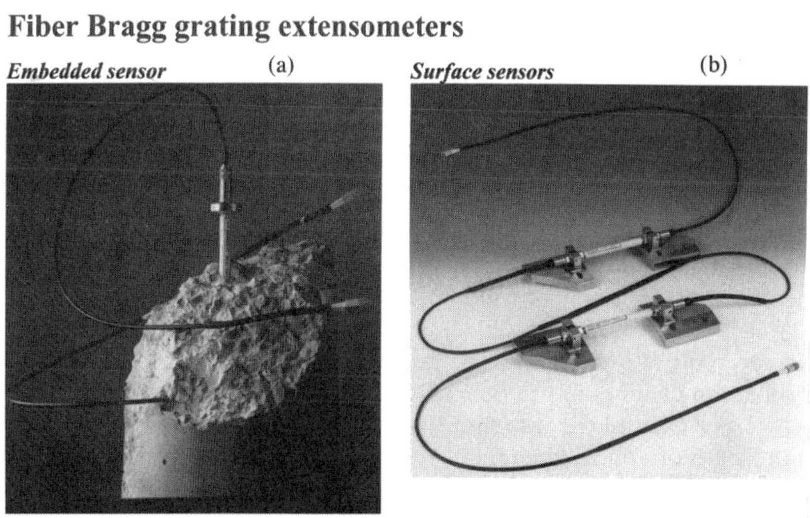

FIGURE 10.2. Photograph showing (a) an embedded fiber optic Bragg grating extensometer and (b) two surface-mounted versions. Courtesy of P. Ferdinand, Cea-Leti, France.

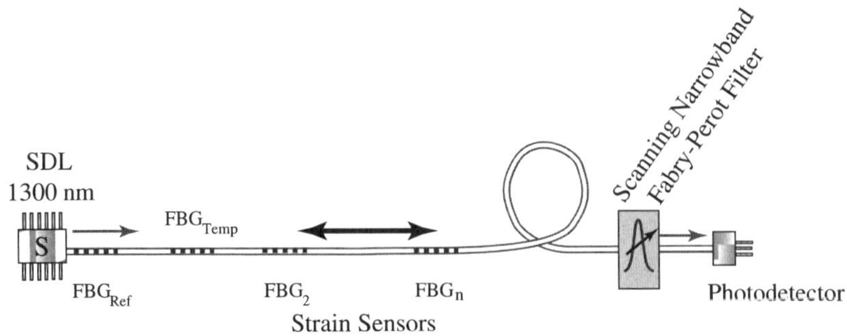

FIGURE 10.3. A fiber Bragg grating based extensometer in which a narrowband tunable Fabry–Perot filter is used to measure the spectrum of the broadband SLD after transmission through the set of FBGs.

10.2.2 Structurally integrated long gauge-length fiber optic strain sensors

The structurally integrated fiber optic long gauge-length (SIFOLG) sensor represents the primary method of undertaking strain measurements for large strutures and can be, quite literally, more flexible than the extensometer approach discussed above. Where the optical fiber is strain-coupled directly to the host structure, it can conform to the surface shape of the structure, or it can be embedded within the structure. In principle, this type of sensor can have an arbitrary gauge length. In practice it is limited by the method of demodulation.

Optical Time Domain Reflectometer. Conceptually, the simplest way to measure the changes in length of a section of an optical fiber is to time the flight of a very short pulse of light between the two selected points (Zimmermann *et al.*, 1989; Zimmermann and Claus, 1993). These could be mirrored splices such as used to create the FFP sensors described in Chapter 9 (Lee and Taylor, 1988; and Hogg *et al.*, 1991a). Unfortunately, this can require extremely short pulse lasers and very fast electronics if small strain changes are to be resolved even for gauge lengths of several meters. For example, if we want to achieve 10 $\mu\varepsilon$ strain resolution with a 10-m length of optical fiber, then we need to be able to measure a changes in length of $\sim$100 μm. If this is to be accomplished by a time-of-flight measurement, a laser pulse of about 0.5 ps duration ($t = nL/c$) is required in an optical fiber with an index of refraction $n = 1.5$. At present this would take a very costly *optical time domain reflectometer* (OTDR; see Chapter 5), or the performance would have to be considerably relaxed.

As we shall see in Chapters 12 and 13, new developments in *mode-locked fiber lasers* and nonlinear detection techniques could lower the cost of this approach.

Low-Coherence Interferometry. Currently, it is much less expensive to measure such small changes in the length of an optical fiber by means of polarimetry, modal interferometry, coherent interferometry, or low-coherence interferometry. The last has an important advantage over the other approaches, in that it enables absolute measurements to be undertaken. Since an absolute measurement is interrupt immune, such measurements can be undertaken at what ever frequency interval is appropriate for checking on the status of the structure, or acquiring its load history. Experience has proven that the *low-coherence Michelson interferometer* is well suited for this task, and a number of studies have demonstrated good results with a this form of long gauge-length fiber optic strain sensor (Inaudi *et al.*, 1994; Elamari *et al.*, 1995).

In Section 3.8, *low coherence interferometry* was introduced in terms of a bulk-optic Michelson interferometer and in Section 9.4 (Fig. 9.99), the low coherence short gauge-length EFPI sensor was discussed. In Fig. 10.4a a simplified illustration of the principle behind the low coherence fiber optic Michelson interferometer is presented in terms of the *coherence wavepacket* (CWP) concept introduced in Chapter 3. If the *Michelson path imbalance*, $2L$, is greater than the *coherence length* of the source, L_c, then the CWPs, S_1 and S_2, reflected from the two end mirrors do not overlap, and no interference is observed at the photodetector. Conversely, if the Michelson path imbalance is *less* than the coherence length of the source, then S_1 and S_2 do overlap, and interference is observed at the photodetector.

In the case of the dual Michelson interferometers (Fig. 10.4b) that constitute the sensing system of Inaudi *et al.* (1994), the path imbalance L_R of the reference Michelson interferometer is adjusted, by means of a translatable mirror, until it just matches the path imbalance, L_S, of the sensing Michelson interferometer, i.e., $L_R = L_S$; see Fig. 10.4b. At this point, interference is observed by the photodetector, and if L_R, is determined from the actuator system that translates the mirror, then the sensing length, L_S, is also known. Note that interference is observed when $L_R = L_S$, as the CWPs A_1R_2 and A_2R_1 overlap.

Smartec, Switzerland, has developed a commercial system, named SOFO, based on this approach, and its systems have been used extensively in that country. A schematic illustration of their system is depicted in Fig. 10.5, and a photograph of an actual system at a field site is shown as Fig. 10.6. The small inset to Fig. 10.5 provides a typical three-peak optical signal resulting from a scan of the mobile mirror in the reference

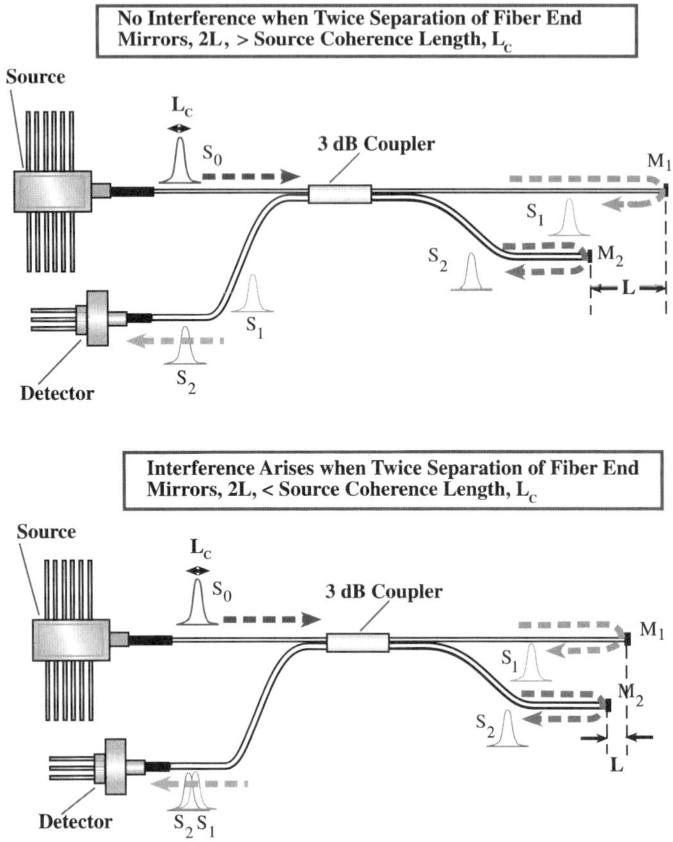

(a)

FIGURE 10.4. Schematic illustrations used to explain low-coherence interferometry in terms of coherence wavepackets for the case of a single fiber optic Michelson interferometer (a) and a pair of Michelson interferometers, (b) where one is used for sensing and the other serves as a reference. The inset displays the variation in the photodetector signal as a function of the path imbalance in the reference interferometer. A peak arises when the path imbalance of the two interferometers matches.

Michelson interferometer. The sensing optical fiber is held under tension between two anchor points within the tube, ensuring that the sensor experiences the same integrated strain as the transduction tube. As indicated in Chapter 6, the Michelson interferometer suffers from a major weakness: The total length of both optical fibers between the coupler and the mirrored ends serves as the interferometer. This means that nonmeasurand influences, such as changes in the shape of either optical fiber, local temperature changes, or transverse pressure points, are

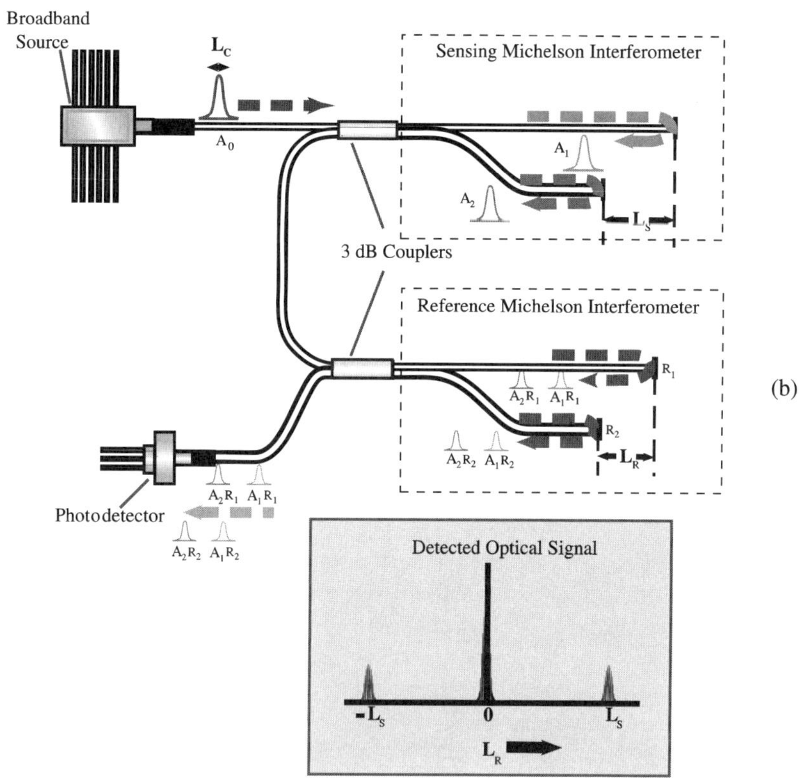

FIGURE 10.4. *(contined)*

indistinguishable from measurand-induced changes. Vurpillot *et al.* (1996) overcame this problem by placing both the sensing and reference optical fibers that constitute the sensing Michelson interferometer and the coupler within a single tube. By strain decoupling the reference optical fiber from this tube, it also provides temperature compensation; see Fig. 10.5. The SOFO system permits the measurement of displacement on the order of 10 μm for gauge lengths of 25 cm to 10 m, with a stability that has been tested over 2 years.

Although the method developed by Inaudi *et al.* (1994) for overcoming the inherent weakness of the Michelson interferometer (i.e., integrating two optical fibers and their coupler within a fairly rigid tube) makes the measurement system quite robust and relatively easy to use in field sites, it also limits the range of application for which this approach can be used. Indeed, this form of SIFOLG sensing system is

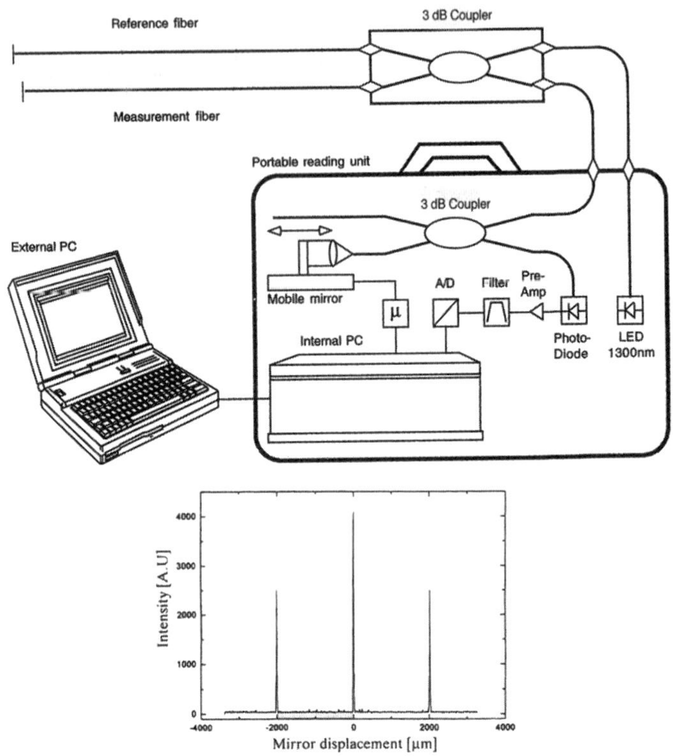

FIGURE 10.5. Basic layout of the SOFO long gauge-length sensor system, which uses a reference Michelson interferometer to interrogate a sensing Michelson interferometer. The path imbalance of the former is controlled by a translatable mirror. The lower autocorrelation spectrum reveals that the path imbalance of the sensing interferometer in this instance was about 2000 μm. From Inaudi *et al.* (1997b). Top part from Inaudi *et al.*, (1988), "Structural Monitoring by Curvature Analysis Using Interferometric Fiber Optic Sensors," *Smart Materials and Structures*, **7**, 1998, 199–208. Lower part from Inaudi *et al.*, (1995b), "Low Coherence Fiber Optic Sensors for Structural Monitoring," *Structural Engineering International* 1/95, **5**(1), 43–47.

restricted in a manner similar to an extensometer in the applications for which it can be used. We shall look at some of the applications of this technology in the next section and see that for measuring the deformation of large concrete structures, this approach works well.

Members of the author's laboratory recognized that *localization* of the long gauge-length sensor would make it more versatile and capable of being used for a wider range of applications, such as the new field of FRP patching and wrapping for the rehabilitation and strengthening of

FIGURE 10.6. Photograph of the SOFO long gauge-length sensor demodulation system at a field site. Courtesy of D. Inaudi, Smartec.

concrete structures. This localization was accomplished by using a Fabry–Perot interferometer, rather than a Michelson, for the sensing system (Fan *et al.*, 1997).

A simplified illustration of the configuration and principle based again on the coherence wavepacket concept is presented as Fig. 10.7. As can be seen from this figure, the reference interferometer took the form of a fiber optic Michelson interferometer, as opposed to the bulk-optic version used by Inaudi *et al.* (1994). This made the demodulation system more compact and rugged, but requires stretching the reference optical fiber, rather than translating a mirror to match the optical paths of the sensing and reference interferometers. In this configuration, stretching the optical fiber in the reference Michelson interferometer varies L_R, and as shown in the inset to Fig. 10.7, interference is observed when $L_R = \pm L_S$, or $L_R = 0$. The latter situation is of little interest, as it conveys no information about the sensor. Note that when the path imbalance in the reference Michelson interferometer matches closely the gauge length of the sensing Fabry–Perot interferometer (i.e., $L_R = L_S$) the CWPs A_2R_1 and A_1R_2, seen in Fig. 10.7, overlap.

The pair of *in-fiber reflectors* required to form a fiber optic sensing Fabry–Perot cavity can take the form of reflective fusion splices (Volanthen *et al.*, 1996d), as in the case of the short gauge-length strain sensor (see Section 9.4.1), or they can be short (0.5-mm) fiber Bragg gratings (Fan *et al.*, 1998). A schematic illustration of the experimental

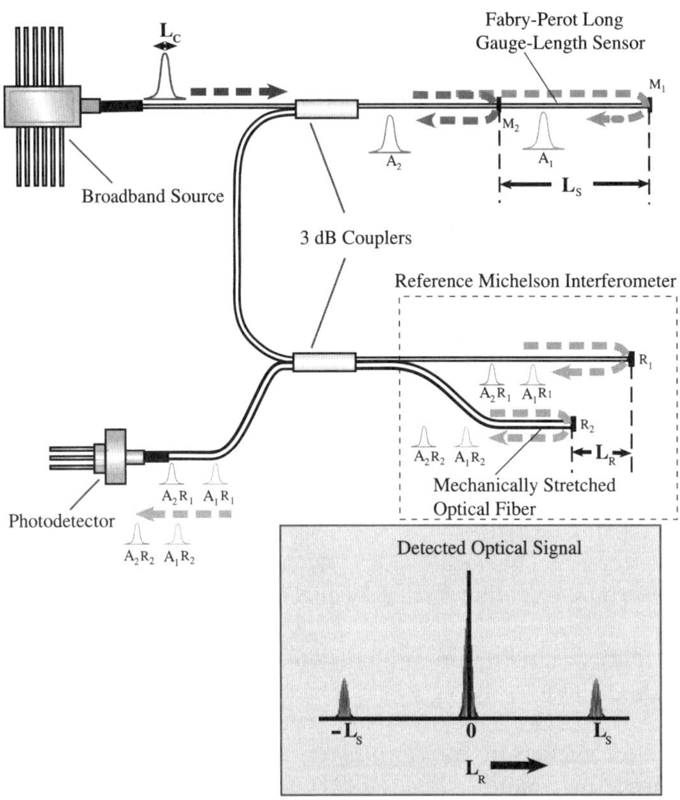

FIGURE 10.7. Schematic illustration of the localized, long gauge-length fiber optic Fabry–Perot strain sensor, using two identical short-length FBGs to define the FP cavity. A reference Michelson interferometer is again used to read the Fabry–Perot sensor by path imbalance matching.

system of Fan *et al.* (1998) is depicted in Fig. 10.8. When two FBGs are used, it is important that they be kept in the same strain and temperature environment. If not, the overlap of their spectra will diminish and the system performance will degrade. In circular geometries this can be avoided by locating the two FBGs very close to each other (see Fig. 10.9). As indicated earlier, a major advantage of the low-coherence approach to reading the Fabry–Perot interferometer is that it provides an *absolute measurement*. This is interrupt immune, so measurements can be undertaken at what ever time interval is relevant. Furthermore, as we shall see in Chapter 11, Sorin and Baney (1995) demonstrated that FBG-based Fabry–Perot interferometric sensors enabled serial multiplexing of a set of such long gauge-length sensors.

Since a broadband LED source was used in the experiment of Fan *et al.* (1998), it is the much narrower spectral width of the FBG that essentially defines the bandwidth of the measurement system. This in turn determines the width of the coherence wavepackets and the precision of measurement. Equation (3.58) indicates that the length of the coherence wavepacket is inversely proportional to the system effective linewidth. The reflection spectrum of the FBG used by Fan *et al.* (1998) is displayed as

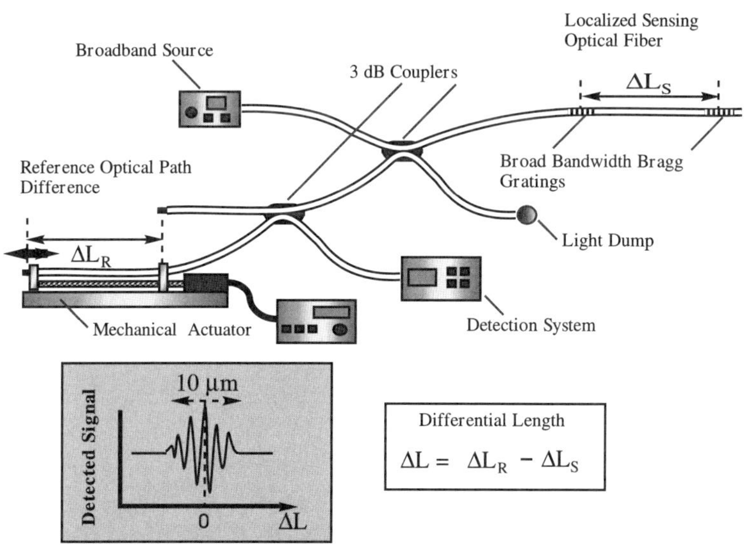

FIGURE 10.8. Localized, long gauge-length Fabry–Perot strain sensor using two broadband FBGs to define the sensor gauge length. The inset indicates the variation in the detected signal as a function of the differential length between the FP cavity length (sensor gauge length) and the imbalance in the Michelson interferometer.

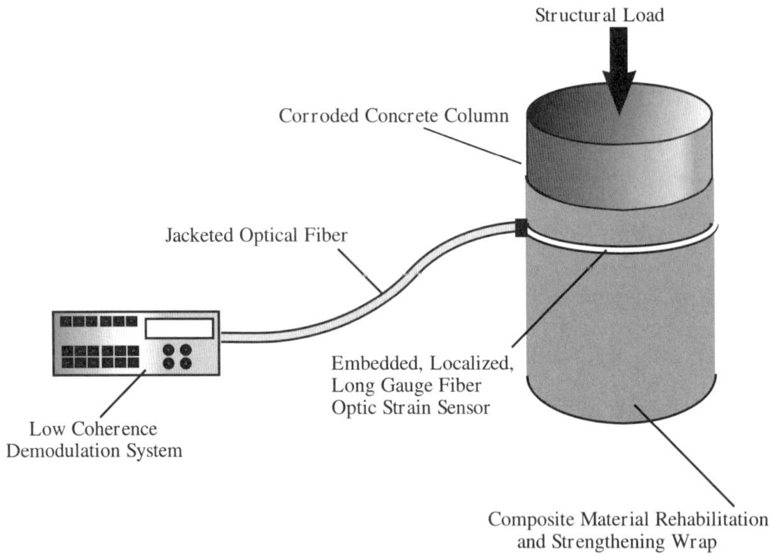

FIGURE 10.9. Schematic illustration of the localized long gauge-length fiber optic strain sensor embedded within a FRP wrap used to rehabilitate and strengthen a concrete structure.

Fig. 10.10. The FBG linewidth can be seen to be approximately 1 nm. The reference optical fiber is mounted on a translational stage and driven by a Newport linear actuator. The actuator is controlled by a central electronics unit, which also evaluates the length change of the reference optical fiber from an encoding system on the actuator. A measurement precision of about ±1.5 μm was achieved by noting the reading of the central electro-

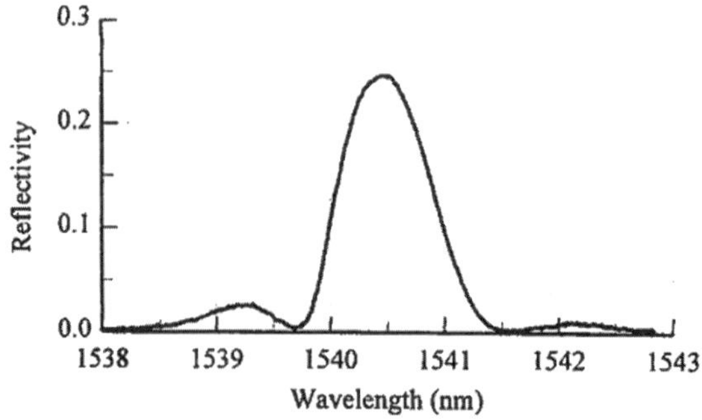

FIGURE 10.10. Reflection spectrum for one of the broadband, short FBGs used as an intrinsic cavity mirror for the long-gauge Fabry–Perot sensor. From Fan N.Y., Huang S., Measures, R., (1998), "Localized Long Gauge Fiber Optic Strain Sensors," *Smart Materials and Structures Journal*, 7, 257–264, with permission from the Institute of Physics Publishing Ltd.

nics unit that corresponds to the peak of the interference during the stretching of the reference optical fiber. The variation of strain, measured by this means, is seen in Fig. 10.11 to be a linear function of the displacement of the optical fiber, as measured by a micrometer.

Tunable Laser Demodulation. Although the experimental system of Fan *et al.* (1998), shown in Fig. 10.8, has been used in a number of applications to be discussed in the next section, it is not as portable or robust as desired by field-site workers, primarily because of the mechanical translational and actuation system. Fan *et al.* (1997) developed and demonstrated a purely optoelectronic approach that avoids the disadvantage of using moving parts and has the potential to be made into a very compact and portable system (Fig. 10.12). In this scheme the wavelength of a tunable laser diode is scanned over a wavelength interval $\Delta\lambda$, and the light reflected from the combination of a fiber optic sensing Fabry–Perot interferometer and reference Michelson interferometer is low-pass filtered and detected, PD2. The change of phase $\Delta\phi$ recorded by the low-pass filtered photodetector, PD2, is directly proportional to the shift in laser wavelength, $\Delta\lambda$, and the difference in the optical path lengths of the reference and sensing interferometers, $|L_R - L_S|$, i.e.,

$$|\Delta\phi| = \frac{4\pi n \Delta\lambda}{\lambda^2} |L_R - L_S|, \tag{10.1}$$

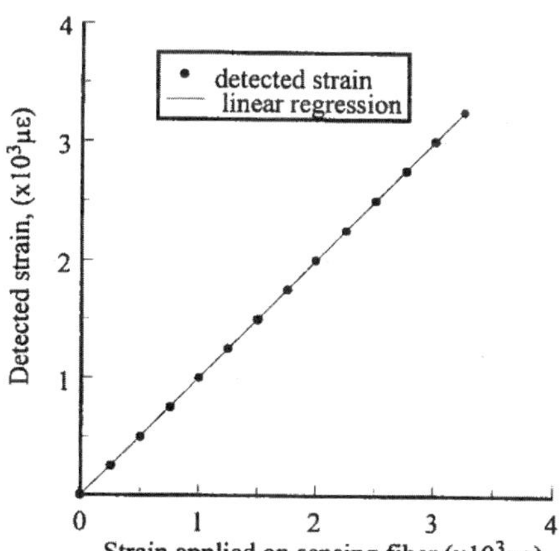

FIGURE 10.11. Variation of the measured strain, using the long-gauge Fabry–Perot sensor, against the applied strain for the experimental system illustrated in Fig. 10.8. From Fan, N.Y., Huang, S., Measures, R., (1998), "Localized Long Gauge Fiber Optic Strain Sensors," *Smart Materials and Structures Journal*, **7**, 257–264, with permission from the Institute of Physics Publishing Ltd.

(a)

(b)

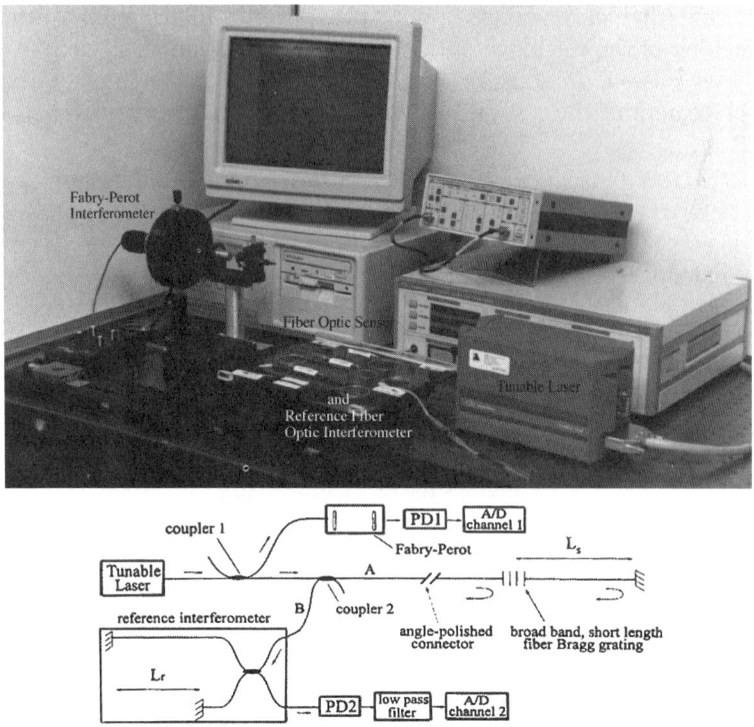

FIGURE 10.12. (a) Photograph of the experimental arrangement used to evaluate the strain in the localized long gauge-length Fabry–Perot sensor based on the change of phase per unit change of laser wavelength. (b) Schematic of system shown in photograph. Adapted from Fan, N.Y., Huang, S.Y., Measures, R.M., (1997), "Localized Long Gauge Length Fiber Optic Sensor Demodulated with Wavelength Tuning Technique," *SPIE* **3042**, 366–371.

where λ is the free-space laser wavelength and n is the index of refraction of the core of the optical fibers, assumed the same. In this experiment a second photodetector, PD1, was used with a high-finesse Fabry–Perot interferometer, comprising one FBG and a mirror end of the optical fiber, to measure the change in laser wavelength. A photograph of the experimental arrangement is presented as Fig. 10.12a and a schematic illustration of the system is displayed as Fig. 10.12b.

Vindication of relation (10.1) is provided in Fig. 10.13, by the linear dependence of the elongation of the sensing optical fiber as measured against the elongation predicted from a resistive foil strain gauge reading. Both the optical fiber sensor and the foil gauge were mounted on a cantilever beam for this experiment. However, as we shall see later in

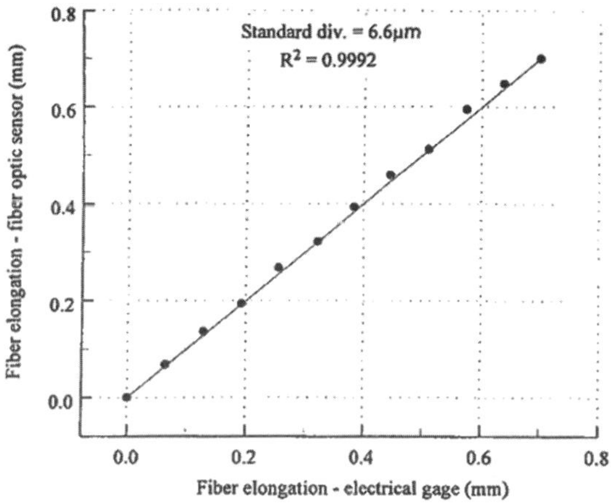

FIGURE 10.13. Comparison of optical fiber elongation measured by means of a resistive foil gauge and the Fabry–Perot sensor based on two FBGs and the tunable laser illustrated in Fig. 10.12. Adapted from Fan, N.Y., Huang, S.Y., Measures, R.M., (1997), "Localized Long Gauge Length Fiber Optic Sensor Demodulated with Wavelength Tuning Technique," *SPIE* **3042**, 366–371.

Chapter 13, Section 13.2, the value of the injection current used to drive the laser diode may be adequate to determine the change in the laser's wavelength, $\Delta\lambda$. This would make a very much simplified and compact demodulation system for the long-gauge fiber optic sensor.

Posey *et al.* (1997) have shown that it is possible to demodulate a Michelson-based long gauge-length fiber optic sensor using a broadband source, a tunable Fabry–Perot filter, and a chirped fiber Bragg grating. In essence, low-coherence interferometry is used with a tandem reference and sensing pair of Michelson fiber optic interferometers. As can be seen from the schematic of the experimental arrangement presented as Fig. 10.14, the *chirped FBG* forms one arm of the reference Michelson interferometer. Since the penetration distance of light within the chirped FBG depends upon its wavelength (see Section 5.9), it is possible to match the path lengths of the two Michelson interferometers by choosing the right wavelength with the tunable narrowband Fabry–Perot filter.

Figure 10.15 displays the normalized reflection profile of the chirped FBG, and Fig. 10.16 the ratio of detector signals versus the tunable Fabry–Perot filter wavelength for sensing fiber strains of 0, 3000, and 6000 $\mu\varepsilon$. The measurement accuracy was estimated to be 50 $\mu\varepsilon$. The electronic system locks onto the peak of the integrated interference pattern (rather than individual fringes) and produces a tracking voltage, which is fed to the

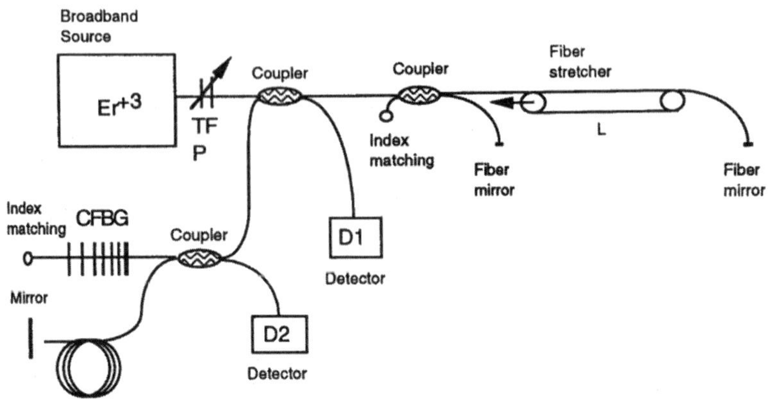

FIGURE 10.14. Experimental arrangement for a long-gauge Fabry–Perot sensor demodulation system based on a broadband source and a chirped FBG. From Posey, R. Jr., Kersey, A.D., LeBlanc, M., and Davis, M., (1997), "Low-Coherence Wavelength-Encoded Addressing for Integrated Long-Gauge Length Fiber Optic Strain Sensing," OSA 12th Int. Conference on Optical Fiber Sensors, Williamsburg, 574–577.

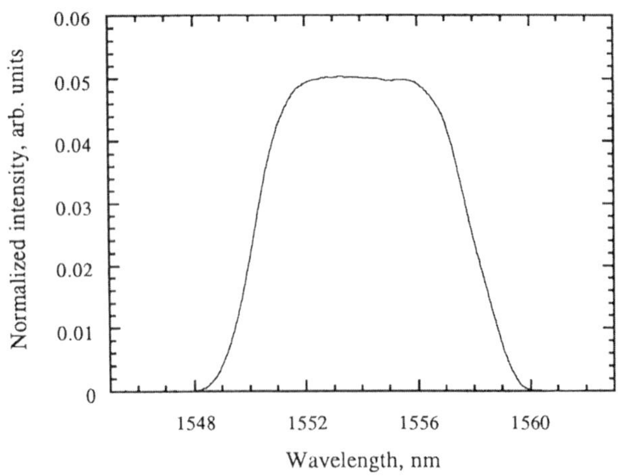

FIGURE 10.15. Normalized reflection spectrum of the chirped FBG used in the demodulation system illustrated in Fig. 10.14. From Posey, R. Jr., Kersey, A.D., LeBlanc, M., and Davis, M., (1997), "Low-Coherence Wavelength-Encoded Addressing for Integrated Long-Gauge Length Fiber Optic Strain Sensing," OSA 12th Int. Conference on Optical Fiber Sensors, Williamsburg, 574–577.

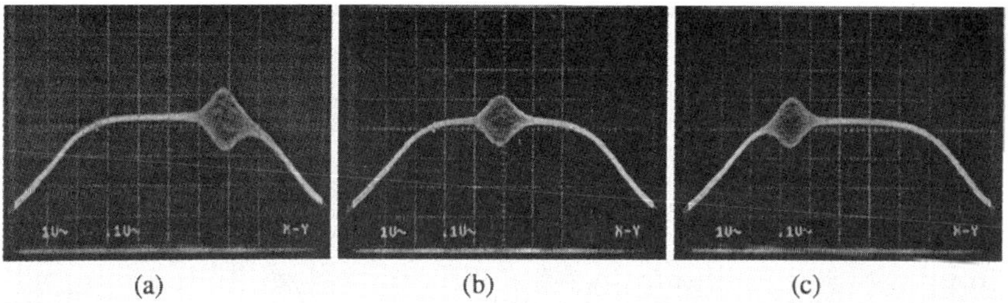

FIGURE 10.16. Ratio of detector signals versus the tunable Fabry–Perot filter wavelength for the measurement of three optical fiber strains0 με (a), 3000 με (b), and 6000 με (c). From Posey, R. Jr., Kersey, A.D., LeBlanc, M., and Davis, M., (1997), "Low-Coherence Wavelength-Encoded Addressing for Integrated Long-Gauge Length Fiber Optic Strain Sensing," OSA 12th Int. Conference on Optical Fiber Sensors, Williamsburg, 574–577.

tunable Fabry–Perot filter to hold the interrogation wavelength corresponding to the optimum optical path matching condition. The linear relationship between the tunable Fabry–Perot filter wavelength and the strain in the sensing optical fiber is seen in Fig. 10.17. This low-coherence wavelength-encoded addressing approach to long gauge-length demodulation avoids the need for a mechanical actuation system and can therefore be lightweight, robust, and compact.

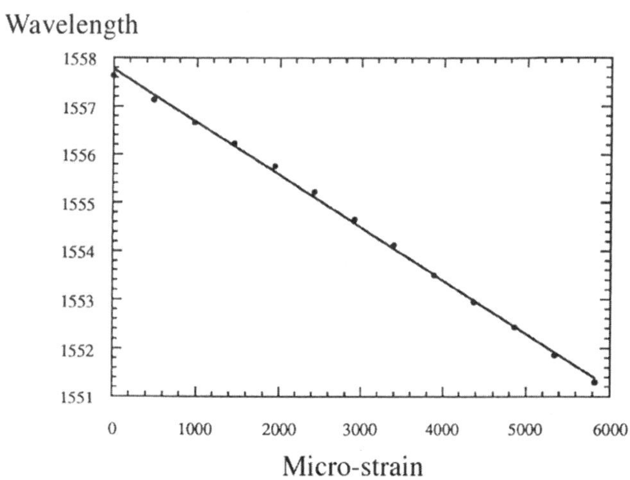

FIGURE 10.17. Variation in the tunable Fabry–Perot filter wavelengths with strain in the sensing optical fiber. From Posey, R. Jr., Kersey, A.D., LeBlanc, M., and Davis, M., (1997), "Low-Coherence Wavelength-Encoded Addressing for Integrated Long-Gauge Length Fiber Optic Strain Sensing," OSA 12th Int. Conference on Optical Fiber Sensors, Williamsburg, 574–577.

10.3 LONG GAUGE-LENGTH SENSOR APPLICATIONS

In many civil structures such as bridges, tunnels, and dams, deformation of the structures represent one of the most important parameters to be monitored. Short gauge-length strain measurements provide only local information about the material behavior. A complete understanding of a structure's response to its loading would require a large array of such short gauge-length sensors. Inaudi *et al.* (1997b) observed that fiber optic deformation sensors with a measurement base of 1 to 10 m can provide valuable structural information both during the construction phase and over the life of the structure. During the period immediately following construction, the long gauge-length sensors can provide valuable data on thermal expansion due to the exothermic setting reaction and the successive thermal and drying shrinkages. In the long term these sensors can also track geometrical deformations and creep of the structure under static loads, especially its dead load.

We will first look at some applications of the fiber optic extensometer developed by Dewynter-Marty *et al.* (1997); then we will review several papers by Inaudi and co-workers as they have spun off a company, called Smartec, which has engineered long gauge-length fiber optic sensing systems and have undertaken a substantial number of field-site demonstrations of this technology. As indicated in the previous section, the use of a Michelson fiber optic configuration limits this technology to measurements undertaken with fairly bulky transduction tubes. Although these are well suited for deformation measurements in large concrete structures, they are not easily adapted for hoop-stress measurements of circular structures, and are too large to embed within FRP composite structures. By contrast, the fiber optic Fabry–Perot-based long gauge-length sensor developed by Fan *et al.* (1998) is very well suited for these types of applications, as this sensor involves a single optical fiber that can readily be bonded to a structure and conform to its shape, or be embedded within FRP composite material structures. Recently, a new company called Fostek has been formed to commercialize this technology.

10.3.1 Applications of fiber optic extensometers

Fiber optic extensometers were introduced in Section 10.2.1 and involve short gauge-length strain sensors, such as FBGs, that are supported under a tensile strain bias within a transduction tube. Two such FBG extensometers (0.2 m in length) were bonded onto the cylindrical surface of concrete columns that were 0.25 m in diameter and 0.5 m long; see Fig. 10.18 (Dewynter-Marty *et al.*, 1997). The concrete column was also

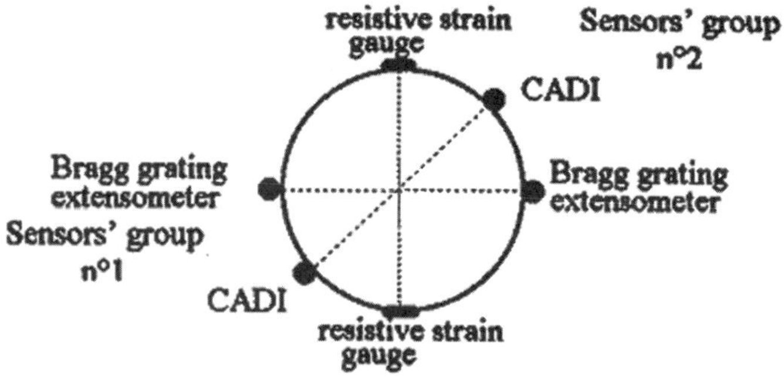

FIGURE 10.18. Siting of two Bragg grating extensometers and other sensors around a concrete column that was axial loaded. From DeWynter-Marty, V., Rougeault, S., Ferdinand, P., (1997), "Concrete Strain Measurements and Crack Detection with Surface-Mounted and Embedded Bragg Grating Extensometers," OSA 12th Int. Conf. on Optical Fiber Sensors, Williamsburg, 600–603.

instrumented with two resistive foil strain gauges and two inductive extensometers (CADI, 0.2 m in length). For the purpose of reporting, the sensors were divided into two groups, as indicated in this figure. The measured strain for each sensor in group 1 is plotted against the compressive axial load applied to the concrete cylinder in Fig. 10.19a, while the results for the second group of sensors are presented as Fig. 10.19b. The FBG extensometers are seen to perform consistently well and in both graphs follow closely the theoretical prediction (the straight line). The repeatability of the FBG measurements was $\pm 4\ \mu\varepsilon$.

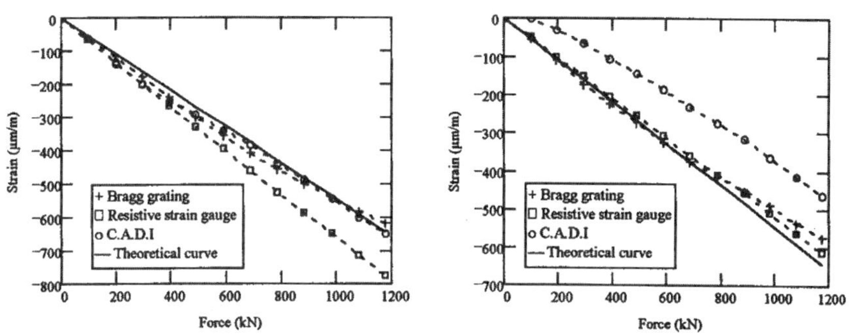

FIGURE 10.19. Variations in strain against applied load measured by several sensors mounted on the cylindrical surface of a concrete column. From DeWynter-Marty, V., Rougeault, S., Ferdinand, P., (1997), "Concrete Strain Measurements and Crack Detection with Surface-Mounted and Embedded Bragg Grating Extensometers," OSA 12th Int. Conf. On Optical Fiber Sensors, Williamsburg, 600–603.

In another set of experiments a concrete floor (1.5 m × 1.0 m × 0.26 m) was constructed with some metal plates included to create weak areas and provide some degree of control over crack formation. One FBG extensometer and two inductive extensometers (CADI) were positioned above the plates, while a second group of sensors was placed over a section of floor that was free of metal plates. The arrangement is depicted in Fig. 10.20 and shown in the accompanying photograph. The floor was then placed in a metal frame and stressed with two hydraulic jacks. Figure 10.21 reveals the dynamic response of the sensors to the loading and formation of cracks. Figure 10.21a presents the results for the region of floor including the metal plate, while Fig. 10.21b show the results for the metal-free part of the floor. The FBG extensometer was observed to have a strain detection threshold of 2 με as compared to ~30 με for the inductive extensometers. The load on the floor was increased in steps of 30 kN and the results indicate that the FBG response is in good agreement with the CADI. These results and others involving additional experiments have led Dewynter-Marty *et al.* (1997) to state that the FBG extensometer represents

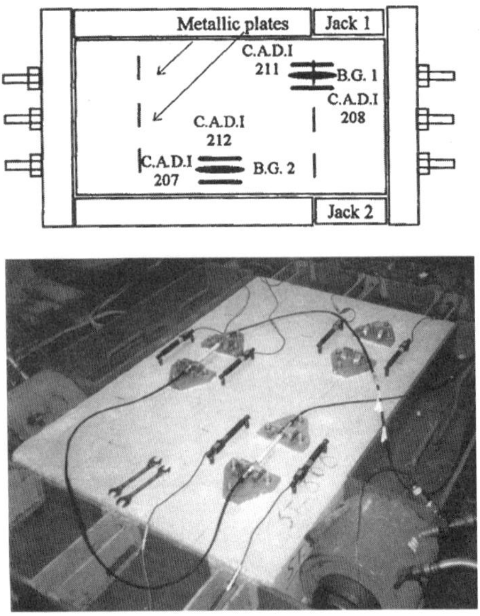

FIGURE 10.20. Photograph and schematic of an array of sensors, including the Bragg grating extensometers, positioned on a concrete test slab. From DeWynter-Marty, V., Rougeault, S., Ferdinand, P., (1997), "Concrete Strain Measurements and Crack Detection with Surface-Mounted and Embedded Bragg Grating Extensometers," OSA 12th Int. Conf. On Optical Fiber Sensors, Williamsburg, 600–603.

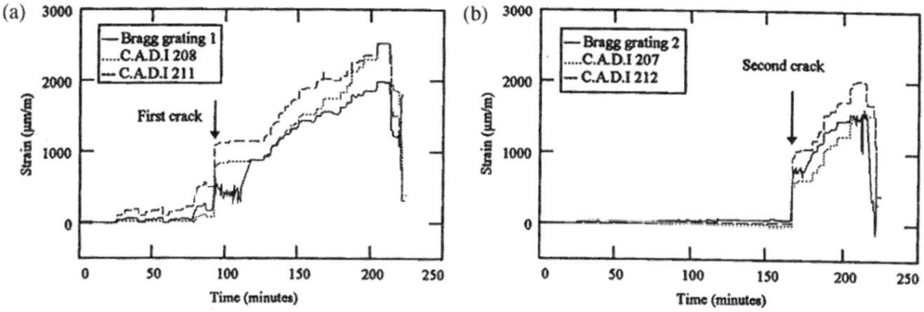

FIGURE 10.21. Comparison of the temporal variation in the strain measured by the different sensors shown in Fig. 10.20 as the test slab is loaded. From DeWynter-Marty, V., Rougeault, S., Ferdinand, P., (1997), "Concrete Strain Measurements and Crack Detection with Surface-Mounted and Embedded Bragg Grating Extensometers," OSA 12th Int. Conf. on Optical Fiber Sensors, Williamsburg, 600–603.

a valuable new tool for measuring compressive and tensile strains, as well as the early detection and monitoring of cracks, in concrete structures.

10.3.2 Applications of the Michelson-based long-gauge sensor

The all-fiber tandem Michelson interferometer developed by Inaudi and co-workers and discussed earlier sometimes goes under the French acronym SOFO (Surveillance d'Ouvrages par Fibres Optiques) for structural monitoring by optical fibers. A schematic illustration of the SOFO system was presented in Fig. 10.5, and a photograph of an actual system at a field site was shown as Fig. 10.6. This structural monitoring system has been used in a substantial number of field sites (the number of sensors installed is indicated in parentheses):

Versoix Highway Bridge	(104)
Venoge Highway Bridge	(40)
Lutrive Highway Bridge	(36)
Lully Highway Viaduct	(12)
Vignes Highway Tunnel	(16)
Luzzone Dam	(13)
Emosson Dam	(2)
EDF Nuclear Power Plant	(20)

Since 1993, almost 600 fiber optic sensors have been installed with a survival rate of $\geq 90\%$ (Inaudi, 1997b). Because the sensors are protected within a tough tube and supplied with a robust and easy-to-use connection, they are easily installed by trained site workers. Special care has to be taken in the design and network cabling used to connect the sensors to the reading unit at the central location. At some sites, such as Versoix and Lutrive bridges and the Luzzone Dam (see Fig. 10.22), an optomechanical switch with a cellular phone and modem are used to automate the connection between the field-site sensors and the distant reading unit.

From the various field-site installations it has been determined that long gauge-length deformation sensors are useful for measuring and monitoring a wide range of phenomena:

- Concrete deformation during setting
- Prestressing
- Neutral axis evolution
- Concrete–steel interaction
- Postseismic damage evaluation
- Spatial displacement (curvature analysis)
- Crack opening
- Creep-flow
- Long-term deformation
- Restoration

The basic design of the SOFO deformation sensor is schematically depicted in Fig. 10.23. It can be directly embedded in concrete, or surface mounted on existing structure (Inaudi *et al.*, 1996a). The active sensing length can be fabricated from 20 cm to 8 m, while the passive leadlength

FIGURE 10.22. Luzzone Dam, Switzerland. Courtesy Smartec, Switzerland.

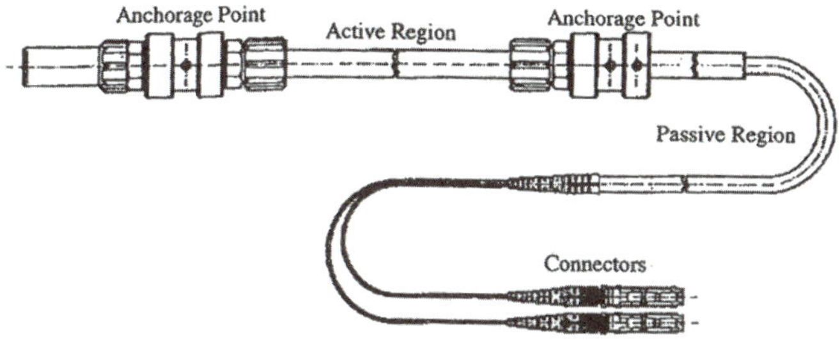

FIGURE 10.23. Schematic of the SOFO deformation sensor as embedded within concrete structures, or mounted on the surface of existing structures. From Inaudi, D., Vurpillot, S., Casanova N., "Bridge monitoring by interferometric deformation sensors" *SPIE Intl. Photonics China Symposium on Laser Optoelectronics & Microphotonics: Fiber Optic Sensors*, Beijing, China, 4–7/11/96, pp. 1–12.

can be up to several kilometers long. Multiplexing of these long gauge-length fiber optic sensors will be touched on briefly in Chapter 11.

1. *Venoge Bridge: Material Testing during Construction and Load Tests.* This four-span highway bridge, near Lausanne, comprises two parallel steel girders (1.0 m by 1.9 m) supporting a 23-cm-thick concrete deck (Fig. 10.24). More than 30 fiber optic sensors, a few wire gauges, and about 24 thermocouples were installed in the deck and on the steel girders to

FIGURE 10.24. Venoge Bridge near Lausanne, Switzerland. Courtesy Smartec, Switzerland.

monitor the behavior of the bridge as it was widened (Inaudi *et al.*, 1996a). While undertaking deformation strain measurements on the Venoge bridge, Vurpillot *et al.* (1996) discovered that it was necessary to use polyimide-coated optical fibers to avoid premature failure when the sensors were loaded to 0.5% strain for about a month. This fracture of the optical fibers arose where the acrylate coating had to be stripped away to provide good strain coupling. The precision of their measurements was also improved to the required value of 0.004% by integrating the coupler into the measurement rod. More details on this technology are provided by Inaudi *et al.* (1996c).

The concrete deck shrinkage measurements were divided into two phases: the first 14 hours, and the first 20 days. The results obtained from eight fiber optic deformation sensors during the first 14 hours followed by the results in the first 20 days after concreting are presented as Fig. 10.25a. The results from two vibrating wire gauges placed in roughly the same location, for the same time periods, are included as Fig. 10.25b. The thermal expansion phase can be seen to give rise to a short-lived

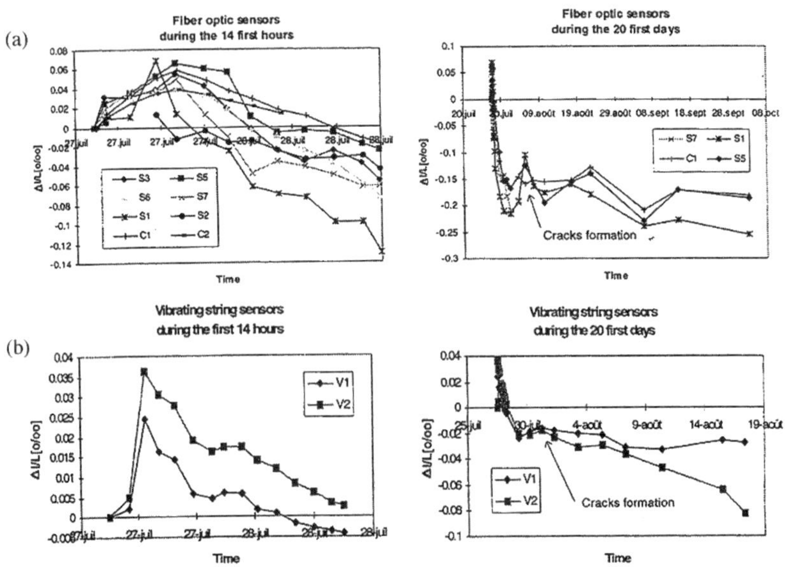

FIGURE 10.25. (a) Variation in the deformation measured over two time periods (first 14 hours, and first 20 days) by a number of long-gauge fiber optic sensors installed into the concrete deck of the Venoge Bridge. (b) Comparable measurements undertaken by vibrating string sensors. Reprinted from *Smart Structures and Materials* 1996, **2719**, "Bridge monitoring by fiber optic deformation sensors: design, emplacement and results," (S. Vurpillot, D. Inaudi and J.M. Ducret) pp. 141–149.

elongation of about 0.06%; then during the subsequent cooling phase the fiber optic sensors register a discontinuity suggesting the formation of cracks about 8 days after concreting (between the 4th and 9th of August). A good correlation was found between the crack-width measured with a magnifying glass and the fiber optic sensor readings. Load tests undertaken with a 25-ton truck verified that the fiber optic sensor deformation readings are repeatable with a error of about 4 μm when compared to an external dial gauge.

2. *Versoix Bridge: Displacement Measurements with More Than 100 Fiber Optic Sensors.* Vurpillot *et al.* (1997) report on the installation of an of array of 104 long gauge-length fiber optic deformation sensors installed in the Versoix Bridge near Geneva (Switzerland). A photograph of this bridge is presented as Fig. 10.26. The desire for monitoring arose from the plan to enlarge this bridge by widening the two prestressed concrete beams and extend the overhangs of this classically designed concrete bridge, see Fig. 10.27a. The engineers involved in this project were interested in the spatial displacements of the bridge and in the effects of concrete shrinkage between the old and new parts of the enlarged bridge. The location of some parts of the fiber optic sensing network is shown in Fig. 10.27b. An overview of the sensor installation in the external over-

FIGURE 10.26. Versoix Bridge near Geneva, Switzerland. Courtesy D. Inaudi, Smartec.

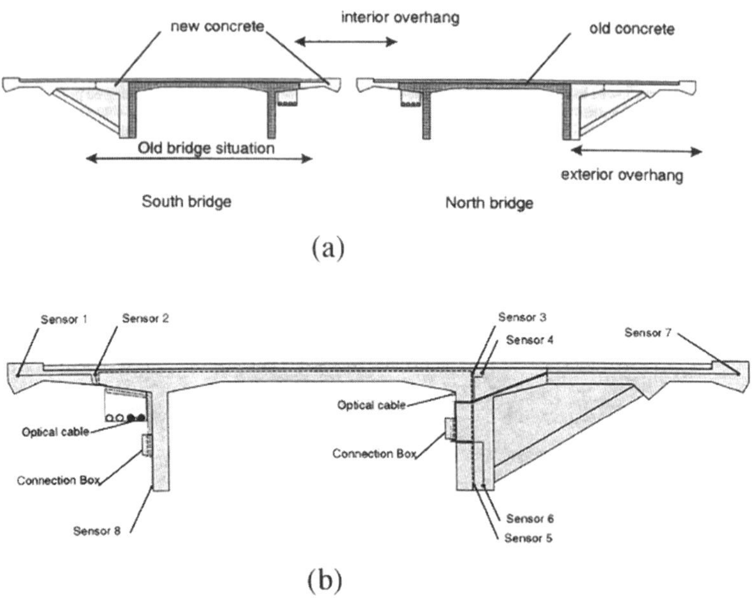

FIGURE 10.27. Schematic diagrams showing the widening extensions to be built on the Versoix Bridge. Reprinted from *Smart Structures and Materials*, 1997, **3043**, "Bridge spatial displacement monitoring with 100 fiber optic sensors deformations: sensors network and preliminary results," (S. Vurpillot, N. Casanova, D. Inaudi, P. Kronenberg) pp. 51–57.

hang section is presented as Fig. 10.28. The inset to this figure shows a fiber optic deformation measurement tube held in place by plastic strings (not fixed) to the steel rebars in this new section of the bridge. The fiber optic deformation sensors used in this project had a gauge length of 4 m and a passive lead length of 2 to 10 m. There were 14 connection boxes that fed the signals to one central box where an optical switch permitted any sensor to be connected to the reading unit. A portable computer and mobile phone are used to connect the system through the Internet.

Figure 10.29 shows some preliminary results for a few of the sensors installed in the interior overhang section. Three stages are discernible in this figure. The first shrinkage results from drying of the concrete; the second corresponds to a stabilization phase; and the third is due to a temperature decrease of 10°C in bridge during November, when the observations were made. Shrinkage of the concrete of about 0.02% is seen to arise in stage 1, while shrinkage of 0.005% was noted in stage 2.

FIGURE 10.28. Photograph of the external extension for Versoix Bridge. *Inset*: A measurement tube including a long-gauge fiber optic deformation sensor placed into the steel rebar frame for this extension. Courtesy of D. Inaudi, Smartec.

3. *Moesa Railway Bridge: Damage Assessment during Construction.* An array of 30 long gauge-length fiber optic strain sensors was used to measure the deformations of the concrete deck of the new Moesa railway bridge (Southern Switzerland) during its setting and shrinkage phases, as well as when it was slid 5 m by hydraulic jacks to replace the old bridge (Inaudi *et al.*, 1997a). This monitoring program avoided the development of excess curvature and resulting cracking in the concrete deck. The 53-cm-thick concrete deck is supported by two continuous steel I-beams that are over 89 m long. These are clearly seen in Fig. 10.30. Because the large mass of concrete and the presence of the steel beams partially hindering the concrete shrinkage, the sensors were important to verify that the

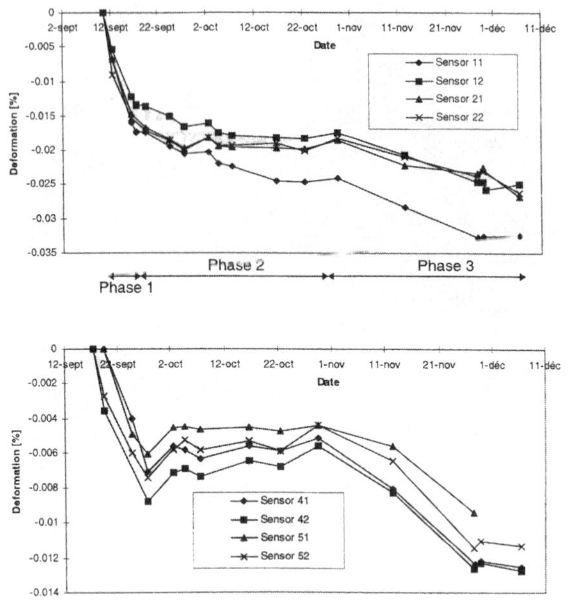

FIGURE 10.29. Fiber optic deformation measurements with time for the interior overhang extension to the Versoix Bridge. Reprinted from *Smart Structures and Materials*, 1997, **3043**, "Bridge spatial displacement monitoring with 100 fiber optic sensors deformations: sensors network and preliminary results," (S. Vurpillot, N. Casanova, D. Inaudi, P. Kronenberg) pp. 51–57.

concrete deck did not suffer cracking during shrinkage and movement of the bridge.

The sensors had gauge lengths of between 2 and 4 m, were placed parallel to the bridge length, and were embedded in the deck, attached to the reinforcement bars, in a manner similar to that shown in Fig. 10.28. Connection boxes, such as the one shown in Fig. 10.31, were used to permit the reading unit to access the sensors. Figure 10.32 shows a typical sensor measurement and reveals the shrinkage experienced by the first concrete section. The jump arising on the 15th of February corresponds to the elongation produced by the additional weight of the concrete used in the second section and the humidity intake due to removal of the external air heating. The sensors were rapidly installed in the rebar cage without any delay in the schedule, and about 90% of the sensors survived the installation and concreting.

The embedded fiber optic sensors, augmented by some surface-mounted sensors, monitored any curvature variations arising from

FIGURE 10.30. Installation of the fiber optic deformation measurement tubes during construction of the concrete deck for the new Moesa railway bridge, Switzerland. From Inaudi, D., Casanova, N., Kronenberg, P. and Vurpilllot, S., (1997a), "Railway Bridge Monitoring During Construction and Sliding. SPIE, *Smart Structures and Materials*, **3043**, San Diego, USA, 58–64.

uneven progression of the four hydraulic jacks during the 6-hour movement of the bridge. The results from two 2-m-long fiber optic sensors placed parallel to the bridge length and close to the second hydraulic jack are presented in Fig. 10.33. The elongation of the bridge is presented with its curvature during the period of pushing the bridge into place (Fig. 10.34; Inaudi *et al.*, 1998). The elongation is primarily due to the heating by the sun and is derived by averaging the left and right sensors (see subsection 6, later), while the curvature is obtained from the difference in the sensor readings divided by the sensor length and their separation. Since the bending was kept below 30% of the value required to crack the concrete deck, no damage was done by sliding the bridge in place.

4. *The Schiffenen Shell-Mass and Emosson Shell Dams.* Monitoring of dams has always been an important part of hydroelectric power management. It is required to guarantee the safety of the structure, optimize utilization, and ensure sound maintenance of the dam. Many existing dams require additional monitoring, either to replace existing systems that no longer provide the structural information required today, or to extend an existing monitoring system to supply information such as crack and joint openings, or displacements of the dam's foundation. There are also new

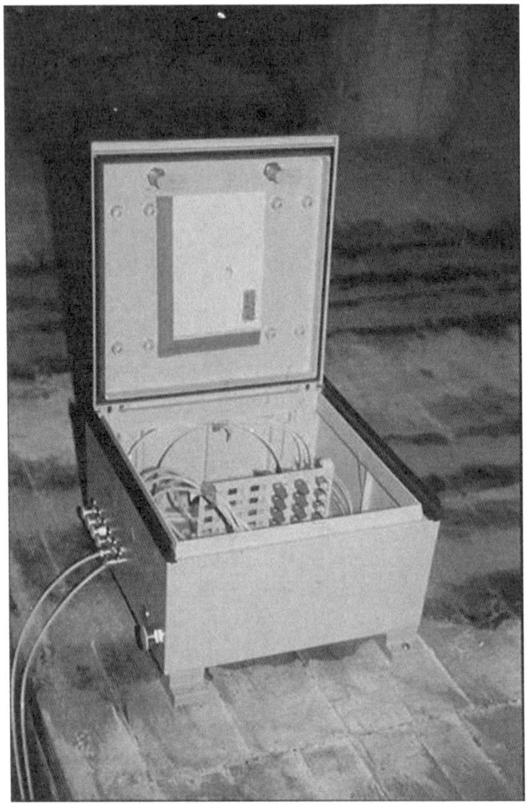

FIGURE 10.31. Connection box for the fiber optic deformation sensing system installed on the Moesa railway bridge. Courtesy of D. Inaudi, Smartec.

structures, or structural upgrades, which require more sophisticated monitoring systems.

Kronenberg *et al.* (1997) report on two major installations of fiber optic sensors in dams. The *Schiffenen Shell-Mass Dam*, near Fribourg in Switzerland, was used for a theoretical study to assess the feasibility of monitoring the internal radial deflection of the structure with a spatial fiber optic deformation sensor network. This dam is 42 m high and 370 m long with a 200-m radius of curvature at the coping. The curvature of the shell was to be evaluated by a pair of standard fiber optic deformation sensors placed at different points. Then, by double integrating this curvature information and introducing the appropriate boundary conditions it is possible to extract the radial deflection of the dam, as was done by Vurpillot *et al.* (1996) in the case of a bridge. Since the Schiffenen dam is an existing

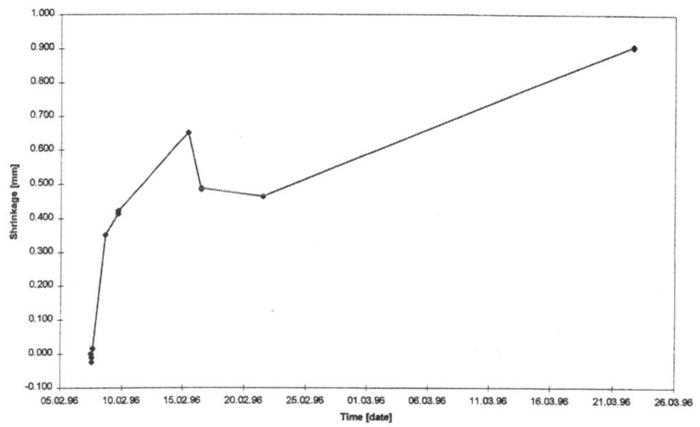

FIGURE 10.32. Measured shrinkage with time, as determined by one of the installed fiber optic sensors, for a concrete section of the Moesa bridge. From Inaudi, D., Casanova, N., Kronenberg, P., and Vurpilllot, S., (1997a), "Railway Bridge Monitoring During Construction and Sliding, SPIE, *Smart Structures and Materials*, **3043**, San Diego, USA, 58–64.

structure, the fiber optic sensor network has to be mounted on the external surface of the concrete. A computer simulation of the dam deflections was used to assess the number of sensors and their best location, with the mounting constraints imposed by the existing dam. The gauge length was determined to be 3 m for the arch measurements and 2 m for the wall. The proposed configuration is illustrated in Fig. 10.35. The reading unit is to be connected to the array of sensors through an optical switch, allowing the complete structure to be monitored without any manual intervention in the dam.

The *Emosson Shell Dam* is a much larger dam that is located in the Swiss Alps. It is 180 m high and close to 550 m long at the coping, where it is 10 m thick. At the base the dam is 35 m thick. The primary goal of this project was to develop two very long fiber optic extensometers with gauge lengths of 30 and 39 m, respectively, to replace a mechanical rockmeter that is anchored into the bedrock within an underground cavity of the dam. A schematic of the fiber optic extensometer and its mounting arrangement is depicted in Fig. 10.36. Within the dam the sensor was designed to be anchored or not by bending a special component. The measurement fiber is in mechanical contact with the structure and given a strain bias ($\sim$0.5%) so that it can follow both elongation and shortening of the structure. The reference fiber is strain decoupled by means of a microtube and positioned along side the measurement fiber to provide temperature compensation.

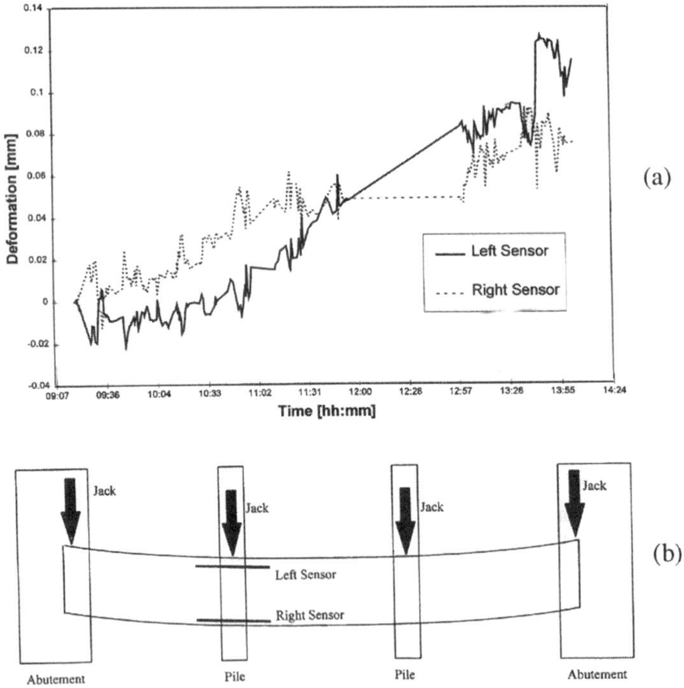

FIGURE 10.33. (a) Variation with time of the deformation measured by 2-m-long fiber optic sensors during the movement of the bridge. (b) Schematic showing relative positions of the two sensors and the hydraulic jacks used to move the bridge. From Inaudi, D., Vurpillot, S., Casanova, N., Kronenberg, P., (1998), "Structural Monitoring by Curvature Analysis Using Interferometric Fiber Optic Sensors," *Smart Materials and Structures Journal*, **7**, 199–208, with permission from the Institute of Physics Publishing Ltd.

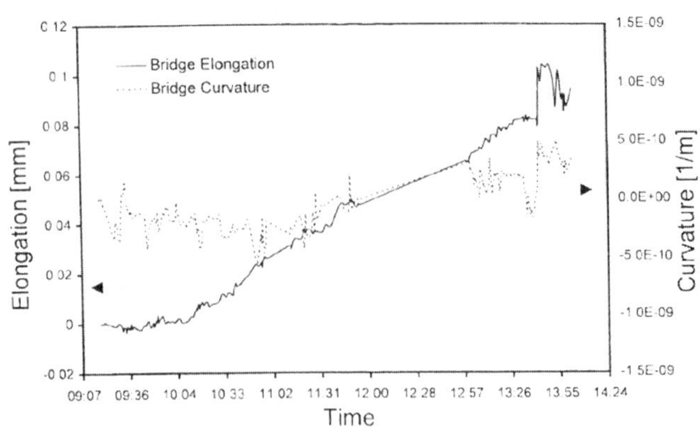

FIGURE 10.34. Fiber optic elongation and curvature measurements of the Moesa bridge while it was being pushed into place with the set of hydraulic jacks (see Fig. 10.33). From Inaudi *et al.*, "Structural Monitoring by Curvature Analysis Using Interferometric Fiber Optic Sensors," *Smart Materials and Structures Journal*, **7**, (1998), pp. 199–208, with permission from the Institute of Physics Publishing Ltd.

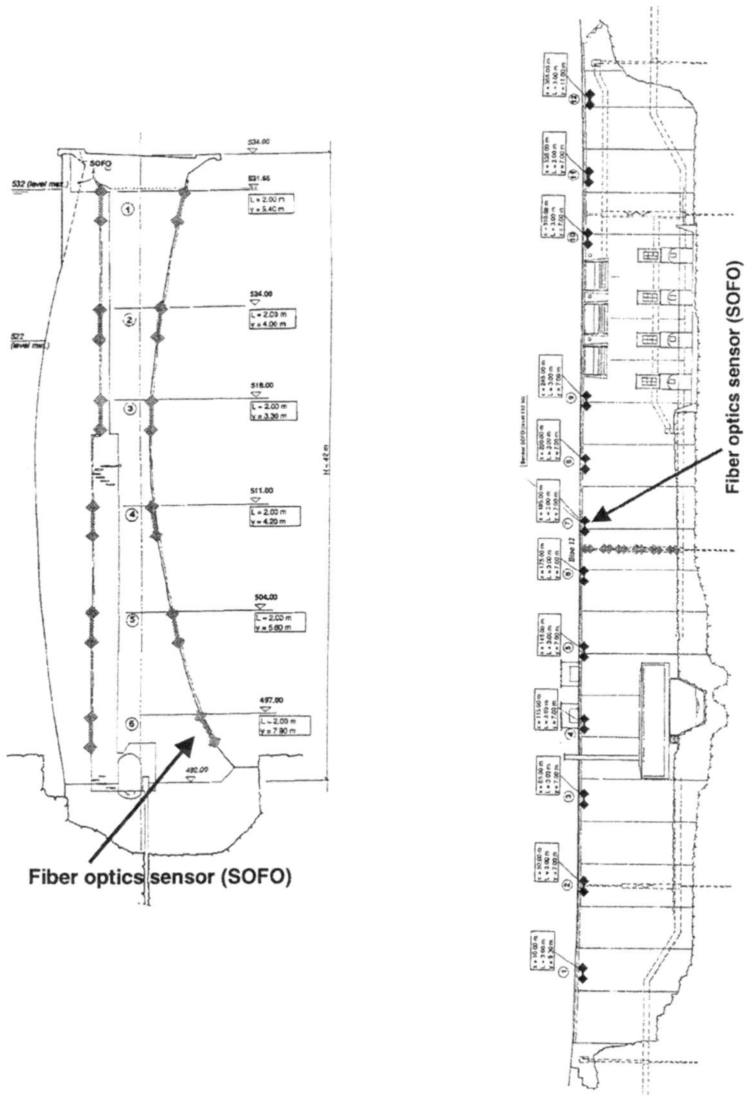

FIGURE 10.35. Proposed placement of long-gauge fiber optic sensors in the Schiffenen Shell-Mass Dam, Switzerland. From Kronenberg *et al.*, (1997), "Dam Monitoring with Fiber Optics Deformation Sensors," *Proc SPIE* **3043**, pp. 2–11, *Smart Structures and Materials* 1997: Smart Systems for Bridges, Structures, and Highways. Norris Stubbs, (Ed).

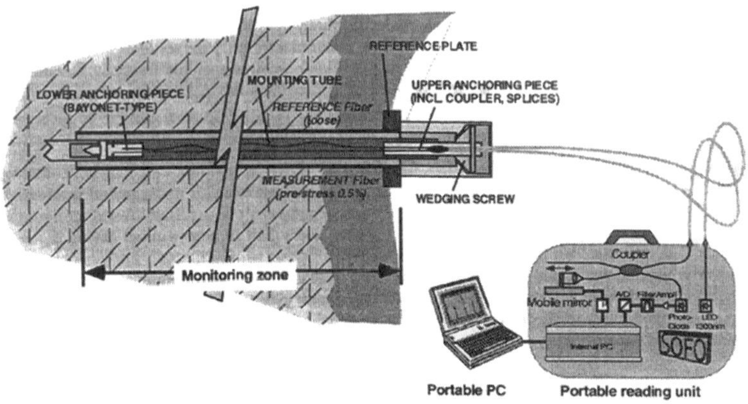

FIGURE 10.36. Schematic of the fiber optic extensometer and its mounting arrangement used in the Emosson Shell Dam, Switzerland. From Kronenberg *et al.*, (1997), "Dam Monitoring with Fiber Optics Deformation Sensors," *Proc SPIE* **3043**, pp. 2–11, *Smart Structures and Materials* 1997: Smart Systems for Bridges, Structures, and Highways. Norris Stubbs, (Ed).

5. *Anchorage Cable Monitoring.* Anchorage cables used to stabilize geological features, rock walls, slurry walls, or soil are often monitored because ground movement can either reduce their tension (and efficiency) or increase it, inducing possible failure. This monitoring is accomplished with load cells installed on the anchorage head. However, these cells are insufficient to provide a full understanding of the interaction behavior between the soil and structure, because different effects can have the same results on the anchor load. A better interpretation can be attained by measuring the actual change in length between the anchor head and the grouted zone. The archorage point between the optical fiber and the rock at the bottom of the anchorage consisted in custom design pieces, in which the optical fiber were glued (Fig. 10.37a, Inaudi *et al.*, 1995). The outer side of these elements were designed in such a manner that after injection of the cables, the hardened grout would ensure sound mechanical contact with the surrounding rock. The second measuring point was installed at the anchor head in one of the unused cable holes (see Fig. 10.37b).

Although most of the optical fibers survived the assembly, installation, injection, and tensioning of the anchors, it became clear that installing both the sensing and reference optical fibers in the same pipe was a poor choice, since complete independence of the two fibers could not be guaranteed. It was observed that pulling on the sensing optical fiber also stressed the reference optical fiber, making deformation measurements

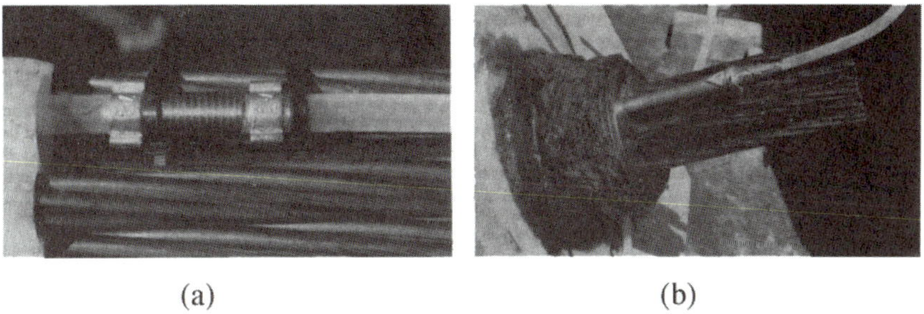

FIGURE 10.37. (a) Close-up view of the fiber optic sensor installation in an anchorage used to stabilize geological structures. (b) View of the optical fiber egress from the anchor head. From Inaudi, D., Vulliet, L., Pflug, L., Vurpillot, S., Wyser, A., "Low Coherence Interferometry for the Monitoring of Underground Works" SPIE on *Smart Structures and Materials,* **2444**, pp. 171–178, San Diego, USA, 1995.

unreliable. The solution to this problem adopted by Inaudi *et al.* (1997) was to place each optical fiber within its own microtube and separate these with a third larger tube. The whole assembly is then protected within a PVC mounting pipe (see Fig. 10.38).

6. *Structural Curvature Measurements with Fiber Optic Sensors.* Long gauge-length fiber optic sensors can be used to acquire deformation information on an entire structure, or substructure, that allows the

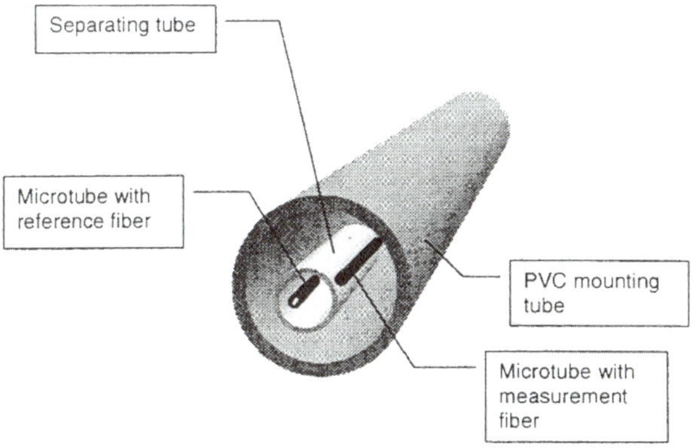

FIGURE 10.38. Arrangement for mounting the sensing, and reference optical fibers within the transduction tube. From Kronenberg *et al.*, (1997), "Dam Monitoring with Fiber Optics Deformation Sensors," *Proc SPIE* **3043**, pp. 2–11, *Smart Structures and Materials* 1997: Smart Systems for Bridges, Structures, and Highways. Norris Stubbs, (Ed).

global behavior of the structure to be better monitored and understood. Deformations can result from changes in the performance or properties of the materials constituting the structure or external forces and environmental conditions. Cracks, creep, shrinkage, and debonding are examples of the first category, while dynamic loads, ground movement, moisture, and changes of temperature are representative of the second. Inaudi *et al.* (1997b) indicate how a structure undergoing complex deformations can be analyzed by dividing the structure into cells, each instrumented with its own array of sensors.

Bending represents a fundamental form of structural deformation and Inaudi *et al.* (1997b) indicate how it is possible to determine the radius of curvature of a structural element from a set of linear displacement measurements derived from an array of long gauge-length fiber optic sensors. The key to this observation is the installation of one or more such sensors at known distances from the neutral axis of the structure. The axis of each sensor is parallel to the neutral axis for that particular structural element (Fig. 10.39a). The equation relating the radius of curvature to the local strain for a beam in simple flexion is obtained from considering Bernoulli's law for an element of the structure (see Fig. 10.39b). We see that if the length of the element, at the neutral axis, is L and that at a distance y from this neutral axis the element has been stretched to a length $(L + \Delta L)$, by virtue of the bend in the structure, then

$$\frac{\Theta}{2} = \frac{\Delta L/2}{y} \tag{10.2}$$

and

$$L = -R(z)\Theta, \tag{10.3}$$

where Θ is approximately the angle subtended by the element of length, L, at the neutral axis, at the center of curvature, located at $-R(z)$ (the *radius of curvature*) from the neutral axis (see Fig. 10.39b). In such a case we can write

$$\frac{1}{R(z)} = -\frac{\Delta L}{yL}. \tag{10.4}$$

Equation (10.4) shows that a displacement gauge placed parallel to the neutral axis, but at a known distance from it, is able to evaluate the radius of curvature, providing that bending is the only cause of strain. If a change of temperature is also possible, a pair of measurements, at

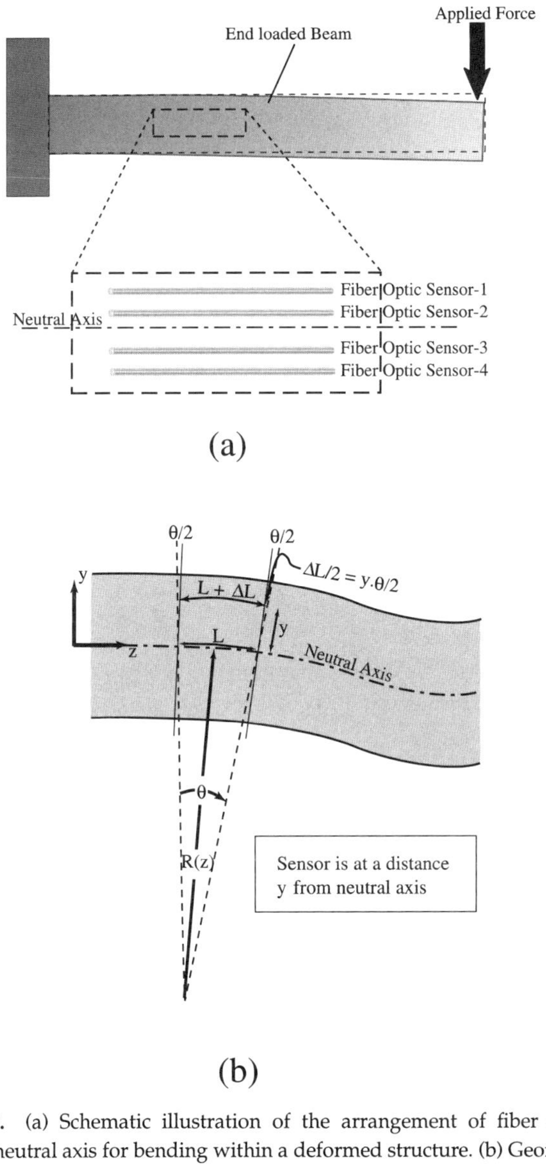

FIGURE 10.39. (a) Schematic illustration of the arrangement of fiber optic sensors in relation to the neutral axis for bending within a deformed structure. (b) Geometrical relations involved in determining curvature from deformation. Adapted from Inaudi, D., Vurpillot, S., Casanova, N., Kronenberg, P., (1998), "Structural Monitoring by Curvature Analysis Using Interferometric Fiber Optic Sensors," *Smart Materials and Structures Journal*, **7**, 199–208, with permission from the Institute of Physics Publishing Ltd.

different distances from the neutral axis, are required to determine the radius of curvature, i.e.,

$$\frac{1}{R(z)} = \frac{\Delta L_2 - \Delta L_1}{\{y_1 - y_2\}L}, \tag{10.5}$$

where ΔL_1 and ΔL_2 are the respective deformations measured by two sensors (of original length L) at distances y_1 and y_2, respectively. This analysis was used for the Moesa bridge (see Fig. 10.34). Furthermore, the displacement of the neutral axis, a_N, for that cell can be determined by the relation (Inaudi *et al.*, 1998)

$$a_N = \frac{y_1 \Delta L_2 - y_2 \Delta L_1}{\Delta L_2 - \Delta L_1}. \tag{10.6}$$

10.3.3 Applications of the Fabry–Perot-based long–gauge sensor

As indicated earlier, the advantages of the fiber Bragg grating–based Fabry–Perot long gauge-length sensor over the Michelson configuration are as follows:

1. A single optical fiber is involved, which makes it practical to incorporate the sensor into FRP composite material structures via pultrusion, filament winding, or prepreg layups.
2. The sensor can also conform to the shape of the host structure when bonded to its surface.
3. Since the sensing length is localized, there is no influence from the extraneous forces acting on the lead fibers.
4. Demodulation is possible by purely optoelectronic means, making a lightweight, low-cost, reliable, highly portable system practical.

In this section we shall show that the mechanical actuation form of the structurally integrated fiber optic long gauge-length (SIFOLG), Fig. 10.8, has been used with filament-wound FRP composite pressure cylinders, a series of corrosion-accelerated concrete columns, and some field-site installations of FRP rehabilitation and strengthening wraps of existing concrete structures.

1. CFRP Composite Cylinder Pressure Monitoring. Environmental and economic forces have made it attractive to consider the development of carbon FRP composite materials in place of steel as *natural gas vehicle* (NGV) tanks. An example of such a filament-wound pressure cylinder is 0.47 m in diameter by 1.24 m in length and is about a third of the weight of its metal counterpart. This reduced weight constitutes a major incentive for their development and makes them very attractive in terms of installation and maintenance. The explosive nature of their contents and the high pressure under which it is stored (service pressure ~3600 psi) require that these fuel tanks be inspected on a regular basis to ensure that they continue to meet their burst-pressure standard. This requires the tank to be removed from the vehicle and tested, which is time consuming and a real inconvenience for the vehicle owner, and represents a strong deterrent to their broad acceptance.

An experimental program was undertaken to explore the prospect of developing a fiber optic structural monitoring system that could be integrated into such CFRP NGV-cylinders during their filament winding fabrication. A CFRP cylinder was instrumented with a long gauge-length Fabry–Perot fiber optic sensor and tested by exposing it to the kinds of abuse it could experience during its life mounted, for example, on a bus in New York City. In this initial work, two long gauge-length fiber Bragg grating–based Fabry–Perot sensors (A and B) were surface-mounted on one of the end-zones of the cylinder, as shown in the schematic of Fig. 10.40. A photograph of the instrumented CFRP cylinder (manufactured at the time by EDO Canada) is displayed as Fig. 10.41.

According to the manufacturer's experience a damaged cylinder should deform abnormally under high pressure, resulting in a rupture at a pressure well below the designed burst pressure of 8000 psi. It was considered that a long gauge-length fiber optic sensor might be better able to detect the onset of structural weakness after the cylinder sustains damage, through a change in the compliance of the cylinder. The gauge lengths of the two sensors, A and B, were 2.63 and 2.13 m, respectively. Each sensor was monitored as the cylinder was pressurized, up to 6000 psi

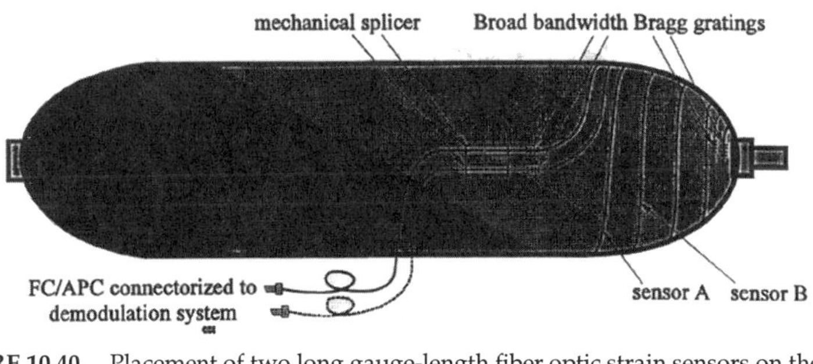

FIGURE 10.40. Placement of two long gauge-length fiber optic strain sensors on the surface of a CRFP natural gas vehicle cylinder. From Fan, N. (1997), "Development of Fiber Optic Sensing Techniques," PhD. Thesis, University of Toronto.

(a)

(b)

FIGURE 10.41. (a) Optical fiber long gauge-length sensor being installed on CFRP-NGV cylinder. (b) Close-up view of the end of CFRP-NGV cylinder showing attached optical fiber sensor. From Fan, N., (1997), "Development of Fiber Optic Sensing Techniques," PhD. Thesis, University of Toronto.

for sensor A and 8000 psi for sensor B. The integrated path length change for each fiber optic sensor can be seen to be very close to a linear function of the pressure in the cylinder (Fig. 10.42). Sensor A is observed to have a higher sensitivity to pressure than sensor B because of its greater gauge length.

To simulate the kind of impact damage likely to be sustained by a CFRP cylinder in service, it was dropped from a height of ~2 m onto a concrete floor with its longitudinal axis at 45° to the ground plane (see inset to Fig. 10.42). The cylinder was then repressurized and the fiber optic sensors monitored its compliance. These results are also included in Fig. 10.42. The cylinder appeared to be unscathed by this fall and attained its design burst pressure of 8000 psi. Further impacts were manually applied by striking the cylinder with a hammer until some visible local delamina-

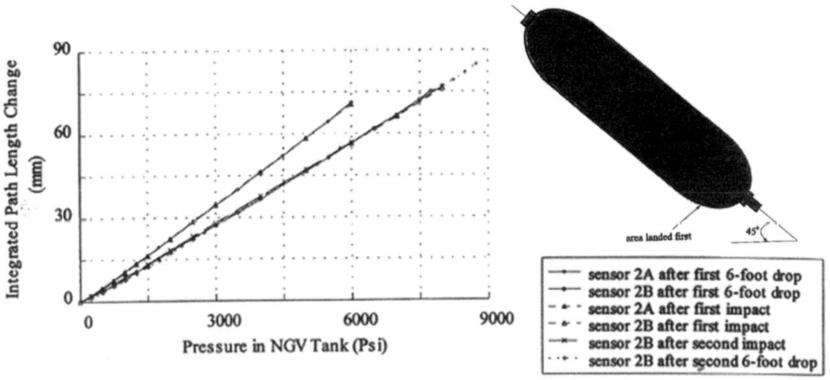

FIGURE 10.42. Variation in the integrated path length around the CFRP-NGV cylinder as a function of internal pressure as measured by both fiber optic long gauge-length sensors after several different kind of impacts were delivered to cylinder. Inset reveals orientation of cylinder during one drop test. From Fan, N., (1997), "Development of Fiber Optic Sensing Techniques," PhD. Thesis, University of Toronto.

tion was visible. Since, subsequent measurements still indicated no substantial damage, the cylinder was again struck, and unfortunately sensor A was destroyed in this procedure (Fan, 1997). Measurements by sensor B revealed no discernible weakness in the cylinder as it was pressurized once again to 8000 psi. The cylinder was then dropped once more from a height of 2 m and retested. The cylinder was this time pressurized until it burst. The last reading from sensor B was at 8750 psi.

These results confirmed the rugged nature of the CFRP NGV cylinder and that of the fiber optic sensing system, especially if one recognizes that any practical monitoring system would be integrated into the cylinder, as oppose to being surface-mounted. To this end tests were done, some time earlier, to prove that the optical fiber could be filament-wound into the cylinder.

2. Long Gauge Fiber Optic Monitoring of Concrete Columns. The deterioration of reinforced concrete infrastructure due to corrosion is a major problem worldwide, especially in cold climates where extensive use is made of deicing salts. Key bridge components, such as the supporting columns, are especially vulnerable to corrosion attack. Traditional repair methods have demonstrated only limited success, as they are often costly and labor-intensive. Furthermore, the repaired structure is vulnerable to continuation of the same deterioration mechanisms. The timely emergence of FRP composite materials could provide innovative, cost-effective rehabilitation and strengthening of existing structures.

It is well known that confinement increases the strength and ductility of wrapped or jacketed concrete members. Epoxy-bonded steel jackets, and more recently, FRP wraps and patches have been used extensively for structural retrofits in seismic regions. In recent years, this technology has been tested for use in the rehabilitation and strengthening of corrosion-damaged concrete columns (see Fig. 2.21). In addition to enhancing the mechanical behavior, such repairs can also decrease postrepair corrosion rates by (i) serving as a diffusion barrier to inhibit the ingress of: chlorides, oxygen, and moisture through the cover; and (ii) stifling the formation of expansive corrosion products. Because these FRP composite materials are extremely lightweight, the relatively high material costs are offset by significant reductions in installation time and labor costs. By far the most desirable advantage, however, is the immunity of these materials to corrosion attack.

Growing alongside the demand for innovative repairs is the need for robust instrumentation that can provide valuable information on the status of the repaired structure. Fiber optic long-gauge fiber Bragg grating sensors, unlike the traditional strain gauges, allow the integrated strain over the gauge length to be determined. In combination with the advanced composite warps, these open up the possibility of creating "smart" repairs, with the capacity to collect and transmit information regarding the condition of the column, such as state of strain, temperature, and internal pressure. As part of a study aimed at clarifying and quantifying the relationship between expansion and corrosion within reinforced concrete members, research is underway to relate column expansion to column integrity so that the condition of a corroded concrete column can be inferred simply from the expansion data.

The use of long gauge-length Bragg grating fiber optic sensors could play a key role in this regard, so several were surface-mounted 50 mm below the center line of special steel reinforced concrete columns (Fig. 10.43). The optical fiber sensors were wrapped around the concrete columns and protected from accidental damage with a silicon caulking. Also seen in this photograph is the more conventional mechanical expansion collar used to track the column expansion. It might be noted that a number of these collars broke during the test period, as seen by the breaks in their column expansion data presented as Fig. 10.44. These concrete columns were 305 mm in diameter and 1016 mm in height, and in order to discourage early corrosion in the end regions the steel reinforcement cage was epoxy-coated in those end zones. Corrosion of the steel within these columns was accelerated by driving a small current though

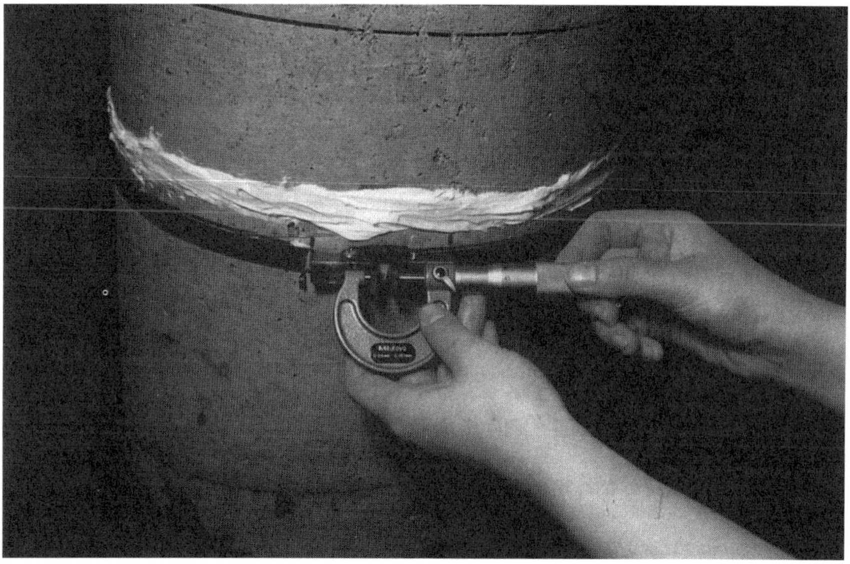

FIGURE 10.43. Close-up view of concrete column instrument with a long gauge-length fiber optic strain sensor (protected by white silicon caulking) and a conventional mechanical expansion collar. From MASc Thesis, C. Lee, 1998, Civil Engineering, University of Toronto, Canada.

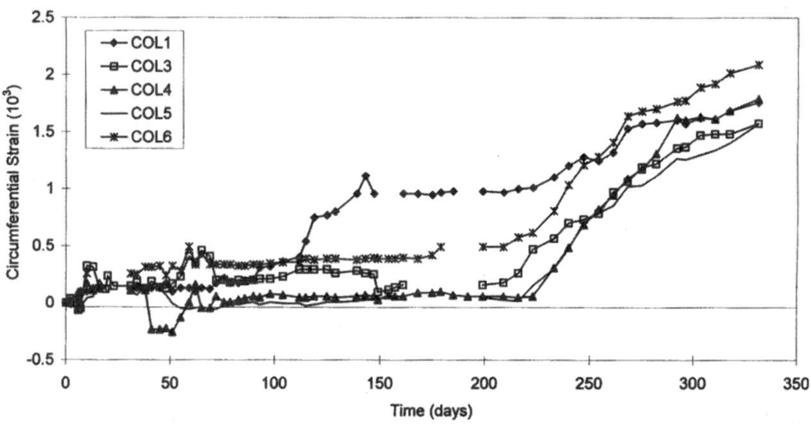

FIGURE 10.44. Variation of circumferential strain with time for several concrete columns subject to accelerated corrosion, as measured by the mechanical expansion collars. From MASc Thesis, C. Lee, 1998, Civil Engineering, University of Toronto, Canada.

the metal cage (Lee, 1998). A photograph of the fiber optic experimental setup, including a view of the interference pattern recorded on the oscilloscope from the low-coherence demodulation system, is presented as Fig. 10.45.

The fiber optic expansion data is presented as Fig. 10.46a. Note that two optical fiber sensors on two different columns were initially tracked, but as the mechanical collar on column 5 broke, it fractured the optical fiber lead so close to the sensor that it could not be repaired—clearly, a lesson in protecting optical fibers with a resilient jacket! A comparison of the results for three measurement techniques for column 4 are displayed as Fig. 10.46b. The third technique involves measuring the sum of the widths of all of the cracks in the concrete that are formed as a result of the internal pressure generated by the corrosion process. It is apparent that although there is some difference early in the experiment, the same trend

FIGURE 10.45. View of the localized, long gauge-length fiber optic sensing system, including a test concrete column instrumented with such a sensor. From MASc Thesis, C. Lee, 1998, Civil Engineering, University of Toronto, Canada.

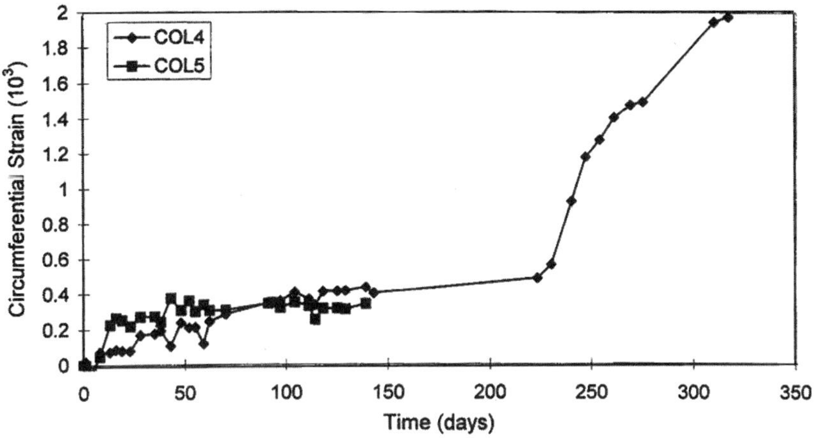

(a)

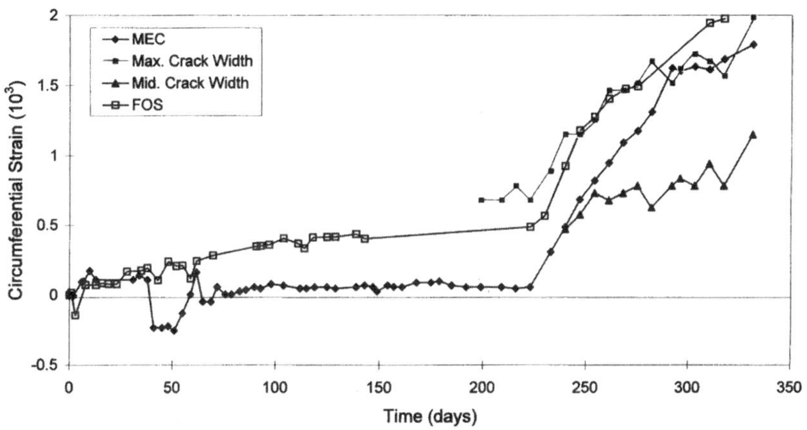

(a)

FIGURE 10.46. (a) Variation in the circumferential strain experienced by two concrete column subject to accelerated corrosion and measured by localized, long gauge-length fiber optic sensors. Note: the fiber optic sensor on column 5 was accidentally broken just prior to 150 days. (b) Comparisons of several measurement techniques used to track the circumferential strain of concrete column 4. From MASc Thesis, C. Lee, 1998, Civil Engineering, University of Toronto, Canada.

is displayed by all three techniques, and in end the mechanical expansion collar (MECCA) and the fiber optic sensor (FOES) data converge to a similar value. The crack width summation technique is known to be less accurate because it fails to account for tensile strain that is below the cracking strain for concrete.

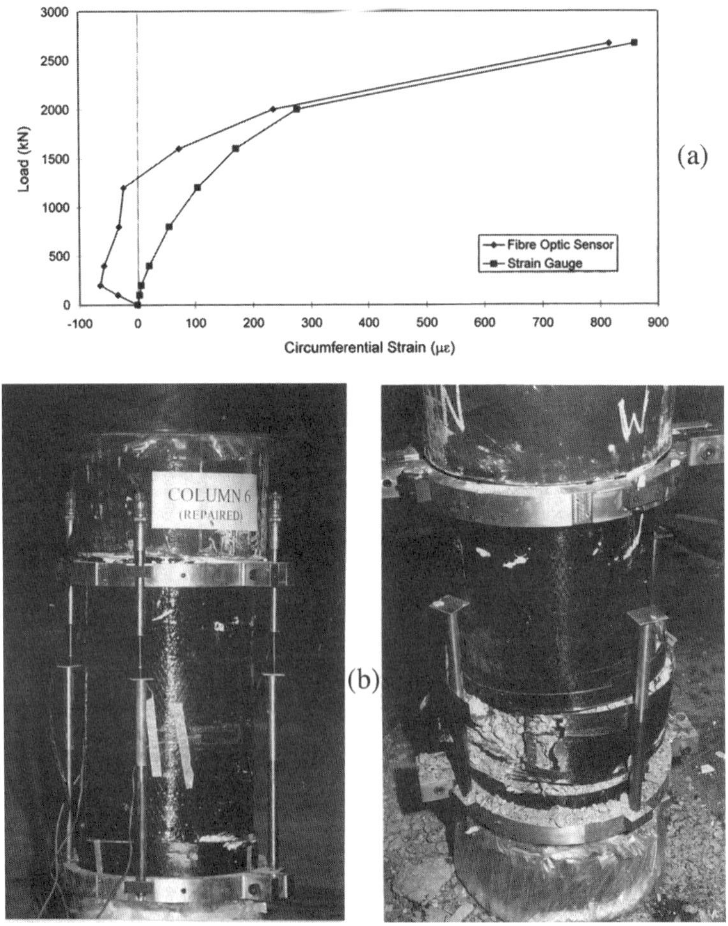

FIGURE 10.47. (a) Comparison of circumferential strain measurements undertaken by a long-gauge fiber optic sensor and a resistive foil gauge during axial loading of a concrete column that was wrapped with CRFP material. (b) Before and after loading to failure of a CFRP wrapped and instrumented concrete column. From MASc Thesis, C. Lee, 1998, Civil Engineering, University of Toronto, Canada.

FIGURE 10.48. Installation of localized, long gauge-length fiber optic sensors into the FRP rehabilitation and strengthening wrap of concrete columns supporting the 401 highway in Toronto.

The fiber optic Fabry–Perot long gauge-length sensors were also tested with CFRP repair wraps for concrete columns that were subject to large axial loads. Figure 10.47 shows a comparison of the circumferential strain values obtained from this sensor and a resistive foil gauge during an axial loading experiment. The inset shows two photographs of the CFRP wrapped and instrumented concrete column, before and after loading. It is anticipated that part of the discrepancy at low loads is due to the point sampling nature of the foil strain gauge compared to the hoop strain measurement of the fiber optic sensor. For high values of axial stress the CFRP wrap, on which the strain gauges are mounted, tends to average out the strain so the two different types of sensor tend to agree.

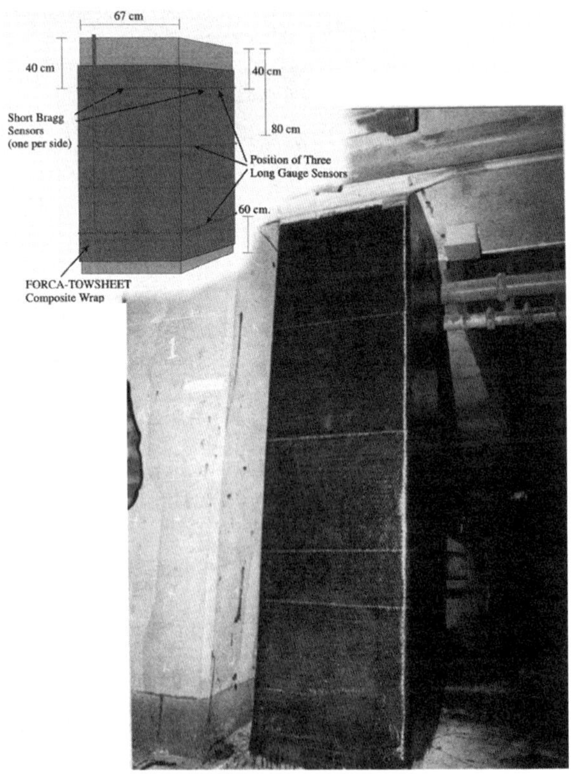

FIGURE 10.49. Installation of both short and long gauge-length fiber optic sensors into the FRP rehabilitation wrap of concrete columns in an underground parking garage, Hull, Quebec. Inset shows the location of both short and long gauge-length sensors within the FRP wrap.

The following conclusions can be drawn from these experiments: long-gauge fiber optic sensors are relatively easy to install directly on existing concrete columns, or embed within CFRP wraps of such columns, and they can be used for monitoring circumferential expansion with great precision (Lee, 1998). This has encouraged their use in field experiments, and several such sensors have been installed within FRP rehabilitation and strengthening wraps of highway concrete columns (Fig. 10.48) and underground parking garage columns (Fig. 10.49). These experiments are being undertaken as part of a collaboration between the University of Toronto Civil Engineering Department, the Ministry of Transportation Ontario, ISIS—Canada, and members of my laboratory.

Clearly, further research and development are required, particularly in terms of producing a very small and low-power demodulation system that could be left at the field site and inspected, possibly in a drive past, or via telephone or satellite links (see Chapter 13). Nevertheless, it is quite apparent from the current work that much valuable information could be gleamed from a broad implementation of this technology. This is especially true in earthquake zones, where a sophisticated array of such sensors could be used to quickly assess the degree and extent of highway overpass and bridge damage sustained immediately after an earthquake. It is even possible that a broad array of such sensitive sensors could turn the highways into a sensitive large-scale antenna for studying ground movements, and eventually maybe serve as an advanced warning detection system for the Big One!

CHAPTER 11

Multiplexed Fiber Optic Structural Sensing

11.1 INTRODUCTION

Fiber optic structural sensors, and in particular those based on fiber Bragg gratings, have a number of advantages over conventional electrical strain gauges. The ease with which they can be serially multiplexed, however, is certainly one of the most compelling. It is not possible to serially address a number of resistive foil, or piezoelectric, strain gauges along a single conductor. Each electrical gauge has to be individually connected to an instrument package. This can be a significant barrier to their use in applications requiring substantial numbers of sensors and certainly rules them out when embedding within FRP composite materials is desired.

As we have indicated in Section 5.10, an array of fiber optic sensors strung out along an optical fiber can be individually addressed and provide *quasi-distributed sensing* (see Fig. 5.57). This raises the question of distinction between the terms *distributed* and *quasi-distributed sensing*. Indeed, some researchers use the term *distributed sensing* when they are discussing serial multiplexed sensor arrays. In this book we distinguish between the terms, using *quasi-distributed sensing* to be synonymous with serial multiplexed sensing and reserve the term *distributed sensing* for a system that can provide information at every point along a specific section of an optical fiber (see Fig. 2.9). Raman and Brillouin scattering, discussed briefly in Chapter 6, are examples of techniques that permit truly distributed sensing. However, the spatial resolution possible with these techniques is currently so large, ≥ 1 m, that it is possible to derive more detailed information on a strain or temperature profile over distances of a meter or so by using a string of short gauge-length fiber optic sensors. It is clear that this could lead to some confusion in the terminology.

optic sensors. It is clear that this could lead to some confusion in the terminology.

In the case of fiber Bragg grating technology, however, the distinction is much clearer. Quasi-distributed sensing is accomplished with a string of FBG sensors that are serially multiplexed, while truly distributed sensing is achieved by "intragrating" measurements—that is to say, measurements undertaken within a single grating. Indeed, a spatial resolution of better than 1 mm has been shown to be possible with *Bragg intragrating* (BIG) sensing, and this topic is discussed in detail in Chapter 12.

In this chapter we focus on serial multiplexing of fiber optic sensors, with a considerable emphasis on FBG sensors. We will first show that it is practical to produce strings of fiber optic sensors; otherwise, the exercise would be rather academic. Then we shall review the different serial multiplexing techniques for demodulating such strings of fiber optic sensors, and finally we shall provide examples of field-site applications of this technology. Kersey (1995, 1997a) has prepared fairly extensive reviews of fiber optic sensor multiplexing.

11.2 FABRICATION OF SERIAL ARRAYS OF FIBER OPTIC SENSORS

Fiber Bragg gratings and Fabry–Perot interferometers appear as the primary choice of fiber optic structural sensors for either short or long gauge-length form. Since the most popular type of short gauge-length Fabry–Perot sensor, the EFPI, has to be manually fabricated, it is clear that there is no reason why several such sensors could not be built along the same optical fiber. There are, however, practical limits as to how close together they can be fabricated (generally speaking, ≥ 10 cm).

Fiber Bragg gratings are produced in an entirely different manner, quite literally with a flash of light. Moreover, this can be done during the manufacture of the actual optical fiber so that writing of FBGs becomes an integral part of the production of the optical fiber. Friebele *et al.* (1995) have demonstrated FBGs can be produced during the drawing of an optical fiber from a preform at a sustained rate of ~2000 per hour.

Two types of FBG can be produced by a single laser pulse (required for in-line production), depending on the energy in the laser pulse. As shown in Fig. 11.1, Type 1 gratings, are made with ~50 to 70 mJ pulse energy and are characterized by narrow linewidths and low reflectivities, ~3 to 5%. By contrast, Type 2 gratings are produced with pulse energies approaching the damage threshold of silica, have reflectivities close to

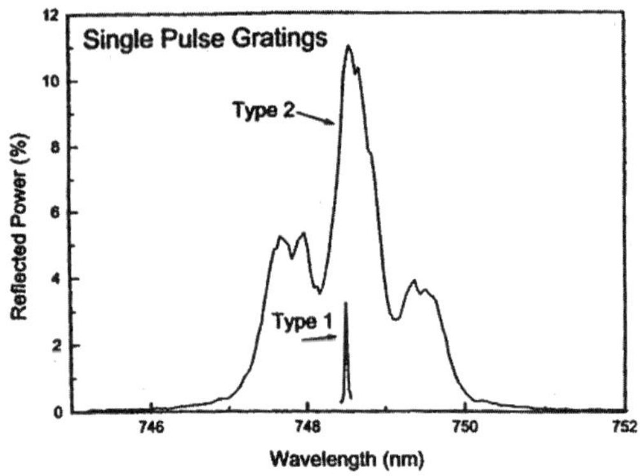

FIGURE 11.1. Reflection spectra for Type 1 and Type 2 fiber Bragg gratings written on the draw tower by single laser pulses. From Friebele *et al.*, (1995), "Fabrication and Application of Low-Cost Optical Fiber Sensor Arrays for Industrial and Commercial Applications." In: *Industrial and Commecial Applications of Smart Structures Technologies*, C.R. Crowe and G.L. Anderson, (Eds.), *SPIE* **2477**, 1995, 305–311.

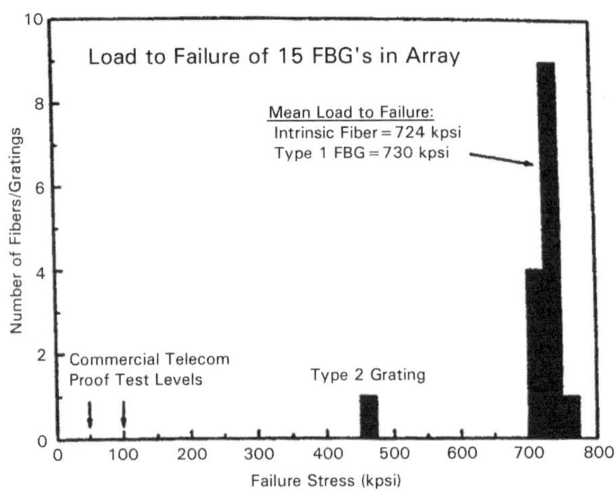

FIGURE 11.2. Distribution of failures as a function of stress loading for Type 1 and Type 2 FBGs. From Friebele *et al.*, (1995), "Fabrication and Application of Low-Cost Optical Fiber Sensor Arrays for Industrial and Commercial Applications." In: *Industrial and Commecial Applications of Smart Structures Technologies*, C.R. Crowe and G.L. Anderson, (Eds.), *SPIE* **2477**, 1995, 305–311.

100%, and are spectrally broad and structured. Since this "in-line" method of production writes the FBG into the pristine optical fiber, which is then immediately coated with a protective polymer jacket, the strength of Type 1 gratings is very close to the value for normal optical fiber (see Fig. 11.2). Indeed, we see from this figure that the mean load to failure of the Type 1 FBGs is about 730 kpsi, as compared to 100 kpsi for off-line manufactured FBGs.

It should be noted that these FBGs had a Bragg wavelength of around 800 nm. Since the maximum reflectivity is related to this wavelength, given by Eq. (3.140), it is apparent that the reflectivity would be even less at 1550 nm. Putnam *et al.* (1995a), however, have shown that peak reflectivities of ~9% should be possible at 1550 nm with an improved quality of the writing laser beam. This magnitude of reflectivity should be fine for serial multiplexed arrays of FBGs.

11.3 SERIAL MULTIPLEXING OF FIBER BRAGG GRATING SENSORS

In this section we review the most important techniques for accomplishing serial multiplexing, with the major emphasis being on FBG sensors. As we saw in Section 5.10, there are a number of multiplexing techniques; however, we shall focus here on the three primary approaches which have been explored at some depth and show the greatest potential:

1. Wavelength division multiplexing (WDM)
2. Time division multiplexing (TDM)
3. Coherence division multiplexing (CDM)

In the case of short gauge-length FBG sensors, either WDM or TDM is seen to be satisfactory for serial multiplexing. However, the former would require a source with considerable bandwidth if many FBG sensors are to be interrogated, while the latter requires short pulse optoelectronic technology, which is fairly expensive. A combination of both techniques is shown to be one way of handling a large number of sensors at modest cost.

A discussion on the use of a tunable laser for serial multiplexing is omitted from this chapter even though it has considerable potential. There are two reasons for this: (i) it is the primary focus of Chapter 12, where a technique for serial multiplexing involving the measurement of the "complex reflection spectrum" for a FBG-sensor is described, and (ii) it is also covered in Chapter 13 in the context of a universal demodulation system for FBG-based sensings.

11.3.1 Wavelength division multiplexing

The fact that FBGs can be produced with a wide range of wavelengths makes wavelength division multiplexing a natural method of addressing individual FBGs in a string of such sensors along a single strand of optical fiber. In Section 9.2.2 we reviewed a number of narrowband filter demodulation techniques for the case of a single FBG. Most of these techniques have also been used to interrogate a string of FBGs, each with a reference Bragg wavelength separated from its neighbor by at least the maximum wavelength interval needed to cover the full measurement range of the sensor. The combination of a Fabry–Perot interferometer to select the waveband specific to one particular FBG out of the string of sensors, and the passive spectral ratiometric technique to measure the actual wavelength of that FBG sensor, was discussed in Chapter 9 and illustrated in Fig. 9.60a.

Scanning Fiber Fabry–Perot Filter. The scanning fiber Fabry–Perot filter technique introduced in Section 9.2.2 is capable of examining the composite spectra from a set of FBGs strung out along a single optical fiber with a suitable spectral separation between them. An example of the transmission spectra for a string of 12 FBGs is presented as Fig. 11.3 (Kersey, 1997a). The scanning fiber Fabry–Perot filter can also be used to measure the set of Bragg wavelengths for such a sensor string in the manner discussed in Section 9.2.2 and illustrated in Fig. 9.11.

A ramp waveform voltage is applied to the piezoelectric drive of the tunable FFP, forcing its passband to sweep over the entire wavelength range of the three gratings. As mentioned in Section 9.2.2, a small dither voltage is added to the FFP filter ramp voltage, modulating the filter's passband wavelength by ~0.01 nm. The photodetector signal is mixed electronically with the dither frequency and low-pass filtered so as to extract the component at the dither frequency. As the FFP filter sweeps over the full wavelength range, the output of the mixer–filter arrangement provides the derivative of the photodetector signal (see Fig. 9.11b). The zero-point crossings of this derivative signal correspond to the peak of each FBG spectrum and consequently give the Bragg wavelengths for the set of FBG sensors.

Kersey *et al.* (1994) demonstrated that this technique could be used for structural shape sensing and vibrational mode analysis. In this experiment three FBGs were surface bonded to an aluminum cantilever beam

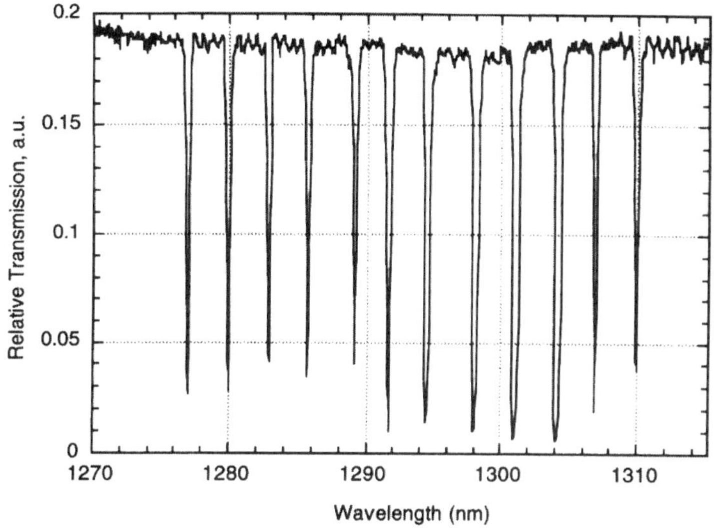

FIGURE 11.3. Transmission spectrum of an optical fiber containing a string of 12 FBGs. From Kersey A. D., 1997a, "Multiplexing Techniques for Fiber Optic Sensors" Reprinted with permission from *Optical Fiber Sensors*, Vol. 4, J. Dakin and B. Culshaw (Eds), Artech House, Inc., Norwood, MA, USA. www.artechhouse.com.

(96 cm × 5.5 m × 1.58), as indicated in Fig. 11.4a and shown in the photograph that constitutes Fig. 11.4b. Also displayed in Fig. 11.4(a) are the first three free-vibration mode shapes [$d_j(x)$ is the beam deflection as a function of x for jth mode] and strain distributions [$\varepsilon_j(x)$]. The FFP tunable filter has a spectral width that is comparable to that of the FBG and a free-spectral range of 27 nm.

Vibrational analysis of the beam was performed by individually interrogating the three FBG sensors. Figure 11.5a shows the close correlation between the FBG sensor located at the root of the beam (top trace) and the resistive foil strain gauge (bottom trace) when the beam was excited in its first two modes. The sensitivity of an individual sensor to a specific vibration mode depends quite strongly on its location. This can be seen by reference to Fig. 11.5b, which displays the frequency spectrum attained from the FBG sensor and the resistive foil strain gauge, when all three vibration modes were excited (Kersey *et al.*, 1994). It is apparent that the second FBG sensor was placed at such a location that it exhibits a ~30 dB suppression of the third vibration mode (24 Hz). By contrast, the placement of the RSG was such that it responded quite well to all three vibration modes. In fact, its sensitivity to the first mode is greater than that of the FBG sensor.

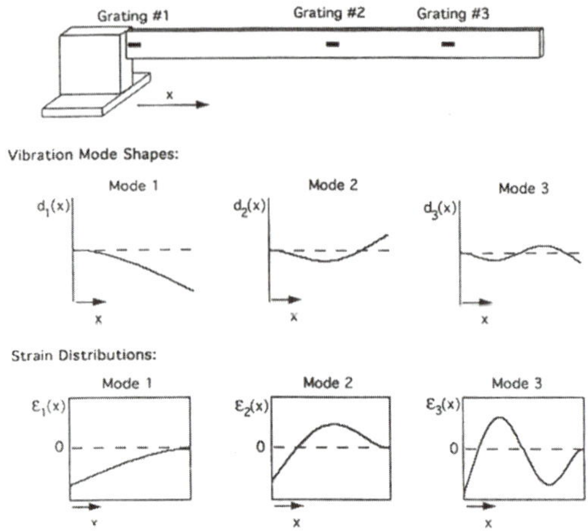

(a)

(b)

FIGURE 11.4. (a) Schematic of a cantilever beam instrumented with three FBGs arranged to measure mode shapes and strain profiles. (b) Photograph of the experimental arrangement showing the cantilever beam. From Davis, M.A., Kersey, A.D., Sirkis, J., and Friebele, E.J., (1994), "Fiber Optic Bragg Grating Array for Shape and Vibration Mode Sensing," *SPIE* **2191**, 94–101.

In the scanning mode of operation, the FFP filter was driven with a 10-Hz ramp signal and a 1-kHz dither. Using optimized trial functions, a comparison between the actual beam shapes (full curves) derived from digitized photographs and the calculated shapes (broken line) using the

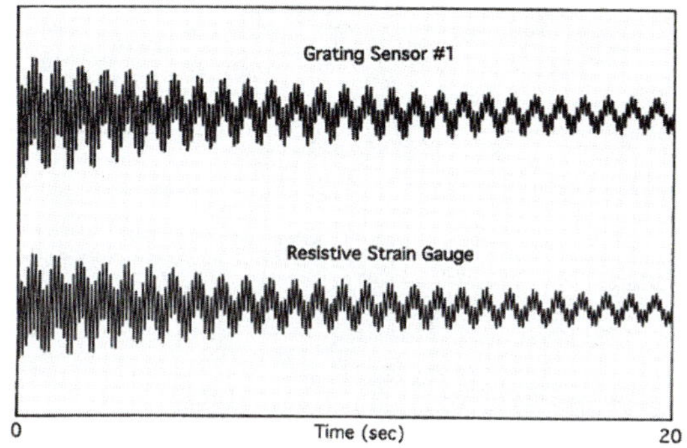

(a)

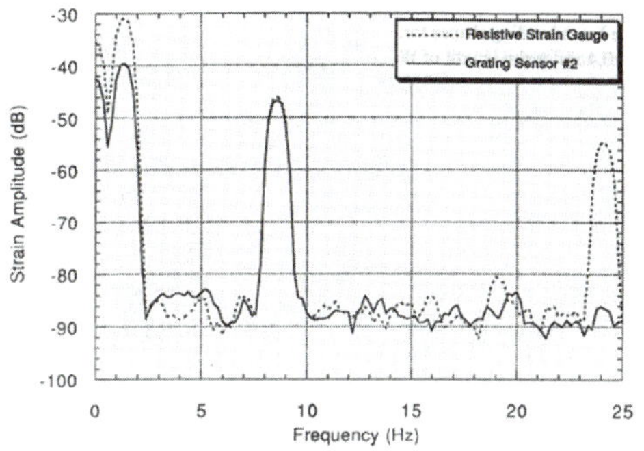

(b)

FIGURE 11.5. (a) Temporal strain variations for vibrating cantilever beam as measured by one of the three sensors shown in Fig. 11.4 and a resistive foil strain gauge. (b) Frequency spectrum as determined by this FBG sensor and the resistive foil strain gauge. From Davis, M.A., Kersey, A.D., Sirkis, J., and Friebele, E.J., (1994), "Fiber Optic Bragg Grating Array for Shape and Vibration Mode Sensing," *SPIE* **2191**, 94–101.

FBG sensor data, is presented as Fig. 11.6. These results show an excellent agreement and indicate the capability of this technique to measure strain distributions and vibration mode shapes.

Acousto-optic Tunable Filter. The acousto-optic tunable filter (AOTF) is another form of electronically controlled tunable narrowband filter (see Sections: 5.8 and 9.2.2) that can be used to interrogate a string of FBG sensors along a single strand of optical fiber. The linewidth of an AOTF is somewhat greater than that of a Fabry–Perot filter (4 nm as compared to 0.3 nm), but its tuning range can exceed 1000 nm, making it possible to interrogate a much larger number of FBGs. An indication of this is

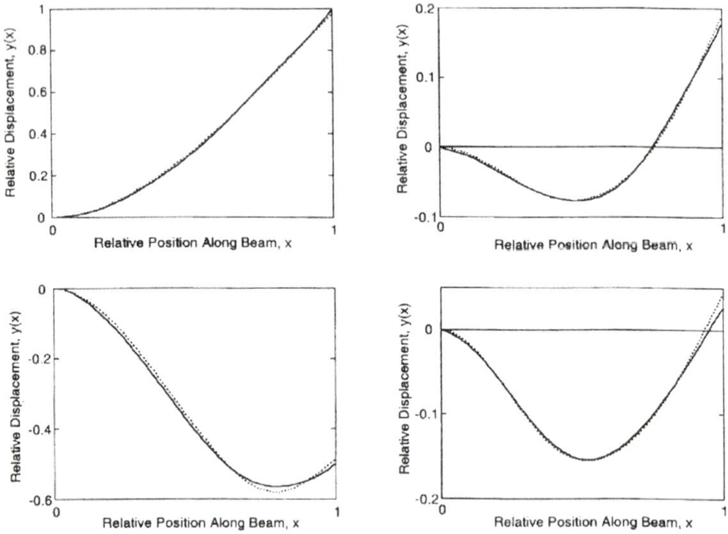

FIGURE 11.6. Comparison of theoretical and FBG measured modal shapes for a cantilever beam at different times. FRomDavis, M.A., Kersey, A.D., Sirkis, J., and Friebele, E.J., (1994), "Fiber Optic Bragg Grating Array for Shape and Vibration Mode Sensing," *SPIE* **2191**, 94–101.

provided in Fig. 11.7, where an AOTF was used to capture the spectral output of a broadband source and four fiber Bragg grating reflections at wavelengths 1299, 1305, 1309, and 1313 nm (Meltz, 1996). Xu *et al.* (1996) have modeled and undertaken a performance analysis of the AOTF demodulation system. They have also tested quasi-static and dynamic

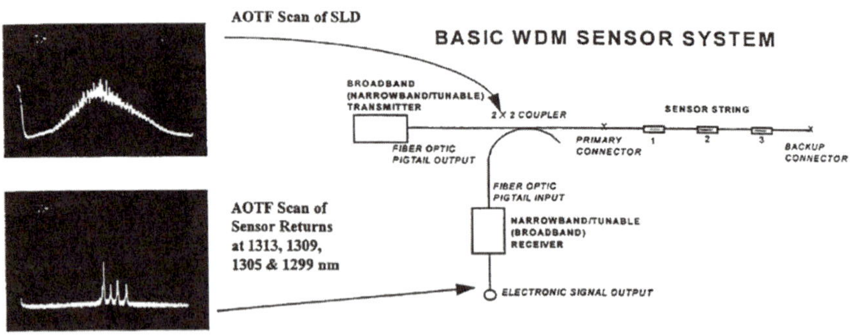

FIGURE 11.7. Schematic of a basic wavelength division multiplexing system based on the use of either a narrowband source or a narrowband detection system. Two insets show the AOTF scans of an SLD source, and the returns from several FBGs. Adapted from Meltz, (1996). "Overview of Fiber Grating-Based Sensors," *SPIE* **2838**, pp. 2–22.

loading of cantilever beams at low levels of strain, and with high levels of static tensile strain. An example of the real-time response of their system is provided in the two insets. This display from a computer shows (a) the spectral scan mode (top left), and (b) the lock-in mode for real time monitoring of a ±50 $\mu\varepsilon$ strain cycle (bottom) with a measurement time of 100 ms.

It is also possible to monitor several FBGs simultaneously, since a number of different transmission windows can be formed at the same time by driving the AOTF with the appropriate set of RF-voltage frequencies and creating the required set of standing waves. An appreciation of the potential for FBG multiplexing by means of AOTF demodulation can be obtained from the simultaneous monitoring of two FBG sensors, with nominal Bragg wavelengths of 1300 and 1550 nm, respectively (Volanthen *et al.*, 1996c). They compared the two possible methods of demultiplexing a string of FBGs with an AOTF: *optical demultiplexing*, in which each wavelength interval is detected by a different photodetector that provides an input to a separate lock-in loop, and *electronic demultiplexing*, in which two lock-in loops are toggled at different frequencies and only one photodetector is required.

Their experimental arrangements are shown in Figs. 11.8a, 11.8b. In the optical demultiplexing approach (a), each loop locks its mean interrogation wavelength to the wavelength of the grating it is monitoring and measures the corresponding mean AOTF drive frequency generated by its own voltage-controlled oscillator (VCO). In the electronic demultiplexing method (b), a simple photodetector provides the amplitude information for both lock-in loops, but at different toggling frequencies. Each loop locks its mean interrogation wavelength to the wavelength of the grating it is monitoring and measures the corresponding mean AOTF drive frequency, generated by its own VCO.

Two ELED sources operating at 1300 and 1550 nm were used with the two FBGs. The stretched FBG had a reference wavelength of 1305.8 nm, a bandwidth of 0.4 nm, and 71% reflectivity. The specifications of the reference FBG were 1549.3 nm, 0.65 nm, and 100%, respectively. The AOTF had a transmission bandwidth of 3 nm and was controlled by two feedback circuits. The results they attained are displayed as Fig. 11.9a (optical) and 11.9b (electronic). Both approaches worked well and provided simultaneous real-time monitoring. The electronic technique requires fewer optical components, but is limited by crosstalk at the toggling frequencies. When large numbers of sensors need to be interrogated, a combination of both techniques would be optimum.

Matched Fiber Bragg Gratings. Fiber Bragg gratings are also tunable narrowband filters, and the concept of matched FBGs was originally

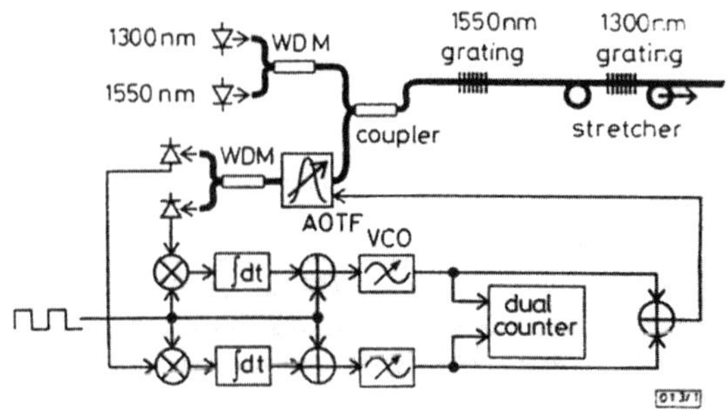

(a)

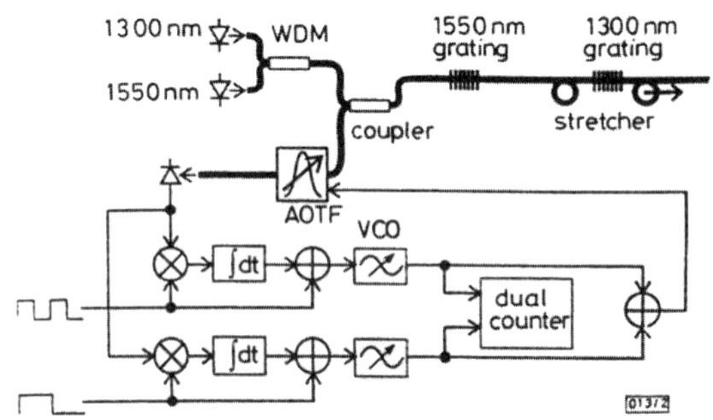

(b)

FIGURE 11.8. Schematics of two demodulation systems for two FBGs based on the use of an AOTF. (a) Optical demultiplexing is achieved by RF-frequency tracking of each FBG with its own photodetector. (b) Electronic demultiplexing is accomplished by toggling the two spectral windows of the AOTF at different frequencies. From Volanthen *et al.*, "Simultaneous Monitoring of Multiple Fibre Gratings with a Single Acousto-Optic Tunable Filter," *Electronics Letters*, **32**, (13), 1228–1229, with permission from IEE Publishing Ltd.

explored by Jackson *et al.* (1993), who showed that a reflective FBG could be used to interrogate a sensing FBG in either a scanning or tracking mode (Fig. 9.18; see section 9.2.2). Davis and Kersey (1995b) improved on this approach by using the demodulating FBGs as narrowband transmission filters (Fig. 9.19). This transmissive configuration eliminates the need for multiple photodetectors and couplers, as required by the reflective configuration, and reduces optical power losses.

The signal from the single photodetector was fed to two lock-in amplifiers, each referenced to the dither voltage signals applied to the piezoelectric elements used to tune the interrogating FBGs. The outputs of the two lock-ins were summed with the dither signals and fed back onto the interrogating FBGs. This feedback loop allowed the filter gratings to

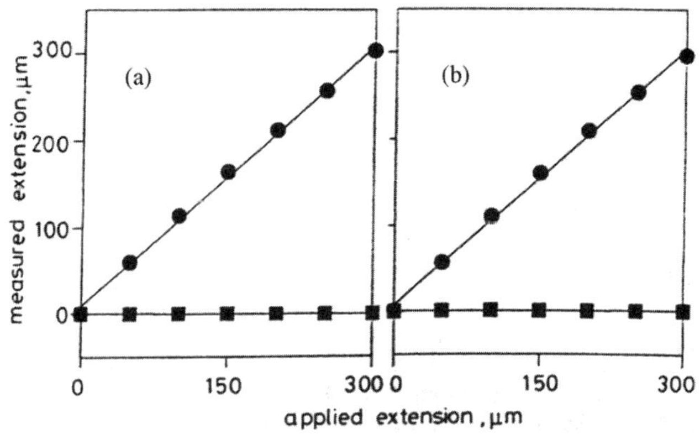

FIGURE 11.9. The measured extension for the two FBGs is plotted against the imposed extension for the demodulation approaches based on the use of AOTFs with (a) optical (b) electronic multiplexing. From Volanthen, M., Geiger, H., Xu, M.G., and Dakin, J.P., "Simultaneous Monitoring Multiple Fibre Gratings with a Single Acousto-Optic Tunable Filter," *Electronics Letters* **32**, (13), 1228–1229, with permission from IEE Publishing Ltd.

accurately track any strain induced wavelength shift experienced by the sensing FBGs. The lock-in output signals providing a voltage directly proportional to the imposed strain. This is borne out in Fig. 11.10, where the signal is seen to be a linear function of the applied strain imposed on the two sensing FBGs over a strain range of ~200 με. Different dither frequencies applied to the interrogating FBGs permitted simultaneous measurement of the strains placed on the sensing gratings. A dynamic strain resolution of 0.01 με $Hz^{-0.5}$ was attained at low frequencies.

Fallon *et al.* (1997) developed a simpler, passive approach that was also based on using interrogation FBGs as transmission filters. The important difference, however, is that identical broadband chirped FBGs are used for sensing and interrogation. In essence the measurand-induced wavelength shift of the sensing chirped grating is determined from the change in the light transmitted by the interrogating identical chirped grating. Clearly, the amount of light detected will diminish with increasing spectral mismatch between the gratings (see Fig. 11.11a). The experimental arrangement is schematically illustrated in Fig. 11.11b, where it is seen that either a coupler or an optical circulator can be used for directing the light to the sensing chirped FBG and the receiving chirped FBG. However, this scheme suffers from the same vulnerability to power fluctuations as all intensity encoded techniques.

The wide (~10 nm) transmission spectra of the matched chirped sensing and receiving FBGs are displayed in Fig. 11.12. This spectral width allows a strain range of about 9000 με, while the broadband

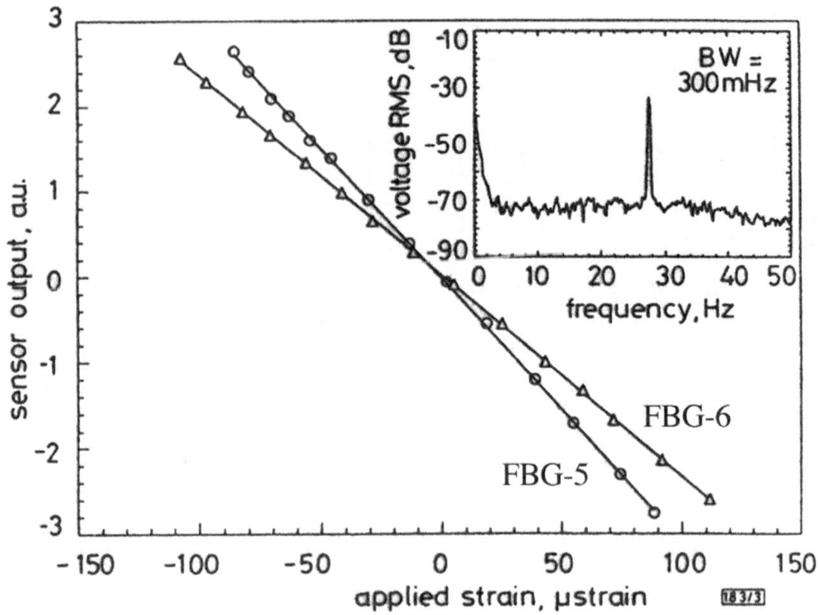

FIGURE 11.10. Linear sensor response to strain imposed on two FBGs that are closed-loop tracked by matched (reading) FBGs. The inset reveals the frequency spectrum of the measurement. From Davis, M. A., and Kersey, A. D., "Application of a Fiber-Fourier Transform Spectrometer to the Detection of Wavelength Encoded Signals from Bragg Grating Sensors," *J Lightwave Technology* **13**, 1289–1295, © 1995, IEEE.

(~50 nm) source, in the form of a 980 nm diode pumped erbium-doped optical fiber, makes it possible to serially multiplex five such sensors. The strain and temperature response of this technique is shown in Fig. 11.13. The inset in the upper graph reveals the transmitted spectra as a function of strain.

To compensate for the variation in the source optical power with wavelength, particularly at the shoulder wavelengths of the source, Zhang *et al.* (1997) introduced a reference photodetector via the second arm of an additional coupler. This ensures that the sensing system is linearized at all wavelengths. The ratio of signals from these two photodetectors can be written in the form

$$\frac{P_{\rm B}}{P_{\rm REF}} = \frac{\int I(\lambda)T_{\rm B}(\lambda)R_{\rm B}(\lambda, \varepsilon)d\lambda}{\int I(\lambda)R_{\rm B}(\lambda, \varepsilon)d\lambda}, \tag{11.1}$$

where P_B is the power received by the photodetector behind the interrogating chirped FBG with a transmission spectrum, $T_{\rm B}(\lambda)$, and $P_{\rm REF}$ is the

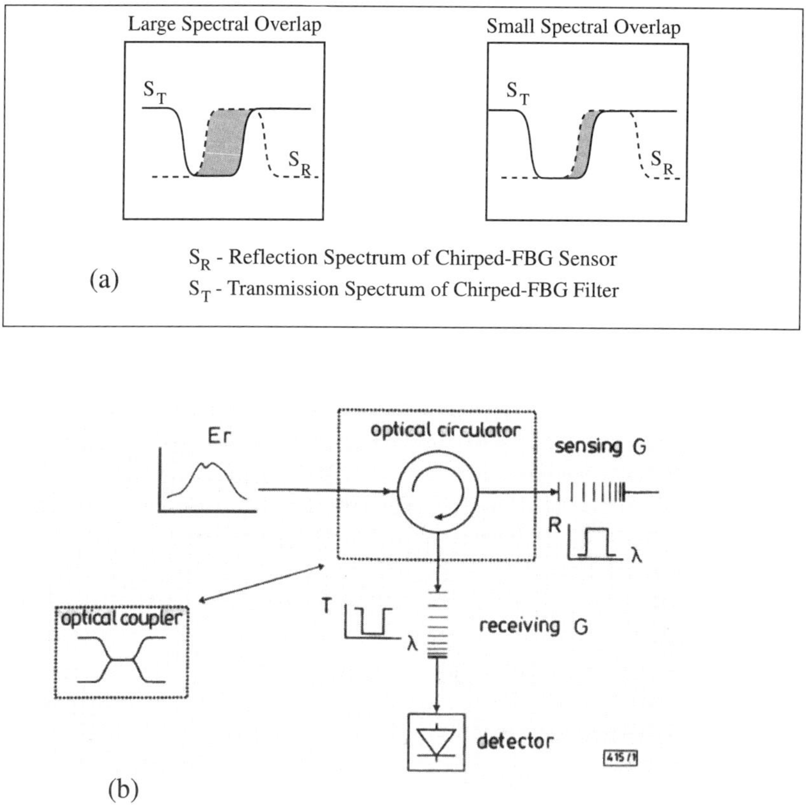

FIGURE 11.11. (a)Schematic illustration of the overlap between a chirped sensing FBG and an interrogating chirped FBG, used in transmission. The greater the spectral overlap, the larger the received signal by the photodetector. (b) Schematic of experimental system based on this form of FBG demodulation. From Fallon, R. W., Zhang, L., Gloag, A. and Bennion, I., "Multiplexed Identical Broadband Chirped Grating Interrogation Technique for Large-Strain Sensing Applications," *IEEE Proceedings*, **9**, 1616–1618, © 1997, IEEE.

power received by the reference photodetector. $I(\lambda)$ represents the incident optical power on the sensing chirped FBG having a reflection spectrum $R_B(\lambda, \varepsilon)$. The sensitivity of the system can be enhanced by introducing a second matched chirped FBG (with a known wavelength shift, λ_a, relative to the first interrogating chirped FBG) ahead of the reference photodetector. Under these circumstances a linearized output is attained from the relation

$$\text{Output} = \frac{P_{B1} - P_{B2}}{P_{B1} + P_{B2}}, \tag{11.2}$$

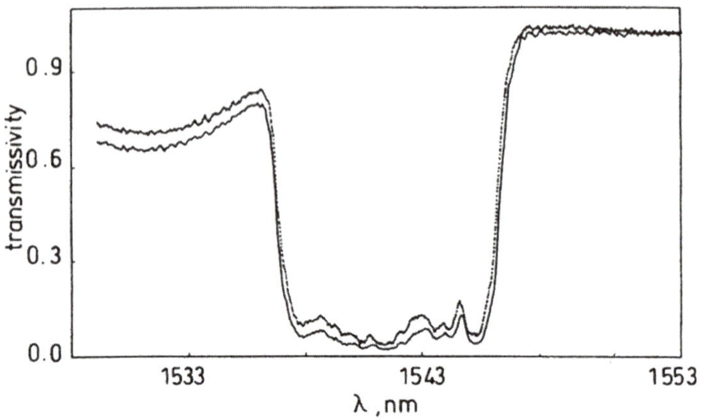

FIGURE 11.12. Transmission spectra of two near-identical FBGs used in a matched FBG demodulation scheme. From Fallon, R. W., Zhang, L., Gloag, A. and Bennion, I., "Multiplexed Identical Broadband Chirped Grating Interrogation Technique for Large-Strain Sensing Applications," *IEEE Proceedings*, **9**, 1616–1618, © 1997, IEEE.

where

$$P_{B1} = \int I(\lambda)T_B(\lambda)R_B(\lambda, \varepsilon)d\lambda \tag{11.3}$$

and

$$P_{B2} = \int I(\lambda)T_B(\lambda - \lambda_a)R_B(\lambda, \varepsilon)d\lambda. \tag{11.4}$$

The use of a spectrally mismatched chirped FBG ahead of the reference photodetector also eliminates the offset seen in Fig. 11.13, but provides a smaller signal. Zhang *et al.* (1997) demonstrated that this technique can be used with serial or parallel multiplexed systems. Figure 11.14 schematically illustrates the kind of architecture proposed for serial multiplexing. In their experiments, however, they used only two sensing gratings and two interrogation gratings.

Direct Spectral Measurements. The development of very compact spectrometers combined with linear charge-coupled detector (CCD) arrays and microelectronics (see, for example, Fig. 5.45) has made possible WDM by direct spectral measurements. The limited number of photodetector elements (pixels) in these arrays, however, requires some trade-off between spectral range and resolution, which is translated into a compromise between strain (or temperature) range and resolution. Intrapixel interpolation by means of digital algorithms has permitted wavelength

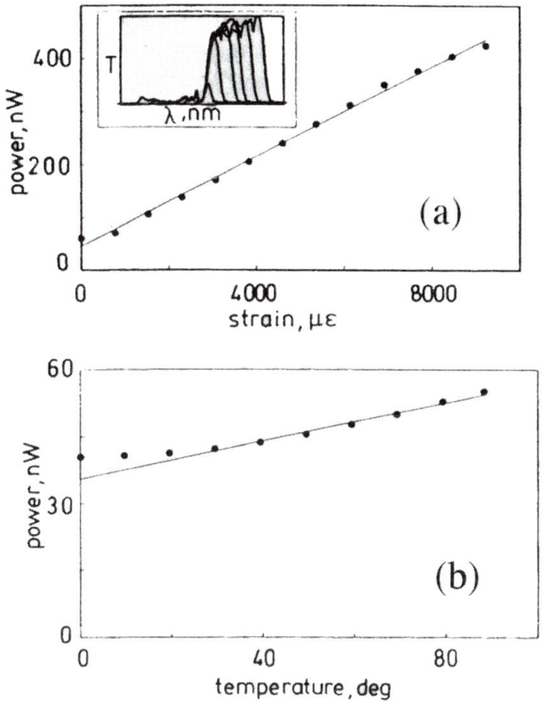

FIGURE 11.13. Variation of optical power received by a photodetector as a function of (a) strain and (b) temperature changes imposed on the sensing FBG. Small inset in (a) displays the transmitted spectra as function of the overlap between the sensing and reading FBGs. From Fallon, R. W., Zhang, L., Gloag, A. and Bennion, I., "Multiplexed Identical Broadband Chirped Grating Interrogation Technique for Large-Strain Sensing Applications," *IEEE Proceedings*, **9**, 1616–1618, © 1997, IEEE.

shifts of 1 pm to be achieved with a spectrometer having a spectral resolution of 0.1 nm (Ezbiri *et al.*, 1997), which translates to a strain resolution of 1 με. For example, the centroid detection algorithm is based on the sum of the intensity-weighted wavelengths divided by the spectrally integrated intensity, i.e.,

$$\lambda_B = \frac{\sum_j \lambda_j i_j}{\sum_j i_j}, \tag{11.5}$$

where i_j and λ_j represent the intensity and center wavelength of the jth-pixel, respectively, and λ_B is the Bragg wavelength.

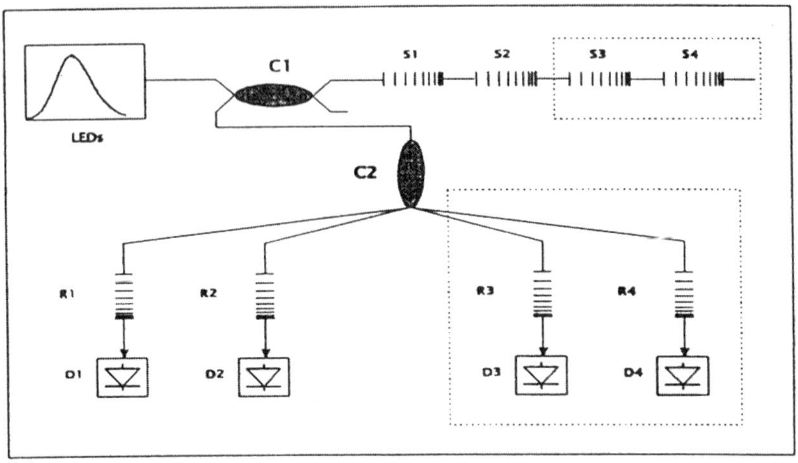

FIGURE 11.14. Schematic of one method of serial multiplexing and demodulating with matched FBGs. From Fallon, R. W., Zhang, L., Gloag, A., and Bennion, I., "Multiplexed Identical Broadband Chirped Grating Interrogration Technique for Large-Strain Sensing Applications," *IEEE Proceedings* **9**, 1616–1618, © 1997 IEEE.

The experimental layout used by Ezbiri *et al.* (1997) is schematically illustrated in Fig. 11.15 with the inset displaying the reflection spectra from five FBGs written into a single optical fiber. Each FBG had a length of 0.4 mm and a 0.6 nm spectral width, which corresponds to ~9 pixels. The FBGs shown had reflectivities of 7.5% at 810.8 nm, 2.5% at 819.3 nm, 12% at 825.3 nm, 2% at 827.5 nm, and 20% at 834.9 nm.

To overcome the wavelength limitation on the number of FBG sensors that can be multiplexed by such WDM schemes alone, Chen *et al.* (1997) combined spatial and wavelength multiplexing techniques. In essence, they wavelength-multiplexed several FBG sensors along each of an array of optical fibers that are focused by an optical system onto a bulk diffraction grating (see Fig. 11.16). The spectrally dispersed light from each of the optical fibers is then focused onto a two-dimensional CCD chip so that the image contains both wavelength and spatial (i.e., which optical fiber) information.

In the actual experimental test system of Chen *et al.* (1997), the light from a broadband source at about 830 nm is directed into two single-mode optical fibers by means of a star coupler. There are three FBGs with different reference center wavelengths (824, 830, and 854 nm) along one optical fiber and two (824 and 850 nm) along the second optical fiber. Figure 11.16b displays the spectral image projected on to the CCD chip.

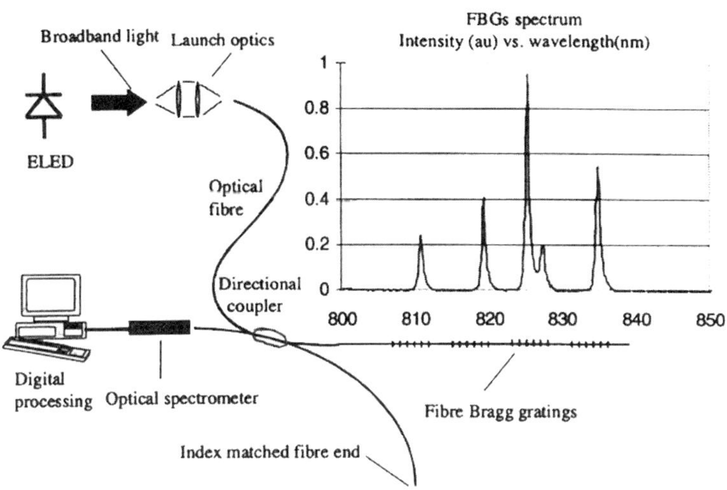

FIGURE 11.15. A demodulation system for a string of FBG sensors based on direct spectral measurement of the back-reflected light. The inset displays the measured spectra from a set of five FBGs with very different reflectivity. From Ezbiri, A., Kanellopoulos, S. E., Handerek, V. A., (1997), "High Resolution Instrumentation System for Demodulation of Bragg Grating Aerospace Sensors," OSA 12th Int. Conf. on Optical Fiber Sensors, Williamsburg, 456–459.

Unfortunately, because the CCD camera was low-cost (512×512 active pixels) and not designed for operation at around 850 nm, the images of the spectrally dispersed light from the core of each optical fiber is blurred and distorted, and not the small spot that could be achieved with optimized equipment. Even so, this system was estimated to be capable of interrogating 125 FBG sensors at a sample rate of 30 Hz using 25 optical fiber channels with five FBGs per optical fiber. Chen *et al.* (1997) also used digital algorithms to achieve subpixel resolution and a strain resolution of 1 $\mu\varepsilon$. In Fig. 11.17, the calculated peak centroid positions for all five FBG sensors is plotted while an increasing load was applied to FBG-2 and FBG-5; the other three FBG sensors were left alone. These results provide a glimmer of the potential of this approach.

Ecke *et al.* (1997) have suggested that changes in the state of polarization (SOP) of light reflected from FBGs can lead to systematic errors for many of the wavelength measurement techniques just described. These errors can arise when the emission (e.g., ELEDs), transmission (e.g., fiber couplers or acousto-optic tunable filters), or detection (e.g., grating spectrometers) components are polarization dependent. Polarization changes can arise when the optical fiber is laterally compressed or bent due to stress-induced birefringence. To avoid polarization effects it is

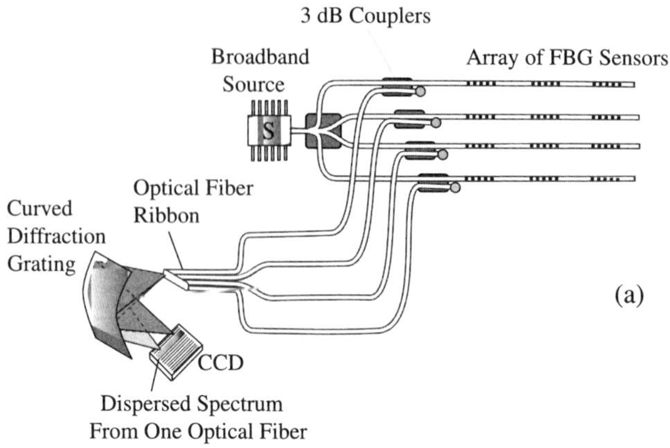

(a)

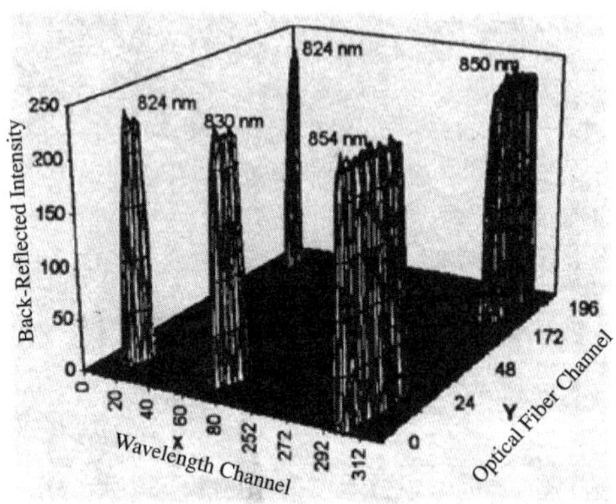

(b)

FIGURE 11.16. (a) A spectral demodulation system for an array of FBG sensors that combines spatial and wavelength multiplexing. (b) A display of the spectral image as projected onto a CCD chip showing the intensity of the back-reflected light as a function of wavelength for the different FBGs. From Chen, S., Hu, Y., Zhang, L., Bennion, I., (1997), "Digital Spatial and Wavelength Domain Multiplexing of Fiber Bragg Grating Based Sensors," OSA 12th Int. Conf. On Optical Fiber Sensors, Williamsburg, 448–451.

desirable to scramble the SOP in either the wavelength or time domains. Ecke *et al.* (1997) demonstrated that a passive fiber version of a Lyot filter is suitable for fast response systems while a mechanical system based on different lengths of highly birefringent optical fiber is suitable for narrow-band systems. If all of the optical components are made polarization-insensitive, then there is no need for any form of polarization scrambler.

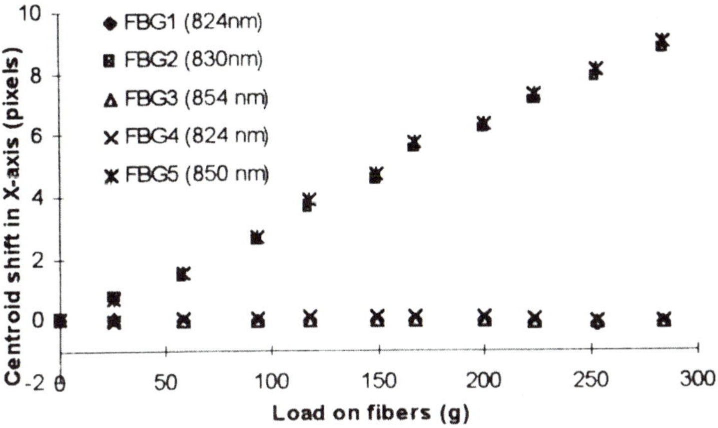

FIGURE 11.17. Variation with load in the wavelength of the five-FGBs portrayed in Fig. 11.17, as measured by the centroid shift in the position of the peak intensity on the CCD. From Chen, S., Hu, Y., Zhang, L., Bennion, I., (1997), "Digital Spatial and Wavelength Domain Multiplexing of Fiber Bragg Grating Based Sensors," OSA 12th Int. Conf. on Optical Fiber Sensors, Williamsburg, 448–451.

Twin FBG-Based Fabry–Perot Sensors. For applications such as pipelines, power lines, dams, or suspension bridges, where the structure could be hundreds to thousands of meters in length, WDM is not practical because of source bandwidth limitations. The possibility of creating Fabry–Perot interferometric sensors with twin fiber Bragg gratings has been explored from the perspective of serial multiplexing large numbers of FBGs.

In this approach, an identical pair of fiber Bragg gratings are formed in an optical fiber with a predetermined separation (Fig. 11.18a). Under these circumstances the reflection spectrum comprises a cosine wave modulated by the FBG spectrum (Fig. 11.18b). The frequency of the

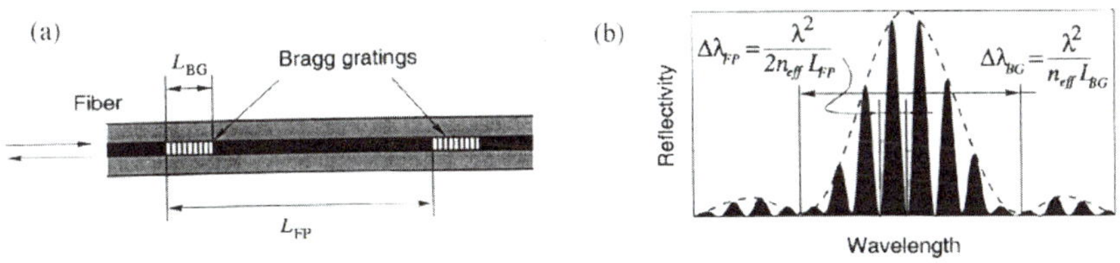

FIGURE 11.18. (a) Schematic of a fiber Fabry-Perot interferometric sensor formed by a pair of identical FBGs. (b) Reflection spectrum of the FBG-based Fabry–Perot sensor. From Shlyagin, M. G., Miridonov, S. V., Tentori, D., Castillo, J., (1997), "Twin Grating-Based Interferometric Fiber Sensor," OSA 12th Int. Conf. on Optical Fiber Sensors, Williamsburg, 472–474.

cosine wave is determined by the separation of the pair of identical FBGs and can be used to identify the sensors from their peaks in the frequency domain of the Fourier transform. Changes in the measurand (strain or temperature) shift the peak of the Bragg spectrum and can be determined from the phase shifts of the corresponding frequency components. As indicated above, the advantages of this approach are that the number of sensors is not restricted by the bandwidth of the light source, as is the case in pure WDM systems.

Figure 11.19 displays schematically how this approach permits both wavelength and frequency multiplexing. As can be seen, the wavelength of the reflection spectrum for a fixed separation of FBGs is controlled by the Bragg wavelengths of the gratings used to form the sensor. Changing the physical separation of the FBG pair changes the modulation frequency within the FBG envelope. In the lower part of Fig. 11.19, Shlyagin *et al.* (1997) have suggested a possible approach at detection and demodulation of such a set of fiber Bragg grating Fabry–Perot (FBG-FP) sensors.

If a string of such FBG-FP sensors are formed within an optical fiber, crosstalk arising from multiple reflections can be avoided if the reflectivity of the FBGs is kept very small. Shlyagin *et al.* (1997) have suggested that

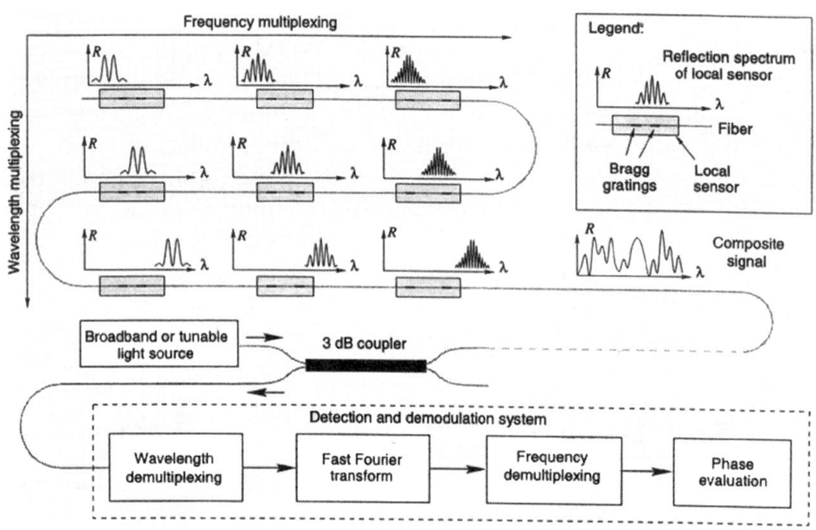

FIGURE 11.19. Demodulation system FFP sensors formed by pairs of identical FBGs. The sensors are distinguished by their FP-frequency spectrum and the wavelength of their FBGs. From Shlyagin, M. G., Miridonov, S. V., Tentori, D., Castillo, J., (1997), "Twin Grating-Based Interferometric Fiber Sensor," OSA 12th Int. Conf. on Optical Fiber Sensors, Williamsburg, 472–474.

the reflectivity must be less than 0.13% if 30 such FBG-FP sensors (formed from 60 FBGs) are to be used along a single strand of optical fiber. The low reflectivity of the FBGs ensures that the Fabry–Perot interferometer has a low finesse and a reflection spectrum of the form

$$R_{\mathrm{FP}}(\lambda) = R_{\mathrm{FBG}}(\lambda)\left[1 + \cos\left\{\frac{4\pi n_{\mathrm{eff}} L_{\mathrm{FP}}}{\lambda}\right\}\right], \qquad (11.6)$$

where $R_{\mathrm{FP}}(\lambda)$ is the FBG-FP reflection spectrum, $R_{\mathrm{FBG}}(\lambda)$ is the FBG reflection spectrum, n_{eff} is the effective index of refraction of the optical fiber, λ is the free-space wavelength, and L_{FP} is the Fabry–Perot cavity length, i.e., the FBG separation (see Fig. 11.18). An example of the reflection spectrum and its Fourier transform for a FBG-FP made from two short (1-mm-long) FBGs separated by 7.5 mm is presented as Fig. 11.20. The measured reflection efficiency of the gratings was about 0.5%.

Measures *et al.* (1998) have shown how Fourier transform analysis can be used to identify each FBG-FP. If $f(\lambda)$ describes the reflection (or transmission) intensity spectrum for a given pair of FBGs, acquired by scanning the wavelength of a laser over the spectral interval $(\lambda_0, \lambda_0 + \Delta\lambda)$, then the corresponding frequency spectrum (of wavelength oscillations) for this pair of FBGs can be derived from the Fourier transform of their intensity spectrum $f(\lambda)$;

$$F(\sigma) \approx \int_{\lambda_0}^{\lambda_0+\Delta\lambda} f(\lambda) e^{-2\pi\lambda\sigma} d\lambda, \qquad (11.7)$$

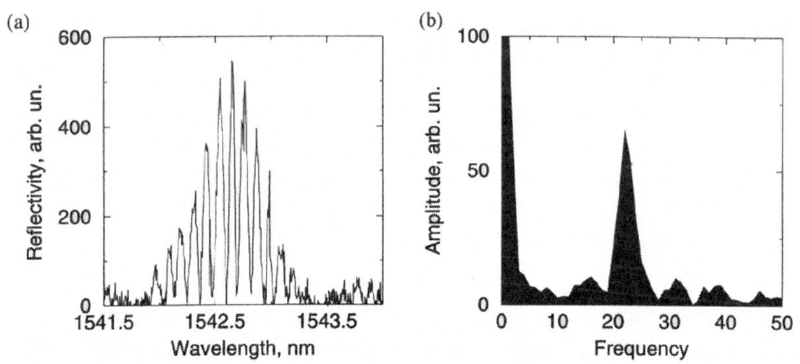

FIGURE 11.20. (a) Reflection spectrum, and (b) Fourier transform for an FBG-FP sensor made of twin FBGs separated by ~ 7.5 mm and operating at a wavelength of ~ 1542 mm. From Shlyagin, M. G., Miridonov, S. V., Tentori, D., Castillo, J., (1997), "Twin Grating-Based Interferometric Fiber Sensor," OSA 12th Int. Conf. on Optical Fiber Sensors, Williamsburg, 472–474.

where σ represents the frequency of wavelength oscillations in the intensity spectrum and corresponds to the inverse of the wavelength separation of the Fabry–Perot modes, $\delta\lambda$, formed by the pair of FBGs, i.e.,

$$\sigma \equiv \frac{1}{\delta\lambda}, \tag{11.8}$$

with

$$\delta\lambda \equiv \frac{\lambda^2}{2nL}. \tag{11.9}$$

Since $\delta\lambda$ is inversely proportional to the Fabry–Perot cavity length L, we see that the Fourier transform of the intensity spectrum gives directly the separation of the FBG pair.

A schematic of the experimental arrangement used by Measures *et al.* (1998) in a preliminary test of this concept is presented as Fig. 11.21. Five pairs of FBGs were produced in germanium-doped photosensitive optical fiber with measured grating pair separations plus penetration depths (i.e., Fabry–Perot cavity lengths) of $L_{\mathrm{I}} = 9.3$ cm; $L_{\mathrm{II}} = 10.0$ cm; $L_{\mathrm{III}} = 11.1$ cm; $L_{\mathrm{IV}} = 11.9$ cm, and $L_{\mathrm{V}} = 12.6$ cm. Each grating was about 1 cm long and each pair of gratings was separated from its neighbors by approximately 3 m of Corning SMF-28 optical fiber. This 3 m of fiber loop ensured that each FBG pair was out of coherence with the adjacent FBG pair so that interference was restricted to the pairs of FBGs and little crosstalk occurred. Each FBG-FP sensor was subject to its own strain state as each segment of optical fiber containing a FBG pair is bonded at one end to a stationary block while the other end is bonded to a traveling stage controlled by a micrometer (Fig. 11.22).

The reflection and transmission spectra of five pairs of FBGs are displayed in Fig. 11.22, when a strain of about 380 $\mu\varepsilon$ is applied to all of the FBGs. The spectral walkoff (as seen in transmission) for the reflection spectrum of FBG pairs I, II, and IV when a strain of approximately 380 $\mu\varepsilon$ is applied to each FBG-FP sensor, in turn, can be seen in Fig. 11.23. Although it is not possible to tell which FBG-FP had been stressed from the reflection spectra, the Fourier transform of the wavelength displaced reflection spectrum indicates clearly which pair of FBG has been subject to loading (see Fig. 11.24). Each FBG-FP sensor can be seen to have a distinctive frequency spike in the Fourier transform frequency spectrum corresponding to the resonance frequency of its Fabry–Perot cavity length.

In these experiments a strain of about 60 $\mu\varepsilon$ applied to one FBG-FP sensor appears to be adequate to separate its spectrum from the composite spectrum of the remaining FBG-FP sensors. Furthermore, a difference of just 1 cm in the separation of the two gratings constituting a FBG-FP

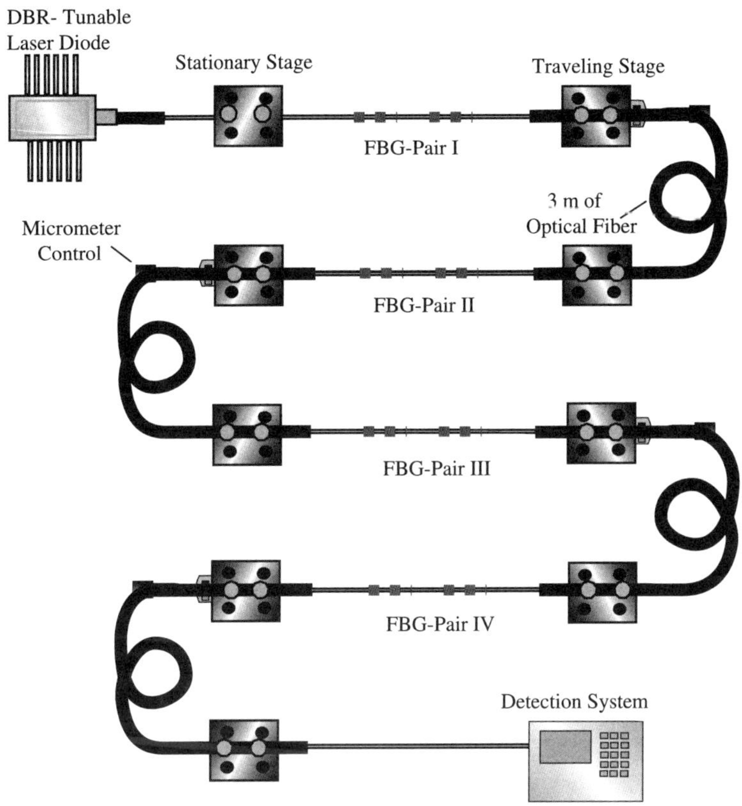

FIGURE 11.21. An experimental arrangement used to demonstrate that Fourier transform analysis can demodulate a string of FBG-FP sensors. The tunable DBR laser wavelength is scanned to produce the reflection spectrum for each FBG-FP sensor. Micrometers on the traveling stages permit each sensor to be precisely strained. From Measures, R. M., Ohn, M. M., Huang, S. Y., Bigure, J., Fan, N. Y., (1998), "Tunable Laser Demodulation of Various Fiber Bragg Grating Sensing Modalities," *Smart Materials and Structures Journal*, **7**, 237–247, with permission from the Institute of Physics Publishing Ltd.

sensor is sufficient for the Fourier transform analysis to identify which FBG pair is involved. Even though these experiments are preliminary, they suggest that the technique is worthy of further consideration, as it may be possible to greatly improve on these figures with a less noisy laser.

Dakin *et al.* (1997) have proposed an alternative method of interrogating such a string of FBG-FP sensors using low-coherence interferometry. A schematic illustration of their suggested system is depicted in Fig. 11.25a. The Bragg wavelength for each FBG-FP sensor can be determined from

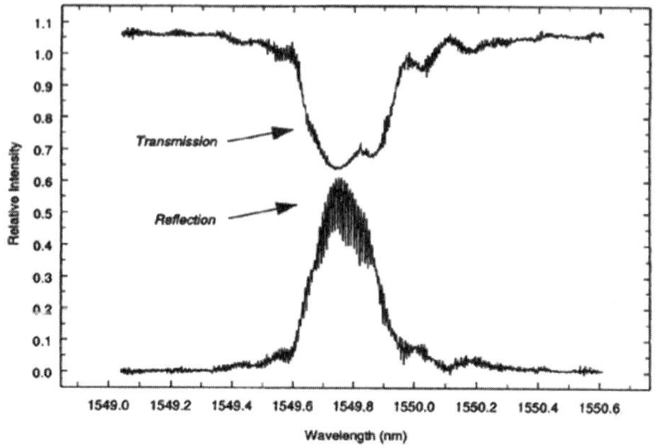

FIGURE 11.22. Combined reflection and transmission spectra for the set of paired FBGs forming FFP sensors, when a strain of 380 με is applied to all of them. From Measures, R. M., Ohn, M. M., Huang, S. Y., Bigure, J., Fan, N. Y., (1998), "Tunable Laser Demodulation of Various Fiber Bragg Grating Sensing Modalities," *Smart Materials and Structures Journal*, **7**, 237–247, with permission from the Institute of Physics Publishing Ltd.

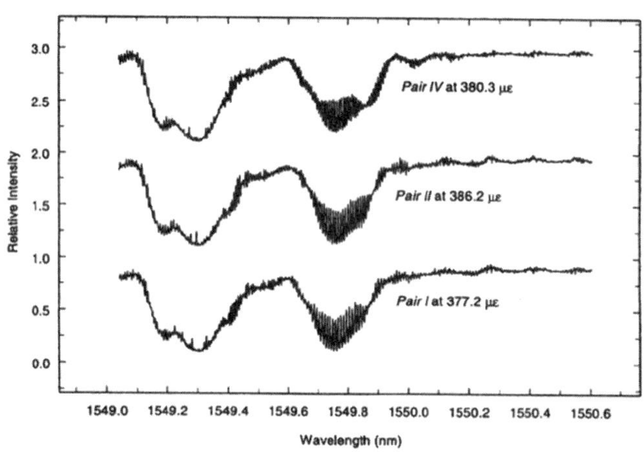

FIGURE 11.23. Spectral walkoff when about 380 με is applied to each of the FBG-pairs (I, II, and IV) in turn. From Measures, R. M., Ohn, M. M., Huang, S. Y., Bigure, J., Fan, N. Y., (1998), "Tunable Laser Demodulation of Various Fiber Bragg Grating Sensing Modalities," *Smart Materials and Structures Journal*, **7**, 237–247, with permission from the Institute of Physics Publishing Ltd.

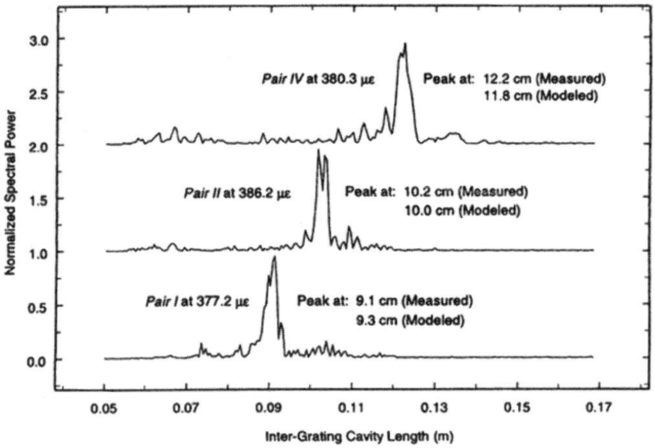

FIGURE 11.24. Fourier transforms of the reflection spectra for FBG pairs I, II, and IV, revealing the measured spacing of each FBG pair compared to the modeled value. From Measures, R. M., Ohn, M. M., Huang, S. Y., Bigure, J., Fan, N. Y., (1998), "Tunable Laser Demodulation of Various Fiber Bragg Grating Sensing Modalities," *Smart Materials and Structures Journal*, **7**, 237–247, with permission from the Institute of Physics Publishing Ltd.

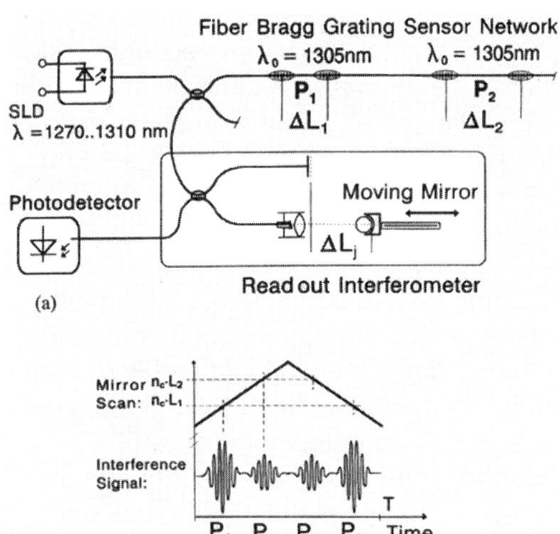

FIGURE 11.25. (a) Experimental arrangement for using low-coherence interferometry to interrogate a string of FBG-FP sensors. (b) Interferometric fringe patterns produced by the scan of the moving mirror in the readout Michelson interferometer. From Dakin, J. P., Ecke, W., Rothardt, M., Schauer, Usbeck, K., and Willsch, R., (1997), "New Multiplexing Scheme for Monitoring Fiber Optic Bragg Grating Sensors in the Coherence Domain," OSA 12th Int. Conf. on Optical Fiber Sensors, Williamsburg, 31–34.

the temporal spacing of each pattern of interference fringes, formed sequentially as the readout-Michelson interferometer's path length is scanned (by moving the mirror), and matches that of each grating pair in turn (see Fig. 11.25b). Which sensor is selected is of course known from its place in the sequence, as the separation of the FBG pairs is designed to increase with position along the optical fiber. The readout interferometer essentially separates the returns from each FBG-FP in the coherence domain.

Using this technique a strain resolution of $\sim$1 $\mu\varepsilon$ is expected in a measurement time $\sim$1 s. Unfortunately, in an initial experiment Dakin *et al.* (1997) observed that the residual FP modes of the SLD source forced the spacing of the grating pairs to be greater than 20 mm because of the large coherence of the source. Use of a broadband source with a smoother spectral distribution could reduce this to the coherence length corresponding to the 0.5-nm bandwidth of the FBGs. This would permit sensor lengths of less than 5 mm.

11.3.2 Time division multiplexing

In time division multiplexing, the position of each sensor along an optical fiber is determined from the time it takes a very short duration laser pulse to travel out to the sensor and return to the photodetector (see Fig. 5.58). Repetitive pulsing of the system allows each sensor to be addressed in turn by simple selective time gating of the photodetector. The term *optical time domain reflectometry* (OTDR) is often used to describe this approach to serial multiplexing.

One of the simplest forms of serially multiplexed fiber optic sensors uses a set of microbend sensors strung out along an optical fiber (Kersey, 1997b). The attenuation imposed on the light traveling through the section of optical fiber under the influence of the microbend sensor increases with the transverse pressure applied to the device. The degree of this attenuation and its location can be determined by an OTDR (see Fig. 11.26). This approach suffers a major flaw, however. The loss of light is cumulative, so the signal-to-noise ratio attained from those sensors that are furthest away becomes very poor, especially if any intervening sensor imposes a large loss.

The use of partial reflectors along a length of optical fiber permits in-line serial multiplexing of interferometric sensors, such as low-finesse fiber Fabry–Perot sensors (Lee and Taylor, 1988). Interrogation with a pulsed light source is practical, providing the duration of the optical pulse, τ, is less than the *round-trip delay* T_d ($= 2nL/c$), between reflectors

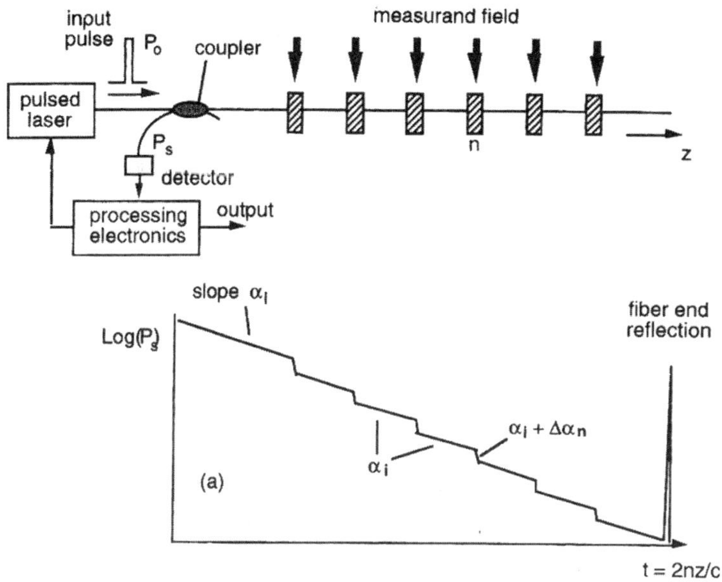

FIGURE 11.26. Schematic illustration of a serial multiplexed array of microbend sensors, using a short pulse laser and fast detection electronics. From Kersey, A. D., (1997a), "Multiplexing Techniques for Fiber Optic Sensors," *Optical Fiber Sensors*, Vol. 4, J. Dakin and B. Culshaw (Eds). Artech House, Inc., Norwood, MA, USA.

separated by a distance L. An interferometric signal from each sensor in the array can be generated in time sequence by applying a double-input pulse interrogation signal, with the two input pulses separated by an interval equal to the round-trip delay T_d, such that optical pulses reflected from consecutive partial reflectors in the array overlap and undergo interference. If the optical frequencies of the two input optical pulses are set to be different, the overlapping pulses generate a *beat frequency*. This concept, termed *differential delay heterodyne* interrogation, was the scheme originally used with this multiplexing arrangement (Kersey, 1997b).

Unfortunately, this reflective topology gives rise to multiple pulse interactions and crosstalk, the magnitude of which is proportional to the mirror reflectivity, suggesting the need for low reflectivities. The fabrication of low-reflectivity splices with consistent reflectance was originally problematic, but the advent of fiber Bragg gratings has provided an ideal solution to this problem. As discussed earlier, the use of fiber Bragg gratings also provides wavelength selectivity to such reflectors.

If spatial resolution of a few centimeters is desired for strain or temperature profile mapping by means of serial multiplexing in the time domain, then very short optical pulses are required. A spatial resolution of ΔL requires that the optical light pulse duration be

$$\tau \leq 2n\Delta L/c, \tag{11.10}$$

where n is the index of refraction of the optical fiber and c the free-space speed of light. Thus, if we need a spatial resolution of 1 cm, then the optical pulse duration should be less than $2 \times 15 \times 1\ \text{cm}/3 \times 10^{10}\ \text{cm s}^{-1} = 10^{-10}\ \text{s} = 0.1\ \text{ns}$ or 100 ps. This involves extremely fast electronics but is achieved with the commercial OTDR of Opto-Electronics Incorporated, Ontario (see Fig. 6.6). However, for most structural monitoring applications this solution is rather expensive.

Mode-Locked Lasers. In addition to requiring very fast electronics, time division multiplexing with good spatial resolution also suffers from low-intensity signals and corresponding poor signal-to-noise ratios. One way around this difficulty is to use a mode-locked laser as the source of very short-duration optical pulses. This type of laser can generate extremely intense light pulses of subpicosecond duration. The problem with this approach is that in the past such lasers were large solid-state lasers that could not be developed into portable field-site demodulation systems.

This situation is on the verge of changing with the development of *passively mode-locked ring fiber lasers* (Tamura *et al.*, 1994). In this type of laser, light builds up within a length of erbium–yittrium-doped fiber amplifier (EYDFA) as it is pumped by an external laser. The fiber laser

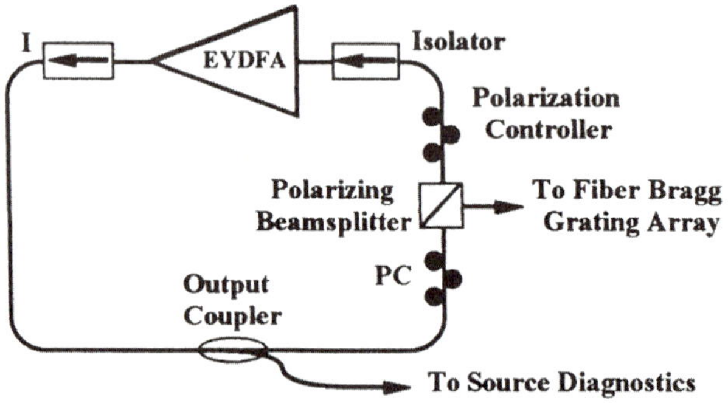

FIGURE 11.27. Schematic of mode-locked fiber laser arrangement to be used for OTDR. From Putnam, M. A., Dennis, M. L, Kang, J. U., Tsai, T. E., Duling, I. N., Friebele, E. J., (1997), "Sensor Grating Demodulation Using a Passively Mode Locked Fiber Laser," *SPIE*, **3042**, 352–357.

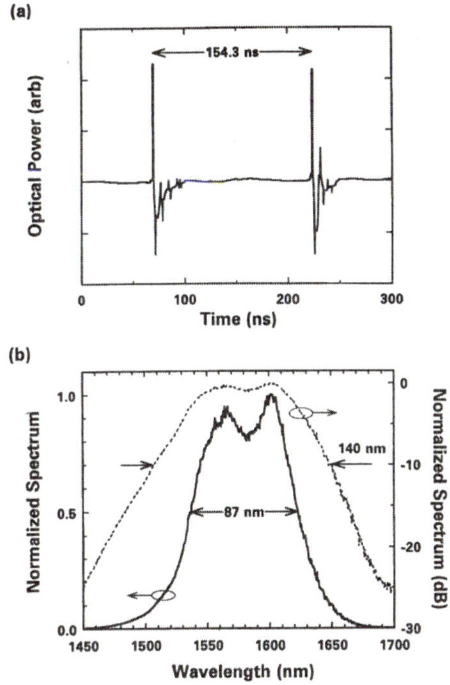

FIGURE 11.28. (a) Pair of representative laser pulses from the mode-locked fiber laser illustrated in Fig. 11.28. (b) Normalized spectrum of mode-locked laser pulse presented on both linear and logarithmic scales. From Dennis, M., Putnam, M., Kang, J., Tsai, T., Duling, I., Friebele, E. J., (1997), "Grating Sensor Array Demodulation by Use of a Passively Mode-Locked Fiber Laser" *Optics Letters*, **22**, 1362–1364.

light is effectively trapped within the optical fiber loop schematically depicted in Fig. 11.27. Nonlinear birefringence locks the large number of frequency modes within the broad gain profile of this highly doped optical fiber, causing a single, broadband laser pulse to be generated per round trip around the ring fiber laser cavity. Two representative laser pulses, as captured by a sampling oscilloscope, are displayed in Fig. 11.28a, while their broadband spectrum is shown as Fig. 11.28b.

Greater details of this laser can be found in the paper by Dennis *et al.* (1997). Note that the 31.4-m length of the laser cavity gives rise to a pulse repetition rate of 6.48 MHz, which in turn limits the length of the string of FBG sensors to less than 20 m for TDM interrogation. If the separation of the first and last FBG sensor was greater than 20 m there would be overlap in the pulse return. This limitation was overcome by the inclusion of 500 m of dispersion-shifted fiber into the passively mode-locked fiber laser. This improves the performance of the laser by increasing the nonlinear interaction length and cavity round-trip time. The latter effect decreases the pulse repetition rate, allowing a string of FBG sensors up to 250 m in length to be interrogated.

Since nonlinear birefringence is used to mode-lock the laser, the greater interaction length lowers the pump power threshold for switching and allows a lower-power, inexpensive laser diode to be used as the pump laser. An additional merit of this demodulation scheme is that the high spectral brightness of the source permits the use of low-reflectivity FBG arrays produced on the draw tower and fabricated at relatively low cost (Askins *et al.*, 1994). A demonstration of this is provided in Fig. 11.29, where the reflected signals from one 5% FBG and two 2% FBGs, as recorded by a sampling oscilloscope and a 125-MHz high-gain InGaAs detector, are displayed (Putnam *et al.*, 1997).

Dennis *et al.* (1997) developed a novel method of interrogating a string of FBG sensors by converting the strain- or temperature-induced wavelength shifts they experience into changes in the arrival time of the ultrashort pulses reflected from the gratings. This dispersive wavelength–time conversion interrogation technique is schematically illustrated in Fig. 11.30 and is made possible by virtue of the extremely short duration of the laser pulses. A 3.25-km length of standard dispersion compensation fiber (DCF) is introduced before the FBG array and performs this transformation by means of chromatic dispersion. A representative example of this effect is displayed in Fig. 11.31. The temporal

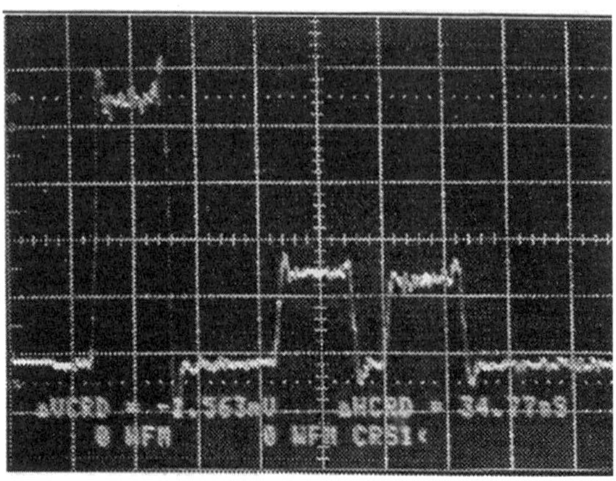

FIGURE 11.29. Reflection of mode-locked laser pulses from low-reflectivity FBGs produced on-line on the draw tower. From Putnam, M. A., Dennis, M. L., Kang, J. U., Tsai, T. E., Duling, I. N., Friebele, E. J., (1997), "Sensor Grating Demodulation Using a Passively Mode Locked Fiber Laser" *SPIE*, **3042**, 352–357.

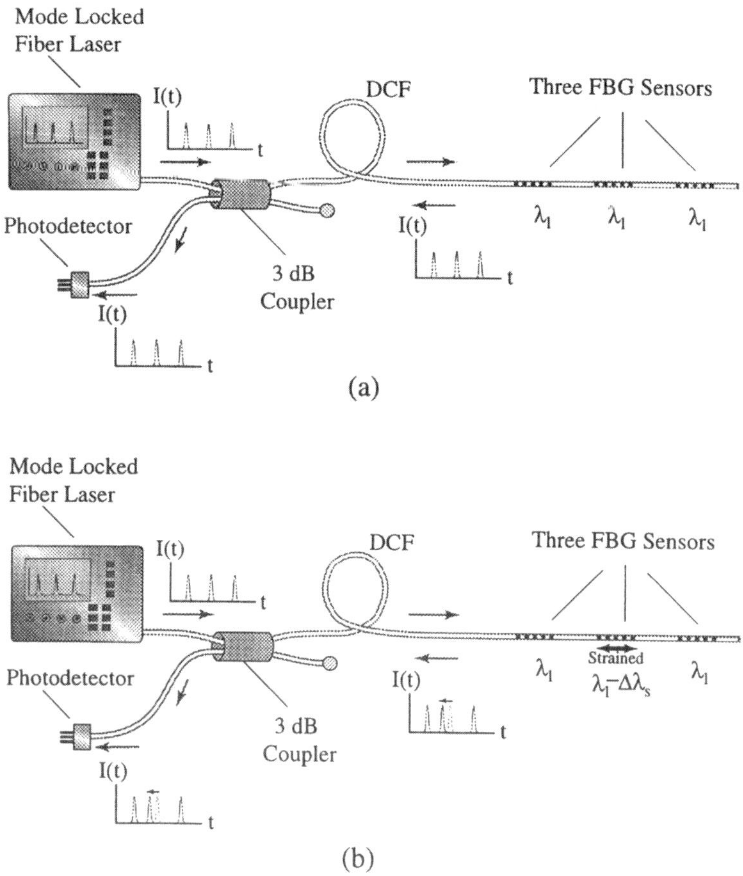

FIGURE 11.30. Schematic of a mode-locked fiber laser system for demodulating a string of FBG sensors based on dispersive wavelength–time conversion. A comparison of the upper and lower sketches reveals that the light pulse reflected from the middle FBG has its time of arrival shifted commensurate with the wavelength shift experience by this FBG.

delay introduced by the DCF is seen to be a linear function of the applied strain for this system in Fig. 11.32. In this approach the 20-ps time resolution of the photodetection system corresponds to about 20-$\mu\varepsilon$ strain resolution.

When standard wavelength-domain demodulation is used, the maximum strain that can be measured by a FBG sensor in an array is limited by the spectral separation between adjacent gratings. This restriction is removed in this approach. Since only the measurand-induced wavelength shift (i.e., temporal delay) is measured, the wavelengths of different

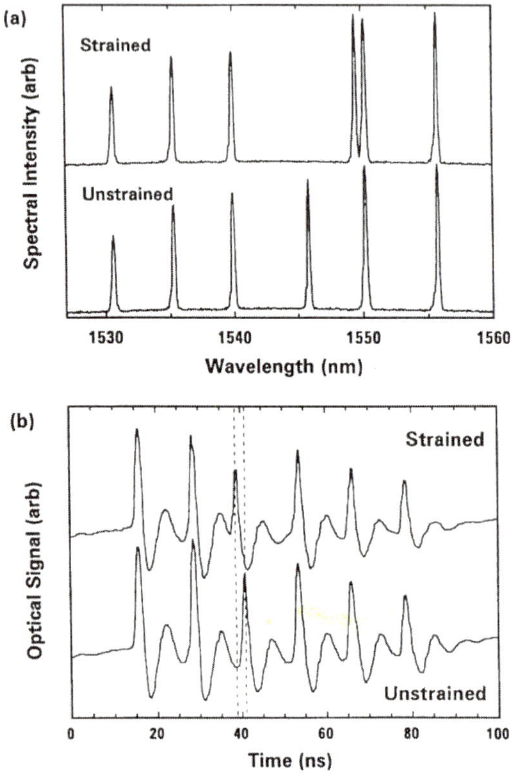

FIGURE 11.31. (a) Experimental set of reflection spectra from a string of FBGs corresponding to the straining and unstraining of the FBG at about 1545 nm. (b) Corresponding temporal signals reflected by this set of FBGs. It is clear that the pulse reflected from the FBG, experiencing a shift in its center wavelength, is also time shifted. From Dennis, M., Putnam, M., Kang, J., Tsai, T., Duling, I., Friebele, E. J., (1997), "Grating Sensor Array Demodulation by Use of a Passively Mode-Locked Fiber Laser," *Optics Letters*, **22**, 1362–1364.

gratings can be permitted to shift past one another, or even overlap spectrally, so long as sufficient light is returned from each FBG for determination of the arrival time of its reflection. This yields a greater dynamic range of operation or increases the number of gratings per spectral bandwidth in an array. It is also worth noting that the localized strain over the length of the array can be obtained simply from a comparison of the direct temporal interrogation (analogous to standard optical time-domain reflectometry) with this dispersive wavelength–time conversion interrogation technique (Putnam *et al.*, 1997).

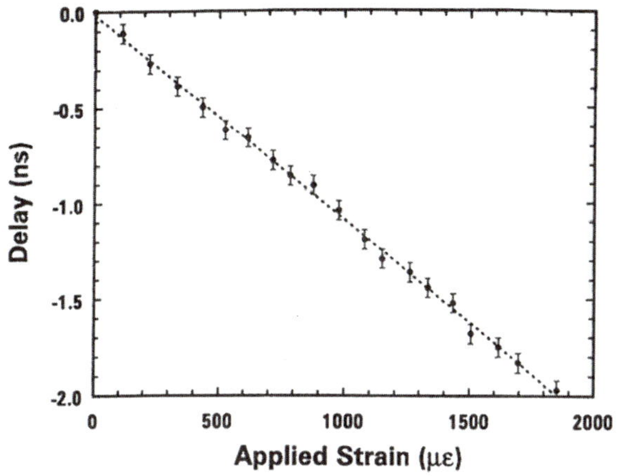

FIGURE 11.32. Variation in the measured delay time of the mode-locked laser pulse reflected from a strain FBG. From Dennis, M., Putnam, M., Kang, J., Tsai, T., Duling, I., Friebele, E. J., (1997), "Grating Sensor Array Demodulation by Use of a Passively Mode-Locked Fiber Laser," *Optics Letters*, **22**, 1362–1364.

Interferometric and TDM Multiplexing. We saw in Section 9.1.3 that an unbalanced Mach–Zehnder interferometer can be used as a wavelength-shift detector, translating strain (or temperature)-induced changes in the Bragg wavelength reflected from a FBG sensor into variations in the phase difference of the interferometer; see Fig. 9.20. As suggested by Kersey *et al.* (1992a), this approach can also be used for serial multiplexing of a string of FBG sensors using time-division addressing as depicted in Fig. 11.33. Pulsed light from a broadband source is launched into an optical fiber containing N fiber Bragg grating sensors along its length. Each FBG has a different peak reflection wavelength. The optical pulse duration is arranged to be equal to, or less than, the round-trip time between two gratings. The reflected pulses are directed through the Mach–Zehnder (MZ) interferometer that acts as the wavelength-shift detector. The output of the MZ interferometer is then electronically time-gated to address any FBG sensor in that array.

The information contained in the phase modulated output of the interferometer can be extracted using a pseudo-heterodyne technique in which a resetting electrical ramp (serrodyne) signal is applied to the piezoelectric fiber stretcher (PZ in Fig. 11.33) that produces a linearly changing phase shift of $\sim 2\pi$ in one arm of the MZ interferometer. After bandpass filtering (BPF) at the fundamental frequency of the serrodyne signal, the output signals are of the form

$$S(\lambda_j) = A\cos\{\omega_0 t + \Delta\phi(\lambda_j)\}, \qquad (11.11)$$

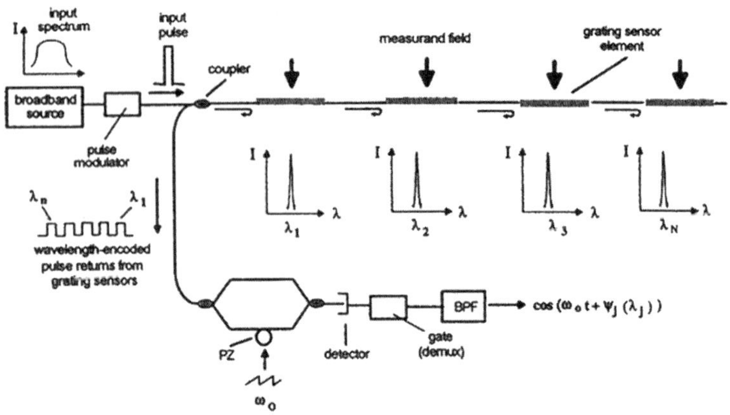

FIGURE 11.33. Experimental layout of an interferometric demodulation scheme for serially demultiplexing a string of FGB sensors. From Weis, R.S., "A Four-Element Fiber Grating Sensor Array with Phase-Sensitive Detection," *IEEE Photonics Technology Letters*, **6**, 1469–1472, © 1994, IEEE.

where λ_j is the peak reflected wavelength from the jth FBG sensor, ω_0 is the angular frequency of the voltage signal applied to the PZ, and $\Delta\phi(\lambda_j)$ represents the phase change produced in the MZ interferometer resulting from the shift in the Bragg wavelength of the jth FBG sensor; see Eq. (9.6). Thus, the FBG sensor wavelength shifts are linearly transposed into phase shifts of temporally separate low-frequency electrical carrier signals.

Using this approach, Weis (1994) demonstrated that four-FBG sensors along an optical fiber could be time-division multiplexed. They used an ELED with a center wavelength of 1306 nm (FWHM of 24 nm) and a pulse duration of 50 ns with a 1 : 5 duty cycle. The FBG sensors were spaced 5 m apart (to keep the cost of electronics reasonable) and had Bragg wavelengths between 1284 and 1290 nm, with reflectivities that ranged from 60 to 99% and spectral widths in the range 0.3 to 0.5 nm. A fiber MZ interferometer with a optical path difference OPD = 1 mm, refractive index $n = 1.46$, a normalized strain responsivity (or what we have defined as the strain-gauge factor G_ε, in Section 7.3) of the FBG of $0.74 \times 10^{-6}\ \mu\varepsilon^{-1}$, and a free-space wavelength $\lambda_0 = 1300$ nm yields a strain-induced phase shift responsivity of 5.2 mrad $\mu\varepsilon^{-1}$.

The high-resolution, dynamic phase shift detection capability (typically about 1 μrad $\mu\varepsilon^{-1}$) achievable with interferometric sensors coupled with this phase shift responsivity gives a possible dynamic strain resolution of ~200 pε Hz$^{-0.5}$. Figure 11.34 displays a set of multiplexed signals resulting from the phase modulation of the MZ interferometer. From the

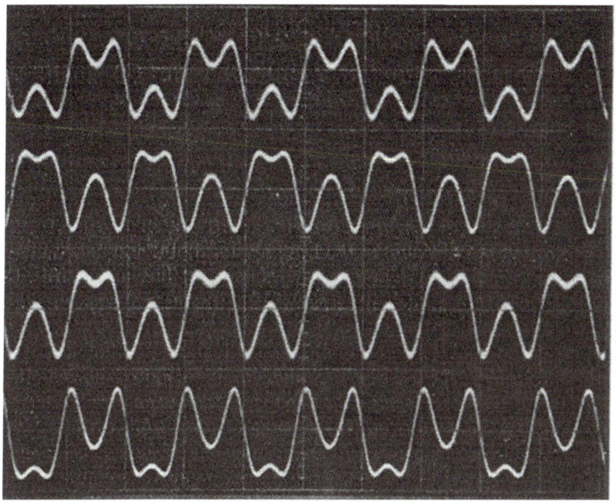

FIGURE 11.34. Set of multiplexed signals resulting from phase modulation of the Mach–Zehnder interferometer shown in Fig. 11.33. From Weis, R.S., "A Four-Element Fiber Grating Sensor Array with Phase-Sensitive Detection," *IEEE Photonics Technology Letters*, **6**, 1469–1472, © 1994, IEEE.

signal-to-noise measurements of the gated signals , the minimum detectable strain was determined to be about 2 nε Hz$^{-0.5}$.

Combination of WDM and TDM. The limited bandwidth of light sources compared to the possible wavelength excursions required for sensing restricts the number of FBG sensors that can be wavelength multiplexed. The combination of TDM and WDM offers a practical method of addressing a large number of FBG sensors. This concept is illustrated in Fig. 11.35 and has been demonstrated using a 3×3 linear array of FBG sensors by Berkoff *et al.* (1995) (Fig. 11.36). The gratings used were of low reflectivity (5 to 10%) and written at three different nominal wavelengths: $\sim$1530 nm (λ_1), 1535 nm (λ_2), and 1540 nm (λ_3). The pulsed broadband source comprised a pair of erbium-doped optical fiber amplifiers separated by an integrated optic switch driven by a pulse generator. This source produced 100 ns, 1.5 mW optical pulses at a 600-ns repetition rate. The voltage pulse applied to the source with the corresponding three photodetected return signals (each including the three wavelength-encoded FBG sensor signals) is presented as Fig. 11.37.

Detection of the wavelength shifts for the three FBG sensors in one of the three subarrays of FBGs was accomplished by using a scanning Fabry–Perot filter and a given electronic delay for the time-gated photo-

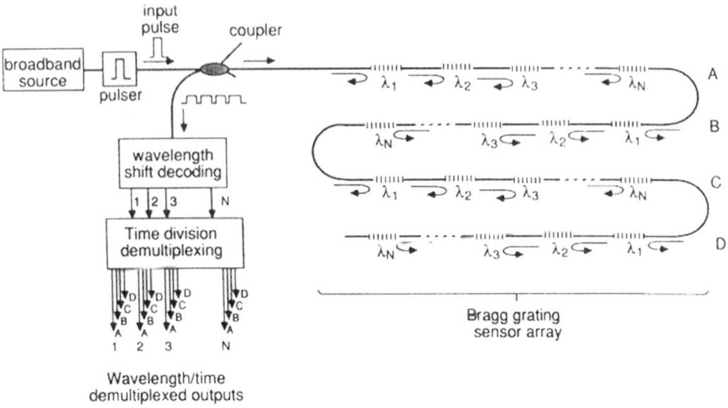

FIGURE 11.35. Schematic of combined time and wavelength multiplexing concept for a large number of FBGs along an optical fiber. From Kersey, A. D., "Fiber Optic Sensor Multiplexing Techniques" In "Fiber Optic Smart Structures" Eric J. Udd, Ed. ©1995 Wiley Interscience. Reprinted by permission of John Wiley and Sons, Inc.

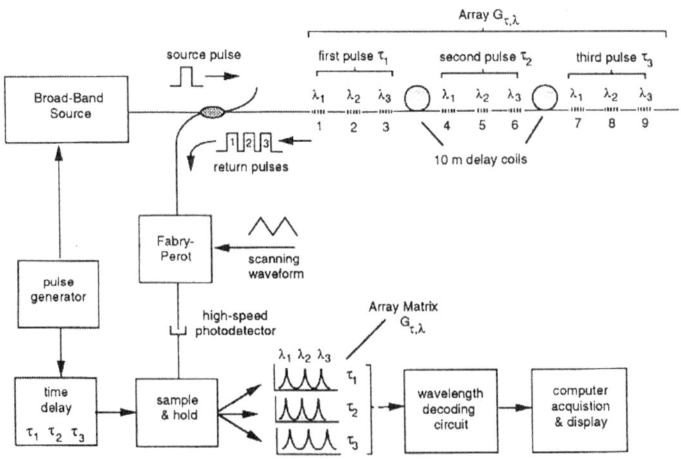

FIGURE 11.36. An experimental arrangement designed to demodulate a string of FBG sensors using a combination of wavelength division multiplexing and time gating of the photodetection system. From Berkoff, T.A, Davis, M.A., Bellemore, D.G., Kersey, A.D., Williams G.M., and Putnam M.A., (1995), "Hybrid Time and Wavelength Division Multiplexed Fiber Bragg Grating Sensor Array," *SPIE*, **2444**, 288–294.

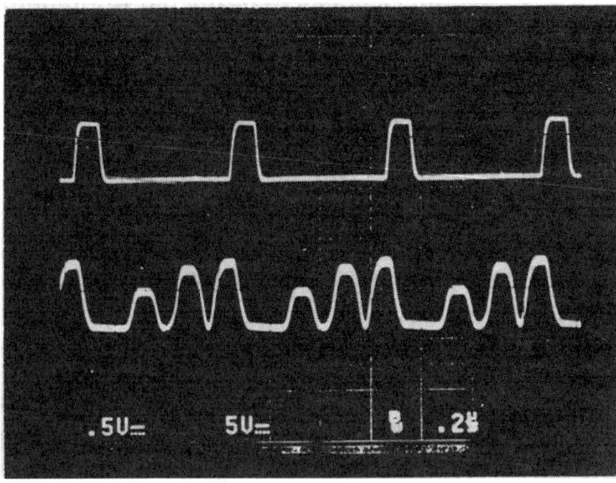

FIGURE 11.37. Upper trace displays a set of broadband pulses launched into the optical fiber with the string of FBG sensors, while the lower trace shows the set of back-reflected optical signals from each of the three FBGs. From Berkoff, T.A, Davis, M.A., Bellemore, D.G., Kersey, A.D., Williams G.M., and Putnam M.A., (1995), "Hybrid Time and Wavelength Division Multiplexed Fiber Bragg Grating Sensor Array," *SPIE*, **2444**, 288–294.

detector so it was only able to accept light reflected from that group of FBGs. To provide enough optical delay for the cost of the electronics to be reasonable, the three subarrays were connected in series using 10 m of optical fiber; see Fig. 11.36. A series of outputs obtained from the sample and hold circuits when the time delay is adjusted to the three different time channels of the system are displayed as Fig. 11.38. The fully demultiplexed and demodulated strain signals obtained for all nine FBG sensors, when a 1-Hz sinusoidal 100 $\mu\varepsilon$ signal was applied to the ninth sensor in the array, is presented as Fig. 11.39.

11.3.3 Coherence division multiplexing

One of the main advantages of low-coherence (or white-light) interferometry over high-coherence (or single-frequency) interferometry is that the former yields an absolute measurement and is therefore interrupt immune; see Sections 3.8 and 10.2.2. *Coherence division multiplexing* refers to the technique for interrogating an array of fiber optic sensors based on low-coherence interferometry. In one of the more popular forms of this technique, a reference Michelson interferometer is used to interrogate an array of low-coherence sensors by matching its path imbalance

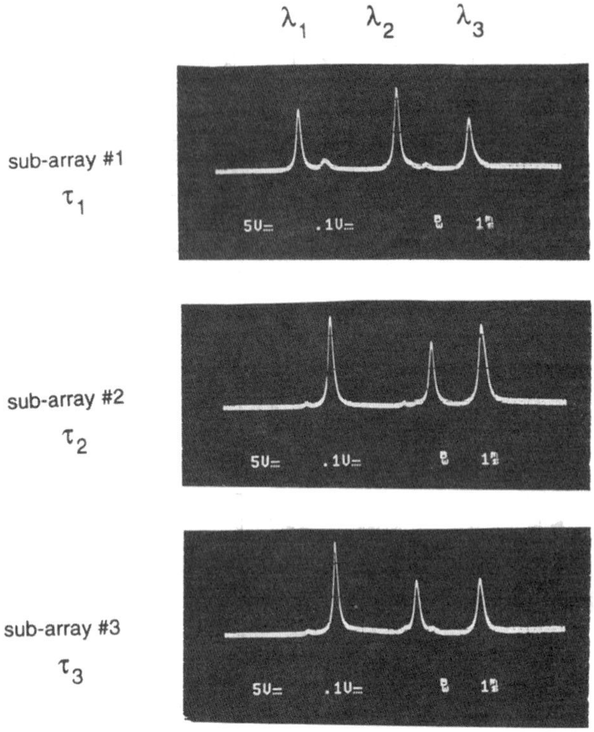

FIGURE 11.38. Set of three spectral returns from the three FGBs, corresponding to three different time gate settings. From Berkoff, T.A, Davis, M.A., Bellemore, D.G., Kersey, A.D., Williams G.M., and Putnam M.A., (1995), "Hybrid Time and Wavelength Division Multiplexed Fiber Bragg Grating Sensor Array," *SPIE*, **2444**, 288–294.

L_R to the path difference L_{sj} of each sensing interferometer in turn. The subscript j is used here to designate the jth interferometric sensor, if the nominal path difference of each sensing interferometer needs to be sufficiently different from all of the other sensing interferometers such that there can be no ambiguity no matter how much it is strained. In Fig. 10.4b a simplified illustration of the principle behind the low-coherence fiber optic Michelson interferometer is presented in terms of the coherence wavepacket (CWP) concept introduced in Chapter 3.

If ε_{max} is the maximum strain likely to be encountered and L is taken as the base length for the sensing interferometers, then the nominal path length of the second sensing interferometer is at least $L + 2L\varepsilon_{max}$, while that of the third sensing interferometer is at least $L + 4L\varepsilon_{max}$ and the nominal path length of the jth sensing interferometer is at least

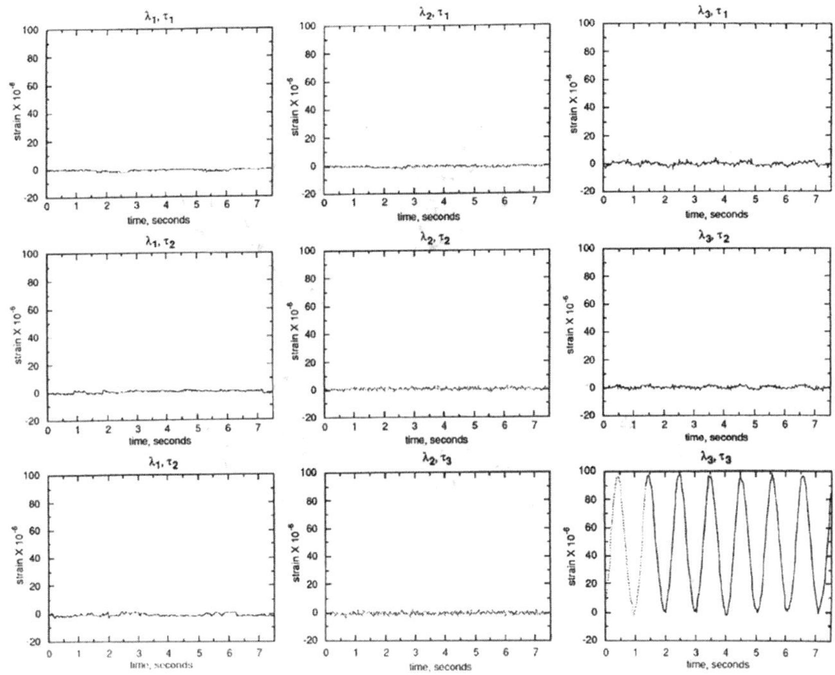

FIGURE 11.39. A set of nine temporal signals corresponding to three wavelength intervals and three time gates for the experimental setup shown in Fig. 11.36. From Berkoff, T.A, Davis, M.A., Bellemore, D.G., Kersey, A.D., Williams G.M., and Putnam M.A., (1995), "Hybrid Time and Wavelength Division Multiplexed Fiber Bragg Grating Sensor Array," *SPIE*, **2444**, 288–294.

$L + 2jL\varepsilon_{max}$. In such a case, if there are N sensing interferometers, the maximum change likely to arise in the path imbalance of the reference interferometer $\Delta L_R = 2(N-1)L\varepsilon_{max}$, and this corresponds to the maximum translation required of the movable mirror in the reference Michelson interferometer.

Sorin and Baney (1995) demonstrated that low-coherence interferometry could be used for precise length measurements for a number of long-gauge interferometric fiber optic sensors located at large distances from the interrogation system. Their experimental arrangement is shown in Fig. 11.40. Seven separate sensing lengths of optical fiber were created within a single optical fiber by means of telecommunication patch cords terminated in FC–PC connectors. A laser pumped erbium-doped optical fiber operating at 1550 nm with a spectral width of 25 nm served as the broadband source. A Hewlett-Packard precision reflectometer was used to

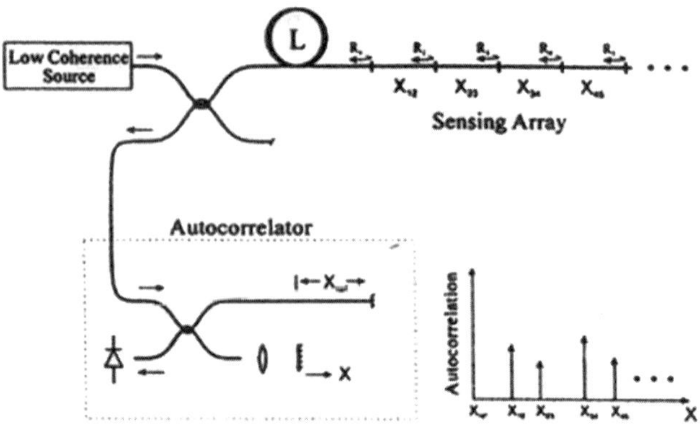

FIGURE 11.40. Schematic of a low-coherence system for interrogating a string of long gauge-length sensors formed by a set of fiber optic patch cords terminating in FC–PC connectors. From Sorin, W. V. and Baney, D. M., "Multiplexed Sensing Using Optical Low – Coherence Reflectometry", *IEEE Photon Technology Letters*, **7**, 917–919, © 1995, IEEE.

perform the autocorrelation and a 6-m fiber optic patchcord was used in its reference arm.

The reflection peaks corresponding to the seven sensing optical fibers located at the end of a 6-km spool of optical fiber is displayed in Fig. 11.41. These experiments and others demonstrated that low-coherence interferometry could be used for length measurements to an accuracy of 10 μm for an interferometric fiber optic sensor with a path length of 6 m situated many kilometers from the interrogation system.

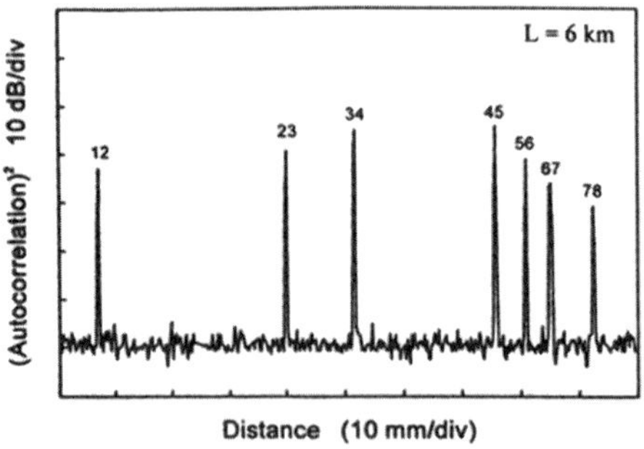

FIGURE 11.41. Autocorrelation signal against position for the set of fiber optic sensors interrogated by the system shown in Fig. 11.41. From Sorin, W. V. and Baney, D. M., "Multiplexed Sensing Using Optical Low –Coherence Reflectometry", *IEEE Photon Technology Letters*, **7**, 917–919, © 1995, IEEE.

11.4 SERIAL MULTIPLEXED FIBER BRAGG GRATING APPLICATIONS

In this section we examine a number of major applications that demonstrate field-site use of serially multiplexed fiber Bragg grating sensors. The FBG sensor interrogation schemes used in these applications have, by and large, been discussed in the previous section, so descriptions of the technology will be brief and we shall focus on the application and the results stemming from this innovative research.

11.4.1 Shape measurements with two-dimensional FBG sensor array

Accurate deformation information is required for many structures, including bridges, aircraft, space structures and submersible vessels. Jones *et al.* (1998) used a two-dimensional array of 16 fiber optic Bragg grating sensors to map the out-of-plane deformation of a cantilever honeycomb plate subject to single-point loading. A least-square strain-fitting algorithm was developed to determine the full deformation field from the strain information provided by the array of FBG sensors. The experimental arrangement is displayed in Fig. 11.42, (Jones *et al.*, 1996). The sensors were interrogated using a wavelength division multiplexing scheme involving two ELEDs and a scanned Fabry–Perot filter. The output of the tunable FP filter is electronically processed to obtain the derivative of the signal from which the zero-point crossings and Bragg wavelengths were obtained.

Finite-element comparisons were performed to test the accuracy of the strain–displacement algorithm. The ability of a single strand of optical fiber to map the strain field of a structural component with a simple array of FBG sensors distributed along its length is a significant improvement over conventional technology, which would require two leads to every strain gauge. This simplification could make structural monitoring much easier to use and greatly increase the frequency with which it is undertaken.

In order to compare the performance of the least-squares shape determination algorithm with the actual shape of the experimental specimen, a two-dimensional translational stage was constructed to allow movement of an Omega *linear variable differential transformer* (LVDT) to any point on the plate. An example of the raw and fitted displacement measurement data as a function of in-plane plate location is presented in Fig. 11.43a, while Fig. 11.43b, displays a comparison of the LVDT

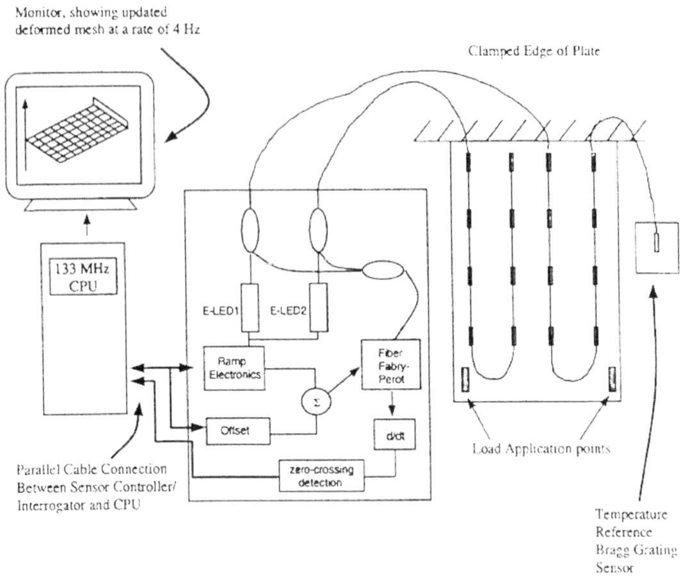

FIGURE 11.42. Experimental arrangement involving a scanning FP filter for interrogating an array of FBG sensors distributed over a loaded cantilever plate. From Jones, R. T., Berkoff, T. A., Bellemore, G., Early, D. A., Sirkis, J. S., Putnam, M. A., Friebele, E. J. and Kersey, A. D., (1996), "Cantilever Plate Deformation Monitoring Using Wavelength Division Multiplexed Fiber Bragg Grating Sensors," *SPIE*, **2718**, 258–267.

displacement measurements and the least-squares estimation for single point loading. It is apparent from this figure that the point of maximum difference (~13%) corresponds to the point of load application.

11.4.2 High strain monitoring of CFRP-wrapped concrete columns

Davis *et al.* (1996b), describe the use of an FBG sensing system based on a scanning Fabry–Perot filter (FPF), to measure the strong tensile and compressive loads experienced by carbon fiber composite wraps of concrete columns subject to test-to-failure loading. The FBG sensors were embedded in both the axial and circumferential (hoop) directions. Under axial compressive loading differential strains as high as 5% resulted in wavelength shifts of >50 nm. Figure 11.44 schematically illustrates the FBG sensing system, revealing that a fiber optic switch was used to interrogate either the hoop or axial FBG sensors. Separate

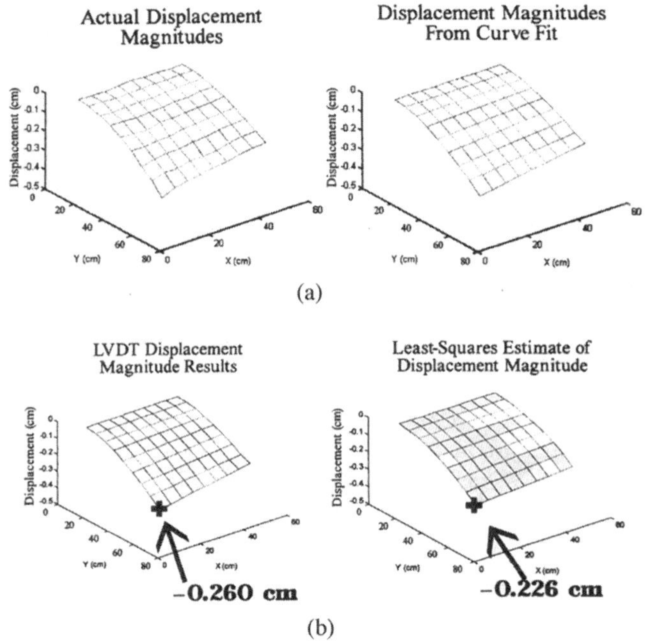

FIGURE 11.43. Comparison of experimental and computed displacement maps for a cantilever plate subject to point loading. From Jones, R. T., Berkoff, T. A., Bellemore, G., Early, D. A., Sirkis, J. S., Putnam, M. A., Friebele, E. J. and Kersey, A. D., (1996), "Cantilever Plate Deformation Monitoring Using Wavelength Division Multiplexed Fiber Bragg Grating Sensors," *SPIE*, **2718**, 258–267.

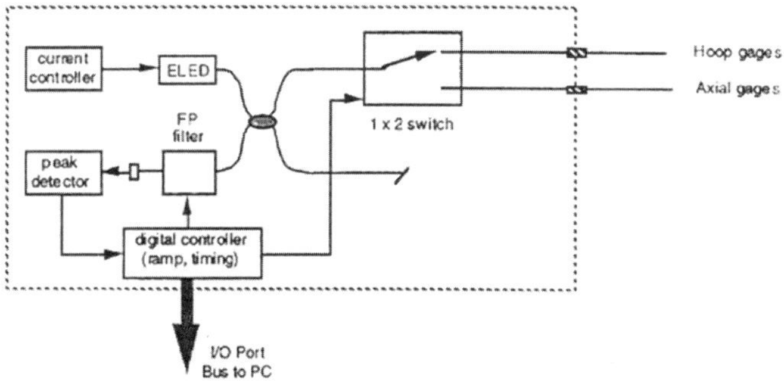

FIGURE 11.44. Demodulation system based on scanning FP filter for interrogating either hoop or axial FBG sensors mounted on a concrete cylinder. From Davis, M. A., Bellemore, D. G., Putnam, M. A. and Kersey, A. D., (1996b), "High Strain Monitoring in Composite-Wrapped Concrete Cylinders Using Embedded Fiber Bragg Grating Arrays," *SPIE* **2721**, 149–154.

interrogation of the hoop and axial FBG sensors allowed the bias operating point of the FP filter to be set differently. This was dictated by the large spectral shifts expected in this experiment. The scanning Fabry–Perot filter had a free spectral range of 40 nm, permitting a maximum range of about 40 000 $\mu\varepsilon$ for a single FBG sensor.

For the unstrained arrays, the FBG were written at wavelengths grouped toward either the higher (compressive axial array) or the lower (tensile hoop array) end of the tunable FPF scanning range; see Fig. 11.45. Under high loads the FPF operating bias is switched to independently track the two arrays, and an allowance is made for this bias voltage in calculating the system response. Figure 11.46a schematically illustrates the instrumentation of two CFRP composite material wrapped concrete columns, specimens B and C, while Fig. 11.46b displays (at left) the axial and hoop strains recorded by the FBG sensors for specimen C under loading, and (at right) the tensile (hoop) and compressive (axial) strains recorded by the FBG sensors during the testing of specimen B up to an axial load of 3 million lb.

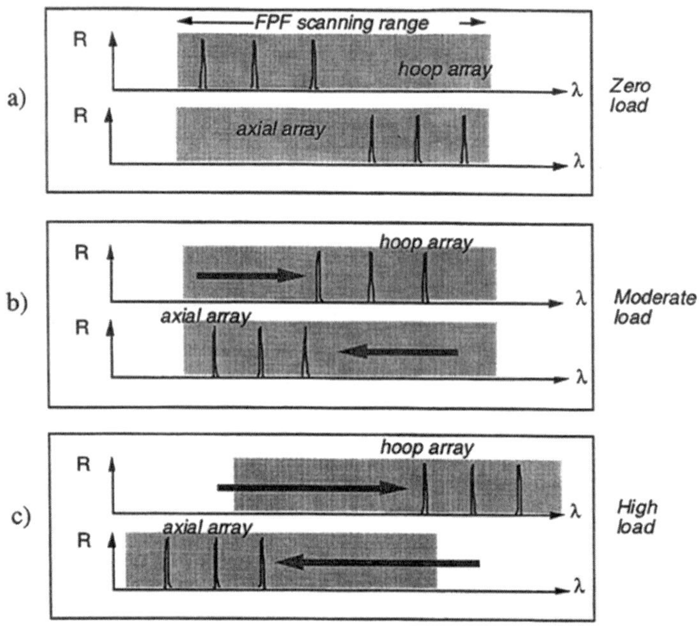

FIGURE 11.45. Spectral returns from hoop and axial-mounted FBG sensors under conditions of (a) zero load, (b) moderate load, and (c) high load. From Davis, M. A., Bellemore, D. G., Putnam, M. A. and Kersey, A. D., (1996b), "High Strain Monitoring in Composite-Wrapped Concrete Cylinders Using Embedded Fiber Bragg Grating Arrays," *SPIE* **2721**, 149–154.

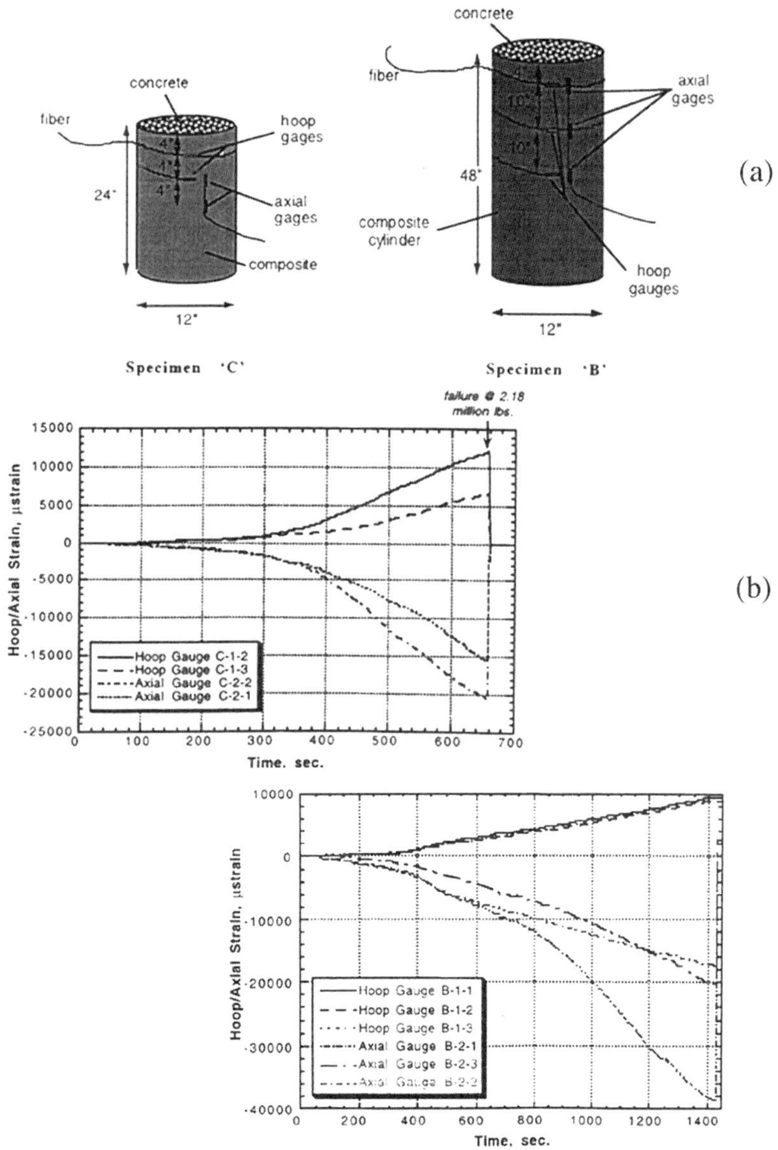

FIGURE 11.46. (a) FBG sensor configurations used on the two concrete cylinders loaded to failure. (b) Experimental measurements of hoop and axial strain for each of the two concrete specimens. From Davis, M. A., Bellemore, D. G., Putnam, M. A. and Kersey, A. D., (1996b), "High Strain Monitoring in Composite-Wrapped Concrete Cylinders Using Embedded Fiber Bragg Grating Arrays," *SPIE* **2721**, 149–154.

FIGURE 11.47. Four frames of a high-speed video showing explosive failure of a CFRP-wrapped concrete cylinder to high axial loading. From Davis, M. A., Bellemore, D. G., Putnam, M. A. and Kersey, A. D., (1996b), "High Strain Monitoring in Composite-Wrapped Concrete Cylinders Using Embedded Fiber Bragg Grating Arrays," *SPIE* **2721**, 149–154.

The tensile and compressive strains are seen to reach 21 000 $\mu\varepsilon$ and 12 000 $\mu\varepsilon$, respectively, for specimen C just prior to failure of the column, while the compressive and tensile strains reached over 38 000 $\mu\varepsilon$ and 10 000 $\mu\varepsilon$, respectively, at an axial load of 3 million lb for specimen B. Failure was subsequently induced in B at 2.3 million lb after drilling holes into the CFRP composite wrap. Figure 11.47 shows four frames of a video that captures the explosive failure following rupture of the CFRP composite wrap. The FBG sensors operated up to, and in some cases after, the destructive failure of the specimen. It is also worth reiterating that the CFRP-embedded FBG sensors survived compressive strains in excess of 3.8% and tensile strain up to 1.2%.

11.4.3 FBG sensor arrays for CFRP-reinforced concrete beams

Davis *et al.* (1997a) describe the use of FGB sensors to measure the strain within CFRP composite strengthened concrete beams loaded to failure. The FBG demodulation system is very similar to those previously described using a ramp waveform voltage for the piezoelectric drive to sweep the FP filter's transmission window over the wavelength range occupied by the FBG sensors. A schematic of the demodulation system is displayed as Fig. 11.48. The two ELED sources are driven on/off synchro-

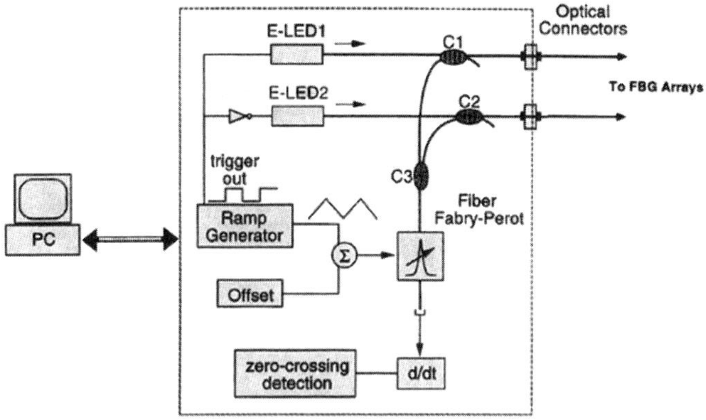

FIGURE 11.48. Demodulation system for FBG sensor array using two ELED sources and a scanning fiber FP filter. Reprinted from Davis, M. A., Bellemore, D. G., Kersey, A. D., (1997), "Distributed Fiber Bragg Grating Strain Sensing in Reinforced Concrete Structural Components," *Cement and Concrete Components*, **19**, 45–57, with permission from Elsevier Science.

nously with the ramp voltage applied to the FPF, but in antiphase with each other. In this manner only one set of back-reflected signals from the two arrays of sensors passes through the FP filter at any instant, since ELED-1 is powered only during the up scan in wavelength and ELED-2 is powered during the down scan. The output of the filter is then detected and the derivative electronically extracted. By referencing any shifts in the zero-point crossings from each FBG sensor to the voltage applied to the FPF, the strain for each can be determined. A resolution of better than 1 $\mu\varepsilon$ was achieved at 150 Hz.

The three oscilloscope traces displayed in Fig. 11.49 help to explain the operation of the demodulation system. The top trace shows the ramp voltage applied to the Fabry–Perot filter. The second trace displays the derivative signal, which was created from the light back-reflected by the five FBG sensors. The derivative of five signal peaks is seen during the ramp up phase, while only four peak derivative signals are seen on the down ramp. It is worth noting that the zero-crossing measurement is insensitive to the shape or magnitude of the back-reflected signals from the FBG sensors provided they are single-peaked. The lower trace shows the pulses created from the derivative signal and fed to the computer. The excellent linearity in the response of this system to strains as high as 3% can be seen by reference to Fig. 11.50.

To test the performance of the fiber optic monitoring system, concrete beams and deck panels were instrumented, then loaded to failure. The FBG sensors were attached to a prepared section of the steel rebars with

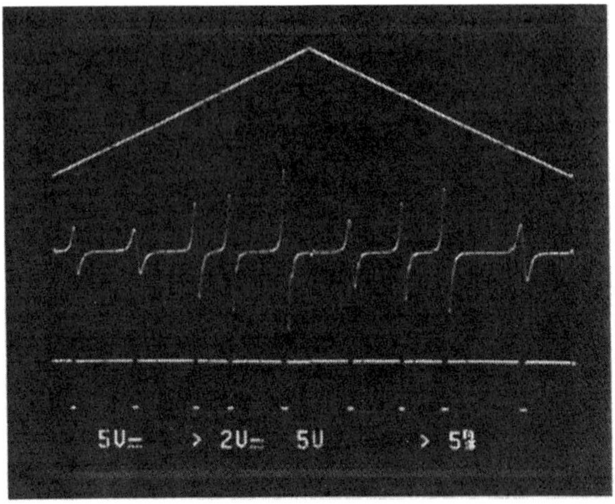

FIGURE 11.49. Upper trace shows ramp voltage used to scan the fiber FP filter. The middle trace displays the derivative of the return signal from the FBG sensors, while the low trace indicates the zero-crossings identified with the center wavelengths of the FBG spectra. Reprinted from Davis, M. A., Bellemore, D. G., Kersey, A. D., (1997), "Distributed Fiber Bragg Grating Strain Sensing in Reinforced Concrete Structural Components," *Cement and Concrete Components*, **19**, 45–57, with permission from Elsevier Science.

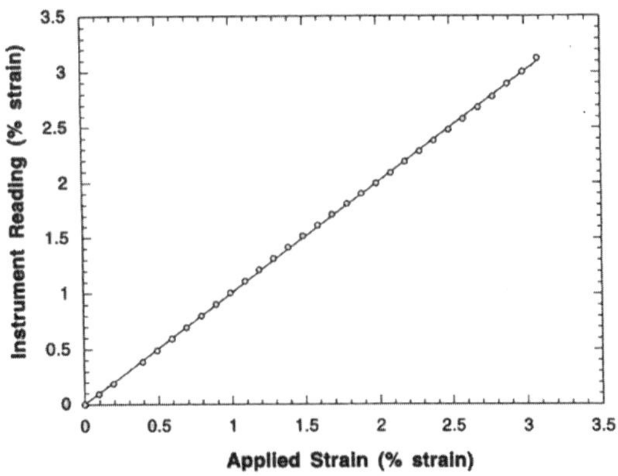

FIGURE 11.50. Linear response of the FBG sensor demodulation system to applied strain. Reprinted from Davis, M. A., Bellemore, D. G., Kersey, A. D., (1997), "Distributed Fiber Bragg Grating Strain Sensing in Reinforced Concrete Structural Components," *Cement and Concrete Components*, **19**, 45–57, with permission from Elsevier Science.

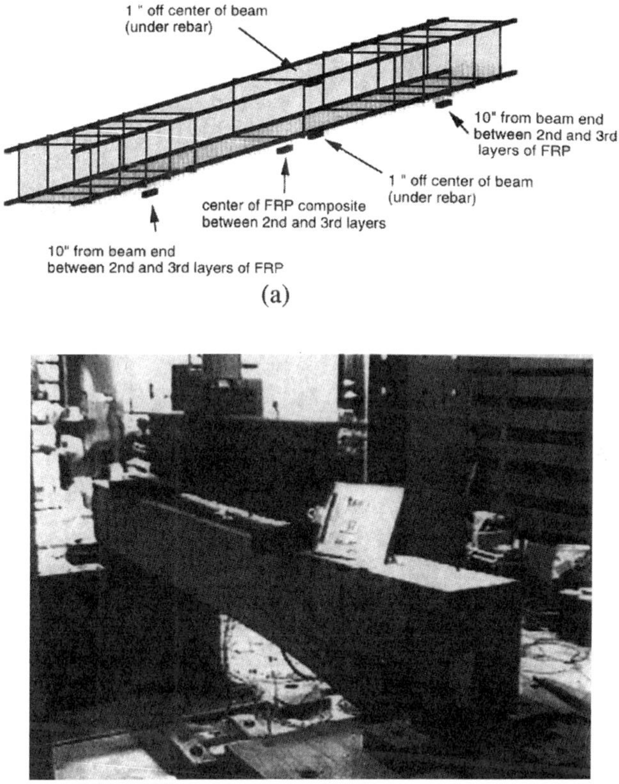

(b)

FIGURE 11.51. (a) Distribution of the FBG sensors in relation to the 3-m-long concrete beam subject to be four-point loaded. (b) Picture of beam after failure. Reprinted from Davis, M. A., Bellemore, D. G., Kersey, A. D., (1997), "Distributed Fiber Bragg Grating Strain Sensing in Reinforced Concrete Structural Components," *Cement and Concrete Components*, **19**, 45–57, with permission from Elsevier Science.

typical foil gauge adhesive (e.g., M-Bond 200) and coated with epoxy for protection. The locations of the FBG sensors in relation to the ~3 m long concrete beam is shown in Fig. 11.51a, while a picture of the beam after four-point loading to failure is presented as Fig. 11.51b. A second array of FBG sensors was placed between the second and third layers of the FRP composite material bonded to the underside of the concrete beams for additional strength.

Figure 11.52 displays the sensor data obtained as the beam was loaded. All the sensors survived beam failure, which occurred 21 minutes

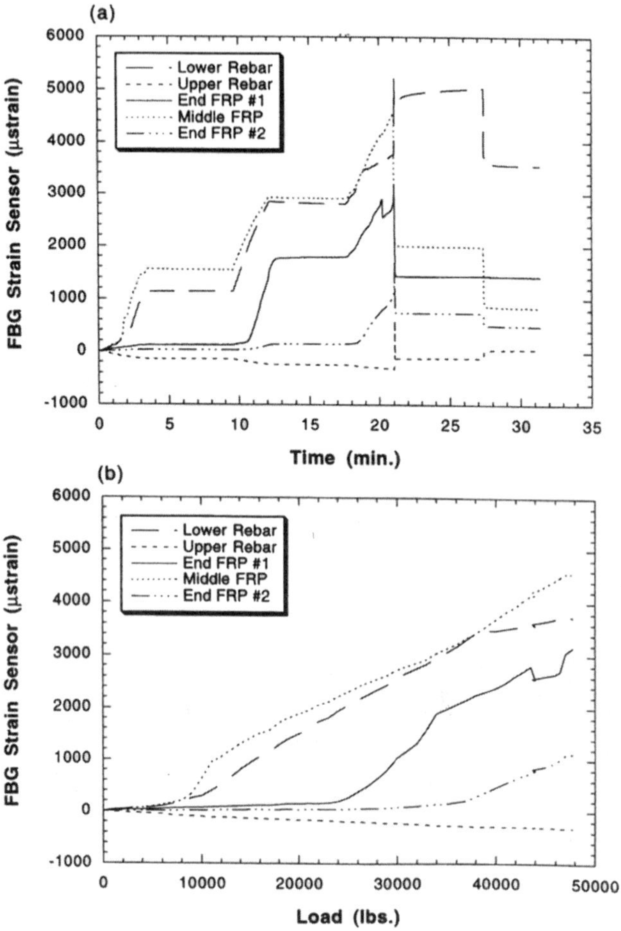

FIGURE 11.52. (a) FBG strain readings with time for loaded concrete beam. (b) FBG strain readintgs as function of load for the concrete beam. Reprinted from Davis, M. A., Bellemore, D. G., Kersey, A. D., (1997), "Distributed Fiber Bragg Grating Strain Sensing in Reinforced Concrete Structural Components," *Cement and Concrete Components*, **19**, 45–57, with permission from Elsevier Science.

into the test at 47 500 lb, and when the load was released at the 27-minute mark the sensors were able to detect residual strains as seen in Fig. 11.52. The sharp increase in strain measured by one end FRP embedded sensor at 25 000 lb is noteworthy. The location of this sensor correlates with the end failure point of the beam. It is believed that under high load, cracks weaken the concrete-rebar section so that the FRP material starts to support the beam, as seen from the increasing strain recorded by the

FBG sensors embedded within the FRP layers. This behavior was observed with other concrete beams that were loaded and instrumented in a similar manner. For each beam, failure arose at one end with the FRP composite material breaking free of the main body of the beam. In each case the failure location was predicted by a large increase in the strain detected by the embedded sensors prior to beam failure.

In the case of the concrete deck panels, suitably placed FBG sensors were able to detect movement of the neutral axis prior to failure. In all, 35 FBG sensors were embedded and tested in either concrete beams or deck panels. These devices functioned to, and beyond, failure of the structures and demonstrated their potential to predict the failure region for heavily loaded structures.

11.4.4 Network of FBG sensors for test bridge damage assessment

Kodindouma *et al.* (1996) described a parallel and serial multiplexed network of 48 FBG sensors that was used to measure the strain distribution through out a quarter-scale laboratory test bridge with a 12.25 m (40 ft)-span. The bridge deck was in the form of a 3.35 m (11 ft) wide and 0.15 m (6 in) thick, steel-reinforced concrete lab supported by three simple W21 × 62 steel beams that were 12.25 m long; see the schematic presented as Fig. 11.53. The bridge was statically loaded and monitored in its pristine condition and as it was progressively damaged. The damage inflicted on the bridge was in the form of a series of cuts made into the east steel girder near its midspan to simulate fatigue cracking.

The FBG sensor demodulation system made use of a scanning fiber Fabry–Perot filter position just ahead of the photodetector (see Section 11.3.1). Interrogation of the 48 FBG sensors was achieved by a hybrid time and wavelength multiplexing scheme. Four optical fibers, each containing 12 FBG sensors, are read by a single Fabry–Perot filter utilizing a fiber optic switch. The system is similar to that shown in Fig. 11.44, except that more channels are connected through the switch. The sampling rate was about 150 Hz. An ELED operating at 1.3 μm is used to excite a string of 12 FBG sensors at a time, and the 45 nm free spectral range of the FPF permits this interrogation with a spectral separation for the FBG sensors of ~3 nm. As previously described (Kersey *et al.*, 1993), the wavelength measurement scheme involves identifying the zero crossings as the Bragg wavelengths from the electronic differentiated signal obtained in each scan of the FPF. The processing time for the array of sensors is ~2.5 s. Alternatively, the optical switch can be set to continuously interrogate a

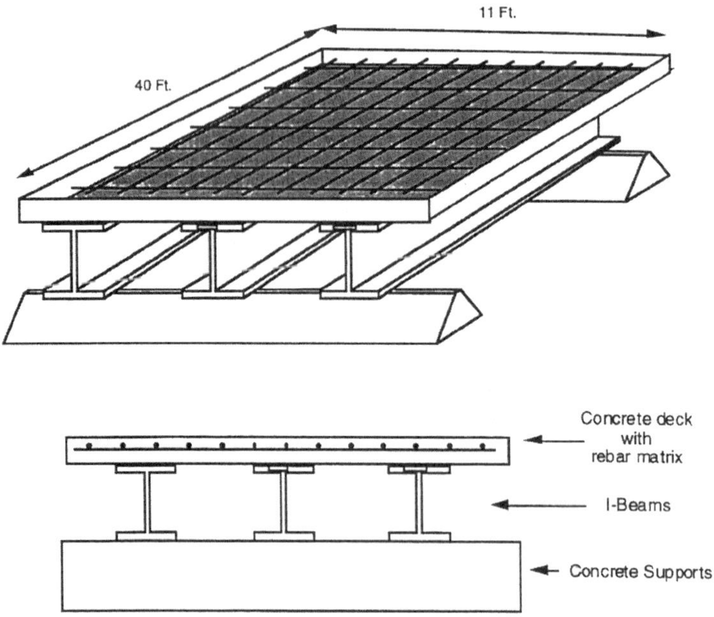

FIGURE 11.53. Schematic of a laboratory model of a bridge deck to be instrumented with FBG sensors for testing their damage assessment capability. From Davis, M. A., Bellemore, D. G., Kersey, A. D., Putnam, M. A., Friebele, E. J., Idriss, R. L. and Kodinduma, M., (1996a), "High Sensor-Count Bragg Grating Instrumentation System for Large-Scale Structure Monitoring Applications," *SPIE* **2718**, 303–309.

single FBG sensor array, leading to a sampling rate of 71 Hz for the 12 sensors (Davis *et al.*, 1996a).

For long-term monitoring, one FBG sensor (C-7) was used to provide reference (temperature) information to stabilize the strain readings throughout the experiment; the remaining FBG sensors were used to gather strain information. Thirty FBG sensors were attached to rebar elements running in both longitudinal and transverse directions through the concrete deck (see Fig. 11.54). The sensor designations in this schematic refer to the four arrays of FBG sensor (A, B, C, and D) and each individual FBG sensor (1–12) in the array. The longitudinal gratings were installed in three sets of five directly over the I-beam deck supports, while the transverse gratings were attached to the rebar near each longitudinal sensor. Fifteen FBG sensors were attached longitudinally to the underside of the steel I-beams, as indicated in Fig. 11.54, while two FBG sensors (D-1 and D-9) were not attached and served as further temperature monitors.

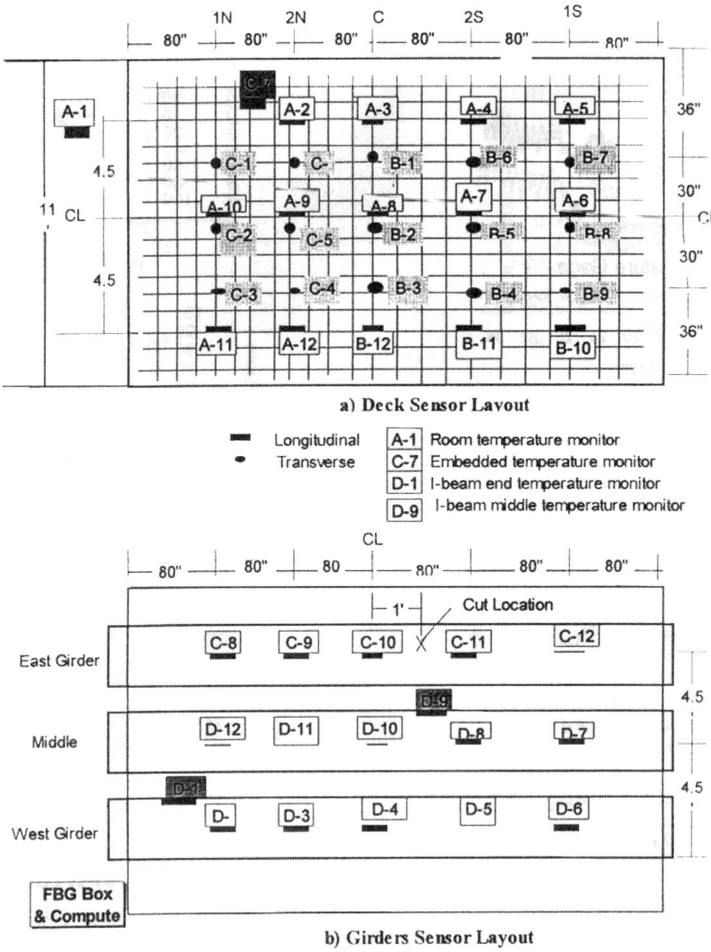

FIGURE 11.54. (a) Map of FBG sensor array and conventional instrumentation on laboratory bridge deck. (b) FBG sensor layout for girder. From Idriss, R. L., White, K. R., Pater, J. W., Vohra, S. T., Chang, C. C., Danver, B. A., Davis, M. A., (1998), "Monitoring and Evaluation of an Interstate Highway Bridge Using a Network of Optical Fiber Sensors," *Fiber Optic Sensors for Construction Materials and Bridges*, F. Ansari (Ed). pp.159–167, Technomic Publishing Co. Inc.

A finite element analysis of the bridge was undertaken to predict first the load redistribution likely after damage to one of the girders, and second the most suitable locations for the 48 FBG sensors (see Fig. 11.54). A resistive foil strain gauge was placed beside each fiber optic strain sensor. The SAP90 analysis also predicted the minimum and maximum strains for a given loading and was used in designing the strain range for the FBG sensing system. A maximum strain of 500 $\mu\varepsilon$ was expected to occur after fracture at the midspan of the middle girder under dead load and live load.

The FBG sensor interrogation system was set to sample the sensors once an hour during the 38 days following the concrete pour to monitor the dead load of the bridge during the concrete cure cycle. A sample of the strain data monitored from the three girders during this period is displayed as Fig. 11.55. The strain is leveled off after a few days but started to increase after the 26th day, when the forms were removed from the concrete, and the full weight of the concrete had to be taken up by the girders (Idriss *et al.*, 1998).

Strain measurements were collected under dead-load conditions and when a live load was provided by a hydraulic jack. Although some redistribution of strain was observed when a cut was inflicted (with an arc torch) on the bottom flange and lower web of the east girder at the midspan position, a major redistribution of strain was observed when the flange was cut through, leaving a half-depth fracture at the midspan of the girder. This redistribution of strain is seen in Fig. 11.56. Under dead-load conditions the damaged (east) girder carried about 66% less load, while the middle girder picked up a 30% increase in load at the midspan. This research lends support to the idea of detecting the onset of damage in structures such as bridges by means of an appropriate system of fiber optic strain monitoring.

11.4.5 Interstate highway bridge monitoring by FBG sensor network

A major test of the potential for retrofitting in-service bridges with fiber optic structural monitoring systems was undertaken on the I-10 bridge over Las Cruces, New Mexico. This bridge consists of 36-m-long spans comprising a concrete deck supported by welded steel plate girders (see Fig. 11.57). In this section we describe the results of a field test carried out using a wavelength division multiplexed array of 32 FBG sensors attached at various locations along the girders of the I-10 bridge. As illustrated in Fig. 11.58a, the FBG sensors were bonded to girders 1, 2, 3, and 4, at the pier, 1/8-span, and 1/2-span locations.

Three FBG sensors, ranging from the lowest to the highest wavelength in a given subarray, are surface mounted on the top flange, upper web, and bottom flange of the girder, as shown in Fig. 11.58b. At each FBG sensor location, a small area of the girder was filed smooth and prepared

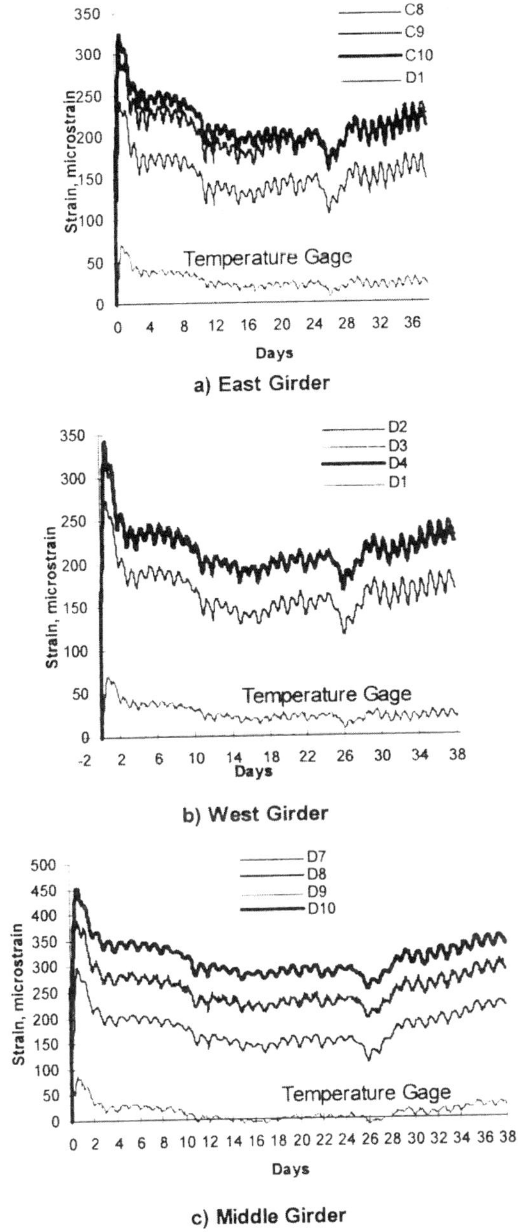

FIGURE 11.55. Experimental strain variation with time, measured by various sensors for (a) east girder, (b) west girder, and (c) middle girder. From Idriss, R. L., White, K. R., Pater, J. W., Vohra, S. T., Chang, C. C., Danver, B. A., Davis, M. A., (1998), "Monitoring and Evaluation of an Interstate Highway Bridge Using a Network of Optical Fiber Sensors," *Fiber Optic Sensors for Construction Materials and Bridges*, F. Ansari (Ed). pp.159–167, Technomic Publishing Co. Inc.

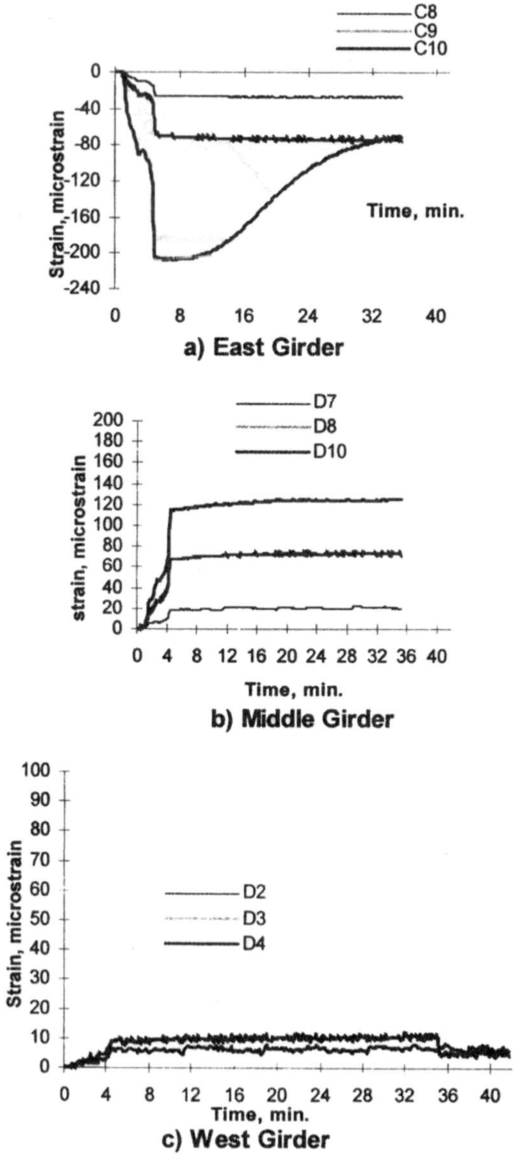

FIGURE 11.56. Strain redistribution following the cutting of the east girder, as determined from the sensor array. From Idriss, R. L., White, K. R., Pater, J. W., Vohra, S. T., Chang, C. C., Danver, B. A., Davis, M. A., (1998), "Monitoring and Evaluation of an Interstate Highway Bridge Using a Network of Optical Fiber Sensors," *Fiber Optic Sensors for Construction Materials and Bridges*, F. Ansari (Ed). pp.159–167, Technomic Publishing Co. Inc.

FIGURE 11.57. Photograph of I-10 Bridge over Las Cruces. From Idriss, R. L., White, K. R., Pater, J. W., Vohra, S. T., Chang, C. C., Danver, B. A., Davis, M. A., (1998), "Monitoring and Evaluation of an Interstate Highway Bridge Using a Network of Optical Fiber Sensors", *Fiber Optic Sensors for Construction Materials and Bridges,* F. Ansari (Ed). pp.159–167, © 1998, Technomic Publishing Co. Inc.

by following the installation tips for the conventional resistance strain gauge and the FBGs were bonded using M-Bond 200 adhesive. For the first FBG array (also known as the 1/2-span array), the temperature reference sensor, T, is installed inside of the FBG sensor demodulation instrumentation located on one of the piers and is used to temperature-compensate the remaining 31 FBG sensors.

The FBG sensor responses are interrogated by means of a scanning fiber Fabry–Perot filter, described earlier, and is designed to have a sampling rate of 45 Hz. The sensor demodulation system is schematically illustrated in Fig. 11.59 and was located on one of the bridge piers. The choice of 16-bit resolution for the FP filter ramp voltage and a free spectral range of 45 nm produces a minimum detectable wavelength shift of 0.7 pm, or an equivalent strain resolution of less that 1 $\mu\varepsilon$. However, the actual minimum detectable strain was observed to be closer to 10 $\mu\varepsilon$ because of electronic noise in the system. The 45 nm free spectral range of the FPF permits no more than 16 FBG sensors, having a linewidth of ~0.2 nm and spectrally separated by approximately 2.7 nm, to be interrogated per scan of the FP filter. This wavelength spacing is sufficient to allow a strain range of ~1300 $\mu\varepsilon$ for each FBG sensor.

The data from the FBG arrays is stored onto a computer hard disk using a threshold trigger method that allows storage of data only if a certain strain threshold is crossed. The software was designed to store sensing data only from significant load events, defined as ≥ 20 $\mu\varepsilon$

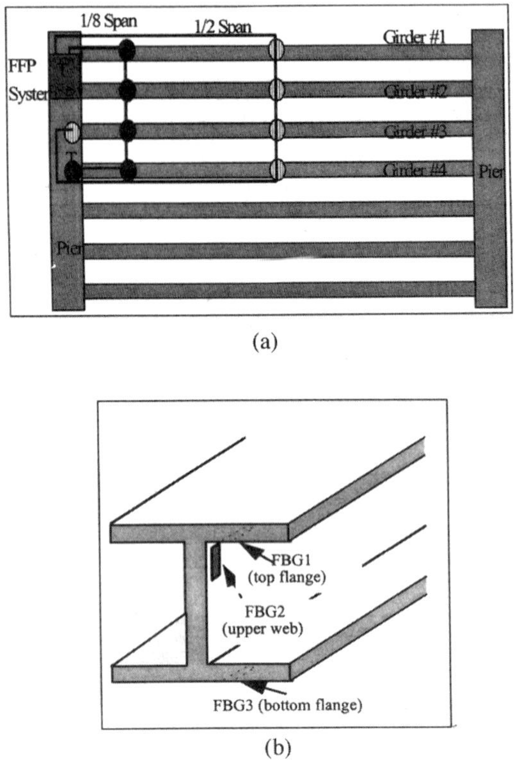

FIGURE 11.58. (a) Schematic of the I-10 Bridge girder layout. (b) Distribution of FBG sensors on one girder. From Vohra, S. T., Chang, C. C., Danver, B. A., Althouse, B., Davis, M. A., and Idriss, R., (1998), "Preliminary Results on the Monitoring of an In-Service Bridge Using a 32–Channel Fiber Bragg Grating Sensor System", *Fiber Optic Sensors for Construction Materials and Bridges*, F. Ansari, (Ed), pp.148–158, © 1998, Technomic Publishing Co. Inc.

experienced by the FBG sensor located on the bottom flange of girder 3. This approach allows judicious storage of data on the computer even though the demodulation system is continuously monitoring the sensor arrays.

The response of the FBG sensors located at half-span on girder-3 of the bridge to a vehicle load event is shown in Fig. 11.60. Data from FBG sensors located on the top flange, the web, and the bottom flange are depicted in this figure. The data clearly shows a large strain event (a truck) followed by a smaller strain event (a car) arising over a 30-s period. All three FBG sensors located on girder-3 recorded the event. The truck, seen as the large peak, produces a peak negative strain of about 40 $\mu\varepsilon$ at the top flange location, peak negative strain of about 100 $\mu\varepsilon$ at the web

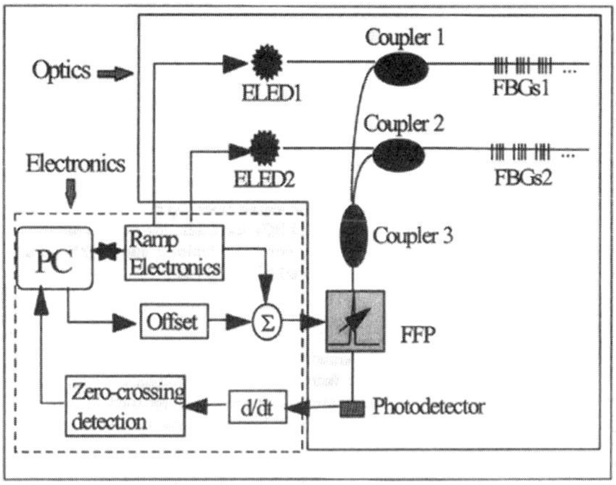

FIGURE 11.59. Demodulation system used to interrogate the 32 FBG sensors distributed over the I-10 Bridge. From Vohra, S. T., Chang, C. C., Danver, B. A., Althouse, B., Davis, M. A., and Idriss, R., (1998), "Preliminary Results on the Monitoring of an In-Service Bridge Using a 32–Channel Fiber Bragg Grating Sensor System", *Fiber Optic Sensors for Construction Materials and Bridges*, F. Ansari, (Ed), pp.148–158, © 1998, Technomic Publishing Co. Inc.

location, and peak positive strain of about 80 $\mu\varepsilon$ at the bottom flange location. Modal excitations of the bridge induced by the transient loading event can be seen in Fig. 11.60.

Since the top flange sensor is on the opposite side of the neutral axis of the I-beam with respect to the bottom flange sensor, the sign of measured strain between the top and bottom flange is reversed for the same loading event at that location. The web sensor is located significantly closer to the top flange of the I-beam. The web sensor shows a large compressive strain signal during the actual transit of the truck directly over the sensor location. This is preceded by a slight positive strain at the web sensor for the same reasons as described earlier. The second event in Fig. 11.60, near the 23-s mark, can be attributed to a smaller vehicle, a car, passing over that location producing a peak strain of about 25 $\mu\varepsilon$. The bottom flange undergoes tension and the top flange undergoes compression, resulting in positive and negative strains at those locations due to the same vehicle. It might be noted that, unlike the truck, the car does not induce noticeable bridge vibrations. An important observation drawn from this work, and indicated in Fig. 11.60, is the absence of slow strain drifts, indicating that the temperature reference sensor compensates the thermal drifts for the strain monitoring FBG sensors.

In Fig. 11.61, a standard data file reveals the response of the four instrumented girders (midspan sensors) during the passage of a truck.

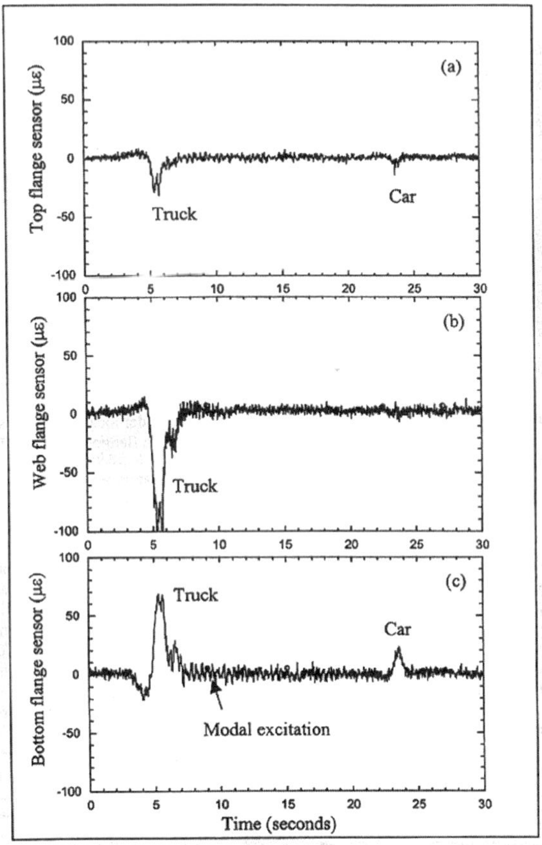

FIGURE 11.60. Dynamic response of three FBG sensors mounted on the (a) top flange, (b) web flange, and (c) bottom flange of a girder of the I-10 bridge, when subject to traffic. From Vohra, S. T., Chang, C. C., Danver, B. A., Althouse, B., Davis, M. A., and Idriss, R., (1998), "Preliminary Results on the Monitoring of an In-Service Bridge Using a 32–Channel Fiber Bragg Grating Sensor System", *Fiber Optic Sensors for Construction Materials and Bridges*, F. Ansari, (Ed), pp.148–158, © 1998, Technomic Publishing Co. Inc.

The girder bottom flanges show the same pattern of behavior: induced compression when the truck is still in the first span at around 5 s, followed by peak tension when the truck passes over the midspan at about 7.5 s on this time scale. From the data collected, detailed information can be generated pertaining to the structure itself, as well as statistical information on the traffic loading of the bridge.

In conclusion the network of 32 FBG sensors was successfully retrofitted to an in-service interstate highway bridge, and preliminary data

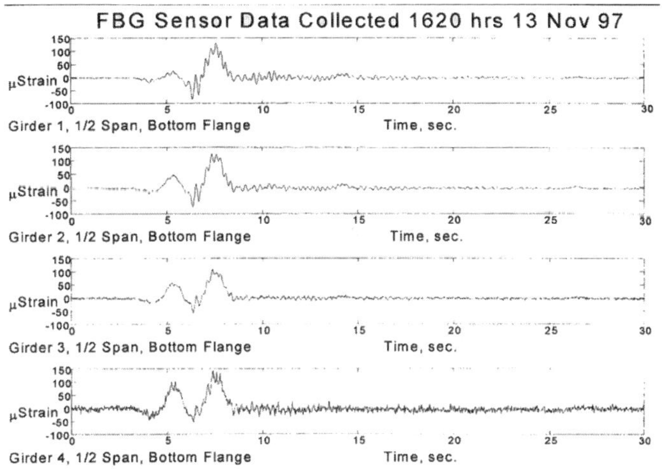

FIGURE 11.61. Dynamic responses of four FBG sensor instrumented girders of the I-10 bridge to the passage of a specific truck. From Idriss, R. L., White, K. R., Pater, J. W., Vohra, S. T., Chang, C. C., Danver, B. A., Davis, M. A., (1998), "Monitoring and Evaluation of an Interstate Highway Bridge Using a Network of Optical Fiber Sensors", *Fiber Optic Sensors for Construction Materials and Bridges*, F. Ansari (Ed). pp.159–167, © 1998, Technomic Publishing Co. Inc.

shows that such a monitoring system can prove to be a very effective tool for the bridge engineer. Traffic information such as vehicle weight and distribution, as well as vehicle speed, can be extracted from the strain signature. Furthermore, a statistical pattern of strain response for the bridge can be developed. Changes from that baseline behavior can be monitored and interpreted as possible defects in the structure. Ultimately, the system can be used to remotely monitor critical structures, giving the bridge engineer a warning of any abnormal conditions as well as providing a record of the loading history of the structure, which could be valuable from a maintenance point of view.

11.4.6 Serial multiplexing by parallel spectral detection

Recent progress toward practical and economic FBG-array fabrication via computer-controlled in-line writing during the drawing of the optical fiber makes necessary the development of serial multiplexed demodulation systems that are capable of working with the low-reflectivity FBGs (typically less than 5%) produced in this manner. Such gratings are classified as Type 1 and are generally narrow-line (<0.2 nm). This instrumentation must make efficient use of the reflected light from the FBG sensors, especially when interrogating a large number of sensors with a minimum of instrumentation and lead-ins. FBG sensors can be produced during the draw of an optical fiber at a rate of thousands per hour and proof-test levels are consistently in excess of 700 ksi.

Although the 1 to 4% reflectivity of such gratings may be adequate for time division multiplexing, there is a question whether they can be used in wavelength division multiplexing schemes, particularly when interrogated by a weak narrow-line source, such as created by a Fabry–Perot filter and a broadband source. For a given FBG reflectivity and linewidth, the strength of the reflected optical signal depends upon the spectral brightness of the source at the Bragg wavelength, lack of crosstalk with other FBGs (i.e., sufficient wavelength spacing), the time over which the signal may be sampled, and the efficiency of the discriminating optics and detection system. In order that many FBG sensors may be wavelength multiplexed, the source must distribute its power over a large spectral window. Maximum channel density and measurement sensitivity, however, necessitates spectrally narrow reflectors.

An integrating-array spectrometer has the potential to satisfy the demands of rapidly extracting many high-resolution strain measurements on many wavelength channels from very little available optical power. A schematic based on the system developed by Chen *et al.* (1997) was presented as Fig. 11.16. Askins *et al.* (1995) demonstrated the merits of this spectral dispersion and parallel detection approach for WDM by comparing it directly with serial detection using narrow pass-band filters, or interferometry. In this comparison, it is assumed that generic FBG sensors, with 2% reflectivity and 0.04-nm bandwidth, are to be illuminated with a superluminescent diode, delivering 300 μW over a spectral interval of 30 nm (assumed flat spectrum). It is further assumed that a wavelength-shift sensitivity of one part per million (ppm) is required on each (1-kHz) scan of several reflectors spaced over a 30-nm wavelength range.

For serial detection, instrument resolution improves (for a sufficiently bright source) as the discriminator passband approaches that of the grating; the example's effective passband width is 0.1 nm. Before losses and efficiencies are considered, the amount of energy available per scan for each grating may be shown not to exceed 7 femtojoules, or 2 nanowatts for <3 μs. Photon noise at this level fundamentally limits SNR to below ~200 : 1. A 100% quantum efficient detector would generate a peak current of around 2 nanoamps and would need a frequency response of several megahertz. If this signal is to be converted into a processable voltage range (say, 100 mV) current-to-voltage amplification is required on the order of 5.0×10^7, again with a frequency response of several megahertz.

Currently available high-quality amplifiers operating under these conditions of gain and bandwidth would essentially obscure the target signal under power-equivalent noise. Resolving a 1-ppm wavelength shift on each scan requires determining the position of the 0.1-nm feature to less that 1/100th of its width. The SNR must exceed 100 : 1 to achieve this result. Serial detection cannot meet the goals set forth without greater FBG reflectivity, higher launched power, or a much reduced scan rate.

Using values from the preceding serial detection example, each FBG signal is assumed to return 2 nW to the instrument. In the parallel detection approach a dispersive instrument, such as a spectrometer, is designed to project a chromatically dispersed image of the FBG reflected signals onto a linear detector array. Because each pixel of the array may integrate the optical signal over the entire scan period, more than 1.0 pJ (equivalent to 2 nW at 1 kHz with 60% efficiency) is available for detection. Commercial silicon charge coupled detectors (CCD) require less than 1 pJ for complete saturation and have dark currents around 1.0×10^{-15} amps. Signal-to-noise at saturation exceeds 2000 : 1, expressed as a readily digitizable signal of >1 V at full scale. The amount of light available per grating is at least four times that needed for a SNR of 2000 : 1, in this example. Gratings distributed over the entire 30-nm wavelength range are simultaneously registered with this high-contrast signal.

The experimental system of Askins *et al.* (1995) included an autocollimating spectrometer, in which the core of the input single-mode optical fiber functions as the input slit. The expanding beam of light from the optical fiber is collimated by a suitable lens to illuminate a 1200-line-mm diffraction grating that spectrally disperses the beam so that different wavelengths are focused on different parts of the linear, 1024-element CCD array placed in the focal plane of the lens (see Fig. 11.62). With this spectrometric detection system, the narrow spectral image formed by the light reflected from a FBG sensor subject to ~130 με shifts by just one pixel position. To accomplish the goal of 1 με sensitivity, intrapixel resolution of better than 0.01 pixel is required. This objective can be attained by slight defocusing of the spectrometer so that the spot of light from the FBG sensor is spread over 3 or 4 pixels, and using a centroid calculation algorithm [Askins *et al.*, 1995].

Three spectral measurements of FBG arrays fabricated on the optical fiber draw tower are presented in Fig. 11.63. The overlap of the sensor spectra (solid lines) with that of the source (broken lines) reveals that low signal strength is always observed on the shorter-wavelength peaks. The wavelength spacings, corresponding to strains of 1900, 4800, and 900 με,

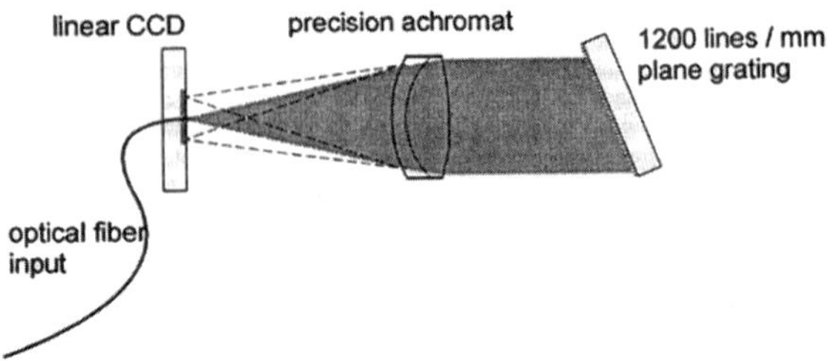

FIGURE 11.62. Dynamic responses of four FBG sensor instrumented girders of the I-10 bridge to the passage of a specific truck. From Askins, C. G., Putnam M. A., and Friebele, E. J., "Instrumentation for Interrogating Many-Element Fiber-Bragg Grating Arrays," *SPIE* **2444**, 257–266.

respectively, for a though c, in this figure dictate the overlap-free strain range. FBG sensor array b in Fig. 11.63 was embedded within a heavy-duty carbon/glass fiber-epoxy composite channel beam fabricated by pultrusion (see Fig. 11.64).

The optical fiber with the string of FBG sensors was inserted beneath the surface ply of the composite without interrupting the commercial manufacture of the channel beam. Weak loads were applied to the channel beam in the form of people stepping on and off of it near gratings C and D, while the beam was supported near gratings A and F. The lower part of Fig. 11.64 reveals the time traces from the demodulation system showing

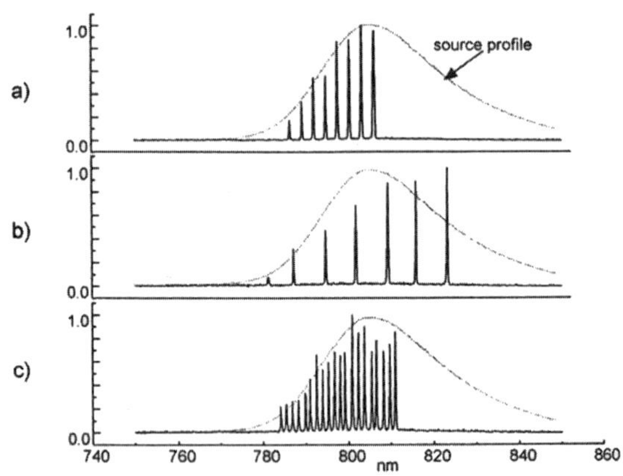

FIGURE 11.63. Three reflection spectra from a string of FBG sensors fabricated on the draw tower. The wavelength spacings correspond to strains of (a) 1990 $\mu\varepsilon$, (b) 4800 $\mu\varepsilon$, and (c) 900 $\mu\varepsilon$. From Askins, C. G., Putnam M. A., and Friebele, E. J., "Instrumentation for Interrogating Many-Element Fiber-Bragg Grating Arrays," *SPIE* **2444**, 257–266.

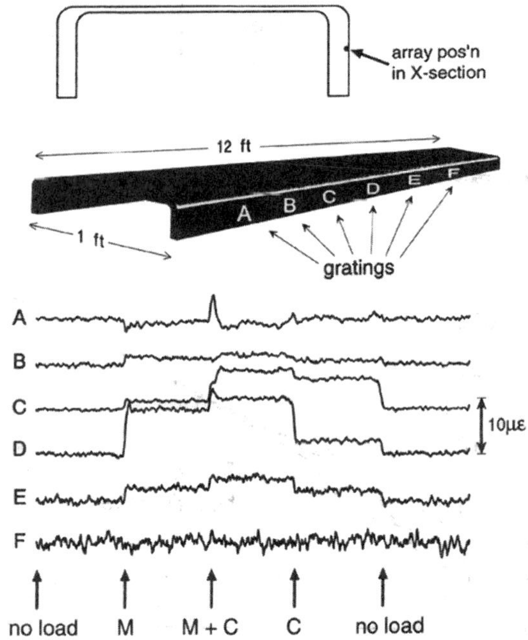

FIGURE 11.64. Schematic of a pultruded channel beam containing a string of FBG sensors embedded during the pultrusion process. Locations of the FBG sensors are indicated as A to F, and their response to step loading of the channel is presented below. From Askins, C. G., Putnam M. A., and Friebele, E. J., "Instrumentation for Interrogating Many-Element Fiber-Bragg Grating Arrays," *SPIE* **2444**, 257–266.

the FBG sensor responses to this transient loading. It is apparent that a strain resolution of about 1 με was achieved with most of the sensors. The noise level of sensor F was higher than the others, as its Bragg wavelength was located toward the spectral edge of the source.

In a second demonstration of this fiber-CCD spectrometric demodulation system, 20 FBG sensors strung out along a single optical fiber were attached to the surface of a panel in the form of a 4 × 5 grid, with all the sensors aligned in the same direction (see Fig. 11.65). The 20 simultaneous time traces emanating from this demodulation system correspond to the strain outputs from this grid of sensors as the panel was flexed, bent, and three-point loaded. Additional measurements have demonstrated that the strain response is linear over three orders of magnitude.

Eight FBG sensors, with 2-nm spectral width and 0.4-m separation, were also pultruded between the 9th and 10th cloth layers of a 10-ply glass fiber, vinyl-ester resin composite panel (Friebele *et al.*, 1994). Egress of the optical fiber from the ~5-cm-wide panel was accomplished by

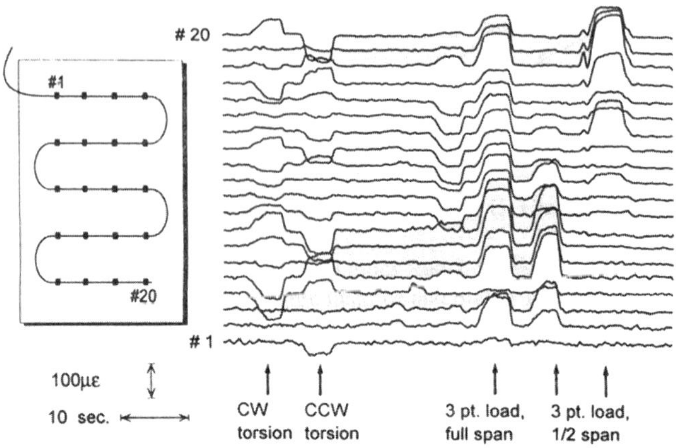

FIGURE 11.65. Layout of a string of FBG sensors bonded to the surface of a plate that is then subject to various loads. The corresponding readings from each of the FBG sensors is presented at right in the figure. From Askins, C. G., Putnam M. A., and Friebele, E. J., "Instrumentation for Interrogating Many-Element Fiber-Bragg Grating Arrays," *SPIE* **2444**, 257–266.

leaving a coiled pigtail within the composite panel in a tape-sealed delamination packet at the ends of each 3-m section. The eighth FBG sensor was included in this pigtail to serve as a reference for the other seven-strain sensors. The spectrum of one such FBG array, measured prior and subsequent to embedding, is shown in Fig. 11.66. The lower trace (a) corresponds to the spectrum as the optical fiber was drawn, and (b) the upper trace to its spectrum after embedding. The Bragg wavelength of the reference FBG (A) at 808 nm is seen to remain fixed, while that of the remaining FBGs are increased because of inadvertent tensile strain ($>1\%$ for G and H) imposed on them from the unequal feed rates on the optical fiber and the fabric as they enter the DIE of a Continuous Resin Transfer Molding (CRTM) machine; see Fig. 11.67. This differential stress was then locked in by the resin cure.

The panel was then supported on rollers located approximately under two of the FBGs (C and I), then loaded by three-point bending. The spectrum of the embedded FBG array was measured, and each of the gratings between the pivot points was seen to suffer a decrease in its Bragg wavelength commensurate with the compressive strain imposed upon it (see Fig. 11.68). It is suggested by Friebele *et al.* (1994) that the difference between the experimental points and the calculated values comes from a variation of about 1/2 ply (0.276 mm) thickness in the position of the sensor visa via the neutral axis of the panel. Further, tests

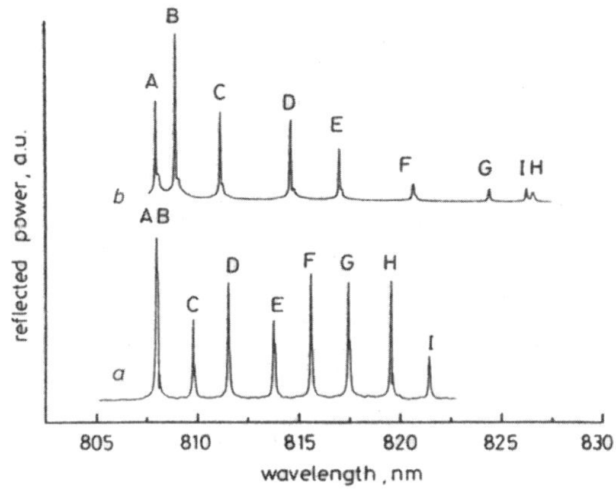

FIGURE 11.66. Reflection spectra from a set of FBG sensors that are pultruded into a 10-ply glass fiber panel. The lower trace (a) displays the spectrum after the optical fiber was drawn, while the upper trace (b) corresponds to the spectrum after embeddment.

and measurements of this demodulation system for arrays of FBG sensors embedded within flat panels and C-channels fabricated by the CRTM process were reported by Friebele *et al.* (1996).

As discussed in Section 9.2.6, Bronnimann *et al.* (1998) have also used a form of fiber-CCD spectrometric demodulation system (see Fig. 9.71), to interrogate the strings of FBG sensors they used to instrument Storck's Bridge in Winterthur, Switzerland. This was the first stay-cable bridge in the world to use CFRP cables in place of steel in two of its 35-m cables and was shown earlier as Fig. 2.7. Each of the CFRP cables was monitored by a sensor array comprising seven-FBGs. Three of the seven FBGs (BG1, BG4, and BG7) were adhered to loaded 5-mm-diameter wires at the circumfer-

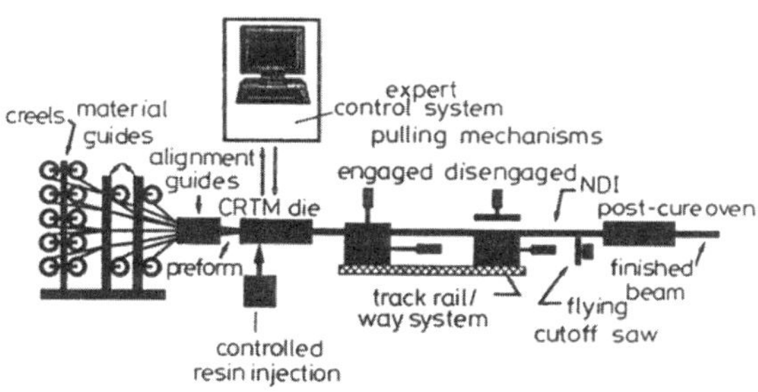

FIGURE 11.67. Pultrusion system used to embed strings of FBG sensors into various structures.

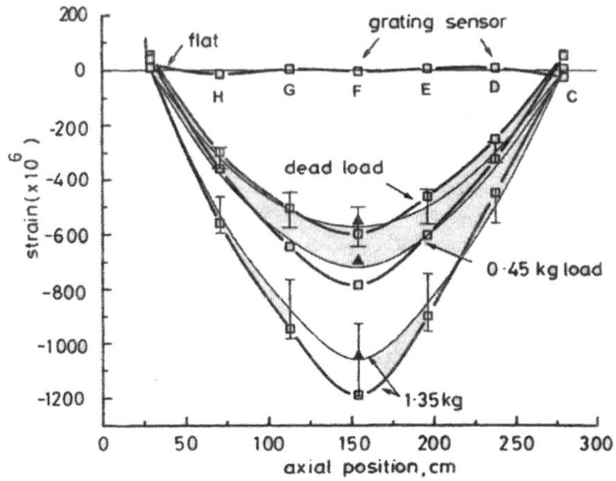

FIGURE 11.68. Strain distribution measured by a set of FBG sensors pultruded within a panel that is subject to increasing three-point bending.

ence of the cable (see Fig. 9.70). The remaining four-FBGs were used as dummies for temperature compensation and for monitoring long-term drifts. The FBGs used were written during fabrication of the optical fiber on a draw tower (Askins *et al.*, 1992). A reference spectrum from an argon lamp was used for absolute wavelength calibration in their system (see Fig. 9.72).

In summary, a number of fiber-CCD spectrometric demodulation systems have been developed and shown to be capable of working with the kind of large serial arrays of weak reflectivity FBG sensors produced during the manufacture of the host optical fiber. This technique is inherently insensitive to fluctuations in optical power and provides an extremely linear response over more than three orders of magnitude of variation in strain. It has been used with a number of serial arrays of embedded FBG sensors fabricated in the manufacture of the optical fiber and used in and with various FRP composite structures. Askins *et al.* (1995) have indicated that a broadband source, emitting less than 1 mW over a 30-nm bandwidth, can provide a strain resolution of better than 1 $\mu\varepsilon$ when used to interrogate 20 FBG sensors simultaneously with a 1-kHz frequency response. It should be recognized that this approach at demodulation requires a more skilled operator and specialized software to extract the strain information from the output of the CCD array. It is anticipated that these performance parameters could be considerably improved with state-of-the-art technology and that new software should make it easier to use.

CHAPTER 12

Distributed Strain and Temperature Sensing

12.1 INTRODUCTION

Truly distributed strain and temperature sensing has been the "Holy Grail" of fiber optic sensing ever since it was appreciated that it was possible to receive information from any point along an optical fiber through Rayleigh scattering. The challenge has been to find a mechanism that would allow the key structural parameters of strain and temperature to be determined at any point along an optical fiber with an appropriate sensitivity and spatial resolution. Unfortunately, the term *distributed sensing* is sometimes attributed to the technique of making a series of point measurements along an optical fiber by a set of short gauge-length sensors, such as FBGs (see Fig. 2.9). The more correct term for this type of measurement is *quasi-distributed sensing*.

It is perfectly true that if the distance between sensors is much smaller than the scale length for change in the variable of interest, then the distinction is academic, and quasi-distributed sensing can provide as much information as a distributed sensing system. Indeed, it is possible to have a situation where the spatial resolution of a distributed sensing system is inferior to that of a quasi-distributed sensing system. An example of this is to be found in distributed temperature sensing via Raman scattering (Dakin, 1995). The spatial resolution of this technique is ≥ 1 m, while it is certainly possible to achieve a much finer spatial resolution using a set of short gauge-length sensors. Nevertheless, in this book we have tried to maintain the distinction between distributed and quasi-distributed sensing, and in this chapter we focus on the former.

We claimed earlier that "fiber optic Bragg grating technology offers unrivaled sensor versatility." One of the strongest vindications of this claim is provided by the recently discovered observation that truly distributed strain and temperature sensing can be achieved through *Bragg intragrating* (BIG) measurements (Huang *et al.*, 1998). It is true that Brillouin scattering can also be used for distributed strain and temperature sensing (Brown *et al.*, 1998), but the spatial resolution actually attained to date with this technique has been limited to ≥ 1 m. This limitation, combined with the large physical size and high cost of the system, makes it impractical for most structural monitoring applications.

Fiber Bragg gratings have been shown to provide strain measurements with a spatial resolution of 1 mm (Huang *et al.*, 1995; Duck, 2000). Current technology permits the fabrication of FBGs with lengths that are limited to about 1 m. Unfortunately, grating lengths in excess of 10 cm are quite expensive at the moment, but considerable research is underway to reduce this cost. It is possible, however, to create a number of extended FBGs in series along an optical fiber so that the effective distributed sensing length is not necessarily limited to that of a single grating.

In this chapter we concentrate fairly exclusively on the exciting new prospect of distributed strain sensing with fiber optic Bragg grating technology. In some sense one could claim that the concept of distributed strain sensing with 1 mm spatial resolution is so new and different from what was previously possible that it is "a solution looking for a problem." But this is exactly what was said of lasers soon after they were invented in 1960, and just look at the incredible number of problems for which lasers have provided solutions over the intervening period. Indeed, in many cases we did not even know of the problem, prior to the invention of the laser.

So it is possible that this new measurement tool could eventually find all kinds of engineering applications once a full appreciation of its capability is widely known. Strain profile measurements might improve the design of FRP cable anchorage systems used by the construction industry, chart the growth of cracks in airplanes, ships or bridges, or warn of weakening in structurally bonded joints. There is little doubt that recognition of partial debonding of a FBG strain sensor from its host structure represents a valuable application, since this is a primary mode of failure for strain sensors. It is possible that the scale of model building might be reduced and made less expensive with the use of this type of *high-resolution strain mapping*. Finally, we speculate that the fine spatial resolution possible with a single strand of optical fiber containing a FBG could permit edge discrimination in tactile sensing for robotic systems.

12.2 FIBER BRAGG INTRAGRATING DISTRIBUTED SENSING CONCEPT

In this section we shall try to explain how *Bragg intragrating* (BIG) measurements permit distributed strain sensing. When we refer to intragrating sensing we mean measurements undertaken *within* a single grating. In essence, any FBG can be thought to comprise a string of very short, closely spaced, subgrating elements. When a uniform strain profile is imposed on the grating, each subgrating experiences the same strain and its Bragg wavelength therefore shifts the same amount. Consequently, all the small optical signals from each of the subgrating elements add together, creating one strong reflection at the Bragg wavelength appropriate for this strain.

If, on the other hand, the strain is nonuniform and varies along the FBG, then each subgrating experiences a slightly different strain and reflects light at a different Bragg wavelength so that the reflection spectrum becomes relatively broad and multiple peaked (see Fig. 12.1). When the grating period or the modulation in the index of refraction are functions of position along the FBG, the grating is said to be "chirped." So a non-uniform strain profile leads to a *strain-chirped FBG*. To illustrate this further we have shown in Fig. 12.2 how the reflection spectrum of a FBG

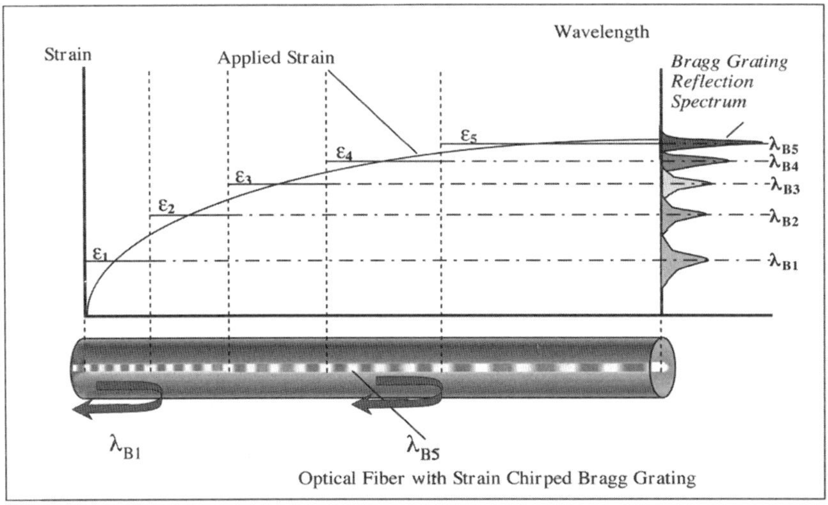

FIGURE 12.1. Strain-induced chirp of an FBG and resulant influence on its reflection spectrum. Note that in the case of chirped grating, each wavelength component is reflected from a different position along the grating.

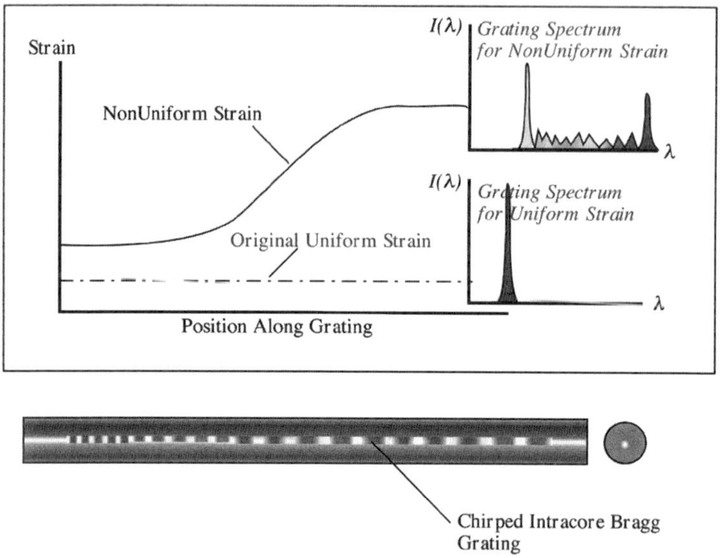

FIGURE 12.2. Schematic illustration of the change in the nature of the reflection spectrum as the strain field goes from uniform to nonuniform.

would change if the strain to which it is exposed changed from being uniform to having a steplike profile. Note that the two flat segments of the strain profile contribute two fairly narrow peaks to the overall spectrum.

The term chirp is used to account for the nonuniformity in the *optical period* (defined as $n\Lambda$) of the FBG index of refraction modulation. Although gratings can be produced with a "built-in chirp" due to a spatial variation in either the mean index of refraction or the index modulation period, we shall be primarily concerned with FBG chirps that result from the imposition of a nonuniform strain field. Associated with the *chirp* of a FBG is the observation that the location from which reflection at a given wavelength occurs becomes dependent upon that wavelength. For a *uniform FBG* the reflection spectrum comprises a *primary peak* (see Section 3.10) centered at the Bragg wavelength,

$$\lambda_B = 2n_0\Lambda_0, \tag{12.1}$$

and from experiments in which the FBG serves as one mirror of a Michelson interferometer, this reflection appears to arise from the *midpoint of the grating*.

If the FBG is chirped, then the optical period becomes a function of axial position z, and so we can write

$$\lambda(z) = 2n(z)\Lambda(z) = 2[n_0\Lambda_0 + \Delta\{n\Lambda\}], \tag{12.2}$$

which indicates that the axial position along the FBG from where a given wavelength appears to be reflected now depends upon the *local strain* $\varepsilon(z)$, and *strain-gauge factor* G_ε, (introduced in Section 7.3); i.e.,

$$\lambda(z) = 2n_0\Lambda_0\{1 + G_\varepsilon(z)\}, \tag{12.3}$$

where $n_0\Lambda_0$ represents the optical period of the unstrained FBG. Equation (12.3) follows directly from the recognition that Eq. (7.24) allows us to write:

$$\Delta\{n\Lambda\} = n_0\Lambda_0 G_\varepsilon \varepsilon. \tag{12.4}$$

Huang *et al.* (1995a) demonstrated that fiber Bragg gratings with non-uniform optical periods, refractive index modulation depths, and mean indices of refraction can be analyzed by the *T-matrix* formalism (Haus, 1984).

12.3 T-MATRIX FORMALISM FOR NONUNIFORM FIBER BRAGG GRATINGS

The behavior of a fiber Bragg grating can be described in terms of two counter propagating plane waves confined to the core of the optical fiber containing the FBG. Initially we assume that the grating is uniform, so its optical period $n\Lambda$ and depth of refractive index modulation Δn_0 are the same throughout the grating. The electric fields of the forward and backward waves in the grating are assumed to be described by the expressions

$$E_f(z, t) = a_f(z)e^{[i\{wt-\beta z\}]} \tag{12.5}$$

and

$$E_b(z, t) = a_b(z)e^{[i\{wt-\beta z\}]}, \tag{12.6}$$

respectively, with $a_f(z)$ and $a_b(z)$ representing the complex amplitudes, ω the angular frequency, and β the propagation constant of the two waves; see Section 3.10. According to Yariv and Nakamura (1977), these complex amplitudes obey the *coupled-mode equations*

$$\frac{da_f(z)}{dz} = -i\kappa^* a_b(z)e^{2iz\Delta\beta} \tag{12.7}$$

and

$$\frac{da_b(z)}{dz} = i\kappa a_f(z)e^{-2iz\Delta\beta} \tag{12.8}$$

over $0 < z < L_G$, where L_G is the length of the grating and we introduce the differential propagation constant,

$$\Delta\beta \equiv \frac{2n\pi}{\lambda} - \frac{\pi}{\Lambda}, \tag{12.9}$$

associated with detuning from the Bragg condition ($\lambda = 2n\Lambda$), so

$$p \equiv \pi/\Lambda. \tag{12.10}$$

For a uniform grating with sinusoidal modulation of the refractive index,

$$n(z) = n + \Delta n_0 \cos\{2\pi z/\Lambda\}, \tag{12.11}$$

and the coupling coefficient

$$\kappa \equiv \sqrt{\left\{\frac{\pi\Delta n_0}{\lambda_B}\right\}^2 - \Delta\beta^2}, \tag{12.12}$$

for wavelengths close to the phase match condition, i.e., $\beta \approx p$ and

$$\kappa \approx \frac{\pi\Delta n_0}{\lambda_B},$$

see Eq. (3.131).

Huang *et al.* (1994) showed that for a uniform subgrating of length, ΔL, a *T-matrix* relation can be used in conjunction with the boundary conditions $a_f(0) = a_0$ and $a_b(\Delta L) = a_{\Delta L}$ to relate the left-side **E**-field complex amplitudes for the subgrating to those of the right-side; see Fig. 12.3a, i.e.,

$$\begin{bmatrix} a_f(0) \\ a_b(0) \end{bmatrix} = \begin{bmatrix} T_{11} & T_{12} \\ T_{21} & T_{22} \end{bmatrix} \begin{bmatrix} a_f(\Delta L) \\ a_b(\Delta L) \end{bmatrix}, \tag{12.13}$$

where

$$T_{11} = T_{22}^* = \left[\frac{\Delta\beta \sinh\{\kappa\Delta L\} + i\kappa \cosh\{\kappa\Delta l\}}{i\kappa}\right] e^{ip\Delta L}, \tag{12.14}$$

and

$$T_{12} = T_{21}^* = \left[\frac{\pi\Delta n_0 \sinh\{\kappa\Delta L\}}{i\kappa\lambda_B}\right] e^{ip\Delta L}. \tag{12.15}$$

This *T*-matrix analysis can be adapted to describe the propagation of light through a chirped FBG by dividing the chirped grating into *N* subgrat-

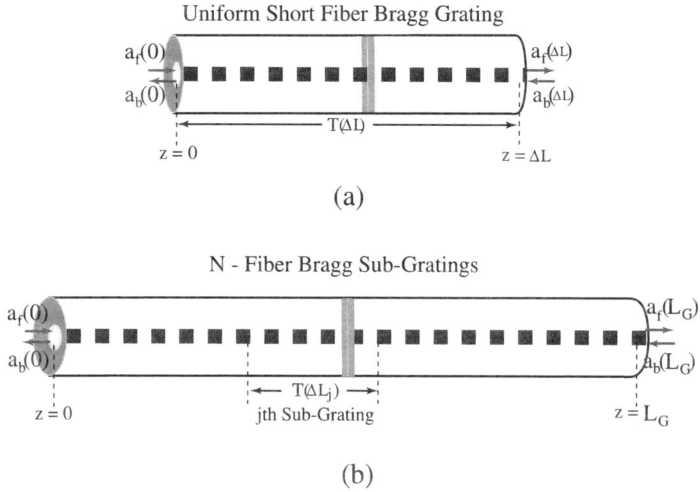

FIGURE 12.3. (a) A uniform short length of a fiber Bragg grating and its boundary conditions. (b) Representation of a fiber Bragg grating by a series of subgrating elements.

ings, each being treated as a uniform grating element with its own T-matrix, and slightly different values of Λ and Δn_0 from those of its neighbors see Fig. 12.3b. In such a case we can write

$$\begin{bmatrix} a_{\text{f}}(0) \\ a_{\text{b}}(0) \end{bmatrix} = [T(\Delta L_1)][T(\Delta L_2)]\cdots[T(\Delta L_N)]\begin{bmatrix} a_{\text{f}}(L_{\text{G}}) \\ a_{\text{b}}(L_{\text{G}}) \end{bmatrix}, \tag{12.16}$$

where

$$L_{\text{G}} = \Delta L_1 + \Delta L_2 + \cdots + \Delta L_N. \tag{12.17}$$

Because there is no backward input wave to the FBG, Huang *et al.* (1994) began with the most right-side elements, that is to say those corresponding to the Nth subgrating, and set $a_{\text{b}}(L_{\text{G}}) = 0$ and $a_{\text{f}}(L_{\text{G}}) = F_N$. After calculating the product of all T-matrices, they obtained the final left-side vector elements $a_{\text{b}}(0)$ and $a_{\text{f}}(0)$ and because F_N turns out to be a common factor for both $a_{\text{b}}(0)$ and $a_{\text{f}}(0)$, it can be set to any nonzero value in calculating the *amplitude spectral reflectivity* of the chirped grating:

$$r(\lambda) \equiv \frac{a_{\text{b}}(0, \lambda)}{a_{\text{f}}(0, \lambda)}, \tag{12.18}$$

where the *intensity spectral reflectivity* $R(\lambda) = |r(\lambda)|^2$. Huang *et al.* (1995a) showed that in the T-matrix calculation for a grating the length selection of each subgrating element is arbitrary and can be as short as one period. To avoid unnecessarily lengthy calculation, the subgrating length is generally set to include a number of periods. The determination of the number of periods depends on the nonuniformity of the grating and the required accuracy. Indeed, different subgrating lengths may be chosen to include different numbers of periods in the same grating.

Huang *et al.* (1994) used the *T*-matrix formalism to determine $R(\lambda)$ for several nonuniform distributions of grating period length $\Lambda(z)$ and the refractive-index modulation depth $\Delta n_0(z)$. They assumed in their calculations that the refractive index is sinuosoidally modulated and used Eq. (12.12) with the phase matched condition (i.e., $\Delta\beta = 0$) to define κ. The

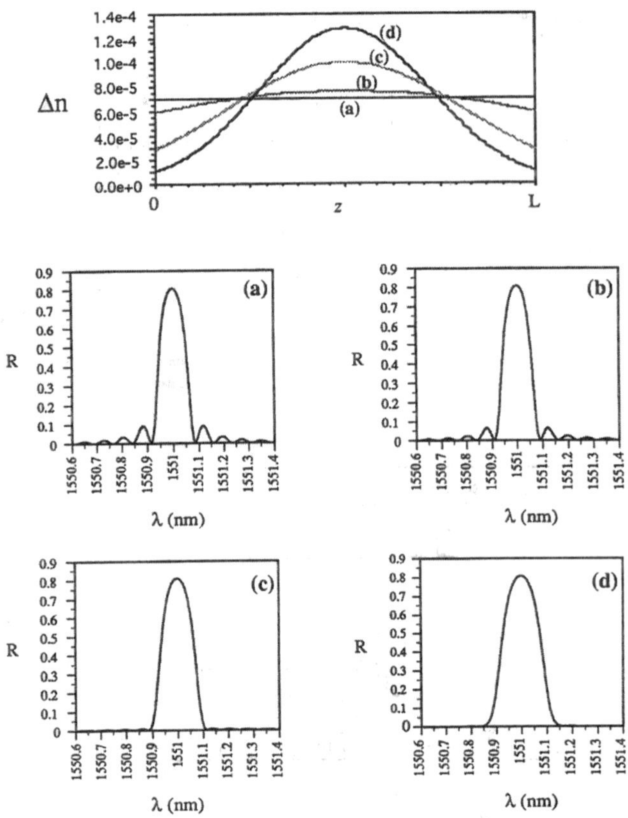

FIGURE 12.4. Four axial profiles for the depth of index modulation, $\Delta n(z)$, and their resulant reflection spectra, $R(\lambda <)$. From Huang, S., LeBlanc, M., Ohn, M.M., and Measures, R.M., (1995a), "Bragg Intra-Grating Structural Sensing," *Applied Optics* **34**, 5003–5009.

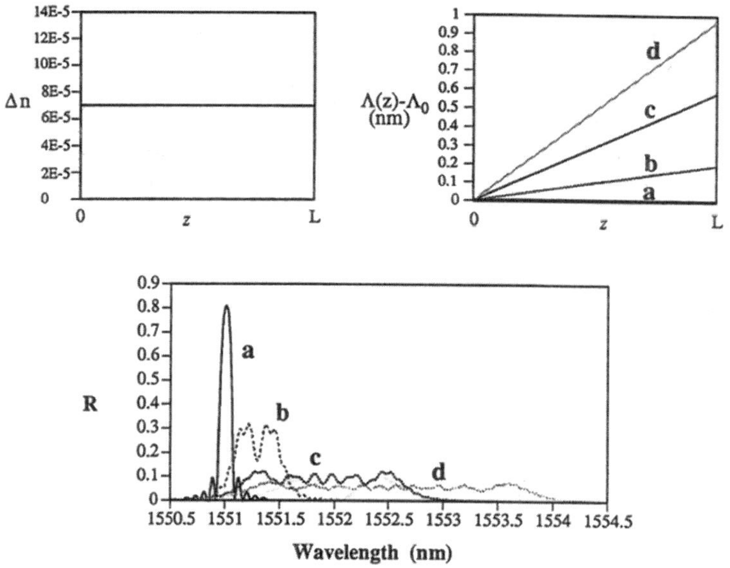

FIGURE 12.5. The influence of varying linear strain gradients on the reflection spectrum of a FBG with a uniform depth of index modulation. From Huang, S., LeBlanc, M., Ohn, M.M., and Measures, R.M., (1995a), "Bragg Intra-Grating Structural Sensing," *Applied Optics* **34**, 5003–5009.

length of the grating, $L_G = 1$ cm, the average refractive index n was $\sim$1.5, and the wavelength was $\sim$1550 nm. Some representative examples are presented next.

For an apodized FBG with a Gaussian-distributed modulation depth of the refractive index, Fig. 12.4 shows the calculated $R(\lambda)$ for gratings with the four different values of peak modulation depth and effective grating length, indicated in the top chart, but constant period length and mean index of refraction. When the optical period of the grating is not uniform, its reflection spectrum becomes wider and is accompanied with a decrease of reflectivity. For a FBG that is subject to a linear strain gradient, the intensity spectral reflectivity, $R(\lambda)$, broadens and comprises multiple peaks, as shown in Fig. 12.5. If a FBG is subject to both a Gaussian-distributed index modulation depth and a linear chirp in the optical period, the resulting intensity spectral reflectivity curves are relatively smooth and the fall-off in peak height is less than the case of the uniform grating; see Fig. 12.6. Experimental verification of this analysis was undertaken by Huang *et al.* (1995a,b) and presented earlier as Fig. 6.25.

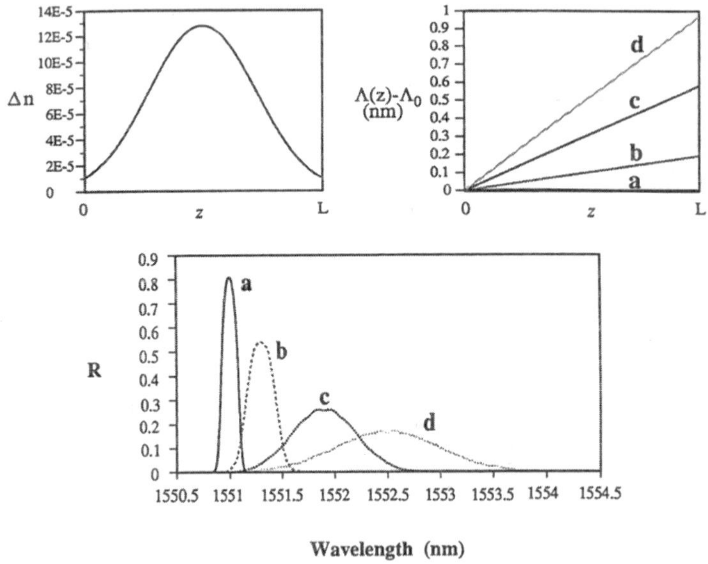

FIGURE 12.6. The influence of varying linear strain gradients on the reflection spectrum of an FBG with a Gaussian-shaped distribution in the depth of index modulation. From Huang, S., LeBlanc, M., Ohn, M.M., and Measures, R.M., (1995a), "Bragg Intra-Grating Structural Sensing," *Applied Optics* **34**, 5003–5009.

12.4 INTENSITY REFLECTION SPECTRUM FOR DISTRIBUTED STRAIN SENSING

The question arises as to how to determine, from the reflected light, the strain profile chirping an FBG sensor. The first approach followed from a simple model of the interaction. Consider an unstrained, uniform FBG that is excited by a broadband source. Since the FBG can be thought of as comprising an array of subgrating elements, each having an identical reflection spectrum, clearly the overall FBG spectrum would simply be the sum of the spectra from all of the sub-gratings. If a nonuniform strain is now imposed on this FBG, each of the subgratings subject to roughly the same strain will contribute the same spectra. For those subgratings exposed to a higher tensile strain, their spectra will be redshifted an amount that is proportional to the difference in strain from its reference value.

We can form a simple model of a strain-chirped FBG by assuming that the strain profile imposed on the FBG is monotonic so that the wavelength of reflection progressively increases for tensile strain (or decreases for compressive strain) the further the light penetrates into the FBG; see Fig. 12.7. We assume that the contribution to the grating intensity spectral reflectivity, $R(\lambda)$, from the m_j (an integer) subgratings subject to the strain corresponding to a reflection wavelength λ_j can be written as the sum over the different wavelength components from the minimum value, λ_{min} (assuming a tensile strain profile) to the wavelength λ_n being reflected from the position of interest, z; i.e.,

$$\sum_{j=1}^{j=n} R(\lambda_j) = \alpha \sum_{j=1}^{j=n} m_j \Delta z, \tag{12.19}$$

where α, is the proportionality constant, and Δz is the length of a subgrating. We see from Fig. 12.7 that relatively high, narrow spectral peaks arise from regions of the grating subject to flat variations of strain as many subgratings contribute to this part of the spectrum. In contrast, regions of the FBG subject to steep strain gradients give rise to small, spectrally wide, multiple-peaked parts of the reflection spectrum.

If we convert the sums into integrals, we can write

$$\int_{\lambda_{min}}^{\lambda} R(\lambda)d\lambda = \int_{\lambda_{min}}^{\lambda} \alpha dz = \alpha z(\lambda), \tag{12.20}$$

and if we integrate along the entire length of the FBG, L_G, we have

$$\int_{\lambda_{min}}^{\lambda_{min}} R(\lambda)d\lambda = \int_{\lambda_{min}}^{\lambda_{max}} \alpha dz = \alpha L_G. \tag{12.21}$$

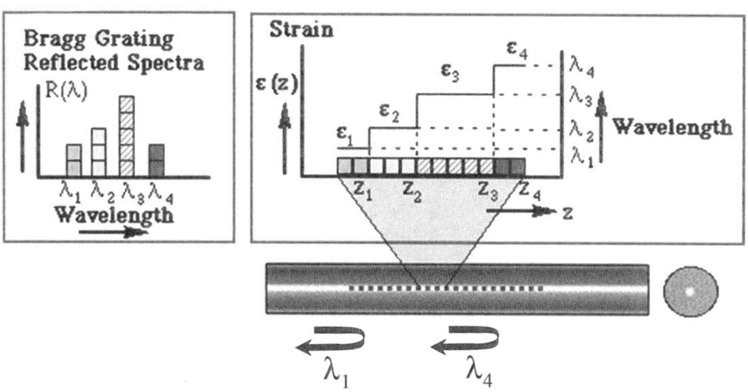

FIGURE 12.7. Simple model for relating the penetration depth for a given wavelength within an FBG to the integrated spectral power to that wavelength, normalized by the total power in reflected.

Combining (12.20) with (12.21) yields a relation for the penetration depth, $z(\lambda)$, commensurate with the wavelength, λ

$$z(\lambda) = L_G \frac{\int_{\lambda_{\min}}^{\lambda} R(\lambda)d\lambda}{\int_{\lambda_{\min}}^{\lambda_{\max}} R(\lambda)d\lambda}. \tag{12.22}$$

We see that Eq. (12.22) enables the distance into the FBG, $z(\lambda)$, that light of wavelength λ penetrates before it suffers reflection to be determined from the spectrally integrated intensity to wavelength λ, normalized by the total intensity reflected. Since this wavelength can be related to the strain at that location via Eq. (12.3), it is clear that the strain profile along the FBG can be evaluated, provided that the strain continually increases or decreases and does not reverse direction. Thus, this method is only applicable to a monotonic strain profile.

Although this simplistic approach provided some insight into the phenomenon, it was not found to be particularly accurate and some improvement was obtained by drawing upon the early work of Matsuhara *et al.* (1975), who described a theoretical procedure for constructing an optical filter with an arbitrary reflection spectrum, $R(\lambda)$, from an aperiodic Bragg grating. LeBlanc *et al.* (1996) adapted this theory to the task of extracting the strain profile from a strain-chirped FBG and showed that the relation

$$-\int_{\lambda(z=0)}^{\lambda(z)} \ln[1 - R(\lambda^*)]d\lambda^* = \pm\frac{\pi^2}{2}\int_{z=0}^{z} \frac{\Delta n^2(z^*)}{n(z^*)} dz^*, \tag{12.23}$$

can be applied, provided the strain distribution is a monotonic function of distance along the grating. In Eq. (12.23), $\Delta n(z)$ is the modulation depth for the index of refraction of the grating as a function of axial position. In regard to the ($\pm$) on the right side of (12.23) the ($-$) sign is used if the slope of $\varepsilon(z)$ is known to be positive, while the ($+$) is used if it is negative. The relationship for $\lambda(z)$ [which is related to $\varepsilon(z)$ through Eq. (12.3)] is obtained by equating the integral on the left-side (λ-generated) with the integral on the right-side (z-generated in the range $0 \le z \le L_G$) using an iterative procedure. Ohn *et al.* (1996) assumed that the index of refraction and its modulation depth are relatively weak functions of axial position. This permits a simplification that makes it possible to directly relate the axial location, z, along the FBG at which the wavelength, λ, is reflected, to the integral on the left side of (12.23), i.e.,

$$z(\lambda) = -\frac{2n}{\{\pi\Delta n\}^2}\int_{\lambda_{\min}}^{\lambda} \ln[1 - R(\lambda^*)]d\lambda^* \qquad \{0 \le z \le L_G\}. \tag{12.24}$$

Once $z(\lambda)$ is obtained from (12.24), the axial strain, ε, at that location (where the wavelength for reflection is λ) is determined using Eq. (12.3) in the form

$$\varepsilon = \frac{\lambda - \lambda_0}{G_\varepsilon \lambda_0}, \tag{12.25}$$

where λ_0 corresponds to the initial wavelength of the unstrained FBG. The accuracy of (12.24) is influenced by the effective interaction length, L_{eff}, introduced by Hill (1974) to account for the actual length of the FBG that contributes to the reflection of light at wavelength λ at the z location

$$L_{\text{eff}}(z) = \sqrt{\frac{\Lambda(z)}{G_\varepsilon |d\varepsilon(z)/dz|}}, \tag{12.26}$$

where $|d\varepsilon(z)/dz|$ represents the local strain gradient. As a result the local strain $\varepsilon(z)$ determined by this spectral integration technique is really the average strain over L_{eff}. Relation (12.26) is easy to appreciate, for the larger the local strain gradient, the fewer the grating periods (i.e., the shorter the length of the subgrating) that can contribute to reflecting a given wavelength of light. This has implications regarding the magnitude of the reflection coefficient at a given wavelength and the minimum strain gradient that can be accurately evaluated by this technique.

Clearly, if the strain gradient drops to a value that makes $L_{\text{eff}} > L_G$, an approximation assumed in deriving Eq. (12.24) becomes invalid and accuracy is lost. Thus, regions of low strain gradient give rise to large errors. In a demonstration of this spectral integration technique for BIG sensing, a three-section tunable DBR laser diode was scanned over a 3-nm range while the reflected optical signal from three special FBGs was recorded by a photodetector (Ohn *et al.*, 1996); see Fig. 12.8. The wavelength of the laser was determined from the injection current used to tune it. The gratings used in this experiment were fabricated at the Institute for Research in Stockholm with lengths 20, 50 and 100 mm, and a $\sim 10^{-3}\%\ \text{mm}^{-1}$ linear chirp rate for the grating periods (pitchs). The reflection spectra for these three FBGs are presented as Fig. 12.9a and the corresponding variation in optical period, calculated by Eq. (12.24), for each of the gratings is displayed as Fig. 12.9b. Also shown in the lower figure, by the dotted line, is the specified chirp.

It is quite apparent that the spectral integration technique was able to evaluate the chirp in the gratings with reasonable accuracy, except toward the ends of each grating. The large errors that rapidly build in the endzones of the gratings can be understood in terms of the breakdown of the condition that L_{eff} be smaller than the length of grating remaining

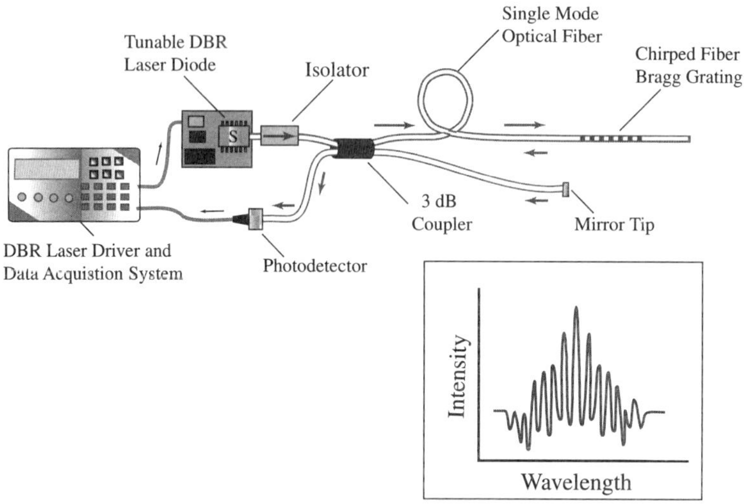

FIGURE 12.8. Experimental arrangement for measuring the linear chirp of a FBG based on the intensity reflection spectrum. From Ohn, M.M., Huang, S.Y., LeBlanc, M. and Measures R., (1996), "Distributed Strain Sensing Using Long Intracore Fiber Bragg Gratings," *SPIE* **2838**, 66–75.

between the current position and the end of the grating. This problem can be alleviated to some extent by the use of a prechirped FBG sensor see Section 12.10.2. This is equivalent to including a strain gradient bias to the grating that prevents any low-gradient region from arising along the sensor. With sufficient bias it is even possible to accommodate some degree of strain reversal so that some deviation from a monotonic strain profile can be evaluated (see Fig. 12.10).

Another demonstration of the spectral integration technique for distributed strain sensing was also described by Ohn *et al.* (1996). In this experiment a 5-mm long FBG sensor (which was originally uniform) was bonded within a groove in an aluminum plate in close proximity to a hole, as indicated in Fig. 12.11. This groove technique was found to minimize unknown strains introduced by surface bonding (LeBlanc *et al.*, 1994). When this plate is loaded under tension, a strain profile (that can be calculated by elasticity theory) is imposed on the FBG sensor and the concomitant change in the reflection spectrum of the FBG measured (see Fig. 12.12a). The experimentally determined strain profiles are compared with those predicted from strain theory in Fig. 12.12b. Although reasonable agreement is obtained over most of the grating length, once again considerable errors are evident toward the ends of the gratings.

(a)

(b)

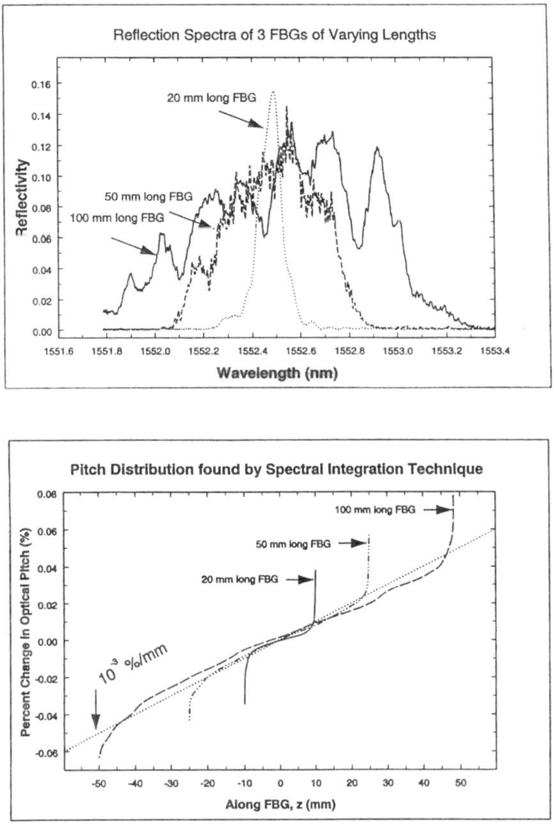

FIGURE 12.9. (a) Reflection spectra for three FBGs of different lengths but the same linear chirp ($10^{-3}\%\mathrm{mm}^{-12}$). (b) Chirp of the three different length gratings as determined by the intensity reflection spectrum technique and the imposed linear chirp. From Ohn, M.M., Huang, S.Y., LeBlanc, M. and Measures, R., (1996), "Distributed Strain Sensing Using Long Intracore Fiber Bragg Gratings," *SPIE* **2838**, 66–75.

Ohn *et al.* (1996) and Huang *et al.* (1998) describe an *interferometric technique* that can be used instead of the *spectral integration*, technique. However, it is also limited to monotonic strain profiles, and if one is prepared for the added complexity of a phase measurement combined with the derivative method of extracting the strain profile, it is more sensible to combine the phase and intensity techniques and be free of these limitations, which brings us to the next section.

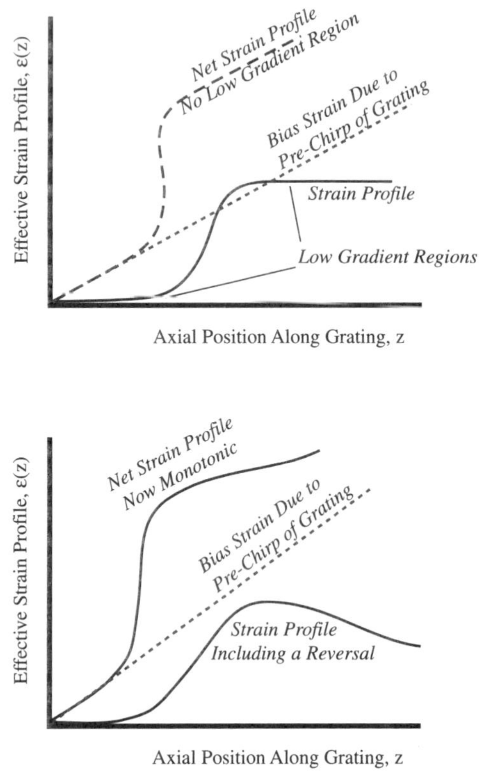

FIGURE 12.10. Schematic illustration of a prechirped FBG and its ability to eliminate the problem of small gradients and make a nonmonotonic strain profile appear as monotonic.

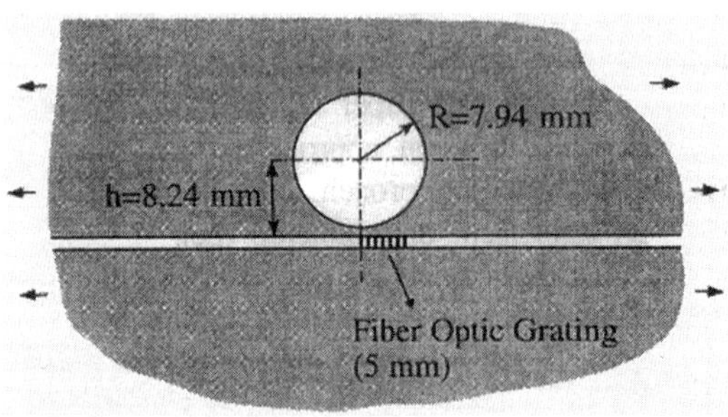

FIGURE 12.11. Experimental arrangement involving mounting a short-gauge FBG sensor adjacent to a hole in a plate under tensile load. From LeBlanc, M., Huang, S., Ohn, M. M., Measures, R.M., Guemes, A. and Orthonos, A., (1996), "Distributed Strain Measurement Based on a Fiber Bragg Grating and its Reflection Spectrum Analysis," *Optics Letters* **21**, 1405–1407.

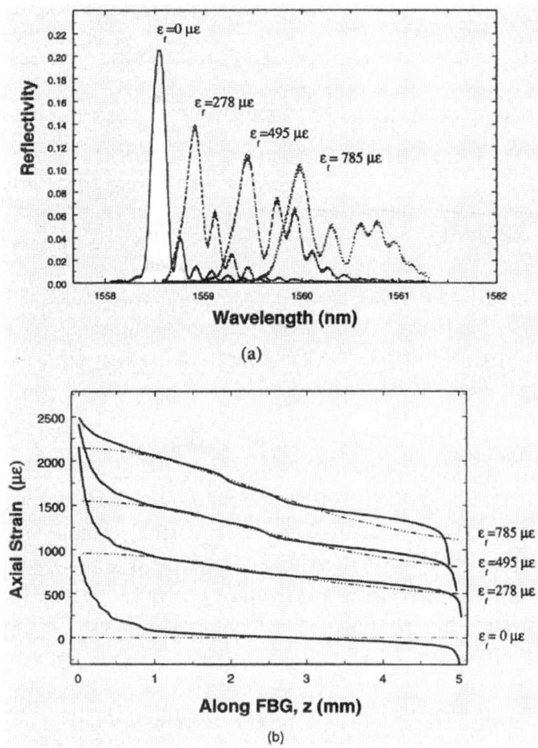

FIGURE 12.12. (a) Set of reflection spectra from the FBG sensor near a hole (see Fig. 12.11) under increasing tensile loading. (b) Experimentally determined strain profile (solid surve) along the FBG, using the intensity reflection spectrum technique, compared to the theoretically predicted profile (broken curve).

12.5 DISTRIBUTED STRAIN SENSING BASED ON FOURIER TRANSFORMS

As was first proposed by Ohn *et al.* (1997), it is possible to undertake "truly" distributed strain sensing of arbitrary strain profiles with a single fiber Bragg grating using a combination of intensity and phase spectral measurements combined with Fourier transform analysis. To appreciate how this is achieved, we assume that within a single-mode optical fiber containing a Bragg grating there exist two waves: a forward wave with a complex amplitude $a_f(z)$, and a backward wave with a complex amplitude $a_b(z)$. Kogelnik (1976) has shown that the relation between these waves for a chirped grating can be written in terms of the grating amplitude reflection coefficient,

$$r(z) = a_{\mathrm{b}}(z)/a_{\mathrm{f}}(z), \tag{12.27}$$

and a first-order differential equation of the form

$$\frac{dr(z)}{dz} = i\kappa(z)e^{-i\{2z\Delta\beta(z)-\phi(z)\}} + i\kappa^{*}(z)r(z)^{2}e^{i\{2z\Delta\beta(z)-\phi(z)\}}, \tag{12.28}$$

where $\kappa(z)$ is the real part of the coupling coefficient for the grating, $\phi(z)$ is the incremental grating phase due to perturbations, and $\Delta\beta(z)$ is the detuning in the propagation constant from its value that is commensurate with the unperturbed Bragg wavelength, i.e.,

$$\Delta\beta(z) = \beta(z) - \pi/\Lambda_0. \tag{12.29}$$

Here $\beta(z)$ is the propagation constant associated with the wavelength that is reflected from location z, and Λ_0 is the unperturbed grating period. Kogelnik's definition of the incremental grating phase $\phi(z)$ is the phase shift arising from the accumulated changes in the period of the grating, between $z = 0$ and $z = z$, relative to the unperturbed state, where $z = 0$ corresponds to the grating end closest to the source. For the case of non-uniform strain-induced perturbations, this must be generalized to account for the strain optic effect, leading to the general relation

$$\phi(z) = -\int_{z=0}^{z} \frac{2\pi}{\Lambda_0}\frac{\Delta\lambda(z)}{\lambda_0}dz = -\frac{2\pi}{\Lambda_0}\int_{z=0}^{z} G_{\varepsilon}\varepsilon(z)dz, \tag{12.30}$$

where the grating strain gauge factor,

$$G_{\varepsilon} = [1 - n_0^2\{p_{12} - \nu(p_{11} + p_{12})\}/2],$$

was introduced in Chapter 7, Eq. (7.25), and $\varepsilon(z)$ is the local strain in the grating, with p_{11} and p_{12} representing the strain optic coefficients for the optical fiber, and ν its Poisson ratio.

For a low-reflectivity grating ($r(z) \ll 1$) the second term in Eq. (12.28) can be neglected. This permits the equation to be integrated along the Bragg grating to yield:

$$r(L_{\mathrm{G}}) - r(0) = i\int_{z=0}^{z=L_{\mathrm{G}}} [\kappa(z)e^{i\phi(z)}]e^{-i2z\Delta\beta(z)}dz, \tag{12.31}$$

where $z = L_{\mathrm{G}}$ corresponds to the end of the grating furthest from the source. Since we can assume $r(L_{\mathrm{G}}) = 0$, as $a_{\mathrm{b}}(L_{\mathrm{G}}) = 0$, and that the complex reflection coefficient is also a function of $\Delta\beta$, Eq. (12.31) takes the form:

$$r(\Delta\beta) = -i\int_{z=0}^{z=L_G} [\kappa(z)e^{i\phi(z)}]e^{-2z\Delta\beta(z)}dz, \tag{12.32}$$

where we have introduced $r(\Delta\beta)$ to represent $r(z=0, \Delta\beta)$, for brevity. Equation (12.32) shows that $r(\Delta\beta)$ is the Fourier transform of the complex coupling coefficient, $(\kappa(z)e^{i\phi(z)})$.

In such a case, the inverse Fourier transform takes the form

$$\kappa(z)e^{i\phi(z)} = \frac{i}{\pi}\int_{-\infty}^{\infty} r(\Delta\beta)e^{i2z\Delta\beta}d(\Delta\beta). \tag{12.33}$$

The complex reflection coefficient, $r(\Delta\beta)$, is also a function of wavelength as

$$\Delta\beta(z) = \frac{2\pi n(z)}{\lambda(z)} - \frac{2\pi n_0}{\lambda_0} \approx -\frac{2\pi n_0}{\lambda_0^2}\Delta\lambda, \tag{12.34}$$

with $\Delta\lambda = (\lambda(z) - \lambda_0)$ and λ_0 representing the Bragg wavelength for the unperturbed grating. We assume that the deviations in n and λ produced by the strain perturbations are quite small. The complex amplitude reflection coefficient can also be expressed in the form

$$R(\Delta\lambda) = \sqrt{R(\Delta\lambda)}e^{-i\psi(\Delta\lambda)}, \tag{12.35}$$

where $R(\Delta\lambda)$ and $\psi(\Delta\lambda)$ represent the intensity and phase spectra, respectively, for the grating measured at $z = 0$. Under these circumstances the inverse Fourier Transform, Eq. (12.33), takes the form

$$\kappa(z)e^{i\phi(z)} = -\frac{2in_0}{\lambda_0^2}\int_{-\infty}^{\infty} \sqrt{R(\Delta\lambda)}e^{-\psi(\Delta\lambda)}e^{-4i\pi n_0 z\Delta\lambda/\lambda_0^2}d(\Delta\lambda). \tag{12.36}$$

Using Eq. (12.36) and the experimental intensity and phase spectra, Measures *et al.* (1998) computed both the amplitude $\kappa(z)$ and the argument $\phi(z)$ of the complex coupling coefficient. The former describes the axial variation in the coupling coefficient along the grating. The relation between the incremental grating phase, $\phi(z)$, and the strain distribution, $\varepsilon(z)$, is given in Eq. (12.30) and can be put in the form

$$\varepsilon(z) = -\frac{\Lambda_0}{2\pi G_\varepsilon}\frac{d\phi(z)}{dz}. \tag{12.37}$$

We see that the strain profile $\varepsilon(z)$ can be derived from the measured intensity and phase spectra of the grating using the inverse Fourier transform as given in Eq. (12.37). Note that for the sake of brevity we shall use $R(\lambda)$ and $\psi(\lambda)$ in place of $R(\Delta\lambda)$ and $\psi(\Delta\lambda)$ in the remainder of this chapter.

12.6 EXPERIMENTAL FOURIER TRANSFORM DISTRIBUTED STRAIN SENSING

The preliminary experiment arrangement used by Huang *et al.* (1998) to test the Fourier transform method of determining the arbitrary strain profile imposed upon a fiber Bragg grating is shown schematically in Fig. 12.13a. A tunable DBR laser operating at about 1552 nm was scanned over the wavelength interval (approximately 2 nm) sufficient to cover the reflection spectrum (intensity and phase) of the fiber Bragg grating. The laser power launched through the optical fiber into the grating is monitored by the reference photodetector as a function of wavelength. The intensity photodetector monitors the light reflected from the grating as a function of wavelength.

The ratio of the two signals from this pair of photodetectors, as a function of wavelength, is proportional to the intensity reflection spectrum, $R(\lambda)$. A calibration of the system, accounting for the various non-common losses and the different spectral responses of the two photodetectors permits $R(\lambda)$ to be determined. A phase photodetector and a fiber optic Michelson interferometric system were used to monitor the phase spectrum, $\psi(\lambda)$. In this first experiment a second coupler and a second optical fiber with a mirrored end were used for this purpose. A Pentium PC computer was configured to control the laser wavelength and collect the data from the three photodetectors.

A 86-mm long FBG was epoxied into a small groove in the top of a specially shaped aluminum cantilever beam (Fig. 12.13b). The beam was clamped at one end and a force applied at the other end. As stated earlier (Section 12.4), this groove technique avoids the creation of unknown strains that can accompany surface bondin (LeBlanc *et al.*, 1994). A small resistive foil strain gauge was adhered to the underside of the beam, directly beneath the near end of the fiber grating. Figure 12.14a and 12.14b, display the measured intensity and phase spectra, $R(\lambda)$ and $\psi(\lambda)$, respectively, recorded for one loading of the beam. The resistive foil gauge registered a strain at the near end of the grating of about 280 $\mu\varepsilon$ for this loading. Figure 12.14c, displays the variation in the coupling coefficient along the FBG as measured by this inverse Fourier transform technique (Huang *et al.*, 1998). The computed strain profile based on pure beam bending theory is shown as the dotted curve in Fig. 12.14d, while the solid

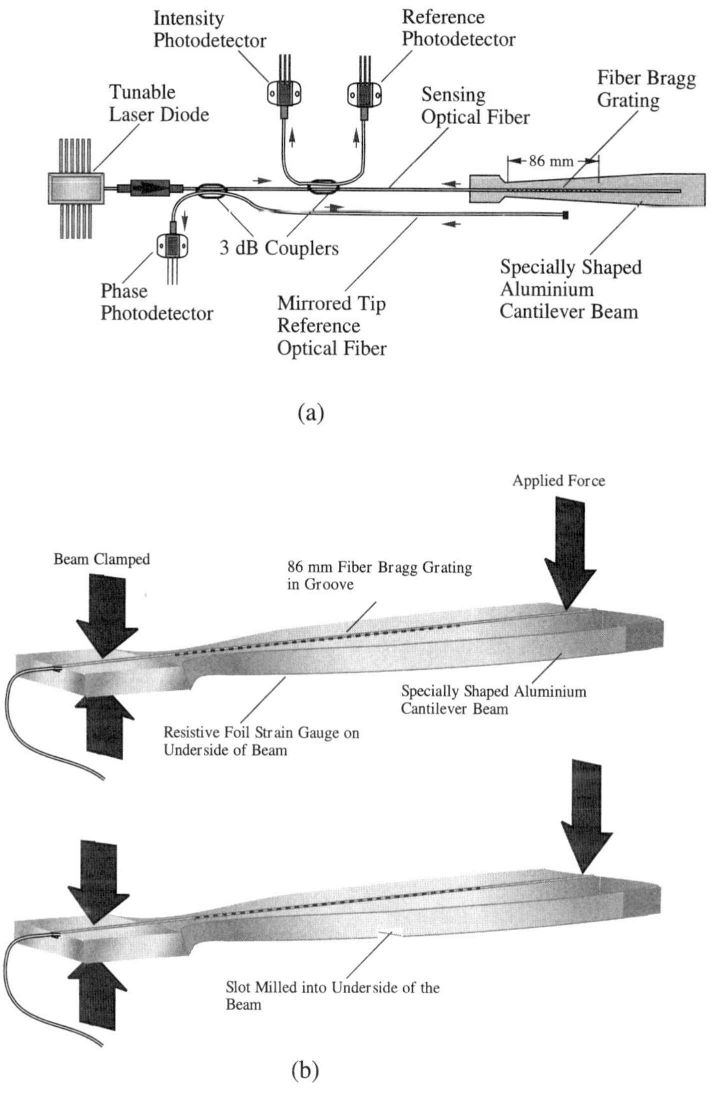

FIGURE 12.13. (a) Experimental arrangement for undertaking both intensity and phase spectral measurements from an 86-mm-long FBG mounted on a specially designed cantilever beam. (b) FBG sensor placement within a groove in the specially shaped cantilever beam, with and without a slot cut into the beam.

curve represents the strain profile deduced from the intensity and phase spectra using the inverse Fourier transform method. Although the experimental curve fits the theoretical curve reasonable well, for an initial experiment, there is considerable noise in the data. Some of the deviations from the smooth theoretical curve could be real and attributed to

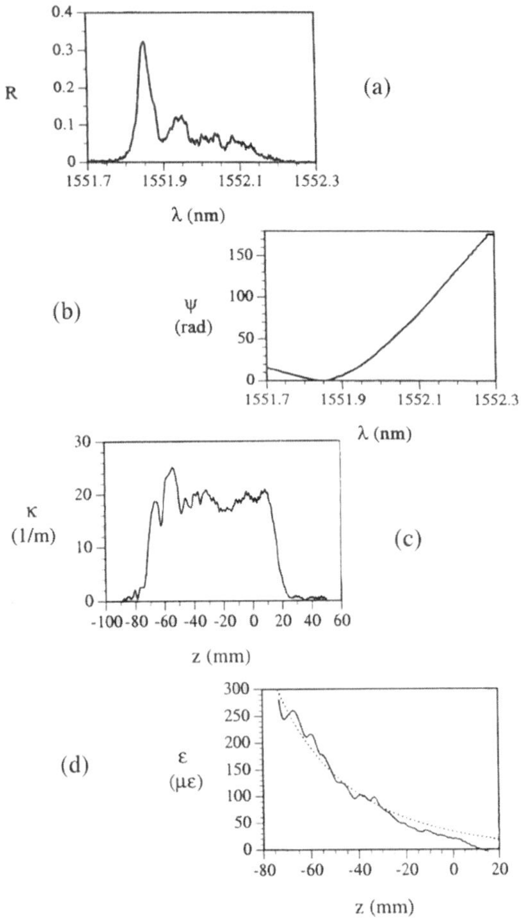

FIGURE 12.14. Intensity spectra for one beam loading of the FBG sensor measured by the tunable laser system illustrated in Figure 12.13. (b) Corresponding measured phase spectra. (c) Coupling coefficient's spatial profile along the FBG determined from the inverse Fourier transform of the intensity and phase spectra (a) and (b). (d) Experimental strain profile (solid curve) derived from the inverse Fourier transform of the intensity and phase spectra compared to the theoretically predicted strain profile (broken curve) for the beam, before the slots were cut into the beam.

imperfections in the grating, or nonuniformity in the bonding of the grating to its host structure.

In an initial attempt to demonstrate the spatial resolution potential of this approach, two slots were cut into the aluminum cantilever beam just described. One slot is indicated in the lower part of Fig. 12.13b, while both are shown more precisely in Fig. 12.15a. The beam was then loaded so that the extended FBG sensor was subject to a tensile and then a

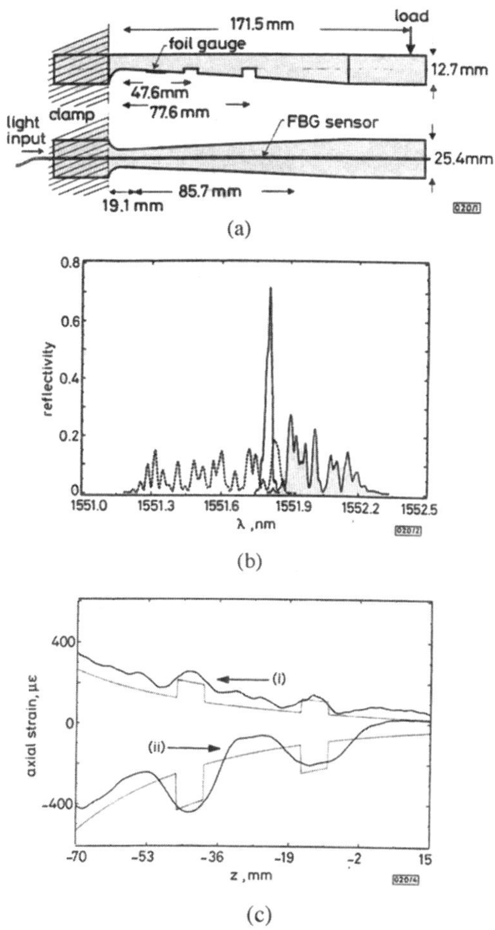

FIGURE 12.15. (a) Detailed design of the cantilever beam showing the two slots cut into its bottom surface. (b) Intensity reflection spectra when the beam is loaded to create both tensile and compressive strain profiles along the FBG sensor. (c) Experimentally determined strain profiles and the guessed profiles as modified by the presence of the two slots.

compressive strain distribution (Ohn *et al.*, 1997). The resulting three intensity reflection spectra for the: unstrained, tension, and compression states are displayed as Fig. 12.15b, while the deduced strain distributions for the tension and compression states (full lines) are presented with the expected strain profiles (dotted lines) in Fig. 12.15c. It is suspected that milling the slots created a residual strain in the cantilever beam and this might account for the net upward displacement of the experimentally derived strain distribution relative to the expected profiles.

More definitive experiments are currently underway and demonstrate a much better performance of the inverse Fourier transform approach when used in conjunction with the rapid scan gain-coupled DFB laser,

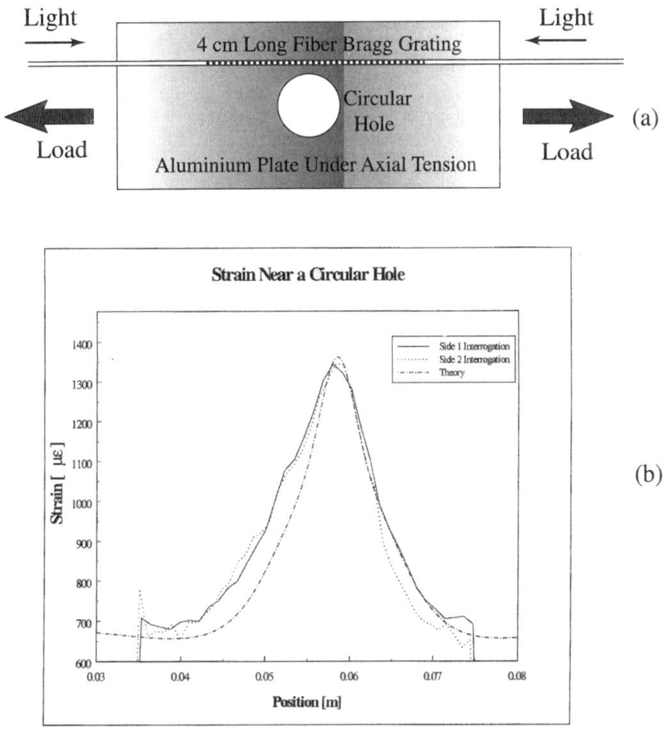

FIGURE 12.16. (a) The experimental arrangement involving a 4-cm-long FBG sensor positioned adjacent to a hole in a tensile loaded plate. The resulting strain profile determined from the inverse Fourier transform of the intensity and phase spectra (using the same tunable laser system shown in Figure 12.13) for light traveling in both directions along the optical fiber is compared with the theoretically predicted strain profile. From Duck, G., (2000), "Distributed Strain Sensing with Fiber Bragg Gratings," University of Toronto, PhD Thesis.

described in Chapter 13. By way of example, Graham Duck (1998) has returned to the classic experiment of the strain profile associated with the hole in the loaded plate. He has succeeded in mapping the strain profile in the vicinity of this circular cutout in the plate under tensile loading. As indicated in Fig. 12.16a, a 4-cm-long fiber Bragg grating was bonded to the beam and light sent both ways through the optical fiber. The two corresponding experimentally deduced strain profiles are compared with the theoretically predicted strain profile. Light was sent both ways through the FBG sensor to see if the slight deviation between the measured profile and the theoretical profile was a function of the direction of light propagation. It can be seen that the results shown in Fig. 12.16b are much better than those presented in Fig. 12 12b. It is clear that the strain profile obtained with the Fourier transform approach represents a considerable improvement over the profile derived from the spectral integration technique. In particular, there is excellent agreement between the experimental peak and quite reasonable agreement with the minimal strain values as compared with those predicted theoretically.

12.7 FOURIER TRANSFORM FOR SERIAL MULTIPLEXED FIBER GRATING SENSORS

Since a string of Bragg gratings along an optical fiber can be thought of as an extended FBG with sections of zero reflectivity between a number of grating elements, it is reasonable to anticipate that the Fourier transform approach might also be used to evaluate the strain distribution from such a string of individual FBG sensors. This represents a novel method of serial multiplexing a linear array of FBG strain sensors, without recourse to the wide spectral width source required for wavelength division multiplexing, or the fast pulse optoelectronics of time division multiplexing.

The use of a Fourier transform approach to demodulate a set of FBGs was first demonstrated by Davis and Kersey (1995c), but their system was rather complex. Measures *et al.* (1998) have tested this concept with both computer simulation and a somewhat simpler experimental setup. The strain field assumed imposed on three FBG sensors in a string along an optical fiber is depicted in Fig. 12.17a. The three simulated intensity reflection spectra concomitant with this strain distribution are presented in Fig. 12.17b, and the comparison between the expected strain readings according to this simulation and the input strain distribution is portrayed as Fig. 12.17c. We can see that there is reasonable agreement if one ignores

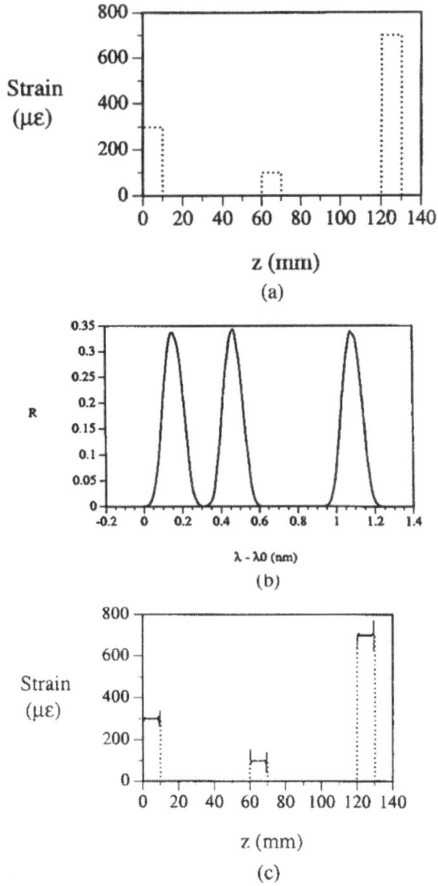

FIGURE 12.17. (a) Theoretical strain imposed on three identical FBG sensors mounted at different positions along a cantilever beam. (b) *T*-matrix calculated intensity reflection spectra for the three FBGs subject to the strains indicated in (a). (c) The strain readings predicted for the three FBGs based on the inverse Fourier transform of their *T*-matrix simulated reflection and phase spectra. From Measures, R.M., Ohn, M.M., Huang, S.Y., Bigure, J. and Fan, N.Y., (1998), "Tunable Laser Demodulation of Various Fiber Bragg Grating Sensing Modalities," *Smart Materials and Structures Journal*, 7, 237–247, with permission from the Institute of Physics Publishing Ltd.

the small oscillatory signals in the vicinity of the grating ends. Measures *et al.* (1998) suggested that these oscillatory signals are attributed to the finite spectral resolution of the system and pointed out that the Fourier transform analysis generates large false oscillatory signals in regions of zero reflection, i.e., where no grating exists. Since no light originates from such regions of the optical fiber they omitted these ghost features in Fig. 12.17c. More complete use of the Fourier transform information provides a means of checking where zero reflection arises, and this could be incorporated into the computer program to avoid the appearance of such false data in the strain profile.

In the initial experiment to test the use of the Fourier transform approach for multiplexing FBG sensors, Measures *et al.* (1998) bonded an optical fiber containing three short FBGs to a specially shaped cantilever beam. They also used the same tunable laser and the same intensity and phase spectrum measurement system used for the distributed sensing work, clearly demonstrating the universality of this demodulation system. Figure 12.18a displays the overlapped intensity spectrum of the three Bragg gratings recorded prior to loading the beam. The spectrum is

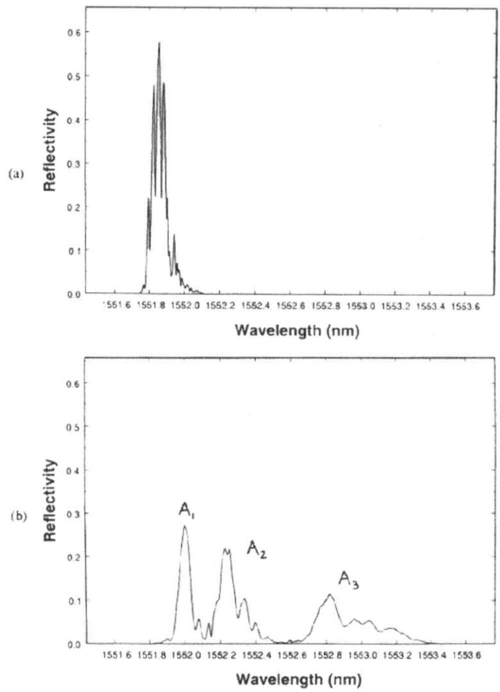

FIGURE 12.18. (a) Complex intensity reflection spectra of three near-identical FBGs placed along an unstrained cantilever beam. (b) Separation of the three spectra from the FBGs due to loading the beam. From Measures, R.M., Ohn, M.M., Huang, S.Y., Bigure, J. and Fan, N.Y., (1998), "Tunable Laser Demodulation of Various Fiber Bragg Grating Sensing Modalities," *Smart Materials and Structures Journal*, **7**, 237–247, with permission from the Institute of Physics Publishing Ltd.

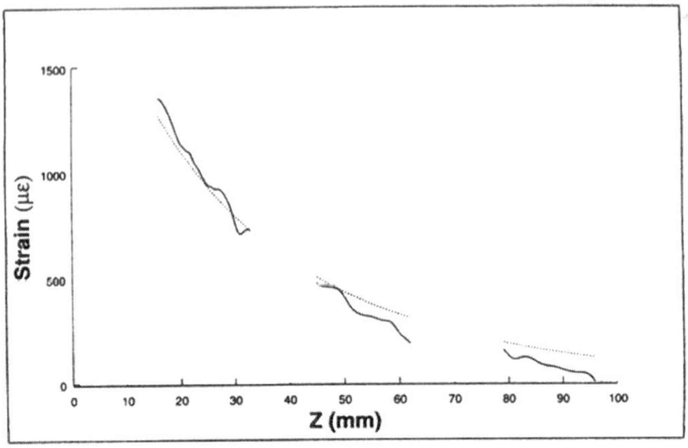

FIGURE 12.19. Comparison of actual strain measurements (solid curves) for the three FBGs, derived from the inverse Fourier transform of their intensity and phase spectra, with the theoretical strain profile (broken curves). From Measures, R.M., Ohn, M.M., Huang, S.Y., Bigure, J. and Fan, N.Y., (1998), "Tunable Laser Demodulation of Various Fiber Bragg Grating Sensing Modalities," *Smart Materials and Structures Journal*, **7**, 237–247, with permission from the Institute of Physics Publishing Ltd.

degenerate and complex in nature resulting from interference between the three FBGs. As can be seen in Fig. 12.18b, loading the beam results in the removal of this spectral degeneracy and a separation of the spectra from the three FBGs. The strain profile estimated from beam bending theory for the loaded beam is presented as the broken curve in Fig. 12.19. Superposed on this curve are the three strain profiles derived for each of the three FBGs from the Fourier transform of their intensity and phase spectra. As can be seen, each sensor provides a part of the overall strain profile, and although the trend is correct, clearly the precision of this technique needs to be improved.

Unfortunately, this Fourier transform approach at reading a linear array of FBGs is limited to lengths of a few meters because of the high frequencies involved in the intensity and phase spectral measurements. Nevertheless, this extends the range over which strain profiles can be measured from the limits of a single extended grating (currently limited to about 10 cm by cost considerations) to a few meters, which could satisfy most practical needs. It should be noted that this distributed strain sensing technique avoids the need for large bandwidth sources and a set of gratings with different reference wavelengths, as required for WDM. It also avoids the high cost of high-resolution OTDR needed by the TDM approach for evaluating strain profiles with a string of gratings spread over a few meters.

12.8 LOW COHERENCE TECHNIQUES FOR DISTRIBUTED SENSING

A number of alternate approaches at extracting arbitrary strain profiles from FBG sensors have been proposed based on low coherence interferometry. Volanthen *et al.* (1996b) demonstrated that this can be accomplished using the experimental arrangement depicted in Fig. 12.20. The strain-chirped FBG (length 3 cm, reflectivity 40%, center wavelength 1531 nm, and chirp of 3 nm) was used as one arm of a single-mode optical fiber Michelson interferometer. A narrowband reference FBG (spectral width 0.18 nm and reflectivity 92%) was part of the other arm of the interferometer and provided a wavelength reflection that was tuned through a load applied to a post around which the end of the optical fiber was wound. Between the 3-dB coupler and this reference FBG, the optical fiber was stretched by means of strain applied to one of two posts separated by 1.7 m and around which 50 m of the optical fiber was wrapped. The piezoelectric cylinder, around which the optical fiber is wrapped, additionally modulated the reference arm optical path length by 2 μm, at 100 Hz to allow several fringes to be observed.

The interferometer was illuminated by a broadband (110 nm) ELED with a peak wavelength at 1515 nm. The distance along the interrogated FBG from which light of a wavelength equal to that of the narrowband reference FBG was reflected is determined by balancing the path lengths

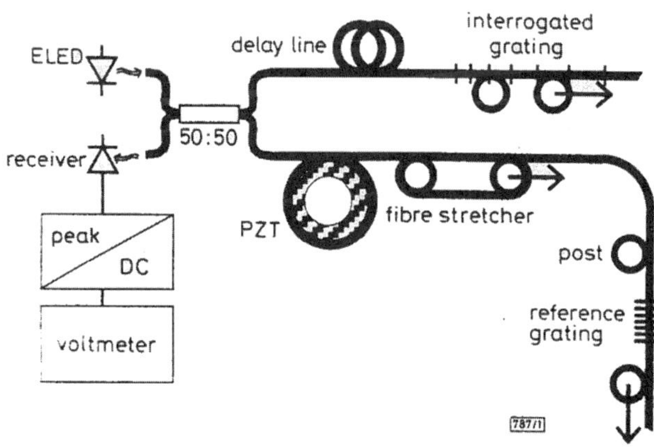

FIGURE 12.20. Experimental arrangement for interrogating a strain-chirped FBG by means of low-coherence interferometry and a narrowband reference FBG. From Volanthen, M., Geiger, H., Cole, M.J., Dakin, J.P., (1996), "Measurement of Arbitrary Strain Profiles Within Fibre Gratings," *Electronics Letters* **33**(11), pp. 1028–1029, with permission of IEE Publishing Ltd.

of the two arms of the interferometer. The strain at that location along the interrogated grating is known from the load needed to be applied to the reference FBG to match wavelengths. Figure 12.21 displays a strain profile measured in this manner. The measured strain accuracy was estimated to be ~50 με.

The preceding approach has been improved by introducing an acousto-optical tunable filter (AOTF) to measure the Bragg wavelength of the interrogated FBG, rather than using a strained reference FBG (Volanthen *et al.*, 1997). Two revised experimental arrangements are schematically depicted in Fig. 12.22. When the wavelength of the AOTF is tuned (by a voltage-controlled oscillator, VCO) toward the wavelength reflected from the location on the interrogated FBG selected by the path matched Michelson interferometer (determined by the stretch applied to the reference arm of the interferometer) optical interference is observed at the photodetector. As with the previous arrangement the piezoelectric transducer (PZT) is used to modulate the reference path length over a few microns so as to excite several fringes. When the amplitude of the fringes is a maximum, the peak filter wavelength matches the local Bragg wavelength.

Both of the experimental arrangements presented in Fig. 12.22 interrogated wavelength profiles along the FBG. The open-loop system (Fig. 12.22a), indicated that a spatial resolution of ~146 μm is possible, while the closed-loop system (Fig. 12.22b), was used to measure spectral

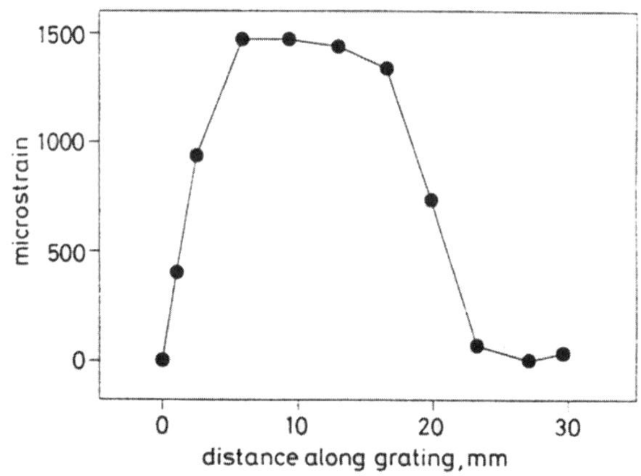

FIGURE 12.21. Strain profile along an FBG determined by the system illustrated in Fig. 12.20. From Volanthen, M., Geiger, H., Cole, M.J., Dakin, J.P., (1996), "Measurement of Arbitrary Strain Profiles Within Fibre Gratings," *Electronics Letters*, **33**(11), pp. 1028–1029, with permission IEE Publishing Ltd.

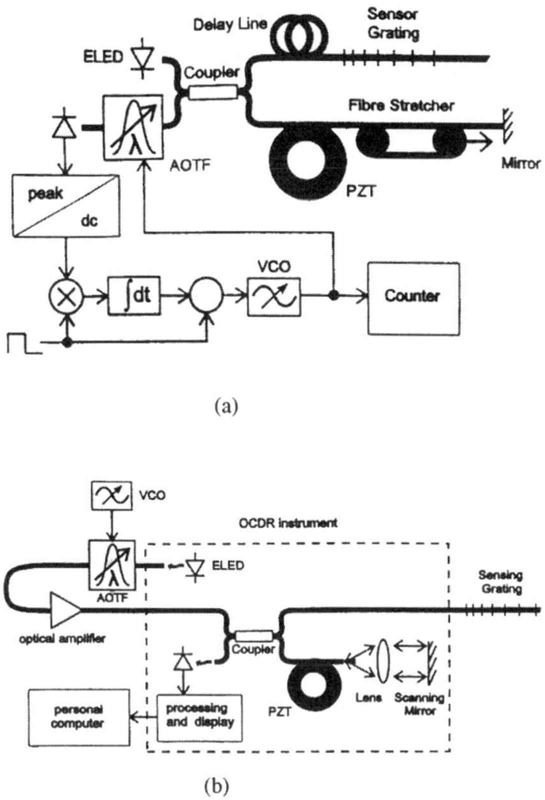

FIGURE 12.22. Two AOTF approaches at measuring arbitrary strain profiles along an FBG sensor based on low-coherence interferometry combined with (a) an open-loop narrowband tunable detection system, and (b) a closed-loop narrowband tunable source. From Volanthen, M., Geiger, H., Dakin, J.P. (1997), "Distributed Grating Sensors: An Alternative to Multiplex Grating Arrays," *SPIE* **3042**, 372–382.

accuracy, by means of continuously scanning the AOFT wavelength windows. An example of the reflectivity variation measured along the FBG, while it was heated at two locations, is presented as Fig. 12.23. In this measurement the wavelength of the AOTF was offset by 4 nm from the unheated local wavelength.

LeBlanc and Kersey (1997) used an experimental arrangement that was almost identical to the closed-loop setup used by Volanthen *et al.* (1997) and shown as Fig. 12.22b. One of the important differences, however, was that LeBlanc and Kersey (1997) used a tunable Fabry–Perot filter as opposed to the AOTF used by Volanthen *et al.* (1997). Figure

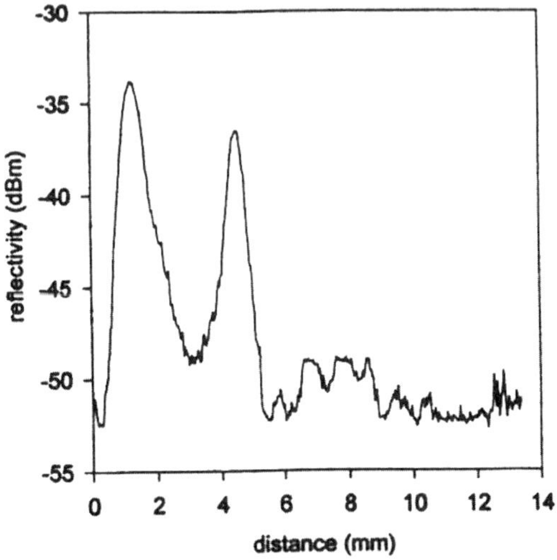

FIGURE 12.23. Reflectivity spatial profile, measured by low coherence interferometry for a FBG heated at two points. From Volanthen, M., Geiger, H., Dakin, J.P. (1997), "Distributed Grating Sensors: An Alternative to Multiplex Grating Arrays," *SPIE* **3042**, 372–382.

12.24a displays the wavelength reflectivity profile, $\lambda(z)$, for the 34 mm long FBG used by LeBlanc and Kersey (1997) and reveals its (as-written) chirp. It is apparent that the bandwidth of the grating was ~13 nm and its center wavelength was 1554.5 nm. Figure 12.24b shows the measured change in this wavelength reflectivity profile as heat was progressively applied to the midsection of the FBG sensor. The calculated temperature profile, corresponding to the 105°C setting of the thermoelectric Peltier block, is determined by subtracting out the original wavelength reflectivity profile. The result is presented as Fig. 12.24c. The temperature and spatial resolutions were deduced to be ~1.4°C and 4.3 nm, respectively.

12.9 DISTRIBUTED SENSING UNDER SIMULATED BRILLOUIN SCATTERING

We saw in Chapter 6, Section 6.6.1, that Brillouin scattering involves inelastic scattering of light by acoustically generated optical inhomogeneinties and this leads to both an upshift and a downshift in the frequency of the light at the Brillouin frequency, v_B (see Fig. 6.19). Stimulated Brillouin scattering (SBS) involves the amplification of light at the *Brillouin Stokes frequency* $v_{BS}(\equiv v_p - v_B)$ propagating counter to the intense pump

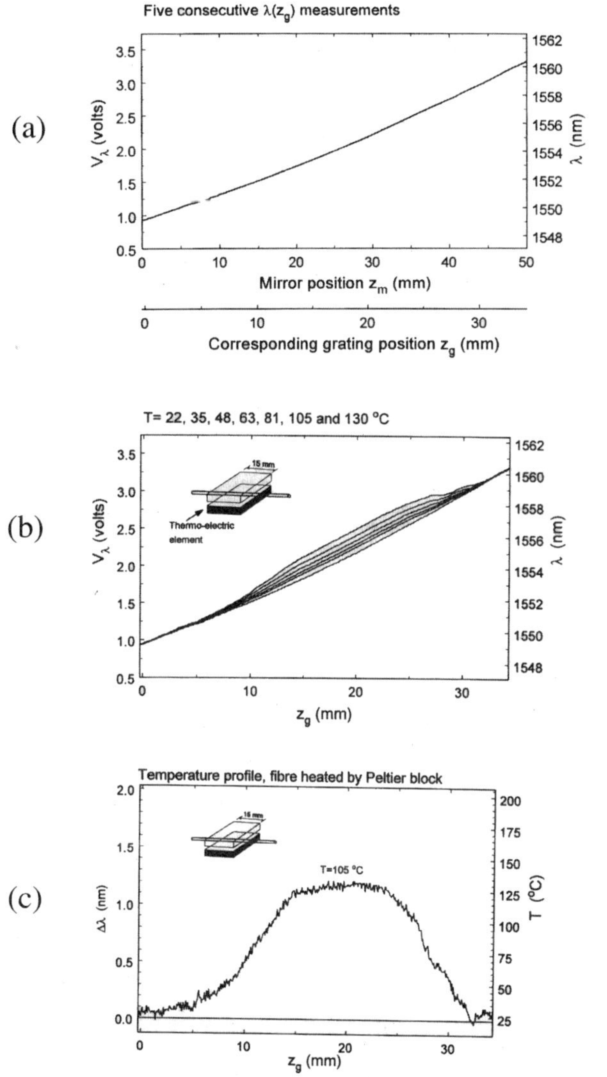

FIGURE 12.24. Meausurements of: (a) the (as-written) wavelength reflectivity profile, $\lambda(z)$; (b) the wavelength reflectivity profiles for seven different temperatures of a thermoelectric Peltier heater element, positioined as shown; (c) the calculated temperature profile for the 105°C case, for a 34-mm-long fiber Bragg grating determined by low-coherence interferometry and a tunable narrowband source. From LeBlanc, M., and Kersey, A.D., (1997), "Distributed Intra-Grating Sensing by Fabry-Perot Wavelength Tuned Low-Coherence Interferometry," OSA 12th Int. Conf. on Optical Fiber Sensors, Williamsburg.

pulse of light at frequency υ_p. *SBS* has the lowest threshold among the nonlinear processes observed in optical fibers and is strongly dependent on local physical parameters of the optical fiber. The Brillouin frequency is proportional to the acoustic velocity within the optical fiber, and since this velocity is a function of temperature and strain, *SBS* has the potential to provide a means of undertaking distributed measurement of these parameters (Bao *et al.*, 1993; Nikles *et al.*, 1994; Fellay *et al.*, 1997; Brown *et al.*, 1998). As indicated in Section 6.6.1, the measured Brillouin strain and temperature frequency shifts are 58 kHz $\mu\varepsilon^{-1}$ and 2 MHz K^{-1} respectively (Horiguchi *et al.*, 1992).

Fellay *et al.* (1997) have examined the spatial resolution limit achievable for distributed sensing based on SBS. Their experimental configuration for studying distributed Brillouin sensing is schematically depicted in Fig. 12.25. It uses a single Nd:YAG laser (150 mW at 1319 nm) and relies on a Mach–Zehnder electro-optic modulator (EOM) to generate the pump and probe pulses of light. The intensity of the pump pulse at 1550 nm is further boosted by means of an Erbium-doped fiber (EDF) amplifier. Amplification of the probe Stokes pulse occurs when pump and probe overlap, provided that the difference between the two frequencies lies within the Brillouin gain spectrum (BGS). This BGS shows a Lorentzian distribution centered on the Brillouin frequency shift υ_B, the quantity to be determined.

If $g_B(\upsilon)$ is the Brillouin gain spectrum, the net amplification of the signal pulse after interaction with the pump pulse (Fellay *et al.*, 1997) is given by

$$I_s = I_0 e^{g_B(\upsilon) L I_P}, \tag{12.38}$$

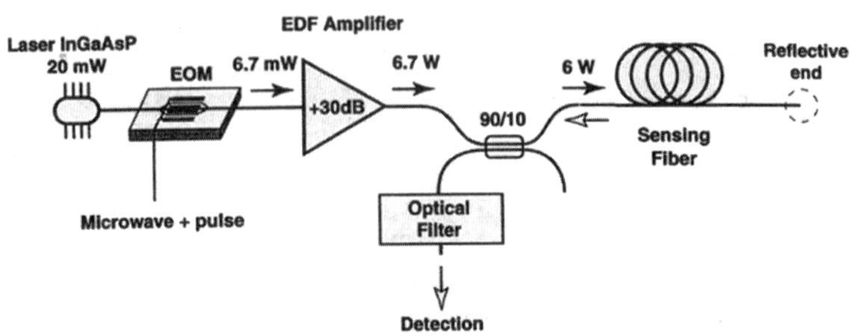

FIGURE 12.25. Experimental configuration used for undertaking distributed sensing along an optical fiber by means of Brillouin scattering. From Fellay, A., Thevenaz, L., Facchini, M., Nikles, M. and Robert, P., (1997), "Distributed Sensing Using Stimulated Brillouin Scattering: Towards Ultimate Resolution," 12th Int Conf. on Optical Fiber Sensors, Williamsburg.

where the intensities are I_0 for the incident signal wave, I_s for the amplified signal wave and I_p for the pump wave, respectively. The equivalent pump pulse length $L = \tau v_g$, where τ is the pump pulse duration and v_g is the corresponding group velocity. The spatial resolution for distributed measurements is directly related to the pump pulse length, L. As shown in Eq. (12.38), the amplification of the Stokes wave decreases exponentially as the pump pulse length becomes shorter, as required for better spatial resolution. To compensate for this decrease in amplification, the pump intensity can be increased so that the product I_pL remains constant, leaving sufficient intensity for the measurement.

The first measurements performed using this novel single laser configuration are presented in Fig. 12.26 and show a 1.5% elongation experienced by an 80-cm length of optical fiber. The measurements shown in this figure were performed using a 1-m long laser pulse. Fellay *et al.* (1997) indicate that any attempt to achieve a significant improvement in this spatial resolution is confronted with two obstacles. The first problem is relatively trivial: the pulse narrowing results in an exponential decrease of the measured signal, as can be seen by reference to Eq. (12.38). Increasing the detector sensitivity and using a more sophisticated data processing system can compensate for this effect to some extent.

The second limitation is more fundamental and results from the pulsed nature of the pump wave. It essentially determines the ultimate spatial resolution attainable with Brillouin sensing. In Eq. (12.38), it was assumed that the Brillouin gain spectrum $g_B(v)$ does not depend upon the length, L, of the pump pulse. This is no longer true for short pulses. As the pulse narrows, its spectrum broadens (see Chapter 3), so that the effective BGS experienced by the signal wave is given by the convolution of the

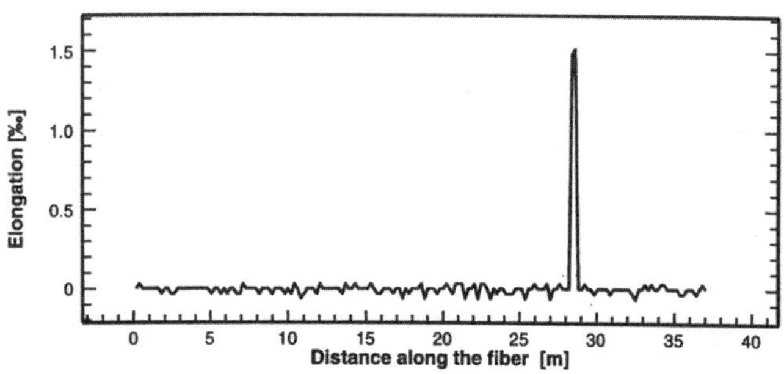

FIGURE 12.26. Elongation measurement along an optical fiber derived from Brillouin scattering. From Fellay, A., Thevenaz, L., Facchini, M., Nikles, M. and Robert, P., (1997), "Distributed Sensing Using Stimulated Brillouin Scattering: Towards Ultimate Resolution," 12th Int Conf. on Optical Fiber Sensors, Williamsburg.

original $g_B(v)$ and the spectrum of the pump pulse. This effect can be seen in Fig. 12.27, which shows the measured effective BGS for three laser pulse widths. The curves spread and become broader the shorter the laser pulse, resulting in a dramatic decrease of the peak gain, and consequently in the measured signal. In addition, this effect is detrimental to the frequency resolution of the measurements, the central frequency of such a spread distribution being determined with less accuracy.

Fellay *et al.* (1997) estimated the ultimate spatial resolution of a distributed Brillouin sensor as a function of the optical fiber length. They showed it is the minimal pulse length giving a sufficient amplification, say 2%, when the spreading of the gain spectrum and the effect of absorption are taken into account. Calculation yields the results shown in Fig. 12.28 for their system specification. It is seen from this figure that the optimal spatial resolution for short optical fibers is just under 1 m.

Brown *et al.* (1998) have suggested that Brillouin optical time domain analysis (BOTDA) might be capable of attaining a spatial resolution of 50 cm and a strain resolution of ~40 με. Their experimental arrangement launches laser radiation from two narrowband Nd : YAG lasers operating at 1320 nm into the opposite ends of a length of optical fiber, (see Fig. 12.29). An electro-optic modulator produces pump pulses (≥1 ns duration, with a 100 ps rise time) from the more powerful laser while the other laser provides the continuous wave (CW) probe beam. The part of the CW

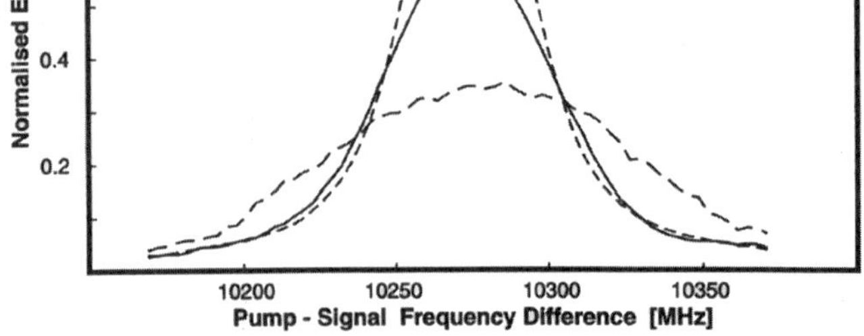

FIGURE 12.27. Three measured Brillouin gain spectra, corresponding to three laser pulse widths ($L = 1, 2$, and 5 m). From Fellay, A., Thevenaz, L., Facchini, M., Nikles, M. and Robert, P., (1997), "Distributed Sensing Using Stimulated Brillouin Scattering: Towards Ultimate Resolution," 12th Int Conf. on Optical Fiber Sensors, Williamsburg.

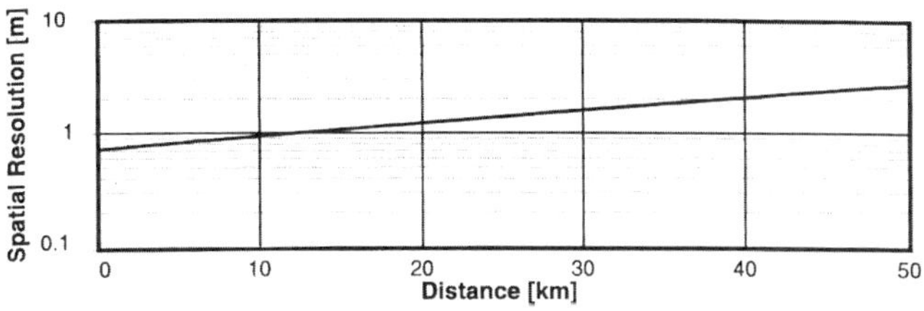

FIGURE 12.28. Variation of the predicted spatial resolution for a distributed Brillouin sensing system as a function of the optical fiber length, based on a resolvable limit of 2% amplification. From Fellay, A., Thevenaz, L., Facchini, M., Nikles, M. and Robert, P., (1997), "Distributed Sensing Using Stimulated Brillouin Scattering: Towards Ultimate Resolution," 12th Int Conf. on Optical Fiber Sensors, Williamsburg.

laser beam arriving at the photodetector at a time t after the launch of the laser pump pulse will have passed through this pump pulse at a time $t/2$ at a distance $ct/2n$ from the photodetector, (see Fig. 12.30). If the CW beam has the higher frequency and the difference in the frequencies of the two laser beams falls within the Brillouin loss spectrum (BLS) then optical power will be transferred from the CW beam to the pump pulse. Since the center frequency of this BLS is linearly dependent on the strain, measuring it will lead to an evaluation of the strain existing at the location along the optical fiber where the two laser beams interacted, i.e., $ct/2n$, from the photodetector.

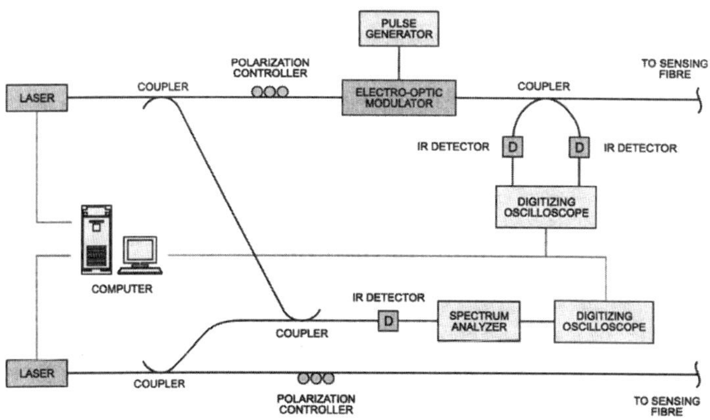

FIGURE 12.29. Experimental layout for Brillouin optical time domain analysis. From Brown, A.W., DeMerchant, M.D., Bao, X., and Bremner, T.W., (1998), "Advances in Distributed Sensing Using Brillouin Scattering" *SPIE* Smart Structures and Materials Conference, San Diego.

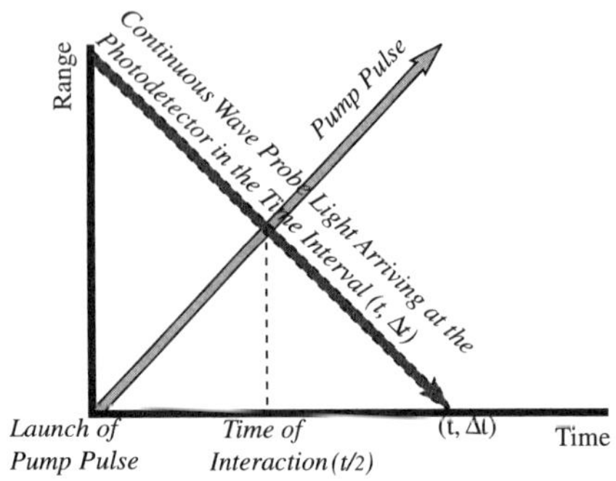

FIGURE 12.30. Range-time schematic of the Brillouin interaciton of a pump pulse that is assumed to be launched into the optical fiber at the time $t = 0$, and that part of a CW probe light beam arriving at the photodetector in the time interval between $t = t$ and $T = t + \Delta t$, i.e., $(t, \Delta t)$.

In order to verify the relation between the load applied to the optical fiber and the Brillouin frequency, Brown *et al.* (1998) placed various masses at a number of locations along a 100-m length of optical fiber, as schematically illustrated in Fig. 12.31. These masses were placed on either 50-cm or 5-m loops of optical fiber. An example of the change in the spectra of the beat frequency (between the two laser frequencies) produced by increasing the load on one of the 5-m loops of the optical fiber is presented as Fig. 12.32a. The resulting linear variation in the peak beat frequency with load is displayed as Fig. 12.32b. This initial work suggested a strain resolution of 17 $\mu\varepsilon$ and a spatial resolution of 1 m. Under conditions of uniform strain, the entire section of optical fiber

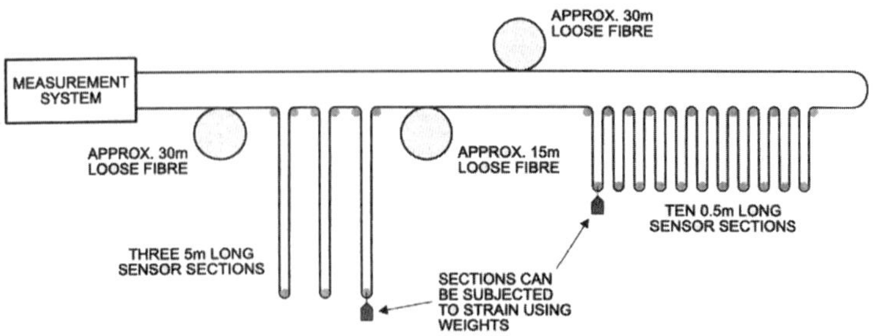

FIGURE 12.31. Optical fiber arrangement for undertaking distributed strain experiments based on Brillouin scattering. From Brown, A.W., DeMerchant, M.D., Bao, X., and Bremner, T.W., (1998), "Advances in Distributed Sensing Using Brillouin Scattering," *SPIE* Smart Structures and Materials Conference, San Diego.

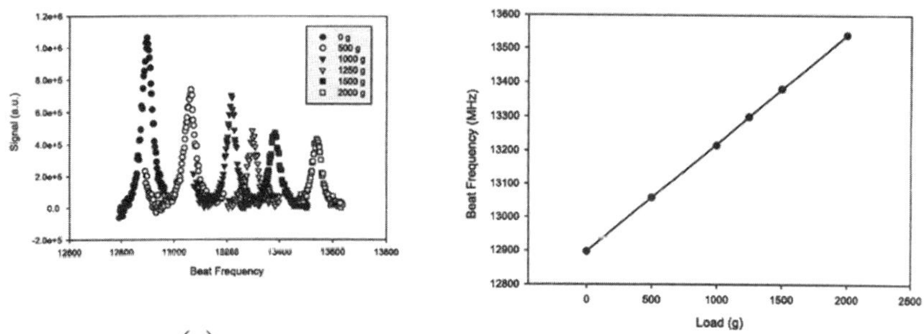

(a) (b)

FIGURE 12.32. (a) Change in the laser beat frequency spectraum observed with increasing load on the optical fiber. (b) Linear variation of the center beat frequency with load. From Brown, A.W., DeMerchant, M.D., Bao, X., and Bremner, T.W., (1998), "Advances in Distributed Sensing Using Brillouin Scattering," *SPIE* Smart Structures and Materials Conference, San Diego.

simultaneously illuminated by both the CW beam and the pump pulse will produce a single peaked spectral profile, as shown in Fig. 12.33a. If the strain varies over this illuminated length of optical fiber, the Brillouin loss spectra will be broadened, or even become multiple-peaked.

Brown *et al.* (1998) suggested that improved spatial resolution could be achieved by using a pulse width sufficient to illuminate two 50-cm loops of optical fiber and resolving the compound beat frequency spectra, (see Fig. 12.33b). Since the loops were of 50-cm length, Brown *et al.* (1998) offer these results as evidence that their experiment demonstrate that a spatial resolution of 50 cm is attainable with a less favorable 40-$\mu\varepsilon$ strain resolution. Obviously, monitoring the magnitude of the received CW beam as a function of time (after the launch of each pump pulse) and the Brillouin beat frequency provides information for evaluating the strain profile along an optical fiber. However, a more definitive set of experiments would be necessary to substantiate the 50-cm claim on spatial resolution.

12.10 DISTRIBUTED STRAIN SENSING APPLICATIONS

As indicated earlier, truly distributed strain measurements could find application in the design of FRP cable anchorage systems used by the construction industry or in the development of technology for monitoring the growth of cracks in airplanes, ships, or bridges, or warning of

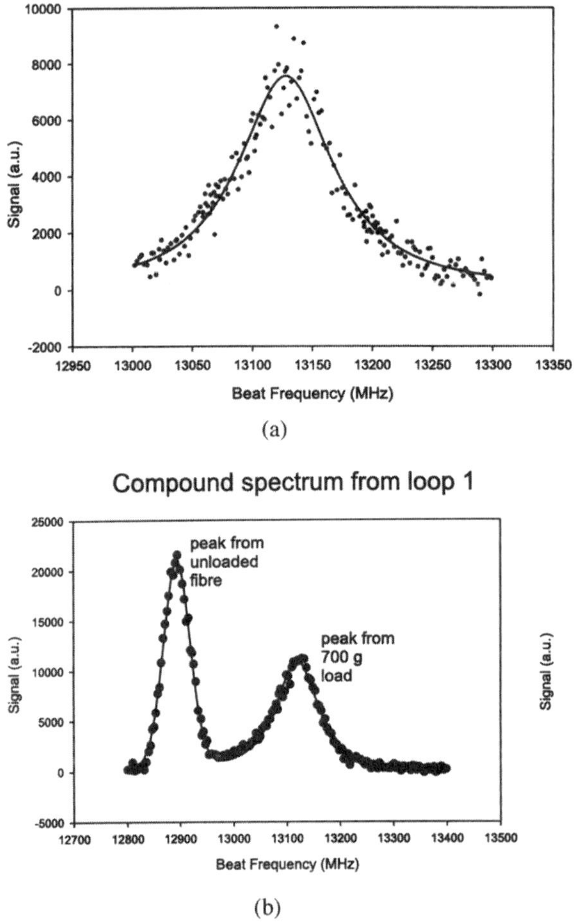

FIGURE 12.33. (a) Beat frequency spectrum corresponding to a uniform load on the optical fiber. (b) Compound beat frequency spectrum when two 50-cm sections of the optical fiber are subject to the laser pump pulse. From Brown, A.W., DeMerchant, M.D., Bao, X., and Bremner, T.W., (1998), "Advances in Distributed Sensing Using Brillouin Scattering," *SPIE* Smart Structures and Materials Conference, San Diego.

weakening in structurally bonded joints. Another possible application for high-resolution distributed strain sensing is studying free-edge effects in FRP composite materials, a condition that often leads to delamination. The most important application, however, might be the development of techniques for detecting the partial debond of a sensor from its host structure (LeBlanc, 1998). This advance might allow a recalibration of such a defective FBG sensor and would constitute a unique method of extending the performance life of structurally integrated FBG sensor technology.

12.10.1 FRP anchorage application

Fiber-reinforced polymer composite materials are finding growing application in the construction industry, particularly in the form of ropes or rods for long-span cable-stayed bridges, and prestressing and post-tensioning of concrete structures. In many of these applications it is necessary to anchor the FRP material. In the case of the more flexible wedge type of anchorage system, this task is made more challenging than for metals because of the large difference between the axial and transverse strength of the FRP material. This difference arises because of the disparity in the properties of the FRP materials. The axial strength is determined by the strong carbon or aramid fibers, while the transverse strength primarily comes from the more brittle resin material. As a consequence of the low interlaminar shear strength of the resin matrix, the anchorage length for FRP pre-stressing tendons has to be much longer than for their steel counterparts. From a cost and convenience perspective, it is often desirable to make the anchorage as short as possible.

Erki and Rizkalla (1993b) have indicated that an optimal solution would involve a continuous variation in the anchorage stiffness, but finding the right design will depend upon the form of the FRP prestressing tendon. An extended FBG sensor that is integrated within the FRP tendon should be capable of providing the strain distribution along the tendon when loaded by an anchor and this information might permit optimization in the design and later provide a warning if there is some degradation during the service life of the system.

A preliminary test of the possible use of the spectral integration technique for measuring monotonic strain distributions within anchorages was undertaken using a specially designed anchorage emulator. This apparatus was designed to impose on a plastic beam strain profiles representative of those present along a tendon within an anchorage. An early version, shown in Fig. 12.34, resulted in a very high transverse stress being applied to the plastic beam at the tip of the screws, so the aluminum clamp was introduced and used in the experiments. As illustrated in Fig. 12.35a, the plastic beam can be clamped at any one of the six possible positions (A, B, . . . , D) through this movable clamp. The other end of the plastic beam is attached to an aluminum block that is longitudinally translatable under the action of an applied force, creating a longitudinal strain along the beam.

FIGURE 12.34. Picture of the anchorage emulator built to test the use of distributed strain sensing based on Bragg intragrating measurements. Note the clamp in the foreground. From Huang, S., LeBlanc, M., Lowery, M., Maaskant, R., and Measures, R.M., (1996), "Distributed Fiber Optic Strain Sensing for Anchorages and Other Applications," *Advanced Composite Materials in Bridges and Structures* E-Badry, M.M., (Ed.), 991–998.

A 9 cm-long fiber optic Bragg grating was bonded on the side of this beam within the anchorage region. In order to simulate different extents of debonding of a tendon from its anchorage, the plastic beam was clamped at various locations. The FBG sensor monitored the corresponding strain distributions using the spectral integration technique. Figure 12.35b reveals the measured strain distributions along the Bragg grating with the beam clamped at the E, D, or C locations, respectively. This corresponds to a debond extending to positions E, D, or C, respectively. The turning point of each strain distribution shown in Fig. 12.35b, clearly indicates where stress transfer between the tendon and the anchorage occured. To ensure that the simulation was realistic, care was taken to make sure that the clamping was not too tight, or Poisson-induced axial strain in the beam would be too localized to be representative of what takes place in a real anchorage.

12.10.2 Detection of debonding in aircraft structurally bonded joints

Aircraft suffer from cracks due to fatigue, bird strikes, and environmental deterioration (Baker and Jones, 1988). Patches of either aluminum or

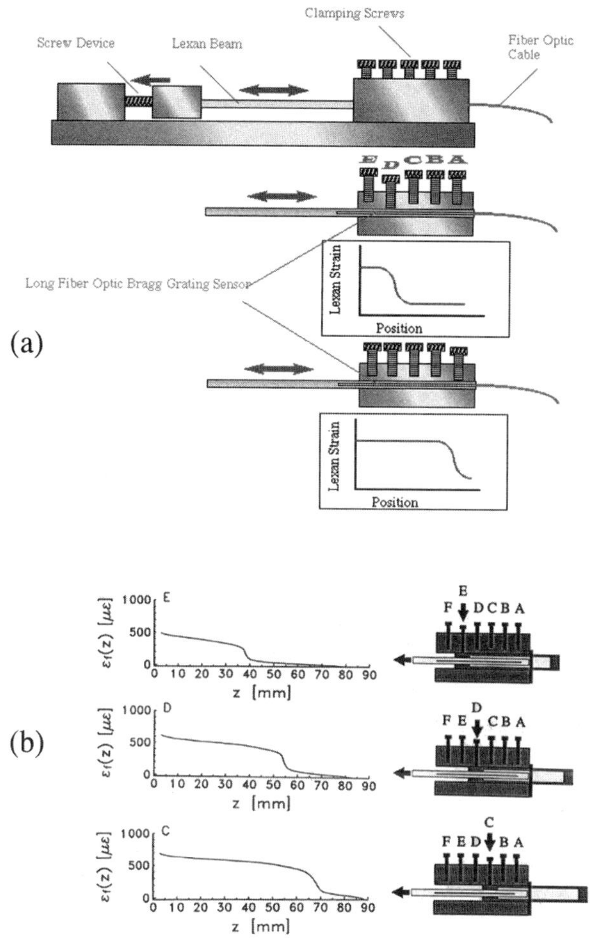

FIGURE 12.35. (a) Early design of the anchorage emulator used to test the distributed strain sensing capability of an extended FBG sensor attached to the side of the plastic beam. (b) Measured strain profiles, corresponding to three clamping positions, as measured by the FBG sensor and the Intensity Reflection Spectrum technique. From Huang, S., LeBlanc, M., Lowery, M., Maaskant, R., and Measures, R.M., (1996), "Distributed Fiber Optic Strain Sensing for Anchorages and Other Applications," *Advanced Composite Materials in Bridges and Structures* E-Badry, M.M., (Ed.), 991–998.

boron fiber reinforced materials can be used to repair this damage. Adhesive bonding is in many ways superior to fastening these patches by rivets because it provides a smoother load transfer across the crack, results in long joint service life; especially for cyclic loading, and offers the

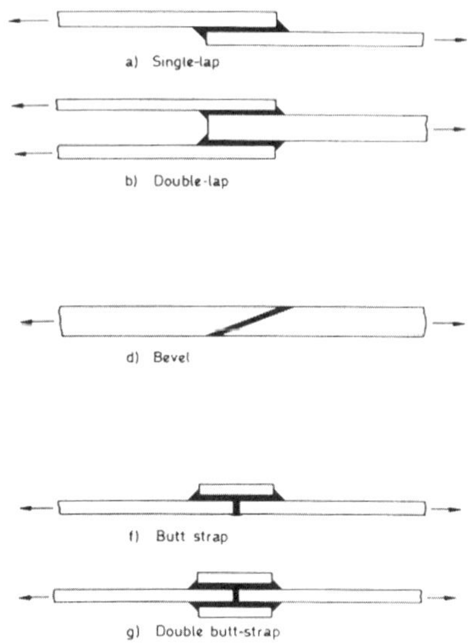

FIGURE 12.36. Schematic of several forms of adhesive joints used in engineering.

benefits of sealing the damaged region from further exposure to the environment (Munns, 1995). A number of common engineering adhesive joints are sketched in Fig. 12.36. The integrity of an adhesive bond in such a joint can be quite insensitive, at least initially, to relatively large manufacturing flaws. Unfortunately, this fact makes it difficult to detect many early defects, since there is no discernible loss of joint strength. However, a defective joint can rapidly weaken. Since joint failures in aircraft can have severe consequences, bond defects must be detected early and tracked if adhesives are to be used more widely by the aerospace industry.

Stringent tests must be performed during the manufacturing process of adhesive joints. However, current test techniques are not as reliable as desired, which leads to large factors of safety and nonoptimal joint design. The common types of defects include voids, porosity, cracking and poor cure, each of which reduces the cohesive strength of the bond. In contrast, a disbond affects the adhesive strength of a joint. Although several methods exist for detecting cohesive defects, there remains some question about detecting adhesive defects reliably. Any current NDT test requires special inspections, complicated procedures, and significant aircraft down-time. It is this lack of any in-service monitoring that prevents the widespread usage of adhesive bonded-structures. A sensor that could detect the onset of debonding would be beneficial by reducing or even eliminating inspection time. Such a sensor could be used in a smart patch that could continually monitor its own integrity and perhaps even provide real-time information on the local strain state of the host structure.

Measures *et al.* (1996) explored the possibility of detecting the presence of a region of debond in an adhesive joint from changes in strain profile using the spectral integration technique. Figurc 12.37 illustrates the aluminum double-lap joint used in this experiment. The joining plates simulate patches covering a cracked aluminum plate. A 5-cm-long fiber optic Bragg grating was bonded to the surface of one of the connecting plates. This grating covered half the length of the plate, from the edge to the central line. Three conventional strain gauges were also bonded on the same surface for reference purposes. When the test specimen was loaded under tension as indicated in Fig. 12.37, the longitudinal strain distribution determined by the FBG provided some insight into the bonding situation. Figure 12.38a shows the FBG spectrum and the recovered strain distribution for a perfectly bonded sample that is unloaded, then loaded. In order to experimentally simulate debonding, a preformed 5-mm edge debond was created in a similar specimen and the results presented in Fig. 12.38b. For comparison, the theoretical strain distribution based on simple elastic bonded-joint theory for both cases is also presented (dotted curves) in Fig. 12.38. As can be seen from Fig. 12.38b, the 5-mm debond is clearly indicated by the $\sim$4 mm strain-free region observed at the edge of the patch.

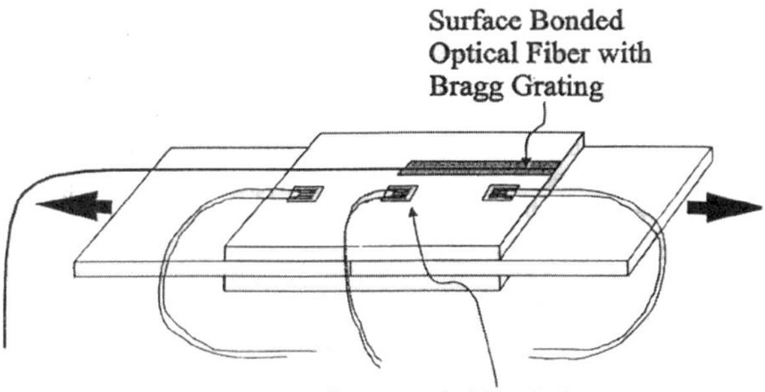

FIGURE 12.37. Aluminum double-lap joint incremented with an extended FBG sensor and three resistive foil strain gauges. From Huang, S., LeBlanc, M., Lowery, M., Maaskant, R., and Measures, R.M., (1996), "Distributed Fiber Optic Strain Sensing for Anchorages and Other Applications," *Advanced Composite Materials in Bridges and Structures* E-Badry, M.M., (Ed.), 991–998.

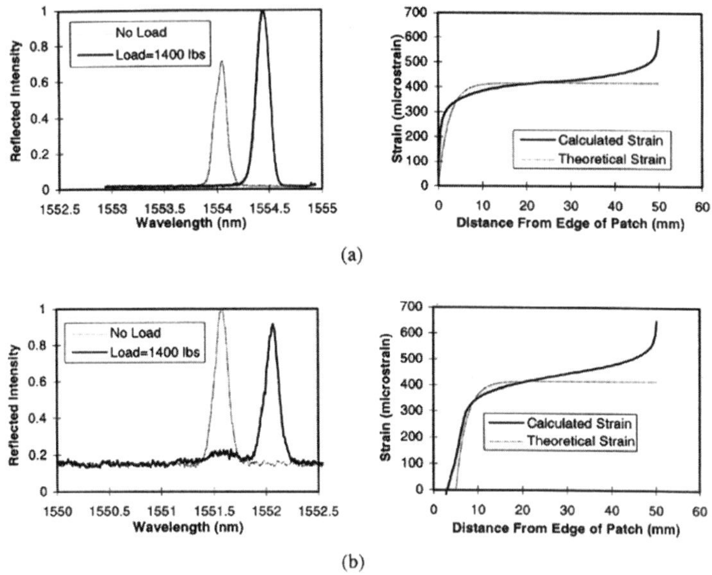

FIGURE 12.38. (a) The left side shows the reflection spectrum for FBG sensor corresponding to no load and a 1400-lb tensile load imposed on the double-lap joint shown in Figure 12.37. The right side shows the strain profiles measured using the intensity reflection spectrum technique for these two cases. (b) Same as (a), except that a 5-mm debond was produced at the edge of the double-lap joint. From Huang, S., LeBlanc, M., Lowery, M., Maaskant, R., and Measures, R.M., (1996), "Distributed Fiber Optic Strain Sensing for Anchorages and Other Applications," *Advanced Composite Materials in Bridges and Structures* E-Badry, M.M., (Ed.), 991–998.

As discussed in Section 12.4, the spectral integration technique for evaluating strain distribution suffers from large errors when the strain profile is flat, for then the effective interaction length (12.26),

$$L_{\mathrm{eff}}(z) = \sqrt{\frac{\Lambda(z)}{G_{\varepsilon}|d\varepsilon(z)/dz|}},$$

can exceed the length of the grating, L_G, and approximations assumed in the theory become invalid. To overcome this weakness, Huang *et al.* (1996) used a prechirped FBG of length 4.5 mm in a similar test. The prechirp in the grating was created by bonding it to the patch reinforcement while the reinforcement was placed in bending compression. After the bonding adhesive was completely cured, the reinforcement was released and then attached to the host to complete the sample. In this sample, we introduced a 6-mm preformed edge debond. The intensity reflection spectra for this grating are shown in Fig. 12.39a while the converted strain profiles are

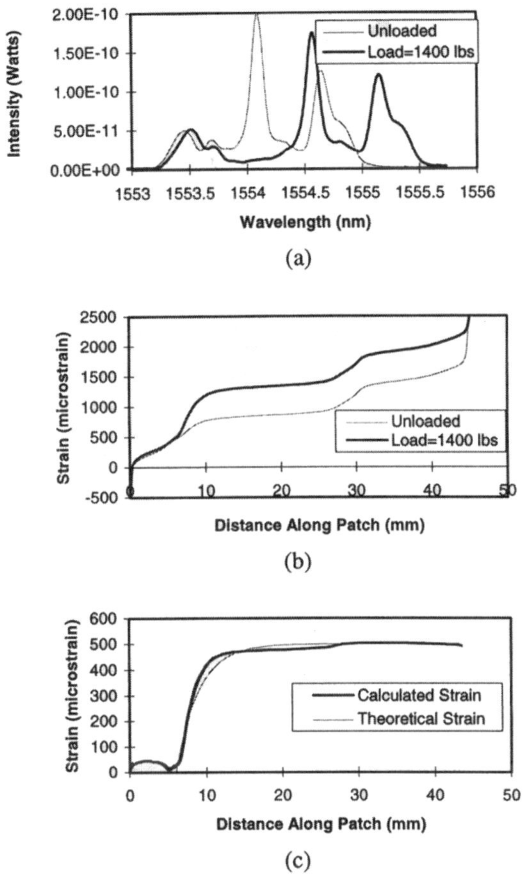

FIGURE 12.39. (a) Reflection spectrum for a "prechirped" FBG sensor used in the double-lap joint experiment illustrated in Fig. 12.37. (b) Strain profile derived for FBG sensor using the intensity reflection spectrum technique for the loaded and unloaded runs. (c) Prechirp strain bias is subtracted from the strain profiles of (b) to yield true strain profiles for the loaded and unloaded cases.

shown in Fig. 12.39b. The strain profile with zero load in Fig. 12.39b represents the prechirp built into the grating. This profile indicates that the grating prechirp averaged around 40 $\mu\varepsilon\ \text{mm}^{-1}$, but is not perfectly linear. It should be noted that the linearity of the prechirp is not critical as long as the profile is known so that it can be subtracted later. Figure 12.39c, shows the subtracted strain profile (i.e., the real strain field) and compares it with the theoretical profile.

It is obvious from this comparison that the prechirp of the FBG sensor served to alleviate many of the difficulties associated with measuring low strain gradients at the endpoints of the grating. Furthermore, the agreement in the size of the strain-free region at the edge of the specimen patch is quite encouraging from the standpoint of detecting a region of debond. Considerable research will be needed to establish the validity and reliability of this technique, but an important start has been made.

12.10.3 Fiber Bragg grating sensor self-diagnosis

We have also demonstrated that distributed strain sensing permits some degree of sensor self-diagnosis in which debonding or yielding of the bond between the sensor and its host structure can be detected. In Fig. 12.40a we present three reflection spectra from a fiber Bragg grating

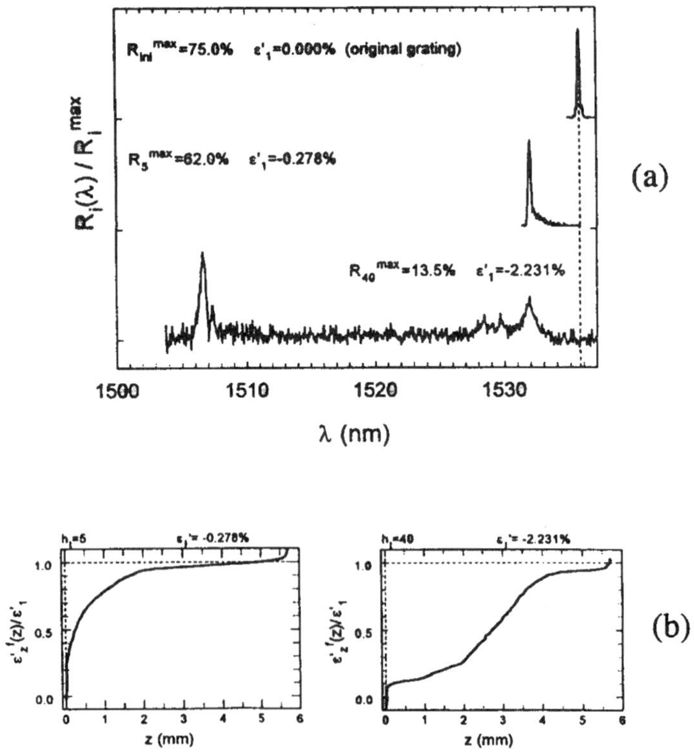

FIGURE 12.40. (a) Fiber optic Bragg grating reflection spectra for different degrees of compressive loading. (b) Strain profiles along the grating sensors as derived from these spectra using the spectral integration technique. From M. Le Blanc, M., (1999), "Interaction Mechanics of Embedded Single-Ended Fibre Sensors Using Novel In-Situ Measurement Techniques," Ph.D, Thesis, University of Toronto, Institute of Aerospace Studies.

sensor that is embedded within a fine groove in an aluminum beam and filled with epoxy to simulate the resin of a host structure (LeBlanc, 1998). The top spectrum is for the FBG before it is embedded in the groove. The middle spectrum is obtained when the the aluminum beam is compressed so the strain is about (−0.278%) and the lower spectrum corresponds to a strain of (−2.231%). Also shown, as Fig. 12.40b, are the associated strain profiles along the grating derived from the resulting back-reflected spectra. In both strained cases the grating is chirped and the spectrum degenerated from the single spike seen for the unstrained FBG. When the FBG is subject to a modest compressive strain of −0.278%, the grating experiences a small chirp and the spectrum develops a long wavelength tail associated with the shearlag strain profile seen on the left side of Fig. 12.40b. When the strain reaches the value of −2.231%, the FBG is strongly chirped and the spectrum degrades considerably, as seen at the bottom of Fig. 12.40a. The strain profile deduced from this spectrum is shown on the right side of Fig. 12.40b and predicts that the optical fiber suffers radial debonding towards its tip and yielding of the epoxy in the central part of the grating. This unrivaled ability of the Bragg grating strain sensor to determine if any part of it has debonded from the host structure (the most common mode of failure of strain sensors) makes this technology unique and could encourage broad implementation of fiber optic structural sensing.

CHAPTER 13

Future Prospects and Summary

13.1 OVERVIEW

There is little doubt that even a conservative engineer would agree that structural monitoring technology should be integrated into new structures if it can meet the following conditions. First, the inclusion of this technology must be practically transparent to those building the structures so it does not slow construction or add appreciably to its cost. This is true no matter whether these structures are sleek new aircraft or sculptured bridges. Second, it must be cost-effective in terms of delivering information that is valuable to the builder and subsequently the owner of the structure. In essence, it should lead to long-term cost savings and enhanced safety. Although significant progress has been made in the past decade, much remains to be done before these ambitious objectives are satisfied.

In this chapter we examine how these objectives might be satisfied and the prospects of attaining this goal. Measures *et al.* (1992b) identified five challenges that must be addressed if structural monitoring with fiber optic technology is to be broadly implemented in the aerospace industry. Although not all applications are as demanding as those in aerospace, nevertheless these challenges serve as a guide to what is generally required of the technology.

The first challenge is to develop a fiber optic sensor that can undertake all of the tasks required of an integrated structural monitoring system. These include the measurement of strain, load, shape change, vibration modes, and cure status for concrete and FRP composite materials. These measurements should provide an evaluation of the status and health of the structure. The fiber optic sensors must be reliable, rugged, manufactured at low cost, and capable of being embedded within complex structural components, especially those made from FRP composite materials. The sensors must also be demodulated in a manner that provides adequate measurand resolution and sufficient time response and that covers the full measurement range. Since fiber optic sensors are temperature sensitive and there will invariably be a difference in the thermal expansion coefficient between the optical fibers and their host material, it is necessary that the temperature be monitored and some form of temperature compensation be applied to any strain measurement.

The second challenge is to design sensing architectures that are *flexible* so as to accommodate a variety of complex structural shapes and layups. It will be necessary to allow for regions of high strain concentration, cutouts, and edges. The sensing architecture should allow full coverage of the structure and display a reasonable degree of damage tolerance so that graceful degradation of the sensing system is possible in the event of damage to the structure.

The third challenge is the development of a *user-friendly interface* that can extract the wealth of optical sensing signals from the instrumented structure, yet be rugged, reliable, immune to vibration, very compact, and as unobtrusive as possible. It should minimize the leads that emerge from the structure, and in the case of an aircraft structure, a relatively unskilled technician should be able to disassemble and reassemble the instrumented host structural component from the main body of the aircraft, even in a harsh environment.

The fourth challenge is concerned with the development of relatively low-cost, rugged, portable *demodulation systems* that can interrogate multiplexed arrays of fiber optic sensors in either parallel or serial configurations. The fifth, and final challenge involves the development of *intelligent processing systems* that can interpret and prioritize the sensing information so it is available for real-time decisions. In certain instances, it may be necessary to combine the real-time sensing information with other pertinent information, such as baseline signatures. Subsequently, some selected fraction of this information will have be downloaded for permanent storage and possibly more detailed analysis. This might be the case where maintenance and inspection decisions are influenced by load history, for example.

As we have seen from the preceding chapters, a considerable amount of work has been done in the intervening period since this paper was written in 1992. Much of it has demonstrated that sensing technology based on fiber Bragg gratings can accommodate the wide range of measurements necessary for structural monitoring. We have also seen that techniques have been developed for addressing the issue of temperature compensation, and FBG sensors have been shown to survive a number of harsh environments, including concrete construction sites, FRP composite cures, and pultrusion.

Serial and parallel multiplexing of FRB sensor arrays have been demonstrated, and since there is no problem producing an FBG sensor at the tip of an optical fiber, this sensing technology can be used in almost any form of complex geometry, including close to cutouts and within the edge of a structure. The successful development of portable, rugged, and reasonably low-cost optoelectronic demodulation systems is borne out by the existence of several commercial systems. Strings of FBG sensors have been produced in an automated manner during the manufacture of the optical fiber. This advance is crucial for serial multiplexed sensing systems and promises low-cost sensors with consistent properties and the strength of the intrinsic optical fiber.

I believe a major challenge for the next decade will be microminiaturization of the demodulation systems and the reduction of their cost to the point where they can be built into the host structure, rather than requiring optical connections and fiber optic leads emerging from the structure. In parallel with this advance it will be valuable to develop a noncontact (wireless, or more appropriately "fiberless") remote means of communicating with the optoelectronic microchip demodulation system, by either radio waves or infrared technology (much like the TV remote control). It will also be necessary to develop methods of delivering power for such a built-in demodulation system. These advances would address the issue of the user-friendly interface with minimal lead egress from the sensor-instrumented host structure.

It is of interest to speculate how the fiber optic structural monitoring technology might evolve in the next millennium. It is anticipated that very small, low-cost demodulation systems will be developed and positioned at sites on an instrumented host structure where they can interrogate a local cluster of sensors. This might involve anything from a single sensor to dozens of sensors distributed in serial, parallel, or mixed arrays. Where several such clusters are well separated over the structure, they may be linked with optical fibers or wires, depending on whether they have a common demodulation system or each cluster has its own demodulation system. In the case of a large bridge, it may be more practical to link the different monitoring sites with a wireless, cellular-phone-type communication network and have that network connected by a phone line, or a cellular link, to some central monitoring station (see Fig. 13.1). For a major structure that is some distance from a population center, such as a dam or an isolated bridge, a satellite link might be more appropriate.

In the case of aircraft or similar structures, networks of sensors might eventually be developed with an adaptive architecture that permits any pattern of sensors to be interrogated at any instant, and if some sensors malfunction or fail, they can be ignored and alternative, nearby sensors

FIGURE 13.1. Possible future fiber optic structural monitoring system, transparent except for a number of small optoelectronic units that communicate remotely with a central unit linked to an off-site monitoring station, as suggested in Fig. 1.1.

used to extract the relevant information. Such redundancy would require inexpensive sensor arrays. One approach might involve forming the pattern of sensors and sensory pathways on some form of polymeric sheet, by means of a photolithographic-type process (Hornack, 1992). In the case of aircraft, for example, such sensing layers could be incorporated into their host structures during fabrication. Structural monitoring patches, possibly also based on polymeric sensor sheets placed within a tough FRP patch, might be retrofitted to the surface of existing structures, such as bridges or aircraft.

New information technologies that could be relevant to this fiber optic structural monitoring field are neural networks and fuzzy logic. The former involves adaptive pattern recognition, while the latter allows modeling based on imprecise or uncertain information. Eventually, optical switching could be possible through soliton interactions (Saleh and Teich, 1991) and the formation of such photonic-controlled pathways might form the basis of *optical neural networks* that would learn from experience and create *optical memory states* that correspond to specific physical states of the structure—a form of *photonic self-awareness*.

In the twenty-first century it may be regarded as unacceptable engineering to permit mechanical failures, or fires, in any major structure to go undetected to the point where they could cause death, injury, an environmental incident, or significant financial loss.

13.2 FIBER BRAGG GRATING UNIVERSAL DEMODULATION SYSTEM

In this section we show that at least a start has been made in the direction of developing versatile demodulation systems that could eventually be microminiaturized and built on an optoelectronic microchip. Although this is only one of many current advances, I have singled it out for two reasons: First I am familiar with it, and second, I believe it has the right elements to form the basis of the future technology.

As we have seen in this book, fiber optic Bragg gratings can be used as short gauge-length, long gauge-length, multiplexed, and distributed sensors. This versatility is unrivaled by any other technology, and it would be fitting if a "universal" demodulation system could be developed that could accommodate this full range of FBG sensing. Members of my research laboratory and I have developed a design concept for such a fiber optic Bragg grating "universal" demodulation system. This, with some of its generic features, is schematically portrayed in Fig. 13.2. What makes it especially exciting is the potential to microminiaturize the system onto an optoelectronic microchip.

At the heart of this system is a rapidly tunable laser diode and an isolator. Where parallel multiplexing is required, a $(1 \times n)$ integrated optic splitter is used to distribute the laser radiation among the n sensing optical fibers. In Fig. 13.2, n was taken to be 4, but it is not limited to this number. Each sensing optical fiber could contain a single FBG sensor; or a linear array of FBG sensors; or a long-gauge sensor; or an extended FBG to serve as a distributed strain sensor; or some combination of these. What makes this demodulation approach so versatile is the *detection module* (DM) that fits between the $(1 \times n)$ splitter and each sensing optical fiber. Each DM has at least a pair of low insertion loss and low reflection connectors, one for interfacing to the $(1 \times n)$ splitter, and the other for interfacing to the sensing optical fiber.

For a single FBG sensor, the DM is very simple and comprises a pair of photodetectors and a 3-dB coupler, as shown in Fig. 13.3a. One photodetector, PD1, is used to monitor the optical power reflected from the FBG, while the second, PD2, is used to measure the laser power sent to

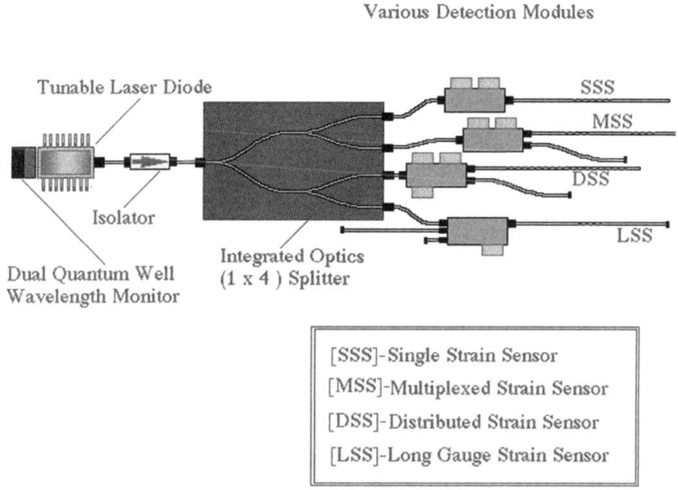

FIGURE 13.2. Generic features of a proposed "universal" fiber optic Bragg grating structural monitoring demodulation system based on a fast-scan laser diode.

the FBG and serves as a reference. The wavelength of the laser during its scans could be monitored with a pair of detectors having some form of differential wavelength dependence (Measures *et al.* 1992). We have demonstrated how this could be accomplished using a quantum-well electroabsorption device (Coroy and Measures, 1996) that could be mounted very close to the laser, or even integrated on the laser chip using *selected area epitaxy* or deposition of a silicon diode layer to ensure different bandgap energies in the laser and detection sections of the chip (Mascher *et al.*, 1998). We have also shown it is possible to develop a FBG short gauge-length sensor demodulation system based on a quantum-well electroabsorption device (Coroy and Measures, 1998); see Section 9.5.2.

For the multiplexed and distributed sensing modalities, a linear array of FBG sensors or an extended FBG distributed sensor is used, and the DM contains three photodetectors and two couplers as shown in Fig. 13.3b. In this case two of the photodetectors, PD1 and PD2, are used to extract the intensity reflection spectrum, $R(\lambda)$, while the third, PD3, is used to measure the phase spectrum, $\phi(\lambda)$. This type of DM also has a third low insertion-loss and low-reflection connector for the mirrored tip reference optical fiber of the Michelson interferometer. In the case of the long gauge-length sensor, the DM would comprise two couplers and a single photodetector connected through a low-insertion-loss and low-reflection connector to a reference Michelson interferometer, as indicated in Fig. 13.3c.

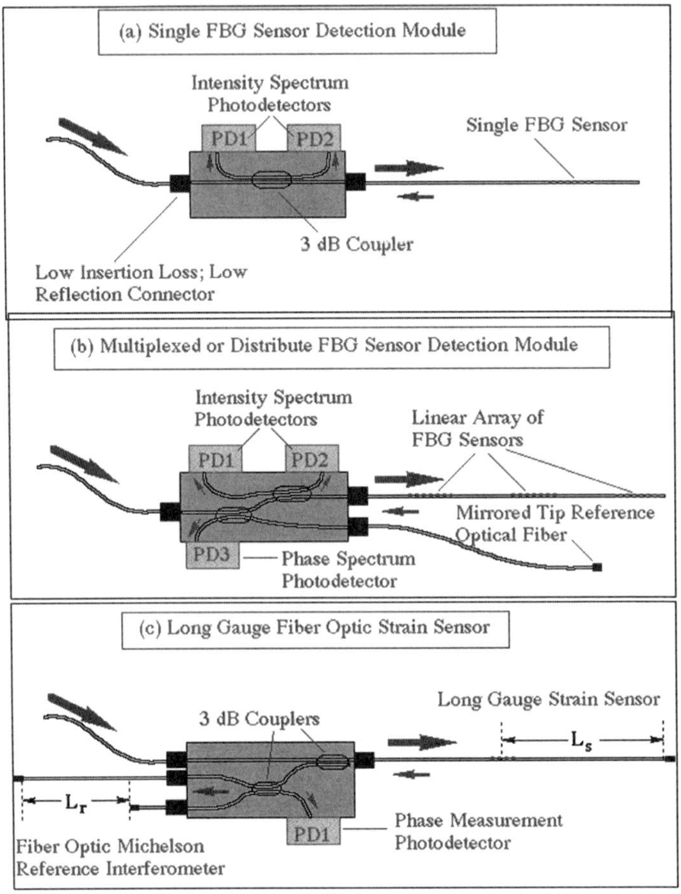

FIGURE 13.3. Proposed integrated optoelectronic circuits for the three detection modules used with the universal demodulation system portrayed in Fig. 13.2.

The beauty of this system lies in its modular nature. This permits great flexibility and adaptability. The tunable laser unit can be used on certain occasions to work with a $(1 \times n)$ splitter and a set of n simple DMs to interrogate n single FBG sensors, while on other occasions it could be used with, say, a $(1 \times n)$ splitter and a set of n interferometric DMs to interrogate a mix of distributed, long-gauge, and multiplexed FBG sensors. In principle, if there is sufficient demand, the tunable laser, its wavelength measurements system, the splitter, the couplers, and the set of photodetectors could be formed into an optoelectronic integrated circuit module. This would greatly reduce the cost of the system, ensure consistent manufacture and reliability, and avoid much of the pigtailing needed with the current fiber optic coupler-based system. It would also lead to a very compact and rugged system with reduced power requirements. Figure 13.4 schematically illustrates an example system involving an eight-channel short gauge-length fiber optic Bragg grating sensing system made of optoelectronic integrated circuit modules.

In an initial test of this concept we have used a Nortel *gain-coupled distributed feedback* (GC-DFB) tunable laser diode, which has the capability of scanning from 1536.5 to 1544.5 nm in about 1 ms. The output of the laser varied from 0.5 to 6.25 mW during such a scan, but since the sensing information is wavelength encoded, this has little consequence. Indeed, this variation in laser power is only a problem if it falls to a point where the signal-to-noise ratio in the detected signal becomes small. The simplest form of the system is schematically illustrated in Fig. 13.5. A standard peak-detection system was used, with three point quadratic curve fitting to interpolate the data, to track the wavelength of a single FBG sensor over a $\pm 1000\ \mu\varepsilon$ range (see Fig. 13.6; Coroy *et al.*, 1997). The strain response is seen to be quite linear with a strain resolution of about $\pm 0.76\ \mu\varepsilon$. Since the sample period for this measurement was 10 ms, the corresponding bandwidth normalized strain resolution was $\pm 0.076\ \mu\varepsilon\ \mathrm{Hz}^{-0.5}$.

A more advanced version of this system was then built to demonstrate the parallel and serial multiplexing potential of this approach. The

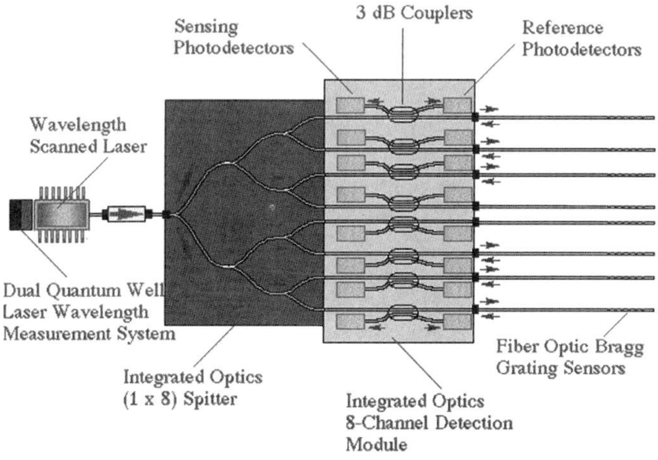

FIGURE 13.4. Schematic projection of an integrated optoelectronic demodulation system that could be used to "simultaneously" interrogate eight optical fibers with as many as five FBG sensors per fiber.

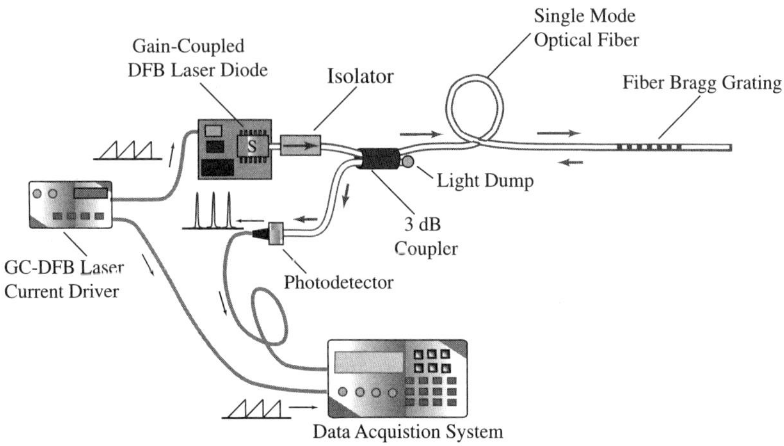

FIGURE 13.5. Fiber Bragg grating demodulation system based on the use of a rapidly scanned Nortel gain-coupled DFB laser diode. From Coroy T., Chappell L.M., Guillermo N.J., Huang S.Y., and Measures R.M.. (1997), "Peak Detection Demodulation of a Bragg Fiber Optic Sensor Using a Gain-Coupled Distributed Feedback Tunable Laser," OSA 12th Int. Conf. on Optical Fiber Sensors, Williamsburg, 210–212.

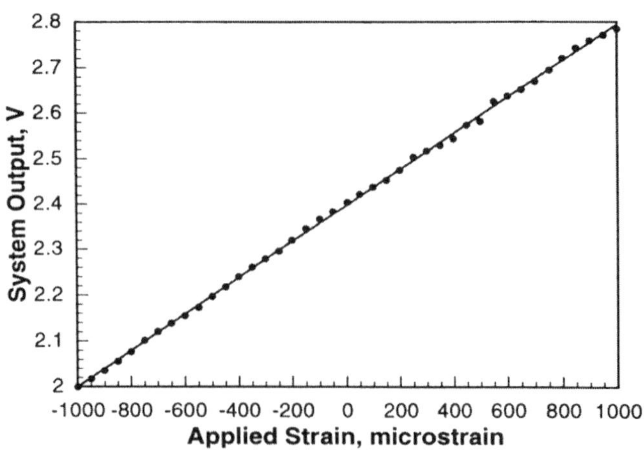

FIGURE 13.6. Linear variation of the system (of Fig. 13.5) response with applied strain on a test FBG sensor. From Chappell, L. (1998). "Development of a Tunable Laser Based Bragg Grating Demodulation System", M.A.Sc. Thesis, University of Toronto.

experimental arrangement is schematically illustrated in Fig. 13.7. A (1 × 4) splitter was used to permit the GC-DFB tunable laser diode to illuminate four optical fibers, each containing three FBGs. Two of the optical fibers had FBG with unstrained Bragg wavelengths of approximately 1537, 1541, and 1543 nm, while the other two optical fibers had unstrained Bragg wavelengths of approximately 1541, 1543, and 1546 nm. The responses from the four sensing photodetectors (PD1 to PD4), as well as the reference photodetector response, were routed, along with a signal from the laser driver used to determine the laser wavelength, to an analog-to-digital board in a computer.

An example of the system's operation is presented as Fig. 13.8, where the real-time variations of the peak wavelength (proportional to strain) for all 12 FBG sensors are displayed simultaneously. Six of the sensors demonstrate their dynamic response, while the remaining six sensors exhibit static signals. The dynamic signals were produced by FBG sensors that were mounted on oscillator cantilever beams, and this explains their

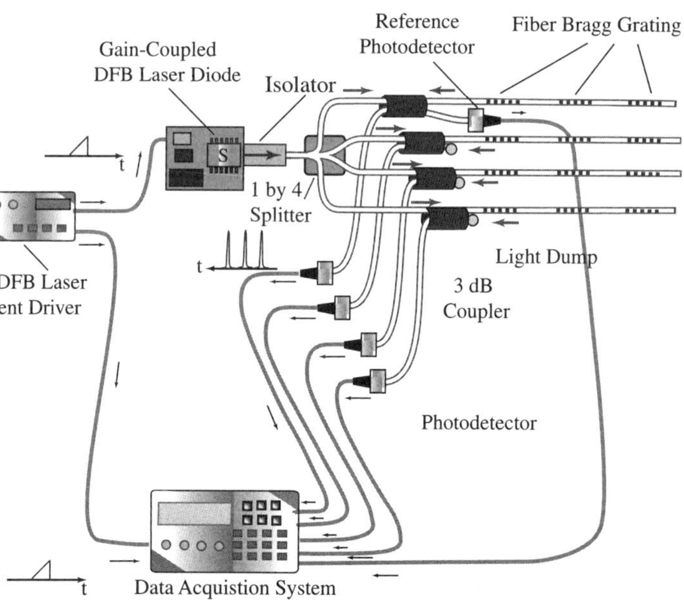

FIGURE 13.7. Simultaneous demodulation of 12 FBG sensors using the rapidly scanned Nortel gain-coupled DFB laser diode. Fiber optic forerunner of the multichannel universal demodulation system proposed in Fig. 13.4.

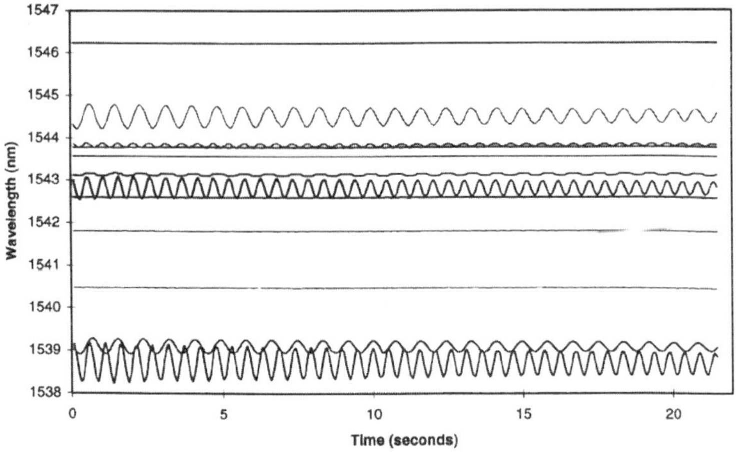

FIGURE 13.8. Simultaneous strain sensor output for the 12 channels of the system illustrated in Fig. 13.7.

amplitude decay with time. The beams were fabricated from various materials and were of different lengths, which accounts for the variations in the signals. The theoretical attenuation level where peak spectral locations can be acceptably determined has been found to be approximately 30 dB. This corresponds to splitting the light 1000 times and suggests that it should be practical to interrogate more than 200 FBGs with one laser, allowing for splitter and coupler losses and assuming an FBG reflectivity of about 80%.

To help the reader appreciate why the tunable laser approach is seen to be very favorable for development into a microchip universal demodulation system, let us consider for a moment an alternative approach that has considerable merit in its own right. I am referring to the compact CCD-spectrometer technique demonstrated by Chen *et al.* (1997) and Ezbiri *et al.* (1997). This approach can certainly be used to multiplex a fairly large number of FBG sensors. The problem is that, because it uses a broadband source, it is limited in terms of the number of sensors it can excite with sufficient signal-to-noise. Furthermore, it cannot provide the phase spectral measurement required for distributed strain sensing, and most important, its physical size cannot be reduced appreciably from its current value because its operating principle involves spectral dispersion. Thus, this approach, as it stands, could not be microminiaturized onto a microchip, or developed into a universal FBG demodulation system.

13.3 SUMMARY

In this book we attempt to demonstrate that fiber optic structural monitoring can be used in many diverse engineering applications, including practical field sites ranging from bridges to ships. We have also indicated that certain aspects of the technology have matured to the point where they are commercially available, and fiber optic structural monitoring systems have been installed on a routine basis and delivered reliable performance.

In Chapters 3 through 5 we provide an introduction to fiber optic technology that is helpful for those wishing to acquire a deeper appreciation of the subject. In Chapter 6 we provide a brief overview of fiber optic sensing and indicate that fiber optic Bragg gratings (FBGs) have emerged as one of the most important components in this new technology.

We show that FBGs can be used as short gauge-length sensors, and since there is no other structural element involved, such as a supporting microcapillary tube, the local perturbation to any host structure is the same as that of the optical fiber alone. The ease of multiplexing FBG sensors, combined with their truly distributed sensing capability and the fact that FBGs are ideal for making arbitrary long gauge-length sensors, gives this technology a potential dominance that is unrivaled. This versatility is indicated in Fig. 13.9. In addition, fabrication of FBGs at a rate of 2000 per hour on the optical fiber draw tower, in an automated process, promises low cost, consistent performance, and pristine strength.

At present there are a number of other fiber optic structural sensors that have been used quite successfully. The transverse sensitivity of the FBG sensors, albeit small, could cause some ambiguity for an embedded sensor that is subject to large transverse loads. However, under such circumstances, it is probably important to measure more than the axial component of strain, and several suggestions have been made in the book as to how this might be undertaken with fiber Bragg gratings. Temperature compensation is an important consideration, especially when the fiber optic sensors are embedded within FRP composite materials. In Chapter 7, a number of approaches have been shown to be capable of addressing this issue.

The most suitable for any specific application depends strongly upon the circumstances. For example, consider a concrete girder, instrumented with a string of FBG sensors pultruded with one of the FRP prestressing tendons. If a single temperature measurement will suffice, this could be accomplished by strain decoupling the last FBG sensor in the string, or simply incorporating a separate temperature sensor in, or on, the girder. Alternatively, the girder may be exposed to substantial temperature

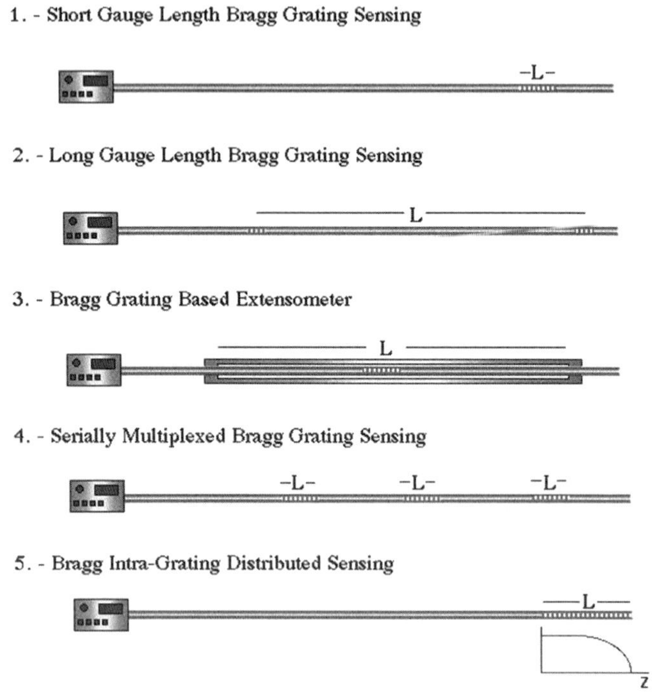

FIGURE 13.9. Different sensing modalities possible with fiber Bragg grating technology.

variations that change during the day as different sections of the girder fall into the shade. Under these circumstances it would be necessary to design into the system the capability of simultaneously monitoring the strain and temperature along the girder. This can be done, for example, by using pairs of gratings, or collocated gratings, with very different Bragg wavelengths. Cross-sensitivity of strain and temperature terms has been shown to be small for many practical situations involving structural monitoring. However, that such complexities may have to be considered if large excursions of strain or temperature are encountered, especially in the case of long lengths of optical fiber.

Chapter 8 indicates that there is substantial evidence that optical fibers with diameters close to 100 μm, and embedded collinear with the local reinforcing fibers of the host FRP composite material cause minimal perturbation and negligible loss of strength. This is especially true if the optical fiber coating is fairly compliant. If the optical fiber cannot be aligned with the adjacent ply direction of the FRP material, then the angle between the optical fiber axis and the ply direction should be kept to a minimum (preferably less than 15°) and a compliant coating is essential.

Where possible, the sensing section of an optical fiber should be positioned out of any strain transfer region within a host structure. Under such circumstances, debonding of the sensor should not affect its ability to monitor the host axial strain. Installation procedures are described in Chapter 8, and it is shown that optical fiber sensors can be embedded within FRP prepreg-laminated, filament-wound, and pultruded structures. Fiber Bragg grating sensors have been found to survive the harsh environment of pultrusion and have little influence on the properties of the pultruded structures, provided only one small optical fiber, with a diameter $<150\ \mu m$ is involved.

Optical connectivity to embedded sensors is another area that requires development, especially in the case of pultruded sensors. In certain applications, it might be necessary to bring an array of embedded optical sensors to one point of egress from the host structure. It is important that the location of this egress point not be in a heavily loaded region of the structure, as it would certainly represent a local perturbation.

Chapter 9 describes a wide range of sensor demodulation techniques and reviews a number of fairly substantial application projects involving short gauge-length fiber optic sensors. As vindication of the maturity of this technology, a number of these applications involved the use of commercial fiber optic structural monitoring systems. The fiber Bragg grating was the sensor of choice for most of the applications, which ranged from instrumenting bridges to ships. Although many techniques for demodulating FBGs were described, those that are capable of handling large parallel arrays of sensors are probably the most important for practical applications, since it is rare indeed that just one or two structural fiber optic structural sensors would be used any field-site location. A number of major test sites for the technology have already been established in Canada, where several large bridges have hundreds of FBG sensors built right into their structure. The Taylor Bridge stands as a state-of-the-art example, where engineers at any off-site location can dial up and, with the right equipment and software, monitor the array of 66 FBG sensors that are integrated into many of the key elements of this bridge.

The sensor demodulation technique most likely to see broad application will be one that is reliable, compact, rugged, easy to use and, of course, low cost. Scanned, narrowband source, or photodetection approaches look quite promising, as do the systems based on spectrometric CCD arrays. The next few years should see a shakedown of these techniques, with probably one emerging as the clear leader. A bit of a dark horse in this race is the tunable laser technique arising from my own laboratory, as it has the potential to be microminiaturized and provide a means of reading all of the different sensing modalities possible with FBGs. This was discussed in Section 13.2.

Chapter 10 reviews the long gauge-length (typically $> 5\,\text{cm}$) fiber optic sensor and shows that it is in a class by itself in terms of measurement capability. This has certainly been demonstrated in Switzerland, where hundreds of such sensors have been installed in bridges, dams, and other structures to measure structural deformation and loading. The greatest application potential for the long-gauge sensor, however, probably lies in monitoring concrete structures that have been rehabilitated and strengthened with FRP composite material wraps. This is especially so for concrete columns and girders damaged by corrosion or earthquakes. These applications will often require the sensor to accommodate the shape of the structure, and one based on FBG technology has been shown to be well suited for the task.

Chapters 11 and 12 focus on the ability of FBGs to undertake distributed sensing and map the strain profile of a structure, whether it be at the tip of a crack under an aircraft repair patch, or along a power line laden with ice. This unique capability stems primarily from the intrinsic nature of the FBG, allowing the optical fiber to serve as both sensor and conduit of sensing data. This capability also permits the function and performance of a sensing array to be checked, a task unmatched by conventional technology. It also makes practical the installation of a large number of sensors in many applications, where the complexity and weight of wiring for an electrically based counterpart would be impractical, no matter how desirable.

In this summary I have tried to touch on just a few of the interesting and important points covered in the book. If, in doing so, I have answered at least some readers' questions and stimulated their interest in looking further at the book, I am pleased.

References

Abdelrahman, A. A., Tadros, G., and Rizkalla, S. H. (1995). "Test model for the First Canadian Smart Highway Bridge," *ACI Struct. J.* **92**, 451–458.

Agnes, G. S. and Silva, K. (1992). "Aircraft Smart Structures Research in the USAF Wright Laboratory," AGARD, Smart Structures for Aircraft and Spacecraft Meeting, Lindau, Germany, 5–7 Oct., pp. 27–1 to 27–9.

Ahmad, A., Crowson, A., Rogers, C. A., and Aizawa, M. (1990). *US–Japan Workshop on Smart/Intelligent Materials and Systems*, Technomic Publishing.

Aktan, A. E., Helmicki, A. J., and Hunt, V. J. (1995). "Instrumentation and Intelligence Issues in Bridge Health Monitoring," *SPIE* **2446**, 106–115.

Alavie, A. T. (1993). "Technique for an Interrupt-Immune Two-Mode Fiber-Optic Strain Sensor," *Optics Lett.* **18**, 1007–1009.

Alavie, A. T., Karr, S. E., Othonos, A., and Measures, R. M. (1993). "A Multiplexed Bragg Grating Fiber Laser Sensor System," *IEEE Photonics Techn. Let.* **5**, 1112–1114.

Alavie, A. T., Maaskant, R., Ohn, M. M. Rizkalla, S., and Measures, R. M. (1994a). "Application and Characterization of Intracore Grating Sensors in a CFRP Prestressed Concrete Girder," *SPIE* **2191**, 102–110.

Alavie, A. T., Maaskant, R., and Measures, R. M. (1994b). "Bragg Grating Laser Sensing System for Smart Structures," Proc. 8th CIMTEC- World Ceramics Congress and Formum on New Materials, Session SV1-3 : L03, Intelligent Optics, Florence, Italy.

Alavie, A. T., Maaskant, R., Ohn, M. M., Glennie, D., and Measures, R. M. (1994c). "Application and Characterisation of Intracore Grating Sensors in a CFRP Prestressed Concrete Girder," *SPIE* **2191**, 13–18.

Alavie, A. T., Maaskant, R., Stubbe, R., Othonos, A., Ohn, M., Sahlgren, B., and Measures, R. M. (1995). "Characteristics of Fiber Grating Sensors and Their Relation to Manufacturing Techniques," *SPIE* **2444**, 528–535.

Albert, J., *et al.* (1995). "Comparison of One-Photon and 2-Photon Effects in the Photosensitivity of Germanium-Doped Silica Optical Fibers Exposed to Intense ArF Excimeter-Laser Pulses," *Appl. Phys. Lett.*, **67**, 3529–3531.

Alexander, J., and Cheng. R. (1996). "Field Application and Studies of using CFRP Sheets to Strengthen Concrete Bridge Girders," in *Advanced Composite Materials in Bridges & Structures*, (El-Badry, M.M., Ed.) pp. 465–472.

Anderson, I. (1992). "Earthquake Tension Builds Up in California," *Science*, April 11, p. 15.
Anderson, I. (1997). "Blowing Hot," *New Scientist*, July, p. 22.
Ansari, F. (1993). "Strain and Temperature Measurement Using Optical Fiber," in *Applications of Fiber Optic Sensors in Engineering Mechanics*, 236–250.
Ansari, F., Ed. (1998). *Fiber Optic Sensors for Construction Materials and Bridges*, Technomic Publishing.
Archambault, J. L., Reekie, L., and Russell, P. (1993). "High Reflectivity and Narrow Bandwidth Fiber Grating Written by Single Excimer Pulse," *Electronic Lett.* **29**, 28–29.
Askins, C. (1994). "Fiberoptic Sensor Arrays Emerge Ready-made from Top Draw Tower," *Laser Focus World*, May, 237–238.
Askins, C. G., Tsai, T. E., Williams, G. M., Putman, M. A., Bashkansky, M., and Friebele, E. J. (1992). "Fiber Bragg Reflectors Prepared by a Single Excimer Pulse," *Opt. Lett.* **17**, 833–835.
Askins, C. G., Putnam, M. A., Williams, G. M., and Friebele, E. J. (1994). "Stepped-Wavelength Optical-Fiber Bragg Grating Arrays Fabricated in Line on a Draw Tower," *Opt. Lett.* **19**, 147–149.
Askins, C. G., Putnam, M. A., and Friebele, E. J. (1995). "Instrumentation for Interrogating Many-Element Fiber Bragg Grating Arrays," *SPIE* **2444**, 257–266.
Askins, C. G., Putnam, M. A. Patrick, H. J., and Friebele, E. J. (1997). "Fiber Strength Unaffected by On-Line Writing of Single-Pulse Bragg Gratings," *Electronics Lett.* **33**, 1333–1334.
Asseh, A., Storoy, H., Sahlgren, B., Asandgren, S., and Stubbe, R. (1997). "A Grating Writing Technique for Long Superstructural Fiber Bragg Gratings," in *Fibre Bragg Gratings and Fibre Optic Structural Strain Sensing*, Doctoral Thesis of H. Storoy, Published by Norwegian University of Science and Technology.
Atkins, R. M., Mizrahi, V., and Erdogan, T. (1993). "248 nm Induced Vacuum UV Spectral Changes in Optical Fibre Preform Cores: Support for a Colour Centre Model of Photosensitivity" *Electronic Lett.* **29**, 385–387.
Austin, F., Knowles, G. J., Jung, W. G., Tung, C. C., and Sheedy, E. M. (1992). "Adaptive/Conformal Wings Design for Future Aircraft," *SPIE* **1777**, 1st Europe Conf. On Smart Structures and Materials, Glasgow, 387–390.
Baker, A. A., and Jones, R. (1988). *Bonded Repair of Aircraft Structures*, Martinus Nijhoff Publishers, Boston.
Balestra, C. L. (195). "Implementation of Integrated Optic Systems with Active Waveguides, Passive Waveguides, Optical Fibers," in *Applications of Photonic Technology* (George A. Lampropoulos, Jacek Chrostowski, and Raymond M. Measures, Eds.), Laurin Publishing, pp. 393–411.
Balestra, C. L., and Shanley, J. F. (1994). "Making the Switch in Photonics," *Photonics Spectra*, Sept., 143–150.
Ball, G. A., and Morey, W. W. (1992). "Continuously Tunable Single-Mode Eribium Fiber Laser," *Opt. Lett.* **17**, 420–422.
Ball, G. A., and Morey, W. W. (1994). "Compression-Tuned Single Frequency Bragg Grating Fiber Laser," *Opt. Lett.* **19**, 1979–1981.

Ball, G. A., Morey, W. W., and Cheo, P. K. (1993). "Single and Multipoint Fiber-Laser Sensors," *IEEE Photonics Tech. Lett.* **5**, 267–270.

Ball, G. A., Morey, W. W., and Cheo, P. K., (1994), "Fiber Laser Source/Analyzer for Bragg Grating Sensors in the Characterization of Scaled Marine Vehicle Models," *J. Lightwave Technol., 700–703.*

Bao, X., Webb, D. J., and Jackson, D. A. (1993a). "32-km Distributed Temperature Sensor Based on Brillouin Loss in an Optical Fiber," *Opt. Lett.* **18**, 1561–1563.

Bao, X., Webb, D. J., and Jackson, D. A. (1993b). "Characteristics of Brillouin Gain Based Distributed Temperature Sensors," *Electron. Lett.* **29**, 1543–1544.

Bao, X., Dhliwayo, Webb, D. J., Jackson, D. A., (1994), "Combined Distributed Temperature and Strain Sensor Based on Brillouin Loss in an Optical Fiber," *Opt. Lett.* **18**, 1561–1563.

Bao, X., Dhliwayo, J., Heron, N., Webb, D. J., and Jackson, D. A. (1995), "Experimental and Theoretical Studies on a Distributed Temperature Sensor Based on Brillouin Scattering," *J. Lightwave Technol.* **13**, 1340–1348.

Barker, A. J., and Balasundaram, V. (1987). "Compression Testing of Carbon Fibre-Reinforced Plastics Exposed to Humid Environments," *Composites* **18**, 217–226.

Barrett, R., and Gross, R. S. (1996). "Super-Active Shape-Memory Alloy Composites," *Smart Mater. Struct.* **5**, 2555–2560.

Barrett, R., and Stutts, J., (1997), "Design and Testing of a 1/12th-Scale Solid State Adaptive Rotor," *Smart Mater. Structures* **6**, 491–497.

Beard, J. (1995). "A Watchful Eye on Bridges," *New Scientist*, Feb. 5, pp. 24–25.

Beauchamp, C. H., Nadolink, R. H., Dickinson, S. C., and Dean, L. M. (1992). "Shape Memory Alloy Adjustable Chamber (SMAAC) Control Surfaces," 1st Europ. Conf. Smart. Struct. & Mater. Glasgow, *SPIE* **1777**, 189–192.

Belleville, C., and Duplain, G. (1993). "White-Light Interferometric Multimode Fiber-Optic Strain Sensor," *Opt. Lett.* **18**, 78–80.

Belleville, C., Morin, A., Caron, S., and Edgecombe, M. (1995). "Smart Skins—A Step towards a Practical Fiber-Optic Sensor," in *Applicaitonis of Photonic Technology* (George A. Lampropoulos, Jacek Chrostowski, and Raymond M. Measures, Eds.), Plenum Press, NY, pp. 329–334.

Bennett, J. D., and McLaughlin, L. R. (1995). "Monitoriing of Corrosion in Steel Structures Using Optical Fiber Sensors," *SPIE* **2446**, 48–59.

Berkoff, T. A., Kersey, A. D., and Friebele, E. J. (1991). "Absolute Strain Sensing Using Fiber Interferometry," in *Optical Fiber Sensor-Based Smart Materials and Structures*, Technomic Publishing, pp. 21–28.

Berkoff, T. A., Davis, M. A., and Kersey, A. D., (1994), "Fiber Optic Sensors for Distributed Vibration Monitoring," Vibration Monitoring and Control Conference, *SPIE* **2264**, 148–154.

Berkoff, T. A., Davis, M. A., Bellemore, D. G., Kersey, A. D., Williams, G. M., and Putnam, M. A. (1995). "Hybrid Time and Wavelength Division Multiplexed Fiber Bragg Grating Sensor Array," *SPIE* **2444**, 288–294.

Berkoff, T. A., and Kersey, A. D., (1996), "Experimental Demonstration of a Fiber Bragg Grating Accelerometer," *IEEE Photonics Tech. Lett.* **8**, 1677–1679.

Berthold, J. W. (1997). "Sensors in Industrial Systems," in *Optical Fiber Sensors*, Vol. 4 (Dakin, J., and Cushaw, B, Eds.), pp. 261–308, Artech House.

Bertholds, A., and Dandliker, R., (1988) "Determination of the Individual Strain–Optic Coefficients in Single-Mode Optical Fibers", *J. Lightwave Tech.* **6**, 17–20.
Bhagavatula, V. A., Holmes, G. T., Nolan, D. A., Jansen, R., and McCourt, M., (1997), "Planar Technology Enhances Coupler Performance," *Laser Focus World*, 155–160.
Bhatia, V., Jones, M. E., Grace, J. L., Murphy, K. A., and Claus, R. O. (1994). "Applicatioins of 'Absolute' Fiber Optic Sensors to Smart Materials and Structures," 10th Int. Conf. Optical Fibre Sensors, Glasgow, Scotland, *SPIE* **2360**, 171–174.
Bhatia, V., Campbell, D., Vries, M. J., Sherr, D., D'Alberto, T., Arya, V., and Claus, R. O. (1997a). "Grating-Based Optical Fiber Sensors for Structural Analysis," *SPIE* **3042**, 390–399.
Bhatia, V., D'Alberto, T., Zabaronick, N., and Claus, R. O. (1997b). "Temperature-Insensitive Long-Period Gratings for Strain and Refractive Index Sensing," *SPIE* **3042**, 194–202.
Biswas, M., Pandey, A. K., and Samman, M. M. (1990). "Modal Technology for Damage Detection of Bridges," in *NATO Advanced Research Workshop on Bridge Evaluation, Repair, and Rehabilitation* (Nowak, A., Ed.), Kluwer Academic Publishers, pp. 161–174.
Bjerkan, L., and Johannessen, K. (1997). "Measurements of Bragg Grating Birefringence due to Transverse Compressive Forces," OSA, 12th Int. Conf. on Optical Fiber Sensors, Williamsburg, 60–63.
Bjerkan, L., Johannessen, K., and Storoy, H. (1995). "Experiments with Fiberoptic Smart Skin Sensors Applied to Marine Vehicles," in *Applications of Photonic Technology* (George A. Lampropoulos, Jacek Chrostowski, and Raymond M. Measures, Eds.), Plenum Press, NY, pp. 325–329.
Bjerkan, L., Johannessen, K., and Storoy, H. (1997). "Experiments with Fiber Optic Smart Skin Sensors Applied to Marine Vehicles," *Fibre Bragg Gratings and Fibre Optic Structural Strain Sensing*, Doctoral Thesis of H. Storoy, Published by Norwegian University of Science and Technology.
Blagojevic, B., Tsaw, W., McEwen, K., and Measures, R. M. (1990). "The Influence of Embedded Optical Fibers on the Interlaminar Fracture Toughness of Composite Materials," in *Review of Progress in Quantitative NDE*, Vol. 9, 1213–1217.
Blake, J. N., Huang, S. Y., and Kim, B. Y. (1987a), "Elliptical Core Two-Mode Fiber Strain Gauge," Fiber Optic and Laser Sensors V, *SPIE* **838**, 332.
Blake, J. N., Huang, S. Y., Kim, B. Y., and Shaw, H. J. (1987b). "Strain Effects on Highly Elliptical Core Two-Mode Fibers," *Opt. Lett.* **12**, 732.
Bock, W. J., (1989), "High Hydrostatic Pressure Effects of Highly Birefringent Optical Fibers," *J. Lightwave Tech.* **7**, 1279.
Bock, W. J., and Wolinski, T. R. (1990). "Temperature-Compensated Fiber-Optic Strain Sensor Based on Polarization-Rotated Reflection," Fiber Optic Smart Structures and Skins III, *SPIE* **1370**, 189–196.
Bogatyriov, V. A., Bubnov, M. M., Dianov, E. M., Rumyantzev, S. D., and Semjonov, S. L. (1991). "Mechanical Reliability of Polymer-Coated and Hermetically Coated Optical Fibers Based on Proof Testing," *Optical Engineering* **30**, 690–699.

Boller, C., and Dilger, R. (1992). "In-Flight Aircraft Structure Health Monitoring Based on Smart Structures Technology," AGARD Conf. Proc. 531, Smart Structures for Aircraft and Spacecraft, FRG.17–1 to 17–19.
Boller, C., Honlinger, H., and Sensburg, O. (1992). "Technological Challenge with Smart Structures in German Aircraft Industry," 1st Europe Conf. on Smart Structures and Materials, Glasgow, *SPIE* **1777**, 289–292.
Bonacci, J. F. (1996). "Strength, Failure Mode and Deformability of Concrete Beams Strengthened Externally with Advanced Composites," in *Advanced Composite Materials in Bridges & Structures*, (El-Badry, M.M., Ed.), The Canadian Society of Civil Engineers, pp. 419–426.
Born, M., and Wolfe, E. (1975). *Principles of Optics*, 5th ed., Pergamon Press, New York.
Borrelli, N. F., and Miller, R. A. (1968). "Determination of the Individual Strain-Optic Coefficients of Glass by an Ultrasonic Technique," *Appl. Optics* **7**, 745–750.
Bronnimann, R., Nellen, P., and Sennhauser, U. (1998). "Application and Reliability of a Fiber Optical Surveillance System for a Stay Cable Bridge," *Smart Mater. Struct.* **7**, 229–236.
Brown, A. W., DeMerchant, M. D., Bao, X., Bremner, T. W., (1998). "Advances in Distributed Sensing Using Brillouin Scattering," *SPIE* Smart Struct. and Mat. Conf. San Diego.
Bullock, D. E., Dunphy, J. R., and Hufstetler, G. H. (1992). "Embedded Bragg Grating Fiber Optic Sensor for Composite Flexbeams," Fiber Optic Smart Structures & Skins V, Boston, *SPIE* **1798**, 253–261.
Butter, C. D., and Hocker, G. P. (1978). "Fiber Optics Strain Gauge," *Appl. Opt.* **17**, 2867–2869.
Carman, G. P., and Averill, R. C. (1992, "Analytical Modeling of Micromechanical Stress Variations in Continuous Fiber-Reinforced Composites," *Proc. IUMTAM Conf.*, 27–61.
Carman, G. P., Averill, R. C., Reifsnider, K. L., and Reddy, J. L., (1993), "Optimization of Coatings to Minimize Micromechanical Stress Concentrations in Composites," *J. Composite Materials*, **27**, 589–613.
Carman, G. P., and Mitrovic, M. (1994). "Health Monitoring Techniques for Composite Materials Employing Thermal Parameters and Fiber Optic Sensors," *SPIE* **2191**, 244–256.
Carman, G. P., and Reifsnider, K. L. (1992). "Analytical Minimization of the Obtrusive Behavior Indicative of Embedded Sensors and Actuators," in *Recent Advances in Adaptive and Sensory Materials and Their Applications* (Craig Rogers, Ed.), Technomic Publishing, Lancaster, pp. 314–331.
Carman, G. P., and Sendeckyi, G. P. (1995). "Review of the Mechanics of Fiber Optic Sensors," *J. Compos. Technol. Res.* **17** 183–193.
Carman, G. P., Case, S. W., Lesco, J. J., and Fogg, B. R. (1992). "Implementation and Evaluation of an Embedded Extrinsic Fabry–Perot Fiber Optic Strain Rosette Sensor," Fiber Optic Smart Structures & Skins V, Boston, *SPIE* **1798**, 237–246.
Carman, G. P., Paul, C. A., and Sendeckyj, G. P. (1993). "Transverse Strength of Composites Containing Optical Fibers," Smart Structures and Intelligent Systems, *SPIE* **1917**, 307–316.

Case, S. W., and Carman, G. P. (1994). "Compression Strength of Composites Containing Embedded Sensors or Actuators," *J. Int. Mat. Syst. & Struct.* **5**, 4–11.
Caussignac, J. M., Barbachi, M., and Chabert, A. (1996). "Bridge Bearings Equipped with Optical Fiber Sensor for Measuring Vertical Load through the Support," *SPIE* **2719**, 220–228.
Cavaleiro, P. M., Ribeiro, A. B. L., and Santos, J. L. (1995). "Referencing Techniques for Intensity-Based Sensors Using Fiber Optic Bragg Gratings," *Electronics Lett.* **31**, 392–394.
Cawley, P., and Adams, R. D. (1979). "The Location of Defects in Structures from Measurements of Natural Frequencies," *J. Strain Analysis* **14**, 49–57.
Chakravorty, K. K. (1993). "Photogeneration of Refractive-Index Patterns in Doped Polyimide Films," *Appl. Optics* **32**, 2331–2338.
Chang, C. C., and Kersey, A. (1997). "Development of Fiber Bragg Grating Sensor Based Load Transducers," 12th Optical Fiber Sensors Conference, October 26–28, Williamsburg.
Chappell, L. (1998). "Development of a Tunable Laser Based Bragg Grating Demodulation System," M.A.Sc. Thesis, University of Toronto.
Chen, Z., and Ansari, F. (1998). "Pulse-Echo Fiber Optic Sensor for Nondestructive Evaluation of Concrete Bridges," in *Fiber Optic Sensors for Constructional Materials and Bridges*, (Farhad Ansari, Ed.), 136–147. Technomic Publishing Comp., Lancaster.
Chen, S., Palmer, A. W., Grattan, K. T. V., and Meggitt, B. T. (1992). "Fringe Order Identification in Electronically-Scanned Optical-Fiber White-Light Interferometry: A Novel Method," 8th Opt. Fiber Sensor Conference, Monterey, Jan.
Chen, X., Ansari, F., and Ding, H. (1996). "Embedded Fiber Optic Displacement Sensor for Concrete Elements," *Proceedings of the 11th Conference on Engineering Mechanics, (Y. K. Lin and T. C. Su, Eds.), 359–365. American Society of Civil Engineers, New York.*
Chen, S., Hu, Y., Zhang, L., and Bennion, I. (1997). "Digital Spatial and Wavelength Domain Multiplexing of Fiber Bragg Grating Based Sensors," OSA, 12th Int. Conf. On Optical Fiber Sensors, Williamsburg, 448–451.
Chen, Z., Mendez, A., Li, Q. and Ansari, F. (1988), "Fiber-Optic White Light Distributed Sensor for Condition Monitoring of Civil Structures," in *Fiber Optic Sensors for Constructional Materials and Bridges* (Farhad Ansari, Ed.), 101–114. Technomic Publishing Comp., Lancaster.
Cheo, P. K. (1985). *Fiber Optics Devices and Systems*, Prentice-Hall, Englewood Cliffs, NJ.
Chefas, J. (1990). "Stretching the Point," *Science* **247**, 630.
Chong, K. P., Dillon, O. W., Scalzi, J. B., and Spitzig, W. A. (1994). "Engineering Research in Composite and Smart Structures," *Composites Engineering* **4**, 829–852.
Chu, K. W., (1991), "Optical Coherence Multiplexing for Interprocessor Communications," *Opt. Eng.* **30**, 337–344.
Clarke, A. (1989). "How Green Is the Wind," *New Scientist*, 27 May, 62–65.
Claus, R. O., and Cantrell, J. H. (1980). "Detection of Ultrasonic Waves in Solids by an Optical Fiber Interferometer," *Proc. IEEE Ultrasonics Symposium*, Vol. 2, 719.

Claus, R. O., Gunther, M. F., Wang, A., and Murphy, K. A. (1992). "Extrinisic Fabry–Perot Sensor for Strain and Crack Operating Displacement Measurements from −200–900 Degrees C," *Smart Mat. & Struct.* **1**, 237–242.

Claus, R. O., Gunther, M. F., Wang, A., Murphy, K. A., and Sun, D. (1993). "Extrinsic Fabry–Perot Sensors for Structural Evaluations of Fibre Optic Sensors in Engineering Mechanics (Ansari, F. Ed.), Ny. 60–70, Technomic Publishing Comp., Lancaster.

Coghlan, A. (1992). "Smart Ways to Treat Materials," *New Scientist*, 4 July, 27–29.

Corke, M., Kersey, A. D., Liu, K., and Jackson, D. A. (1984). "Remote Temperature Sensing Using Polarization Preserving," *Electron. Lett.* **20**, 67–69.

Coroy, T., and Measures, R. M. (1995). "Active Wavelength Demodulation for Bragg Grating Strain Sensors," *Applications of Photonic Technology* (George A. Lampropoulos, Jacek Chrostowski, and Raymond M. Measures, Eds.), 343–347. Plenum Press, N.Y.

Coroy, T., and Measures, R. M. (1996). "Active Wavelength Demodulation of a Bragg Grating Fiber Optic Strain Sensor Using a Quantum Well Electroabsorption Filtering Detector," *Elect. Lett.* **32**, 1811–1812.

Coroy, T., and Measures, R. M. (1998). "Demodulation System for Bragg Grating Fiber Optic Sensors Using a Quantum Well Electroabsorption Filtering Detector," *Smart. Mater. & Struct.* **7**, 265–271.

Coroy, T., Measures, R. M., and Wood, T. H. (1996). "Active Wavelength Measurement System Using an InGaAs–InP Quantum-Well Electroabsorption Filtering Detector," *IEEE Photon. Tech. Lett.* **8**, 1686–1688.

Coroy, T., Chappell, L. M., Guillermo, N. J., Huang, S. Y., and Measures, R. M. (1997). "Peak Detection Demodulation of a Bragg Filter Optic Sensor Using a Gain-Coupled Distributed Feedback Tunable Laser," OSA, 12th Int. Conf. on Optical Fiber Sensors, Williamsburg, 210–212.

Crane, R. M., and Gagorik, J. (1984). "Fiber Optics for a Damage Assessment System for Fiber Reinforced Plastic Composite Structures," *Quantitative NDE* **28**, 1419–1430.

Crawley, E. F. (1992). "Intelligent Structures—A Technology Overview and Assessment," AGARD. Conf. Proc. 531, Smart Structures for Aircraft and Spacecraft, FRG.

Cross, L. E. (1992). "Recent Developments in Piezoelectric and Electrostrictive Sensors and Actuators for Smart Structures," AGARD. Conf. Proc. 531, Smart Structures for Aircraft and Spacecraft, Lindau, FRG.

Culshaw, B., and Dakin, J. (1996). *Optical Fiber Sensors: Components and Subsystems*, Vol. 3, Artech House, Inc., Norwood.

Culshaw, B,. and Gardiner, P. T., (1993), "Smart Structures—The role of Fibre Optics," *Fiber and Integrated Optics*, **2**, 353–373.

Czarnek, R., Guo, Y. F., Bennett, K. D., and Claus, R. O. (1989). "Interferometric Measurements of Strain Concentrations Induced by an Optical Fiber Embedded in a Fiber Reinforced Composite," *Proc. SPIE–Int. Soc. Opt. Eng.* **986**, 120–129.

Dakin, J. P. (1989). "Distributed Optical Fiber Sensor Systems," Chap. 15, 575–598, in *Optical Fiber Sensors* (B. Culshaw and J. Dakin, Eds.), Artech House, Norwood.

Dakin, J. P. (1995). "Distributed Optical Fiber Sensors," in *Fiber Optic Smart Structures* (Eric Udd, Ed.), 373–408.

Dakin, J., and Culshaw, B. (1988). *Optical Fiber Sensors: Principles and Components*, Artech House, Norwood.

Dakin, J., and Culshaw, B. (1997). *Optical Fiber Sensors: Applications, Analysis, and Future Trends*, Artech House, Norwood.

Dakin, J. P., Pratt, D. J., Bibby, G. W., and Ross, J. N. (1985). "Distributed Optical Fibre Raman Temperature Sensor Using a Semiconductor Light Source and Detector," *Electron. Lett.* **21**, 569–570.

Dakin, J. P., Ecke, W., Rothardt, M., Schauer, Usbeck, K., and Willsch, R. (1997). "New Multiplexing Scheme for Monitoring Fiber Optic Bragg Grating Sensors in the Coherence Domain," 12 Int. Conf. on Optical Fiber Sensors, Williamsburg, 31–34.

Daniel, I. M., and Liber, T. (1997). "Effects of Laminate Construction on Residual Stresses in Graphite/Polyimide Composites," *Experimental Mechanics* **17**, 21–25.

Danisch, L. A. (1992). "Bend-Enchanced Fiber Optic Sensors", SPIE Fiber Optic and Laser Sensors X, *SPIE* **1795**, Boston, 8–9 Sept.

Dasgupta, A., and Sirkis, J. S. (1992), "The Importance of Coatings to Structurally Embedded Optical Fiber Sensors in Smart Structures," *AIAA*, **30**, 1137–1343.

Dasgupta, A., Wan, Y., Sirkis, J. S., and Singh, H. (1990). "Micro-Mechanical Investigation of an Optical Fiber Embedded in a Laminated Composite," *Fiber Optic Smart Struct. & Skins III* **1370**, 119–128.

Dasgupta, A., Wan, Y., and Sirkis, J. S. (1992). "Prediction of Resin Pocket Geometry for Stress Analysis of Optical Fibers Embedded in Laminated Composites," *Smart Mater. Struct I*, 101–107.

Davidson, R., and Roberts, S. S. J. (1992). "Finite Element Analysis of Composite Laminates Containing Transversely Embedded Optical Fiber Sensors," 1st Europe Conf. on Smart Structures and Materials, Glasgow, *SPIE* **1777**, 115–122.

Davidson, R., Bowen, D. H., and Roberts, S. J. (1990). "Composite Materials Monitoring Through Embedded Fibre Optics," *Int. J. Optoelectronics* **5**, 397–404.

Davis, M. A., and Kersey, A. D. (1994). "All-Fibre Bragg Grating Strain-Sensor Demodulation Technique Using a Wavelength Division Coupler," *Electron. Lett.* **30**, 75–77.

Davis, M. A., and Kersey, A. D. (1995a). "Matched-Filter Interrogation Technique for Fiber Bragg Grating Arrays," *Electronics Lett.* **31**, 822–823.

Davis, M. A., and Kersey, A. D. (1995b). "Serially Configured Matched Filter Interrogation Technique for Bragg Grating Arrays," *SPIE* **2444**, 295–312.

Davis, M. A., and Kersey, A. D. (1995c). "Application of a Fiber Fourier Transform Spectrometer to the Detection of Wavelength Encoded Signals from Bragg Grating Sensors," *J. Lightwave Techn.* **13**, 1289–1295.

Davis, M. A., and Kersey, A. D. (1995d). "Application of a Fiber Fourier Transform Spectrometer to the Detection of Wavelength-Encoded Signals from Bragg Grating Sensors," *J. Lightweight Tech.* **13**, 1289–1295.

Davis, M. A., and Kersey, A. D. (1996a). "Separating the Temperature and Strain Effects on Fiber Bragg Grating Sensors Using Stimulated Brillouin Scattering," *SPIE* **2718**, 270–278.

Davis, M. A., and Kersey, A. D. (1996b). "Simultaneous Measurement of Temperature and Strain Using Fiber Bragg Gratings and Brillouin Scattering," *SPIE* **2838**, 114–123.
Davis, M. A., Bellemore, D. G., Berkoff, T. A., and Kersey, A. D. (1995), "Design and Performance of a Fiber Bragg Grating Distributed Strain Sensor System," *SPIE* **2446**, 227–235.
Davis, M. A., Bellemore, D. G., Kersey, A. D., Putnam, M. A., Friebele, E. J., Idriss, R. L. and Kodinduma, M. (1996a). "High Sensor-Count Bragg Grating Instrumentation System For Large-Scale Structure Monitoring Applications," *SPIE* **2718**, 303–309.
Davis, M. A., Bellemore, D. G., Putnam, M. A., and Kersey, A. D. (1996b). "High Strain Monitoring in Composite-Wrapped Concrete Cylinders Using Embedded Fiber Bragg Grating Arrays," *SPIE* **2721**, 149–154.
Davis, M. A., Kersey, A. D., Serkis, J., and Friebele, E. J. (1994). "Fiber Optic Bragg Grating Array for Shape and Vibration Mode Sensing" *SPIE* **2191**, 94–101.
Davis, M. A., Kersey, A. D., Sirkis, J., and Friebele, E. J. (1996c). "Shape and Vibration Mode Sensing Using a Fiber Optic Bragg Grating Array," *Smart Materials and Structures* **5**, 759–765.
Davis, M. A., Kersey, A. D., Sirkis, J., and Friebele, E. J., (1996d), "Shape and Vibration Mode Sensing Using a Fiber Optic Bragg Grating Array," *Smart Materials and Structures*, **5**, 759–765.
Davis, M. A., Bellemore, D. G., and Kersey, A. D. (1997a). "Distributed Fiber Bragg Grating Strain Sensing in Reinforced Concrete Structural Components," in *Cement and Concrete Components*, Vol. 19, 45–57, Elsevier Science Limited.
Davis, M. A., Kersey, A. D., Berkoff, T. A. Jones, R. T., Idris, R. L., and Kodinduma, M. (1997b). "Dynamic Strain Monitoring of an In-Use Interstate Bridge Using Fiber Bragg Grating Sensors," *SPIE* **3043**, 87–107.
De Vries, M. J., Murphy, K. A., Goris, J., and Claus, R. O. (1995). "Optical Fiber Sensors for Monitoring Strain on Rebar-Type and Cable-Typed Bolts," *SPIE* **2446**, 236–241.
Demers, N., Gaughierr, M., Labossiere, P., and Neale, K. W. (1996). "The Strengthening of Structural Concrete with an Aramid Woven Fibre/Epoxy Resin Composite," in *Advanced Composite Materials in Bridges & Structures*, (El-Badry, M. M., Ed.), 435–442. The Canadian Society of Civil Engineers, Montreal.
Dennis, M., Putnam, M., Kang, J., Tsai, T., Duling, I., and Friebele, J. (1997). "Grating Sensor Array Demodulation by Use of a Passively Mode-Locked Fiber Laser," *Opt. Lett.* **22**, 1362–1364.
Dewynter-Marty, V., Rougeault, S., and Ferdinand, P. (1997). "Concrete Strain Measurements and Crack Detection with Surface-Mounted and Embedded Bragg Grating Extensometers," OSA, 12th Int. Conf. on Optical Fiber Sensors, Williamsburg, Oct. 28–31, 600–603.
DiFrancia, C., and Clause, R. O. (1989). "Structure/Property Correlation of Several Polyimide Optical Fiber Coatings for Embedding in an Epoxy/Matrix," Fiber Optic Smart Structures and Skins II, *SPIE* **1170**, 505–512.
DiFrancia, C., Claus, R. O., Hellgeth, J. W., and Ward, T. C. (1991). "Discussion of Pullout Tests of Polyimide-Coated Optical Fibers Embedded in Neat Resin," in

Optical Fiber Sensor-Based Smart Materials and Structures, (Clause, R., Ed.), Technomic Publ., 70–83.

Digonnet, M. J. F., and Shaw, H. J. (1983). "Wavelength Multiplexing in Single-Mode Fiber Couplers," *Appl. Optics* **22**, 484–491.

Dong, L., Archambault, J. L., Reekie, L., Russell, P., and Payne, D. N. (1993a). "Single Pulse Bragg Grating Written during Fiber Drawing," *Electron. Lett.* **29**, 1577–1578.

Dong, L., Archambault, J. L., Reekie, L., Russell, P., and Payne, D. N. (1993b). "Bragg Gratings in $Ce3^{+}$-Doped Fibers Written by a Single Excimer Pulse," *Optics Lett.* **18**, 861–863.

Dotto, L. (1992). "Smart Systems," *Challenges*, Winter, 16–21.

Douay, M. (1993). "Thermal Hysteresis of Bragg Wavelengths of Intra-Core Fiber Gratings," *IEEE Photonics Tech. Lett.* **5**, 1331–1334.

Doyle, C., and Fernando, G. (1997). "Condition Monitoring Engineering Materials with an Optical Fiber Variation Sensor System," *SPIE* **3042**, 310–318.

Dry, C., and McMillan, W. (1996). "Three-Part Methylethacrylate Adhesive System as an Internal Delivery System for Resonsive Concrete," *Smart Mat. & Struct.* **5**, 297–300.

Duck, G. (2000). "Distributed Strain Sensing with Fiber Bragg Gratings," University of Toronto Ph.D. Thesis.

Dunker, D. F., and Rabbat, B. G. (1993). "Why America's Bridges Are Crumbling," *Scientific American*, March, 66–72.

Dunphy, J. R., (1996), "Feasibility Study Concerning Optical Fiber Sensor Vibration Monitoring Subsystem," *SPIE* **2721**, 483–492.

Dunphy, J., Ball, G., D'Amato, Ferraro, P., Inserra, S., Vannucci, A., and Varasi, M. (1993). "Instrumentation Development in Support of Fiber Grating Sensor Arrays," Fiber Opt. Smar. Struct., SPIE, Boston.

Dunphy, J. R., Meltz, G., and Morey, W. W., (1995a). "Optical Fiber Bragg Grating Sensors: A Candidate for Smart Structure Applications," in *Fiber Optic Smart Structures* (Eric Udd, Ed.), 271–286, J. Wiley & Sons, N.Y.

Dunphy, J. R., Meltz, G., Varasi, M., Vannucci, A., Signorazzi, M., Ferraro, P., Imparato, S. I., and Voto, C. (1995b). "Embedded Optical Sensor Capable of Strain and Temperature Measurement Using a Single Diffraction Grating," United States Patent #5,399,854.

Eastham, T. R. (1995). "High-Speed Ground Transportation Development outside the United States," *J. Transportation Eng.*, Sept./Oct., 411–416.

Eccleston, D. (1991). "In The Loop, It's Fiber's Mechanical, Not Optical Performance That Counts," *Photonics Spectra*, Feb., 123–128.

Ecke, W., Schauer, J., Usbeck, K., and Willsch, R. (1997). "Improvement of the Stability of Fiber Grating Interrogation Systems, Using Active and Passive Polarization Scrambling Devices," OSA, 12th Int. Conf. on Optical Fiber Sensors, Williamsburg, 484–487.

Elamari, A., Breguet, J., Gisin, N., Inaudi, D., Pflug, D., and Vurpillot, S. (1995). "Low Coherence Fiber Optic Sensors for Structural Monitoring," *Structural Engineering International* **1**, 43–48.

Erdogan, T. (1997). "Fiber Grating Spectra," *J. Lightweight Tech.* **15**, 1277–1287.

Erdogan, T., and Mizrahi, V. (1994). "Fiber Phase Gratings Reflect Advances in Lightwave Technology," *Laser Focus World*, Feb., 73–78.
Erdogan, T., Mizrahi, Lemaire, P. J., and Monroe, D. (1994). "Decay of Ultraviolet-Induced Fiber Bragg Gratings," *J. Appl. Phys.* **76**, 73–80.
Erki, M. A., and Rizkalla, S. H. (1993a). "FRP Reinforcement for Concrete Structures," ACI *Concrete International*, June, 48–53.
Erki, M. A., and Rizkalla, S. H. (1993b). "Anchorages for FRP," ACI *Concrete International*, June, 54–59.
Escobar, P., Gusmeroli, V., Martinelli, M., and Morabito, P. (1992). "Fiber-Optic Interferometric Sensors for Concrete Structures," 1st Euro. Conf. on Smart Structures and Materials, Glasgow, 215–218.
Espindola, R. P., Atkins, R. M., Wang, N. P. Simoff, D. A., Paczkowski, M. A., Windeler, R. S., Brownlow, D. L., Shenk, D. S., Glodis, P. A., Strasser, DeMarco, J. J., and Chandonnet, P. J. (1997). "40 dB Fiber Bragg Grating Written through the Fiber Coating at 257 nm," OSA, 12th Optical Fiber Sensors Conference, October 26–28, Williamsburg.
Ezbiri, A., Kanellopoulos, S. E., and Handerek, V. A. (1997). "High Resolution Instrumentation System for Demodulation of Bragg Grating Aerospace Sensors," OSA, 12th Int. Conf. on Optical Fiber Sensors, Williamsburg, 456–459.
Fallon, R. W., Zhang, L., Gloag, A., and Bennion, I. (1997). "Multiplexed Identical Broadband Chirped Grating Interrogation Technique for Large-Strain Sensing Applications," *IEEE* **9**, 1616–1618.
Fan, N. (1997), "Development of Fiber Optic Sensing Techniques," University of Toronto, Ph.D. Thesis.
Fan, N. Y., Huang, S., Alavie, A. T., and Measures, R. M. (1995). "Rare Earth Doped Fibre for Structural Damage Assessment," *Smart Mater. Struct.* **4**, 179–185.
Fan, N. Y., Huang, S. Y., and Measures, R. M. (1997). "Localized Long Gauge Length Fiber Optic Sensor Demodulated with Wavelength Tuning Technique," *SPIE* **3042**, 366–371.
Fan, N. Y., Huang, S., and Measures, R. M. (1998). "Localized Long Gauge Fiber Optic Strain Sensors," *Smart Materials and Structures J.* **7**, 257–264.
Farahi, F. (1993). "Strain and Temperature Measurement Using Optical Fiber," in *Fiber Optic Sensors in Engineering Mechanics* (Ansari, F., Ed.), ASCE, 236–250.
Farahi, F. (1996). "Simultaneous Measurements of Strain and Temperature Using Fiber Grating Sensors," 11th Conference on Engineering Mechanics (Y. K. Lin and T. C. Su, Eds.), 351–354.
Farahi, F., Webb, D. J., Jones, J. D. C., and Jackson, D. A. (1990). "Simultaneous Measurement of Temperature and Strain: Cross-Sensitivity Considerations," *J. Lightwave Tech.* **8**, 138–142.
Farries, M. C., Sugden, K., Reid, D. C. J., Bennion, I. Molony, A., and Goodman, M. J. (1994). "Very Broad Reflection Bandwidth (44 nm) Chirped Fibre Gratings and Narrow Bandpass Filters Produced by the Use of an Amplitude Mask," *Electronic Lett.* **30**, 891–892.
Favre, F., and Le Guen, D. (1991). "82 nm of Continuous Tunability for an External Cavity Semiconductor Laser," *Electronic Lett.* **27**, 6–7.

Feced, R., Roe-Edwards, M. P., Kannellopoulos, S. E., Taylor, N. H., and Handerick, V. A. (1997). "Mechanical Strength Degradation of UV Exposed Optical Fibers," *Electronic Lett.* **33**, 157–159.

Fellay, A., Thevenaz, L., Facchini, M., Nikles, M., and Robert, P. (1997). "Distributed Sensing Using Stimulated Brillouin Scattering: Towards Ultimate Resolution," 12th Optical Fiber Sensors Conference, October 26–28, Williamsburg.

Ferdinand, P., *et al.* (1995). "Mine Operating Accurate Stability Control with Optical Fiber Sensing and Bragg Grating Technology: The European BRITE/EURAM STASBILOS Project," *J. Lightwave Tech.* **13**, 1303–1313.

Ferdinand, P., Gaucher, J. C., Magne, S., Martinez, C., Marty, V., Pichon, L., Roy, O., and Rougeault, S. (1996). "Potential Applications for Optical Fiber Sensors and Networks within the Nuclear Power Industry," 11 ursos de Verano, Sensores Opticos, 1–23.

Ferdinand, P., Magne, S., Dewynter-Marty, V., Martinez, C., Rougeault, S., and Bugaud, M. (1997). "Applications of Bragg Grating Sensors in Europe," 12th Optical Fiber Sensors Conference, October 26–28, Williamsburg.

Ferreira, L. A., Lobo Ribeiro, A. B., Santos, J. L., and Farahi, F. (1998). "Simultaneous Displacement and Temperature Sensing Using a White Light Intgerrogated Low Finesse Cavity in Line with a Fiber Bragg Grating," *Smart Mater. Struct.* **7**, 189–198.

Figueroa, L., Hong, C. S., Huggins, R. W., Miller, G. E., Popoff, A. A., Porter, C. R., Smith, D. K., and Deventer, B. V. (1991). "Fiber Optic for Military Aircraft Flight Systems," *IEEE LCS*, 52–65.

Fossum, E. R. (1993). "Active-Pixel Sensors Challenge CCDs," *Laser World*, June, 83–87.

Fowles, G. R. (1975). Introduction to Modern Optics, Holt, Rinehart and Winston, New York.

Friebele, E. J., and Kersey, A. D. (1994). "Fiber Optic Sensors Measure Up for Smart Structures," *Laser Focus World*, May, 165–171.

Friebele, E. J., Askins, C. G., Putnam, M. A., Fosha, A. A., Jr, Florio, J. Jr., Donti, R. P., and Blosser, R. G. (1994). "Distributed Strain Sensing with Fibre Bragg Grating Arrays Embedded in CTRM^{TM} Composites, *Electronic Lett.* **30**, 1783–1784.

Friebele, E. J., Askins, C. G., Putnam, M. A. Williams, G. M., and Kersey, A. D. (1995). "Fabrication and Application of Low-Cost Optical Fiber Sensor Arrays for Industrial and Commercial Applications," *SPIE* **2447**, 305–311.

Friebele, E. J., Askins, C. G., and Putnam, M. A. (1996). "Demonstration of Distributed Strain Sensing in Production Scale Instrumented Structures," *SPIE* **2721**, 118–124.

Friend, C. (1996). "Even Aircraft Have Feelings," *New Scientist, Feb. 3, 32–35.*

Fuhr, P. L., and Huston, D. R. (1993). "Multiplexed Fiber Optic Pressure and Vibration Sensors for Hydroelectric Dam Monitoring," *Smart Materials & Structures* **2**, 260–263.

Fuhr, P. L., and Huston, D. R. (1998). "Corrosion Detection in Reinforced Concrete Roadways and Bridges via Embedded Fiber Optic Sensors," *Smart Materials & Structures* **7**, 217–228.

Fuhr, P. L., Huston, D. R., Kajenski, P. J., and Ambrose, T. P. (1991). "Performance and Health Monitoring of the Stafford Medical Building Using Embedded Sensors," *Smart Materials and Structures*, **1**, 63–68.
Fuhr, P. L., Huston, and D. R., and Ambrose, T. (1993). "Interrogation of Multiple Embedded Fiber Sensors in Civil Structures Using Radio Telemetry," *Smart Materials & Structures* **2**, 264–269.
Fuhr, P. L., Huston, D. R., Ambrose, T. P., and Baker, D. A. (1994). "Embedded Sensors Results from the Winooski One Hydroelectric Dam," *SPIE* **2191**, 446–449.
Fuhr, P. L., Huston, D. R., Ambrose, T. P., and McPadden, A. (1995a). "Large Deflection Measurements in Structures Using Diffraction Grating," *SPIE* **2446**, 293–300.
Fuhr, P. L., Huston, D. R., Ambrose, T. P., and Mowat, E. F., (1995b). "An Internet Observatory: Remote Monitoring of Instrumented Civil Structures Using the Information Superhighway," *Smart Materials & Structures*, **4**, 14–19.
Furstenau, N., Janzen, D. D., and Schmidt, W. (1993). "In-Flight Strain Measurements on Structurally Integrated Composite Plates Using Fiber-Optic Interferometric Strain Gauges," *Smart Materials & Structures* **2**, 147–156.
Gambling, W. A. (1992). "Optical Fibres, Lasers and Amplifiers," *Endeavour*, New Series, **16**, 17–22.
Gambling, W. A., and Poole, S. B. (1988). "Optical Fibers for Sensors," in *Optical Fiber Sensors* (Dakin, J., and Culshaw, B., Eds.), Artech, House, 249–276.
Gao, P., Alavie, A. T., Mason, B., and Measures, R. M. (1992). "Effects of Delamination on the Natural Frequencies of Composite Beam as Measured by Fiber Optic Fabry–Perot Sensors," Fiber Optic Smart Structures and Skins V, *SPIE* **1798.**
Garcia, G., and Stubbs, N. (1995). "The Effect of Damage Size and Location on the Stiffness of a Rectangular Beam," *SPIE* **2446**, 151–160.
Gartside, III, C. H., Patel, P. P., and Santana, M. R. (1988). "Optical Fiber Cables," in *Optical Fiber Telecommunications II* (Miller, S. E., and Kaminow, I. P., Eds.), 217–263.
Gauthier, R. R., Farahi, F., and Dahi, N. (1994). "Fiber-Optic White-Light Interferometry: Lead Sensitivity Considerations," *Opt. Lett.* **19**, 138–140.
Geiger, H., Xu, M. G., and Dakin, J. P. (1995). "Multiplexed Measurements of Strain Using Short and Long Gauge Length Sensors," *SPIE* **2507**, 25–34.
Ghatak, A. K., and Thyagarajan, K. (1989). Optical Electronics, University of Cambridge, Cambridge.
Glennie, D., Alavie, T., Liu, K., and Measures, R. M. (1993). "The Applicability of Fiber Optic Sensors to the Detection of Surface Acoustic Waves on Metals," *Smart Structures and Materials, SPIE* **1918**, 97–109.
Gloge, D. (1971). "Dispersion in Weakly Guiding Fibers," *Appl. Opt.* **10**, 2442.
Gloge, D. (1975). "Propagation Effects in Optical Fibers," *IEEE Trans.* **MTT-23**, 106–120.
Glossop, N. D., Lymer, J. D., Measures, R. M. and Tennyson, R. C. (1989). "Image Enhanced Backlighting: A New Method of Non-Destructive Evaluation in Translucent Composite Materials," *J. Nondestructive Evaluation* **8**, 181–193.
Glossop, N., Dubois, S., Tsaw, W., LeBlanc, M., Lymer, J., Measures, R. M., and Tennyson, R. C. (1990). "Optical Fiber Damage Detection for an Aircraft Composite Leading Edge," *Composites* **21**, 71–80.

Gowar, J. (1984). *Optical Communication Systems,* Prentice-Hall International, London.
Grattan, K. T. V., and Meggitt, B. T. (1995), *Optical Fiber Sensor Technology,* Chapman and Hall, London.
Greene, J. A., Murphy, K. A., Fogg, B. R., Claus, R. O., and Vengsarkar, A. M. (1992). "Optical Fiber, Vibration-Mode Filters Incorporating Photoinduced Refractive-Index Gratings," *Smart Materials & Structures* **1**, 243–249.
Greene, J. A., Jones, M. E., Tran, T. A., and Murphy, K. A. (1996). "Grating-Based Optical Fiber-Based Corrosion Sensors," *SPIE* **2718**, 170–174.
Greene, J. A., Jones, M. E., Duncan, P. G., Kozikowski, C. L., Bailey, T. A., May, R. G., and Murphy, K. A. (1997). "Grating-Based Optical Fiber Corrosion Sensors," *SPIE* **3042**, 260–266.
Gregory, S., A., and Silva, K. (1992). "Aircraft Smart Structures Research the USAF Wright Laboratory," NATO AGARD Conference, Nov., 27–1, 27–29.
Gudmunson, P. (1982). "Eigenfrequency Changes of Structures due to Cracks, Notches or Other Geometrical Changes," *J. Mech. Phys. Solids* **30**, 339–353.
Hable, W. R., and Hillemeier, B. (1995). "Results in Monitoring and Assessment of Damages in Large Steel and Concrete Structures by Means of Fiber Optic Sensors," *SPIE* **2446**, 25–36.
Habel, W., Hopcke, M., Basedau, F., and Polster, H., (1994), "The Influence of Concrete and Alkaline Solutions on Different Surfaces of Optical Fiber Sensors" *2nd Europ. Conf. on Smart Structures & Materials,* IOP, Briston. 168–171.
Habel, W. R., and Polster, H. (1994). "The Influence of Cementitious Building Materials on Polymeric Surfaces of Embedded Optical Fibers for Sensors," OFS'10, Conference, 1–9.
Habel, W. R., Hofmann, D., Hillemeier, B., and Basedau, F. (1996). "Fiber Sensors for Damage Detection on Large Structures and for Assessment of Deformation Behavior of Cementitious Materials," *Engineerinig Mechanics Proceedings of the 11th Conference* (Y. K. Lin and T. C. Su, Eds.), 355–358.
Hadfield, P. (1995). "Disaster Quake Wins Grim Place in Record Books," *New Scientist,* Feb., 5.
Hale, K. F. (1992). "An Optical-Fibre Fatigue Crack-Detection and Monitoring System" 1st Europe Conf. on Smart Structures and Materials, Glasgow, *SPIE* **1777**, 147–150.
Hale, K. F., Hockenhall, B. S., and Christodoulou, G. (1980). "The Application of Optical Fibres as Witness Devices for the Detection of Elastic Strain and Cracking," National Maritime Inst. Report, NM1R72m) T-R-8006.
Hamilton, G. (1996). "Making Smarter Stuff," *Equinox,* March/April, 57–63.
Hansen, J. (1995). "Introduction to Advanced Composite Materials," in *Fiber Optic Smart Structures* (E. Udd, Ed.), 23–59. J. Wiley & Sons, Inc., New York.
Hartog, A. H., Leach, A. P., and Gold, M. P. (1985). "Distributed Temperature Sensing in Solid-Core Fibres," *Electron. Lett.* **21**, 1061–1062.
Haslach, H. W., and Sirkis, J. S. (1989). "Stress and Strain Separation in Surface Mounted Optical Fiber Strain Sensors," Proc. Fiber Optic Smart Structures and Skins II, *SPIE* **1170**, 452–461.

Haus, H. A. (1984). "Mirrors and Interferometers," in *Waves and Fields in Optoelectronics* (Holonyak, N., Ed.), Chap. 3, 55–80. Prentice-Hall, Englewood Cliffs.
Haus, H. A., and Lai, Y. (1991). "Narrow-Band Distributed Feedback Reflector Design," *J. Lightwave Tech.* **9**, 754–760.
He, Q. B., Yeh, P., and Gu, C. (1992). "Analysis of Photorefractive Fabry–Perot Etalons: A Novel Device," *Optics Lett.* **17**, 664–666.
Hearn, N. Private communication, 1998.
Hecht, J. (1996). "Fibers Are a Printer's Best Friend," *New Scientisit* Oct., 24.
Hecht, J. (1999). "Diode-Laser Performance Rises as Structures Shrink," *Laser Focus World*, May, 127–143.
Hecht, E., Zajac, A,. (1974), *Optics*, Addison-Wesley.
Henderson, P. J., Rao, Y. J., Jackson, D. A., Zhang, L., and Bennion, I. (1997). "Simultaneous Dynamic-Strain and Temperature Monitoring Using a Wavelength Multiplexed Fibre-Fabry–Perot Array With Low-Coherence Interrogation," OSA, 12th Optical Fiber Sensors Conference, October 26–28, Williamsburg.
Hickman, G. A., Gerardi, J. J., and Feng, Y. (1991). "Application of Smart Structure to Aircraft Health Monitoring," *J. Intel. Mat. Syst. and Struct.* **2**, 411–430.
Hill, K. O. (1974). "A Periodic Distributed-Parameter Waveguide for Integrated Optics," *Applied Optics* **13**, 1853.
Hill, K. O., and Meltz, G. (1997). "Fiber Bragg Grating Technology Fundamentals and Overview," *J. Lightwave Technology* **15**, 1263–1276.
Hill, K. O., Fujii, Y., Johnson, D. C., and Kawasaki, B. S. (1978). "Photosensitivity in Optical Fiber Waveguides: Application to Reflection Filter Fabrication," *Appl. Phys. Lett.* **32**, 647–649.
Hill, K. O., Malo, B., Bilodeau, F., Johnson, D. C., and Albert, J. (1993). "Bragg Gratings Fabricated in Monomode Photosensitive Optical Fiber by UV Exposure through a Phase Mask," *Appl. Phys. Lett.* **62**, 1035–1037.
Hjelme, D. R., Bakke, B., Rambech, J. S., and Neegard, S. (1996). "Multiplexed Fiber Optic Bragg-Grating Strain Sensor System for Use in Marine Vehicle Testing," *SPIE* **2838**, 40–51.
Hjelme, D. R., Bjerkan, L., Neegard, S., Rambech, J. S., and Aarsnes, J. V. (1997). "Application of Bragg Grating Sensors in the Characterization of Scaled Marine Vehicle Models," *Applied Optics* **36**, 328–336.
Hofer, B. (1987). "Fibre Optic Damage in Composite Structures," *Composites* **18**, 309–316.
Hogg, D., Turner, R. D., and Measures, R. M. (1990). "Polarimetric Fiber Optic Structural Strain Characterization," *SPIE* **1170**, 542–550.
Hogg, D., Janzen, D., Valis, T., and Measures, R. M. (1991a). "Development of a Fiber Fabry–Perot Strain Gauge," *SPIE* **1588**, Fiber Optic Smart Structures and Skins IV, 300–307.
Hogg, D., Janzen, D., Mason, B., Valis, T., and Measures, R. M. (1991b). "Development of a Fiber Fabry–Perot (FFP) Strain Gauge with High Reflectivity Mirrors," International Symposium on Active Materials & Adaptive Structures, Alexandria, Virginia, 4–8 November.

Holl, M., and Boyd, S. (1993). "The Effect of Embedded Fiber Optics on the Mechanical Properties of a Composite Host Material," Smart Materials, *SPIE* **1916**, 109–117.

Holmes, B. (1995), "Big One Threatens the Big Apple," *New Scientist*, March 4, 11.

Horiguchi, T., Kurashima, T., and Koyamada, Y. K. (1992). "Measurement of Temperature and Strain Distribution by Brillouin Frequency Shift in Silica Optical Fibers," *Proc. SPIE—Int. Soc. Opt. Eng.*, **1797.**

Horiguchi, T., Kurashima, T., Tateda, M., Ishirara, K., and Wakui, Y. (1992b). "Brillouin Characterization of Fiber Strain in Bent Slot-Type Optical-Fiber Cables," *J. Lightwave Tech.* **10**, 1196–1201.

Horiguchi, T., Rogers, A., Michie, W. C., Stewart, G., and Culshaw, B. (1997). "Distributed Sensors: Recent Developments," in *Optical Fiber Sensors: Applications, Analysis, and Future Trends*, Vol. 4, Artech House, Norwood, 309–368.

Hornack, L. A., Ed. (1992). *Polymers for Lightwave and Integrated Optics—Technology and Applications*, Marcel Dekker, Inc.

Housner, G. W., Masri, S. F., and Soong, T. T. (1992). "Recent Development in Active Structural Control Research in the USA," 1st Europe Conf. on Smart Structures and Materials, Glasgow, *SPIE* **1777**, 201–206.

Huang, S. Y., Blake, J. N., and Kim, B. Y. (1990). "Perturbation Effects on Mode Propagation in Highly Elliptical Core Two-Mode Fibers," *J. Lightwave Techn.* **8**, 23–33.

Huang, S., Ohm, M. M., Le Blanc, M., Lee, R., and Measures, R. M. (1994). "Fiber Optic Intra-Grating Distributed Strain Sensing," *SPIE* **2294**, 81–92.

Huang, S., LeBlanc, M., Ohn, M. M., and Measures, R. M. (1995a). "Bragg Intra-Grating Structural Sensing," *Applied Optics* **34**, 5003–5009.

Huang, S., LeBlanc, M., Ohn, M. M., and Measures, R. M., (1995b). "Bragg Intra-Grating Structural Sensing," in *Applications of Photonic Technology* (G. Lampropoulos, J. Chrostowski, and R. M. Measures, Eds.), 317–320. Plenum Press, New York.

Huang, S., Ohn, M. M., LeBlanc, M., Lee, R., and Measures, R. M. (1994). "Fiber Optic Intra-Grating Distributed Strain Sensing," *SPIE* **2294**, 81–92.

Huang, S., LeBlanc, M., Lowery, M., Maaskant, R., and Measures, R. M. (1996). "Distributed Fiber Optic Strain Sensing for Anchorages and Other Applications," in *Advanced Composite Materials in Bridges & Structures* (El-Badry, M. M., Eds.), 991–998. The Canadian Socieity of Civil Engineers, Montreal.

Huang, S., Ohn, M. M., LeBlanc, M., and Measures, R. M. (1998). "Continuous Arbitrary Strain Profile Measurements with Fiber Bragg Gratings," *Smart Mater. and Struct.* **7**, 248–256.

Huston, D. R., Fuhr, P. L., and Beliveau, J.-G. (1991). "Structural Member Vibration Using a Fiber Optic Sensor," *J. Sound & Vibrations* **149**, 348–353.

Idriss, R. L., Kodindouma, M. B., Kersey, A., and Davis, M. (1997). "Multiplexed Bragg Grating Optical Fiber Sensors for Damage Evaluation in Highway Bridges," *Smart Material and Structures J.* **7**, 209–216.

Idriss, R. L., White, K. R., Pater, J. W,. Vohra, S. T., Chang, C. C., Danver, B. A., and Davis, M. A. (1998). "Monitoring and Evaluation of an Interstate Highway Bridge Using a Network of Optical Fiber Sensors," *Fiber Optic Sensors for Construction Materials and Bridges* (Ansari, F., Ed.), 159–167.

Iizuka, K. (1983). *Optical Sciences: Engineering Optics*, Springer–Verlag, Berlin.

Ikegami, T. (1989). "Longitudinal Mode Control in Laser Diodes," in *Optoelectronic Technology and Lightwave Communications Systems* (Lin, C., Ed.), Van Nostrand Reinhold, 264–298.

Imai, Y., Tgan, K., and Inoue, H. (1997). "Transient Index Grating Fabricated in Er-Doped Optical Fibers," OSA, 12th Int. Conf. on Optical Fiber Sensors, Williamsburg, 225–228.

Inaudi, D. (1997a). "Field Testing and Application of Fiber Optic Displacement Sensors in Civil Structures," OSA, 12th Conf. on Optical Fiber Sensors, Williamsburg, 596–599.

Inaudi, D. (1997b). "Fiber Optic Smart Sensing," in *Optical Measurement Techniques and Applications* (Rastogi, P. K., Ed.), Artech House, Boston, 255–274.

Inaudi, D., Elamari, A., Pflug, L., Gisin, N., Breguet, J., and Vurpillot, S. (1994). "Low-Coherence Deformation Sensors for the Monitoring of Civil-Engineering Structures," *Sensor and Actuators* **44**, 125–130.

Inaudi, D., Vulliet, L., Pflug, L., Vurpillot, S., and Wyser, A. (1995). "Low-Coherence Interferometric Sensors for the Monitoring of Underground Works," *SPIE* **2444**, 170–178.

Inaudi, D., Cananova, N., Kronenberg, P., and Vurpillot, S. (1997a). "Railway Bridge Monitoring during Construction and Sliding," *SPIE* **3043**, 58–64.

Inaudi, D., Casanova, N., Kronenberg, P., Marazzi, S., and Vurpillot, S. (1997b). "Embedded and Surface Mounted Fiber Optic Sensors for Civil Structural Monitoring," SPIE Smart Structures and Materials, San Diego, 1–8.

Inaudi, D., Vurpillot, S., and Casanova, N. (1996a). "Bridge Monitoring by Interferometric Deformation Sensors," SPIE International Photonics China Symposium on Laser Optoelectronics and Microphotonics, Fiber Optic Sensors, 4–7 November, Beijing, China, 1–12.

Inaudi, D., Vurpillot, S., Casanova, N., and Osa-Wyser, A. (1996b), "Development and Field Test of Deformation Sensors for Concrete Embedding," *SPIE* **2721**, 138–148.

Inaudi, D., Vurpillot, S., and Loret, S. (1996c). "In-Line Coherence Multiplexing of Displacement Sensors, A Fiber Optic Extensometer," *SPIE* **2718**, 251–257.

Inaudi, D., Vurpillot, Cassanova, N., and Kronenberg, P. (1998). "Structural Monitoring by Curvature Analysis Using Interferometric Fiber Optic Sensors," *Smart Materials and Structures J.* **7**, 199–208.

Inaudi, D., *et al.* (1995b). Low coherence fiber optic sensors for structural monitoring. *Structural Engineering International* 1/95, **5**, 43–47.

Jackson, D. A. (1981). "A Prototype Digital Phase Tracker for Fiber Interferometer," *J. Phys. E: Sci. Instrum.* **14**, 1274–1278.

Jackson, D. A., and Jones, J. D. C., (1986). "Fibre Optic Sensors," *Optica Acta* **33**, 1469–1503.

Jackson, D. A., and Jones, J. D. C. (1989). "Interferometers," in *Optical Fiber Sensors: Systems and Applications*, Vol. 2 (Culshaw, B., and Dakin, J., Eds.), Artech House, 329–380.

Jackson, D. A., Kersey, A. D., Cork, M., and Jones, J. D. C. (1982). "Pseudo-Heterodyne Detection Scheme for Optical Interferometers," *Electron. Lett.* **18**, 1081–1083.

Jackson, D. A., Lobo, R., Reekie, L., and Archambault, J. L. (1993). "Simple Multiplexing Scheme for a Fiber-Optic Grating Sensor Network," *Optic. Lett.* **18**, 1192–1194.

James, S. W., Dockney, M. L., and Tatam, R. P. (1996). "Demodulation of In Fiber Bragg Grating Sensors Using Volume Holograms," *SPIE* **2838**, 52–57.

Jayaraman, V., Heimbuch, M. E., Coldren, L. A., and Denbars, S. P. (1994). "Widely Turnable Continuous-Wave InGaAsP/InP Sampled Gratings Lasers," *Electron. Lett.* **30**, 1492–1494.

Jensen, D. W., and Sirkis, J. S. (1995). "Integrityy of Composite Structures with Embedded Optical Fibers," in *Fiber Optic Smart Structures* (Eric Udd, Ed.), 109–120.

Jensen, D. W., Pascual, J., and August, J. A. (1992a). "Performance of Graphite/Bismaleimide Laminates with Embedded Optical Fibers Part II: Uniaxial Tension," *Smart Materials and Structures* **1**, 31–35.

Jensen, D. W., Pascual, J., and August, J. A. (1992b). "Performance of Graphite/Bismaleimide Laminates with Embedded Optical Fibers Part I: Uniaxial Tension," *Smart Mat. & Struct.* **1**, 24–30.

Jeunhomme, L. B. (1983). *Single-Mode Fiber Optics Principles and Applications*, Marcel Dekker, New York.

Jewell, J. L., and Olbright, G. R. (1992). "Surface-Emitting Lasers Emerge from the Laboratory," *Laser Focus World*, May, 217–223.

Jewell, J., Askins, C., and Aggarwal, I. D. (1991). "Interferometric Method for Concurrent Measurement of Thermo-Optic and Thermal Expansion Coefficients," *Appl. Optics* **30**, 3656–3660.

Jewell, J., Olbright, G. R., Bryan, R. P., and Scherer, A. (1992). "Surface-Emitting Lasers Break the Resistance Barrier," *Photonics Spectra*, Nov., 126–130.

Jin, X. D., and Sirkis, J. S. (1997). "Optical Fiber Sensors for Simultaneous Measurement of Strain and Temperature," *SPIE* **3042**, 120–127.

Jones, J. D. C. (1997). "Review of Fibre Sensor Techniques for Temperature–Strain Discrimination," OSA, 12th Int. Conf. on Optical Fiber Sensors, Williamsburg, 20–23.

Jones, J. D. C., and Barton, J. S. (1997). "Fiber-Optic Sensors for Condition Monitoring and Engineering, Diagnostics", in *Optical Fiber Sensors*, Vol. 4 (Dakin, J., and Cushaw, B., Eds.), Artech House, 207–259.

Jones, M. E., Greene, A., Bhatia, V., Murphy, K. A., Claus, R. O., Miller, A. E., and Vengsarkar, A. M. (1995). "In-Line Fiber Demodulator for Interrogation of Bragg Grating Sensors," *SPIE* **2444**, 248–256.

Jones, R. T., Bellemore, D. G., Berkoff, T. A., Sirkis, J. S., Davis, M. A., Putnam, M. A., Friebele, E. J., and Kersey, A. D. (1998). "Determination of Cantilever Plate Shapes Using Wavelength Division Multiplexed Fiber Bragg Grating Sensors and a Least-Squares Strain-Fitting Algorithm," *Smart Materials and Structures* **7**, 178–188.

Jones, R. T., Berkoff, T. A., Bellemore, G., Early, D. A., Sirkis, J. S., Putnum, M. A. Friebele, E. J., and Kersey, A. D. (1996). "Cantilever Plate Deformation Monitoring Using Wavelength Division Multiplexed Fiber Bragg Grating Sensors," *SPIE* **2718**, 258–267.

Judge, R. C. B., Scott, R. H., and Gill, P. A. T. (1990). "Strain and Bond Stress Distributions in Tension Lap Joints in Reinforced Concrete," *Mag. Concrete Res.* **42**, 5–14.
Kalamkarov, A. L., MacDonald, D. O., and Westhaver, P. A. D. (1997). "Pultrusion of Smart FRP Composites," *SPIE* **3042**, 400–409.
Kalli, K., Brady, G., Webb, J. D., Reekie, L., Archambault, J. L., and Jackson, D. A. (1994). "Possible Approach for the Simultaneous Measurement of Temperature and Strain Via First and Second Order Diffraction from Bragg Grating Sensors," 10th Optical Fibre Sensors Conference, Glasglow, Scotland, Oct.
Kannellopoulos, S. E., Handerek, V. A., and Rogers, A. J. (1995). "Simultaneous Strain and Temperature Sensing with Photogenerated In-Fiber Gratings," *Opt. Soci. Am.* **20**, 333–335.
Kano, F., Ishii, H., Tohmori, Y., Yamamoto, M., and Yoshikuni, Y. (1993). "Broad Range Wavelengths Switching in Superstructure Grating Distributed Bragg Reflector Lasers," *Electronics Lett.* **29**, 1091–1092.
Kao, K. C., and Hockham, G. A. (1966). "Dielectric-Fiber Surface Waveguides for Optical Frequencies," *Proc. Inst. Electr. Eng.* **113**, 1151–1158.
Kapron, F. P., and Yuce, H. H. (1991). "Theory Measurement for Predicting Stressed Fiber Lifetime," *Optical Eng.* **30**, 700–708.
Kapteyn, K. L., and Miskioglu, I. (1995). "Fully Distributed Fiber Optic Sensing through the Use of the Kerr Nonlinear Optical Effect," *SPIE* **2444**, 94–105.
Kashyap, R. (1994). "Photosensitive Optical Fibers: Devices and Applications," *Optical Fiber Technol.* **1**, 17–24.
Kaufmann, K. (1994). "Detectors Cover the Spectrum of Instrument Applications," *Laser Focus Worlds*, 99–106.
Kersey, A. D. (1995). "Fiber Optic Sensor Multiplexing Techniques," in *Fiber Optic Smart Structures* (Eric Udd, Ed.), 409–444.
Kersey, A. D. (1997a). "Multiplexing Techniques for Fiber-Optic Sensors," in *Optical Fiber Sensors*, Vol. 4 (Dakin, J., and Cushaw, B., Eds.), Artech House, 369–407.
Kersey, A. D. (1997b). "Optical Fiber Sensors," in *Optical Measurement Techniques and Applications* (Rastogi, P. K., Ed.), Artech House, Boston, 217–254.
Kersey, A. D., and Berkoff, T. A. (1993). "Fiber-Grating Based Strain Sensor with Drift-Compensated High-Resolution Interferometric Wavelength-Shift Detection," *Optics Lett.* **18**, 72–74.
Kersey, A. D., and Dandridge, A. (1988). "Distributed and Multiplexed Fiber Optic Sensors," OFS'88, New Orleans, 60–71.
Kersey, A. D., and Morey, W. W. (1993a). "Multiplexed Bragg Grating Fiber-Laser Strain-Sensor System with Mode-Locked Interrogation," *Electronic Lett.* **29**, 112–114.
Kersey, A. D., and Morey, W. W. (1993b). "Multi-Element Bragg-Grating Based Fibre-Laser Strain Sensor," *Electronic Lett.* **29**, 964–966.
Kersey, A. D., Lewin, A. C., and Jackson, D. A. (1984a). "Pseudo-Heterodyne Detection Scheme for the Gyroscope," *Electronic Lett.* **20**, 368–370.
Kersey, A. D., Corke, M., and Jackson, D. A. (1984b). "Linearized Remote Sensing Using a Momode Fibre Polarimetric Sensor," *SPIE* **514**, Optical Fibre Sensors '84, 247–250.

Kersey, A. D., Berkoff, T. A., and Morey, W. W. (1992a). "Fiber-Grating Based Strain Sensor with Phase Sensitive Detection," 1st Europe Conf. on Smart Structures and Materials, Glasgow, *SPIE* **1777**, 61–67.

Kersey, A. D., Berkoff, T. A., and Morey, W. W. (1992b). "High-Resolution Fiber-Grating Based Strain Sensor with Interferometric Wavelength-Shift Detection," *Electronic Lett.* **28**, 236–238.

Kersey, A. D., Berkoff, T. A., and Morey, W. W. (1992c). "Two-Channel Fiber Bragg-Grating Strain Sensor with High-Resolution Interferometric Wavelength-Shift Detection," Fiber Optic Smart Structures & Skins V, Boston, *SPIE* **1798**, 48–55.

Kersey, A. D., Friebele, E. J., and Weis, R. S. (1992d). "Erbium-Doped Fiber Ring Laser Strain Sensor," Fiber Optic Smart Structures & Skins V, Boston, *SPIE* **1798**, 280–285.

Kersey, A. D., Berkoff, T. A., and Morey, W. W. (1993). "Multiplexed Fiber Bragg Grating Strain Sensor-System with a Fiber Fabry–Perot Wavelength Filter," *Optics Lett.* **18**, 1370–1372.

Kersey, A. D., Koo, K. P., and Davis, M. A. (1994). "Fiber Optic Bragg Grating Laser Sensors," *SPIE* **2292**, Fiber Optic and Laser Sensors XII, San Diego, 102–112.

Kersey, A. D., Davis, M. A., Berkoff, T. A., Dandridge, A., Jones, R. T., Tsai, T., Cogdell, G., Wang, G., Havsgaard, G. B., Pran, K., and Knudsen, S. (1997a). "Transient Load Monitoring on a Composite Hull Ship Using Distributed Fiber Optic Bragg Grating Sensors," *SPIE* **3042**, 421–430.

Kersey, A. D., Davis, M. A., Patrick, H. J., LeBlanc, M., Koo, K. P., Askins, C. G., Putnam, M. A., and Friebele, E. J. (1997b). "Fiber Grating Sensors," *J. Lightwave Technol.* **15**, 1442–1463.

Kikuya, Y., Hirano, M., Koyabu, K., and Ohira, F. (1993). "Alignment of Optical Axes by Using Electrostatic Force," *Optics Lett.* **18**, 864–865.

Kim, B. Y., Blake, J. N., Huang, S. Y., and Shaw, H. J. (1987). "Use of Highly Elliptical Core Fibers for Two-Mode Fiber Devices," *Opt. Lett.* **12**, 729–731.

Kim-J.T. (1995), "Robust Damage Localization Algorithm for Highway Plate-Girder Bridges," *Proc. SPIE, Smart Struct. & Mat.*, Smart Systems for Bridges, Structures and Highways.

Kim, K. S., Ismail, Y., and Springer, G. S. (1993). "Measurements of Strain and Temperature with Embedded Intrinsic Fabry–Perot Optical Fiber Sensors," *J. Composite Materials.*

Kirby, G. C., Lindner, D. K., Davis, M. A., and Kersey, A. D. (1995). "Optimal Sensor Layout for Shape Estimation from Strain Sensors," *SPIE* **2444**, 367–376.

Kodindouma, M. B., and Idriss, R. L. (1996). "An Integrated Sensing System for Highway Bridge Monitoring," Smart Structures & Materials, Smart Systems for Bridges, Structures and Highways, 28–29 February 1996, San Diego.

Kodindouma, M. B., Idriss, R. L., Kersey, A. D., Davis, M. A., Bellemore, D. G., Friebele, E. J., and Putnam, M. A. (1996). "Damage Assessment of a Full Scale Bridge Using an Optical Fiber Monitoring System," *SPIE* **2719**, 265–275.

Kogelnik, H. (1976). "Filter Response of Non-Uniform Almost Periodic Structures," *Bell System Tech. J.* **55**, 109–126.

Koo, K. P., LeBlanc, M., Chang, C. C., and Kersey, A. D. (1997). "A Wavelength Encoded Fiber Fabry–Perot Sensor With Multiple Addressable Free Spectral Ranges," OSA, 12th Int. Conf. on Optical Fiber Sensors, Williamsburg, 504–507.
Koob, C. E., Murphy, K. A., Plante, A. J., Vengsarklar, A. M., and Claus, R. O. (1991). "High Temperature Fabry–Perot Based Strain Sensor," in *Optical Fiber Sensor-Based Smart Materials and Structures*, Technomic Publ., 188–193.
Koteles, E. S. (1995). "Techniques for Monolithically Fabricating Photonic Integrated Circuits," in *Applications of Photonic Technology* (G. A. Lampropoulos, J. Chrostowski, and R. M. Measures, Eds.), 413–418. Plenum Press, New York.
Kringlebotn, J. T., Archambault, J.-L., Reekie, L., Townsend, J. E., Vienne, G. G., and Payne, D. N. (1994). "Highly-Efficient, Low-Noise Grating-Feedback Er^{3+}:Yb^{3+} Copdoped Fibre Laser," *Electron. Lett.* **30**, 972–973.
Krishnamoorthy, R. K., Belarbi, A., Chandrashekhara, K., and Watkins, S. E. (1997). "Hybrid Composite Rebars for Smart Concrete Structures," *SPIE* **3043**, 65–71.
Krohn, D. A. (1988). *Fiber Optic Sensors Fundamentals and Applications*, Instrument Society of America.
Kronenberg, P., Casanova, N., Inaudi, D., and Vurpillot, S. (1997). "Dam Monitoring with Fiber Optic Deformation Sensors," *SPIE* **3043**, 2–11.
Kruschwitz, B., Claus, R. O., Murphy, A., May, R. G., and Gunther, M. F. (1992). "Optical Fiber Sensors for the Quantitative Measurement of Strain in Concrete Structures," 1st Europ. Conf. Smart Structures & Materials, Glasgow, *SPIE* **1777**, 241–244.
Kudva, J. H., Munir, N., and Tan, P. W. (1992). "Damage Detection in Smart Structures Using Neural Networks and Finite-Element Analyses," *Smart Mater. & Struct.* **1**, 108–112.
Kurashima, T. (1990). "Distributed Temperature Sensing Using Stimulated Brillouin Scattering Optical Silica Fibers," *Optics Lett.* **15**, 1038.
Kurashima, T., Horigucjho, T., Yoshizawa, N., Tada, H., and Tateda, M. (1991). "Measurement of Distributed Strain due to Laying and Recovery of Submarine Optical Fiber Cable," *Appl. Optics* **30**, 334–337.
Kurashima, T., Tateda, M., Horiguchi, T., and Koyamada (1997). "Performance Improvement of a Combined OTDR for Distributed Strain and Loss Measurement by Randomizing the Reference Light Polarization State," *IEEE Photonics Technology Lett.* **9**, 360–362.
Kurkjian, C. R., and Inniss, D. (1991). "Understanding Mechanical Properties of Lightguides: A Commentary," *Optical Engineering* **30**, 681–689.
Kurkjian, C., and Yuce, H. (1991). "Optical Fiber Reliability," *Opt. Eng.* **30**, 679–680.
Kurkijian, C. R., Krause, J. T., and Matthewson, M. J. (1989). "Strength and Fatigue of Silica Optical Fibers," *J. Lightwave Technol.* **7**, 1360–1370.
Lam, D. D. W., and Garside, B. (1981). "Characterization of Single Mode Optical Fibers," *Appl. Opt.* **20**, 440–445.
Lambelet, P., Fonjallaz, P. Y., Limberger, H. G., Salathé, R. P. Zimmer, C., and Gilgen, H. H. (1993). "Bragg Grating Characterization by Optical Low-Coherence Reflectrometry," *IEEE Photonics Tech. Lett.* **5**, 565–567.
Laming, R. I., and Payne, D. N. (1989). "Electric Current Sensors Employing Spun Highly Birefringent Optical Fibers," *J. Lightwave Tech.* **7**, 2084–2094.

Lawrence, C. M., Nelson, D. V., Spingarm, J. R., and Bennett, T. E. (1996). "Measurement of Process-Induced Strains in Composite Materials using Embedded Fiber Optic Sensors," *SPIE* **2718**, 60–68.

Lawrence, C., Nelson, C., and Udd. E. (1997a). "Development of a Three Axis Strain and Temperature Fiber Optic Grating Sensor," *SPIE* **3042**, 229–236.

Lawrence, C. M., Nelson, D. V., and Udd. E. (1997b). "Measurement of Transverse Strains with Fiber Bragg Gratings," *SPIE* **3042**, 218–228.

Lawrence, C. M., Nelson, D. V., Bennett, T. E., and Spingarn, J. R. (1997c). "Determination of Process-Induced Residual Stress in Composite Materials using Embedded Fiber Optic Sensors," *SPIE*, **3042**, 154–167.

LeBlanc, M. (1998). "Interaction Mechanics of Embedded Single-Ended Fibre Sensors Using Novel In-Situ Measurement Techniques," Ph.D. Thesis, University of Toronto.

LeBlanc, M., and Kersey, A. D. (1997). "Distributued, Intra-Grating Sensing By Fabry–Perot Wavelength Tuned Low-Coherence Interferometry," 12th Optical Fiber Sensors Conference, October 26–28, Williamsburg.

LeBlanc, M., and Measures, R. M. (1992). "Impact Damage Assessment in Composite Materials with Embedded Fiber Optic Sensors," *J. Composite Engineering* **2**, 573–596.

LeBlanc, M., and Measures, R. M. (1993). "Micromechanical Considerations for Embedded Single-Ended Sensors," Smart Structures and Materials Conference, Albuquerque, *SPIE* **1918**, 215.

LeBlanc, M., and Measures, R. M. (1995). "Fiber Optic Damage Assessment," in *Fiber Optic Smart Structures*, (Eric Udd, Ed.), 581–613.

LeBlanc, M., Huang, S., Ohn, M., and Measures, R. M. (1994). "Tunable Chirping of a Fiber Bragg Grating Using a Tapered Cantilever Beam," *Electron. Lett.* **30**, 2163–2165.

LeBlanc, M., Huang, S., and Measures, R. M. (1995). "Fiber Optic Bragg Intra-Grating Strain Gradient Sensing," *SPIE* **2444**, 136–147.

LeBlanc, M., Huang, S., Ohn, M., Measures, R. M., Guemes, A., and Orthonos, A. (1996). "Distributed Strain Measurement Based on a Fiber Grating and its Reflection Spectrum Analysis," *Opt. Lett.* **21**, 1405–1407.

Lecot, C., Lequime, M., Vanotti, P., Lang, D., Windisch, A., Davidson, R., Turpin, M., Chazelas, J., Culshaw, B., Michie, C., and Papadopoulos, P. (1993). "An Introduction to the BRITE-EURAM II Osmos Project," *SPIE* **2075B-43**, Boston, 7–10 Sept.

Lee, C. (1998). M.A.Sc. Thesis, Civil Engineering, University of Toronto, Canada.

Lee, C. E., and Taylor, H. F. (1988). "Interferometer Fiber Optic Sensors using Internal Mirrors," *Electron Lett.* **24**, 193–194.

Lee, C. E., and Taylor, H. F., (1990). "Interferometer Fiber Optic Temperature Sensor using a Low Coherence Light Source," *SPIE* **1370**, Fiber Optic Smart Structures and Skins III, 356–364.

Lee, C. E., and Taylor, H. F. (1992). "In-Line Fiber Fabry–Perot Interferometer with High Reflectance Internal Mirrors," Proc. 8th Int. Conf. Optical Fiber Sensors, OPS'92, 105.

Lee, C. E., and Taylor, H. F. (1995). "Sensors for Smart Structures Based on the Fabry–Perot Interferometer," in *Fiber Optic Smart Structures* (Eric Udd, Ed.), 249–270.

Lee, C. E., Alcoz, J. J., Gibler, W., Atkins, R. A., and Taylor, H. F. (1991). "Method for Embedding Optical Fibers and Optical Fiber Sensors in Metal parts and Structures," *SPIE* **1588**, Fiber Optic Smart Structures and Skins IV, 110–116.

Lee, D. L., (1986), *Electromagnetic Principles of Integrated Optics*, J. Wiley & Sons.

Lee, W., Henderson, C., Taylor, H. F., James, R., Lee, C. E., Swenson, V., Gibler, W. N., Atkins, R. A., and Gemeiner, W. G. (1997), "Railroad Bridge Instrumentation with Fiber Optic Sensors," OSA, 12th Int. Conf. on Optical Fiber Sensors, Williamsburg, 412–415.

Lemaire, P. J., Atkins, R. M., Vizrahi, V., and Reed, W. A. (1993). "High Pressure H_2 Loading as a Technique for Achieving Ultrahigh UV Photosensitivity and Thermal Sensitivity in GeO_2 Doped Optical Fibres," *Electron. Lett.* **29**, 1191–1193.

Lerner, E. J. (1996a). "Charge-Coupled Devices Capture Image Information," *Laser Focus World*, Aug., 103–112.

Lerner, E. J. (1996b). "Avalanche Photodiodes Can Count the Photons," *Laser Focus World*, Oct., 93–102.

Lerner, E. J. (1996c). "Detectors Span Spectrum of Sensing Applications," *Laser Focus World*, May, 93–100.

Lerner, E. J. (1996d). "Infrared Detectors Offer High Sensitivity," *Laser Focus World*, June, 155–164.

Lerner, E. J. (1998). "Diode Lasers Light Up Disks, Communications, and Printers," *Laser World*, 123–132.

Lesko, J., Carman, G., Reifsnider, K., Vengsarkar, A., Miller, B., Fogg, B., and Claus, R. (1991). "Application of Fabry–Perot Fiber Optic Sensors in Composite Macro Models," in *Optical Fiber Sensor-Based Smart Materials and Structures*, Technomic Publ., 65–69.

Lesko, J. J., Carman, G. P., Fogg, B. R., Miller III, W. V., Vengsarkar, A. M., Reifsnider, K. L., and Claus, R. O. (1992). "Embedded Fabry–Perot Fiber Optic Strain Sensors in the Macromodel Composites," *Opt. Eng.* **31**, 13–20.

Leung, C. K. Y., Elvin, N., and Olson, N. (1997). "Optical Fiber Crack Sensor for Concrete Structures," *SPIE* **3042**, 283–292.

Levi, L. (1980). *Applied Optics*, Vol. 2, John Wiley and Sons, New York.

Lewotsky, K. (1996). "Piloting Clever Cars down Smart Highways," *Laser World*, June, 96.

Lewotsky, K. (1999). "Self Calibrated Fiber Optic Sensor Measures Strain," *Laser Focus Word*, December 33–35.

Li, G. P., Makino, T., Moore, R., and Puetz, N., (1992). "1.55 μm Index/Gain Coupled DFB Laser with Strained-Layer Multi-Quantum Well Active Grating," *Electronics Lett.* **28**, 1726–1727.

Li, M.-J., and Najaffi, S. I. (1993). "Polarization Dependence of Grating-Assisted Waveguide Bragg Reflectors," *Appl. Optics* **32**, 4517–4521.

Lin, C. (1989). *Optoelectronic Technology and Lightwave Communications Systems*, Van Nostrand Reinhold, New York.

Lisboat, O., and Kent, C. K. (1994). "An Optical-Fiber Bending Sensor Using Two-Mode Fibers with an Off-Center Core," *Smart Mater. Struct.* **3**, 164–170.

Liu, K., and Measures, R. M. (1992). "Signal Processing Techniques for Localized Interferometric Fiber-Optic Strain Sensors," *J. Intell. Mat. Syst. Struct.* **3**, 432–461.

Liu, K., and Measures, R. M. (1993). "Analysis of Macro-Strain Transfer and Complete Strain State Measurement with Embedded Fiber-Optic Sensors," *Smart Mater. Struct.* **2**, 66–70.

Liu, K., Ferguson, S. M., and Measures, R. M. (1989)." Damage Detection in Composites with Embedded Fiber Optic Interferometric Sensors, "*SPIE* **1170**, Fiber Optic Smart Structures and Skins II, 205–210.

Liu, K., Ferguson, S., McEwen, K., Tapanes, E., and Measures, R. M. (1990)." Acoustic Emission Detection for Composite Using Embedded Ordinary Single-Mode Fiber Optic Interferometric Sensors," *SPIE* **1370**: Fiber Optic Smart Structures and Skins III, 316–323.

Liu, T., Brooks, D., Martin, A., Badcock, R., and Fernando, G. F. (1996). "Design, Fabrication and Evaluation of an Optical Fiber Sensor for Tensile & Compressive Strain Measurements via the Use of White Light Interferometry," *SPIE* **2718**, 408–416.

Liu, T., Fernando, G. F., Rao, Y. J., Zhang, L., Bennion, I., and Jackson, D. A. (1997a). "Simultaneous Strain and Temperature Measurements in Composites Using a Multiplexed Fiber Bragg Grating Sensor and an Extrinsic Fabry–Perot Sensor," *SPIE* **3042**, 203–212.

Liu, T., Fernando, G. F., Zhang, L, Bennion, I., Rao, Y. J., and Jackson, D. A. (1997). "Simultaneous Strain and Temperature Measurements Using a Combined Fibre Bragg Grating/Extrinsic Fabry–Perot Sensor," OSA, 12th Int. Conf. on Optical Fiber Sensors, Williamsburg, 20–23.

Liu, T., Wu, M., Fernando, G. F., Rao, Y., and Jackson, D. A. (1997c). "A Multiplexed Extrinsic Fabry–Perot Sensor System," *Smart Mater. Struct.* **6**, 464–469.

Liu, T., Wu, M., Fernando, G. F., Rao, Y., and Jackson, D. A. (1997d). "A Multiplexed Extrinsic Fabry–Perot Sensor System for In-Situ Strain and Impact Damage Detection in Composites," *SPIE* **3042**, 293–303.

Lo, Y. L., and Sirkis, J. S. (1997). "Simple Method to Measure Temperature and Axial Strain Simultaneously Using One In-Fiber Bragg-Grating Sensor," *Smart Struct. Mat. 1997, SPIE* **3042**, 237–243.

Loken, D. (1990). "Effect of Fiber Straightness on Fatigue of Aligned Continuous CFRP Composites," M.A.Sc. Thesis, University of Toronto (Chemical Eng.).

Lou, K. A., Yaniv, G., Hardtmann, D., Ma, G., and Zimmermann, B. (1995). "Fiber Optic Strain Monitoring of Bridge Column Retrofitted with Composite Jacket under Flexural Loads," *SPIE* **2446**, 16–24.

Lovell, P. A., and Pines D. J. (1997). "A Remote Wireless Damage Detection System for Monitoring the Health of Large Civil Structures," *SPIE* **3043**, 12–22.

Lu, Z. J., and Blaha, F. A. (1989). "A Fiber Optic Strain and Impact Sensor System for Composite Materials," *SPIE* **1170**, Fiber Optic Smart Structures and Skins II, 239–242.

Lu, Z. J., and Blaha, F. A. (1991). "Application Issues of Fiber Optics in Aircraft Structures," *SPIE* **1588**, Fiber Optic Smart Structures and Skins IV, 276–281.

Ma, J., Tang, W., and Zhou, W. (1997). "Second-Order Sensitivity Effects on Optical Fiber Sensor for Simultaneous Measurement of Strain and Temperature," *SPIE* **3042**, 213–217.

Maaskant, R., Alavie, T., Measures, R. M., Ohn, M., Karr, S., Glennie, D., Wade, C., Tadros, G., and Rizkalla, S. (1994). "Fiber Optic Bragg Grating Sensor Network Installed in a Concrete Road Bridge," *SPIE* **2191**, 457–465.

Maaskant, R., Alavie, T., Measures, R. M., Tadross, G., Rizkalla, S. H., and Guha-Thakurta, A. (1997). "Fiber-Optic Bragg Grating Sensors for Bridge Monitoring," *Cement and Concrete Composites* **19**, 21–33.

Maaskant, R., Alavie, A. T., and Measures, R. M. (1998). "A Recent Experience in Bridge Strain Monitoring with Fiber Grating Sensors," in *Fiber Optic Sensors for Constructional Materials and Bridges* (Farhad Ansari, Ed.), 129–135.

Magne, S., Rougeault, S., Vilela, M., and Ferdinand, P. (1997). "State-of-Strain Evaluation with Fiber Bragg Grating Rosettes: Application to Discrimination between Strain and Temperature Effects in Fiber Sensors," *Applied Optics* **36**, 9437–9447.

Maher, M. H. and Nawy, E. G., (1993). "Evaluation of Fiber Optic Bragg Grating Strain Sensor in High Strength Concrete Beams," in *Applications of Fiberoptic Sensors in Engineering Mechanics* (Farhad Ansari, Ed.), 120–133.

Malo, B., Johnson, D. C., Bilodeau, F., Albert, J., and Hill, K. O. (1993). "Single-Excimer-Pulse Writing of Fiber Grating by Use of a Zero-Order Nulled Phase Mask: Grating Spectral Response and Visualization of Index Perturbations," *Opt. Lett.* **1**, 1277–1279.

Malo, B., Theriault, S., Johnson, D. C., Bilodeau, F., Albert, J., and Hill, K. O. (1995). "Apodised In-Fiber Bragg Grating Reflectors Photoimprinted Using a Phase Mask With Variable Diffraction Efficiency," *Electron. Lett.* **31**, 222–223.

Marantidis, C., Gentry, J., and Kudva, J. N. (1993). "Sensor and Sensing Technologies for Structural Health Monitoring of Aircraft," *SPIE* **1917**, Smart Materials and Structures, Albuquerque, Feb. 1–4.

Marcuse, D. (1982). "Influence of Curvature on the Losses of Doubly Clad Fibers," *Appl. Optics* **21**, 4208–4213.

Marcuse, D. (1989). "Fiber-Coupled Short Fabry–Perot Resonators," *J. Lightwave Tech.* **7**, 869–876.

Martin, J., and Ouellette, F. (1994). "Novel Writing Technique of Long and Highly Reflective In-Fiber Gratings," *Electron. Lett.* **30**, 811–812.

Martin, A. R., Hayes, S. A., Fernando, G. F., and Hale, K. F. (1995). "Impact Damage Detection in Filament Wound Tubes Utilizing Embedded Optical Fibers," *SPIE* **2444**, 409–501.

Martin, A., Badcock, R., Nightingale, C., and Fernando, G. F. (1997). "A Novel Optical Fiber-Based Strain Sensor," *Photonics Technology Lett.* **9**, 982–984.

Mascher, P., Boudreayu, M. G., Wallace, S. G., Murugkar, S., Balcaitis, G., Wettlaufer, Ch., and Kaugen, H. K. (1998). "Optical Coatings for Improved Semiconductor Diode Laser Performance," Electrochem. Soc. Symp. On Light Emitting Devices for Optoelec. Appl., San Diego.

Mason, B., Hogg, D., and Measures, R. M. (1992). "Fiber Optic Strain Sensing for Smart Adaptive Structures," 1st European Conference on Smart Structures and Materials, Glasgow, 12–14 May, *SPIE* **1777**, 135–138.

Mason, B., Valis, T., and Hogg, D. (1993). "Parallel Multiplexing and Demodulation of Fiber-Optics Strain Gauges," *Smart Structures & Materials* '93, Alburquerque, *SPIE* **1918**, 174–186.

Mathews, C. T., and Sirkis, J. S. (1990). "The Interaction of Interferometric Optical Fiber Sensors Embedded in a Monolithic Structure" *SPIE* **1370**, Fiber Optic Smart Structures and Skins III, San Jose, 142–153.

Matsuhara, M., Hill, K. O., and Watanabe, A. (1975). "Optical Waveguide Filters: Synthesis," *J. Opt. Soc. Am.* **65**, 804.

Mazur, C. J., Sendeckji, G. P., and Stevens, D. M. (1988). "Air Force Smart Structures/Skins Program Overview," *SPIE* **986**, Fiber Optic Smart. Struct. & Skins, 19–29.

McGowan, J. (1993). "America Reaps the Wind Harvest," *New Scientist*, Aug. 21, 30–33.

Measures, R. M. (1989). "Smart Structures with Nerves of Glass," *Prog. Aerospace Sci.* **26**, 289–351.

Measures, R. M., (1990), Invited paper, "Fiber Optics in Composite Materials—Materials with Nerves of Glass," Int. Congress of Optical Sciences and Engineering, The Hague, *SPIE* **1267**, 241–256.

Measures, R. M. (1991). "Fiber Optic Sensor Considerations and Developments for Smart Structures," *SPIE* **1588**, Fiber Optic Smart Structures and Skins IV, 76–85.

Measures, R. M. (1992a). "Smart Composite Structures with Embedded Sensors," *J. Composites Eng.* **2**, 597–618.

Measures, R. M. (1992b). "Smart Structures, A Revolution in Civil Engineering," Keynote Address, Advanced Composite Materials in Bridges and Structures, K. W. Neale and P. Labossière, Eds., The Canadian Society for Civil Engineers, Montreal, 31–59.

Measures, R. M. (1993). "Fiber Optic Sensing for Composite Smart Structures," *Composite Eng.* **3**, 715–750.

Measures, R. M. (1995). "Fiber Optic Strain Sensing," in *Fiber Optic Smart Structures* (Eric Udd, Ed.), 171–247.

Measures, R. M., Hogg, D., Turner, R. D., Vallis, T., and Giliberto, M. J. (1988). "Structurally Integrated Fiber Strain Rosette," *SPIE* **986**, Fiber Optic Structures & Skins, 32–42.

Measures, R. M., Glossop, N. D., Lymer, J., LeBlanc, M., West, J., Dubois, S., Tsaw, W., and Tennyson, R. (1989). "Structurally Integrated Fiber Optic Damage Assessment Systems for Composite Materials," *Appl. Optics*, **28**, 2626–2633.

Measures, R. M., Liu, K. Davis, A., and Ohn, M. (1991). "Composite Cure Monitoring with Embedded Optical Fiber Sensors," *SPIE* **1489**, Structures Sensing and Control, Orlando, April 1–5.

Measures, R. M., Alavie, A. T., Liu, K., and Melle, S. M. (1992a). "Optoelectronic Smart Structures Interface: A Key Development for Practical Smart Structures," Fiber Optic Smart Structures & Skins V, Boston, *SPIE* **1798** 154–166.

Measures, R. M., Melle, S. M., and Liu, K. (1992b). "Wavelength Demodulated Bragg Grating Fiber Optic Sensing Systems for Addressing Smart Structure Critical Issues," *J. Smart Materials and Structures* **1**, 36–44.

Measures, R. M., Alavie, T., Karr, S., and Coroy, T. (1993), "Smart Structure Interface Issues and Their Resolution: Bragg Grating Laser Sensor and the 'Optical Synapse'," *SPIE* **1918**, Smart Structures and Materials Conference, Albuquerque, February 1–4.

Measures, R. M., Huang, S., LeBlanc, M., Ohn, M., and Alavie, A. T. (1994). "Bragg Intra-Grating Sensing, Implications for Smart Structures," *SPIE* **2191**, 436–445.

Measures, R. M., Alavie, A. T., Maaskant, R., Ohn, M., Karr, S., and Huang, S. (1995). "A Structurally Integrated Bragg Grating Laser Sensing System for a Carbon Fiber Prestressed Concrete Highway Bridge," *Smart Materials & Structures* **4**, 20–30.

Measures, R. M., Huang, S., LeBlanc, M., Lowery, M., Ohn, M., and Maaskant, R. (1996). "Distributed Fiber Optic Strain Sensing Based on Spectral Integration of Bragg Grating Reflection," *SPIE* **2838**, 31–39.

Measures, R. M., Ohn, M. M., Huang, S. Y., Bigure, J., and Fan, N. Y. (1998). "Tunable Laser Demodulation of Various Fiber Bragg Grating Sensing Modalities," *Smart Materials and Structures* **7**, 237–247.

Meier, U., Deuring, M. Meier, H., and Schwegler, G. (1992). "Strengthening of Structures with CFRP Laminates: Research and Applications in Switzerland," in *Advanced Composite Materials in Bridges and Structures* (Neale, K. W., and Labossiere, P., Eds.), Canadian Society for Civil Engineering, 243–252.

Meier, U., Deuring, M., Meier, H., and Schwegler, G. (1993). "Strengthening of Structures with Advanced Composites," reprint from *Alternative Materials for the Reinforcement and Prestressing of Concrete* (J. L. Clarke, Ed.), Chapman & Hall, Glasgow.

Meissner, J., Nowak, W., Slowik, V., and Klin, T. (1997). "Strain Monitoring at a Prestressed Concrete Bridge", OSA, 12th Int. Conf. on Optical Fiber Sensors, Williamsburg, 408–411.

Melle, S. M. (1994). "Today's Sensors for Tomorrow's Structures," *Photonics Spectra*, 88–94.

Melle, S. M., Liu, K., and Measures, R. M. (1991). "Strain Sensing Using a Fiber Optic Bragg Grating," *SPIE* **1588**, Fiber Optic Smart Structures and Skins IV, Boston, 155–263.

Melle, S. M., Liu, K., and Measures, R. M., (1992a). "A Passive Wavelength Demodulation System for Guided-Wave Bragg Grating Sensors," *IEEE Phot. Tech. Lett.* **4**, 516–518.

Melle, S. M., Coroy, T. G., Karr, S., Alavie, A. T., Measures, R. M. (1992b). "Structural Sensing Using a Fiber Laser Strain Sensor," Fiber Optic Smart Structures & Skins V, Boston, *SPIE* **1798**, 227–236.

Melle, S., Alavie, T., Karr, S., Coroy, T., Liu, K. and Measures, R. M. (1993a). "A Bragg Grating-Tuned Fiber Laser Strain Sensor System," *Photonics Technology Lett.* **5**, 263–266.

Melle, S. M., Liu, K., and Measures, R. M., (1993b), "Practical Fiber Optic Bragg Grating Strain Gauge System," *Applied Optics* **32**, 3601–3609.

Melloni, A., Gusmeroli, A., Tonini, A., Sala, G., and Martinelli, M. (1997). "Development and Characterization of Pultruded Fiber-Optic Ribbon," 12th Optical Fiber Sensors Conference, October 26–28, Williamsburg, 213–216.

Meltz, G. (1996). "Overview of Fiber Grating-Based Sensors," *SPIE* **2838**, 2–22.

Meltz, G., Morey, W. W., and Glam, W. H. (1989). "Formation of Bragg Grating in Optical Fibers by a Transverse Holographic Method," *Opt. Lett.* **14**, 823–825.
Melvin, L. D., Rogowski, R. S., Holben, M. S., Namkung, J. S., Kahn, K., and Sirkis, J., (1992). "Evaluation of Acrylate and Polyimide Coated Optical Fibers as Strain Sensors in Polymer Composites," in *Active Materials and Adaptive Structures*, (G. D. Knowles Ed.), Institute of Physics Publishing, Briston, 801–804.
Méndez, A. (1993). "Applications of Embedded Optical Fiber Sensors for Non-destructive Testing of Concrete Elements and Structures," in *Applications of Fiber-optic Sensors in Engineering Mechanics* (Farhad Ansari, Ed.), 144–158.
Méndez, A., and Morse, T. F. (1992). "Overview of Optical Fiber Sensors Embedded in Concrete," Fiber Optic Smart Structures & Skins V, Boston, *SPIE* **1798**, 205–216.
Méndez, A., Morse, T. F., and Médez, F., (1989). "Applications of Embedded Optical Fiber Sensors in Reinforced Concrete Buildings and Structures," *SPIE* Proc. **1170**, Fiber Optic Smart Struct. & Skins II, 60–69.
Méndez, A., Morse, T. F., and Reinhart, L. J. (1993). "Experimental Results on Embedded Optical Fiber Sensors in Concrete," *SPIEE Proc.*, **1918**, Smart Materials and Skins, Albuquerque, NM.
Menn, C., and Billington, D. P. (1995). "Breaking Barriers of Scale: A Concept for Extremely Long Span Bridges," *Structural Engineering International*, Jan., 48–50.
Merzbacher, C. I., Kersey, A. D., and Friebele, E. T. (1996). "Fiber Optic Sensing in Concrete Structures: A Review," *Smart Mater. Struct.* **5**, 196–208.
Mestel, R. (1995). "Tension Mounts as the Big One Stalks LA," *New Scientist*, Jan. 21, 16.
Michalske, T. (1994). "Fiber Coatings Keeps Water Out, Strength Up," *Photonics Spectra*, May, 52.
Michie, W. C., Culshaw, B., Roberts, S. S. J., and Davidson, R. (1991). "Fibre Optic Technique for Simultaneous Measurement of Strain and Temperature Variations in Composite Materials," *SPIE* **1588**, Fiber Optic Smart Structures and Skins IV, 342–356.
Michie, W. C., Thursby, G., Johnstone, W., and Culshaw, B. (1992). "Optical Techniques for Determination of the State of Cure of Epoxy-Resin-Based Systems," Fiber Optic Smart Structrues & Skins V, Boston, *SPIE* **1798**.
Michie, C., Culshaw, B., Thursby, G., Jin, W., and Konstantaki, M. (1996a). "Optical Sensors for Temperature and Strain Measurement," *SPIE* **2718**, 134–146.
Michie, C., Culshaw, B., Thursby, G., Konstantaki, M., Turpin, M., Lecot, C., Noharet, B., Lequime, M., Vannotti, P., Vanucci, A., Lang, D., Kalb, P., Davidson, R., Roberts, S., Papadopoulos, P., Ikiadis, A., Chazelas, J., Bonniau, P., Herbst, T., and Windisch, A. (1996b). "Optic Fiber Sensors for Monitoring of Structures (OSMOS)," *SPIE* **2718**, 385–397.
Midwinter, J. E., and Guo, Y. L. (1992). *Optoelectronics Lightwave Technology*, Wiley, New York.
Miller, S. E., and Kaminow, I. P. (1988). *Optical Fiber Telecommunications II*, Academic Press, New York.
Miller, C. M., Mettler, S. C., and White, I. A. (1986). *Optical Fiber Splices and Connectors: Theory and Methods*, Marcel Dekker, New York.

Miller, M. S., Case, S. W., Carman, G. P., Schmid, C. A., May, R. G., and Clause, R. O. (1992). "Validation of Axial Strain Transfer from a Composite Laminate to Embedded Optical Fiber Sensors," Fiber Optic Smart Structures & Skins V, Boston, *SPIE* **1798**, 19–30.
Morey, W. W., Meltz, G,. and Glenn, W. H. (1989). "Fiber Optic Bragg Grating Sensors," *SPIE* **1169**, Fiber Optic & Laser Sensors VII, 98–107.
Morey, W. W., Dunphy, J. R., and Meltz, G. (1991). "Multiplexed Fiber Bragg Grating Sensors," Proc. Distributed and Multiplexed Fiber Optic Sensors, *SPIE* **1586**, 216–224.
Morey, W. W., Ball, G. A., and Meltz, G. (1994). "Photoinduced Bragg Gratings in Optical Fibers," *Optics and Photonics News*, Feb., 7–14.
Morin, A., Caron, S., Van Neste, R., and Edgecombe, M. H. (1996). "Field Monitoring of the Ice Load of an Icebreaker Propeller Blade Using Fiber Optic Strain Gauges," *SPIE* **2718**, 427–438.
Moss, D., and Chatenoud, F. (1991) "Laser-Compatible Wave-Guide Electro-Absorption Modulators," *Can. J. Phys.* **69**, 497–507.
Moulton, P. (1992). "Tunable Solid-State Lasers," *IEEE* **80**, 348.
Mufti, A. A., Erki, M.-A., and Jaeger, L. G., Eds. (1991). *Advanced Composite Materials with Application to Bridges*, "Anchorages for Prestressing Tendons and Cables," 238–239, Canadian Society for Civil Engineering, Montreal.
Mufti, A. A., Erki, M. A., and Jaeger, L. G., Eds. (1992). *Advanced Composite Materials in Bridges and Structures in Japan*, Canadian Society for Civil Engineering, Montreal.
Muhs, J. D. (1992). "Silicone Rubber Fiber Optic Sensors," *Photonics Spectra*, July, 98–383.
Mulvihill, P. (1997). "Manufacturing Optical Fibre Bragg Grating Strain Sensors with an Excimer Laser for High-Strain, Multiplexed Embedded Applications," M.A.Sc. Thesis, University of Toronto.
Munns, I. (1995). "Adhesive Bond Inspection Using Nondestructive Testing," *Materials World* **3**, 527.
Murphy, K. A., Miller, III, M. S., Vengsarkar, A. M., and Claus, R. O. (1989). "Surface Acoustic Wave Detection Using Michelson-Type Fiber Optic Interferometer," *SPIE* **1170**, 558–565.
Murphy, K. A., Fogg, B. R., Wang, G. Z., Vengsarkar, A. M., and Claus, R. O. (1991a). "Sapphite Fiber Interferometer for Microdisplacement Measurements at High Temperatures," *SPIE* **1588**, Fiber Optic Smart Structures and Skins IV, 117–124.
Murphy, K. A., Gunther, M. F., Vengsarkar, A. M., and Claus, R. O. (1991b). "Fabry–Perot Fiber Optic Sensors in Full-Scale Fatigue Testing on an F-15 Aircraft," *SPIE* **1588**, Fiber Optic Smart Structures and Skins IV, 134–144.
Murphy, K. A., Gunther, M. F., Vengsarkar, A. M., and Claus, R. O. (1991c). "Quadrature Phase-Shifted, Extrinsic Fabry–Perot Optical Fiber Sensors," *Optics Lett.* **16**, 273–275.
Murphy, K. A., Gunther, M. F., Wang, A., and Claus, R. O. (1992). "Extrinsic Fabry–Perot Optical Fiber Sensor," 8th Optical Fiber Sensors Conference, Monterey, 193–200.
Murphy, K. A., Gunther, M. F., Claus, R. O., Tran, T. A., and Miller, M. S. (1993). "Optical Fiber Sensors for Measurement of Strain and Acoustic Waves," Smart Structures & Materials '93, Albuquerque, *SPIE* **1918**, 110–120.

Murphy, K. A., Gunther, M. F., de Vries, M., and Claus, R. O. (1994a). "Characterization and Testing of Materials, Using Multichannel Fiber Sensor Systems," *SPIE* **2294**, 34–42.

Murphy, K. A., Schmid, C. A., Tran, T. A., Carman, G., Wang, A., and Claus, R. O. (1994b). "Delamination Detection in Composites Using Optical Fiber Techniques," *SPIE* **2191**, 227–231.

Nagasawa, S., Yokoyama, Y., Ashiya, F., and Satake, T. (1991). "A High Performance Single-Mode Multifiber Connector using Oblique and Direct Endface Contact between Multiple Fibers Arranged in a Plastic Ferrule," *IEEE Photonics Tech. Lett.* **3**, 937–939.

Najafi, F. T., and Nassar, F. E. (1996). "Comparision of High-Speed Rail and Maglev Systems," *J. Transportation Eng.*, July/August, 276–281.

Nanni, A., Yang, A., Pan, C. C., Wang, K. J., and Michael, R. (1991). "Fiber Optic Sensors for Concrete Strain/Stress Measurement," *ACI Mat. J.* **88**, 257–264.

Narendran, N., Letgcher, S. V., Shukla, A., and Singh, R. P. (1992). "Embedded Fiber Optic Acoustic Sensor for Flaw Detection," Fiber Optic Smart Structures & Skins V, Boston, *SPIE* **1798**, 124–133.

Nath, D. K., Nelson, G. W., Griffin, S. E., Harrington, C. T., He, Y., Reinhart, L. J., Paine, D. C., and Morse, T. F. (1991). "Polyimide Coated Embedded Optical Fiber Sensors," Structures Sensing and Control, *SPIE* **1489**, 17–32.

Natke, H. G., and Yao, J. T. P. (1990). "Systems Identification Methods for Fault Detection and Diagnosis," Int. Conf. on Structural Safety and Reliability, ASCE, New York, 1387–1393.

Nehrich, Jr., R. B., Voran, G. I., and Dessel, N. F., (1967), *Atomic Light Lasers,* Sterling Pub. NY.

Nellen, P. M., Anderegg, P., Bronnimann, R., and Sennhauser, U. (1997). "Application of Fiber Optical and Resistance Strain Gauges for Long-Term Surveillance of Civil Engineering Structures," *SPIE* **3043**, 77–86.

Nellen, Ph. M., Frank, A. Mauron, P., and Sennhauser, U. (1998). "Lifetime and Reliability of Embedded Optical Sensor Fibers," in *Fiber Optic Sensors for Constructional Materials and Bridges* (Farhad Ansari, Ed.), 183–193.

Neumann, E. G., (1989). *Single-Mode Fibers*, Springer-Verlag, Berlin.

Newhook, J. P., and Mufti, A. A., (1996), "A Reinforcing Steel-Free Concrete Deck Slab for the Salmon River Bridge," *Concrete International*, **18**, 30–34.

Newland, D. (1990). "Roads to Ruin," *New Scientist*, Dec., 37.

Newson, T. P., Farahi, F., Jones, J. D. C., and Jackson, D. A. (1989). "Reduction of Semiconductors Laser Diode Phase and Amplitude Noise in Interferometric Fiber Optic Sensors," *Apo. Opt.* **28**, 4210–4215.

Nikles, M., Thevenaz, L., and Robert, P. A. (1994). "Simple Distributed Temperature Sensor Based on Brillouin Gain Spectrum Analysis," 10th Int. Conf. Optical Fibre Sensors, Glasgow, Scotland, *SPIE* **2360**, 138–141.

Normandin, R., Létourneau, S., Chatenoud, F., and Williams, R. L. (1991). "Monolithic, Surface-Emitting, Semiconductor Visible Lasers and Spectrometers for WDM Fiber Communication Systems," *IEEE J. Quantum Electron.* **27**, 1520–1530.

Normile, D., and Vizard, F. (1998). "A Bridge So Far," *Popular Science*, March, 48–53.
Nye, S. F. (1954). *Physical Properties of Crystals*. Oxford Press, London, 235–259.
O'Neill, B. (1993a). "Beating the Bullet Train," *New Scientist*, Oct., 36.
O'Neill, B. (1993b). "Longer Yet and Longer...," *New Scientist*, October 2, 32–34.
Ogawa, Y., Iwasaki, J., and Nakamura, K. (1997). "A Multiplexing Load Monitoring System of Power Transmission Lines Using Fiber Bragg Grating," 12th Optical Fiber Sensors Conference, October 26–28, Williamsburg.
Ohn, M., Davis, A., Liu, K., and Measures, R. M. (1992a). "Embedded Fiber Optic Detection of Ultrasound and its Application to Cure Monitoring," Optical Fiber Sensor-Based Smart Materials and Structures Conferences, Blacksburg, Virginia, April.
Ohn, M. M. Davis, A., Liu, K., and Melle, S. M., (1992b). "Embedded Fiber Optic Detection of Ultrasound and Its Application to Cure Monitoring," Fiber Optic Smart Structures & Skins V, Boston, *SPIE* **1798**, 134–143.
Ohn, M., Sandgren, S., Huang, S., Maaskant, R., Stubbe, R., Sahlgren, B., Measures, R., and Storoy, H. (1995). "Phase Based Bragg Intra-Grating Sensing of Strain Gradients," *SPIE* **2444**, 127–135.
Ohn, M. M., Huang, S. Y., LeBlanc, M., and Measures, R. M. (1996). "Distributed Strain Sensing Using Long Intracore Fiber Bragg Gratings," *SPIE* **2838**, 66–75.
Ohn, M. M., Huang, S. Y., Measures, R. M., and Chwang, J. (1997). "Arbitrary Strain Profile Measurement within Fiber Grating Using Interferometric Fourier Transform Technique," *Electronics Lett.* **33**, 1242–1243.
Othonos, A., and Lee, X. (1995). "Novel and Improved Methods of Writing Bragg Gratings with Phase Masks," *IEEE Photon. Tech. Lett.* **7**, 1183–1185.
Othonos, A., Alavie, T., Melle, S., Karr, S., and Measures, R. M. (1993). "Fiber Bragg Grating Laser Sensors," *Optical Engineering* **32**, 2841–2846.
Othonos, A., Lee, X., and Measures, R. M. (1994). "Superimposed Multiple Bragg Gratings," *Electron. Lett.* **30**, 1972–1973.
Ouellett, F. (1987). "Dispersion Cancellation Using Linearly Chirped Bragg Grating Filters in Optical Waveguides," *Opt. Lett.* **12**, 847–849.
Ouellett, F. (1991). "All-Fiber Filter for Efficient Dispersion Compensation", *Opt. Lett.* **16**, 303–305.
Paek, E. G., Liao, P. F., and Gharavi, H. (1992). "Derivation of Neural Network Models and Their Computational Circuits for Associative Memory," *Opt. Eng.* **31**, 986–994.
Page, D., (1996). "Bridge over Troubled Infrastructure," *Popular Science* **7**, 40.
Pak, Y. E. (1992). "Longitudinal Shear Transfer in Fiber Optic Sensors," *Smart Maerials and Structures* **1**, 57–62.
Pan, J. J., Shih, M., and Shik, K. (1993). "High Performaqnce Fiberoptic Systems Rely on Special Isolators," *Laser Focus World*, June, 167–170.
Pantazopoulou, S. J., Bonacci, J. F., Hearn, N., Sheikh, S., and Thomas, M. D. (1996). "Repair of Corrosion-Damaged Concrete Using ACM," in *Advanced Composite*

Materials in Bridges & Structures, (El-Badry, M. M., Ed.), 457–464. The Canadian Society of Civil Engineeers, Montreal.

Parker, T. R., Farhadiroushan, M., Handerek, V. A., and Rogers, A. J. (1997). "A Fully Distributed Simultaneous Strain and Temperature Sensor using Spontaneous Brillouin Backscatter," *Photonics Technology Lett.* **9**, 979–981.

Patrick, H. J., Kersey, A. D., Pedrazzani, J. R., and Vengsarkar, A. M. (1996a). "Fiber Bragg Grating Sensor Demodulation System Using In-Fiber Long Period Grating Filters," *SPIE* **2838**, 60–65.

Patrick, H. J., Williams, G. M., Kersey, A. D., Pedrazzani, J. R., and Vengsarkar, A. M. (1996b). "Hybrid Fiber Bragg Grating/Long Period Fiber Grating Sensor for Strain/Temperature Discrimination," *IEEE Photon, Tech. Lett.* **8**, 1223–1225.

Paul, C. A. (1993). "Optimization of Optical Time Domain Reflectometry (OTDR) Based Strain Sensing," Soc. Exper. Mechanics (SEM) Conf., 7–9 June, Dearborn, Michigan.

Paul, C. A., and Schoeppner, G. A. (1993). "Static Test Study of Composites with Embedded Optical Fibers Using Fractional Factorial Methods," Soc. Exper. Mechanics (SEM) Conf. 7–9 June, Dearborn, Michigan.

Paul, C. A., Sendeckyj, G. P., and Carman, G. P. (1993a). "Detection of the Onset of Damage Using An Extrinsic Fabry–Perot Interferometric Strain-Sensor (EFPI-SS)," Smart Structures & Materials '93, Albuquerque, *SPIE* **1918**, 154–164.

Pierce, S. G., Philp, W. R., Culshaw, B., Gachagan, A., McNab, A., and Haywardm G. (1996). "Ultrasonic Inspection of CFRP Plates Using Surface Bonded Optical Fiber Sensors," *SPIE* **2718**, 345–354.

Poole, S. B., Paye, D. N., and Fermann, M. E. (1985). "Fabrication of Low-Loss Optical Fibres Containing Rare-Earth Ions," *Electron. Lett.* **21**, 737–738.

Posey, R. Jr., Kersey, A. D., LeBlanc, M., and Davis, M. (1997). "Low-Coherence, Wavelength-Encoded Addressing for Integrated Long-Gauge Length Fiber Optic Strain Sensing," OSA, 12th Int. Conf. on Optical Fiber Sensors, Williamsburg, 574–577.

Prohaska, J. D., Snitzer, E., Chen, B., Maher, M. H., Nawy, E. G. (1992). "Fiber Optic Bragg Grating Strain Sensor in Large-Scale Concrete Structures," Fiber Optic Smart Structures & Skins V, Boston, *SPIE* **1798**, 286–294.

Przemieniecki, J. S. 1985) *Theory of Matrix Structural Analysis*, Dover Publications.

Putnam, M. A., Askins, C. G., Williams, G. M., Friebele, E. J., Baskansky, M., and Reintjies, J., (1995a). "Single Pulse Fabrication of Fiber Bragg Gratings Using a Phased-Conjugated kRF Excimer Laser," *SPIE* **2444**, 403–409.

Putnam, M. A., Williams, G. M., and Friebele, E. J. (1995b). "Fabrication of Tapered, Strain-Gradient Chirped Fibre Bragg Gratings," *Electron Lett.* **31**, 309–310.

Putnam, M. A., Dennis, M. L., Kang, J. U., Tsai, T. E., Duling, I. N., and Friebele, E. J. (1997). "Sensor Grating Demodulation Using a Passively Mode Locked Fiber Laser," *SPIE* **3042**, 352–357.

Rao, Y. J., and Jackson, D. A. (1993a). "Prototype Fiber-Optic Based Pressure Probe with Built-In Temperature Compensation with Signal Recovery by Coherence Reading," *Appl. Optics* **32**, 7110–7113.

Rao, Y. J., and Jackson, D. A. (1993b). "Prototype Fiber-Optic Based Ultrahigh Pressure Remote Sensor Using Dual-Wavelength Coherence Reading," *Appl. Optics* **29**, 2142–2143.

Rao, Y. J., and Jackson, D. A. (1994). "Improved Synthesised Source for White Light Interferometry," *Electron. Lett.* **30**, 1440–1441.

Rao, Y. J., Jackson, D. A., Jones, D., and Shannon, C. (1994). "Development of Prototype Fiber Optic Based Fizeau Pressure Sensors with Temperature Compensation and Signal Recovery by Coherence Reading," *J. Lightwave Techn.* **12**, 1685–1695.

Rao, Y. J., Ribeiro, A. B. L., and Jackson, D. A. (1996). "Simultaneously Spatial-Time and Wavelength-Division-Multiplexed In-Fiber Bragg Grating Sensor Network," *SPIE* **2838**, 23–30.

Rao, Y. J., Hurle, B., Web, D. J., and Jackson, D. A. (1997). "In-Situ Temperature Monitoring in NMR Machines With a Prototype In-Fibre Bragg Grating Sensor System," OSA, 12th Int. Conf. on Optical Fiber Sensors, Williamsburg, 646–649.

Rausch, H. (1995). "Fiber Optic Sensors: Rx for the Infrastructure," *Photonic Spectra*, March, 80–89.

Reismann, H., and Pawlik, P. S. (1980). *Elasticity Theory and Applications*, Wiley Interscience, New York.

Rizkalla, S. and Tadros, G. (1994). "First Smart Bridge in Canada," *ACI Concrete Internat.* **16**, 42–44.

Rizkalla, S., Shehata, E., Abdelrahman, A., and Tadros, G. (1997). "Headingley Smart Bridge: A New Generation of Civil Engineering Structures," US–Canada–Europe Workshop on Bridge Engineering, July, Zurich, Switzerland.

Roberts, S. S., and Davidson, R. (1991). "Mechanical Properties of Composites Materials Containing Embedded Fiber Optic Sensors," Fiber Optic Smart Structures and Skins IV, *SPIE* **1588**, 326–341.

Roberts, S. J., and Davidson, R. (1992). "Short Term Fatigue Behavior of Composite Materials Containing Embedded Fiber Optic Sensors and Actuators," 1st Europ. Conf. Smart Structures & Materials, Glasgow, *SPIE* **1777**, 255–262.

Roberts, D. R., Cuellar, E., Kennedy, M. T., and Ritter, J. E. Jr. (1991). "Calculation of Static Fatigue Lifetime of Optical Fiber," *Opt. Eng.* **30**, 716–727.

Rogers, C. A. (1993). "Intelligent Material Systems—The Dawn of a New Materials Age," *J. Intell. Mat. Syst. & Structures* **4**, 4–12.

Russell, P. St. J., and Archambault, J. L., (1996), "Fiber Gratings," in *Optical Fiber Sensors Components and Subsystems*, (Culshaw, B., and Dakin J., Ed.), Artech House, 9–67.

Russell, P. St. J., and Archambault, J. L. (1997a). "Fiber Gratings," in *Optical Fiber Sensors* (Culshaw, B., and Dakin, J., Eds.), 9–67.

Russell, P. St. J., and Archambault, J. L. (1997b). "Fiber Gratings," Chapter 2 in *Optical Fiber Sensors Components and Subsystems* (Culshaw, B., and Dakin, J., Eds.), Vol. 3., Artech House, 9–67.

Rutherford, P., Ikegami, R., and Shrader, J. (1996). "Novel, NDE Fiber Optic Corrosion Sensor," *SPIE* **2718**, 158–169.

Salane, H. J., and Baldwin, J. W. (1990). "Identification of Modal Properties of Bridges," *ASCE J. Struct. Eng.* **116** (7), 2008–2021.

Saleh, B. E. A., and Teich, M. C. (1991). *Fundamentals of Photonics*, J. Wiley and Sons, New York.

Salehi, A., Tay, A., Wilson, D., and Smith, D. (1989). "Strain Concentrations Around Embedded Optical Fibers, by FEM and Moiré Interferometry," 5th Annu. ASM/ESD Adv. Compos. Conf./Exspos., Dearborn MI, 11–19.

Sansonetti, P., Guerin, J. J., Lequime, M. and Debrie, J., (1991). "Parallel Coherence Receiver for Quasi-Distributed Optical Sensor," SPIE Fiber Optic Smart Structures and Skins IV, Boston, 5–6 Sept.

Sansonetti, P., Guerin, J. J., Viton, D., Gouffier, C., and Engrand, D. (1992). "Unidirectional Glass Reinforced Plastic Composite Monitoring with White Light Quasi-Distributed Polarimetric Sensing Network," 1st European Conference on Smart Structures and Materials, Glasgow, 77–80.

Sarrazin, H., Kim, B., Ahn, S., and Springer, G. S. (1995). "Effects of Processing Temperature and Lay-up on Springback," *J. Composite Mater.* **29**, 1278–1294.

Schmidt, W., and Boller, C. (1992). "Smart Structures: a Technology for Next Generation Aircraft," AGARD. Conf. Proc. 531, Smart Structures for Aircraft and Spacecraft, FRG, 1–1 to 1–14.

Schoess, J. N., and Seifert, G. (1997). "Smart Fastener for KC-135 Structural Integrity Monitoring," *SPIE* **3042**, 278–282.

Schulz, W. L., Udd, E., Seim, J. M., Laylor, H. M., and McGill, G. E. (1998). "Single and Multiaxis Fiber Grating Based Strain Sensors for Civil Structure Applications," *Proc. SPIE* **3489**, 71–78.

Scott, R. H. (1991). "A Preliminary Investigation of Strains in Reinforced Concrete Beam–Column Connections due to Seismic Loading," *Mag. Concrete Res.* No. 43, Mar., 59–64.

Scott, R. H., and Gill, P. A. T. (1987). "Short-Term Distributions of Strain and Bond Stress along Tension Reinforcement," *The Structural Engineer* **65B**, 39–48.

Searle, I., Ziola, S., and Seidel, B., (1997), "Crack Detection on a Full Scale Aircraft Fatigue Test," *SPIE* **3042**, 267–277.

Senior, J. M. (1985). *Optical Fiber Communications Principles and Practice*, Prentice-Hall International.

Sennhauser, U., Bronnimann, R., Mauron, P., and Nellel, Oh. M. (1998). "Reliability of Optical Fibers and Bragg Grating Sensors for Bridge Monitoring," in *Fiber Optic Sensors for Constructional Materials and Bridges* (F. Ansari Ed.), 117–128.

Sheem, S. K., Giallorenzi, T. G., and Koo (1982). "Optical Techniques to Solve the Signal Fading Problem in Fiber Interferometers," *Appl. Optics* **21**, 689.

Shelley, T. (1989). "Fibre Optic Nerves Give Life to Composites," *Eureka*, July, 74–77.

Shlyagin, M. G., Miridonov, S. V., Tentori, D., and Castillo, J. (1997). "Twin Grating-Based Interferometric Fiber Sensor," OSA 12th Int. Conf. on Optical Fiber Sensors, Williamsburg, Oct., 472–474.

Shukla, A., Singh, R., Letcher, S. V., and Narendran, N. (1993). "Fiber-Optic Sensors and Fracture Mechanics," in *Applications of Fiberoptic Sensors in Engineering Mechanics* (Farhad Ansari, Ed.), 177–191.

Shuldiner, H., and Retseck, G. (1993). "Rebuilding the World Trade Center," *Popular Mechanics*, July, 41–44.

Siegman, A. E. (1986). *Lasers*, University Science Books.

Simonsen, H. D., Paetsch, R., and Dunphy, J. R. (1992). "Fiber Bragg Grating Sensor Demonstration in Glass-Fiber Reinforced Polyester Composite," 1st Europ. Conf. Smart. Struct. & Mater., Glasgow, *SPIE* **1777**, 3–76.

Singh, H., and Sirkis, J. S. (1991). "Micromechanics of Laminated Composites with Embedded Optical Fibers," in *Optical Fiber Sensor-Based Smart Materials and Structures*, Technomic Publ. 55–64.

Singh, H., Sirkis, J. S., and Dasgupta, A. (1991). "Microinteraction of Optical Fibers Embedded in Laminated Composites," *SPIE* **1588**, Fiber Optic Smart Structures and Skins IV, 76–85.

Sirkis, J. S. (1991). "Phase–Strain–Temperature Model for Structurally Embedded Interferometric Optical Fiber Strain Sensors with Applications," *SPIE* **1588**, Fiber Optic Smart Structures and Skins IV, 26–43.

Sirkis, J. S. (1993a). "A Unified Approach to Phase–Strain–Temperature Models for Smart Structure Interferometric Optical Fiber Sensors: Part I—Development," *Opt. Engineering* **32**, 752–763.

Sirkis, J. S. (1993b). "A Unified Approach to Phase–Strain–Temperature Models for Smart Structure Interferometric Optical Fiber Sensors: Part II—Applications," *Opt. Eng.* **32**(4), 763–773.

Sirkis, J. S. (1993c). "Interpretation of Embedded Optical Fiber Sensor Signals," in *Applications of Fiberoptic Sensors in Engineering Mechanics* (Farhad Ansari, Ed.), 85–99.

Sirkis, J. S. (1993d), "Optical and Mechanical Isotrophies in Embedded Fiber Optic Sensors," *Smart Materials & Structures* **2**, 255–259.

Sirkis, J. S. (1998). "Using Bragg Grating Sensor Systems in Construction Materials and Bridges: Perspectives and Challenges," in *Fiber Optic Sensors for Constructional Materials and Bridges* (F. Ansari, Ed.), 44–61.

Sirkis, J. S., and Dasgupta, A. (1990). "Optimal Coatings for Intelligent Structure Fiber Optic Sensors," *SPIE* **1370**, Fiber Optic Smart Structures & Skins III, San José, 129–140.

Sirkis, J. S., and Dasgupta, A. (1991a). "Thermal-Plastic Metal Coatings on Optical Fiber Sensors for Damage Detection," Int. Symposium on Active Materials and Adaptive Structures, Nov. 4–8, Alexandria, Virginia.

Sirkis, J. S., and Dasgupta, A. (1991b). "Thermal Plastic Metal Coatings on Optical Fiber Sensors," *SPIE* **1588**, Fiber Optic Smart Structures and Skins IV, 88–99.

Sirkis, J. S., and Dasgupta, A. (1992). "What Do Embedded Optical Fibers Really Measure?" 1st Europ Conf. Smart. Struct. & Mater., Glasgow, *SPIE* **1777**, 69–72.

Sirkis, J. S., and Dasgupta, A. (1993). "The Role of Local Interaction Mechanics in Fiber Optic Smart Structures," *J. Intell. Mat. Systems & Struct.* **4**, 260–271.

Sirkis, J. S., and Dasgupta, A. (1995). "Optical Fiber/Composite Interaction Mechanics," in *Fiber Optic Smart Structures* (E. Udd, Ed.), Wiley, 61–107.

Sirkis, J. S., and Haslach, H. W., Jr. (1990). "Full Phase–Strain Relation for Structurally Embedded Interferometric Optical Fiber Sensors," *SPIE* **1370**, Fiber Optic Smart Structures and Skins III, San Jose, 248–259.

Sirkis, J. S., and Haslach, H. W., Jr. (1991). "Complete Phase-Strain Model for Structurally Embedded Interferometric Optical Fiber Sensors," *J. Intell. Mater. Syst. Struct.* **2**, 3–24.

Sirkis, J. S., and Lo, Y.-L. (1994). "Simultaneous Measurement of Two Strain Components Using 3×3 and 2×2 Coupler-Based Passive Demodulation of Optical Fiber Sensors," *J. Lightwave Tech.*, **12**, 2153–2161.
Sirkis, J. S., and Matthews, C. T., (1993), "Experimental Investigation of Phase–Strain–Temperature Models for Structurally Embedded Interferometric Fiber Optic Sensors," Experimental Mechanics March. 26–31.
Sirkis, J., Putnam, M. A., Berkoff, T. A., Kersey, A. D., Friebele, E. J., Jones, R. T., and Brennan, D. (1994a). "In-Line Fiber Etalon (ILEE) for Internal Strain Measurement," *SPIE* **2191**, 136–147.
Sirkis, J. S., Lo, Y. L., and Nielson, P. L. (1994b), "Phase-Strain Model for Polarimetric Strain Sensors Based on Fictitious Residual Strains," *J. Intell. Mat. Systems & Struct.* **5**, 494–500.
Snitzer, E. (1961). "Cylindrical Dielectric Waveguide Modes," *J. Opt. Soc. Am.* **51**, 491–498.
Sorin, W. V., and Baney, D. M. (1995). "Multiplexed Sensing Using Optical Low-Coherence Reflectometry," *IEEE Photonics Tech. Lett.* **7**, 917–919.
Spillman, W. B., Jr. (1992). "The Evolution of Smart/Materials," 1st European Conf. on Smart Struct. & Mat. Glasglow, 97–113.
Spillman, W. B., Jr. (1993). "Instrumentation Architecture Development for Smart Structures," Smart Structures & Materials '93, Albuquerque, *SPIE* **1918**, 165–171.
Spillman, W. B., and Lord, J. R. (1987). "Self-Referencing Multiplexiing Technique for Fiber-Optic Intensity Sensors," *J. Lightwave Tech.*, 865.
Spillman, W. B., Jr., and Lord, J. R. (1995). "Methods of Fiber Optic Ingress/Egress for Smart Structures," in *Fiber Optic Smart Structures* (Eric Udd, Ed.), 121–153, J. Wiley & Sons, New York.
Starodubov, D. S., Grubsky, V., and Feinberg, J. (1997a). "Efficient Bragg Grating Fabrication in a Fiber through Its Polymer Jacket Using Near-UV Light," *Electron. Lett.* **33**, 1331–1333.
Starodubov, D. S., Grubsky, V., Feinberg, J., Dianov, E. M., Semjonov, S. L., Guryanov, A. N., and Vechkanov, N. N. (1997). "Fiber Bragg Grating with Reflectivity $> 97\%$ Fabricated through Polymer Jacket Using Near-UV Light," OSA, 12th Optical Fiber Sensors Conference, October 26–28, Williamsburg.
Storoy, H., and Saether, J. (1997). "Fiber Optic Condition Monitoring during a Full Scale Destructive Bridge Test," Fibre Bragg Gratings and Fibre Optic Structural Strain Sensing, Doctoral Thesis of Storoy, H., Published by Norwegian University of Science and Technology.
Sudo, M., Nakai, M., Himeno, K., Suzaki, S., Wada, A., and Yamauchi, R. (1997). "Simultaneous Measurement of Temperature and Strain Using PANDA Fiber Grating," OSA, 12th Int. Conf. on Optical Fiber Sensors, Williamsburg, 170–173.
Sugden, K., Bennion, I., Molony, A., and Copner, N. J. (1994). "Chirped Gratings Produced In Photosensitive Optical Fibers by Fiber Deformation during Exposure," *Electron. Lett.* **30**, 440–442.
Sugita, M., Yanagida, H., and Muto, N. (1995). "Materials Design for Self-Diagnosis of Fracture in GFRP Composite Reinforcement," *Smart Materials & Structures* **4**, Suppl. 1A, A52–A57.

Sumida, M., and Takemoto, K. (1984). "Lens Coupling of Laser Diodes to Single Mode Fibers," *J. Lightwave Technol.* **Lt-2**, 305–311.

Takagi, T. (1992). "A Concept of Intelligent Materials and the Current Activities of Intelligent Materials in Japan," 1st Europe Conf. on Smart Structures and Materials, Glasgow, 1–6, *SPIE* **1777**, 13–18.

Tamari, Y. (1997). "Custom Silicon Photodiodes Offer Design Flexibility," *Laser Focus World*, Jan., 123–126.

Tamir, T. (1990). *Guided-Wave Optoelectronics*, Springer–Verlag, Berlin.

Tamura, K., Doerr, C. R., Nelson, L. E., Haus, H. A., and Ippen, E. P. (1994). "Techniques for Obtaining High-Energy Ultra-short pulses from Additive-Pulse Mode-Locked Erbium-Doped Fiber Ring Lasers," *Opt. Lett.* **19**, 46.

Tay, A. K. (1992). "Evaluation of an Optical Signal Response and Failure Mechanisms Induced by the Transverse Shear Deformation in Smart Structures," 1st Europ. Conf. on Smart Structures and Materials, Glasgow, *SPIE* **1777**, 277–280.

Tay, A., Wilson, D. A., and Wood, A. (1989). "Strain Analysis of Optical Fibers Embedded in Composite Materials Using Finite Element Modelling," *SPIE* **1170**; Fiber Optic Smart Structures, and Skins 11, 521–533.

Teral, S. R. (1996). "Fiber Optic Weigh-in-Motion Sensor: Correlation between Modeling and Practical Characterization," *SPIE* **2718**, 417–426.

Thomas, J. J., Luong, M. P., and Semblat, J. F. (1995). "Vibratory Signature of Electric Pylons under Impulse Testing," *SPIE* **2446**, 258–267.

Todd, M. D., Nau, G. M., Danver, A., Tveten, A. B., and Vohra, S. T. (1997). "A Low-Frequency Fiber Optic Accelerometer Array for Mechanical Motion Detection," OSA, 12th Int. Conf. on Optical Fiber Sensors, Williamsburg, 20–23.

Tohmori, Y., Yoshikuni, Y., Ishii, H., Kano, F., Tamamura, T., and Kondo, Y. (1993). "Over 100 nm Wavelength Tuning in Superstructure Grating (SSG) DBR Lasers," *Electron. Lett.* **29**, 352–354.

Troy, C. T. (1997). "Fiber Optic Smart Structures: A Technology Ahead of Its Time Is Finally Winning Acceptance," *Photonics Spectra*, 112–128.

Tsai, S. W., and Hahn, H. T. (1980). *Introduction to Composite Materials*, Technomic Publishing, 414–416.

Tulley, A. (1994). "The Age of Wind Power Dawns in Alberta," *Canadian Geographic*, March/April, 12.

Tulloch, M. H. (1992). "Luminescent Paint Reveals Aircraft Pressure Points," *Photonics Spectra*, April, 20.

Tullock, M. T. (1995). "Fiber Optic Strain Gauge May Influence Oil Tanker Design," *Photonics Spectra*, Jan., 18.

Turner, R. D., Valis, T., Hogg, W. D., and Measures, R. M. (1989). "Fiber Optic Strain Sensors for 'Smart' Structures," *J. Intelligent Material Systems and Structures* **1**, 26–49.

Turner, R. D., Valis, T., Hogg, W. D., and Measures, R. M. (1990). "Fiber-Optic Sensors for Smart Structures," *J. Intell. Mater. Syst. Struct.* 26–49.

Tutton, P. A., and Underwood, F. M. (1992). "Structural Health Monitoring Using Embedded Fibre Optic Sensing," AGARD. Conf. Proc. 531, Smart Structures for Aircraft and Spacecraft, FRG. 18–1 to –10.

Udd, E. (1988). 1st Fiber Optic Smart Structures and Skins Conference, *SPIE* **986**, Boston.
Udd. E. (1991). *Fiber Optic Sensors: An Introduction for Engineers and Scientists*, John Wiley & Sons, New York.
Udd, E., Ed. (1995). *Fiber Optic Smart Structures*, J. Wiley & Sons, New York.
Udd, E. (1996a). "Distributed Fiber Optic Strain Sensor Based on the Sagnac and Michelson Interferometers," *SPIE* **2719**, 210–212.
Udd. E. (1996b). "Three Axis Strain and Temperature Fiber Optic Grating Sensor," *SPIE* **2718**, 104–100.
Udd, E., Lawrence, C., and Nelson, D. (1997). Development of a Three Axis Strain and Temperature Fiber Optic Grating Sensor. *Proc SPIE* **3042**, 229–236.
Udd, E., Michal, R. J., Higlry, S. E., Theriault, P., LeChong, P., and Holin, D. A. (1987). Fiber Optic Sensor System for Aerospace Applications," *SPIE* **838**, 162–168.
Udd, E., Schulz, W., Seim, J., McGill, G., and Laylor, H. M. (1998). "Distributed Multiaxis Fiber Grating Strain Sensor Applications for Bridges," in *Fiber Optic Sensors for Constructional Materials and Bridges* (Farhad Ansari, Ed.), 168–180.
Uffen, D. R., Scholaert, H., and Schmid, G. (1992). "Smart Structures at Aastra Corporation," AGARD, Smart Structures for Aircraft and Spacecraft Meeting, Lindau, Germany, 5–7 October.
Valery, N. (1995). "Earthquake Engineering, " *The Economist* April, 3, 12.
Valis, T., and Measures, R. M. (1992). "Far Field In-Plane Stress Measurement with an Embedded Eigenaxis-Flipped Fiber Fabry–Perot Strain Gauge," Proc. of Conference on Fiber Optic Sensor Based Smart Structures, Blacksburg, Virginia, 23–29.
Valis, T., Hogg, D., and Measures, R. M. (1989a). "Localized Fiber Optic Strain Sensors Embedded in Composite Materials," *SPIE* **1170**, Fiber Optic Smart Structures and Skins 11, 495–504.
Valis, T., Turner, R., and Measures, R. M. (1989b). "Distributed Fiber Optic Sensing Based on Counterpropagating Waves," *Appl. Optics* **28**, 1984–1990.
Valis, T., Hogg, D., and Measures, R. M. (1990a). "Composite Material Embedded Fiber Optic Fabry–Perot Strain Rosette," *SPIE* **1370**, Fiber Optic Smart Structures and Skins III, San Jose, 154–161.
Valis, T., Hogg, D. E., and Measures, R. M. (1990b). "Fiber Optic Fabry–Perot Strain Gauge," *IEEE Photonics Tech. Lett.* **2**, 227–228.
Valis, T., Tapanes, E., Liu, K., and Measures, R. M. (1991). "Passive-Quadrature Demodulated Localized-Michelson Fiber-Optic Strain Sensor Embedded in Composite Materials," *J. Lightwave Tech.* **9**, 535–544.
Valis, T., Hogg, D., and Measures, R. M. (1992). "Fiber-Optic Fabry–Perot Strain Rosettes," *Smart Mat. & Struct.* **1**, 227–232.
Van Steenkiste, R. J., and Springer, G. S. (1997). *Strain and Temperature Measurement with Fiber Optic Sensors*, Technomic Publishing, Lancaster.
Varadan, V. V., Bao, X., Ramanathan, S., and Piscotty, D. (1997). "Wireless Passive IDT Sytrain Microsensor," *Smart. Mat. Struct.* **6**, 745–751.
Varadan, V. V., Chin, L.-C., and Varadan, V. K. (1992). "Modelling Integrated Sensor/Actuator Functions in Realistic Environments," 1st Europ. Conf. on Smart Structures and Materials, Glasgow, *SPIE* **1777**, 1–6.

Varadan, V. V., Varadan, V. K., Varelas, D., Limberger, H. G., and Salathe, R. P. (1997). "Enhanced Mechanical Performance of Single Mode Optical Fibers Irradiated with a cw UV Laser," *Electron. Lett.* **33**, 704–705.

Varelas, D., Limberger, H. G., and Salathé, R. P. (1997). "Enhanced Mechanical Performance of Singlemode Optical Fibres Irradiated by a CW UV Laser, *Electron. Lett.* **33**, 704–705.

Varnham, M. P., Barlow, A. J., Payne, D. N., and Okamoto, K. (1983). "Polarimetric Strain Gauges Using High Birefringence Fibres," *Electron Lett.* **19**, 699–700.

Vengsarkar, A. M. (1996). "Long-Period Fiber Grating Shape Optical Spectra," *Laser Focus World*, June, 243–247.

Vengsarkar, A. M., Michie, W. C., Jankovic, L., Culshaw, B., and Clause, R. O. (1990). "Fiber Optic Sensor for Simultaneous Measurement of Strain and Temperature," *SPIE* **1367**, Fiber Optic and Lasers VIII, 249–260.

Vengsarkar, A. M., Zhong, Q., Inniss, D., Reed, W. A., Lemaire, P. J., and Kosinski, S. G., (1994). "Birefringence Reduction in Side-Written Photoinduced Fiber Devices by a Dual-Exposure Method," *Optics Lett.* **19**, 1260–1262.

Vensarkar, A. M., Lemaire, P. J., Judkins, J. B., Bhatia, V., Erdogan, T., and Sipe, J. E. (1996). "Long-Period Fiber Gratings as Band-Rejection Filters," *IEEE J. Lightwave Technol.* **14**, 58–64.

Vohra, S. T., Chang, C. C., Danver, B. A., Althouse, B., Davis, M. A., and Idriss, R. (1998). "Preliminary Results on the Monitoring of an In-Service Bridge Using a 32-Channel Fiber Bragg Grating Sensor System" in *Fiber Optic Sensors for Constructional Materials and Bridges* (Farhad Ansari, Ed.), 148–158.

Volanthen, M., Geiger, H., Cole, M. J., and Dakin, J. P. (1996a). "Measurement of Arbitrary Strain Profiles within Fiber Gratings," *Electron. Lett.* **32**, 1028–1029.

Volanthen, M., Geiger, H., Cole, M. J., Laming, R. I., and Dakin, J. P. (1996b). "Low Coherence Technique to Characterize Reflectivity and Time Delay as a Function of Wavelength within a Long Fiber Grating," *Electron. Lett.* **32**, 757–758.

Volanthen, M., Geiger, H., Xu, M. G., and Dakin, J. P. (1996c). "Simultaneous Monitoring of Multiple Fiber Gratings with a Single Acousto-Optic Tunable Filter," *Electron. Lett.* **32**, 1228–1229.

Volanthen, M., Geiger, H., and Dakin, J. P. (1997). "Distributed Grating Sensors: An Alternative to Multiplex Grating Arrays," *SPIE* **3042**, 372–382.

Volanthen, M., Geiger, H, Xu, M. G., and Dakin, J. P. (1996d). "Lock-In Techniques for Interrogation of Long and Short Gauge Length Optical Fibre Sensor Arrays," *SPIE* **2838**, 87–97.

Vries, M. de, Nasta, M., Bhatia, V., Trant, T., Greene, J., Claus, R. O., and Masri, S. (1995). "Performance of Embedded Short–Gage–Length Optical Fiber Sensors in a Fatigue–Loaded Reinforced Concrete Specimen," *Smart Materials & Structures* **4**, Suppl. 1A, A107–A113.

Vurpillot, S., Casanova, N., Inaudi, D., and Kronenberg, P. (1997). "Bridge Spatial Displacement Monitoring with 100 Fiber Optic Sensors Deformations: Sensors Network and Preliminary Results," *SPIE* **3043**, 51–57.

Vurpillot, S., Inaudi, D., and Ducret, J. M. "Bridge Monitoring by Fiber Optic Deformation Sensors: Design, Emplacement and Results," *SPIE* **2719**, 141–149.

Wada, B. K. (1990). "Adaptive Structures: An Overview," *J. Spacecraft* **27**, 330–337.

Wada, O. (1982). "Performance and Reliability of High Radiance InGaAsP/InP DH LEDs Operating in the 1.15–15 μm Wavelength Region," *IEEE J. of Quantum Elect.* **QE-18**, 368–374.
Wagreich, R., and Sirkis, J. S. (1997). "Distinguishing Fiber Bragg Grating Strain Effects," OSA, 12th Int. Conf. on Optical Fiber Sensors, Williamsburg, 20–23.
Wagreich, R. B. (1996). "Effects of Diametric Load on Fiber Bragg Gratings Fabricated in Low Birefringent Fiber," *Electron. Lett.* **32**, 1223–1224.
Waite, S. R., and Sage, G. N. (1988). "The Failure of Optical Fibers Embedded in Composite Materials," *Composites* **19**, 288–294.
Wang, G., Havsgard, Urnes, E., Pran, K., and Knudsen, S. (1997). "Digital Demodulation and Signal Processing Applied to Fiber Bragg Grating Strain Sensor Arrays in Monitoring Transient Loading Effects on Ship Hulls," OSA, 12th Int. Conf. on Optical Fiber Sensors, Williamsburg, 612–615.
Warkentin, D. J., Crawley, E. F., and Senturia, S. D. (1992). "The Feasibility of Embedded Electronics for Intelligent Structures," *J. Intell. Mater. Syst. and Struct.* **3**, 462–482.
Weis, R. S. (1994). "A Four-Element Fiber Grating Sensor Array with Phase-Sensitive Detection," *IEEE Photonics Tech. Lett.* **6**, 1469–1472.
Weis, R. S., and Beadle, B. M. (1997). "MWD Telemetry System for Coiled-Tubing Drilling Using Optical Fiber Grating Modulators Downhole," OSA, 12th Int. Conf. on Optical Fiber Sensors, Williamsburg, 416–419.
White, S. R., and Hahn, H. T. (1992). "Process Modeling of Composite Materials: Residual Stress Development During Cure. Part II. Experimental Validations," *J. Composite Mater.* **26**, 2423–2453.
Williams, III, G. (1993). "Smart Materials," *Omni* Magazine, April, 43–91.
Williams, D. L., Ainslie, B. J., Armitage, J. R., Kashyap, R., and Campbell, R. (1993). "Enhanced UV Photosensitivity in Boron Codoped Germanosilicate Fibers," *Electron. Lett.* **29**, 45–47.
Wilson, A., James, S. W., and Tatam, R. P. (1997). "Time-Division-Multiplexing of In-Fibre Bragg Gratings Using a Pulsed Laser Diode Source," OSA, 12th Int. Conf. on Optical Fiber Sensors, Williamsburg, 475–478.
Wilson, J. (1997). "Glass Bridges," *Popular Mechanics*, Dec., 74–77.
Wilson, J., and Hawkes, J. F. B. (1989). *Optoelectronics: an Introduction*, Prentice Hall.
Wolff, R., and Miesseler, H. J. (1992). "Monitoring of Prestressed Concrete Structures with Optical Fiber Sensors," 1st Europ. Conf. Smart Structures and Materials, Glasgow, *SPIE* **1777**, 23–29.
Worden, K., Ball, A. D., and Tomlinson, G. R. (1993). "Fault Location in a Framework Structure Using Neural Networks," *Smart Mat. & Struct.* **2**, 189–200.
Wosinski, L., Bétend-Bond, J. P. Breidne, M., Sahigren, B., and Stubbe, R. (1992). "Quasi-Distributed Fiber-Optic Sensor for Simultaneous Absolute Measurement of Strain and Temperature," 1st Europ. Conf. on Smart Structures and Materials, Glasgow, *SPIE* **1777**, 53–56.
Wright, H. D., and Lloyd, R. M. (1992). "Possibilities for the Use of Strain Gauged Reinforcement in Smart Structures," 1st Europ. Conf. on Smart Structures and Materials, Glasgow, *SPIE* **1777**, 219–222.

Wu, C., Rolland, C., Shepherd, F., Larocque, C., Puetz, N., Chik, K. D., and Xu, J. M. (1993). "InGaAsP/InP Vertical Directional Coupler Filter with Optimally Designed Wavelength Tunability," *IEEE Photo. Tech. Lett.* **4**, 457–459.
Wyatt, T. (1994). "How to Build on Disasters," *New Scientist*, July, 37.
Xu, M. G., Johnson, M., Farhadiroushan, M., and Dakin, J. P. (1993a). "Novel Polarimetric Fibre Device for Interrogating 'White-Light' Interferometers," *Electron. Lett.* **29**, 378–379.
Xu, M. G., Geiger, H., Archambault, J.-L. Reekie, L,. and Dakin, J. P. (1993b). "Novel Interrogating System for Fibre Bragg Grating Sensors Using an Acousto-Optic Tunable Filter," *Electron. Lett.* **29**, 1510–1511.
Xu, M. G., Reekie, L., Chow, Y. T., and Dakin, J. P. (1993c). "Optical in-Fibre Grating High Pressure Sensor," *Electron. Lett.* **29**, 398–399.
Xu, M. G., Archambault, J.-L, Reekie, L., and Dakin, J. P. (1994a). "Discrimination between Strain and Temperature Effects Using Dual-Wavelength Fibre Grating Sensors," *Electron. Lett.* **30**, 1085–1087.
Xu, M. G., Geiger, H., and Dakin, J. P. (1994b). "Multiplexed Point and Stepwise-Continuous Fibre-Grating Based Sensors: Practical Sensor for Structural Monitoring?" *SPIE* **2294**, Distributed and Multiplexed Fiber Optic Sensors IV, 69–80.
Xu, M. G., Archambault, J.-L. Reekie, L, and Dakin, J. P. (1994c). "Structural Bending Sensor Using Fiber Gratings," *SPIE* **2292**, 407–413.
Xu, M. G., Dong, L., Reekie, L, Tucknott, J. A., and Cruz, J. L., (1995a), "Temperature–Independent Strain Sensor Using a Chirped Bragg Grating in a Tapered Optical Fiber," *Electron. Lett.* **31**, 823–825.
Xu, M. G., Geiger, H., and Dakin, J. P. (1995b). "Interrogation of Fiber-Optic Interferometric Sensors Using Acousto-optic Tunable Filter," *Electron. Lett.* **31**, 1487–1488.
Xu, M. G., Geiger, H., and Dakin, J. P. (1996). "Modeling and Performance Analysis of a Fiber Bragg Grating Interrogation System Using an Acousto-optic Tunable Filter," *J. Lightwave Tech.* **14**, 391–396.
Yarcho, W. B. (1989). "In-House Study to Determine the Effects of Embedded Optical Glass Fibers on the Strength of Graphite Composite Material," Materials & Processes Report MoP-89-Jan-FDBCC-006. Wright Laboratory, Wright–Patterson AFB.
Yariv, A., and Nakamura, M. (1977). "Periodic Structures for Integrated Optics," *IEEE Quantum Electron.* **4**, 233–253.
Young, W. (1991). "Introduction to Reliability-Related Problems in Optical Fiber Connnectors," *Opt. Eng.* **30**, 821–823.
Young, W. C. (1989). "Optical Fiber Connectors, Splices, and Jointing Technology," in *Optoelectronic Technology and Lightwave Communications Systems* (Lin, C., Ed.), Van Nostrand Reinhold, 155–174.
Yu-Lung, L, and Sirkis, J. S. (1997). "Simple Method to Measure Temperature and Axial Strain Simultaneously Using One In-Fiber Bragg–Grating Sensor," *SPIE* **3042**, 237–243.
Yurek, A. M., and Dandridge, A. (1988). "Optical Sources," in *Optical Fiber Sensors: Principles and Components*, (Culshaw, B. and Dakin, J., Ed.), Artech House, 151–187.

Zabaronick, N., Sherrer, D., Claus, R., Murphy, K., Duncan, P., and Shinpaugh, K. (1996). "Remote Optical Interrogation of Embedded Optical Fiber Sensors," *SPIE* **2718**, 234–238.

Zhang, L., Fallon, R. W., Gloag, A., Bennion, I., Haran, F. M., and Foote, P. (1997). "Spatial and Wavelength Multiplexing Architectures for Extreme Strain Monitoring Systems Using Identical-Chirped-Grating-Interrogation Technique," OSA, 12th Int. Conf. on Optical Fiber Sensors, Williamsburg, 452–455.

Zimmermann, B., and Claus, R. (1993). "Spatially Multiplexed Optical Fiber Time Domain Sensors for Civil Engineering Applications," *Applications of Fiber Optic Sensors in Engineering Mechanics* (Ansari, F., Ed.), 280–287.

Zimmermann, B., Claus, R., Kapp, D., and Murphy, K. (1989). "Optical Time Domain Reflectometry for Local Strain Measurements," *SPIE* **1170**, 534–541.

Zuliani, G., Hogg, D., Liu, K., and Measures, R. M. (1991). "Demodulation of a Fiber Fabry–Perot Strain Sensor Using White Light Interferometry," *SPIE* **1588**, Fiber Optic Smart Structures and Skins IV, 308–313.

Index

94.76